SpringerWienNewYork

Die Herausgabe erfolgt mit Unterstützung des
Friederike und Oskar Ermann Fonds

Naturhistorisches Museum Wien

Maria Teschler-Nicola (Ed.)

Early Modern Humans at the Moravian Gate

The Mladeč Caves and their Remains

SpringerWienNewYork

a.o. Univ.-Prof. Dr. Maria Teschler-Nicola
Anthropologische Abteilung, Naturhistorisches Museum Wien, Österreich

SpringerWienNewYork is part of Springer Science+Business Media
springer.com

Typesetting and Printing: Theiss GmbH, 9431 St. Stefan, Austria

Printed on acid-free and chlorine-free bleached paper
SPIN: 11339250

With numerous (partly coloured) Figures

Library of Congress Control Number: 2006927801

ISBN-10 3-211-23588-4 SpringerWienNewYork
ISBN-13 978-3-211-23588-1 SpringerWienNewYork

Dedicated to

Jan Jelínek
(6.2.1926–3.10.2004)

and

Emil Breitinger
(15.10.1904–1.5.2004)

PREFACE

The early Upper Palaeolithic human fossils from the Mladeč (Lautsch) cave (German *Fürst Johann's Höhle*) are among the most valuable inventories of the Vienna's Natural History Museum (Naturhistorisches Museum Wien). They are closely associated with the early history of the museum, the Anthropological Department in particular, and remain at the centre of scientific discussion on the biological and cultural evolution of early modern humans.

The history of the retrieval and preservation of these finds is marked by circumstances and events that are both fortunate and tragical. Geologist Ferdinand von Hochstetter and his assistant Josef Szombathy played a key role in the discovery. Very successful in a variety of positions, and acting for a number of institutions, Hochstetter also played a specific role in establishing the "science of man" in Vienna. President of the Geological Society and First Director of the newly created imperial-royal Court Museum, he founded the Anthropological-Ethnographical Department and became its Director. The fact that he was also a Real Member of the Academy of Sciences and chairman of the *Prehistorical Commission of the imperial Academy of Sciences,* founded in 1878, was also of particular importance in the context of early speleological research. This Commission was entrusted the task of initiating and promoting speleological investigations and "palaeo-ethnographical" studies and excavations on Austrian territory and of preventing the "unscientific exploitation of major sites for private purposes". A number of caves in the Moravian karst, whose wealth of diluvial bones had long been known from historical sources and which were exposed to massive pilferage for industrial purposes in the middle of the 19th century, also faced particular danger. From 1879, the Viennese scientists, who were familiar with contemporary geological and topographical literature on the margraviate of Moravia, first concentrated on Výpustek and other minor caves situated on territory belonging to John II, the Prince of Liechtenstein. Incidentally, the Prehistorical Commission of the Academy of Sciences also sponsored Maška's excavations in the Šipka cave, and Hochstetter visited the cave and its scarce finds in July 1879. From his personal contacts with the Liechtenstein administrators, Hochstetter was well aware of the potential inherent in that region. In any event, in the summers of 1881 and 1882, Josef Szombathy complied with a request to include the Mladeč Cave in the study. Despite successful campaigns, culminating in sensational finds, the Viennese scientists did not continue excavations in the Mladeč Cave. As Szombathy's central concern – proving the contemporaneity of "man and the raindeer" – had been answered by the results of the 1882 excavations, the focus was shifted to other caves in the Moravian karst.

It is due to some fortunate circumstances that practically all the Mladeč finds donated to the Anthropological-Ethnographical Department of the imperial-royal Court Museum of Natural History survived two World Wars without suffering any damage; on the other hand, those recovered a few years later – most of which were preserved at the Moravian Museum Brno (Moravské zemské muzeum of Brno) – were unfortunately destroyed during their attempted removal at the end of World War II, except for a male calvarium and a few other minor finds. As a result, these finds were lost for science forever.

Josef Szombathy first presented and interpreted the finds more than forty years after being recovered. Although some aspects were singled out for re-interpretation in the eighties and nineties of the past century, that presentation continues to be the most comprehensive one. But again it was to take almost (another) fifty years until this important assemblage of finds was to come into the focus of scientific debate again. It was realised that this assemblage was characterised by great variability and distinctive sexual dimorphism and that there was evidence of a Neanderthal heritage for the early Upper Paleolithic. Though in the seventies and eighties, Jan Jelínek and American palaeo-anthropologists Milford Wolpoff, David Frayer and others reinvestigated the human fossils stored at the Naturhistorisches Museum Wien and at the Brno Moravské zemské muzeum, the studies – to be conducted in association with the Museum of Brno – for various reasons were never presented as planned. In the late nineties, renewed efforts were made, now from Vienna, to arrange for the printing of the finalised principal manuscripts at the Museum of Natural History in Vienna. But despite numerous attempts at its timely completion, and against all expectations, the project progressed rather slowly and eventually took a completely different course: It seems likely that by including a number of additional topical issues, the project gained a momentum of its own that some of the authors considered rather difficult to comprehend.

The book was to be enhanced by a few additions concerning the interpretation of the cave and its use that had hitherto not been dealt with in a satisfactory way or not been dealt with at all, including an analysis of the faunal remains stored at various museums and the taphonomic changes of human skeletal remains. Another intention was to subject historical perspectives to critical review, such as Szombathy's diary and other written documentary sources capable of shedding light on the cultural historical context and the evolution of specific scientific issues and museum developments. In parallel to the above, a pilot study conducted by collaborators of the Max Planck Institute for Evolutionary Anthropology had succeeded in providing positive evidence of sufficiently well-preserved organic components in some randomly selected human and animal skeletal remains from Mladeč. It seemed that absolute dating, which is crucial for the interpretation of the finds and had hitherto invariably failed, became feasible for the first time – even though concurrently, applying such invasive methods of study would confront us with specific issues of curatorial responsibility. Finally, these processes and considerations ultimately prompted us to redesign the content of our initial concept and to extend our timeframe for the finalisation of this monograph.

Nevertheless, our original intention in writing this volume was unaltered: This presentation still aims to introduce to the discussion a concise set of new data on a find that must be considered "old" in the historical sense; in so doing, to convey different views and interpretations and to open up our minds to diverging opinions, thereby contributing, *inter alia*, to a discussion of the models developed to describe the origin of anatomically modern human beings.

The printing of this volume has been made possible by generous funding from the Natural History Museum, in particular the "Friederike und Oskar Ermann Fonds". Therefore I am, first of all, particularly indebted to professor Bernhard Lötsch, the director-general of the Museum, and to Herbert Kritscher, his deputy.

I also thank the Springer publishing house in Vienna, in particular Messrs. Raimund Petri-Wider and Wolfgang Dollhäubl, for their friendly collaboration, their support in conceptual issues and their comprehension and assistance in the graphic design.

I also wish to express my gratitude to all the collaborators of the Anthropological Department, who I could rely on for all the technical, taxidermical and photographical work; in particular to Mr. Wolfgang Reichmann, who entered all objects in a photographic documentation and reproduced a number of copies. He was also in charge of the photoshop editing and layout of the plates, provided competent support to authors and editors, and is co-responsible for the high-quality photography of this volume; as well as to Mr. Ronald Mühl, who did all the sampling, made the histological thin sections and recorded many findings.

I am particularly grateful to Mrs. Michaela Zwölfer (LanguageWorks) for her precision in translating some of the texts and her many linguistic suggestions and friendly support; as well as to

Mrs. Louise Chantale Blundell, Mr. Thomas-Bence Viola and Mrs. Doris Schamall for their editing of some texts that were available in English.

I wish to extend particular acknowledgement to Mrs. Bettina Voglsinger, my assistant, who accompanied the making of this book as it evolved through its different stages over the past two years. I owe my thanks to her for all her assistance relating to organisational tasks, her many rounds of meticulous proof-reading, her painstaking verification of sources and her invariably ready physical and mental "emergency programme for busy coordinators and authors". This service was also extended to my family, who have been kind enough to tolerate my style of work for many years.

Vienna, May 2006 Maria Teschler-Nicola

FOREWORD

Ever since its initial excavation in the early 1880s by Josef Szombathy, the Mladeč cave has played a pivotal role in discussions of the prehistory of central Europe. Perhaps the most significant early contribution was Szombathy's presentation in Paris in 1900, at the Twelfth International Congress of Anthropology and Archaeology, of the first of many *Homo sapiens* skulls eventually recovered from the site. This specimen, Szombathy argued, was unambiguously associated not only with blade tools, an animal-bone dagger, and perforated teeth of beaver and reindeer, but also with the bones of ancient animals. This was among the few occasions up to that point on which human skeletal remains had ever been claimed to be as old as the fossilized remains of animals that were agreed to be extinct. For, prior to Szombathy's presentation of Mladeč to the anthropological community, the history of human paleontology records little more than Fuhlrott's largely unsuccessful argument in 1859 for the great antiquity of the Feldhofer Grotto Neanderthal remains, Lartet and Christy's association during the 1860s of the Cro-Magnon remains with extinct mammals, and Fraipont and Lohest's representation in the late 1880s of the Spy Neanderthals as truly ancient humans. In light of this scant history, it is hardly surprising that Szombathy initially faced some scepticism in making the case for the Pleistocene age of the Mladeč remains. The task was further complicated by the fact that the first and second human specimens found were not only fragmentary - a condition Szombathy rather implausibly attributed to cannibalism or some other form of human activity - but had been recovered very close to the surface of the cave deposits.

Eventually Szombathy's, Janda's and subsequent excavations of the Mladeč Cave yielded a treasure trove of specimens - faunal, artifactual, and human. But while the early excavations at Mladeč were apparently quite meticulously conducted for their period, they were rather poorly recorded and published. The principal result of this unfortunate circumstance is that, while the singular significance of the Mladeč fossils is widely acknowledged, they have figured principally as elements in broader analyses rather than as subjects of study in their own right. This volume is a sorely-needed and highly successful attempt to rectify this situation. It is well known that many of the jewels of the Mladeč human collection were tragically destroyed in 1945, in a paleoanthropological disaster exceeded only by the loss of the Peking Man collection four years earlier, also in connection with the events of the Second World War. But what many readers of this volume may be surprised to learn is that almost four dozen original human specimens survive. These fossils are inventoried here, along with those lost, by Wolpoff and colleagues, who also provide a splendid series of photographic plates. The specimens consist of variably preserved adult and immature craniodental and postcranial elements, of both sexes, that provide a wealth of morphological information. All in all, the Mladeč human fossil collection still offers palaeoanthropologists and prehistorians a rich resource that presents a unique insight into the nature of Aurignacian human populations in Central Europe.

Given the rather fragmentary nature of the information on the site and its excavation contained in Szombathy's diary and correspondence, it is gratifying that many chapters in this volume strive as far as possible to fill in the blanks. Antl-Weiser begins with a review of Szombathy's notes, and Svoboda continues the "sleuthing" with a computer reconstruction of the cave, including the section known as the "Dome of the Dead" from which the human remains came. Svoboda concludes that a vertical chimney had afforded a possible but difficult entrance into the cave from the outside, and that - *contra* Oliva's interpretation later in the volume - the site could not easily have been a ritual area that had been regularly visited by humans. Indeed, Svoboda favours the notion that the human remains, like those of the other animals found at Mladeč, fell into the cave through an opening or openings to the surface, although he leaves open the question of whether the presence of the human remains in the cave was somehow related to human activity of some kind outside. In contrast, on the basis of his reanalysis of Szombathy's notes, Oliva concludes that although Mladeč was clearly not a habitation site, the locality was visited occasionally by humans, most likely in ritual contexts. Because of the reported presence of hearths, and because in one instance human remains and artifacts were found together in a niche in the northwestern wall of the cave, Oliva rejects the notion that the human remains were introduced into the cave via a chimney connecting it to the surface. Rather, he hypothesizes the existence during the Upper Paleolithic of a horizontal entrance to the cave that would have permitted access by humans.

Clearly, the issue of how the human remains came to be in the Mladeč deposits, and how the site may or may not have been used by *Homo sapiens* during the Aurignacian, is not an open-and-shut case. Teschler-Nicola makes this clear in her taphonomic study, from which she concludes that the "charcoal" that suggested domesticated fire was more likely deposits of manganese. Further, in her view the fragmentary state of the human bones may well reflect carnivore or scavenger activity – which she observes may not preclude their introduction into the cave via a vertical chimney. Teschler-Nicola's interpretation gains support from Pacher's study of site formation based on analysis of the faunal remains. This analysis confirms that the Late Pleistocene faunal assemblage most likely consists of individuals that fell into a natural trap. Overall the Mladeč thanatocoenose, which is dominated by bovids followed by reindeer, horse and wolf, is markedly different from those found in penecontemporaneous human occupation sites in the region. Like the other mammal bones, the human remains appear to be related to the debris cone below the chimney, and none of the animal bones shows any sign of human activity, although there is some minor evidence of gnawing by hyenas.

Among the many significant outcomes of this collective endeavor is Wild et al.'s ^{14}C dating of samples derived from animal and human skeletal remains. While carbonate samples yielded a minimum age of 35-34 ^{14}C kyr BP, analysis of the animal bones produced a wide range, from ~8.5 to ~42.5 ^{14}C kyr BP (which might reflect the irregular depositional history of the site). Direct dating of four human samples resulted in dates ranging from ca. 32 to 30.5 ^{14}C kyr BP (uncalibrated), while two others (possibly contaminated) were dated as 3 and 4 kyr younger. The preponderant age of the Mladeč human specimens puts them solidly within the frame of the middle to late Aurignacian of Central Europe, and makes them the most complete early human remains from the region. In the region only the Romanian site of Peştera cu Oase appears to contain older *Homo sapiens* fossils, at about 35 kyr.

Following Wolpoff et al.'s inventory come two substantial chapters on the human remains. The first of these deals with the Mladeč craniodental specimens (including ones that were destroyed) that are considered by Frayer et al. to be male, and the second focuses on those identified by Wolpoff et al. as female. The first of these chapters also provides a detailed summary of the history of the discovery of the human remains, and of their archaeological context. In both contributions the authors provide information on the morphology of individual specimens, which makes these chapters useful even to those who do not agree with the controversial Multiregional model within which these authors situate the Mladeč specimens. This is particularly important since the Mladeč *Homo sapiens* have frequently been cited – in our view, erroneously – as retaining certain characters, particularly of the brow and cranial rear, that indicate some kind of contact or affinity with the Neanderthals who had earlier occupied Central Europe. In this regard, we note that although Frayer et al. identify in the nasal cavity of Mladeč 8 what they believe is the base of a medial projection, a feature we have described as a Neanderthal apomorphy, the structure in question actually appears to be part of the conchal crest, a feature that is primitively retained in *Homo sapiens*. Clearly, the debate over continuity vs replacement in Central Europe is set to continue, with the Mladeč hominids as pivotal participants; and these chapters will endure as indispensable documentation of these unique fossils.

Against this basic background, this volume then subjects the Mladeč crania to closer scrutiny than simple traditional description and measurement of their outer surfaces. Prossinger and Teschler-Nicola use analysis of CT scans of the Mladeč 1 cranium to "rid" this fragile and unique specimen of foreign material, and to reveal for the first time its true preserved bony morphology both inside and out. This approach will certainly provoke new avenues of research on this and other specimens. Through a different use of computer technology, Minugh-Purvis et al. reconstruct the general shape of the braincase of the Mladeč 3 infant. Although the interpretation of this reconstruction is steeped in a "non-evo-devo" paradigm that envisions an amalgam of Neanderthal and modern human gene pools, this contribution contains valuable information allowing comparison of this specimen with modern and ancient juveniles of *Homo sapiens* and other hominid taxa. While Minugh-Purvis and colleagues emphasize continuity, a few chapters later in the volume Weber et al. morphometrically compare the external geometry of skull shape between the three better preserved Mladec specimens, "anatomically modern humans" and Neanderthals, and conclude that the Mladeč fossils group firmly with the modern sample.

Turning to the postcranial skeleton, Trinkaus et al. present an overview comparable in morphological and metric detail to that provided for the crania. Although the postcranial remains representing juveniles, adolescents, and adults were not found (or at least kept) associated when excavated, Trinkaus et al. make strong cases for assigning bones of similar developmental status or robusticity to particular individuals. They conclude that in general the Mladeč humans were relatively robust and in many ways morphologically similar to Upper Paleolithic humans from other sites, while leaving open the question of whether their postcranial reflects "Neanderthal ancestry" or merely Upper Paleolithic behavioral patterns.

The final three chapters in this volume deal, successively, with trauma and disease, mtDNA, and strontium isotope ratios in the Mladeč hominid sample. In the first of them, Teschler-Nicola et al. use CT scanning and radiography, in addition to surficial visual assessment, to reveal a array of ailments or traumatic events that affected the Mladeč humans. These include a malformed cochlear-vestibular region and extruded 7th cranial nerve, healed traumatic lesions, blunt traumatic and other depressions of the skull, and even a small osteoma. The chapter by Serre et al. on mtDNA deals mainly with the more general question of whether "Neanderthal-like mtDNA sequences" can be identified in early modern human remains from Europe. The authors conclude that no Neanderthal-like mtDNA sequences are present in the latter, and specifically find no support for Neanderthal elements in the two Mladeč samples from which mtDNA fragments were retrieved. Finally, Prohaska et al. used noninvasive techniques to provide preliminary information on $^{87}S/^{86}S$ isotope ratios in various Mladeč specimens in the hope of providing insight into possible migratory patterns, since strontium ratios are locally specific. Although the authors are very tentative about their results, it is interesting that the Sr isotope ratio pattern in the dental enamel of Mladeč 1 and Mladeč 2 suggests that these individuals were not local but had instead migrated to this region of Moravia. This in and of itself raises a host of questions and should whet one's intellectual appetite.

This volume successfully represents a broad collaboration that will serve as a model for future similar endeavors. It is made all the more exciting by housing a substantial diversity of opinion within its covers, a diversity that reflects the central position of the Mladeč human fossils in a variety of ongoing paleoanthropological debates. The editor and authors are to be congratulated for rising so effectively to the challenge of definitively monographing this highly significant but until now woefully under-documented group of early European modern humans.

April 2006

Jeffrey H. Schwartz
University of Pittsburgh

Ian Tattersall
American Museum of Natural History

CONTENTS

Chapter 1

SZOMBATHY'S EXCAVATIONS IN THE MLADEČ CAVE AND THE FIRST PRESENTATIONS OF THE RESULTS

Walpurga Antl-Weiser

When Szombathy began studies in the Moravian karsts, prehistory was far from being its own discipline at the Vienna University. People who were dealing with prehistoric subjects, therefore, came from various backgrounds - geology, paleontology, art history or classical archaeology. If we want to understand the way Szombathy was doing his studies, it is important to have a closer look at his scientific background. After a solid scientific education at the Polytechnic Institute of Vienna in botanic, paleontology, mineralogy and geology, J. Szombathy became an assistant to Ferdinand v. Hochstetter, professor and director of the Technical University, and at the same time, he attended lectures at the University of Vienna concerning physical geography and the drawing of maps. He first made contact with archaeology when he joined an excavation with Ferdinand v. Hochstetter at the Hallstatt cemetery in 1877. When Hochstetter became director of the Imperial Mineralogical Cabinet, Szombathy followed him as his assistant. Szombathy undertook excavations in order to enlarge the paleontological and prehistoric collection (Heinrich, 2003). In 1880, he excavated in the caves of Southern Moravia on behalf of the Austrian Academy of Sciences. Due to his education, he made exact cave surveys and plans, sieved the sediments and took samples for later scientific research, which was rather unusual for archaeological excavations in those days.

The original documents of the Prehistoric Department

From Szombathy's excavations, we mainly have his diary and some letters he exchanged with the foresters of Prince Liechtenstein, with Janda who completed his research in 1882 and with Dr. Smyčka who continued the research in the Mladeč Cave.

His diary is the only record we have about the excavations themselves. For the Mladeč Cave, we have descriptions from the years 1881, 1882, 1904 and 1925. The first two years are his excavations, while the two other entries are later visits to the cave.

The diary includes a series of very accurate plans of different parts of the cave and detailed descriptions of the structure of the different layers in a trench in area "a". Yet these plans are not suitable to identify the position of the animal bones and the artifacts. There is no information left with the bones either. Therefore, only a thorough analysis of the original documents and the first publications may offer a solution.

As for the doubts concerning the preservation of the original prehistoric surface, only the exact analysis of all the facts can help to elucidate the problem.

Description of the cave at Szombathy's first visit

Szombathy started the excavation in June and was supported by the local forester caring for the forest of Prince Liechtenstein. According to the diary, it seems that Szombathy first took down only some

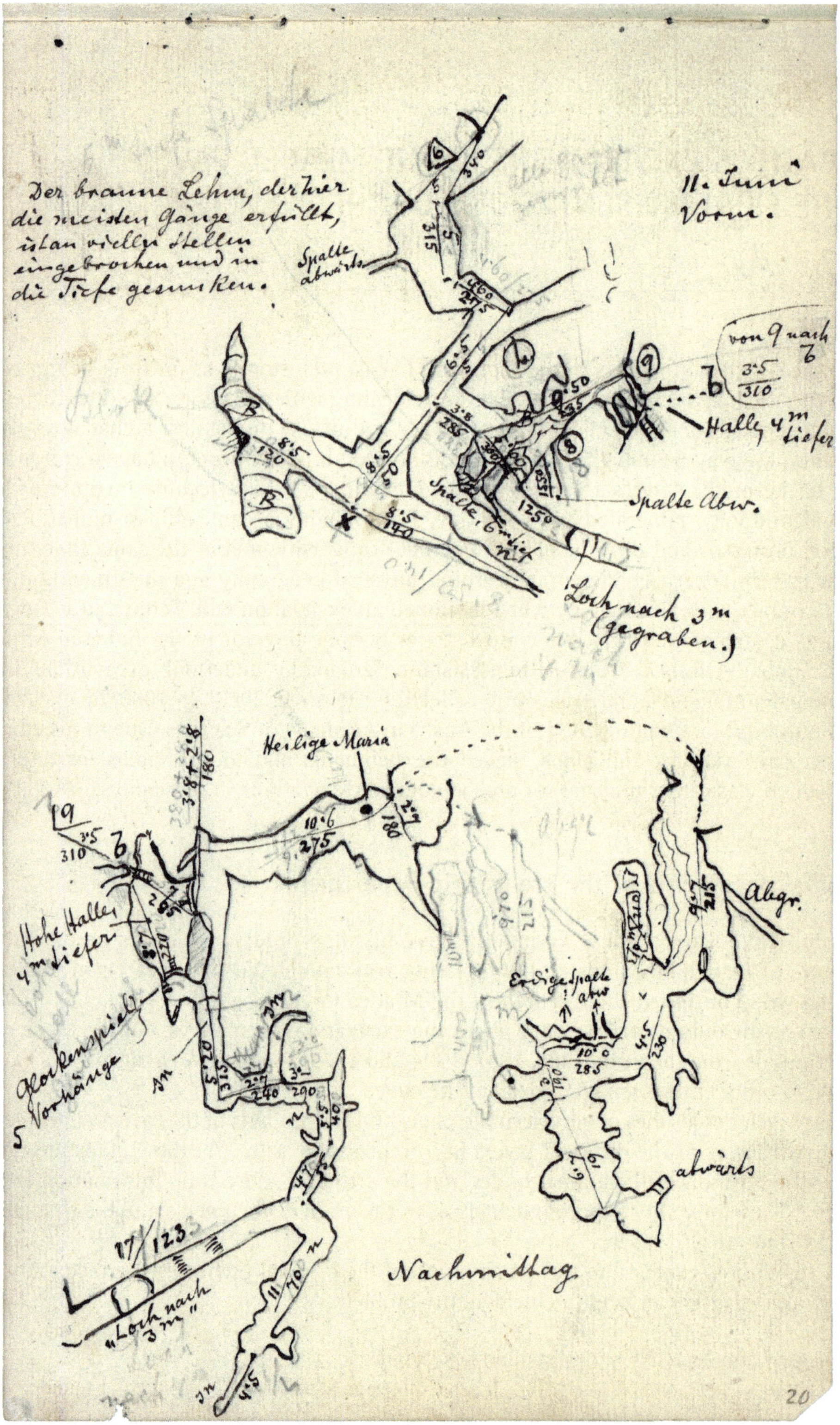

Fig. 1. Szombathy's diary, 11th June 1881, p. 20

short notes and produced a longer description shortly afterwards because his first impressions were written with a pencil and the other in ink. The beginning of both versions is the same, therefore, he probably wrote the version in ink shortly after his first notes - in the evening or some days later.

The entrance to the cave was outside the village Mladeč (Lautsch). This was not the original entrance because the cave was detected while blasting for a quarry. Through this new entrance Szombathy entered the cave. In front of the entrance was a loamy embankment falling steeply to the entrance hall that was 2 m high and 6 m in diameter. From this hall to the right were two small gaps and to the left a small passage and then another higher entrance to the cave. In Szombathy's plan, the entrance hall is "A" and the next hall is "B". First of all, there is a small hall with two side passages, after that a bigger one with rocks and on the left a passage that leads to the surface. Straight ahead the way continues horizontally, to the right the way leads down through a small room with rocks and to the left there is a hall with an even floor. In this Hall (probably D) there are many side passages. Szombathy supposed that it would be best to start excavations there or in the small hall.

The mentioned passage straight ahead was ten meters long with stalactites and rocks. Szombathy supposed that there had to be a passage on the right-hand side to Hall I but he was not sure about the exact location. The left passage from the bigger hall with rocks to the surface was covered with brown loamy sediments. Szombathy continued his way through a hole using a ladder and then moved further on horizontally over rocks. From there the left way leads up and the right one down to a chapel with deep fissures and pillars. Szombathy mentioned a possibility to wind oneself through rocks on the left. Straight away there was a passage with disintegrated rocks and from there a small passage downwards that Szombathy called "chicken's staircase", then to the left over rocks to a hall with big lumps of rock and many side passages. After that he found an elongated hall with deep fissures falling 50° to the right. It was not possible to get to the end of this hall. To the left, there was a room with beautiful stalactites. The upper way led to this room as well. From there on the cave was a labyrinth with passages, holes and precipices. The floor was covered with brown loamy clay that was produced by the weathering of limestone. The small stalagmites have been taken away. The wall was eroded partly by running, partly by stagnant water. There were many stalactites with small curtains on the ceiling but there were big parts where this surface was cracked off. On the floor there were big rocks. He saw big lumps of rock hanging from the ceiling in constant danger of falling down at one stroke of a hammer. On the walls there were beautiful contrasts of color: weathered limestone in bluish gray, brown traces of loam coming down with the water from the ceiling, white and milky white stalactites and sinter similar to chalcedony on various places of the wall. When Szombathy visited the cave, there were already stairs and ladders to get through the cave more easily. Szombathy mentions older descriptions were people had been held back by the water in this area. When Szombathy visited the cave there was no water there[1]. The fact that stalagmites had been taken away

1 Szombathy's diary 1881

p. 10a

„Hinter dem letzten Haus von Lautsch ist der Eingang der Höhle. Vor dem Eingang ist eine Schutthalde und die [...] Felsen vor dem Eingang zeigen, dass sie vor kurzhem blosgelegt wurden. Man kam beim Versuch einen Steinbruch anzulegen auf den Eingang.

Von dem Wall geht ein niederer Eingangsbogen in die Vorhalle hinab. Diese hat 6m im dm, von ihr gehen zwei niedere Gänge fort, von denen einer nach oben, der andere horizontal [...] geht.

Nach oben geht es fort durch eine kleine Halle mit 2 Seitengängen – l in einen großen mit Blöcken auf dem Boden. Von da geht links ein kleiner Schlupf gegen Tag, geradeaus geht es horizontal weiter und r geht es abwärts.

Wir gehen horizontal durch einen niederen Gang mit Stalagtiten weiter."

p. 11

„Eingang im W v. Lautsch hinter dem letzten Haus. Er zeigt sich als eine niedere, hinter einer bewachsenen Schutthalde versteckte Öffnung an einem felsigen Anschnitte der Thallehne. Diese Felsenblöhse zeigt die Kalkbänke zertrüm-

in the part of the "labyrinth", shows how far people got into the cave even before Szombathy's first visit.

The excavation of 1881

On June 7th, Szombathy started excavating in Hall I (=D). The floor of this chamber was covered with loam. They began to dig in the southeast of the hall. Twenty centimeters beneath the surface they found a skull and a femur without epiphyses in a scattered position. The skull was covered with sinter. The bones were gray and relatively hard. The skull is described as "hypsi-dolichocephal". The

---

mert und mit Thonmassen durchsetzt, ihre Oberflächen Verwitterung ist so gering, daß man vermuthen kann, sie seien erst vor wenigen Dezennien blosgelegt worden.

Der Lehmwall ist nach innen steil geböscht und bildet den Boden einer im Durchschnitt 2 m hohen Vorhalle von 6x Durchmesser. Nach r 2 kl. Spalten, nach l zuerst ein niederer, daneben ein höherer Eingang in die weitere Höhle. Zuerst eine ganz kleine Halle mit zwei Seitengängen l., dann eine größere mit Blöcken auf dem Boden und l. ein Gang gegen Tag. Geradeaus geht es horizontal weiter, nach r geht es abwärts durch einen kleineren Raum über Geröll gegen l in eine Halle mit ebenem Boden. Nach allen Seiten Nebengänge. (Halle I). Hier wäre zu graben, oder etwas weiter vorne in der Kl. Halle.

Horizontal geht es durch einen niederen, vielleicht 10 m langen Gang mit Stalaktiten und Felstrümmern weiter. Rechts geht es wieder zu der Halle I, (aber ich werde mir nicht klar, auf welchem Wege) nach l ein Gang mit braunem Lehm nach aufw.

Nun durch ein Loch mit einer kleinen Leiter hinab und über Trümmer horizontal weiter. Hier theilt sich der Weg. l oben, r unten.

p. 11a

Am Ende dieses etwa 10 m langen Ganges hat man (Links durchgestrichen) r unten die Räume, in welche man von der vorigen Halle r abw. gelangen kann. Links einen Gang mit braunem Lehm, der zu Tage führt

Nach r über eine Leiter in den unteren Raum, eine Halle mit arg zerklüfteten Decke mit mehren Pfeilern und Nebengängen.

Von der Leiter weiter durch einen zerfallenen Gang über Trümmer (Hühnersteige) in eine Trümmerhalle mit vielen Seitentheilen. Dann eine Spalten Halle mit Gefälle nach l (r oben schöne Stal.)

p. 12

Abw nach r. über eine Leiter hinab in eine allerseits zerklüftete Kapelle mit Pfeilern und Nebengängen, von denen man dann nach l. sich zwischen Blöcken durchwinden könnte.

Dann wieder von der Leiter gerade aus (recte l.) weiter durch einen zerfallenen Gang über schauerliches Trümmerwerk und über eine Hühnersteige (rechts wieder ein Seitengang) dann nach l über Blöcke hinab in eine Trümmerhalle mit vielen Seitentheilen.

Dann folgt eine längliche Spaltenhalle, welche im Allgem. 50° nach r einfällt. Man kann nicht bis ganz hinab. L. oben ist ein Raum mit schönen Stalaktiten, zu welchen man auch auf dem oberen Wege gelangen kann. Weiterhin folgt ein Labyrinth von Gängen, Löchern und Abgründen, das aller Beschreibung spottet. Der Boden der Gänge wird von braunem plastischem Lehm gebildet, der vielfach mit einer häufig wieder zerbrochenen Sinterdecke belegt ist und an vielen Stellen abgebrochen und in Abgründe gerutscht ist. Dieser Lehm ist der Verwitterungsrückstand des Kalkes und leer. Die kleineren Stalakmiten sind alle weggenommen und nur einige größere sind erhalten.

Die Felsen zeigen vielfach denudierte Flächen, theils von fliessendem Wasser, theils von stehendem.

p. 13

Die Decke ist an vielen Stellen mit Stal. besetzt, von welchen auch kleinere und auch mehrere tönende Vorhänge erhalten sind. Aber an vielen Stellen liegen die verwitterten Zerklüftungs und Bankungsflächen der Decke zu Tage und sehen auf die niedergestürzten Blöcke herab; an gewissen Punkten hängen tonnenschwere Blöcke in so losem Zusammenhange an der Decke, dass einige Hammerschläge genügen müssten, um sie herab zu bringen. Zeichen für das fortwährende In-sichzusammenfallen der Trümmermassen, über welche man schreitet, fallen einem zu Hunderten in die Augen.

An den Wänden hat man oft schöne Farbengegensätze: Der blaugraue, abgewitterte Kalk, die braunen, aus den Deckenspalten mit dem Wasser niedergehenden Lehmstriemen, die hellbraunen und die weissen Stalaktiten und milchweisse, Chalcedon ähnliche Sinterüberzüge an vielen Wandstellen.

teeth show that the skull belongs to a young individual. In Hall I, only a few bone fragments were left in the loamy ground. Thirty-five centimeters below was a small layer of charcoal. After that Szombathy tried a second place and found bones of mammals and a mandible from a ruminant possibly a reindeer[2].

The draft written with pencil concerning the excavation is incomplete compared to the description of the cave. It mentions only the charcoal layer. This is the last entry written with a pencil. The excavation continues on the 9th of June. There were mainly bones of reindeer and cave bear.

Szombathy instructed one of the workers to start at a third place where a large number of bones from cattle and a supra-occipital of human had been found. There were also finds of reindeer[3].

Together with these finds Szombathy quotes an older description of the cave from the year 1839 that he does not seem to have known about before. According to this description mentioning several sections and domes, only few stalagmites and stalactites and bones of prehistoric animals, it is evident that the whole cave Szombathy visited at the beginning of his excavation had been known to the public for a very long time. This old description also mentions that the cave was already detected in 1826[4]. The point is we do not know the number of animal and human bones that were car-

---

Raschendorfer kennt die Höhle gut und hat sie mit Leitern, Steigen usw. in hohem Grade gangbar gemacht. Die Stelle, an welcher die Besucher vor 20 Jahren durch Wasser aufgehalten worden sein sollen, kennt er nicht. Mir ist auch keine solche Stelle aufgefallen.

2 Szombathy's diary

p. 13a
In einer Tiefe von 35 cm eine (horizontale Kohlenschichte) an benachbarten Stellen mehrere kl. Knöchelchen

p. 14
Versuchsgrabung in der Halle I, vollkommen lehmiger Boden. Wir schlugen im SÖ Theile ein und fanden 1 Calvarium und einen Femur ohne Epiphysen in zerstreuter Lagerung, 20 cm unter der Oberfläche. Am Schädel lag eine Sinterschicht an.
Die Knochen sind durch und durch grau, aber ziemlich fest. Der Schädel ist hypsi-dolichocephal und hat eine wohl entwickelte Stirne. Den Zähnen nach ist er von einem jungen Individuum.
8. Juni. Regen
Heintz reist ab
In der Halle I nur mehr einige ganz kleine Knochenfragmente und in ca. 35 cm Tiefe etwas Kohle in einer horizontalen Schicht gefunden.
An einem 2. Platze angegangen. Dort fanden sich Säugethier-Knochen. Wiederkäuer-Kiefer Geweihstück von Renn?

3 Szombathy's diary

p. 15
9. Juni Herr Alois Kroker u. Herr Apotheker Scholda graben wieder den ganzen Tag. Kr. findet Renn- u. U. spel. Reste. Scholda setzte ich an einer 3. Stelle an, wo er bald auf eine Masse von Rinderknochen kam und auch ein Supra occipital von Mensch fand.
Am Abend brachten wir 3 Butten voll Knochen, meist von Rind und Rennthier nach Hause. Schädelfragmente von Mensch.
Ich begann mit der Aufnahme einer Planskizze.
10. Juni, Freitag
Vormittags Knochen eingepackt, 2 Kisten: in der großen die oberste Lage Renn, das andere Rind; in der kleineren alles, bes. Mensch.

4 Gregor Wolny, Die Markgrafschaft Mähren, topographisch, statistisch und historisch geschildert von Brünn, 1839, p. 166 Lautsch [...] hat in dem nahen Übergangskalkgebirge eine erst im Jahre 1826 zufällig entdeckte Höhle, die bei 210 Klafter lang ist, verschiedene Abteilungen und einige hohe Kuppeln aber nur sehr wenig Stalagmiten und Stalaktiten enthält. Die Höhle, worin man auch einige Thierknochen aus der Urzeit gefunden, steht jenen bei Sloup und Adamsthal befindlichen in jeder Beziehung weit nach und ist etwas beschwerlich zu besehen.

ried out of the cave in these 54 years, and to what extent the layers of the cave had been disturbed before Szombathy came to the Mladeč Cave. At the end of the 1881 excavation, Szombathy drew a series of plans of the whole cave (Fig.1). He gave no details about the position of the finds and therefore, it is difficult to identify special pieces. Szombathy left Mladeč on June 12th[5]. The complete length of the cave that Szombathy studied was more than 100 m. After the excavation, he decided to name the cave "Fürst Johanns Höhle" after the landlord Prince Johann of Liechtenstein.

The excavation of 1882

The 1882 excavation started on July 13th.

"Traveled to Neuschloss at 6 o' clock in the morning. Haunold shows the bone fragments he had found in the ditch in front of the cave entrance: atlas of mammoth, two molars of equus and an antler of deer. [...] The entrance of the cave closed by heavy bars is completely free now. The way to the cave has been prepared. All bone fragments came from the inner side of the ditch that had blocked the entrance. Within the cave the way to Hall D is in good condition. In the morning debris is still being cleared out of "D". In the afternoon [...] I start with a trench at the finding place of the cranium in "D". The earth is thrown to the rocks between "C" and "D". [...] July 14th – the trench is continued up to 2 m in clay. At a depth of 1.5 m there were remains of a big rodent that unfortunately had been broken. Multiple discordant layers of clay show deposition by running water.

Two workers are brought to site "b" to continue last year's excavations and to move westwards. In the trench there are rocks at 1.75 m that seem to build a gap with the other side. In the SW in a corner between the rocks there was a slightly darker earthy loam with small bones. I reserve a series of baskets for sieving. Passage "d" is completely ploughed through although I excavated only a part of it last year. Raschendorfer denied having excavated there and yet it is certain. In some places of the cave there are a few bones on the surface.

July 15th – In the whole part of passage "d" spared last year we didn't find a single bone. The surface seems to be untouched but during the excavation it became clear that this part had previously been dug up. Raschendorfer denies. Research continues in this place only to complete the picture.

In the trench we reached fluvial sands with debris of schist. The gap is becoming smaller so we think we have reached the ground. At 2.5 m we found a black charcoal layer on a stone that was 1 cm thick and at 2.6 m another one directly in the loamy sediment. We stopped excavating at 3 m. The limestone in this area is deeply weathered, smooth and white. The black layer was lying above this white layer.

The vicinity of last year's excavations was again dug up and leveled again. Having drunk half a liter of spirits Raschendorfer confessed that some men from Littau carried out excavations in the cave in spring. [...] These villains destroyed the sites that were most important for the cave. [...] Hall D has been excavated up to 0.75 m because most of the bones can be found in the dark sandy clay. The

5 Szombathy's diary

p. 19
11. Juni Vormittag
Vermessung der Höhle
Reduction der gestern vermessenen Partien. Aufnahmearbeit bis 1h.
Nachm. Schluß der Aufnahme der Höhle um 9h abends
Der Gang von "6" nach N war dem Rasch. noch nicht bekannt und da sich in demselben überhaupt keine Spuren von Menschen entdecken liessen, so glaube ich ihn entdeckt zu haben

p. 22
Rückfahrt 12. Juni, Frühmorgens die letzten Knochen gepackt.

layers below 0.6 were empty except this single concentration of bones. But for a concentration like this we cannot excavate the whole area up to 2 m and remove so much material out of the cave. We continue to excavate the western part of "D".[6] In 1882, Szombathy seems to have gotten to know a few more details about the first opening of the cave in 1826. The description of giant bones in the first Hall (A) let us suppose that they found bones of a big mammal possibly the skeleton of a big

6 Szombathy's diary

p. 37
Morgens um 6h nach Neuschloß gefahren. Haunold zeigt mir die Knochenfragmente, welche in dem Schuttkegel am Eingange der Höhle gefunden wurden: Atlas von Mammut, 2 Mol. von Equus, 1 Geweihfragment von Cervus.
Haunold zeigt die Knochen dem Fürsten – ist interessiert
Der Eingang d. Höhle ist nun ganz freigemacht, mit einem starken Gitter abgeschlossen und vor ihm ist ein schönes Plateau hergerichtet. Auch der Zugang bis zur Höhle ist sehr praktikabel hergerichtet worden.
Die Knochen fanden sich an der inneren Böschung des Walles, welcher den Eingang verlegt hatte.

p. 38
In der Höhle ist auch der Weg bis in die Abt. D. sehr gut hergestellt, [...] vormittags wird aus D noch einiger Schutt ausgeräumt
Nachm. mit H. zur Höhle. 3 Arbeiter und der Raschendorfer. Vor allem beginne ich mit einem Schacht an der Stelle des Schädelfundes in D. Die gewonnene Erde wird in den Felssturz zwischen C und D geschüttet. [...]
14. Juli 6 Arbeiter
Der Schacht geht nun schon 2. M. durch Thon. Bei 1 1/2 m an der S. Wand Reste eines großen Nagethieres, Sch(ädel) und Extremitäten, die leider ganz zerbrochen wurden. Die Beschaffenheit des Thones wechselt in falscher Schichtung, u. zwar so häufig, so daß man seine Ablagerung durch fließendes Wasser erkennen kann.

p. 39
2 Arbeiter werden an der Stelle b angestellt, um die vorjährigen Grabungen fortzusetzen und nach W herauszugehen.
Im Schacht stoßen wir bei 1 3/4 m auf den von der Seite hereinkommenden Felsen, der mit der Gegenseite in eine Spalte zusammenzulaufen scheint.
Bei 2 m tritt in der SW Ecke in einem Felsenwinkel ein dunklerer erdiger Lehm mit kleinen Knöchelchen auf, von welchem ich mehrere Körbe voll zum Schlämmen reservire.
Der Gang d ist ganz umgewühlt, obwohl ich im Vorjahre nur einen Theil davon durchgraben konnte. Raschendorfer läugnet, hier gegraben zu haben und doch ist es sicher.
An einigen Stellen der Höhle werden oberflächlich einige Knochen gefunden.
15. Juli, Samstag 4 Arbeiter
Im ganzen, im Vorjahr geschonten Theile des Ganges b wird kein Knöchelchen gefunden. Die Oberfläche scheint intact und doch sieht man beim Graben, daß auch hier schon früher gegraben wurde. Raschendorfer läugnet. Ich lasse die Durchgrabung nur mehr der Vollständigkeit halber vollenden.

p. 40
Im Schacht sind wir nun auf Flußsand mit Geschieben von Grauwackenschiefer gestoßen. Die Spalte verengt sich, so daß wir annehmen können, auf dem Grunde zu sein.
Bei 2.50 ist auf einem der eingelagerten flachen Steine eine 1 cm starke kohlige Schicht aufgedeckt worden und bei 2.60 eine ähnliche direct im Lehm
Bei 3 m hören wir auf zu graben.
Aller an dem Lehm angrenzender Kalk ist bis zu 1 cm tief zersetzt, mürbe und weißlich. Die obige Kohlenschichte lag separat auf der weißlichen Schichte.

p. 41
Die ganze Nachbarschaft meiner vorjährigen Grabungen erwies sich als neuerdings durchgegraben und wieder planiert.
Nach 1/2 Liter Schnaps beichtet Raschendorfer, daß die Herren von Littau im Frühjahr in der Höhle gegraben hätten. [...] Diese Schurken haben mir gerade jene Fundorte zerstört, die für die ganze Höhle am wichtigsten waren. [...]
Nun wird die ganze Abtheilung D bis 3/4 m durchgegraben, weil sich bisher gezeigt hat, daß es vorzüglich der braune, etwas sandige Thon ist, welcher die Knochen enthält. Die Schichten von 0.60 abwärts waren leer bis auf das Knochennest und dem zu Liebe kann man doch nicht 2 m Erde umgraben und aus der Höhle schaffen.

cave bear because it is not very likely that they found a more or less complete skeleton of a mammoth. Szombathy mentions the story elderly people had told him: "Pekar's father-in-law, 72 year old Walloch, worked in the quarry when the entrance of the cave was found. In the first small hall to the right there was the skeleton of a giant lying with out-stretched arms. He had teeth bigger than a thumb. The parish priest of Mierotein took the bones and buried them. Raschendorfer's mother, now 70 years old, was a girl of 16 years when the cave was opened. The head of the giant was so big that a man could have put it over his head."[7] The story of the giant found in the cave made even scientists think that in 1826 a large amount of human remains was removed from the cave together with lots of animal bones. The original description of this incident doesn't mention any human bones except the "giant". It was Maška who first concluded that human remains had been brought out of the cave (Maška, 1886, 50–56) followed by Hoernes, Obermaier and even Bayer (Hoernes, 1903, 75 and 171; Obermaier, 1912, 308–309; Bayer, 1922, 178). As the priest buried only the "giant" that was more likely an ice age animal, we have no real evidence of human remains brought out of the cave before Szombathy's excavation although there might have been various possibilities.

Szombathy's excavation further on was concentrated in Hall D. Before he could finish his studies there, he was ordered to go back to Vienna. "Finds in the west of "D" are increasing constantly. Bones can be found within the first 50 cm. Beneath this layer there is a lighter plastic loam that was rather crumbly during excavation. The layer with bones was darker and earthier, containing pieces of limestone and dripstones. Raschendorfer again denies excavations within the cave but the other workers confirmed his earlier report.

July 18th – The box is packed with bones from Hall D, the few pieces from "b" and "d" can be recognized by their brighter color."[8] Szombathy excavated in the west and east of "D" until he was told to return to Vienna.

Janda's excavation in August 1882

Szombathy, who was told to return to Vienna in July 1882, seems to have initiated further research in those places he could not have a look at. From this excavation, we have a report from Janda who was in charge of this work:

---

p. 42
Es wird mit dem Durchgraben des westlichen Theiles von D fortgefahren [...]

7 Der Schwieger-Vater des Pekar: der 72 jährige Walloch war als Steinbrecher dabei, als sich der Eingang der Höhle zeigte, sowie ich ihn noch im vorigen Jahre fand. In der 1 Kl. Halle lag rechts das Skelet eines „Riesen", welcher die Arme von sich wegstreckte. Er hatte Zähne, größer als ein Daumen. Der Pfarrer von Mierotein kam, nahm die Knochen mit und ließ sie beerdigen.
Die Mutter des Raschendorfer, 70 Jahre alt, war als 16 jähriges Mädchen dabei, als die Höhle aufgemacht wurde. Der Schädel des "Riesen" war so groß, daß ihn ein Mann hätte ganz über den Kopf stülpen können.

8 Szombathy's diary

p. 43
Die Funde im W. Theile von D. mehren sich ansehnlich. Die Knochen kommen in den oberen 50 cm vor. Unter dieser Schichte folgt lichterer, plastischer, beim Graben etwas bröckeliger Lehm, der leer ist. Die Knochenschichte ist in verschiedenem Grade erdig und enthält Kalksteine und Tropfsteine.
Raschendorfer leugnet heute, daß die Littauer in der Höhle gegraben hätten. Die anderen Arbeiter opponieren ihm und bestättigen seine früheren Angaben.
18. Juli 1882
1. Kiste gepackt. Knochen aus D, die wenigen Fragmente von b und d, kenntlich an der lichteren Farbe.

"Having been instructed recently by the forester to give you a report about the results of the excavation in the "Prince John's cave" I had been in charge of, I hasten to fulfill this pleasant order, Sir.

As you will remember you instructed me to do two things: further research in the stone debris in Hall "D" near "a", and to continue the excavation in the southwest of this hall, where the last trench under your direction had been made, Sir.

I began with this work on August 7th with 7 men and ended on August 12th. Working simultaneously on both places I am free to inform you shortly about these excavations.

In Hall "D" near "a" we cleared the stone debris until we reached pure loam; only then we began to examine the heap of rubble. We tried to go slightly under the rubble as far as the dangerous wedged in rocks allowed it; yet in the whole heap of rubble we found nothing but 2 incisors of beaver and some unimportant bone fragments. Only on top of it, between the big wedged in rocks we found a fragment of antler. In contrast our excavation in the southwest of the hall behind the debris was very interesting and it was there we found all the animal bones. Just behind the trench I let the men dig the earth until – as you wished, Sir – they reached the small layer of plastic clay. From there we continued horizontally. Having continued half a meter we reached a big rock that had fallen from the ceiling and was buried under debris up to now just in the middle of this part of the hall where a pointed rock meets the earth. In front of this rock half a meter to the right we found a circular fireplace 1 1/4 m in diameter that was clearly recognizable and enclosed by stones on one side. These stones were burnt; the upper crust was partly carbonized. The whole extent of the hearth could be recognized by its black burnt earth.

Just behind this hearth – i.e., on the right side of the rock – we found a rather well preserved skull of a bear and other bones in considerable quantity. The second skull we found like most of the bones on the left side of the rock in the wider room between the rock faces.

All the finds from this side chamber lay between 15 and 50 cm, or until 1 m which was the part we usually examined. The highest layer – 0–15 cm – consisted of debris from the cave sinter; the lower layer – 50–75 cm – contained only few finds whereas the lowest layer up to 1 meter consisted of loam with only one small bone fragment. Further to the back in the narrower part of the hall we found only a few bones – the finds decreased until there was nothing at all. Therefore, I ceased excavating there.

On the field above the last houses of Lautsch I looked for finds because the owner had ploughed the field, but I found nothing. The second man, who guided us one day, has not yet ploughed his field, where we found the fragments of the skull. When he will have done so I will have a look; when I find something I will leave it with the forester.

If you should wish to get further information, Sir, please let me know [...]".[9]

9 Fundakten der Prähistorischen Abteilung des Naturhistorischen Museums: Lautscher Höhle
„Nachdem mir Herr Forstmeister dieser Tage mitteilte, ich soll über die Resultate der unter meiner Aufsicht gepflogenen weiteren Nachgrabungen in der „Fürst Johanns Höhle" an euer Wohlgeboren berichten, beeile ich mich diesen angenehmen Auftrage hiemit zu entsprechen.
Wie sich Herr Assistent erinnern werden, haben Sie mir 2 Aufgaben gestellt: Weitere Durchforschung des Steinschuttes in der Halle D bei a, und Fortsetzung der Nachgrabungen in der südwestlichen Abzweigung dieser Halle. Dort wo unter der Leitung des Herrn Assistenten der letzte Schacht ausgehoben wurde.
Ich habe mit 7 Männern diese Arbeit am 7. August begonnen und am 12. August beendet. Gleichzeitig an beiden Punkten arbeitend, erlaube ich mir also in Kürze über diese Nachgrabungen zu berichten.
Bei a der Halle D habe ich den Steinschutt so tief ausheben lassen, bis wir auf reinen Lehm kamen und dann erst gingen wir daran, den Schutthaufen zu durchforschen. Wir drangen – soweit es die Gefährlichkeit der oben eingekeilten Steintrümmer erlaubte – bis kurz unter diese; doch fanden wir in diesem ganzen Steinschutthaufen nichts als 2 Schneidezähne vom Biber und einige unwesentliche Bruchstücke von Thierknochen. Erst ganz oben, also schon zwischen den großen zusammengepressten Steintrümmern fanden wir ein Stück Stange von einem Geweih.
Dagegen waren die Nachgrabungen in der südwestlichen Halle – hinter dem aufgeworfenen Schutte – von vielem In-

Szombathy's visit in 1904

In 1904, Szombathy returned to the cave in August with Dr. Smyčka who showed him the latest results of his excavation. The site of 1904 was in a quarry 55 steps away from the first entrance to the west. There, workers had found three human skulls together with other animal and human bones. Smyčka sent a report to the Central Commission for Art and Historical Monuments with plans from the situation which Szombathy presented in the "Jahrbuch der k. k. Zentralkommission" (Szombathy, 1904, 9–16). The human remains Smyčka had excavated comprised two calvae, fragments of an upper jaw, a lower jaw and fragments of the postcranial skeleton. Three meters below the skeletons there were traces of fire. Szombathy wrote in his diary: "I take samples from the sooty stones, in order to identify organic remains. I advise against deepening the excavation in this place. Eventually the side passage should be cleaned because it could lead to a deeper part of the cave."[10]

---

teresse und haben wir auch dorten die ganzen gesamten Knochenfunde gemacht. Gleich hinter dem besagten Schachte ließ ich vorerst die Erdschichte ebenfalls so tief abgraben, bis wir, wie es Euer Wohlgeboren gewunschen – auf die bedeutete schwache Schichte plastischen Thones kamen. Dann gingen wir in wagrechter Richtung nach vorwärts.
Nachdem wir beiläufig einen 1/2 Met. vorgerückt waren, stießen wir – genau in der Mitte dieses Hallentheiles u. zw. dort, wo der Fels in eine Spitze auslaufend bis zur Erde herabgeht – auf einen jedenfalls von der Felsdecke abgefallenen und jetzt verschüttet gewesenen großen Block. Vor diesem Felsblock, etwa 1/2 Met. tief nach rechts fanden wir (3/4 Met. tief) eine ganz gut kennbare auf einer Seite mit Steinen eingefasste circa 1 1/4 Met. im Durchmesser betragende kreisrunde Feuerstelle. Die einfassenden Steine waren ganz angebrannt und teilweise die obere Kruste verkohlt. Die Weite der Feuerstelle war an der geschwärzten, ausgebrannten Erde erkennbar.
Gleich hinter diesem Feuerplatz – daher auf der rechten Seite des Felsblockes – fanden wir den ziemlich gut erhaltenen Bärenschädel und andere Knochen in ziemlicher Menge. Den zweiten Schädel sowie die meisten Knochen fanden wir jedoch links von dem Felsblock in dem zwischen den Felswänden liegenden breiteren Raum.
Alle in dieser Seitenhalle gemachten Funde liegen in einer Tiefe von 15 bis 50 Centim. beziehungsweise bis 1 Met. – der Normaltiefe, die wir durchforschten. Die oberste Schicht bis 15 cm bestand aus Trümmern der Sinterdecke; die untere Schichte von 50–75 Centim. war sehr schwach an Fundobjekten, während die letzte Tiefschichte bis 100 Centim. nur mehr festen Lehm zeigte, und sie uns [...] ein kleines Knochenfragment ergab.
Weiter nach hinten in die engeren Theile der Halle vorgerückt, fanden wir nur mehr sehr wenig Knochen vor, und verringerte sich der Fund immer mehr, bis er in den schwachen Ausläufern ganz aufhörte. – Und so hörte ich auch auf zu graben.
Auf dem Felde ober den letzten Häusern von Lautsch habe ich, nachdem der Eigenthümer das Feld geackert hatte, nachgeschaut, habe jedoch nichts gefunden. Der zweite Mann, – derselbe der uns damals geführt – hat bis jetzt sein Feld, auf dem Schädelknochen fanden, – noch nicht geackert. Bis er das getan haben wird, werde ich dorten Nachschau halten und sollte ich etwas finden, so werde ich dies beim Herrn Forstmeister deponieren.
Sollten Euer Wohlgeboren irgend eine weitere Auskunft wünschen, so bitte ich Ihren bezüglichen Wunsch mitzutheilen.
Indem ich mich schließlich bestens empfehle, zeichne ich in vorzüglichster Hochachtung
Euer Wohlgeboren
Zu jedem Dienste bereiter
(Robert?) Janda
Lautsch den 24. September 1882.

10 Szombathy's diary

p. 49
Die Stelle mit den Feuerspuren liegt ca. 3 m tiefer als die Fundstellen der Skelette (in Lautsch)
die [...] Tiefengrabung wurde unterstützt durch einen ca. 1 1/2 m br. natürlichen Schacht, der W. neben der Fundstelle der Schädel in die Tiefe ging.

p. 49a
Ich nehme von den angerussten Steinen Proben mit, um sie auf die Anwesenheit organischer Substanz prüfen zu lassen. Weitere Tiefgrabung abzurathen, eventuell den Seitengang ausräumen, ob dieser nicht auf eine tiefgelegene Höhle führt.

Szombathy's visit in 1925

In 1925, Szombathy revisited the cave. His report reveals that enormous masses of sediment had been brought out of the cave between 1904 and 1925 in order to prepare comfortable paths for the general public. This work had not been supervised by archaeologists so an unknown amount of information about the use of the cave by prehistoric man was destroyed. Szombathy gives a detailed account of the state of the cave in 1925: "The Littau museums society has bought the cave together with the adjacent land. They have built a nice two-story house for tourists in front and above the cave entrance with a big garden, swings, a bowling lane, a pianino, veranda, etc. and electric lighting with 160 lamps in the cave. The visitor enters the cave where huge amounts of earth and stone had been removed through this house. The passage from D to S where I expected the old entrance has been cleared and is used as an exit when there are lots of visitors. It had just been blocked by earth. The new entrance was once closed by the rock, which had been removed by quarrying in 1828. After having sunk the floor of the cave by 2 m in Hall D, "d" and "E" have been united to a big hall with many columns. By clearing passages from "E" to the south and southeast new rooms had been opened: Panenská jeskyně (Virgins cave) and Netopiře jeskyně (Bats cave), because many bats were found on the ceiling. The bats left the cave after electric light had been installed. In this part of the cave bones of young cave bears could be found but no human bones. From Panenská jeskyně a narrow crevice crossing beneath the old cave to the northwest and corresponding to passage "g" leads to a water-cave on the level of the river Morawa. In several sections of this cave there are accumulations of water that increase or decrease with the water level of the river. Hall K is enlarged to a Přérodné Chrám, a natural dome; the floor has been sunk by 3 to 4 m and consequently widened.

The big stalagmite = Egyptian mummy

Hall N is widened to the Mordrá jeskyně = Blue grotto also by sinking the floor. From the dome and the blue grotto openings to the north are initiated but not yet cleared. East of the dome new halls had been explored and partly cleared but they cannot yet be visited. They looked for an opening to the north by clearing already existing but totally blocked passages and blasting the rock to create new tunnels. They preferred to bring out the earth and the stones from the backward rooms because they did not want to bring it through the clean passages and the tourist house. The exit is laid out with bricks and is situated near the forester's lodge.

In a place possibly belonging to "e", northeast of a rock pillar the site of a hearth is shown: a charcoal layer surrounded by stones. In its vicinity animal and human bones were found more than 1 m beneath the old cave floor. Above the hearth there was a 70 cm thick deposit and above it a stalactite. From this place to the northwest there were heaps of cattle bones – two skulls, spinal column and complete limbs. The human skeletons were lying near them to the east. As to the position of the small gray fragmented human bones the custodian of the cave Mr. Nowotny could give no details."[11]

---

11 Szombathy's diary

p. 51a

7.5.1925 Besuch der Lautscher Höhle (Bočkova dira, Fürst-Johanns-Höhle)

Der Littauer Museumsverein hat die Höhle mit den zugehörigen Grundstücken angekauft, über und vor dem Eingange ein einstöckiges, hübsches Touristenhaus gebaut mit großem Gastgarten, Schaukeln, Kegelbahn, Pianino, Veranda, etc. elektrische Höhlenbeleuchtung mit mehr als 160 Lichtern.

Man geht durch das Haus in die Höhle, aus der riesige Mengen von Erde und Stein ausgeräumt wurden. Der aus der Abteilung D nach S führende Gang, in dem ich den alten Eingang vermutete, ist nun ausgeräumt und als Ausgang bei stärkerem Besuch in Verwendung. Er war nur durch Erde verlegt. Der neuere Eingang war einstmals durch die Felswand verschlossen, die 1828 durch den Steinbruch abgetragen wurde.

Die Räume D, d und E sind durch Tieferlegung des Bodens um etwa 2 m und Ausräumung von Felsgestein zu einer großen, durch viele Felssäulen unterbrochenen Halle vereinigt. Durch die Ausräumung der von E nach S gegangenen Äste eröffnete man S.O. von E neue Räume: Panenská jeskyně (Jungfrauenhöhle) und Netopiře jeskyně (Fleder-

The first presentations of the human remains and the archaeological finds from Mladeč

The Twelfth International Congress of Anthropology and Archaeology in Paris, 1900

The first presentation of the human remains from Mladeč took place at the Twelfth International Congress of Anthropology and Archaeology in Paris, 1900 (Szombathy, 1900, 133–140). According to the comments in this paper, he had brought the original skull to Paris to show it to the scientific audience. In his report, he mentions loamy cave deposits being covered by a horizon of stalagmites. Within the first loamy layer between 0–30 cm below the floor, a human cranium was found. Like the animal bones from this layer in a niche it was covered with sinter. According to Szombathy's report at the congress simple blades, perforated teeth of beaver and reindeer, and a bone dagger that is now called point type Mladeč, were found within a distance of about one meter from the cranium. They were also covered with sinter; the bones were gray like the other animal and human bones. Szombathy therefore concluded a simultaneity of the finds, which was not contradicted by other members of the congress. As further evidence he mentioned a fragment of a rib from reindeer that was sintered to the right orbit, and a stalagmite that was sintered to the base of the skull (Szombathy, 1900, 134–135). For our judgement of the whole situation it is not only important what he told the scientific audience, it is equally important what he did not tell them. Szombathy did not mention the first opening of the cave in 1826 and the possible disturbance of the original layers. He did not speak about the excavation of some citizens from Littau in the cave between 1881 and 1882. On the other hand, he presented details about the skull and the artifacts he did not mention either in his diary or in his very first report at the Imperial Academy of Sciences: the rib of a reindeer sintered to the skull and the stalagmite sintered to the base. It might also be interesting that this description at the congress in 1900 is the first one concerning the position of the artifacts. In his diary, Szombathy does not speak about artifacts at all[12]. In Hochstetter's report about Szombathy's excavation in Mladeč at

maushöhle), da an ihrer Decke viele Fledermäuse Aufenthalt hatten, die erst durch die elektrische Beleuchtung vertrieben wurden. In der letzteren wurden Knochen junger Höhlenbären, aber nichts vom Menschen gefunden. Von der Panenská jeskyně geht in einem tieferen Niveau, unter der alten Höhle durch nach NW, übereinstimmend mit dem Gange g, eine teilweise sehr enge Spaltenverbindung zu einer im Niveau der March liegenden Wasserhöhle, die in mehreren Abteilungen Wasseransammlungen hat, welche mit dem Wasserstande der March steigen oder fallen.
K ist zu einem Přírodní Chrám = Naturdom ausgeräumt, der Boden 3–4 m tiefer gelegt und damit auch verbreitert.
Der große Stalakmit=Mumie egyptská
Der Raum N ist durch die Tieferlegung ebenfalls erweitert zur Mordrá jeskyne=Blaue Grotte.
Vom Dom und von der blauen Grotte sind gegen N hin Durchbrüche angebahnt, die noch nicht ausgeräumt wurden. Auch O. vom Dom wurden neuerlich größere Räume erkundet und teilweise ausgeräumt, sind aber noch nicht zu begehen. Um das Erd- und Gesteinmaterial aus den rückwärtigen Räumen nicht durch die gepflegten Gänge und das Touristenhaus befördern zu müssen wurde ein Durchbruch nach N gesucht und im Zickzack durch die Ausräumung vorhandener aber gänzlich verlegter Gänge sowie durch Aussprengung von Stollen aus dem festen Felsen ausgeführt. Der Ausgang ist ausgemauert und kommt bei Försterhause heraus.
In dem etwa zu e gehörigen Teile wird NO an einem Felspfeiler die Stelle gezeigt, an der ein „Herd" gefunden wurde, eine mit einigen Steinen umstellte Holzkohlenschichte, neben der Tierknochen und Menschenknochen gefunden wurden, mehr als 1 m unter dem alten Höhlenniveau. An der Stelle hatte sich eine etwa 70 cm dicke Schicht und darauf ein Stalaktit angesetzt. NW davon wurden Haufen von Rinderknochen, 2 Schädel, Wirbelsäule, vollständig erhaltene Gliedmaßen, gefunden. Die menschlichen Skelete lagen weiter O. daneben.
Über die Lage der grauen, zerkleinerten Menschenknochen weiß H. Nowotny, der Höhlenverwalter keine Angaben zu machen.

12 Szombathy's diary

p. 43
Finds in the west of D are increasing constantly. Bones can be found within the first 50 cm

the Imperial Academy of Sciences in Vienna, we can find only one sentence: "In the Prince John's cave the simultaneity of man and reindeer, which was already plausible after the excavation of 1881, could be confirmed with finds of characteristic human artifacts" (Hochstetter, 1883, 169). In the year 1900 in Paris, he mentions there are "teeth of beaver and reindeer which are without any doubt part of a necklace" and "in the neighboring niche, in about 1 meter's distance of the last mentioned finds there was the human cranium."[13] It is obvious that he tried to emphasize the unity of all finds in an undisturbed position which was the only way to prove the age of the human remains in those days. The unity of all finds was of vital interest for Szombathy especially after the critical note of Maška who said he could not exclude that parts of the layers had been destroyed by visitors between 1826 and 1881, although he would not deny the presence of Paleolithic man in the cave (Maška, 1886). Hoernes (1903) and Obermaier (1912) later repeated K. J. Maška's arguments in their judgment of the Mladeč finds. Nevertheless after his lecture in Paris, Szombathy was congratulated for his finds because they were also seen as evidence for the age of the finds from Cro-Magnon. This shows the importance the assemblage had and still has for our knowledge of early *Homo sapiens* in Europe. O. Menghin mentions the skull from Mladeč in his book about the prehistory of Bohemia and Moravia in connection with the Cro-Magnon man (Menghin, 1926, 25): "Among the skeletons from Moravia the skull from the Prince John's cave in Lautsch is assigned to this type, which is often – I think unjustly – doubted."

Bayer 1922

In 1922, Bayer tried to meet the critics by stylistic arguments and attributed the assemblage to the younger Aurignacian period parallel to layers 5–9 of Willendorf II, which are now the Gravettian layers. As original sources for his work, he used Szombathy's diary and the report to the Academy. Even in those days – 40 years after the excavation – no further documents seem to have existed. With a view to the details of the position of the finds we see an increase of information from Szombathy's diary to the report and later on to 1900 and 1925, which perhaps was not based on written documents but on the memory of the excavator Josef Szombathy. Therefore, it is difficult to examine the authenticity of information by going back to the roots. A similar development could be observed concerning the finding of the Venus of Willendorf although we have much more documents in that case (Antl, 2001).

In 1922, Bayer mentions 19 perforated teeth from beaver, reindeer, cave bear, wolf and wild horse "being without any doubt part of a necklace which had been given to one of the dead" (Bayer, 1922, 178). The expression "being without any doubt part of a necklace" is an exact translation from the sentence Szombathy used in Paris in 1900; "which had been given to one of the dead" is Bayer's conclusion. He regarded the finds from Mladeč as disturbed graves. He considered the "necklace" as a grave good for the young individual. Concerning the position of the teeth, his description differs from that given by Szombathy. According to Bayer, a well preserved skull and a femur was found near site "a"; the other human bones among them the fragments of a child's skull were scattered in the area "b" where also the perforated teeth and lots of animal bones had been found (Bayer, 1922, 181)[14]. Following Szombathy (1900), the artifacts were found between "a" and "b" only one meter from "a".

The decisive difference between Szombathy and Bayer was that Bayer was convinced the finds had been submitted to postdepositional disturbance. He did not necessarily mean modern distur-

13 J. Szombathy, 1900, 134 : [...] dents de Castor et de Renne percées, provenant sans doute d'un collier, [...]; 135: Dans la niche avoisinante, à environ 1 metre de distance des trouvailles mentionées en dernier lieu, gisait le crâne humain.

14 see also Szombathy, 1900, 134–135

bance but he could not exclude it because Szombathy found a piece of a rope in area "b". For Obermaier, this was another reason to doubt the unity of the finds and consequently, the age of the human remains (Obermaier, 1912, 308–309). Bayer argued against it with the state of conservation and the comparison with the Cro-Magnon skull that had been accepted by the congress in 1900.

Bayer also noticed the illegal excavation of some citizens from Littau in 1882 and the unknown extent of loss to the archaeological record. In contrast to Szombathy, Bayer listed all arguments that could lead to doubt of the diluvial age of the remains and discussed them on the basis of all the documents and the finds. Unlike Szombathy, Bayer did not believe that the position in the highest layer - even if it was undisturbed - would qualify for dating the assemblage.

Szombathy 1925

Three years later, Szombathy gives a detailed description of the finds in the Mladeč Cave. Even there we cannot find any reference to Jandas's excavation in 1882. Szombathy describes the finding of skull I and for the first time, informs about bones of reindeer and cattle in the vicinity of the skull. Concerning the artifacts, Szombathy opposed the interpretation of Bayer, bringing a new description of their distribution within the layer. "Twenty-two perforated pendants made of the teeth of ruminants such as bear, wolf, horse, reindeer and beaver. All of them had been perforated near the end of the root and certainly belonged to one or more necklaces; they lay scattered over the whole center of the Hall (D)" (Szombathy, 1925, 12–13). In another place of the same publication: "In the center of the chamber we found bones of reindeer and cattle and fragments of a human spinal column, ribs, skulls, pelvis and limbs and an isolated upper jaw, finally bone artifacts and hornstone unevenly spread over an area of about 20 square meters" (Szombathy, 1925, 6). The widespread human remains and artifacts between animal bones, Szombathy interprets as remains of cannibalistic rites.

As a sort of answer to Bayer's suspected damage to the layers by the illegal excavation in 1882, Szombathy now denies - against his own entry in the diary - a bigger impact to the results of his research in the cave caused by these excavations: "The loss of material caused by this operation cannot be very big and certainly can have no influence to the main results of our research" (Szombathy, 1925, 6).

Conclusion

When we try to judge whether the assemblage from Mladeč had been deposited like Szombathy found it or not, it is necessary to consider all facts that could have disturbed the original situation. The documents show that the cracks of the bone fragments - Szombathy only mentions fragments of a human skull - were partly covered with sinter. If we do not conclude with Szombathy that the fragmentation of the human bones was due to cannibalism, we must think of post-depositional changes that took place as long as the original entrance to the cave was open, perhaps already during the Paleolithic. Another point is the rather shallow position of the majority of the finds. We should be aware of possibly rather long periods without sedimentation. In caves like La Garma in Spain, no sedimentation took place after Paleolithic man had left the cave. If this was also the case in Mladeč, bones and artifacts from different periods could have been deposited in the same layer. The position of the finds could have also been changed by animals coming to the cave. Szombathy describes big rocks hanging from the ceiling. There is much debris in the cave that goes back to former collapses of the ceiling. Together with the rocks, material from the surface of the Plavatisko might have come into the cave. According to Szombathy's descriptions of the cave in 1881 and 1925, it seems that at least the finds in area "a" and the centre of "D" and probably also the ones in area "b", were deposited after the big collapse of the ceiling in "E", because Szombathy mentions that there was no direct connection between "a" and "E" when he excavated in the cave. When the Littau museums society cleared the cave to prepare it for visitors, the floor of the cave was sunk by at least 2 m. Today it is possible to go directly from "E" to "a" which probably is a consequence of the clearing of the cave between 1904 and 1925.

Apart from these considerations, there are many possibilities for modern disturbances. The cave was detected in 1826; at least a big number of animal bones were brought out of the cave in those days. We do not know how much this has affected the prehistoric layer where Szombathy found the human remains and the artifacts. From Szombathy's description of the cave in 1881, we learn that people even brought out stalagmites from deep inside the cave. Did they only take stalagmites? Would Szombathy have realized 50 year old slight disturbances in plastic loam given the lack of electric light in the cave when he carried out his excavations? In 1882, Szombathy realized that the layers had been disturbed in some parts of the cave between 1881 and 1882. He got furious - witness the details in his diary - about these clandestine excavations. He mentioned that it would not make any sense to continue research in those devastated parts. In front of a scientific audience in 1900, he did not even mention that fact. In 1925, he spoke of minor losses. Due to these rather controversial statements, the original documents are not really helpful. The entries in the diary are rather concise. Facts we would need to elucidate the situation are not mentioned there. On the other hand, Szombathy presents in his publications many details that are not included in the original records. The rib of a reindeer and the piece of stalagmite sintered to the skull is first mentioned in 1900 when the unity of the material had to be defended. Why was it not worth mentioning when he documented his first impressions? The position of the archaeological finds scattered over 20 square meters is first mentioned in 1925 when he opposed Bayer who interpreted the perforated teeth as a necklace given to one of the burials. The most detailed description of the Mladeč finds was published more than 40 years after the excavations, but there are no written documents except his diary. Even when Bayer wrote his article about the chronological position of the assemblage, he could only refer to the diary, the report to the academy in 1882 and the lecture in Paris in 1900. So if there were no other documents, did Szombathy really remember every detail or did he only think he would remember? We can at least observe a slightly selective argument adjusted to the particular purpose that makes us suspicious about the authenticity of his later and very detailed descriptions.

Another point is the lack of documentation as far as the position of the finds is concerned. Szombathy described different sites and drew various very accurate plans of the cave but he did not leave any note where the bones and artifacts were exactly found. So today - except the skulls - we cannot identify the position of the bones and artifacts.

Taking all these arguments into account, we have to conclude that we cannot prove an integrated whole within the Mladeč material by archaeological means. Therefore, as a start for this publication every piece has to be regarded as a single find that must be individually dated. The dating of all elements will then prove whether the Mladeč assemblage is a real unity or whether only parts of it belong together.

References

Antl, W. (2001) Die Auffindung der Venus von Willendorf - eine unendliche Geschichte. *Mitteilungen der Anthropologischen Gesellschaft Wien* 130/131, 39–58

Bayer, J. (1922) Das Aurignac-Alter der Artefakte und menschlichen Skelettreste aus der „Fürst Johanns-Höhle" bei Lautsch in Mähren. *Mitteilungen der Anthropologischen Gesellschaft Wien* 52, 173–185

Heinrich, A. (2003) Josef Szombathy (1853–1943). *Mitteilungen der Anthropologischen Gesellschaft Wien* 133, 1–45

Hochstetter, F. v. (1883) Sechster Bericht der Prähistorischen Commission der mathematisch-naturwissenschaftlichen Classe der kaiserlichen Akademie der Wissenschaften über die Arbeiten im Jahre 1882. *Sitzungsberichte der kaiserlichen Akademie der Wissenschaften math.-nat. Classe* 87, 168–174

Hoernes, M. (1903) *Der diluviale Mensch in Europa.* Braunschweig: Vieweg, p. 75 und 171

Maška, K. J. (1886) *Der diluviale Mensch in Mähren. Ein Beitrag zur Urgeschichte Mährens.* Programm der mährischen Landes-Oberrealschule in Neutitschein für das Schuljahr 1885/1886. Neutitschein: Selbstverlag des Verfassers, pp. 50–56

Menghin, O. (1926) *Einführung in die Urgeschichte Böhmens und Mährens.* Anstalt für Sudetendeutsche Heimatforschung 1, Reichenberg

Obermaier, H. (1912) *Der Mensch der Vorzeit.* Berlin München Wien: Allgemeine Verlagsgesellschaft, pp. 308–309

Szombathy, J. (1882) Über Ausgrabungen in den mährischen Höhlen im Jahr 1881. In (F. v. Hochstetter) Fünfter Bericht der prähistorischen Commission der mathematisch-naturwissenschaftlichen Classe der kaiserlichen Akademie der Wissenschaften über die Arbeiten im Jahre 1881. *Sitzungsberichte der kaiserlichen Akademie der Wissenschaften math.-nat. Classe* 85, [90]–[107]

Szombathy, J. (1900) Une crâne de la race de Cro-Magnon trouvé en Moravie. *Congrès International d'anthropologie et d'archéologie prehistoriques. Compte rendu de la douzième session.* Paris: Masson et Cie, pp. 133–140

Szombathy, J. (1904) Neue diluviale Funde von Lautsch in Mähren. *Jahrbuch der k. k. Zentralkommission für Kunst und historische Denkmäler* 2, 9–16

Szombathy, J. (1925) Diluviale Menschenreste aus der Fürst-Johanns-Höhle bei Lautsch. *Die Eiszeit* 2, 1–34, 73–95

Wolny, G. (1839) Die Markgrafschaft Mähren, topographisch, statistisch und historisch geschildert (Band 5, Olmützer Kreis). Brünn, p. 166

Chapter 2

IN SEARCH OF PROTOTYPES – HISTORICAL SOFT-TISSUE RECONSTRUCTIONS OF MLADEČ 1

Maria Teschler-Nicola

Introduction

The collections of the Anthropological Department still contain a plaster soft-tissue reconstruction which had been produced on the basis of Mladeč 1. The object was made by the academic sculptor Friedrich Fahrwickel, presumably in the late 1930s. A second sculpture, made as early as the beginning of the thirties by the academic sculptor Egon Grenzer, under scientific instructions by the curator Viktor Lebzelter, is apparently missing. However, this sculpture was used as an early element of anthropological showcasing at the Naturhistorisches Museum Wien and forms a major, historically relevant item in the context of correspondence preserved to this day. The object underscores the motivation of the artists and curators involved as well as the collecting strategies and exhibition policies practiced in the museum during the interwar period.

"We look into the faces to differentiate human races and their variants"[1]

Plastic reproductions in the form of masks and casts of body parts belonged to the methodological "repertoire" of early anthropology. In his first text book, Rudolf Martin (1864–1925) made a point of allowing "relatively large room" for anthropological methods (Martin, 1914, V). In his opinion, "the fate" of anthropology depended on technological developments. Accordingly, his work not only included instructions for standardized measurement of humans, but also very comprehensive descriptions of figurative and plastic reproductions as well as soft-tissue reconstructions of human heads and faces based on bone structures.

Although viewed by him as feasible "with some degree of approximation to reality", he considered the reliability of this method as uncertain, in particular regarding the representation of physiognomic details of the face. On the other hand, many anatomists and artists were convinced of the efficiency of the plastic reconstruction method in anthropology and forensic medicine.

Initially, this technique facilitated a physiognomic comparison of historical personages with cranial remains as well as their subsequent identification; later the reverse route was taken by creating portraits based on existing skulls. This gave rise to a more vivid perception of the appearance of early hominids such as the Neandertals or (pre)historic humans.

Welcker (1883; 1884) and His (1895) were the first anatomists to examine the correlations between the skeleton and soft parts of the face in greater detail and to provide evidence on the presence of relevant laws of regularity. Welcker's method of "craniological diagnosis" was based on a graphic reconstruction of the facial contours. This contour drawing was then fitted into the profile outline of the death masks, which permitted examination of the authenticity of cranial remains (e.g.,

1 Kollmann and Büchly (1898).

the skull of Friedrich Schiller). Toward the end of the 19th century, based on Welcker's data, His went even a step further and commissioned a sculptor to produce the first three-dimensional plastic reconstruction of facial features (in this case, a cranial cast of Johann Sebastian Bach; Stadtmüller, 1922).

Kollmann and Büchly (1898) were the first to apply the reconstruction method to scientific purposes. They believed that this technique could provide an answer to various questions on the origin of Europeans. The sculptures would provide a clue on "the race-specific appearance of the face" (Stadtmüller, 1922, 338). This idea was based on the fundamental assumption of race persisting by inheritance (Kollmann and Büchly, 1898, 331). Therefore – using a "racial skull" – the idea was to materialize race-specific rather than individual portraits (Kollmann and Büchly, 1898, 329). Such a skull should correspond to "one of the basic European forms in every detail" and "reflect the pure form, without any other admixture" (Kollmann and Büchly, 1898, 333). Kollmann and Büchly's whole "racial" reconstruction argument is circular. They apparently proofed what they already believed in terms of racial purity and facial form. This concept was used until the fifties of the 20th century (see, v. Eickstedt, 1925) and was also implicit in attempts to reconstruct the Upper Paleolithic Mladeč 1 skull. Using a modified technique, the Russian anthropologist Gerasimov (1968) was the first to focus on individual physiognomies (Ullrich, 1958).

"Starving out a department affects the reputation of the whole museum"[2]

The Anthropological Collection - part of the institutional structure of "Anthropologisch-ethnographische Abteilung" (covering the disciplines of Ethnography, Prehistory and Anthropology) - established by Ferdinand v. Hochstetter in 1876, developed very slowly. Compared to the two other, extraordinarily prosperous disciplines, the collection was at a disadvantage in the late twenties (after WW I in particular) as far as human resources and space were concerned. The collection was marked by chronic underendowment and, according to Josef Bayer, "neglect in an almost irreparable manner"[3].

Between 1919 and 1928, no objects were purchased for the Anthropological Collection. Therefore, the curators of the department, in particular Viktor Lebzelter, developed strategies to halt the stagnation taking place in the upgrading of the collection, particularly in view of the plans for an exhibition hall dedicated to "human biology"[4]. In 1930, this "first European racial show"[5] was made accessible to the general public and with it, anthropological knowledge was popularized "successfully"[6], possibly giving rise to a new tradition[7]. Essential elements of this popu-

2 Josef Bayer to Hans Rebel (director of the NHM), March 10, 1932. NHM Vienna, Department of Anthropology, correspondence folder 1929–1932.

3 Josef Bayer to Hans Rebel (director of the NHM), March 10, 1932. NHM Vienna, Department of Anthropology, file "History/formation of the collections/exhibitions".

4 Press information, apparently attached to a letter of Josef Bayer, March 8, 1930. NHM Vienna, Department of Anthropology, file "History/formation of the collections/exhibitions".

5 Viktor Lebzelter, contribution to the museum guide (n. d.). NHM Vienna, Department of Anthropology, file "History/formation of the collections/exhibitions".

6 Shortly after the opening ceremony Viktor Lebzelter reported an acquisition (a skull from South America), which resulted from the exhibition; Josef Bayer to first director of the museum, April 22, 1930. NHM Vienna, Department of Anthropology, correspondence folder 1929–1932.

7 A similar presentation, namely the combination of skulls and photographs of living people was made in the permanent exhibition, opened in 1978.

larization were soft-tissue reconstructions and portrait busts which had been produced by external artists since 1923 under instructions by different experts,[8] or acquired by purchase or exchange[9].

Along with other public shows of newly acquired knowledge, the creators used the potential of soft tissue reconstructions and models to disseminate knowledge. Not only because of this programmatic focus on "soft tissue typology", but also with the collection in mind and for budgetary reasons, plastic reconstruction played a key role for Viktor Lebzelter (1886–1936)[10], who had been in charge of the Anthropological Department since 1932 (see Weninger, 1936): Indeed, in his view, these reconstructions were of "limited benefit to the advancement of science"; but they were helpful to "exhibit past forms of race to interested laymen" (Lebzelter, 1933). The plastic reproductions could also be delivered in exchange for original skulls and casts of important paleo-anthropological finds or offered for sale. In the end, trading in these objects not only contributed to stocking up the inventories of the collection but also to improve departmental budgets and enhance international reputation: With such "highly professional reconstruction method" – which obviously became an obsession to Lebzelter – "the Museum could prepare unique reconstructions of all prehistoric racial types, thereby making our museum world-famous in this field"[11].

Under his supervision, soft-tissue reconstructions were now manufactured in a proper Department "studio". For 1934, approx. 30 reconstruction projects were planned to be produced by young sculptors or pupils. These works included the first soft-tissue reconstruction of the "ice age person of Lautsch" as well as numerous reconstructions based on human crania from prehistoric as well as historic time periods (among others a man of the Neolithic, a man of the Bronze-age, a man from the Bell-Beaker period and a Tasmanian; Lebzelter 1932; 1934; 1935a; 1935b). At smaller exhibitions, many of these works were presented as "new acquisitions" immediately after their completion. Viktor Lebzelter's attempts to distinguish himself in the general public area of the museum must presumably also be seen in the context of political developments: At the end of December 1933, the Federal Minister of Education called upon subordinate authorities to support the work of the newly installed commissioner of propaganda. In consultation with the *Vaterländische Front* and by using certain print media, the aim was to specifically strengthen the awareness of the Austrian people for their "homeland and history"[12]. By disseminating i.a., their "racial" findings, Anthropology followed

8 Some of the reconstructions were manufactured by voluntary external artists, e.g., Erna von Engel-Baiersdorf. She ran a studio for scientific and artistic sculpture in Pécs, Hungary, and was a correspondent of the Naturhistorisches Museum in Vienna and other institutions; Engel-Baiersdorf was an enthusiastic sculptor and worked together with experts such as Egon Frhr. v. Eickstedt, Paul Schebesta, Martin Gusinde as well as Viktor Lebzelter (see Engel-Baiersdorf, 1949). Other reconstructions stored at the NHM were made by Hugo Heese and Franz Klinghofer. See NHM Vienna, Department of Anthropology, inventory record "Abgußsammlung".

9 For example, from the Museum of Ethnology in Hamburg, from the "Gipsgießerei Berlin" (a complete body moulding of a San, made by Felix v. Luschan) and from the Viennese Institute for Anthropology; most of them were obviously acquired for the realisation of the first permanent exhibition, which opened in 1930.

10 Viktor Lebzelter studied Anthropology, Zoology, Prehistory and Ethnology; he became collaborator of the Department of Anthropology in 1926, tentative head in 1932 and director in 1934; he carried out an expedition to South Africa between 1926–28, investigated living populations in Austria and Romania, etc. and died in 1936 (a bibliography is in preparation).

11 Viktor Lebzelter to first director of the NHM, January 22, 1934. NHM Vienna, Department of Anthropology, correspondence folder 1933–1934.

12 Federal Minister of Education to provincial governments and other subordinated authorities, December 30, 1933. NHM Vienna, Department of Anthropology, correspondence folder 1933–1934.

such call[13]. "Scientific reconstruction[s] of facial types of extinct and prehistoric races" (Lebzelter, 1936a) played a significant role in this regard[14].

In consideration of Lebzelter's ambitious plans to expand the collection, it is not surprising that in 1935, he set up a "proper laboratory" (Lebzelter, 1936b) for these works at the Naturhistorisches Museum, concentrated the reconstruction work within the Anthropological Department and put it under the direction of the new employee Rosa Koller[15]. Like Viktor Lebzelter, Rosa Koller was a member of the *"Vaterländische Front"* (since 1933) and manager of the "Arbeitsgemeinschaft der vaterländischen und katholischen Gelehrten am NHM"[16] ("Working group of vaterländische and catholic scholars of the NHM"). From then, reconstructions were predominantly made by Rosa Koller[17] and the sculptor Friedrich Fahrwickel[18] (e.g., the second reconstruction from Mladeč 1). Both were said to be "100% Aryan"[19].

The soft-tissue reconstructions of Mladeč 1

Numerous attempts were made to illustrate the physiognomy of prehistoric humans by using different technologies like drawings, pictures or busts (Trinkaus and Shipman, 1993). Most of them were Neandertals, but there was also an interest in reconstructing individuals from more recent time period. It was probably Schaaffhausen who had not only initiated already in 1877 a 3D-reconstruction based on a medieval skull, but also the first sculptor on a Neandertal skull (Schmidt, 1898; see Winkler, 1988). Further attempts to reconstruct Neandertals have been made by Boule (1921), v. Eickstedt (1924), the Field Museum (N. N., 1929), Mollison (1931) and others (among them the Neandertal reconstruction attempts of McGregor, Heberer, Wandel, Coon, Schultz, see Kurth, 1956). Most of them were done by scientific artists under the direction of anatomists and paleontologists. The majority were for museums exhibits (e.g., the Chicago Fields Museum), but many were also produced for magazines and books (see Trinkaus and Shipman, 1993). In contrast to the "Neandertal euphoria", the general interest in soft-tissue reconstruction based on skeletal remains of early modern humans was obviously less keen.

---

13 Viktor Lebzelter to first director of the NHM, April 24, 1934. NHM Vienna, Department of Anthropology, correspondence folder 1933–1934.

14 Viktor Lebzelter, Jannuary 1934, „Eine österreichische Kulturfrage". NHM Vienna, Department of Anthropology, correspondence folder 1933–1934, document 17, p. 4.

15 Viktor Lebzelter to Otto Schlaginhaufen, February 8, 1935. NHM Vienna, Department of Anthropology, correspondence folder 1935–1936.

16 Wolfgang Adensamer to district authority of *Vaterländische Front*, March 25, 1935. NHM Vienna, Department of Anthropology, correspondence folder 1935–36.

17 Rosa Koller was a collaborator at the Department of Anthropology since 1933; her activities included both scientific tasks and artwork. See Josef Wastl, September 20, 1938. NHM Vienna, Department of Anthropology, correspondence folder 1937–1939.

18 Friedrich Fahrwickel acted as a voluntary sculptor since 1935; according to Josef Wastl he was very talented and worked alone or under supervision of a specialist; he also fabricated a bust of Adolf Hitler, which was purchased by Reichsminister Rust. See Josef Wastl, September 20, 1938. NHM Vienna, Department of Anthropology, correspondence folder 1937–1939.

19 Viktor Lebzelter to *Staatliches Museum für Vor- und Frühgeschichte Berlin,* November 21, 1935. NHM Vienna, Department of Anthropology, correspondence folder 1935–36.

At the Naturhistorisches Museum, the first reconstruction using an Upper Paleolithic cranial specimen from the Mladeč Caves (Mladeč 1), was carried out by sculptor Egon Grenzer in 1931[20] (see Figs. 1 and 2; Lebzelter, 1933); by request of Lebzelter, the artist was also supposed to provide a sculpture based on the Předmostí finding, dated to the early Upper Paleolithic as well and still unpublished at that time (the skull used was Předmostí 3, attributed to a male individual, see Matiegka, 1934; Schwartz and Tattersall, 2002, 302). In Lebzelter's view, both findings represented "two quite different racial types" (Lebzelter, 1933, [6]); by showing "successful" reconstructions of these specimens, he believed that his view could be strengthened and made plausible to the untrained viewer. Lebzelter considered the reconstruction of the "Lautsch man" as faultless, assigned the product "without doubt" to the Cro-Magnon group and identified a physiognomy close to the so-called "dalisch" type. In his view, the reconstruction of the Předmostí man was a more difficult task – "more dangerous", because the fossils featured an unusual combination of "australiform" and "europid" morphological characters. In his opinion, even the artist would be lost with "totally unknown types" and left in the dark when "creating the lips, the tips and wing region of the nose" (Lebzelter, 1933, [7]). Despite these problems, Lebzelter was convinced of both reconstructions and immediately used them in the exhibition hall as a topical illustrative material. It is important to know that the differences which Lebzelter noted between the Mladeč and Předmostí specimen were largely based on the sex difference between them; it was not recognized until later that Mladeč 1 is most probably a female.
Figure 3 shows some specimens

Fig. 1. The soft tissue reconstruction of Mladeč 1 in frontal view (original sculpture lost), manufactured by sculptor Egon Grenzer in 1931 under the supervision of Viktor Lebzelter (copy of a lantern slide, photo archive, Department of Anthropology, Inv. no. 7523)

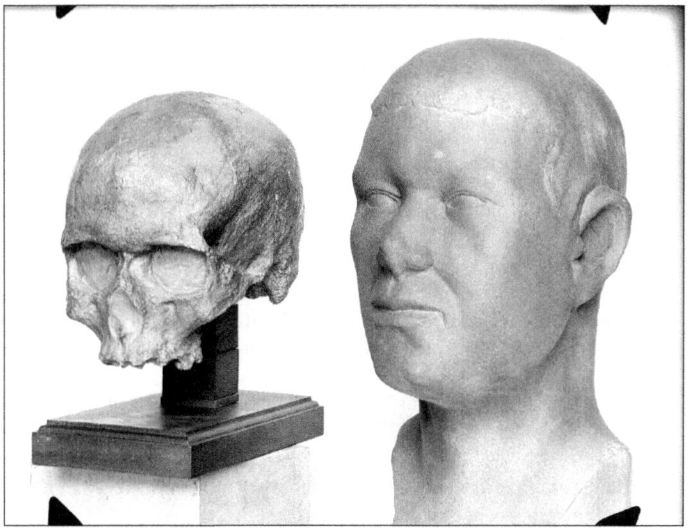

Fig. 2. Cast of the Mladeč 1 cranium and the soft tissue reconstruction in 3/4 profile (original sculpture lost) (copy of a lantern slide, photo archive, Department of Anthropology, Inv. no. 10862)

20 NHM Vienna, Department of Anthropology, inventory of the "Abgußsammlung".

of a special exhibition which was opened on 30th of October 1931 in hall XII of the Naturhistorisches Museum[21]. Below the constitutional types modeled by Erna v. Engel-Baiersdorf[22], the figure shows the two "early-diluvial race types from Moravia" made by Egon Grenzer. At the same time Lebzelter published these works in scientific journals. This strategy was used as a means to push the marketing of the objects - which had already proved "a rather lucrative commercial enterprise" elsewhere[23] - at the Museum in Vienna. In a report to the Federal Ministry in 1933, Lebzelter noted that "the effect the exhibits render in the German museums [...] is largely based on the plentiful use of plaster casts and models". In Berlin, a *Gipsformerei* (a lab where plaster casts were made) was installed, which developed into a prosperous business. By exchanging the casts against originals, German museums could often "enrich their collections in an inexpensive manner"[24].

Fig. 3. Showcase of a temporary exhibit at the NHM Vienna Hall XII (opened end of 1931). Below are the reconstructions of the early Upper Palaeolithic specimens Mladeč 1 and Předmostí (copy from the inventory record of the "Abgußsammlung")

At a time of financial bottlenecks and exceptional economic need, this strategy seemed an excellent method of choice. Therefore, and as a matter of routine, Lebzelter offered the local productions, in particular the Mladeč 1 reconstruction, to many European and non-European institutions (among others to the Danish anthropological committee[25], the Museo Nacional de Arqueología in Mexico[26] and the Greek Embassy[27]) "in exchange for ten recent skulls" or copies of fossils. "According to the practice of the Louvre in Paris"[28], Lebzelter also turned to the association

21 Annual report, Jannuary 22, 1932. NHM Vienna, Department of Anthropology, correspondence folder 1930–1932.

22 See Footnote 8.

23 Landesmuseum Hannover to Viktor Lebzelter, April 30, 1934. NHM Vienna, Department of Anthropology, correspondence folder 1933–1934; see also Viktor Lebzelter to Landesmuseum Hannover, May 17, 1934, ibd.

24 Viktor Lebzelter to Federal Ministry of Education, 1933 (n. d.). NHM Vienna, Department of Anthropology, correspondence folder 1929–1930.

25 Viktor Lebzelter to the Danish anthropological committee, March 9, 1933. NHM Vienna, Department of Anthropology, correspondence folder 1933–1934.

26 Viktor Lebzelter to Rubin de la Borbolla, May 11, 1934. NHM Vienna, Department of Anthropology, correspondence folder 1933–1934.

27 Viktor Lebzelter to Greek Embassador, August 10, 1933. NHM Vienna, Department of Anthropology, correspondence folder 1933–1934.

of friends of the Naturhistorisches Museum to undertake on behalf of the department the selling of "plaster casts of artistic-scientific reconstructions of prehistoric human types" in the entrance hall of the Naturhistorisches Museum[29]. The agreed price was 40,- S; this was about half of the monthly budget for the Anthropological Department (approx. 100,- schillings). The artists agreed to this procedure and took part in the profit.

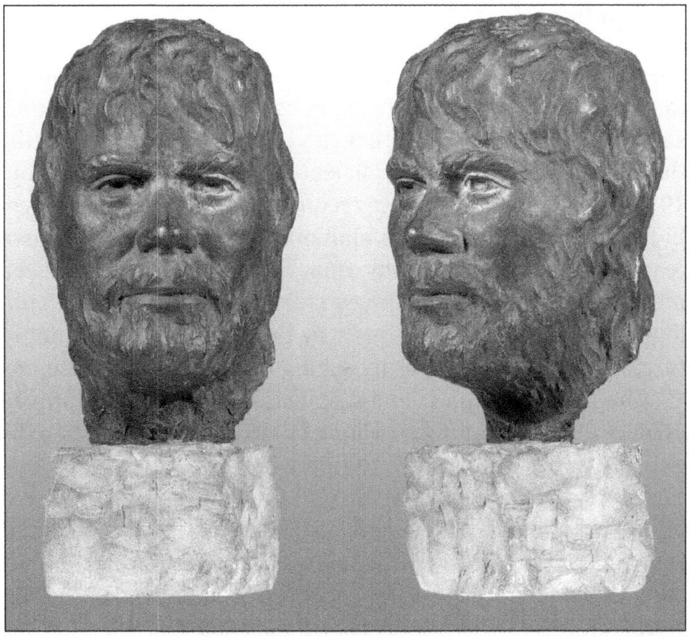

Fig. 4. Soft tissue reconstruction manufactured by Friedrich Fahrwickel (most probably 1937) in frontal and 3/4 profile view (Inv. no. "Abgußsammlung" 21350)

The second reconstruction made on the basis of the skull of Mladeč 1 was carried out by Friedrich Fahrwickel, presumably in 1937 (Inv. no. 21350, see Fig. 4). According to Josef Wastl, who succeeded Viktor Lebzelter in heading the Department (Lebzelter died in 1936), several of the soft tissue reconstructions - among them the reconstructions of Mladeč and Předmostí - were "made by non-Aryans" and had to be redone by Aryans (Erna v. Engel-Baiersdorf was "most probably Jewish"; Egon Grenzer, too)[30]. Fahrwickel's work differs substantially from the reconstruction made by Grenzer, both as far as the physiognomy and execution technique are concerned. Lacking a protocol or a publication of this work, it is difficult to justify the reproduction from the sculptor's or supervisor's point of view. While Grenzer's reconstruction seems to be rather soft and close to the given bony structures - in spite of basically wrong gender allocation - Fahrwickel had subjectively lent much broader artistic scope to his interpretation.

Fahrwickel reinforced a relatively less prominent supra-orbital region by attaching thick eyebrows; the alveolar prognatism was "concealed" by attaching a beard in a way for the vertical shaping to approach a straight line; to a certain degree, this reconstruction suggests ideological motives. According to recent research, this work has never been presented in the general public area of the Naturhistorisches Museum.

28 Viktor Lebzelter to Erna v. Engel-Baiersdorf, December 22, 1932. NHM Vienna, Department of Anthropology, correspondence folder 1933–1934.

29 Viktor Lebzelter to the friends of the NHM Vienna ("Freunde des Naturhistorischen Museums Wien"), Jannuary 16, 1933. NHM Vienna, Department of Anthropology, correspondence folder 1933–1934.

30 Josef Wastl to Dr. Krantz, Rheinisches Mineralien-Kontor Bonn, February 1, 1939. NHM Vienna, Department of Anthropology, correspondence folder 1937–1938, 122; moreover, a request for the Předmostí reconstruction in 1940 was defeated by Wastl, since "it was produced by a Jewish sculptor"; see Josef Wastl to the Reich governor of the district of Oberdonau, May 14, 1940. NHM Vienna, Department of Anthropology, correspondence folder 1937–1938, 249.

Epilogue

Soft tissue reconstructions played a key role in the collection and exhibition policies of the Naturhistorisches Museum in the late twenties and early thirties of the past century. The effort that was made specifically by Viktor Lebzelter, "to build a complete, scientifically accurate record of all the peoples that have ever lived on its ancient soil", has been quoted by the international community (Thone, 1936, 42). To Lebzelter, these reconstructions were considered "racial portraits" rather than being of high scientific value. In his opinion, a "vivid plastic reconstruction can transmit all an expert can read only by laborious study from the analysis of cranial features" (Lebzelter, 1936a). In addition, these objects would also convey political messages to the public in which the staff of the Anthropological Department would in a way begin to take part from the early thirties. Moreover, he expected to increase the inventories by selling or exchanging these novel products.

The Vienna sculptors and scientists developed an improved way to carve likenesses of prehistoric predecessors: According to Thone (1936, 42), they compared skulls "critically with skulls of peoples existing in the world today, to find what modern racial types they most resemble" [...] and than they turned "to measurements made on living possessors of these modern counterparts of ancient skulls" to get an idea on the thickness of the soft parts, which should be reconstructed. As Viktor Lebzelter kept intervening in the production of these sculptures, they not only bear the artist's, but also his own, trademark. He basically conceded artistic freedom to the sculptors, checked and discussed their work, suggested changes or consulted other experts who had more concrete ideas about the physiognomy in question. Nevertheless, some of these works were also subjected to heavy criticism[31].

Today, both of the early reconstructions introduced here are of primarily historical relevance. The fact that in the thirties of the past century, the Mladeč 1 cranium - excavated in 1881 and serving as the bony structure for the reconstruction - was still assigned to a male individual, dramatically underscores the fact that these reproductions only reflect instantaneous images of the status quo. Beyond that, however - and given the fact that both Mladeč 1 soft tissue reconstructions were materialized within a few years from each other, and within the same institution - they also demonstrate that each reproduction is not only influenced by a variable final state of knowledge, but also by subjective judgment and by the ideological background of the sculptor.

Acknowledgement

I thank Erik Trinkaus for critical comments on an earlier version of the manuscript and Wolfgang Reichmann for technical assistance. Translation by Michaela Zwölfer.

References

Boule, M. (1921) *Les hommes fossiles.* Paris: Masson et Cie

Eickstedt, E. v. (1924) Eine Rekonstruktion des Urmenschen. *Die Umschau* 28, 714

Eickstedt, E. v. (1925) Eine Ergänzung der Weichteile auf Schädel- und Oberkörperskelett eines Neanderthalers. *Zeitschrift für Anatomie und Entwicklungsgeschichte* 77, 363–380

Engel-Baiersdorf, E. v. (1949) The method of reconstructing human and animal remains in sculpture and in paintings. *Museum and Art Notes* 1, 5–11

31 Viktor Lebzelter to Erna v. Engel-Baiersdorf, March 9, 1935. NHM Vienna, Department of Anthropology, correspondence folder 1935–1937.

Gerasimov, M. M. (1968) *Ich suchte Gesichter. Schädel erhalten ihr Antlitz zurück - Wissenschaft auf neuen Wegen.* Gütersloh: C. Bertelsmann
His, W. (1895) *Johann Sebastian Bach, Forschungen über dessen Grabstätte, Gebeine und Antlitz. Bericht an den Rath der Stadt Leipzig im Auftrage einer Commission erstattet.* Leipzig: F. C. W. Vogel
Kollmann, J. and Büchly, W. (1898) Die Persistenz der Rassen und die Reconstruction der Physiognomie prähistorischer Schädel. *Archiv für Anthropologie* 25, 329–359
Kurth, G. (1956) Überlegungen zu Neandertalerrekonstruktionen. In (K. Tackenberg, Ed.) *Der Neandertaler und seine Umwelt.* Bonn: Habelt, pp. 36–48
Lebzelter, V. (1932) Rekonstruktion des Kopfes und Gesichtes eines Landmannes aus der Bronzezeit von Gemeinlebarn, Niederösterreich. *Mitteilungen der Anthropologischen Gesellschaft* 62, [26]
Lebzelter, V. (1933) Plastische Rekonstruktionen der Mammuthjäger von Lautsch und Předmost. *Mitteilungen der Anthropologischen Gesellschaft Wien* 63, [5]–[7]
Lebzelter, V. (1934) Plastische Rekonstruktion eines Tasmaniers. *Mitteilungen der Anthropologischen Gesellschaft Wien* 64, [3]–[4]
Lebzelter, V. (1935a) Plastische Rekonstruktion über einen Schädel aus dem 4. Jahrh. n. Chr. aus Tulln (Niederösterreich). *Mitteilungen der Anthropologischen Gesellschaft* 65, [4]
Lebzelter, V. (1935b) Plastische Rekonstruktion über einen Schädel aus einem Kurgan bei Simbirsk. *Mitteilungen der Anthropologischen Gesellschaft* 65, [21]–[22]
Lebzelter, V. (1936a) Anthropological sculpture: Prehistoric types reconstructed. *The Illustrated London News* (11. April 1936), p. 623
Lebzelter, V. (1936b) Vorgeschichtliche Menschenrassen rekonstruiert. *Die Umschau in Wissenschaft und Technik* 40, H 48, 952–955
Martin, R. (1914) *Lehrbuch der Anthropologie in systematischer Darstellung mit besonderer Berücksichtigung der anthropologischen Methoden.* Jena: Gustav Fischer
Matiegka, J. (1934) *Homo předmostensis,* vol. 1 and 2. Prague: Česká Akad Věd Umění
Mollison, Th. (1931) Eine neue Rekonstruktion des Homo primigenius. *Anthropologischer Anzeiger* 7, 285–288
N. N. (1929) Cave man's life resurrected. *Science News Letter* 15 (429), 399
Schmidt, E. (1898) Die Rekonstruktion der Physiognomie aus dem Schädel. *Globus* 74, 307
Schwartz, J. H. and Tattersall, I. (2002) *The human fossil record,* vol. 1. New York: Wiley-Liss
Stadtmüller, F. (1922) Zur Beurteilung der plastischen Rekonstruktionsmethode der Physiognomie auf dem Schädel. *Zeitschrift für Morphologie und Anthropologie* 22, 337–372
Thone, F. (1936) Ancestral portraits. *Science News Letter* 30 (797), 42–44
Trinkaus E. and Shipman P. (1993) *The Neandertals.* London: Jonathan Cape, pp. 384–410
Ullrich, H. (1958) Die methodischen Grundlagen des plastischen Rekonstruktionverfahrens nach Gerasimov. *Zeitschrift für Morphologie und Anthropologie* 49, 245–258
Welcker, H. (1883) *Schillers Schädel und Todtenmaske, nebst Mitteilungen über Schädel und Todtenmaske Kants.* Braunschweig: Vieweg
Welcker, H. (1884) Der Schädel Rafaels und die Rafael-Portraits. Sendschreiben an Geh. Rath Dr. H. Schaaffhausen. *Archiv für Anthropologie* 15, 417–440
Weninger, J. (1936) Viktor Lebzelter †. *Anthropologischer Anzeiger* 13, 305–306
Winkler, E.-M. (1988) Methoden der Weichteilrekonstruktion. In (R. Knußmann, Ed.) *Anthropologie. Handbuch der vergleichenden Biologie des Menschen.* Stuttgart New York: Gustav Fischer, pp. 600–605

Chapter 3

THE STRUCTURE OF THE CAVE, STRATIGRAPHY, AND DEPOSITIONAL CONTEXT

Jiři A. Svoboda

Introduction

The Mladeč Cave system is located in the Devonian limestones of the Konice-Mladeč formation, inside the Třesín Hill (343 m a.s.l.), dominating the Upper Moravian Plain. Previously known as "Bočkova díra" in Czech or "Fürst Johanns-Höhle" in German, this multi-floor karstic system, with mean elevations around 250 m a.s.l., includes two major human fossil sites: Mladeč Ib and Mladeč II, and a possible site Ia (Table 1; Szombathy, 1882, 1904, 1925; Maška, 1905; Knies, 1906; Bayer, 1925; Smyčka, 1922; Skutil, 1938; Jelínek, 1983, 1987; Oliva, 1989, 1993; Valoch, 1993; Svoboda, 2000, 2001, 2002). Other Upper Paleolithic sites at the Třesín Hill include a smaller "Podkova" (Horseshoe) Cave (Mladeč III; Skutil, 1938; Svoboda, 2002), located at 270 m a.s.l. in the northern slope, and an open-air site "Plavatisko" on the top plateau (Valoch, 1981; Oliva, 1996). Typically, the limestone is penetrated by vertical fissures and chimneys interconnecting the horizontal cavities and providing more or less direct access to the surface.

In order to address the depositional context of the human fossils, two lines of evidence are combined: the original records of early excavators and witnesses, and the revision of the actual topographic situation inside the cave. Nevertheless, several issues remain unresolved, not the least of which is the question of whether the Upper Paleolithic people frequented the interior of the cave, or whether their remains fell in through the chimneys. Furthermore, there are several questions related to: the original entrance, the hearths, and finally, possible rock art.

The interpretation of Upper Paleolithic burials is widely influenced by both the personal experience of the excavators and available analogies. Naturally, these viewpoints have altered with the advancement of European Paleolithic research during the last century. As the majority of Upper Paleolithic burials discovered earlier in caves of Italy and France (e.g., Grimaldi, Cro-Magnon) were situated in regularly frequented and settled sites, the first interpretations of Mladeč automatically sought an analogous explanation. Elsewhere in Europe, however, a more specialized type – the funeral cave – was repeatedly encountered, as at Cussac where the human skeletal

Table 1. Structure and numeration of the Upper Paleolithic sites at the Třesín Hill in Mladeč

Site	Years	Excavators	
Ia.	Mladeč - Entrance	1815?, 1826, 1828	
Ib.	"Dome of the Dead"	1881–1882 and later	Szombathy
		1903–1911	Knies
		1922	Fürst, Smyčka
		1958–1962	Jelínek
II.	Side fissure	1904	Knies, Smyčka, Szombathy
III.	Podkova Cave		
IV.	Plavatisko - surface site		

remains were associated with parietal art on the walls (Ajoulat et al., 2001), or at Paviland where a burial was located in a scarcely settled cave on the western periphery of the Upper Paleolithic world (Aldhouse-Green, 2000). In 1950, a funeral cave was discovered in the Zlatý kůň Hill at Koněprusy in the Bohemian karst (Prošek et al., 1952), and it has been used as another type of analogy for Mladeč (Jelínek, 1987; Svoboda, 2000).

The situation after 1815: Site Mladeč Ia

Although the existence of underground cavities within the Třesín Hill was known since earliest times and reflected in local tales, the actual entrance to the cave was not opened until either 1815, 1826 (as reported by G. Wolny) or 1828 (report by J. G. Sommer) by a limestone quarry in the northern slope of the hill.

In the first hall behind this entrance (later labeled "A" by Szombathy), skeletal remains of a large ("giant") individual were found. The following reports by local witnesses were later recorded by Szombathy (1925, 4). Walloch: "In the first small hall on the right lay the skeleton of a giant with arms stretched out. He had teeth larger than a thumb. The priest of Mierotein came, took the bones with him and had them buried in a funeral". Mrs. Raschendorfer: "His skull was so big that a man could place it wholly over his head". In addition, Knies (1905) referred to a carved bone spear-point associated with the body, and Skutil (1938), who explored the entrance part of the cave, reported the discovery of two lithic artifacts.

The skeletal remains were presumably reburied in the nearby village of Měrotín, possibly near the church; there is no report of a more precise location of the burial. Since that time, however, a new and larger cemetery was built near Měrotín and consequently, there is no hope of rediscovering this skeleton. Following the report, the skeleton may have belonged to a large human or to a carnivore, presumably a bear. We have no evidence of the stratigraphy, but the finds were likely right on the surface. If we judge from the artifacts both the undocumented bone point and the two lithics drawn by Skutil (1938) could possibly be Aurignacian.

Actual revision of this part of the cave shows that an almost vertical natural corridor several meters long - the "Entrance chimney" - opens towards the surface behind the modern entrance to the right. Even if passage through this chimney is difficult, it is accessible to humans. A date of 1902 is inscribed on the rock wall in the middle of this pathway. If it was open during the past as it is today, the skeleton and the artifacts could have passed through during any time period. In addition, the "Entrance chimney" could have enabled passage further inside the cave system.

Activities after 1881: Site Mladeč Ib

Situation and stratigraphy

Following the entrance to the cave, several halls are separated by massive limestone pillars, labeled "B–E" by Szombathy (Figs. 1 and 2). Today, after removal of a large portion of the sediments, these parts of the cave form a unique, oblong space called "The Dome of the Dead". During the 19th century, however, it was filled with sediments and large limestone blocks, and sealed with calcite layers. Thus, orientation in this part of the cave was considerably more difficult.

After discovery, this part of the cave was vulnerable to exploitation of the faunal remains and stalactites up until 1881–1882, when Joseph Szombathy first began scientific exploration of the cave. Areas of his excavation are indicated on his map (Fig. 1; Szombathy, 1925, Abb. 1), making this excavation the best documented and most successful fieldwork ever undertaken inside the cave.

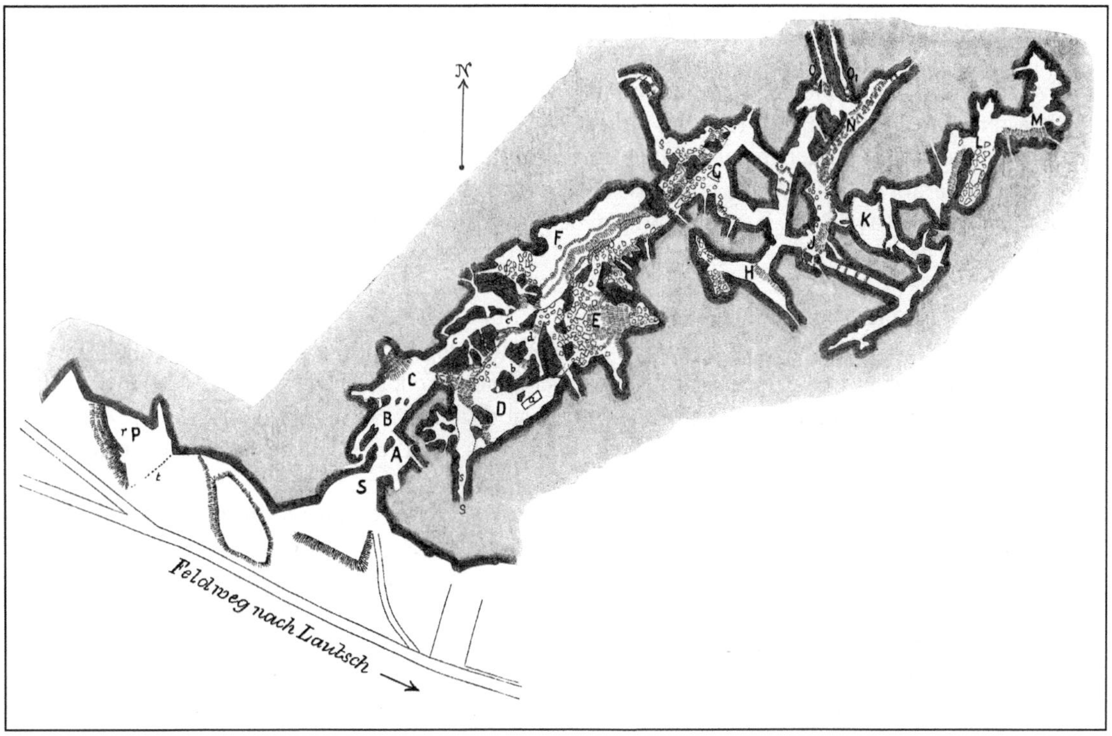

Fig. 1. Historical plan of the caves by J. Szombathy (1925), showing his numeration of the halls (A–E) and findspots (a–e). Scale: 1:800

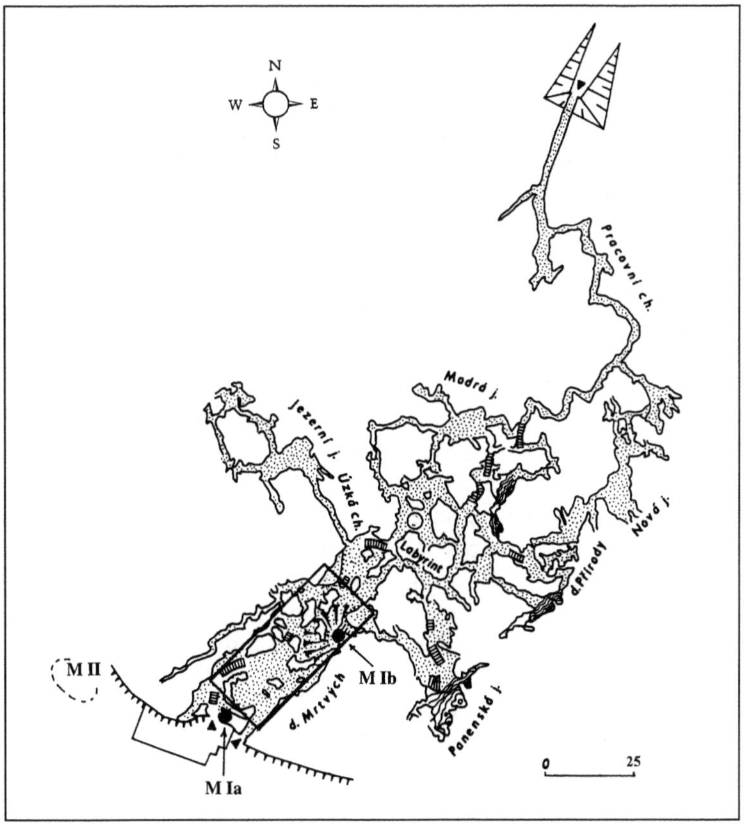

Fig. 2. Actual plan of the caves, showing the position of the sites Ia, Ib and II. The square indicates the analyzed area in the "Hall of the Dead" (site Ib). The points indicate the chimneys

Szombathy did not mention the chimney or the massive cone of underlying deposits in the northeastern part of the hall; instead, he represents this area as a solid rock on his plan, presumably because of the large boulders accumulated on the top of the cone. According to our recent reconstruction of the surface (Fig. 4a, b), his most important trench at locus "a", Hall D, was located at the deepest part of the cave floor. It was about 3 m deep and showed the following stratigraphy (Fig. 3a; Szombathy, 1925, Abb. 2):

(1) A solid calcite (travertine) cover
(2) Brown, plastic, partly sandy clay, 60 cm thick; at a depth of 30 cm it was subdivided into two parts by a continuous layer of calcite; at a depth of 35 cm a lens of charcoal was found, about 50 cm wide; it contained Pleistocene human and faunal remains.
(3) Light clays (with so-called "false stratification") with bands of smaller limestone elements, about 80 cm thick
(4) Dark brown earthy clays with hard, "bituminous" interlayer, 20 cm thick; Pleistocene fauna
(5) Dark brown, "fat" clay with individual limestone blocks, 50 cm thick
(6) Brown sandy clay with gravel of various dimensions ("Grauwackengeschiebe"), 90 cm thick; at the depth of 40 cm in this layer lay a large limestone platform overlain by a charcoal layer; at the depth of 50 cm another charcoal layer was present and contained a "nest" with microfaunal remains
(7) Sandy, partly silty, layer with gravel

At a depth of only 20–50 cm, Szombathy found a large portion of a human skull and a femur, together with faunal remains. Below this was a layer of calcite with charcoal and a series of finely bedded loam interspersed with concentrations of microfaunal remains, reaching a depth of 3 m. Additional human fossils and artifacts were later found in a similar situation in locus "b", and also in other parts of the hall ("d–e") during the 1882 excavations. They were spread over an area altogether not surpassing 20 m^2. Even though Szombathy excavated quite deep, no finds were located beneath a depth of 50–60 cm. This observation is important, because in disputes with K. J. Maška about the authenticity and Pleistocene age of his finds, Szombathy would certainly have favored a lower depth.

The exploration of this hall was continued after 1903 by J. Knies (1905). Knies was the first to describe the large, 6 m high debris cone in the northeastern part of the Dome (between Halls D and E). In this area, he mostly found Pleistocene remains, including a reindeer skeleton. In his letter to K. Maška (dated April, 1904), Knies mentioned "an almost complete reindeer skeleton, only the skull is fragmented. The extremities perfectly preserved! The animal died in the rock fissure into which it fell". It is hard to say what he meant by the "fissure". But based on the taphonomy of this skeleton, the preservation of the bones in several anatomically ordered clusters, and the two observed types of coloration, Knies tried to reconstruct the depositional process. He concluded that the animal would have fallen in through the chimney, and individual parts of the body would have been gradually redeposited together with the gravel in two basic directions.

Since 1911, the sediments have been disturbed by large-scale earth removals in order to accommodate the arrangement of stairs and paved floors for tourists visiting the cave. Another important discovery of human fossils was made by Fürst and Smyčka as late as 1922, but there are not enough contextual data with which to evaluate the finds. According to Smyčka (1922) and Skutil (1938), the fossils lay not far from Szombathy's findspot below "a chimney", but it is not certain to which chimney they are referring. The location of the findspot, as indicated today on the rock wall, was questioned by Jelínek (1987, 58). The finds include both human and animal bones, unusual bone artifacts, and partly or totally pierced micro-fauna and snails.

Compared to that of Szombathy, the stratigraphic description is less complete. The sediments were separated into two layers by a 30–50 cm thick calcite desk, and the human fossils were discovered in the lower layer (Smyčka, 1922). Given the late date of the discovery, it is highly probable that the "upper" layer represented, in fact, redeposited sediments and rubbish from earlier excavations; this coverage may help to explain why the finds were not discovered earlier. If we compare this report with the section of Szombathy, there is one striking similarity: human fossils were found under a solid, rather shallow calcite cover in both instances. With this last find, the Late Pleistocene deposits and their anthropological, paleontological, and archaeological content were, in fact, exhausted.

All subsequent fieldwork performed in this area aimed to record the local topography and revise the stratigraphy. Stratigraphic trenches excavated by Jelínek (1987), and Horáček and Ložek (1984), demonstrated that the majority of the preserved deposits date to the Middle Pleistocene and earlier.

Near the former trench "a", Jelínek reopened and confirmed the lower portion of the classical section of Szombathy to a total depth of 5 m, with the critical upper portion already removed. Based on the diaries of F. Adámek (April 3, 1959), the stratigraphic sequence in trench 3 of Jelínek's excavation (Fig. 3b) was published by K. Valoch (1993; see Fig. 1):

(1) Redeposited sediments below the actual floor
(2) Light-brown loessic earth with schist and limestone debris
(3a) Thin deposit of red-brownish clay with small-sized limestone debris
(3b) White-grayish silt
(4) Brown, compact clay, with lime concretions and fragments, pebbles, and scarce limestone debris
(5) Brownish, fine clay with light ochreous interlayers
(6) Brownish clay including coarse-grained sand, interlayers of pure sand and gravel (and one possible chopper of quartz)
(7) Thin interlayer of white-grayish silt
(8) Sandy loam including sand and gravel interlayers
(9) Coffee-colored, compact loam

Even if it is not visible in Fig. 3b, the description mentions that layers 2, 3a, and 5 were rising towards the debris cone.

In 1996, the Institute of Archaeology (AS CR) documented a section on the rock wall adjacent to Szombathy's findspot "a", with remains from the sediments of the upper part of the section (Fig. 3c):

(1) Solid calcite (travertine) with clayey interlayers
(2) Brownish to grayish clay
(3) Light yellow clay
(4) Grayish clay with reddish-brown spots and particles of limestone detritus
(5) Solid rock wall with remains of clayey coverage

The boundaries between layers 3–5 were unclear.

Finally, during winter 2003/2004, the Institute of Archaeology supervised further surface adaptations inside the Mladeč Caves that were carried out by the Natural Protection Agency. During these efforts, human and animal osteological material (both domesticated and wild), was gathered from redeposited sediments just below the floor (Svoboda and Nývltová-Fišáková, 2004). Even though the finds display various stages of fossilization, they all are of Holocene age, and are in no way related to the Pleistocene fossils discovered earlier.

Based on Jelínek's discoveries, Valoch (1993) suggested a Lower Paleolithic occupation event inside the cave. The pieces (from layer 6, for example) are comprised, however, of ambiguous artifacts and the fact that they are made of non-local materials (chert, limonite, quartz) and thus manuports, is not convincing given the evident transport of the sediments from above. The only convincing artifact, a chopper/core made of a larger quartz pebble, was found in the superficial calcite layer in one of the side corridors, and may well be of a later age.

The Upper Paleolithic artifacts made of organic materials collected by Szombathy, Knies and Smyčka, are consistent with the Aurignacian age of the site, as was first suggested by Bayer (1922). The lithics are rare and culturally undiagnostic. It should be added that Szombathy originally mentioned the find of a torch, but later (1925) critically rejected its Paleolithic age.

Dating

All explorers of the Dome, including Szombathy, Knies and Smyčka, stressed the stratigraphic role of the thick, upper calcite platform and indicated that the human finds were directly related to this horizon or oriented just above or below it. In fact, portions of the calcite are still visible on some of the fossils preserved in the Naturhistorisches Museum Wien. Today, several (two or three) generations of calcite formation are preserved on the walls of the hall.

Adjacent to Szombathy's findspot "a", we collected samples for dating (Fig. 3c). The calcite is partly compact, with curved layering over and around vugs formed either by sediment or water. It is also partly structured into micro-layers separated by air-gaps or interlayers with a clayey appearance. First, the

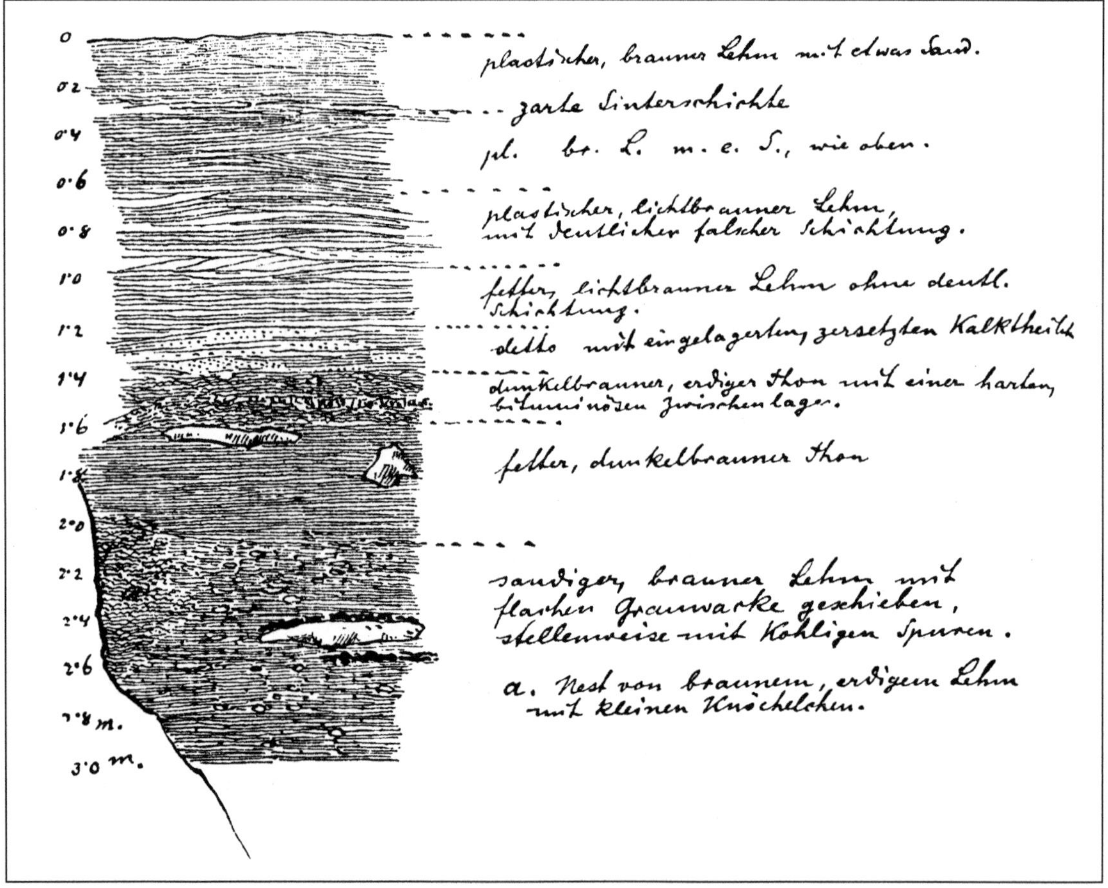

Fig. 3a. Mladeč Ib: The classical section of the trench in findspot "a" in 1882 (after Szombathy, 1925). Description of the layers in the text

samples were submitted for U-series dating at the University of Liverpool, but this was unsuccessful because the samples were clay contaminated and brecciated (A. Latham, pers. comm., August 26, 1994).

Two additional samples, both taken from the top calcite layers and located 5 cm apart, were collected for ^{14}C dating (Svoboda et al., 2002). The results obtained from the carbonate are: 34,160 (+520–490) BP for the upper sample (Gr.N-26333), and 34,930 (+520–490) BP for the lower sample (Gr.N-26334). Based on 13d values, the reservoir ages are ca. 800 years (sample 1), and ca. 1,200 years (sample 2). There is one basic assumption here: the travertine was formed by deposition from more or less streaming, recent water (at the time of formation), and not from fossil age water.

The interval between the two samples documents a rapid formation of the series of the calcite layers, at least at this particular location. Naturally, the deposition of the fossiliferous sediments above and below these layers was a considerably longer process, as indicated by the time-span covered by ^{14}C dates obtained directly from human and animal bones.

Reconstruction

It follows from the history of investigation that two of the hominid finds in the "Dome of the Dead" were of importance, the 1881–82 and the 1922 finds. Only the former, however, has a relatively good documentation. Both were discovered in a relatively shallow position below the surface. The lo-

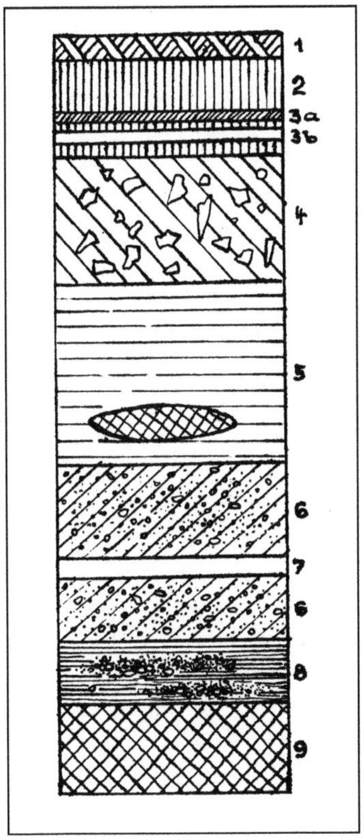

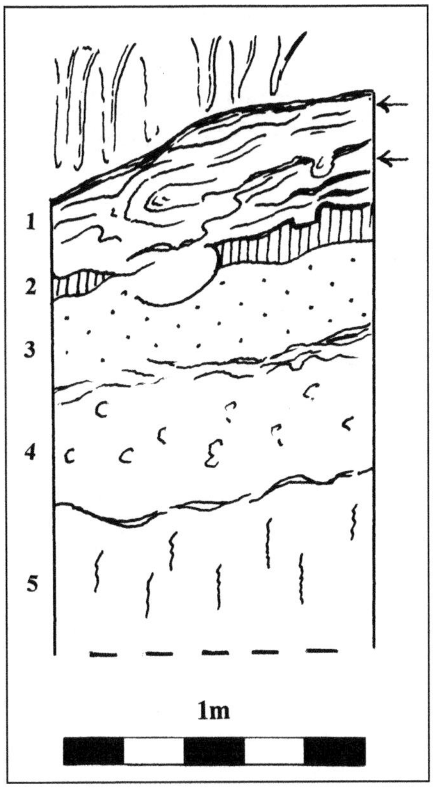

Fig. 3b. Mladeč Ib: Lower part of the sediments, as recorded in trench 3, 1959 (after Valoch, 1993). Description of the layers in the text. Total depth = 5 m

Fig. 3c. Mladeč Ib: Relicts of the upper part of the sediments attached to the cave wall, as recorded in 1996. Description of the layers in the text. The arrows indicate the location of the ^{14}C samples

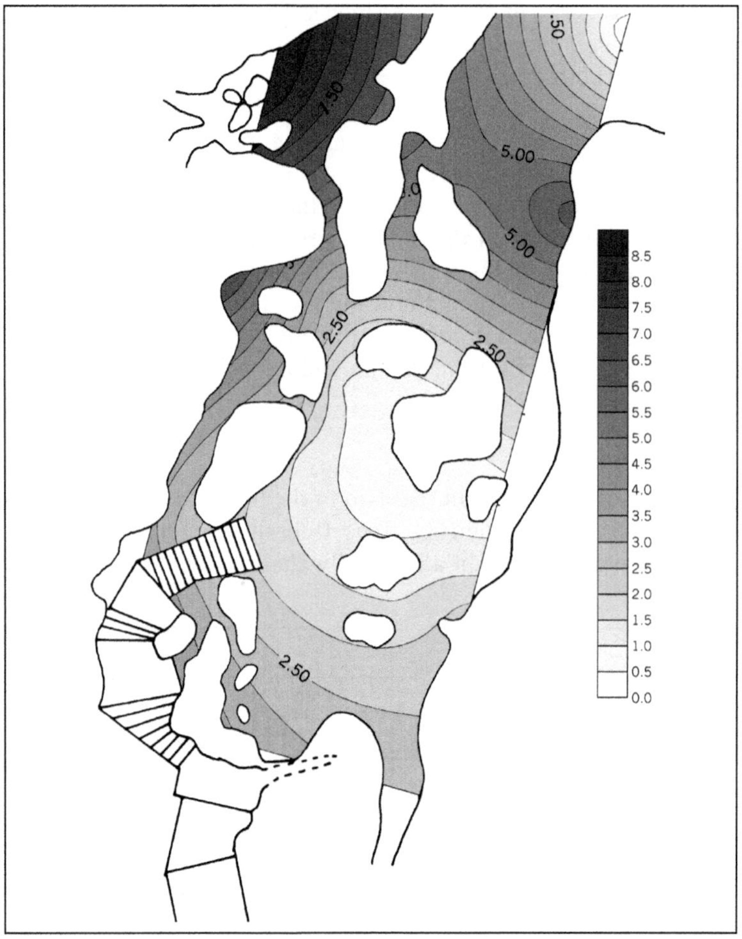

a

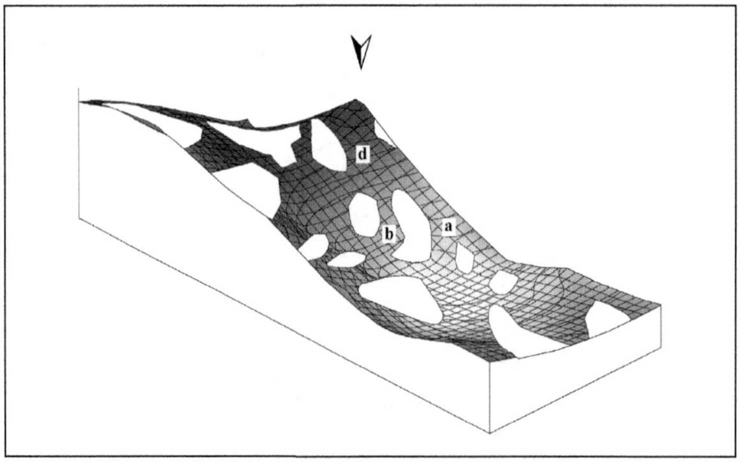

b

Fig. 4a, b. Mladeč Ib: Detailed plan of the "Hall of the Dead" and Surfer reconstruction of the original sedimentary filling. White areas correspond to solid rock cation of Szombathy's trench "a" is visible on his plan (Fig. 1), and the location of the find by Fürst and Smyčka is recorded by an authentic label on the rock wall. Even if the latter location was questioned by Jelínek (1987), we do not have a better indication today.

In order to understand the original situation and the mode of deposition, it was crucial to reconstruct the original surface of the cave floor. Today, the hall is a large rectangular area with several massive pillars. Remaining sediments are preserved especially in the corners of the hall, and traces of removed sediments are also partly visible on the rock walls. In the spring of 1996, we measured the relative elevations of the latest calcite formation wherever it was still preserved on the walls of the hall, and elevations of the existing remains of the cones were also recorded. Based on these data, the Surfer program enabled us to generate a hypothetical reconstruction of the original cave filling (Svoboda, 2000).

The resulting images (Fig. 4a, b) indicate that the filling of the cavity accumulated from at least four source areas in each "corner". The southwest cone corresponds to the current cave entrance where sediments were recently redeposited when the cave system was opened, especially for tourists. Sediments in the southeast corner (Middle Pleistocene and earlier) are cut by the second, parallel entrance, opened in the early 20th century. In the northwest corner, a system of fissures in the ceiling is evidently responsible for the sediment deposits. The most important source of material, both paleontologically and archaeologically, is the chimney in the northeast – the "Chimney of the Dead" – that forms a large and typically developed debris cone. Today, only the central parts of this cone are preserved, and they contain Middle Pleistocene deposits with microfauna (Ložek, pers. comm.), while the Late Pleistocene sediments were almost completely removed in the past.

When plotting Szombathy's finds "a–e" into this plan (Fig. 4b; after Szombathy, 1925, Abb. 1), it became clear that all lay on the surface and periphery of the northeast cone. The most important findspot "a" would have been a distance of about 15 m from the Chimney of the Dead, at the foot of the cone. The findspot of Smyčka (1922), if we trust the label on the rock wall, would have also been about a 15 m distance from the chimney. Following the verbal description of Knies (1905; see also Szombathy, 1925, 9), his finds, though smaller in dimension, were associated with a cone that was very likely the same one. It seems that the larger and rounded bones such as the skulls reached as far as the foot of the cone (this is a taphonomic observation encountered at the Pavlov 1 burial, for example, that was also affected by slope movements), while the others were variously located on the surface.

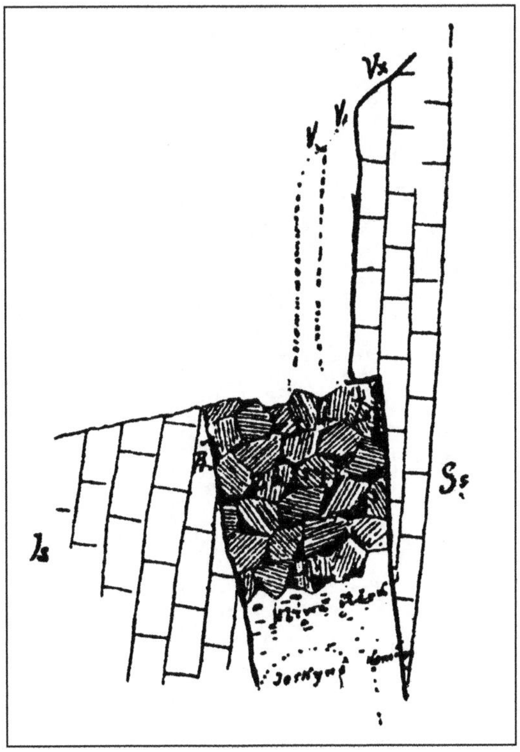

Fig. 5. Mladeč II: Section at the findspot in 1904 (after Knies, 1905)

Activities in 1904: Site Mladeč II

On March 22, 1904, workers in the quarry in the southern part of the Třesín Hill, 43 m west of the present entrance to the large caves, found and destroyed a small cavity. As early as April 1904, Knies mentioned in a letter addressed to Maška that he had "from a certain place a child's mandible, skull fragments, and about 20 bones of extremities, mostly fragmented." Smyčka commented on this discovery in a letter to Szombathy who published it (Szombathy, 1904). Maška also published his observations (1905), but more comprehensive summary reports were reserved to Knies (1905) and Szombathy (1925).

During the archaeological examination, it was observed that the cavity itself was already damaged, and that the finds were dispersed in the vicinity. According to Smyčka, the findspot was located about 10 m below the original surface in solid limestone rock, without an entrance; human finds were concentrated in a cavity with a triangular ground-plan (after Knies), in a space measuring 2 x 3 m (after Smyčka). Following Knies's section sketch (Fig. 5), the cavity would have been the relict of a vertical fissure or chimney leading to deeper parts of the cave system. Stratigraphically, the filling was composed of large limestone blocks, smaller gravel, and "yellow earth", probably loessic sediment.

In order to clarify the situation, excavations in the chimney continued until 7 m below the findspot, where Smyčka believed he would find a prolongation somewhere towards the Main Cave system. Later, Szombathy (1904) continued the excavation to a depth of 13 m, proving the existence of a continuing fissure below the findspot, but without further continuation towards the cave system. In August, Maška (1905) observed the relict of another chimney in the western part of the site, leading to the surface and continuing further north into the hill. In the upper part of the chimney, Maška recorded more Pleistocene faunal remains, but at the time of his visit, it was no longer possible to reconstruct the relationship of this chimney to the human fossil findspot. Today, vertical fissures of this kind are still visible at the remaining rock wall adjacent to the findspot (Fig. 6).

Fig. 6. Mladeč II: One of the vertical fissures as visible at the limestone rock wall of the site

Due to damage to the cavity, it was also difficult to complete the associated bone and artifact inventory, and to establish its contextual relationship. Some pieces were, in fact, collected in the broader vicinity ("from the mud on the road...", etc.). In general, however, the points made of organic material are consistent with the Aurignacian age of the site, whereas the lithics are culturally undiagnostic, and some may not even belong to this assemblage (Oliva, 1989; Svoboda, 2002).

Interpretations

Entered or fallen through?

For more than a century, the situation at the Mladeč sites has evoked discussions amongst several individual archaeologists. Both the shallow stratigraphic location and incompleteness of the skeletal material were recognized as the most remarkable features by the early researchers. Bayer (1925) interpreted this situation as a disturbed burial site, and a settlement regularly accessible by a hypothetical entrance. Szombathy (1925, 8), who had to defend the authenticity and Pleistocene age of the fossils against Maška, opposed any reference to a disturbance, and explained the fragmentary state of the skeletons by Paleolithic anthropophagy. J. Knies was the first to suggest that at least part of the deposits and fossils came in through the chimneys. Following his reconstruction, the cave would have had an open entrance during deposition of the human fossils, but the formation of the debris cone would have occurred during a later period when the cave entrance was closed. Subsequent investigations (Horáček and Ložek, 1984; Jelínek, 1987; Svoboda, 2000) revealed, however, that the real sequence is in fact the opposite: the main body of the cone dates to the Middle Pleistocene or earlier, whereas the fossiliferous Late Pleistocene layers form a relatively shallow coverage on this surface. The idea of bodies and artifacts falling into the cave through the chimneys was published in full by Smyčka (1922) and Jelínek (1987), and is supported by the reconstruction of the original surface by Svoboda (2000).

The idea of a regularly visited ritual area has recently been raised by M. Oliva, who discovered and described simple signs on the walls of the cave (Oliva, 1989, 1993). One of his arguments states that some of the finds were located by the reverse side of the rock pillars. Redeposition from a higher debris cone following the slope movements, however, may well explain this mechanism of transport around a block, ending just on the reverse side (Fig. 4a, b).

At Mladeč II, where the original situation was largely destroyed at the time when the archaeologists came, contradictory explanations were put forward. Maška (1905) interpreted the site as a classical small cave, accessible by a hypothetical entrance from the south. Knies and others stressed the role of redeposition that was apparent from both the morphology of the vertical fissure - a chimney - and from the character of the sediments. This was also clear to Szombathy, but in accord with Maška, he explained the situation as the relict of a destroyed cave.

What about the entrance?

All interpretations suffer from the absence of an original entrance to the "Dome of the Dead". As mentioned, the actual entrance is artificial. Szombathy (1925, 3) looked for another possible entrance in the southeast corner of the "Dome of the Dead", but subsequent stratigraphic investigation revealed that during the Late Pleistocene, this area was already filled by deposits from the Middle Pleistocene and earlier. J. Skutil (1938, 32) turned back to the present entrance in the southwest corner, arguing that the single artifacts he found in this area pointed to human passage there. Given the actual topography of the place, we may add that the "Entrance chimney", opening from this part of the cave directly to the surface, certainly offers a difficult, but accessible entrance (Svoboda, 2000).

Hearths?

The existence of charcoal and/or hearths has been mentioned both at site Ib (Szombathy, 1925) and II (Maška, 1905), and it has been used as a major argument for a regular human occupation inside the caves. Based on the discovery of a charcoal (or, rather, "black") layer with fauna attached to the rock wall 4 m from the original findspot at site II, Maška (1905) hypothesized that the site was originally a small cave accessible by an entrance from the south. For the first time, Szombathy (1904) questioned the existence of charcoal as recorded by Maška, and explained the black coloration by the presence of iron and manganese. Jelínek (1987, 62), who returned to the classical site Ib, addressed and discussed the same problem there. Although it was no longer possible to identify the presumed charcoal lenses and the stone alignment mentioned by Szombathy in his upper layers, the deeper, dark lenses found by Jelínek have, in fact, been shown to be manganese deposits. It should be recalled that even if the presence of charcoal inside these cavities could be proved, it may be redeposited, similar to the other objects.

Rock art?

J. Skutil (1938, 32) mentions that H. Breuil pointed to the Mladeč Caves as "one of the rare sites that would be most suitable for preservation of traces of parietal Upper Quaternary art, hitherto unknown in our region. After this recommendation by Breuil [...] I have surveyed very carefully the walls of the Mladeč Caves, but with no success."

Half a century later, M. Oliva (1989) announced the presence of simple lines and signs, in red color, at several locations inside the "Dome of the Dead" and elsewhere. It is difficult to judge whether Skutil did not observe these signs, whether he did not consider them important, or whether they simply were not yet there at his time. Unfortunately, use of the red color excludes a direct dating. It is suspicious, however, that several alphabet letters are drawn in the same color in one of the adjacent corridors, and that some of the signs are located on the wall below the level of the original sedimentary filling. All this leads to skepticism regarding the Pleistocene age of the signs. Nevertheless, all these signs are certainly worthy of topographic documentation and speleoarchaeological inventory.

Koněprusy Caves: An analogy

The karstic system of the Zlatý kůň Hill (475,9 m a.s.l.) resembles the situation at Mladeč as it is another multi-floor cave site, with skeletal remains of a female individual found in one of the deep cavities (the Prošek Hall) on the surface of a debris cone under a vertical chimney. At this site, the topography of the chimney, the cone and its stratigraphy were all well documented, so that the interpretation raised no doubts (Prošek et al., 1952; Vlček, 1957; Svoboda, 2000; and unpublished documentation at the Institute of Archaeology, AS CR, Brno). Whereas at Mladeč the deposition of human bodies seems to be a repeatedly practiced act, Koněprusy shows a single event. At both sites, human bodies were deposited during terminal stages of the accumulation of the debris cones. This kind of analogy, together with the associated fauna, led researchers to date the time of deposition at both sites to the Early Upper Paleolithic. Supporting evidence for Koněprusy, however, was scarce: stratigraphically, the human remains were deposited on or just below the surface, the associated lithic artifacts were culturally undiagnostic, and the presumed bone projectile fragment (Mladeč-type) later appeared to be just a fragmented bone. In addition, the glacial fauna from the upper layers of the debris cone may be older than the human fossils.

For the purposes of radiocarbon dating, V. Kuželka selected a human bone fragment 4 x 2 mm in size, most likely from the cranial base of the buried individual. The result, 12,870 ± 70 BP (Gr.A-13696), places this site in the Magdalenian (dated to 12,420 ± 470 BP – Ly 1108, at the nearby site of Hostim) and is therefore, considerably younger (Svoboda et al., 2002; 2004). In light of this date, the analogy with the Early Upper Paleolithic site of Mladeč would suggest that deposition of human bodies into deep karstic cavities was a type of mortuary behavior with a longer duration and transcultural significance.

Conclusions

The complex situation inside the Mladeč Caves does not correspond to a settlement or to a frequented site. There is an almost total absence of reliable activity traces, serious doubts remain regarding the "hearths", and lithic implements are rare. On the other hand, there is a relative abundance of decorative objects such as pierced animal teeth, and of projectiles and other artifacts made of organic materials associated with human skeletal remains.

Considering the excavation history and available information on the Mladeč Caves, combined with a three-dimensional reconstruction of the original surface in the "Dome of the Dead", it seems most likely that sites Ia, Ib and II (contrary to site III) do not represent primary foci of human activity. The fossils were obviously in a secondary position, having been redeposited by slope movements during a longer time-span. Sites Ia and II are small, and the finds are directly associated with vertical karstic fissures. The majority of finds from the larger site Ib, were found at a distance of about 15 m from the "Chimney of the Dead", and thus, on the periphery of a large debris cone deposited below it. These situations are best interpreted as accumulations of human remains within karstic cave systems, through chimneys or fissures. Whether the deposition of the human material was associated with activities on the surface or related to intentional placing of the human remains and artifacts within the openings, is uncertain. All in all, it is likely that the Mladeč Caves closely paralleled the later and well documented situation of Prošek's Hall of the Koněprusy (Zlatý kůň) Caves.

Naturally, within subterranean systems as complex as both sites provided above, we cannot exclude the possibility of penetration of living animals and humans inside the cave systems, even though such an access was probably quite difficult (cf. the Entrance chimney). Passages to unknown underground worlds has always provoked the human mind, and especially during the Upper Paleolithic.

References

Ajoulat, N., Geneste, J. M., Archambeau, Ch., Barraud, D., Delluc, M., Duday, H. and Gambier, D. (2001) La grotte ornée de Cussac. *INORA* 30, 3–9

Aldhouse-Green, S. (Ed.) (2000) *Paviland Cave and the 'Red Lady': a definitive report.* Bristol: Western Academic Specialist Press

Bayer, J. (1922) Das Aurignac-Alter der Artefakte und menschlichen Skelettreste aus der Fürst-Johanns-Höhle bei Lautsch in Mähren. *Mitteilungen der Anthropologischen Gesellschaft Wien* 52, 173–185

Bayer, J. (1925) Die ältere Steinzeit in den Sudetenländern. *Sudeta* 1, 21–120

Horáček, I. and Ložek V. (1984). Z výzkumu výplně Mladečské jeskyně u Litovle. *Československý kras* 35, 98–100

Jelínek, J. (1983) The Mladeč finds and their evolutionary importance. *Anthropologie* (Brno) 21, 57–64

Jelínek, J. (1987) Historie, identifikace a význam mladečských antropologických nálezů z počátku mladého paleolitu. In *25 let pavilonu Anthropos 1961–1986.* Brno: Moravské muzeum, pp. 51–70

Knies, J. (1905) Nový nález diluviálního člověka u Mladče na Moravě. *Věstník klubu přírovědeckého Prostějov* 8, 3–19

Knies, J. (1925) Přehled moravského paleolitu. *Obzor prehistorický* 4, 89–116

Maška, K. J. (1905) Poznámky k diluviálním nálezům v jeskyních mladečských a stopám glaciálním na severovýchodní Moravě. *Časopis Moravského muzea zemského* 5 (reprint), 1–3

Oliva, M. (1989) Mladopaleolitické nálezy z Mladečských jeskyní. *Časopis Moravského muzea zemského* 74, 35–54

Oliva, M. (1993) Le contexte archéologique des restes humains dans la grotte de Mladeč. In (J. Pavúk, Ed.) *UISPP, XIIe Congres*, Volume 2. Bratislava: Institut d'Archéologie, pp. 207–216

Oliva, M. (1996) Středomoravská skupina gravettienu: periodizace versus regionalita. *Pravěk* 6, 7–50

Prošek, F. et al. (1952) Výzkum jeskyně Zlatého koně u Koněprus. *Československý kras* 5, 161–179

Skutil, J. (1938) *Pravěké nálezy v Mladči u Litovle na Moravě.* Litovel: Krajinská musejní společnost

Smyčka, J. (1922) Nálezy diluviálního člověka v Mladči u Litovle na Moravě. *Obzor prehistorický* 1, 111–120

Svoboda, J. (2000) The depositional context of the Early Upper Paleolithic human fossils from the Koněprusy (Zlatý kůň) and Mladeč Caves, Czech Republic. *Journal of Human Evolution* 38, 523–536

Svoboda, J. (2001) Mladeč and other caves in the Middle Danube region: early modern humans, late Neandertals, and projectiles. In (J. Zilhao, T. Aubry and A. Faustino Carvalho, Eds.) *Les premiers hommes modernes de la Péninsule Ibérique. Actes du colloque de la Commission VIII de l'UISPP.* Lisboa: Instituto Portugués de Arqueologia, pp. 45–60

Svoboda, J. (2002) Mladeč II. In (J. Svoboda, Ed.) *Prehistorické jeskyně – Prehistoric Caves.* Brno, pp. 344–354

Svoboda, J. and Nývltová-Fišáková, M. (2004) Mladeč. *Přehled výzkumů* 45, 109

Svoboda, J., van der Plicht, J. and Kuželka, V. (2002) Upper Palaeolithic and Mesolithic human fossils from Moravia and Bohemia (Czech Republic): some new ^{14}C dates. *Antiquity* 76, 957–962

Svoboda, J., van der Plicht, J., Vlček, E. and Kuželka, V. (2004) New radiocarbon datings of human fossils from caves and rockshelters in Bohemia (Czech Republic). *Anthropologie* (Brno) 42, 161–166

Szombathy, J. (1882) Über Ausgrabungen in den mährischen Höhlen im Jahre 1881. In (F. v. Hochstetter) Fünfter Bericht der prähistorischen Commission der math.-nat. Classe der kaiserlichen Akademie der Wissenschaften über die Arbeiten im Jahre 1881. *Sitzungsberichte der kaiserlichen Akademie der Wissenschaften* 85, [90]–[107]

Szombathy, J. (1904) Neue diluviale Funde von Lautsch in Mähren. *Jahrbuch der k. k. Zentralkommission für Kunst- und historische Denkmäler* 2, 9–16

Szombathy, J. (1925) Die diluvialen Menschenreste aus der Fürst-Johanns-Höhle bei Lautsch in Mähren. *Eiszeit* 2, 1–34, 73–95

Valoch, K. (1981) Beitrag zur Kenntnis des Pavlovien. *Archeologické rozhledy* 33, 279–298

Valoch, K. (1993) Starý paleolit v Mladečských jeskyních. *Časopis Moravského muzea zemského* 78, 3–9

Vlček, E. (1957) Pleistocénní člověk z jeskyně na Zlatém koni u Koněprus. *Anthropozoikum* 6, 283–311

Chapter 4

THE UPPER PALEOLITHIC FINDS FROM THE MLADEČ CAVE

Martin Oliva

Introduction

In spite of the early date of excavations and the unique importance of the Mladeč archaeological locality, the archaeological material found during the first field campaign was not published until 40 years later (Szombathy, 1925). A complete list of the material was published only 67 years after the last exploration of intact Upper Pleistocene sediments (Oliva, 1989; 1993a, b). This rather peculiar situation was caused by the unsystematic course of the research as well as by the fact that the finds were deposited in three museums in two countries. Furthermore, some objects from amateur excavations or random finds have been lost, without ever having been mentioned in the archaeological literature. Other complications were caused by the mergence of collections of the former hunting museum of the Dukes of Liechtenstein in Úsov with those of the museum in Litovel, and finally by the integration of the two collections to form the Regional museum in Olomouc. This disconsonant state is naturally connected with an utter lack of documentation, causing frequent confusions in the literature. In this paper, I shall try to summarize the available information, correlate the individual sources and overhaul the existing finds.

Revision of the circumstances of the finds

The first group of reliably documented archaeological material comes from J. Szombathy's second season of excavation in 1882. At the center of space E ("The Dome of the Dead") in the upper part of the sediments, there lay scattered 22 perforated animal teeth, a long bone point, several fragments of points or awls, a used lower jaw of *Ursus spelaeus*, and two flint artifacts (Szombathy, 1925, 8). These finds were mixed with bones of reindeer and *Bos* or *Bison*, and with human skeletal fragments. Apparently, the anthropological finds discovered in previous years at points "a" and "b", were not accompanied by any artifacts. Five almost complete bone points were later submitted to the Moravian Museum, along with the collection of J. Knies. It is very difficult to locate the exact place where they were found. Jan Knies concentrated on studying the area adjoining the debris cone between sites D and E (see Knies, 1905 and his research diary IV; Szombathy, 1925, 9), but there is no mention of the discovery of bone points in his records. It appears that he did not acquire them during his own excavations; perhaps he may have obtained them from some of the local people.

In 1904, an *ad hoc* quarry was opened near the entrance to the cave (site P in Szombathy's chart), and the sediments with the archaeological material of the crumbled down portal of the small horizontal cave were disposed of without informing any of the archaeologists (Knies, 1905; Maška, 1905; Smyčka, 1907, 141). Besides important anthropological and paleontological material, some bone and stone were also saved, while others, without doubt, were destroyed. Knies wrote (1905, 12) that there were bones scattered and trampled down along the road leading to the top of the Třesín Hill.

The same author mentioned that for some time phosphate loam was also exploited at this location (Knies, 1928), and it can be presumed that numerous finds may have been destroyed during the quarrying operation.

In 1911, the area became the property of the Museum Society in Litovel, and this was a real turning point in the history of research of the Mladeč Caves. The Society cleared the sediments from all areas now forming part of the route marked out for visitors. Until 1922, little was known about these clearing operations.

In 1922, a group of amateur researchers (J. Fürst, E. Smékal, H. Rohm and others) dug their way to the surface beneath the big debris cone near point "e", and found a great concentration of both animal and human bones. On the basis of the incomplete and fragmentary records made by the participants of these excavations (Fürst, 1922; 1923–24; Smyčka, 1922; 1925), it is impossible to precisely locate the site where the human remains were found. H. Rohm, one of the participants of the event however, marked the site of the find in a sketch of the cave system (Rohm in Weiser, 1928, 281, Point 3), which is in line with the report by Szombathy who located it at point "e". It follows from the above records that the human bones were discovered in the corridor between Szombathy's Dome D and E, north of the debris cone.

Szombathy (1925, 10) informs us about the discovery of two "fireplaces"; at least one of which was characterized by a thin layer of charcoals and lined with a circle of stones. Its discoverers speak of a "stone hearth" (Fürst, 1923–24; Smyčka, 1925). Around it and at the same level, there were numerous crushed animal and human bones, and several stones. About 2 m from this location, there was a heap of fragmented animal bones, bone implements and perhaps also, some human bones. The other fireplace was situated somewhat closer to both the surface and the debris cone. Large quantities of animal bones were found northwest of the fireplaces. Most of these finds consist of non-fragmented skeletal remains of *Bos* or *Bison*, with a very low degree of petrification, and are related to similar finds by Szombathy at point "d". All of these finds come from the upper layers of the trench, roughly 1 m thick. The trench reached down to a depth of 2 m, and the sediments beneath the finds were removed. The lower layers likely also contained Middle Pleistocene fauna. Smyčka (1922, 118) mentions numerous finds of mollusks and "backbones of small animals", probably snakes coming from the lower layers. There is no doubt at all, however, that large amounts of bones were pillaged and unlawfully appropriated by unwelcome visitors (the Moravské zemské muzeum acquired, for example, a collection of worked bones from M. Charvát from Třebič in 1935). Later, numerous items from the collection of the Litovel Museum were lost (e.g., lithics and all bones of Plate III.11, published by Szombathy, 1925, Tafel 3, Abb. 9). The further fate of 51 fragments of "bone awls" (?) from the old exposition near the caves, is also unclear (compare Skutil, 1938, Note 76).

In the following years no important discoveries occurred at the caves; only J. Skutil (1938, Note 77) mentions that J. Novotný found a blade core below the chimney. J. Skutil also discovered Aurignacian stone tools in the loess-loam of entrance 2 (Skutil, 1938, 32, Fig. 60).

The later excavations organized by the Moravské zemské muzeum (1958–1961) did not result in the discovery of further traces of the Upper Paleolithic settlement.

The most recent possible traces of the presence of prehistoric man in the cave were discovered by M. Oliva in 1981, in the form of red ochre signs on the walls of the "Dome of Dead" (D), and also sporadically elsewhere (Oliva, 1987).

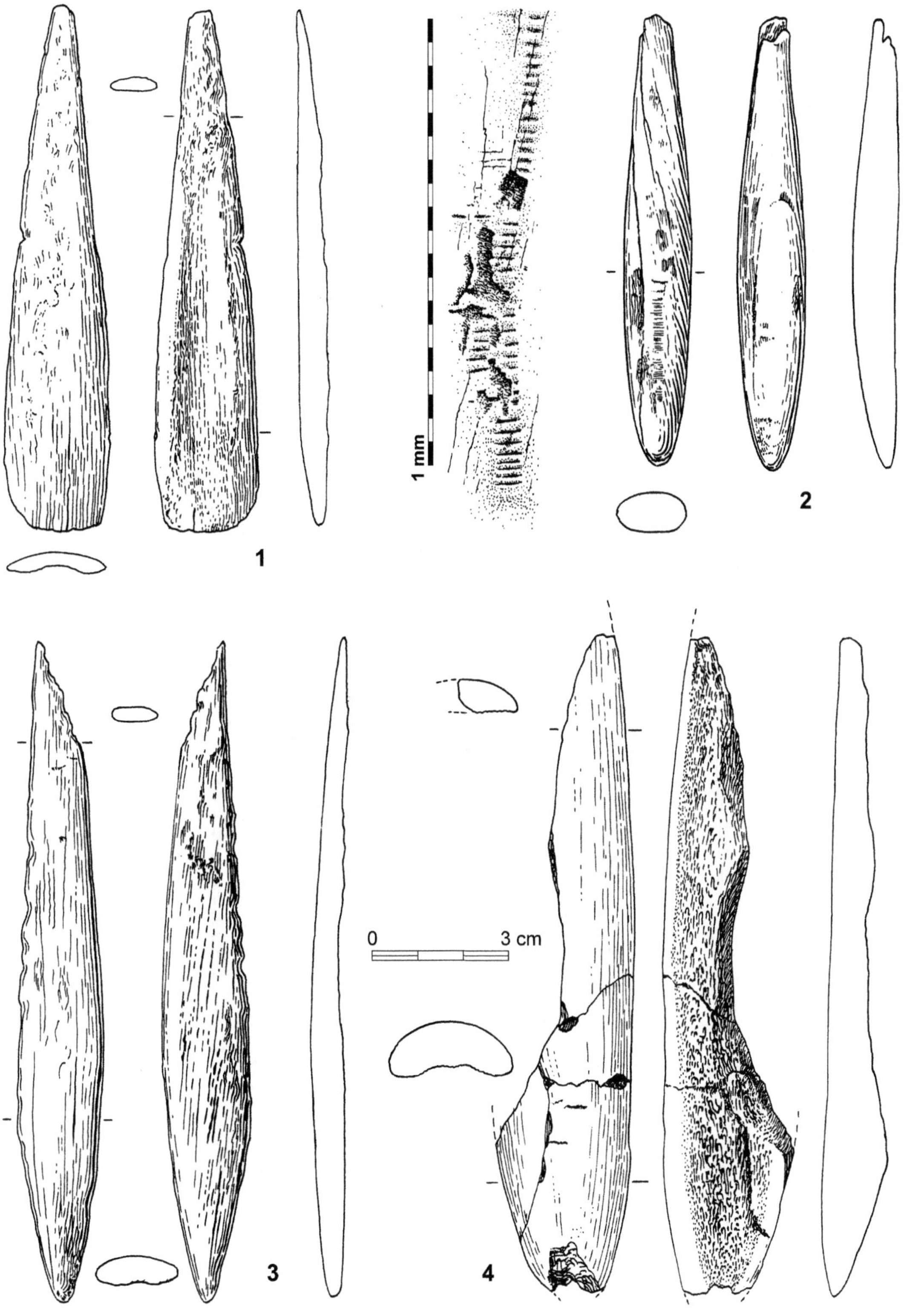

Fig. 1. Mladeč, bone points, (1–3) Main Cave, (4) Small Cave P (Drawings M. Latzmann)

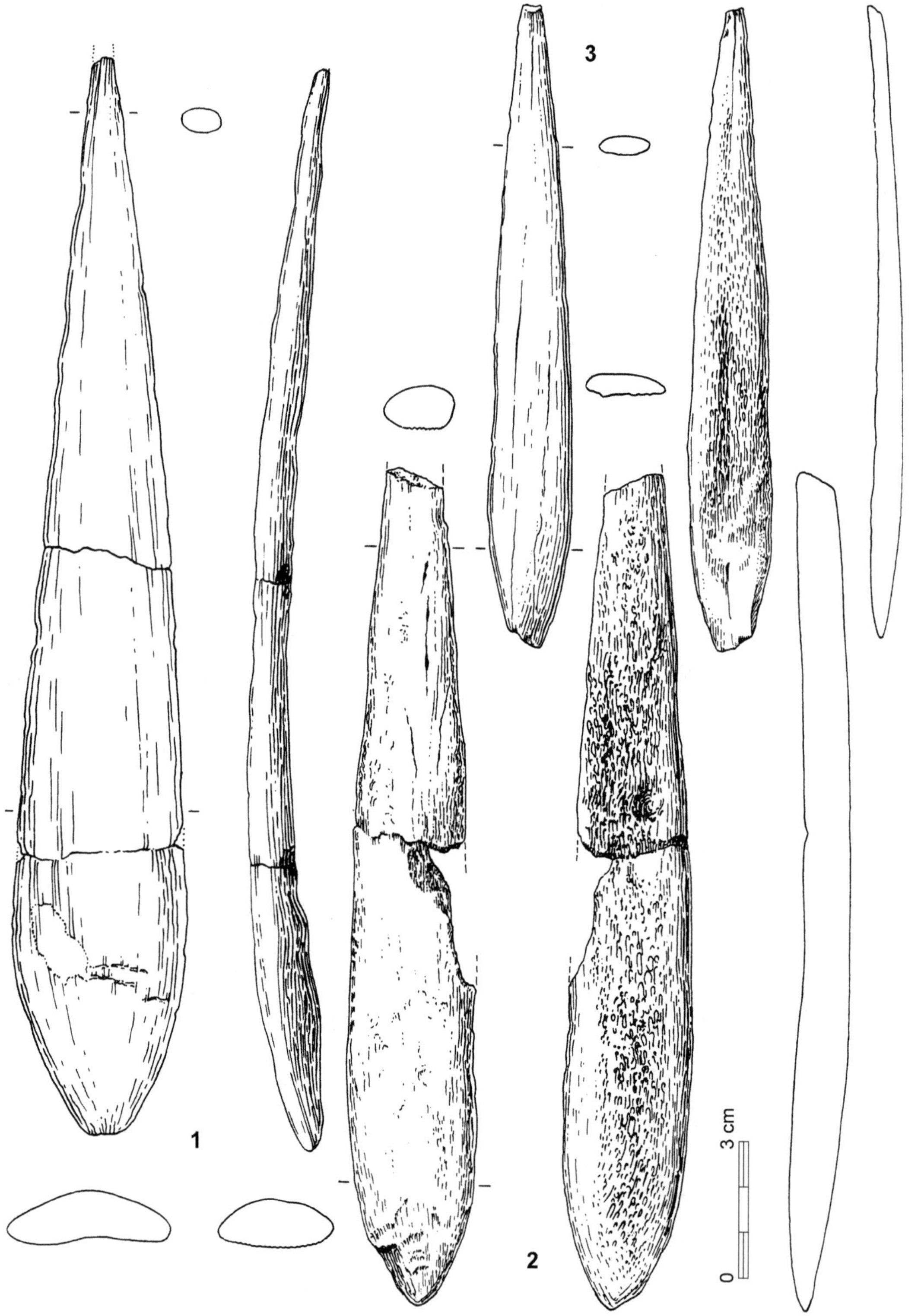

Fig. 2. Mladeč, bone points, **(1–3)** Main Cave (Drawings M. Latzmann)

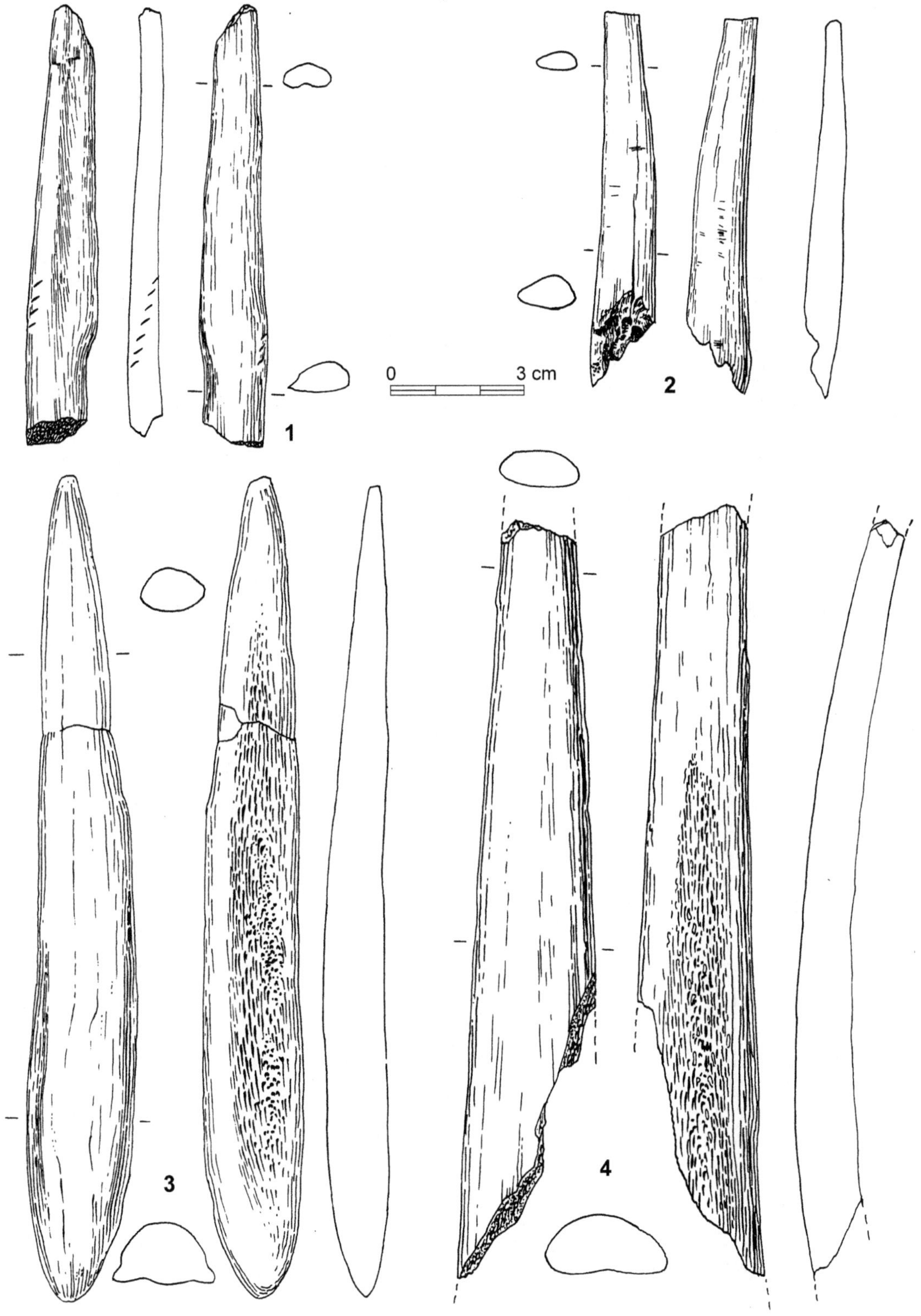

Fig. 3. Mladeč, (1–2) bones with incisions, (3–4) bone points; (2) Main Cave, (1, 3–4) Small Cave P (Drawings M. Latzmann)

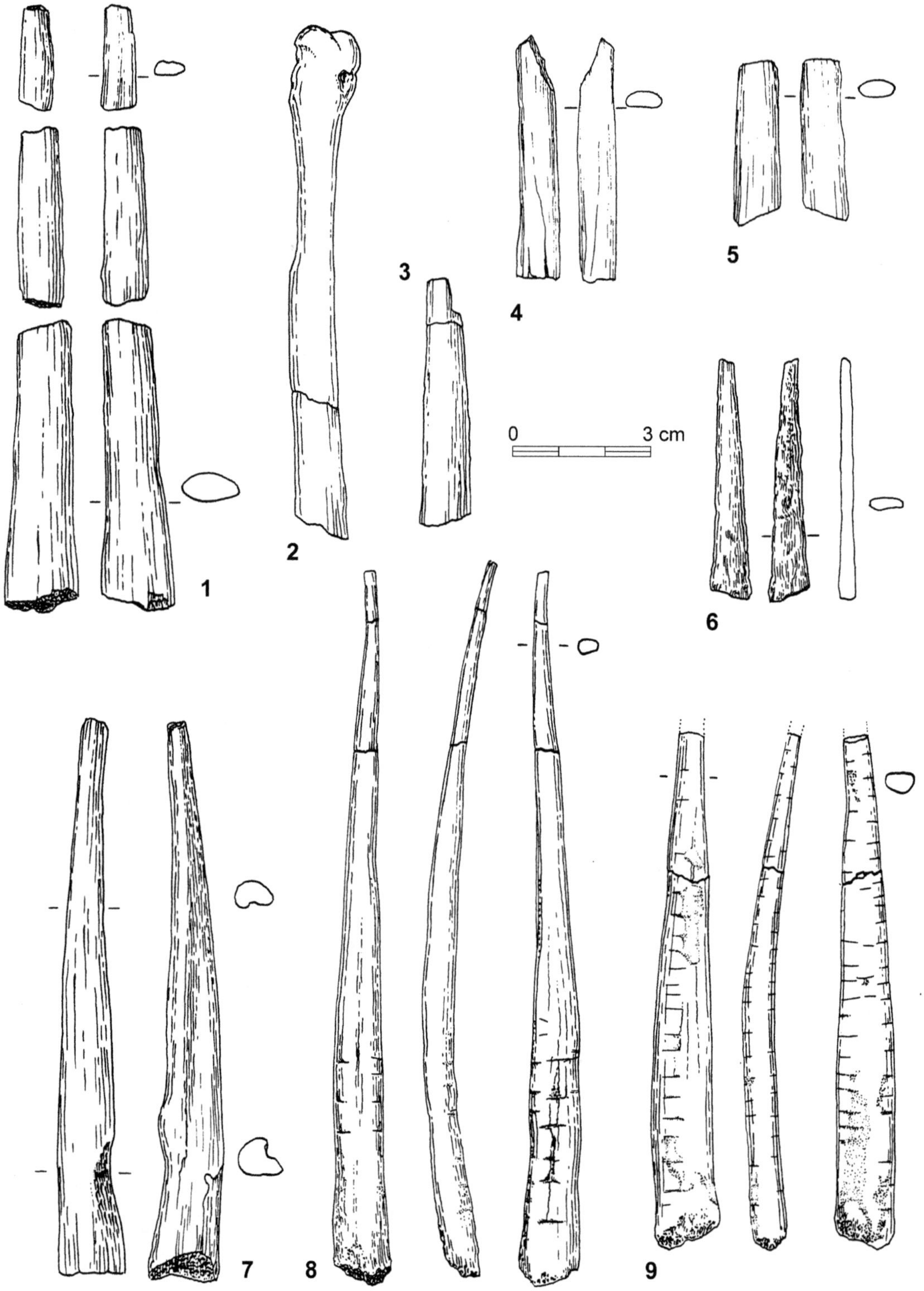

Fig. 4. Mladeč, **(1–7)** partially worked bones, **(8–9)** ornamanted bones; **(1–9)** Main Cave (Drawings M. Latzmann)

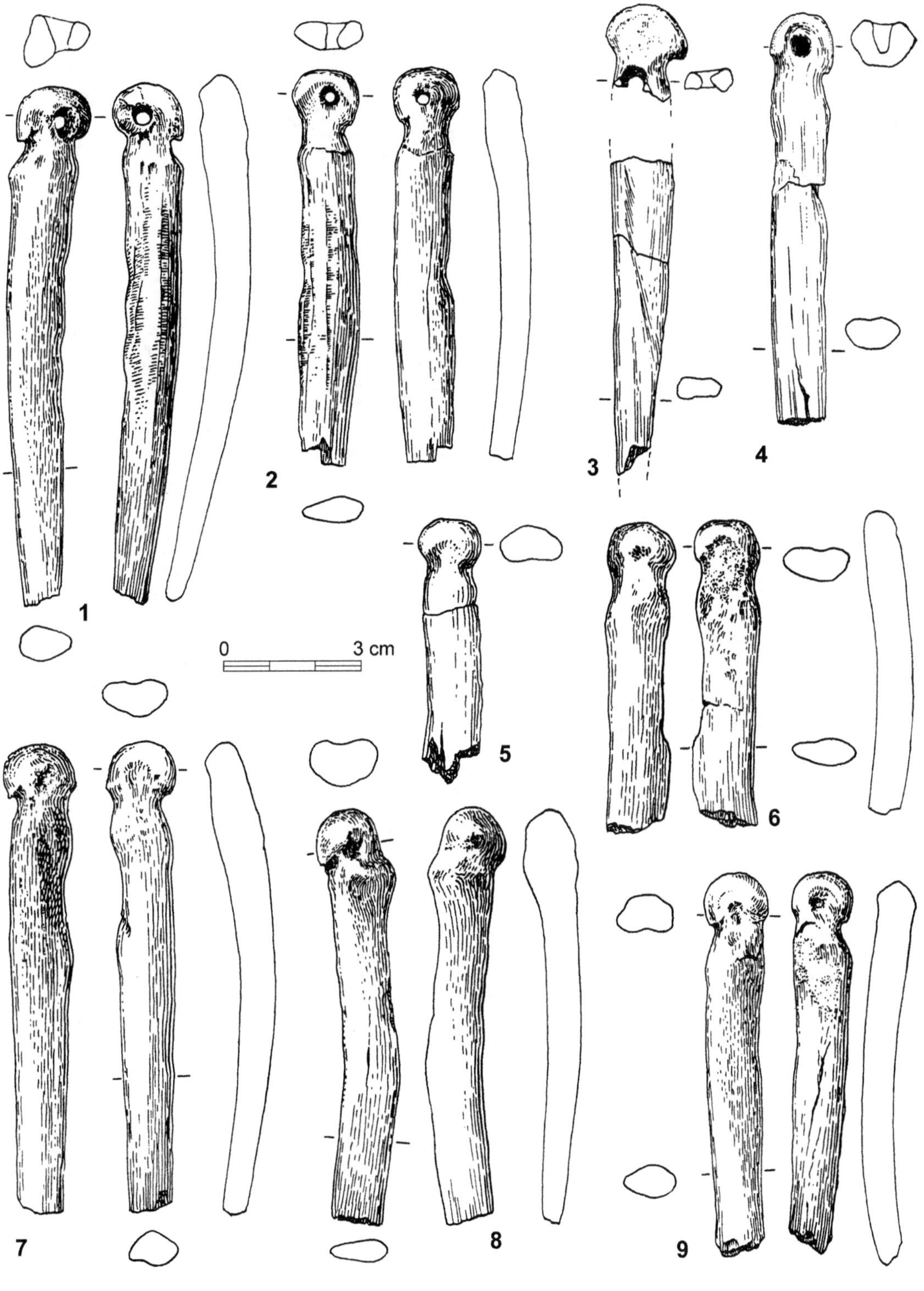

Fig. 5. Mladeč, partially worked metacarpals, **(1–9)** Main Cave (Drawings M. Latzmann)

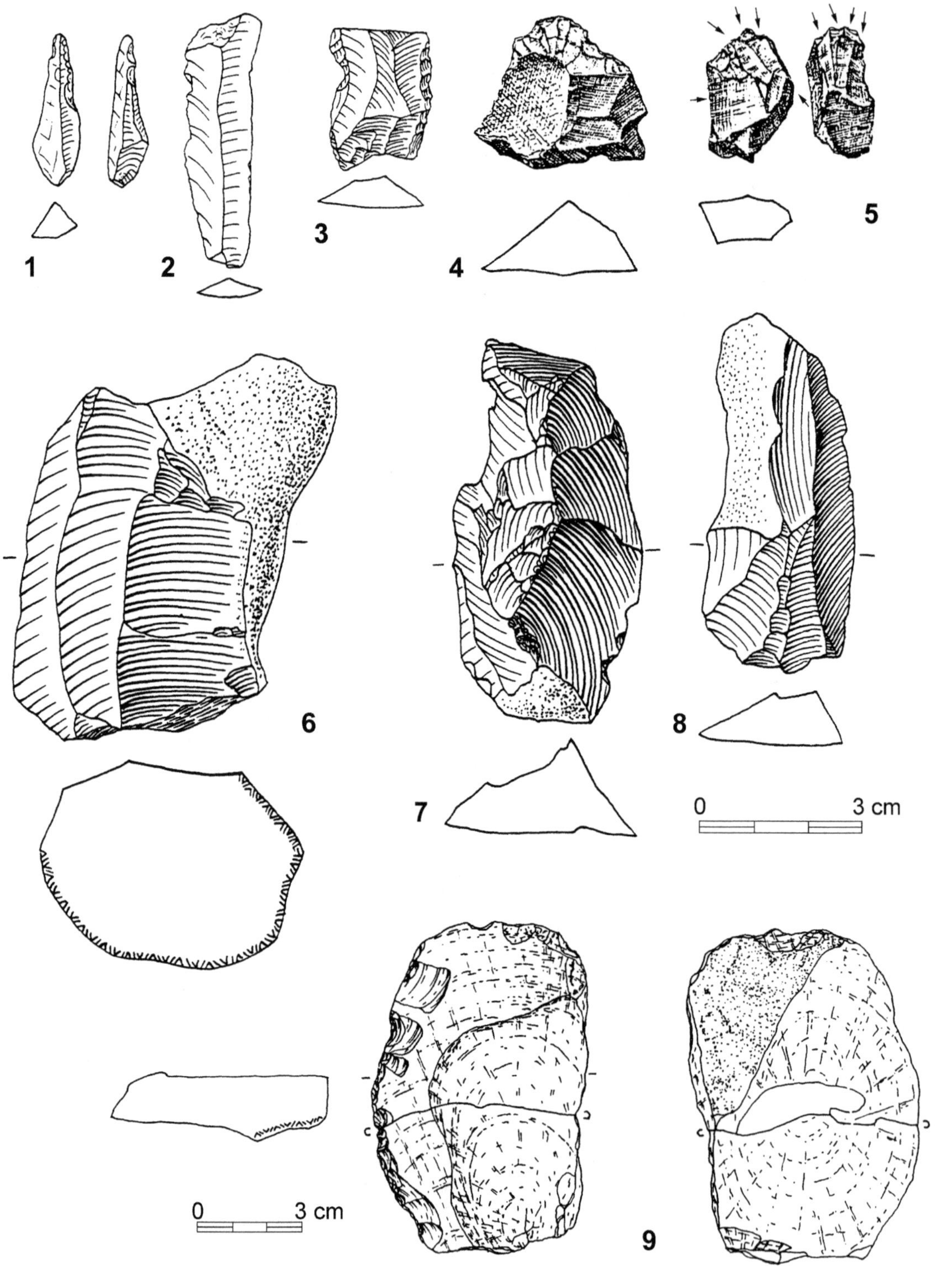

Fig. 6. Mladeč, lithic industry, **(1–6)** Main Cave, **(7–9)** Small Cave P (Drawings M. Latzmann and Z. Nerudová, (4–5) after Skutil, 1938)

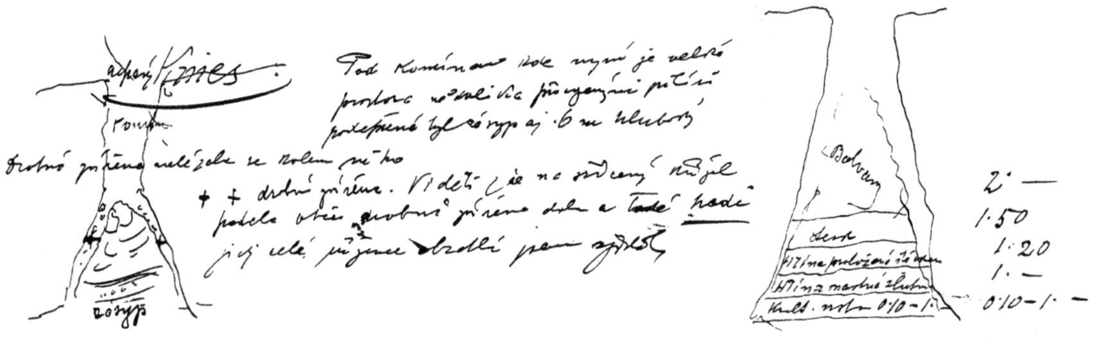

Fig. 7. Big chimney in Hall E with the horizontal deposition of archaeological layers. A sketch from Knies's diary IV, at left before excavations, at right during excavations in 1922

Descriptions of the finds

Bone industry

Bone points

1. (Fig. 2.2, Plate I.6) Reg. Nos. 4529/55 and 4545/55, Moravské zemské muzeum (later MZM): max. length 183 mm (incomplete), max. breadth 28 mm (at the proximal third), max. thickness 12 mm, the tip is broken, the base is obtuse-pointed; surface: laterally there are traces of scraping, the dorsal face is corroded, ventrally there are remnants of spongy bone. The piece has been glued together from two fragments with different alteration of the surface. *Mat.:* antler. *Loc.:* either Dome D or E (coll. J. Knies). *Ref.:* Szombathy, 1925, 13; Albrecht et al., 1972, No. 126, Tafel 8; Hahn, 1977, Tafel 126.5.
2. (Fig. 1.2, Plate I.3) Reg. No. 4530/55, MZM: max. length 96 mm (incomplete), max. breadth 15 mm (at the basal third), max. thickness 10 mm, the tip is broken, the base is obtuse-pointed; the surface is polished. The series of very fine "incisions" at the proximal part is not intentional and represents a consequence of the vibration of the iron knife during cleaning of the object. *Mat.:* ivory. *Loc.:* as above. *Ref.:* Szombathy, 1925, 13; Albrecht et al., 1972, No. 122; Hahn, 1977, Tafel 127.1.
3. (Fig. 1.4, Plate I.4) Reg. No. 4532/55, MZM: max. length 139 mm (incomplete), max. breadth 28 mm (incomplete), max. thickness 12 mm, the tip and base are broken; dorsally and laterally polished, ventrally there is spongy bone (reindeer antler), glued together from 5 fragments. *Loc.:* Small Cave "P" (coll. J. Knies, 1904). *Ref.:* Knies, 1905, 18; Szombathy, 1925, Abb. 12; Bayer, 1929, Tafel 4, bottom right.
4. (Fig. 2.3, Plate I.5) Reg. No. 4535/55, MZM: max. length 140 mm (incomplete), max. breadth 19 mm (at the proximal fourth), max. thickness 7 mm, the tip is narrow and pointed (broken off), the base is round-pointed; the whole surface is polished, ventrally there are remnants of spongy bone. *Mat.:* bone. *Loc.:* Dome D or E (coll. J. Knies). *Ref.:* Szombathy, 1925, 13, Tafel 3, Abb. 8; Albrecht et al., 1972, No. 124, Tafel 8; Hahn, 1977, Tafel 128.2.
5. (Fig. 1.3, Plate I.1) Reg. No. 4534/55, MZM: max. length 139 mm (incomplete), max. breadth 19 mm (at the proximal third), max. thickness 7 mm, the tip was presumably narrow and pointed (damaged), the base is pointed; the whole surface is polished (proximally with striae on the dorsal face), ventrally there are remnants of spongy bone. *Mat.:* antler. *Loc.:* as above. *Ref.:* Szombathy, 1925, 13, Tafel 3, Abb. 8; Albrecht et al., 1972, No. 125, Tafel 8; Hahn, 1977, Tafel 126.1.

6. (Fig. 1.1, Plate I.2) Reg. No. 4534/55, MZM: max. length 112 mm, max. breadth 72 mm (at the proximal fourth), max. thickness 6 mm, the tip is obtuse-pointed, the base is round-even; the whole surface is polished, ventrally there are remnants of spongy bone. *Mat.:* antler. *Loc.:* as above. *Ref.:* Szombathy, 1925, 13, Tafel 3, Abb. 8; Albrecht et al., 1972, No. 123, Tafel 8; Hahn, 1977, Tafel 127.2.
7. (Fig. 2.1, Plate IV), Reg. No. 72171, Prähistorische Abteilung NHM Wien: max. length 291 mm (incomplete), max. breadth 48 mm (at the proximal fourth), max. thickness 14 mm, the tip is pointed, the base is round-pointed; the surface is polished. *Mat.: Mammoth* or *Bison* rib. *Loc.:* central part of Dome D, research by Szombathy, 1882. *Ref.:* Bayer, 1922, Abb. 3, 1929, Tafel 4, bottom left; Breuil, 1925, 289; Szombathy, 1925, 13, Tafel 3, Abb. 6; Albrecht et al., 1972, No. 127, Tafel 9; Hahn, 1977, Tafel 125.
8. Base of point (Plate V.8) Reg. No. 72172, Prähistorische Abteilung, NHM Wien: max. length 48 mm, max. breadth 33 mm, max. thickness 13 mm, cross section: rectangle with arched walls, the base is slightly oblique; the surface is polished. *Loc.:* as above. *Ref.:* Szombathy, 1925, 13, Tafel 2, Abb. 5, bottom left.
9. Point base (Plate V.7) Reg. No. 72173, Prähistorische Abteilung, NHM Wien: max. length 57 mm, max. breadth 20 mm, max. thickness 9 mm, the base is round-pointed, the cross section is oval; ventrally there is spongy bone, the whole surface is finely polished. *Loc.:* as above. *Ref.:* Szombathy, 1925, 13, Tafel 2, Abb. 5, center left.
10. Tip of point? (Plate V.3) Reg. No. 72174, Prähistorische Abteilung NHM Wien: max. length 71 mm, max. breadth 20 mm, max. thickness 7 mm, both extremities are broken off, the cross-section is planoconvex; both sides are polished and finely facetted, ventrally there is spongy bone. *Loc.:* as above. *Ref.:* Szombathy, 1925, 13, Tafel 2, Abb. 5, central line, 2nd from the left.
11. Tip of point (Plate V.2) Reg. No. 72175, Prähistorische Abteilung, NHM Wien: max. length 70 mm, max. breadth 19 mm, max. thickness 8 mm, both extremities are broken off, the cross-section is planoconvex; the surface and sides are polished. *Loc.:* as above. *Ref.:* Szombathy, 1925, Tafel 2, Abb. 5, central line, 3rd from the left.
12. Distal part of point (Plate V.5) Reg. No. 72176, Prähistorische Abteilung, NHM Wien: max. length 104 mm, max. breadth 20 mm, max. thickness 10 mm, both extremities are broken off, the cross-section is planoconvex; the surface and sides are polished, ventrally there is spongy bone and a fine cover of sinter. *Loc.:* as above. *Ref.:* Szombathy, 1925, Tafel 2, Abb. 5, bottom line, 1st from the right.

    Two other objects (Reg. Nos. 72177–72178, here Plate V.1. and V.6) mentioned by Szombathy (1925, Tafel 2, Abb. 5, bottom line, 3rd from the right) as tips – *"Vermutlich einer Lautscher Spitze"* – presumably of a Mladeč point; they show no evidence of modification.
13. (Fig. 3.3, Plate II.1) Reg. No. L 2307/1, Museum Olomouc: max. length 181 mm, max. breadth 23 mm (at the proximal half), max. thickness 12 mm, the tip is pointed, the base is rounded, the cross-section is planoconvex; dorsally there are 2 grooves on the sides, it is polished especially near both extremities and ventro-laterally, ventrally there are remains of spongy bone. *Loc.:* Small Cave "P", 1904 (found by workers and handed over to the Liechtenstein collection, and later to the Museum Litovel). *Ref.:* Bayer, 1922, Abb. 5, bottom left; Szombathy, 1925, 13, Tafel 3, Abb. 10, left; Skutil, 1938, Fig. 65, left.
14. (Fig. 3.4, Plate II.2) Reg. No. L 2307/3, Museum Olomouc: max. length 168 mm (incomplete), max. breadth 28 mm (incomplete), max. thickness 19 mm, both ends are broken off, the cross-section is planoconvex; bent profile, ventrally there are remains of spongy bone. *Loc.:* as above. *Ref.:* Bayer, 1922, Abb. 5, bottom left; Szombathy, 1925, 14, Tafel 3, Abb. 10, center; Skutil, 1938, Fig. 65, center.

A 97 mm long cervid metacarpal (Fig. 3.1, Plate III.3) was presented as a bone point, but one cannot see any traces of processing on it. On its left edge, several fine cuts are visible. The bone was found together with the two above-mentioned points, and was published along with them. The 16 bone fragments (Plate III.11) found near point "e" in 1922–23 and presently unaccounted for, were described by Szombathy (1925, 13, Tafel 3, Abb. 9) as points.

Awls and perforated long bones

The older literature regarded numerous fragments of deer metacarpals as awls. The absence of pointed distal parts indicates that at least in part, fragmented metacarpal material is represented. Traces of longitudinal scraping are only visible on three medial fragments: Reg. Nos. L 2906/21–22 and 15, Museum Olomouc (Fig. 4.3–5, Plate III.6–7), and perhaps also on Reg. No. L 2906/5 (but it is not certain whether the three parts are at all related, Fig. 4.1).

Microscopic examination also revealed longitudinal striae on seven fragments: Reg. No. 4543/55, MZM. The only preserved distal tip is a laterally scraped splinter (see Fig. 4.6). There are several micro-incisions on the fragment in Fig. 3.1, similarly as the joint head in Reg. No. L 2906/7 (Fig. 4.2, Plate III.5) seems to be artificially deepened, as well as the natural reduction on the cervid metacarpal (Fig. 4.7, Plate III.4). All these finds likely come from point "e". Scores of other fragments without macroscopic traces of processing are stored in Brno and Olomouc.

From Szombathy's excavations in the central part of Dome D, six objects with certain traces of modification are deposited in the NHM in Vienna (Reg. Nos. 72176-72181, e.g., the pointed metacarpal of a horse: Szombathy, 1925, Tafel 2, Abb. 5, bottom center). On the reindeer metacarpal fragment, Reg. No. 72178, there is a well discernable proximal cut on the left side, while Reg. No. 72180 exhibits some indistinctive cuts on both sides (Plate V.4).

The most interesting ornamented awls come from the collection of Jan Knies (MZM Reg. Nos. 4535/55 and 4543/55). Both are made of metacarpals of a large deer, perhaps of *Megaceros* (determined by L. Seitl). The larger of the two (Fig. 4.8, Plate III.2) exhibits 6 oblique incisions ventro-proximally at its center, 3 ventral incisions on the left, 2 on the right, and 3 dorsally on the left and 4 on the right. On the smaller artifact (Fig. 4.9, Plate III.1), the incisions are positioned in the following way: 22 dorsally on the right (19 + 3 on the proximal part), about 17 on the left, and at least the same number ventrally on the left half, partially covered with sinter. Both awls show marked traces of lateral scraping (and also, distally on the surface).

Area "e" beneath the big chimney also yielded numerous metacarpals of large deer or elk, with a preserved and often worked joint head. As documented by the fragments connected by sinter (Plate III.8–10), they must have been heaped up in the given area by human hands. On three pieces, one can see complete biconical drill holes in the epiphyses (Reg. Nos. 4538–9/55, L 2906/1, Fig. 5.1–3, Plate II.5–7) and in those very pieces, traces of longitudinal scraping with subsequent polishing in a transverse direction, can be seen most markedly. The drill hole in Fig. 5.4 (Plate II.3) has not been completed, and in other specimens, there are only shallow pits in the joint heads (Fig. 5.7? 5.8, 5.9?). Other alterations consist of deepening the natural reductions below the epiphyses (Fig. 5.5, 5.8, Plate II.4). In view of the missing distal parts, one cannot determine their function. Only two somewhat different (but unfortunately lost) perforated pieces of the old Litovel collection (Plate III.11, 3rd and 5th from the left) resemble awls.

Perforated teeth (all deposited in the Prähistorische Abteilung, NHM Wien)

All 21 perforated teeth lay scattered at the center of space D:

1. Reg. No. 72183 (Plate VI, left), canine of a brown bear, length 74 mm, distally broken, bore-hole on one side damaged, originally likely biconical.

2. Reg. No. 72184 (Plate VI, center), canines of a wolf, length 53 mm, broken in the bore-hole, probably artificially flattened on one side around it.
3. Reg. No. 72185 (Plate VI, right), incisor of a horse, length 73 mm, root slightly damaged. A small irregular hole is visible only on one side, 2–3 mm in diameter, probably not artificial.
4. Reg. No. 72186 (Plate VII), 9 beaver incisors, biconical bore-holes commenced by bilateral whetting (not boring), among them:
   - 3 with a complete bore-hole (max. length 53 mm)
   - 4 with an irregular break in the bore-hole
   - 2 with the entire bore-hole area broken off
5. Reg. Nos. 72187 and 72188 (Plate VIII), 7 elk or Megaceros incisors (Hoernes, 1903, Abb. 73; Szombathy, 1925, Tafel 2, Abb. 5), among them:
   - 1 complete, length 54 mm
   - 6 broken in the bore-hole (1 recently)

All bore-holes are biconical and commenced by easy whetting. In contrast to the data given by Szombathy (1925, Tafel 2, Abb. 5), one of the items is missing in the collections.

Special finds

1. Reg. No. 72182, Prähistorische Abteilung, NHM Wien: left half of a mandible of *Ursus spelaeus* with condyles, molars and canine broken off (Szombathy, 1925, 13, Tafel 3, Abb. 7). The authenticity of this modification is questionable.
2. Posterior part of an aurochs skull, after Szombathy, with traces of hammering (Szombathy, 1925, Tafel 2, Abb. 3). Not found in the NHM collections.
3. Reg. No. 72216, Prähistorische Abteilung, NHM Wien: femur of an aurochs with parts of both epiphyses broken off (artificially?, Szombathy, 1925, Tafel 2, Abb. 4).

Re-examining the osteological material from excavations carried out between 1922–1923, we found minor but very dense striations resembling, at first sight, traces of processing. Detailed microscopic examination revealed that the striae have the same intensity over differently eroded parts of the bone surface. Thus, they may have arisen as a result of movement through sandy sediment (d'Errico and Giacobini, 1986; Cook, 1986). The striae mostly appear on the pelvic bones, and rarely on those of other animals. The most intense striations occur on one calcaneus of a *Bos* or *Bison,* however, they appear on the eroded surface (Plate IX.6). Besides this, some of the objects (ivory point, Reg. No. 4530/55, Fig. 1.2; fragment of bovid pelvis) exhibit a series of very fine and short striae (Plate IX.7) that, in the opinion of N. Praslov, are produced by the vibration of the knife-edge during cleaning of the specimen.

Lithic industry

1. Non-retouched bladelet (Fig. 6.2) Reg. No. 72169, Prähistorische Abteilung, NHM Wien: length 45 mm, erratic flint or Jurassic chert. *Loc.:* central part of Dome D. *Ref.:* Hoernes, 1903, Abb. 73; Szombathy, 1925, 13, Tafel 2, Abb. 5; Hahn, 1977, 111.
2. Burin spall with distal end utilized (Fig. 6.1) Reg. No. 72170, Prähistorische Abteilung, NHM Wien: length 27 mm, erratic flint or Jurassic chert. *Loc.* and *Ref.:* as above.
3. Elongate flake of core crest (Fig. 6.7) Reg. No. 4508/55, MZM Brno: length 69 mm, erratic flint. *Loc:* Small Cave "P" (coll. J. Knies, 1904). *Ref.:* Knies, 1905, 16; Szombathy, 1925, 14, Tafel 3, Abb. 12; Skutil, 1938, Fig. 67.
4. Irregular blade (Fig. 6.8) Reg. No. 4507/55, MZM Brno: length 65 mm, erratic flint. *Loc.* and *Ref.:* as above.

5. Natural fragment of erratic flint with some local trimming on the edge (Fig. 6.9) Reg. No. L 2988/A 61597, Museum Olomouc: length 90 mm. *Loc.:* Small Cave "P" (coll. Smyčka, 1904). *Ref.:* Smyčka, 1907, 148; Szombathy, 1925, 14, Tafel 3, Abb. 11; Skutil, 1938, 46, Fig. 66.
6. Trimmed pebble (pre-core, Plate IX.2) Reg. No. 2985, Museum Olomouc, and broken off fragment Reg. No. 4510, MZM Brno: length 173 mm, stripped chert with black cortex. *Loc.:* as above. *Ref.:* Knies, 1905, 16; Smyčka, 1908; Bayer, 1922, Abb. 5; Szombathy, 1925, 141.
7. Pebble with two flake scars (Plate IX.1) Reg. No. L 2989, Museum Olomouc: length 53 mm, chert with black cortex (Krumlovský les-type). *Loc.:* Small Cave "P". *Ref.:* Szombathy, 1925, 14.
8. Pebble with flat flake scar (Plate IX.5) Reg. No. L 2984, Museum Olomouc: length 173 mm, thickness 48 mm, limestone. *Loc.:* at point "e" (coll. J. Fürst, 1922–23). *Ref.:* Szombathy, 1925, 13; Skutil, 1957, Table 6, on the right.
9. Hammerstone (Plate IX.3) Reg. No. L 2987, Museum Olomouc: length 44 mm, thickness 32 mm, both ends are heavily worn, quartz. *Loc.:* as above. *Ref.:* Szombathy, 1925, 13.
10. Pebble-hammerstone (Plate IX.4) Reg. No. L 2986, Museum Olomouc: length 98 mm, thickness 55 mm, traces of hammering at the end, at places ochre-colored stains, limestone. *Loc.* and *Ref.:* as above. Skutil, 1957, Table 6, left.
11. Pebble-hammerstone, allegedly Museum Olomouc: dimensions 102 x 93 x 33 mm, battered along its circumference, quartzite. *Loc.:* as above. *Ref.:* Skutil, 1957, 415, Table 5, left.
12. Pebble-hammerstone, allegedly Museum Olomouc: dimensions 96 x 87 x 41 mm, its circumference is battered, quartzite. *Loc.:* as above. *Ref.:* Skutil, 1957, 315, Table 5, right.

13–14. Two small chert flakes, lost. *Loc.:* as above. *Ref.:* Fürst, 1928–30, 7 (?); Szombathy, 1925, 13; Blekta, 1932, 29.

15. Core artifact, lost, dimensions 75 x 20 x 18 mm, flint or chert. *Loc.:* as above. *Ref.:* Skutil, 1938, 65, Note 77.
16. Uni-directional blade core (Fig. 6.6) Reg. No. L 2990, Museum Olomouc: dimensions 70 x 50 x 39 mm, blackish silicite (flint?). *Loc.:* cave interior? *Ref.:* Oliva, 1989, Fig. 6.10.
17. Retouched blade, MZM Brno: length 41 mm, chert. *Loc:* cave interior (coll. V. Gross). *Ref.:* Oliva, 1989, 41.
18. Retouched blade fragment with a notch on opposite edge (Fig. 6.3) MZM Brno: length 25 mm, erratic flint. *Loc.:* near point "e". *Ref.:* Oliva, 1989, 41.
19. Thick nose-shaped scraper (Fig. 6.4) originally MZM Brno, lost, cast in NHM Vienna (cf. Hahn, 1977, 111): length 28 mm, flint or chert. *Loc:* Hallway A. Coll. and *Ref.* Skutil, 1938, 32, Fig. 62.
20. Small core or polyhedric burin (Fig. 6.5): length 29 mm, other data as above.

Rope

In his first report, Szombathy (1882, 17) documents his discovery of a carbonized rope beside sintered fragments of a human calvarium:

> *"Daneben fand sich nebst anderen kohligen Spuren ein gänzlich verkohltes Stück von 12 cm Länge, welches wahrscheinlich der Rest eines starken Seiles ist und noch sehr deutlich parallel neben einander liegende gedrehte Schnüre von 1 mm Durchmesser erkennen lässt".*

In Szombathy's final report, however, the rope is no longer mentioned, since it became one of the reasons for doubting the Paleolithic age of the human remains (Obermaier, 1905). Nevertheless, this could obviously be the earliest record of Paleolithic (?) rope-making; a skill that became a much discussed topic in connection with the discovery of Pleistocene textiles (Adovasio et al., 1996). It should be pointed out that remains of a rope were also found at Lascaux (Leroi-Gourhan and Allain, 1979).

Ochre-colored marks on walls

On August 20th, 1981, accompanied by Dr. Přemysl Ryšavý, I visited the Mladeč Caves to examine the possibility of continuing of our research. On that occasion, I noticed several ochre-colored marks on the walls of the "Dome of the Dead" (D), as well as in other spaces. So far, only a preliminary note has been published on these observations (Oliva, 1987; 2003).

1. Double line, irregularly interrupted in places, straight, 22 cm long. Site: the northern wall of the rock column in the middle of Dome D, on the right of the vertical erosion furrow in the wall (Plate X.1), about 2.3 m high above the concrete floor.
2. Double line, blurred in places, converging towards the lower ends, not straight, widened in places, 18 cm long. Site: as above, to the left of the vertical furrow (Plate X.1), about 2.35 m high.
3. A pair of irregular lines, uneven in length, max. length 19 cm, max. distance 2 cm (Plate X.2). Site: not quite 2 m east of No.1, on the edge of the column, about 2.4 m high.
4. "Barbed signs" tapering towards the bottom, left one interrupted by a change in direction, length 16 cm (Plate X.3). Site: on the left, over "Fürst's plate", on the wall north of point "b" in Szombathy's map, 2.5 m high.
5. A row of dots, double on the left (distance between the rows about 5–7 cm), simple on the right, made with pale brick-colored ochre pigment, length 19 cm. In contrast to the preceding marks, the dots are on the very rock surface. Site: 35 cm to the right of No. 4, to the left of the vertical cavity in the wall, 2.6 m high.
6. Approximately six irregular, not blurred small dots, rather only irregular blotches of the pigment. Site: min. 28 cm away from No. 4, 2.5 m high.
7. An irregular arrow-shaped figure pointing obliquely downwards, on the rock surface (easily erased), length 16 cm. Site: on the north side of the column with marks 4 to 6, about 2 m above the stairs, approximately 15 cm lower than No. 4.
8. Some unclear parallel vertical lines, approximately 15 cm long, apparently consolidated in the flowstone. Site: 3.5 m above the stairs between F to E on Szombathy's plan, 4 m from No. 7.
9. Interrupted horizontal line about 2–3 cm broad and about 2.5 m in total length, interrupted at its left part by the erosion furrow in the wall. Below the line, there are indistinct blurred stains of the same rusty color. Site: 10 m east of the preceding group, about 3.5 m above the stairs.
10. Almost vertical line, 1 to 1.5 cm broad and 20 cm long, brownish ochre in color, applied to a surface of warty sinter, readily erased (Plate X.4). Site: to the left of the passageway from Dome E to the Panenská Cave, 60 cm high above a relict sinter plate on the wall, actual height about 3 meters.
11. Several blurred ochre blotches, probably of natural origin, in a smooth stalagmite flowstone. Site: to the right of the passageway from Dome E to the Panenská Cave, under an overhanging rock, about 1.4 m high above the recent floor.
12. Several indistinct, vertically blurred ochre blotches. Site: in the Netopýří Cave (newly discovered south of the H in Szombathy's plan) above the eastern lateral room, about 3 m high above the ground.
13. A compact group of blotches and lines, pale ochre in color, 30–40 cm in length (the so-called "Hedgehog"). Site: above the stairs leading to the Nová (New) Cave (to the east of the K in Szombathy's plan) about 4–5 m above the ground.

A sample of the ochre pigment taken from mark No. 6 was subjected to semi-quantitative analysis (performed by mgr. V. Selucká). The analysis revealed spectral lines of Fe and Cu, and to a smaller extent, Si, Mg and Ca. The question of the age of the marks, however, remains open to debate, and it is even probable that not all of them originated in the same epoch. This is suggested by three letters of D, written in the same ochre pigment on the wall bearing numerous ancient signatures near the stairs leading to the "Upper floor". Above all, the marks applied to the flowstone are the most uncertain. Marks Nos. 1, 2 (Plate X.1) and 3 (Plate X.2), found in the eastern part of the "Dome of the Dead" directly above the finds of human skulls in 1881, however, are very consolidated and covered in place with a hard calcareous coat. Illuminated by an ultraviolet lamp, these old sinters appear black, whereas the recent (active) flowstones and the subrecent inscriptions emit a clear glow. All marks, the same as the cave loam, appeared black. Even the symmetrical location of marks 1 and 2 is conspicuous: they are situated on either side of the vertical erosion furrow, with which they might form a semantic unit.

Comparisons

The only culturally significant lithic artifact is the Aurignacian nose-shaped scraper found in the entrance area. The most striking thing about the other lithics is that in spite of their modest occurrence and the variety of raw materials, they comprise specimens from all production phases of the chipped industry, i.e., roughly prepared cores (Plate IX.2), prismatic cores (Fig. 6.6), blades and retouched tools. The isolated occurrence of the lithic industries and the variety of raw materials used (Moravian Jurassic chert for rough cores, erratic flint for flakes and tools), however, suggest that they were not manufactured locally. It appears as though this is a group of artifacts randomly brought together, without functional relations. The disproportionately large number of hammerstones seems to support the viability of the above conclusion.

On the other hand, at least part of the much richer bone industry was manufactured on the spot, as documented by the heaps of deer metacarpals in various phases of processing. The most important components of the bone tools are the points with massive bases. With their relatively versatile shapes, numerous analogies can be found with various Central European assemblages. The most typical variant is represented by the two elongated points with a pointed base and tip (Nos. 4 and 5), prevailing analogy with the Potočka Cave in Slovenia (Brodar, 1983). But in both cases, their cross-section is extraordinary thin. Less frequent are the thin points of triangular shape, known only from layer VII in the Bockstein Törle and Vogelherd VI in Central Europe (Albrecht et al., 1972, Table 2.7, 4; 52).

But thin triangular-shaped points are quite common amongst points with a split base. The large point No. 7 is extraordinary long, with its original length, no doubt, exceeding 30 cm. Only two specimens of similar dimensions are known from the Mamutowa Cave near Cracow (Kozlowski, 1924), and only one from Wildhaus-Höhle considerably exceeds them both in being 40 cm long (Hahn, 1977, Tafel 75.1). The closest morphological analogy is the large point made of a femur of *Ursus spelaeus* from the Ciutarun Cave in northwestern Italy, published as a chalcolithic dagger (d'Errico and Giacobini, 1985). Point No. 1 (Fig. 1.2) is remarkable with regards to its raw material: it is made of (mammoth) ivory. Mammoth ivory points are rare; they have occurred, for example, in the Mamutowa Cave near Cracow, in the upper layer of the Istállóskö Cave in Hungary, and in some caves in southern Germany. Apparently, they are missing from the largest bone-point collection found in the Potočka Cave (compare Brodar, 1983, 195). Even their shape is somewhat unusual. To the few examples (e.g., from Willendorf II/4, Potočka and Temnata Dupka, comp. Albrecht et al., 1972, Fig. 16), one may also add a surface find of a bone point from Hluchov, about 18 km south of the Mladeč Cave (Oliva, 1987, 44; Valoch, 1993b). The relatively variable assemblage of points from Mladeč does not include a rhomboid type known, for example, from Willendorf, nor does it include one with elongated and narrow distal parts, typical of the Mamutowa Cave.

It is quite surprising that there is not a single point with a split base. The thin cross-sections, typical of Mladeč, would fit in better with split bases (Albrecht et al., 1972, 60).

The medial fragment in Fig. 3.4 can be interpreted as a point or "lissoir" (compare Leroy-Prost, 1975, Fig. 7; Hahn, 1977, 74).

The rest of the bone industries are surprising in their absence of such common utility tools as awls. Two ornamented pieces resembling awls, but likely belonging to the group of non-utilitarian artifacts, are both made of cervid metacarpals, and are unusual in their form of processing and lack of distal tips. The effort of narrowing the heads in Fig. 5.5, 6?, 9?, has its closest analogy with the fragment of rib exhibiting bilateral reductions from the upper layer of the Istállóskö Cave (Vértes, 1955, Plate XLI.8). Bilateral incisions are quite frequent in the Aurignacian, either on the points (namely from the Potočka Zijalka, Brodar, 1983, Tafel 7–12, 14, 16, 21–22 etc.), or on flat bone stylets from Vogelherd (Hahn, 1977, Tafel 34, 36, 58–59), Sirgenstein (67, 9), Peskö (149, 2), Velika Pečina (Malez, 1967, Tafel VI.2), and Potočka (Brodar, 1983, Tafel 22.21). The two ornamented awl-like artifacts from Mladeč are thus, unique in the Central European Aurignacian.

The discovery of polished metacarpals with perforated or at least narrowed points is most unusual. Perforated bones from other localities have a completely different character: holed bear mandibles and long bones from the Potočka Zijalka (Brodar, 1983, 155–158), and long bones from both Lokve (Bayer, 1929, Tafel 5) and the Istállóskö Cave - upper layer (Vértes, 1955, Plate XLIII), are actually considered to be consequences of natural forces. Ornamented bones are only known from Vogelherd (Hahn, 1977, Tafel 37.1, 59.5) and the Geissenklösterle Cave (Hahn and Münzel, 1995; for other such finds see Turk and Kavur, 1997 in Divje Babe cave, d'Errico et al., 2003, 39–48) from which the well-known bone flute derives. In addition, a point with a broken perforated base is also known from the Potočka Zijalka (Brodar, 1983, Tafel 12.75), and perforated proximal parts of awls have also been published from Southwestern France (Leroy-Prost, 1975, Fig. 20.4, 11). The closest analogy to our objects in Fig. 5.1–4, however, can be found with the Kostienki 8 site, layer II, situated in the upper humus layer (Praslov and Rogačev, 1982, Fig. 33.17).

Pendants made of animal teeth are typical Aurignacian ornaments (Hahn, 1972; White, 1993), but their prevalence over the lithic implements at Mladeč is quite exceptional. The regular rows of incisions on bones are quite common in the Aurignacian (Marshack, 1972).

On the chronological position and cultural pertinence of the assemblages with Mladeč points

Bone points, of both the Aurignacian and Mladeč type, occur in Central Europe in assemblages with thick endscrapers, Aurignacian blades, leaf points and backed instruments.

Some prehistorians hold, therefore, that these bone points represent an intercultural phenomenon, more likely connected with hunting activities than with a particular culture (Albrecht et al., 1972, 72). But one should not overlook the fact that non-Aurignacian lithics occur with bone points only in caves, and above all, in those that can be reliably regarded as hunting stations due to the narrow spectrum of tools represented (retouched blades and flakes, points), and the presence of skeletal remains of hunted fauna (with the cave bear prevailing). There is perhaps one exception, an artifact resembling a Mladeč point found at Pavlov (Klima, 1957, Fig. 16). In the outfit of Aurignacian hunters, bone points have the same role as pointed specimens of the leaf points in Szeletian and the backed points in Gravettian. It is logical, therefore, in caves occasionally serving as hunting stations like Mamutowa, Oblazowa, Dzeravá skala, Szeleta, Istállóskö and Vindija, that these tools should occur side by side. It is a meeting of functionally identical components of tools and weapons used by various cultures. The characteristic Aurignacian-type chipped tools, i.e., carinated scrapers and burins, obviously did not form a necessary part of the hunters' outfit and thus, occasionally we do not find them at these localities (missing from Istállóskö, but appearing at Potočka, Brodar, 1983, Tafel 4). Carinated scrapers, mostly without any trace of utilization, are very likely the result of relaxation activities. The occurrence of bone and leaf points in the same stratigraphic context is also due to considerable cryoturbational and biological disturbances of Middle Würmian sediments (Dzeravá skala, Vindija). At such sites, intrusions from older Mousterian layers are also possible (see the confusion in the radiocarbon and other dates at Vindija, Smith et al., 1999; Karavanič, 2000). Thus, the "archaism" and the typological poverty of the assemblages mentioned, may be the result of functional and geological circumstances, and do not necessarily justify the resurgence of the term "Olschewian" to designate a regional cultural unit at the beginning of the Upper Paleolithic in Croatia and Slovenia (Montet-White, 1996; Karavanič, 2000). In Vogelherd, Bockstein, Geissenklösterle, Bacho Kiro and Temnata, where rich lithic industries document continuous occupation, bone points occur in an unequivocally Aurignacian context. A similar situation is found at the Willendorf open-air site.

The overall geographic occurrence of Mladeč points covers practically the extension of the Aurignacian culture, including the Transcaucasian (Kozlowski, 1998, Fig. 14) and Near-Eastern sites (Bar Yosef and Belfer-Cohen, 1996). In the classical Aurignacian region of Southwestern France, however, there is apparently no such characteristic group of hunting stations, and the bone points commonly occur in localities under rockshelters, thus combining both settlement and hunting functions.

In Western and Central Europe, most of the points with a split base are found stratigraphically lower than those with a massive base. The question is whether this phenomenon is of any chronological importance, i.e., whether the points with a split base form a closed, older horizon.

The majority of bone points with a massive base from the Balkans and Central Europe are contemporaneous or even older than the main chronological horizon containing split-base points (i.e., Aurignacian I) in Western Europe (Geissenklösterle layer II: 33,5 to 30,6 kyr BP; Willendorf II/4: 32,0 - 31,7 kyr; Bacho Kiro 8: >33,3 kyr). It is true, however, that the split-base points are still sometimes older in the Balkans and Central Europe (Bacho Kiro 9: ~36 kyr; Istállóskö: 39 kyr?; Peskö: 34,6 kyr; Szeleta: 40 or 32,5 kyr; Divje Babe: 35,3 kyr) as well as exceptionally in the West (El Castillo: 37,7–40 kyr; Arbreda: 37,3 kyr). On the other hand, there are also very late occurrences of this point type (Vogelherd V: 31,9–27,4 kyr; Fumane: 31,7 kyr). Additionally, the top layer from Istállóskö, dated to 31,5–30,9 kyr, has yielded some specimens with a split base (compare Cabrera-Valdez and Bischoff, 1989; Broglio, 1993; Hahn, 1977, 168; 1988, 205, 208; Hahn et al., 1977, 20; Knecht, 1993; Kozlowski and Otte, 2000; Kozlowski, 1982; Montet-White, 1996, 100; Richter et al., 2000; Ringer et al., 1995). At the Croatian localities that have yielded poor assemblages (Velika Pečina, Vindija), the two point types occur in one and the same layer; a fact that may, but by far does not suggest that they were also used at one and the same time period.

While the assemblages with prevailing massive-base points occur separately (Mladeč, Potočka, Dzeravá skala, Mamutova Cave, etc.), the layers dominated by split-base points in Central and Southeastern Europe are always covered with layers containing the Mladeč type points (Vogelherd, Geissenklösterle, Istállóskö, Mokriška jama, Bacho Kiro 8 and 9). Even though the two basic bone point types considerably overlap chronologically on the overall European scale, not a single case is known thus far in which a massive-base point has been found beneath a layer with prevailing split-base points. Chronospatially, one can see here relations of a vertical (chronological) rather than horizontal character. The question is posed whether this is only due to the present status of research (or the imperfection of radiometric dating), or whether it is a manifestation of independent developments of locally bounded populations, returning to a certain cave site over a long period of time.

This development may have also been due to purely technological causes. The more complex and sophisticated shape of the point with a split base was a rapidly spreading novelty (in fact, it was the first bone tool shaped in a complex way), but its functional properties were poor. The point is that the progressive idea of fitting the handle into the split base of the bone tool was dissatisfactory and caused frequent ruptures of the point (Bordes, 1967, 28). Later, therefore, the less thought-out but more effective full-base type prevailed.

Were the caves accessible in the Upper Paleolithic?

All those who have witnessed excavations in original layers of interpleniglacial age agree that caves were visited by man and that therefore all anthropological remains and animal bones were brought into them together with parts of man's prey. Other parts of animal bones may have entered into the caves through a chimney (this pertains especially to those bones that were found directly at its base). Only J. Smyčka advocated the idea that even the human remains were dropped into the caves through a chimney, probably after a ritual. In despite of this, he considered the presence of humans inside the caves as obvious (Smyčka, 1922, 118–119). Following the excavations carried out by the MZM, however, hypotheses appeared stating that during the last Glacial the caves were no longer accessible, and that all Upper Pleistocene sediments had dropped into them from the surface of the Třesín Hill through the chimney over the surface of the Middle Pleistocene talus. The facts supporting this hypothesis were summarized by Jelínek (1987, 62–63, 97 (Discussion)) and followed up by Svoboda (2000; 2001; 2002), on the basis of a comparison with the locality of Koněpruské jeskyně (Zlatý Kůň) in the Bohemian karst.

Having studied all available reports as well as the situation inside the caves, I prefer the original explanation. My opinion is based on the following reasons:

(1) There was no direct connection between the sites of Szombathy's Upper Pleistocene finds and the space below the chimney. Szombathy (1882, 101) reports that there are two bends in the corridor between site D (with finds of undamaged bovid bones) and space E below the chimney. Point "b" with anthropological finds (near the present, erroneously placed "Fürst Memorial Plaque") communicates even at present with the space in front of the talus by a mere curved and narrow corridor that, at the time of excavations performed by the Moravian Museum, was already filled solely with Middle Pleistocene sediments. On Szombathy's map (1882, Tafel II, Fig. 3), however, even this connection is missing. Thus, the two spaces were probably not connected on the level of the Upper Pleistocene sediments. A similar situation is also found in the middle part of the "Dome of the Dead" with finds of human bones; there was no connection between the eastern projection of Dome D (with the find of a human skull near point "a") and space E below the chimney. In Szombathy's original plan, the "Dome of the Dead" ends east of point "a"; in the direction from the talus, it is only crossed by a little corridor that passes on a different level, as distinctly indicated on the plan. Apparently, this corridor corresponds with the small chasm that contained a reindeer skeleton and a human mandible (Knies, 1905, 8; 1929, 485), and joined the eastern part of Dome D, as reported by Szombathy (1925, 9).

(2) Considering the absence of a major transport by water, the translocation of sediments containing finds from the talus to spaces 30 m away (bee line), or even more in the case of an indirect connection, would require a considerable slope. According to the detailed post-war mapping by Alois Štefka, however, the slope is rather contrary, with the bottom of Dome D lying several centimeters higher than the pathway in space E below the talus. Naturally, the thickness of the sediment removed during the major adaptations after 1911, is unknown. In one of his later contributions, Smyčka (1925) reported that the remains from 1922–1923 lay 4 meters deeper (in a "lower storey") than those from 1881–1882. Even if these data appear to be considerably exaggerated, Szombathy's observations indicate that the slope necessary for the transport of sediments was directed elsewhere: the deepest point in space E below the chimney was in the middle of the space, not in the direction towards the "Dome of the Dead". The chasm mentioned above containing the reindeer bones, however, was situated in that very direction towards point "a". Between the foot of the talus and the chimney, there

was a crevice ("Spaltenraum") that largely contained bones of microfauna, above all, abundant skeletons of snakes and mollusks (Szombathy, 1925, 9, 12). On the basis of Knies's record (see his diary IV) of the horizontal course of the Upper Pleistocene layers, Szombathy (1925, 12) considered the finds discovered beneath the talus to be younger and in part, even recent. According to V. Ložek (in Svoboda, 2001, 47), however, the talus contains Middle Pleistocene fauna (it is the snake skeletons that are typical of the fauna in the Mladeč Caves) (Ivanov, 1997; Valoch, 1993a). Bones of bovids that predominate the finds from 1922–1923 are missing among the 24 species enumerated from Knies's investigations (Szombathy, 1925). It is important to note that those excavations are earlier than those of J. Fürst, so that they still comprised the talus coat itself, and not its later core that could incidentally be older still (cf. Svoboda, 2001, 47). Nor do any of the reports mention a marked slope characterizing those layers. In the sketches from the years before and during the excavations of the debris cone area (Knies's diary IV; here Fig. 7), we can clearly see the horizontal position of the archaeological layers. There is no doubt that a part of the overlying rocky filling came from the collapsed ceiling, and that a part of it moved to that place from the higher lying corridor F (Szombathy, 1882, 101). That, too, is why the center of the dome is the lowest.

Before the investigations were launched, a horizontal course of loamy sediments without any larger rocks was noted in the vicinity of point "a" in the "Dome of the Dead" (Szombathy, 1882, 101; 1925, Tafel 2, Abb. 2). On the contrary, rocky sediments had sunk into the northern part of the Dome from space C (Szombathy, 1882, 101). The course of the surface of the sediments, as reconstructed by Svoboda (2001, Fig. 3), cannot be checked. The form of the rock columns, as indicated by this author, corresponds with their present condition as it developed on the one hand, by removing the sediments deep below the level of the Upper Pleistocene layer, and on the other, by breaking off the ceilings. The date obtained from the sinter on the wall 7 m westwards from point "a", is at least older than those obtained later directly from human and animal bones. Hence, they may not have been connected with the original finding layer.

The depth of the layer does not decrease in the direction away from the chimney. According to Szombathy (1925, 5–6), it was 50 cm deep in the western part of the "Dome of the Dead" (i.e., in the most distant place), 60 cm in the middle of the Dome and at point "a" (with the finds being discovered only 20–30 cm deep), and at point "d", lying the nearest to the chimney, bovid bones were found "in geringen Mengen von Höhlenlehm" and "fast zu Tage" (Szombathy, 1882, 106; 1925, 6). The finds were present at greater depths only at point "e", but there, the stratigraphic situation was quite different from that in the "Dome of the Dead": the finding layer was situated beneath a travertine layer 30–50 cm deep and contained stalagmites up to 70 cm tall. Moreover, this layer was covered with another layer of cave loam (Smyčka, 1922, 115–116).

(3) In contrast to uniformly distributed sediments, archaeological finds show a conspicuously unequal distribution. In the "Dome of the Dead", they occur only in the middle of the space, roughly 20 m^2 in area, and only in the upper half of the layer containing the bones. Moreover, their character is very specific: 21 drilled animal teeth, probably from a single necklace; a long and thin point, almost undamaged; fragments of other points, etc. No artifacts have been reported from points "a", "b", "d" or from the western part of the Dome. A large number of bone needles, 4 of which have very specific perforations, come from point "e" beneath the chimney. The needles were deposited there in the form of a depot, as indicated by the fact that the specimens are still connected with sinter (Plate III.8–10). Such needles are missing from the "Dome of the Dead" and conversely, no drilled animal teeth occur near point "e". No report is available on any finds made at the eastern edge of the talus in the so-called Strž. Also, there is a difference in the extent of damage done to the animal bones: in space D, the bovid bones are mostly undamaged (Szombathy, 1925, 6, 10), whereas in other places (i.e., where anthropological and Paleolithic finds occur), they show traces of heavy crushing. Human skulls from point "b" are accompanied by an almost complete selection of reindeer bones coming from a single individual (Szombathy, 1882, 106; 1925, 6).

(4) On the contrary, the presence of a horizontal entrance to the cave during the Upper Pleistocene can be considered very probable. This may not have been, however, the small corridor 's' that Szombathy believed to be the original entrance, since in that place the Middle Pleistocene sediments reached almost up to the ceiling (judging from traces on the walls ascertained by the Moravian Museum excavations). Nevertheless, Upper Paleolithic finds have been documented in the spaces of the present entrance. In antechamber C, the first visitors who discovered the cave encountered a large skeleton (Sova, 1904, 8; Szombathy, 1925, 4), and Szombathy (1925, 6–7) himself found remains of loams containing cave bear bones. In the loess loams on the eastern side of antechamber A, J. Skutil found two Aurignacian stone tools in 1937 (Fig. 6.4–5; Skutil, 1938, 32). If the animal bones show a wide range of dating, this indicates that the caves must have been accessible for a long period of time, but were only occasionally visited by humans. Obviously, the entrance was closed at some later time since Holocene sediment is missing inside the caves. The sediment did not enter even through the chimney because at present, it is impassable even for moisture during the spring thaw. It is uncertain at which time the chimney was closed; none of the reports mention Upper Pleistocene finds in the talus proper.

(5) Finally, the presence of man is inferred by evidence of fireplaces. This does not refer to the two black sites at depths of 2.5 and 2.6 m in Szombathy's trench "a" (1925, 5) that are probably mere manganese precipitates in Middle Pleistocene layers. In the same profile, one can observe, at a depth of 35 cm (i.e., close beneath the sinter on which skull 1 was resting), "eine geringe und etwa 50 cm breite Schicht mit Holzkohlenklein und in ihrer ganzen Ausdehnung vereinzelte diluviale Säugetierknochen" (Szombathy, 1925, 5). The author was aware of the possible danger in mistaking charcoals with the manganese stains, as he had long before pointed out that the manganese precipitate was mistaken for a fireplace by Maška (Szombathy, 1904, 15). Still more marked may have been the first fireplace, discovered in 1922, that was surrounded by rocks the size of a human head.

(6) Also, the Small Cave in the quarry (P) is sometimes described as "chasm-like". All original publications, however, clearly demonstrate that it was a horizontal space accessible from the southeast, with a recently broken off ceiling (Szombathy, 1904; Maška, 1905; Knies, 1905). The original entrance was covered by diluvial sediments, and this was why the Small Cave, still unknown during Szombathy's investigations, was not opened until quarrying for limestone begun. At the time when the above explorers visited the cave, its ceiling had already been removed. Its previous existence, however, was indicated by the astonished eyewitnesses who "were unable to understand how those things could have got under the solid rock" (Knies, 1905, 10). Human remains, together with artifacts, were found in a nook at the northeastern wall. A chimney descending down to a considerable depth was then opened at the northwestern end of the cavern, about a 7 m distance from the find. The idea of throwing human remains into cave chimneys during the Aurignacian lost its last support after a new radiometric date of a skull from the Koněpruské Caves, which placed it in the time span of the Magdalenian (Svoboda et al., 2002).

Conclusions

The artifacts and human skeletal remains from the Mladeč Caves can be placed in the middle or late phase of the Aurignacian. This phase boasts one of the largest concentrations of Aurignacian sites east of the Rhine River (Oliva, 1987; 1993b). It documents not only the extraordinary quantity of localities (about 100 sites in the area west of the Morava River), but also their chronological span, reaching from the Middle Würmian to the Pleniglacial B (Oliva, 1996). Unfortunately, most of them are surface localities that do not allow observations of stratigraphy and fauna.

The position of the Mladeč Caves within the framework of the Central European Aurignacian is quite extraordinary. The low frequency of lithic industries and the position of bone tools deep inside the cave in permanent darkness, suggest that the locality was not permanently occupied. The selection of the fauna with a very low proportion of cave bears, does not say much about the use of the cave as an occasional hunting site, in contrast to other localities with an abundance of bone points. The fragmentary remains of the human skeletons that were presumably originally even more numerous, and their occurrence along with split animal bones suggest that it was perhaps a ritual aggregation site. Some special bone artifacts found at the site may have been connected with this function, and tend to support this hypothesis, for example: the necklace of large teeth, the perforated metacarpals and perhaps even the red ochre signs on the walls of the "Dome of the Dead". If we had excavated and documented the locality in its original state, it would have offered us a more profound insight into the spiritual life of the Early Upper Paleolithic man.

References

Adovasio, J. M., Soffer, O. and Klíma, B. (1996) Upper Paleolithic fibre technology: interlaced woven finds from Pavlov I, Czech Republic, c. 26,000 years ago. *Antiquity* 70, 526–534

Albrecht, G., Hahn, J. and Torke, W. G. (1972) *Merkmalanalyse von Geschoßspitzen des mittleren Jungpleistozäns in Mittel- und Osteuropa.* Stuttgart Berlin Köln Mainz: Kohlhammer

Bar-Yosef, O. and Belfer-Cohen, A. (1996) Another look at the Levantine Aurignacian. In *XIII Int. Congress of Prehistorical and Protohistorical Sciences, Colloquia 6 – The Upper Palaeolithic.* Forli, pp. 139–150

Bayer, J. (1922) Das Aurignac-Alter der Artefakte und menschlichen Skelettreste aus der "Fürst Johanns-Höhle" bei Lautsch in Mähren. *Mitteilungen der Anthropologischen Gesellschaft Wien* 52, 173–185

Bayer, J. (1929) Die Olschewakultur. *Eiszeit und Urgeschichte* 6, 83–101

Blekta, J. (1932) Kras mezi Konicí a Litovlí. *Věstník Přírodovědeckého klubu v Prostějově* 22, 1–48

Bordes, F. (1967) Considérations sur la Typologie et les techniques dans le Paléolithique. *Quartär* 18, 25–55

Bordes, F. (1968) *Le Paléolithique dans le Monde.* Paris: Hachette

Breuil, H. (1925) Notes de voyage paléolithique en Europe centrale 3, Les cavernes de la Moravie. *L'Anthropologie* 35, 271–291

Brodar, S. and Brodar, M. (1983) *Potočka Zijalka.* Ljubljana: Slovenska akad. znanosti i umetnosti

Broglio, A. (1993) L'aurignacien au sud des Alpes. *Actes du XIIe Congrès Int. des Sci. Préh. et Protohistoriques Bratislava* 2, 193–202

Cabrera-Valdez, V. and Bischoff, J. L. (1989) Accelerator ^{14}C dates for Early Upper Palaeolithic (Basal Aurignacian) at El Castilló Cave (Spain). *Journal of Archaeological Science* 16, 577–584

Cook, J. (1986) The application of scanning electron microscopy to taphonomic and archaeological problems. In (D. A. Roe, Ed.) *Studies in the Upper Palaeolithic of Britain and Northwest Europe.* BAR 296, Oxford, pp. 143–163

d'Errico, F. and Giacobini, G. (1985) Approche méthodologique de l'analyse de l'outillage osseux. Un exemple d'étude. *L'Anthropologie* 89, 457–472

d'Errico, F. and Giacobini, G. (1986) L'emploi des répliques en vernis pour l'étude de surface des pseudo instruments en os. In *Outillage peu élaboré en os et en bois de cervidés 2. Artefacts* 3, CEDA, Viroinval, pp. 57–68

d'Errico, F., Henshilwood, Ch., Lawson, G., Vanhaeren, M., Tillier, A-M., Soressi, M., Bresson, F., Maureille, B., Nowell, A., Lakarra, J., Backwell, L. and Julien, M. (2003) Archaeological evidence for the emergence of language, symbolism, and music – an alternative multidisciplinary perspective. *Journal of World Prehistory* 17, 1–70

Fürst, J. (1922) Nové nálezy v Mladečských jeskyních. *Litovelské novinky* 26. 8. 1922

Fürst, J. (1923–24) Nálezy v Mladečských jeskyních. *Vlastivědný sborník střední a severní Moravy* 2, 6–7

Fürst, J. (1928–30) Co nám vypravuje pazourek. *Vlastivědný sborník střední a severní Moravy* 8, příloha (Beilage) Litovelsko, 6–8

Hahn, J. (1972) Aurignacian signs, pendants and art objects in Central and Eastern Europe. *World Archaeology* 3, 252–266

Hahn, J. (1977) *Aurignacien, das ältere Jungpaläolithikum in Mittel- und Osteuropa.* (Fundamanta A/7). Köln: Böhlau Verlag

Hahn, J. (1988) *Die Geissenklösterle-Höhle im Achtal bei Blaubeuren I.* Stuttgart: Konrad Theiss Verlag
Hahn, J., Koenigswald, W. v., Wagner, E. and Willie, E. (1977) Das Geissenklösterle bei Blaubeuren, Alb-Donau-Kreis. Eine altsteinzeitliche Höhlenstation der mittleren Alb. *Fundberichte aus Baden-Würtenberg* 3, 14-37
Hahn, J. and Münzel, S. (1995) Knochenflöten aus dem Aurignacien des Geissenklösterle bei Blaubeuren, Alb-Donau Kreis. *Fundberichte aus Baden-Würtenberg* 20, 1-12
Hoernes, M. (1903) *Der diluviale Mensch in Europa.* Braunschweig: Friedrich Viehweg und Sohn
Horáček, I. and Ložek, V. (1984) *Z výzkumu výplně Mladečské jeskyně u Litovle. Československý kras* 35, 98-100
Ivanov, M. (1997) *Hadi evropského kenozoika.* Mgr-thesis, Masaryk University Brno
Jelínek, J. (1983) The Mladeč finds and its evolutionary importance. *Anthropologie* 21, 57-64
Jelínek, J. (1987) Historie, identifikace a význam mladečských antropologických nálezů z počátku mladého paleolitu. In *Anthropos - 25 let pavilonu Moravskho muzea.* Brno: Moravské museum, pp. 51-69
Karavanič, I. (2000) Olschewian and appearance of bone technology in Croatia and Slovenia. In (J. Orschiedt and G.-C. Weniger, Eds.) *Neanderthals and modern humans - discussing the transition.* Mettmann: Neanderthal-Museum, pp. 159-168
Klíma, B. (1957) Übersicht über die jüngsten paläolithischen Forschungen in Mähren. *Quartär* 9, 85-130
Knecht, H. (1993) Splits and wedges: the techniques and technology of early Aurignacian antler working. In (H. Knecht, A. Pike-Tay and R. White, Eds.) *Before Lascaux. The complex record of the Early Upper Palaeolithic.* Boca Raton: CRC Press, pp. 137-162
Knies, J. (1905) Nový nález diluviálního člověka u Mladče na Moravě. *Věstník klubu přírodovědeckého v Prostějově* 8, 3-19
Knies, J. (1928) K 100. výročí objevení Mladečských jeskyň. *Litovelský kraj* 5, č. 45-49
Knies, J. (1929) Diluviální zvířena kraje olomouckého. *Vlastivěda župy olomoucké* 1, 465-493
Kozlowski, J. K. (Ed.) (1982) *Excavation of the Bacho Kiro Cave. Final report.* Warszawa: PWN
Kozlowski, J. K. (1998) The Middle and the Early Upper Palaeolithic around black Sea. In (T. Akazawa, K. Aoki and O. Bar-Yosef, Eds.) *Neanderthals and modern humans in western Asia.* New York: Plenum Press, pp. 461-482
Kozlowski, J. K. and Otte, M. (2000) La formation de l'Aurignacien en Europe. *L'Anthropologie* 104, 3-15
Kozlowski, L. (1924) Die ältere Steinzeit in Polen. *Die Eiszeit* 1, 112-163
Leroi-Gourhan, A. and Allain, J. (Eds.) (1979) *Lascaux inconnu.* Suppl. à Gallia Préhistoire 12. Paris: CNRS
Leroy-Prost, Ch. (1975) L'industrie osseuse aurignacienne. Essai régional de classification: Poitou, Charente, Périgord. *Gallia Préhistoire* 18, 66-156
Malez, M. (1967) Paleolit Velike Pečine na Ravnoj Gori u sjeverozapadnoj Hrvatskoj. *Archeološki radovi i rasprave* 4/5, 7-68
Marshack, A. (1972) *The roots of civilisation.* New York: McGraw-Hill
Maška, K. J. (1905) Poznímky k diluviálním nálezům v jeskyních mladečských a stopám glaciálním na severovýchodní Moravě. *Časopis moravského muzea zemského* 5, Sept. 1-3
Montet-White, A. (1996) *Le Paléolithique en ancienne Yougoslavie.* Grenoble: Jérôme Millon
Obermaier, H. (1905) Les restes humains quaternaires dans l'Europe Centrale. *L'Anthropologie* 16, 385-410
Oliva, M. (1987) *Aurignacien na Moravě.* Studie Muzea Kroměřížska '87. Kroměříž: Muzeum Kroměřížska, pp. 1-128
Oliva, M. (1989) Mladopaleolitické nálezy z Mladečských jeskyní. *Acta Musei Moraviae - Časopis Moravského muzea v. spol.* 74, 35-54
Oliva, M. (1993a) Le contexte archéologique des restes humains dans la grotte de Mladeč. *Actes du XIIe Congrès Int. des Sci. Préh. et Protohistoriques Bratislava* 2, 207-216
Oliva, M. (1993b) The Aurignacian in Moravia. In (H. Knecht, A. Pike-Tay and R. White, Eds.) *Before Lascaux. The complex record of the Early Upper Palaeolithic.* Boca Raton: CRC Press, pp. 37-55
Oliva, M. (1996) Epiaurignacien en Moravie: le changement économique pendant le deuxième interpleniglaciaire wurmien. In: *XIII Int. Congress of Prehistorical and Protohistorical Sciences, Colloquia, 6 - the Upper Palaeolithic.* Forli: ABACO edizioni, pp. 69-81
Oliva, M. (2003) *Pohřebiště kromaňonců v Mladečských jeskyních.* Olomouc: Archeologické centrum
Praslov, N. D. and Rogačev, A. N. (Eds.) (1982). *Paleolit kostěnkovsko - borševskovo rajona na Donu.* Leningrad: Nauka
Richter, J., Waiblinger J., Rink W. J. and Wagner G. A. (2000) Thermoluminiscence, electron spin resonance and ^{14}C-dating of the Late Middle and Early Upper Palaeolithic site of Geissenklösterle cave in southern Germany. *Journal of Archaeological Science* 27, 71-89

Ringer, A., Kordos, L. and Krolopp, E. (1995) Le complex báboniyen-szeletien en Hongrie du nord-est dans son cadre chronologique et environnemental. In *Les industries à pointes foliacées d'Europe centrale (Paléo-suppl. 1)*. Les Eyzies: Soc. des amis du Musée nat. de préhistoire, pp. 27–30

Skutil, J. (1938) *Pravěké nálezy v Mladči u Litovle*. Litovel: Krajinská musejní společnost

Skutil, J. (1957) Drobné příspěvky k paleolitiku Moravy. *Anthropozoikum* 6, 412–432

Smith, F. H., Trinkaus, E., Pettitt, P. B., Karavanič, I. and Paunovič, M. (1999) Direct radiocarbon dates from Vindija G1 and Velika Pecina Late Pleistocene hominid remains. *Proceedings of the National Academy of Science USA* 96, 12281–12286

Smyčka, J. (1907) Litovel a okolí za pravěku. *Pravěk* 3, 140–150

Smyčka, J. (1922) Nálezy diluviálního člověka v Mladči u Litovle na Moravě. *Obzor praehistorický* 1, 111–120

Smyčka, J. (1925) Kdy přišel první člověk do litovelského kraje. *Vlastivědný sborník střední a severní Moravy* 3, příloha (Beilage) Olomoucko a Litovelsko, 5–7

Sova, J. (1904) *Památnosti jeskyň Mladečských u Litovle*. Litovel

Svoboda, J. (2000) The depositional context of the Early Upper Palaeolithic human fossils from the Koněprusy (Zlatý kůň) and Mladeč caves, Czech Republic. *Journal of Human Evolution* 38, 523–536

Svoboda, J. (2001) Mladeč and other caves in the Middle Danube region: early modern humans, late Neanderthals, and projectiles. In (J. Zilhao, T. Aubry and A. Faustino, Eds.) *Les premier's hommes modernes de la péninsule Ibérique*. Lisboa: Inst. Portuges de Arqueologia, pp. 45–60

Svoboda, J. (2002) Mladeč 2. K problematice funerálních jeskyní. In (J. Svoboda, Ed.) *Prehistorické jeskyně*. Brno: Archeologický ústav AV ČR, pp. 344–354

Svoboda, J., van der Plicht, J. and Kuželka, V. (2002) Upper Palaeolithic and Mesolithic human fossils from Moravia and Bohemia (Czech Republic): Some new ^{14}C dates. *Antiquity* 76, 957-962

Szombathy, J. (1882) Über Ausgrabungen in den mährischen Höhlen im Jahre 1881. In (F. v. Hochstetter) Fünfter Bericht der prähistorischen Commission der math.-nat. Classe der kaiserlichen Akademie der Wissenschaften über die Arbeiten im Jahre 1881. *Sitzungsbericht der kaiserlichen Akademie der Wissenschaften* 85, [90]-[107]

Szombathy, J. (1904) Neue diluviale Funde von Lautsch in Mähren. *Jahrbuch der k. k. Zentralkommission für Kunst- und historische Denkmale* 2, 9–16

Szombathy, J. (1925) Die diluvialen Menschenreste aus der Fürst-Johanns-Höhle bei Lautsch in Mähren. *Die Eiszeit* 2, 1–34, 73–95

Turk, I. and Kavur, B. (1997) Palaeolithic bone flutes - comparable material. In (I. Turk, Ed.) *Mousterian "Bone Flute" and other finds from Divje Babe I cave site in Slovenia*. Ljubljana: ZRC, pp. 179–184

Valoch, K. (1993a) Starý paleolit v Mladečských jeskyních. *Acta Musei Moraviae - Časopis Moravského muzea v. spol.* 78, 3–9

Valoch, K. (1993b) Zwei Knochenspitzen des Aurignacien aus Mähren. *Acta Musei Moraviae - Časopis Moravského muzea v. spol.* 78, 23–27

Vértes, L. (1955) Neuere Ausgrabungen und paläolithische Funde von Istállóskö. *Acta Archaeologica Academia Scientarium Hungaricae* 5, 111–131

Weiser, E. (1928) *Reise und Wanderbuch*. Freudenthal: Schlesische Verlagsanstalt W. Kromer

White, R. (1993) Technological and social dimensions of "Aurignacian-Age" body ornaments across Europe. In (H. Knecht, A. Pike-Tay and R. White, Eds.) *Before Lascaux. The complex record of the Early Upper Palaeolithic*. Boca Raton: TRC Press, pp. 277–299

Plates

Plate I.
Mladeč, bone points, (1–3), (5–6) Main Cave, (4) Small Cave P (Photos V. Gánoczyová)

Plate II.
Mladeč, (1–2) bone points, (3–7) partially worked metacarpals, (1–2) Small Cave P, (3–7) Main Cave (Photos V. Gánoczyová)

Plate III.
Mladeč, (1–2) ornamented awls, (3–7) partially worked bones, (8–10) fragments of (worked?) bones in sediment, (11) lost awls and worked bones after Szombathy (1925), out of scale. (1–2), (4–11) Main Cave, (3) Small Cave P (Photos V. Gánoczyová)

Plate IV.
Mladeč, the largest bone point, Main Cave (50%); left: ventral view, right: dorsal view

Plate V.
Mladeč, possible tips (2–5) and bases (7–8) of bone points, probably unworked bones (1, 6), Main Cave

Plate VI.
Mladeč, perforated canines of bear, wolf and probably unperforated incisor of horse (at right), Main Cave

Plate VII.
Mladeč, perforated incisors of beaver, Main Cave

Plate VIII.
Mladeč, perforated incisors of elk or *Megaceros,* Main Cave

Plate IX.
Mladeč, (1, 2, 5) worked pebbles, (3–4) hammerstones, (6–7) naturally striated bones of Bos or Bison. (1?), (3–7) Main Cave, (2) Small Cave P (Photos V. Gánoczyová, out of scale)

Plate X.
Mladeč (1), ochre double line, partially covered by sinter, Main Cave, Dome D; Mladeč (2), ochre double line, partially covered by sinter, Main Cave, Dome D; Mladeč (3), barbed sign, Main Cave, Dome D; Mladeč (4), simple ochre line, Main Cave

PLATE I

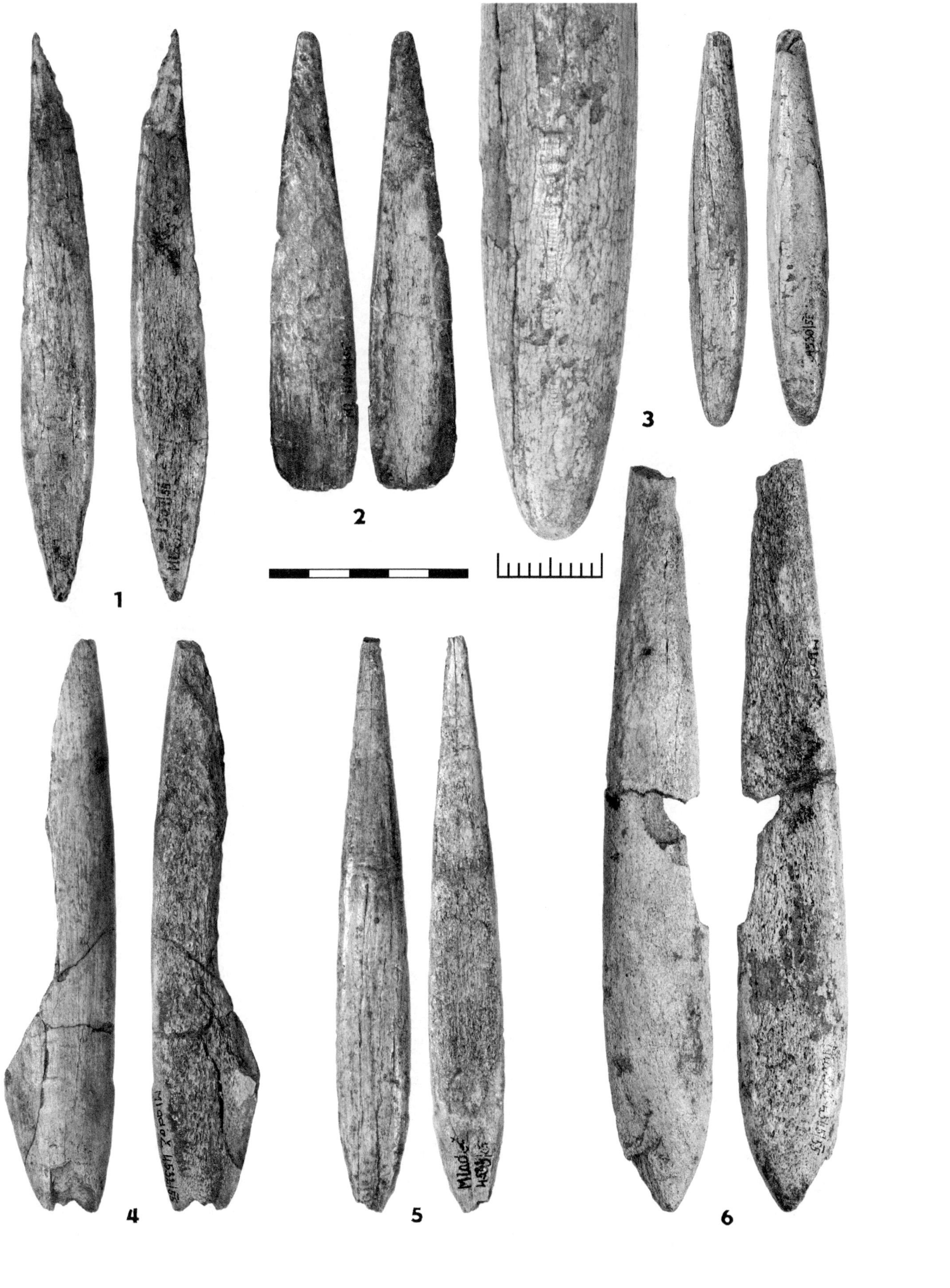

PLATE II

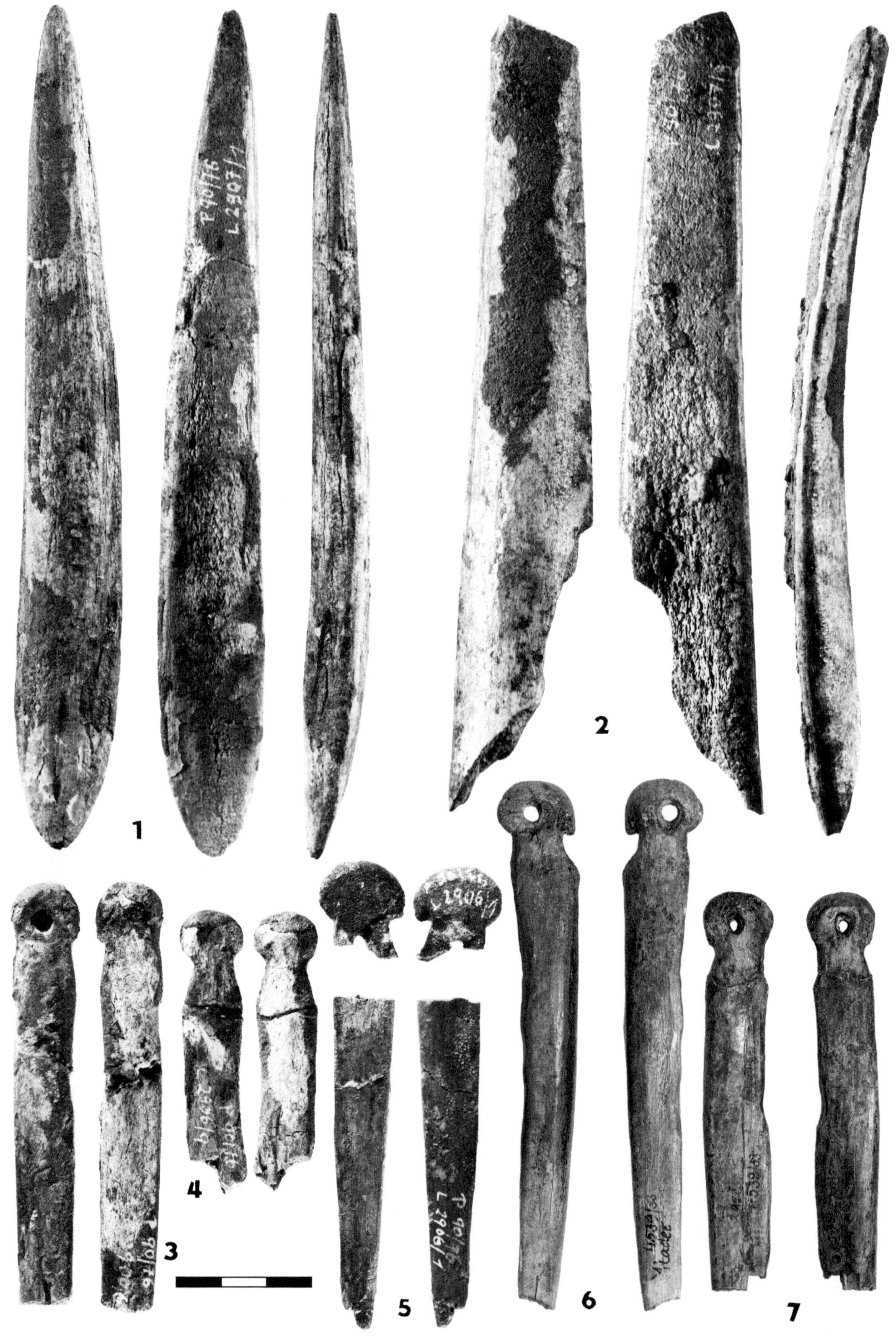

PLATE III

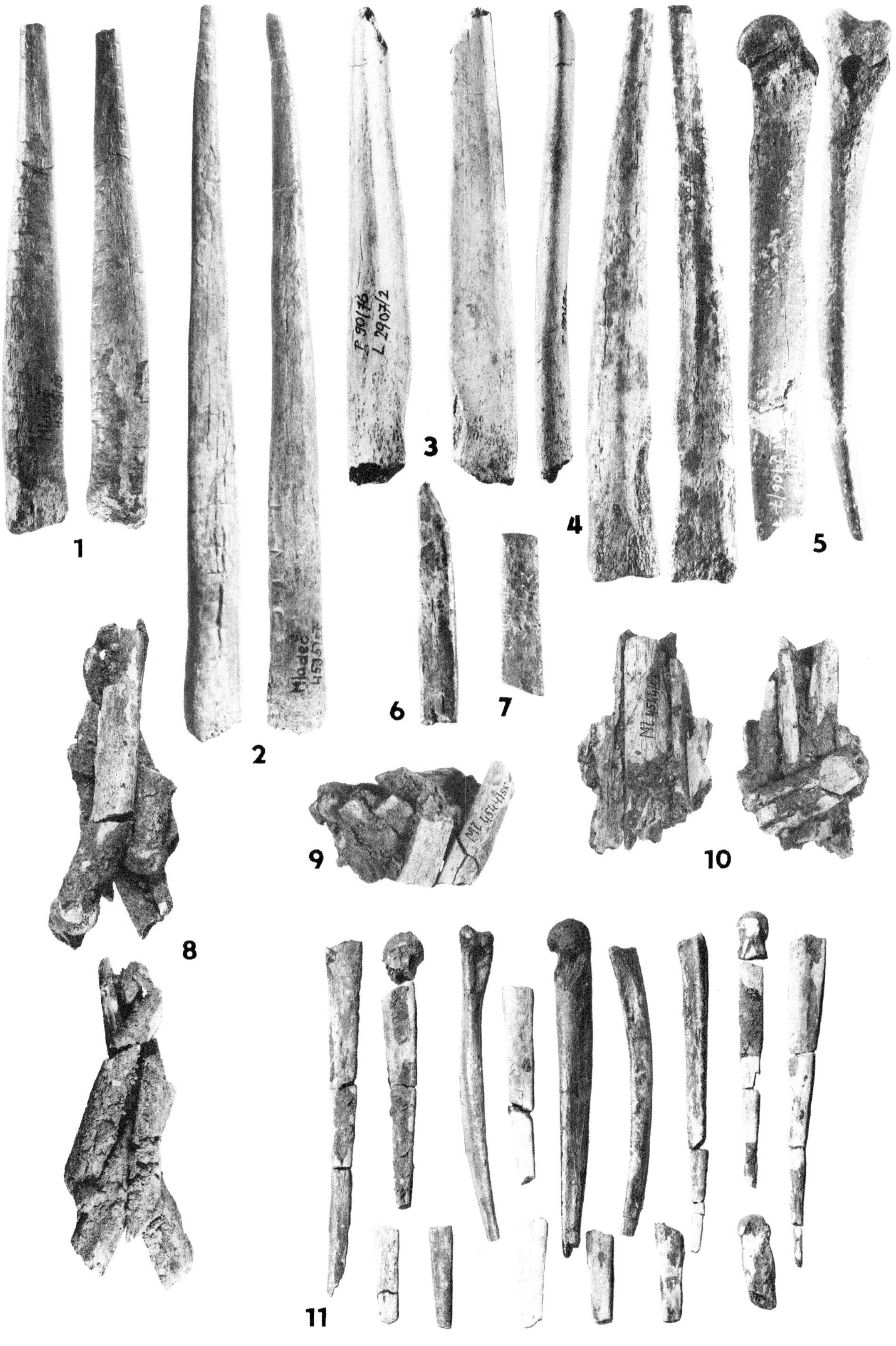

PLATE IV

PLATE V

PLATE VI

1 2 3

PLATE VIII

PLATE IX

1

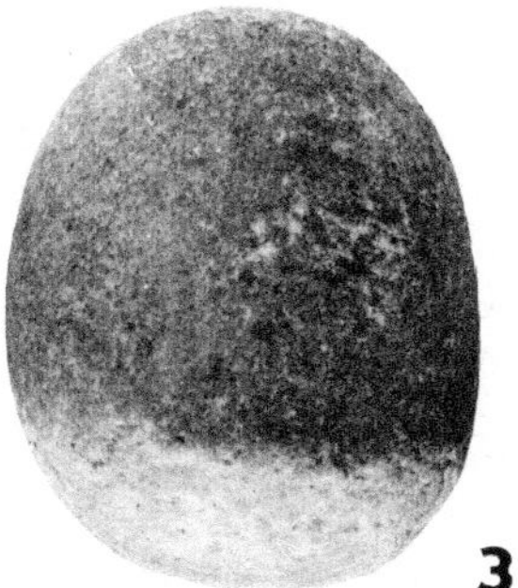

3

4

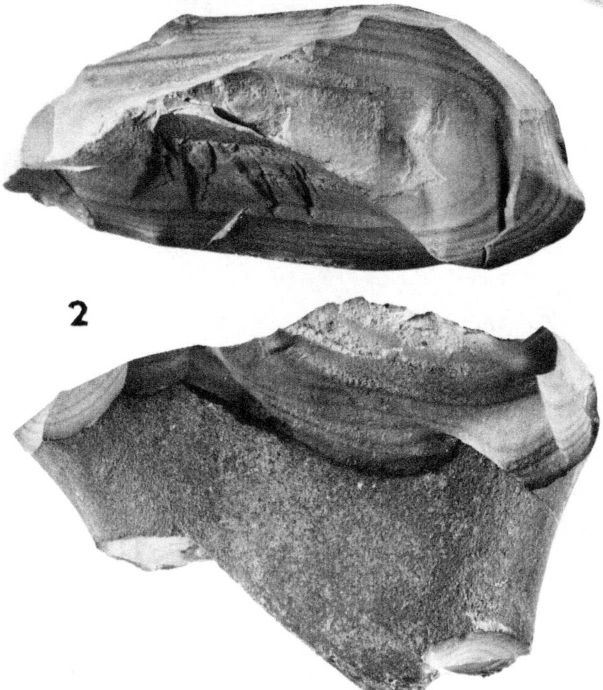

2

5

7

6

PLATE X

1

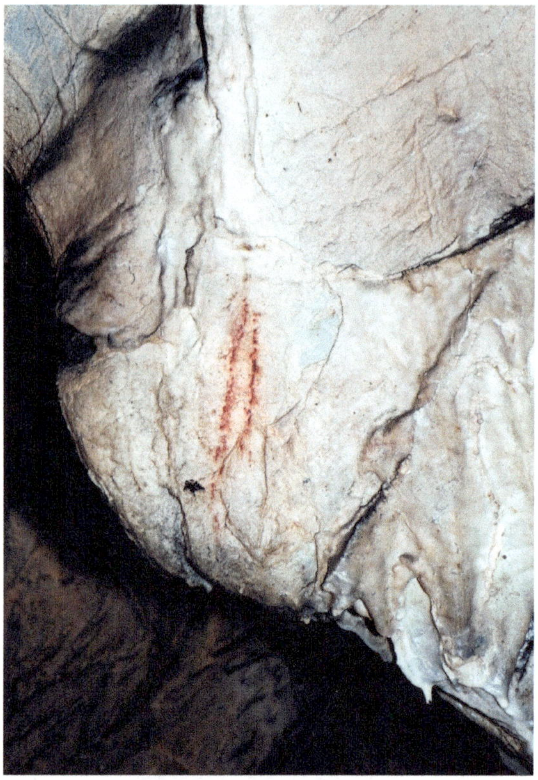

2

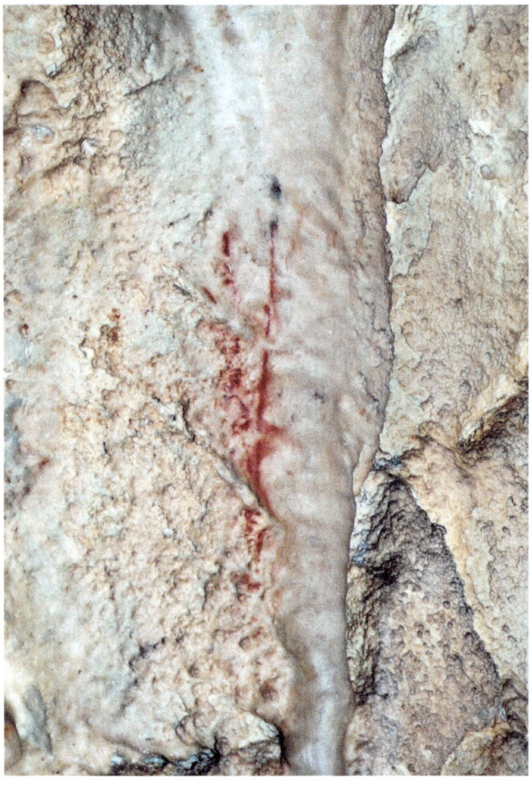

3

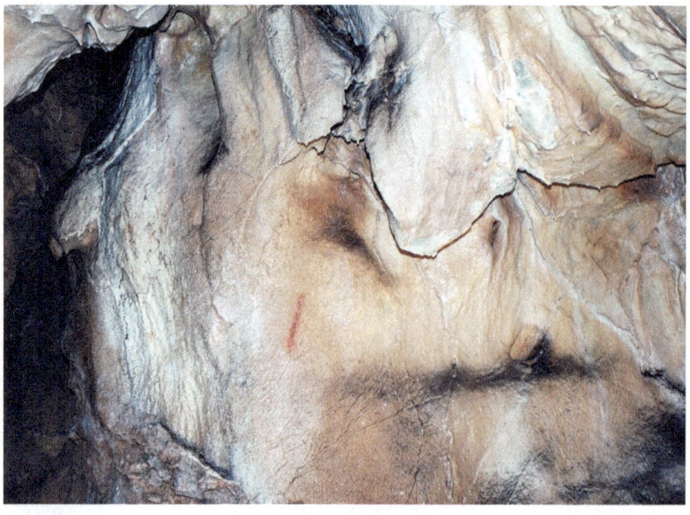

4

Chapter 5

TAPHONOMIC ASPECTS OF THE HUMAN REMAINS FROM THE MLADEČ CAVE

Maria Teschler-Nicola

Introduction

The first excavations to be carried out at Moravian caves by collaborators of the Naturhistorisches Museum Vienna were initiated by the geologist Ferdinand v. Hochstetter[1]. He was not only the first director of the new Museum, but concurrently also chairman of the "Prähistorische Commission" of the Academy of Sciences in Vienna. It was this commission which decided on 5th February 1880 to continue with the excavations at Výpustek cave near Brno, and eventually also to carry out an "exploration of other caves in the Moravian province"[2]. Among them was the Mladeč Cave, whose investigation (which started in 1881 and continued in 1882) was entrusted to Josef Szombathy, Hochstetter's assistant at the Museum since 1878 (Heinrich, 2003).

Although Josef Szombathy spent only five days at the Mladeč site in 1881 (7th–11th June) and just a few days in 1882 (13th–18th July), he recovered a considerable number of pleistocene human and animal bones as well as several archaeological objects (Hochstetter, 1882, 1883; Szombathy, 1900, 1904). But despite the fact that Szombathy carried out the excavation systematically, there is still lack of evidence, in particular concerning the documentation of the position of these finds in situ, their attribution to the sediment layers and the association of the different objects (see Antl-Weiser, this volume, chap. 1). As a consequence, answers to questions such as the purpose of the cave or the deposition of the bones in the site sediments, i.e., whether due to a burial, a ritual (see Oliva, this volume, chap. 4) or an incidental deposition via open chimneys and slots (Svoboda, 2000; Svoboda, this volume, chap. 3; Pacher, this volume, chap. 6) are still insufficient.

Szombathy justified the lack of provided data with time constraints and limited financial resources (he could not carry out a systematic investigation of layers, which would have required laborious evacuation of the soil); as a result, he could only obtain a "quick collection" of bones (Szombathy, 1925, 5). Unfortunately, Szombathy did not continue with the excavation in the following year, instead he concentrated on behalf of Hochstetter on the Výpustek cave and a small cave localized in the Kiriteiner-valey, the Žitny cave (Hochstetter, 1884). However, a significant body of evidence relating to taphonomy of the Mladeč human bone assemblage, in particular the dispersal of the bones is missing due to poorly documented fieldwork. Questions concerning the in-

1 Ferdinand v. Hochstetter, between 1860–1881 professor at the Polytechnicum in Vienna (Technical University), up from 1876 first director of the new Royal-Imperial Natural History Museum and director of the newly established Department of Anthropology-Ethnography. In 1878, he entrusted Josef Szombathy, his assistant at the Polytechnicum, with the responsibility for the Paleontological and Anthropological collections at the Royal-Imperial "Hofmineralienkabinett" and the investigation and excavation of different archaeological sites (see Heinrich, 2003).

2 Protocol of the VIIth meeting of the Prehistoric Commission of the Academy of Sciences, 5th February 1880. AÖAW, Prähistorische Kommission, B4, No. 145/1880.

dividual association of the scattered human remains are difficult to answer under these circumstances. Therefore we shall limit our focus on the macroscopic identification and analysis of the type and distribution of peri- and postmortem surface modifications. Furthermore, the micro-structural changes originating in diagenetic phenomena as well as sediment and site specific contamination of a few selected samples will be highlighted. Post-recovery modifications by museums-curators, such as restorative interference (removal of the matrix, completion of substance loss) and sample take off recently carried out for chemical investigations are addressed shortly (the main finding, Mladeč 1, will be discussed more intensively elsewhere; see Prossinger et al., this volume, chap. 11). The present investigation deals with the Mladeč human fossil assemblage housed at the NHM Vienna, exclusively.

Methods

The taphonomic modifications such as matrix concretion, color and decomposition were investigated macroscopically and by the use of a reflected light microscope. Their type and localization were recorded in a data collection protocol (Buikstra and Ubelaker, 1994; Sorg and Haglund, 2002). Specimens with peri- and postmortem damage and carnivore gnawing marks were documented photographically as well.

Furthermore, invasive techniques were used to analyze selected bone samples for their type and degree of microstructural, diagenetic changes (Schultz, 1988; 2001). For that purpose small bone samples were taken from a rib (Mladeč 20) and long bones (Mladeč 25a, Mladeč 25c, Mladeč 28, Mladeč 102), resin-embedded (BIODUR) and cut to block preparations of approximately 5 x 5 x 5 mm; one side of the preparation was polished and fixed to an object slide. Part of the fixed block preparations was then sawn off, polished, carbon-sputtered and investigated by analytical scanning electron microscopy (JEOL JSM6400 equipped with energy dispersive system). The remaining preparation on the object slide was ground to 80 µm, polished manually and investigated by transmitted light microscope (using normal and polarized light) and photographically documented (Olympus BX 50F).

Results and discussion

Macroscopic investigation of taphonomic changes

The current state of preservation of the Mladeč skeletal remains is, to a certain degree, consistent with Szombathy's description given in 1925. Post-recovery modifications (Holland et al., 1997), in particular restorative arrangements and manipulations, can only be identified in a few specimens (see Table 1): Mladeč 1[3] was partly destroyed during the excavation. As Szombathy reported, the considerable substance losses originate from the extraction of extraneous materials (a stalagmite and a reindeer rib) from the cranium.[4] These lesions were later restored (most probably after 1925) and

---

3 The specimen numbers correspond to the inventory list (Wolpoff et al., this issue, chap. 8)

4 "Der Schädel war in der Höhlenerde unversehrt geborgen. Beschädigt wurde er erst bei der Ausgrabung. [...] Besonders die Schädelbasis ist geradezu verhüllt durch die Sinterkrusten, die sich nicht ohne wesentliche Beschädigung der Knochen ablösen lassen würden. Die oben erwähnten, quer über die Basis gelagerten stalagmitischen Wülste konnte ich soweit wegpräparieren, daß das Basion freigelegt wurde. [...] Der erste linke Prämolar war zur Zeit der Entdeckung vorhanden, seine Krone wurde aber später von unbekannter Hand abge-

Table 1. Taphonomic changes of the early Upper Paleolithic human fossil remains of the Mladeč Cave (Specimen no. according to Wolpoff et al., this volume, chap. 8; Perim. = perimortem alterations; Postm. = postmortem alterations; Scr. = scratches; Gn. = gnawmarks; Post-rec. = post recovery modifications)

Specimen	Color	Sinter	Perim.	Postm.	Scr.	Gn.	Post-rec.
M 1	gray-brown	partly	no	yes	no	no	yes
M 2	gray	few	yes?	yes	yes	no	yes
M 3	beige	partly	yes?	yes	yes	?	no
M 8	gray-brown	complete	yes	yes	–	?	yes
M 9a	white-darkbr.	no	no	yes	no	no	yes
M 9b	yellowish	no	?	no	no	no	no
M 10	yellowish	no	no	yes	no	no	yes
M 11	gray	complete	no	yes	–	no	no
M 12	gray	complete	no	no	–	no	no
M 13	gray-brown	complete	yes	yes	–	no	yes
M 14	gray	complete	no	yes	–	no	no
M 15	gray	complete	yes?	?	–	no	yes
M 16	gray	partly	yes	yes	no	yes	no
M 17	gray	complete	?	yes	–	no	no
M 18	gray	complete	?	yes	–	–	no
M 19	gray	complete	no	?	–	–	yes
M 20a	gray	complete	yes	?	yes (cut?)	?	no
M 20b	gray	complete	no	yes	–	no	yes
M 20c	gray	mostly	no	no	–	no	yes
M 20d	gray	no	yes	?	no	?	yes
M 20e	gray	partly	?	?	no	no	yes
M 21	gray	partly	?	yes	no	?	no
M 22	gray	mostly	yes	yes	no	yes	no
M 23	gray	few	yes	yes?	no	yes	no
M 24	gray	very few	yes	?	no	yes	no
M 25a	gray	complete	?	yes	–	no	yes
M 25b	gray	partly	yes	?	yes	?	no
M 25c	yellowish	partly	?	?	no	yes	yes
M 26	gray	few	yes	?	?	?	no
M 27	gray	complete	yes	yes	–	yes	no
M 28	gray	no	yes	yes	yes (cut?)	yes	no
M 29	gray	complete	?	yes	–	?	no
M 30	gray	mostly	no	yes	–	no	no
M 31	gray	partly	?	yes	no	no	no
M 32	gray	completely	no	no	–	no	no
M 34	yellowish	partly	no	yes	no	no	no
M 102	gray	few	yes	yes	no	no	no

brochen, so daß nur seine Wurzel erübrigt. [...] Die übrigen Zähne sind sicherlich aus dem Kiefer genommen worden, ehe der Schädel in die Erde kam, denn die Innenwände der einzelnen leeren Alveolen sind ganz so wie die übrigen Flächen des Schädels mit der Sinterkruste ausgekleidet, einige auch teilweise damit ausgefüllt." (Szombathy, 1925, 14).

completed with hard plaster (Szombathy, 1925, 14; see Prossinger and Teschler-Nicola, this volume, chap. 11); in the 1980s an isolated preserved maxilla (Mladeč 7) was refitted to the Mladeč 2 skull (see Wolpoff et al., this volume, chap. 10); furthermore, the matrix concretions covering the outer table of Mladeč 3 skull fragments were removed mechanically, leaving a considerable number of small bony erosions and scratches. This applies to the femur specimen of Mladeč 28 and two of the small rib fragments from Mladeč 20 as well.

Moreover, small samples were removed by E. Trinkaus and P. Pettitt in the 1990s in an attempt to date several specimens; very recently, samples were taken in the course of ancient DNA-analyses, AMS radiocarbon dating, strontium isotope ratio measurements and the preparation of thin cross sections for the investigation of micro-decomposition and diagenetic changes (protocols are on hand, see also Serre at al., 2004; Wild et al., 2005; Prohaska et al., this volume, chap. 18).

Szombathy already revealed several aspects of taphonomy (although he did not use this term) that were important for the analysis and interpretation of the Mladeč Cave finds. But some of his diagnoses are questionable. In general, all human and faunal remains recovered from Hall D, find spots "a" and "b" (including the finds of 1882 in the "central 20 m^2"; Szombathy, 1925, 6) show a similar (macroscopic) preservation status (Table 1). According to Szombathy (Szombathy, 1925, 6)[5]

> *"they are completely identical, incompletely fossilized; i.e., lacking the organic tissue component, enriched with lime carbonates, thoroughly colored grayish and nerved by dark dentrites. Other than the teeth, the superficial cortical bone layer is covered by a thin gray-brown crust, made up from lime carbonate and fine loam; this crust is barely half a millimeter thick, but crumbles or shows tubercular swellings in many regions. The crust adheres to a certain, but not uniform extent to the cortical bone areas. Hence, the matrix can be removed from some of them without destroying the superficial layer, while it will not act in others. Moreover, this crust covers in a similar manner also the internal layer of the cranial bones and the old fractured surfaces and margins of all human specimens (there are also some new lines of fractures caused during excavation, where, of course, the bony mass lies bare and does not exhibit matrix aggregation)."*

Apart from his assessment concerning the amount of organic components, we largely agree with the given description concerning the color (most of the specimens exhibit a gray color, only specimens 25c and 34 are a bit brighter and more yellowish; so are the isolated teeth) and the amount of matrix concretions. For example, the right femur diaphysis Mladeč 27 is more or less completely covered by matrix and shows crumbled and nodular-formed sinter in an evidentiary manner; the fracture lines at the distal and proximal shaft ends are also covered by matrix (Plate I, Fig. a–d). This applies to other specimens as well. Due to this matrix concretions at the fracture lines, Szombathy assumed that several skeletal remains had become fragmented, "before they were embedded in the ground" (Szombathy, 1925, 20). In this regard, he specifically pointed to

5 "Sie sind in ganz gleicher Weise unvollständig petrifiziert, d.h. der leimgebenden Substanz beraubt, ein wenig Kalkkarbonat angereichert, durch und durch grau gefärbt und mit feinen, dunklen Dentriten durchzogen. Die Oberfläche, mit Ausnahme der Zähne, ist mit einer dünnen graubraunen, aus Kalkkarbonat und feinem Lehm bestehenden Kruste überzogen, die im allgemeinen kaum einen halben Millimeter dick ist, aber an vielen Stellen krümelige oder warzige Verdickungen hat. Sie haftet mit einer gewissen, nicht an allen Stellen gleichen Festigkeit an der Knochenoberfläche, so daß sie sich von manchen Teilen ohne Schädigung des Knochens abheben läßt, von anderen aber nicht. Diese Kruste überzog in gleichem Maße auch die Innenseite der Schädelknochen und die alten Bruchflächen und Bruchränder aller Knochen. (Es gibt wohl auch neue, beim Ausgraben verursachte Bruchstellen, an denen natürlicherweise keine Sinterkruste haftet und wo die Knochenmasse im Durchschnitt bloßgelegt ist.)" (Szombathy, 1925, 6).

the cranial fragments of Mladeč 2[6] and Mladeč 3, the maxillary fragment Mladeč 8[7], the diaphysis of the right humerus Mladeč 24[8] (Plate III, Fig. b), the right proximal radius fragment Mladeč 25a, the proximal third of the right ulna Mladeč 25c (Plate II, Fig. a) and the proximal portion of the left femur Mladeč 28 (Plate II, Fig. c). However, this list might be supplemented by several other remains whose fracture lines are not only covered by matrix but come along with a typical breakage pattern (spiral form, sharp edges, etc.) such as the left radius diaphysis Mladeč 26 (Plate II, Fig. b), which exhibits a peri-mortally induced spiral fracture at the distal margin and a splitting fracture (also induced in fresh, collagen rich bone) at the proximal end of the diaphysis; the right proximal humerus Mladeč 23 (Plate III, Fig. a); the pelvic remain Mladeč 22; the clavicle Mladeč 13; some of the rib fragments (Mladeč 15, Mladeč 16 and two out of the five Mladeč 20 ribs) and the immature femur portion Mladeč 102.

In very few of the fossils, Szombathy observed a damage-pattern caused by severe postmortem injuries. These included the above mentioned left femur fragment Mladeč 28[9], the diaphysis of the right femur shaft Mladeč 27 (Plate I, Fig. a–d)[10] and the proximal portion of the ulna Mladeč 25c, which was "injured by severe blunts, whose features can be clearly observed on five points" (Szombathy, 1925, 25). Surprisingly, Szombathy did not consider that these alterations could eventually represent animal chewing. He referred that "there is no specimen where a trace of carnivore gnawing is evident" (Szombathy, 1925, 6) and suggested that these alterations signify that humans were extracting resources from human subjects, since "the human bones [...] were treated in the same manner as the animal bones remaining from the cave dweller's meals. They are fragmented, the bigger ones systematically minimized and placed quite homogenously between the faunal remains. I believe that the human bones probably originate from the diluvial cave dweller's meal, representing a case of anthropophagia" (Szombathy, 1925, 8).[11] Although there is a lack of distinct traces induced by human activity (such as "blunt marks" with a characteristic chipping

6 "Die Calva ist aus zehn durchschnittlich handtellergroßen Bruchstücken zusammengesetzt, welche zwischen zahlreichen Rentierknochen verstreut gefunden wurden. Einige solche Tierknochen waren auch mit einzelnen Schädelbruchstücken (auch an der Innenwand derselben) zusammengesintert. Die Stücke sind schon in zerbrochenem Zustand in die Erde und zwischen die Rentierknochen gelangt. Das erkennt man daran, daß ihre Bruchflächen ebenso wie die Außen- und die Innenflächen und die Tierknochen mit der dünnen lehmigen Sinterschicht überzogen sind." (Szombathy, 1925, 20).

7 "Es ist zu erkennen, daß das Stück, nachdem es vom Schädel abgebrochen war, an verschiedenen Enden, auch an der Spina nasalis posterior, abgestoßen oder durch Fußtritte beschädigt worden ist, ehe es zur Ruhe kam." (Szombathy, 1925, 22).

8 "von der schon vor alters das Caput und die distale Epiphyse einschließlich der Epicondylen abgebrochen wurden." (Szombathy 1925, 24)

9 "von dem der Gelenkskopf an der Wurzel des Collum unregelmäßig abgebrochen ist, während auch Trochanter major, Crista intertrochanterica und Trochanter minor stark beschädigt, ja beinahe abgetragen sind. [...] Die Bruchflächen zeigen einige alte Schlagmarken. Auch der ein Stückchen oberhalb der Längenmitte gelegene Abbruch der Diaphyse ist alt." (Szombathy, 1925, 25 und Abb. 31).

10 "Das zweite Stück ist die Diaphyse eines rechten Femur, von dem beide Enden in alter Zeit abgeschlagen wurden. Es fehlen die Epiphysen und ein proximales Stück der Diaphyse bis zum Trochanter minor." (Szombathy, 1925, 25).

11 "Mit Ausnahme des bei a gefundenen Calvariums sind diese ganz ebenso behandelt wie die von den Mahlzeiten der Höhlenbewohner übrig gebliebenen Tierknochen. Sie sind zerteilt, die größeren systematisch verkleinert und ganz gleichwertig zwischen die tierischen Reste hineingeworfen. Ich halte es daher für wahrscheinlich, daß auch die Men-

opposite to the impact point) we can not exclude human manipulations completely. The observed scratches in specimens Mladeč 2, Mladeč 3 and Mladeč 25b originate most probably from post-recovery manipulations, while in two specimens (Mladeč 20a and 28), small cut related features are diagnosable (Plate IV, Figs. a and b): At the rib fragment 20a, two small, v-shaped modifications (length ca. 3 mm) localized at the inferior margin are observable. The even surface of one of the cut-like defects seems to be covered by a very thin layer of grayish-brown matrix. The cause of this modification can not be determined with precision, it may eventually originate in post mortem or post-recovery manipulations. Because SEM investigation of this artificially induced feature is not completed yet, we abstain from further interpretations for the present. The second specimen, in which the reflected light microscopic investigation could only now substantiate a cut-like feature, is represented by the left proximal femur portion Mladeč 28: At the posterior surface of the femoral neck, on the base of the intertrochanteric crest, directly at the inferior margin of the obturator externus groove, a small, 3 mm long cut mark is noticeable (Plate IV, Fig. b). Although the cortical bone is fossilized and very hard, we can not exclude a post recovery genesis completely (it might eventually be caused during the artificial removal of the sinter incrustation). But the localization seems to underline a different origin, since "cutmarks tend to be concentrated at the articular ends of bones" (Andrews and Cook, 1985, 690). In addition, micromorphologies like the v-shaped form and the "double lined" scrape at the base of this "cut" (which is similar to those originating from stone tools, Andrews and Cook, 1985, 687) seems to emphasize that this feature originates in a modification shortly after death – by humans rather than scavengers. But the close proximity between this "cut-mark" and the trochanter, where destruction by animal gnawing is observable, makes a reliable decision on the origin difficult. If the mark is indeed humanly-produced, it could – as it was concluded from the cut-marks at Dolní Věstonice 11 occipital – "correspond to another type of mortuary behavior, with the human remains scattered [...] and some intentional disarticulation" (Trinkaus et al., 2000).

Contrary to Szombathy's assumption, we think there is clear evidence that many of the alterations, in particular at the above mentioned femur Mladeč 28 with the loss of trabecular bone at the greater trochanter leaving a smooth, rounded surface, and the changes observed at the proximal and distal epiphyses of the specimen Mladeč 27, eventually also at the right proximal humerus Mladeč 23, the humerus shaft Mladeč 24 and at the rib fragment Mladeč 16, originate from scavenging (Plate I, Fig. a–d; Plate II, Fig. c). In particular, the greater trochanter of the femur is a common area of bone destruction (Haynes, 1980). Such a pattern – destroyed epipyhses, intact shaft, irregular, "sawtooth-formed" fracture lines and puncture marks – represent the well known picture of gnawing by carnivores, as it could be demonstrated by Milner and Smith (1989) and several other authors (Haglund et al., 1988; Blumenschine, 1995; Merbs, 1997; Andrews and Fernandez-Jalvo, 1997; Teschler-Nicola et al., 1999).

Moreover, also the features detectable at the ulna Mladeč 25c, which were identified by Szombathy as severe blunt marks, undoubtedly represent gnawing marks: Five more or less concentric to oval shaped small impressions (ca. 3–4 mm diameter) at the proximal part of the trochlear notch, the trochlar margin and the tuberosity of the olecranon process (here they are opposite to each other) are viewable (Plate II, Fig. a). Similar artificially caused grooves are detectable in the pelvic bone Mladeč 22, around the acetabular fossa. But as matrix covers these impressions, no clear decision can be made on their origin.

---

schenknochen von den Mahlzeiten der diluvialen Höhlenbewohner herrühren, daß hier ein Fall von Anthropophagie vorliegt. Die weitverstreuten Bestandteile des Halsgeschmeides und die anderen Artefakte können auf verschiedene Art und Weise an Ort und Stelle gelangt sein und ebensowohl von den aktiv wie den passiv an den Mahlzeiten Beteiligten herrühren." (Szombathy, 1925, 8).

Interestingly, the same sort of reasoning (carnivorous gnawing) might be responsible for the damage of the epiphyses of the right femur of *Bison priscus* (Szombathy, 1925, 11, Tafel 2, Abb. 4)[12]– but Szombathy addressed them as features of human interference as well (see Pacher, this volume, chap. 6)! It might be of interest that in one femur specimen (Mladeč 28) where the shaft consists of a thick cortical bone, a pattern of sharp breakage in the area between the middle and the proximal portion was identified. Such kind of damage which requires considerable force to break this area has been reported as being consistent with bear activity (Merbs, 1997).

In general, Szombathy was right to identify perimortal taphonomic changes on several human remains of the site, but it was a moot point to attest the total of the alterations and the assignment to other humans, while disregarding factors such as carnivore damage and transport of bone. By all means, the animal-induced artifacts, the damage and absence of bones seem to underline that the corpses or remains left for some time unprotected to the cave ground (or at a site above the cave). It could also be expected that they were buried partially or in a shallow grave which did not protect them from animal scavenging sufficiently (Haglund, 1997).

Regardless of that, there is an other still unsolved (perhaps unsolvable) question which concerns the individual affiliations of the isolated skeletal elements dispersed within the cave. Unfortunately, Szombathy's descriptions of the find spots of the human bones are rather excursive. There are no details recorded, neither in his diary nor in the publications, which could be used for a positive identification of individual affiliation. Rather traceable are the notes given for find spot "a" in Hall D, where he recovered the cranium (later numbered as Mladeč 1) and the diaphysis of a gracile femur (Mladeč 27). North-eastwards of that locus was find-spot "b", where he collected mammalian bones and "among them many human bones" (such as fragments of two skulls, two maxilla fragments, some pieces of ribs and long bones, see above). All finds from 1881 from loci "a" and "b" are similarly preserved as described above. So are the finds from the 1882 campaign, just reported as having been found in the "middle of Chamber D" (fragments of vertebrae, ribs, skull, pelvic and long bones as well as of an isolated maxilla). They are all well fossilized and grayish colored. According to Szombathy, only the finds made in the north-western part of the main Chamber D, around spots "d" and "b_1" – which were not specified – differ from the others in color and petrification. They are brighter and less fossilized (Szombathy, 1925, 6). Following this statement, there is only one bone fragment among the human fossil record (irrespective of the isolated teeth) which seems to fulfill this criterion, namely the proximal ulna specimen Mladeč 25c. This bone fragment is yellowish and the thin matrix partly covering the cortical bone is red-brown. Interestingly, this is the only specimen among the sample of bones selected for the direct radiocarbon dating procedure carried out recently (a total of five samples was used, see Wild et al., 2005) in which the obtained radiocarbon age seems to underline a younger age. But as stressed by Wild et al. (2005), this result may well be a consequence of a post-mortem contamination, and should not be interpreted further.

Lastly, the sparse dates given by Szombathy, combined with the observed type and preservation status of the isolated bones within the Mladeč fossil assemblage plus our experience on scattered human bones as a consequence of animal gnawing (see Teschler-Nicola et al., 1999) do not allow an

12 "Von den Röhrenknochen der Wiederkäuer sind ziemlich viele aufgeschlagen, wohl behufs Gewinnung des köstlichen Markes. So sind auch die Schädel zur Entnahme des Hirns geöffnet. Es kommt aber auch eine besondere Art der Eröffnung des Markkanals großer Röhrenknochen vor, wie sie z. B. an dem starken rechten Oberschenkel von Bison [...] dargestellt ist. Dieser Knochen, dessen Diaphyse in ihrem Mittelteile vielleicht zu fest war, um zerschlagen werden zu können, wurde an den beiden Enden der Diaphyse, wo die kompakte Knochenrinde dünn und locker ist, aufgebrochen. Man scheint sich nicht mit der einfachen Entleerung der Markhöhle begnügt zu haben, sondern hat auch die markhältige Spongiosa der Endteile des Knochens ausgekratzt, wie man besonders am distalen Ende erkennen kann." (Szombathy, 1925, 11 und Abb. 4).

assumption on their possible individual association. This has to be done - alternatively - by using size, morphological characteristics, and age-of-death features (see Trinkaus et al., this volume, chap. 13; Wolpoff et al., this volume, chap. 10).

Thus, the minimum number of individuals represented by the specimen housed at the Naturhistorisches Museum Wien is five to six.

As early as the first fossils were found, a discussion arose about the circumstances by which such a number of human skeletal remains could be inserted to the cave. Szombathy summed up that "whether by formal burial or other operation is a question on its own" (Szombathy, 1925, 7; Bayer, 1922; Maška, 1886). He had dealt with this subject from the very beginning, when he identified a curious on-site mixture of a complete reindeer and scattered human bone fragments. To verify his find, he continued the excavation in 1882, as mentioned above. Bayer and Szombathy did not share the same opinion, nor do we have a consistent opinion on that topic today (Jelínek, 1983; Oliva, 1989; 1993; Svoboda, 2000; Svoboda, this volume, chap. 3; Oliva, this volume, chap. 4). According to Josef Bayer, the Mladeč Cave represents a huge burial place of the late Aurignacian (Bayer, 1925), perished already by humans, which had used the cave later.[13] Szombathy went beyond that. Although he was convinced about the site as a burial place (Szombathy, 1882, 17; Szombathy, 1925, 7), the pattern he assumed was not as unambiguous as in other caves. The human and faunal remains in the Mladeč Cave seemed to exhibit nearly the same taphonomic changes, namely blunt marks, and he could not assess sediment disturbances. Szombathy preferred the interpretation of Mladeč also being used as a ritual- or living ground, taking other arguments such as some charcoal finds and an identification of two "fire places" into account (Szombathy, 1925, 5).

Diagenetic alterations

The type of fragmentation of the main part of the fossils by macroscopic inspection clearly points to perimortem processes, when organic components were still preserved (such as in the radius Mladeč 26, Plate II, Fig. b). To gain insight to the decomposition processes and the depositional impact on the degree of bone preservation at the Mladeč site (in particularly the amount of collagen, still preserved), we additionally investigated epoxy resin-embedded ground sections of a few selected, very small bone samples.

Although it was not to be expected, the thin cross section of one of the Mladeč 20 ribs (rib 20c, see Wolpoff et al., this volume, chap. 8, Plate XI) shows a very well preserved microstructure with lamellar bone, many lacunae of osteocytes arranged in rows (Plate V, Fig. a) and a significant amount of collagen (using a polarizing filter arrangement; the bluish and yellowish color is an effect of the collagen fiber orientation; Plate V, Fig. b). Moreover, the small trabeculae are covered uniformly by a thin layer of inorganic material, a sinter concretion, which mantles also most of the surface of the bones, see Table 1).

A similar preservation status could be verified in the proximal radius portion Mladeč 25a and the robust ulna Mladeč 25c. The cross section of the radius sample shows the compact bone substance with regular vascular canals, surrounded by lamellar increments of bone (Haversian systems) (Plate V, Fig. c). By the use of polarized light just a few preserved organic bone tissues are noticeable. The brightness relates to the collagen fiber orientation (Plate V, Fig. d). Within the cross section of the ulna Mladeč 25c, secondary osteonal structures and interstitial lamellae can be observed

13 "Die weitverzweigte Fürst-Johanns-Höhle ist ein Bestattungsplatz der Aurignacleute, welche hier eine Anzahl Leichen, anscheinend nur wenige Dezimenter tief beisetzten, darunter eine mit Kollier aus Tierzähnen. Noch im Diluvium fand eine Störung der Grabesruhe statt und bei der Auffindung der Knochen durch Szombathy war von Skeletten in situ keine Rede mehr." (Bayer, 1925, 74).

(Plate V, Fig. e); moreover, a high number of clearly separated circumferential lamellae at the periosteal surface of the bone with well preserved organic substance can be identified by the use of a polarizing filter (Plate V, Fig. f). This finding is in accordance with the successful aDNA analysis of that specimen (Serre et al., 2004).

In contrast, the thin cross section of the proximal femur portion Mladeč 28, though macroscopically well preserved, shows a completely different degree of micro-preservation, namely a conspicuous reduction of organic component (Plate VI, Fig. a–f). The images show just a very few cortical areas, where collagen is preserved. Regardless of that, David Serre could demonstrate that the ppm value of collagen was insufficient in several human and animal bones from the Mladeč assemblage for the purpose of aDNA analysis. From the total of twelve human specimen tested, ten failed, among them Mladeč 3, Mladeč 8 (bone), Mladeč 21, Mladeč 25a, Mladeč 26, Mladeč 27, Mladeč 102 and Mladeč 28, respectively (David Serre, list of results not published, oral information).

Since some of our recent projects deal with the analysis of isotope ratios and the pattern and amount of trace elements, several of the above mentioned specimens investigated by light microscopic techniques were also studied in the scanning electron microscope for their postmortem alterations. Fig. 1a shows the rib Mladeč 20 in BSE-mode. Applying the potential of this technique, diagenetic alterations in form of irregular, uneven small matrix layers on the trabeculae and within the Haversian channels could be clearly identified. Secondary mineralization products are even visible within the osteocyte lacunae. Many of these mineralizations which infill the channels and porosities, could be identified as carbonate (calcite, Fig. 1b), most probably exogenous $CaCO_3$, as it was suggested by Jackes et al. (2001). On the other hand, siliciclastics, e.g., feldspar and quartz (SiO_2), could be verified along small old fracture cracks and in the lumen of the Haversian channels in the specimens Mladeč 25a (Fig. 2a, b; 3a, b) and Mladeč 28 (Fig. 4). All of them have been reported as typical components of the Mladeč Cave sediments (see Prohaska et al., this volume, chap. 18). Such site specific postmortem alterations were also identified at the sample of Mladeč 102 (Figs. 6a and b).

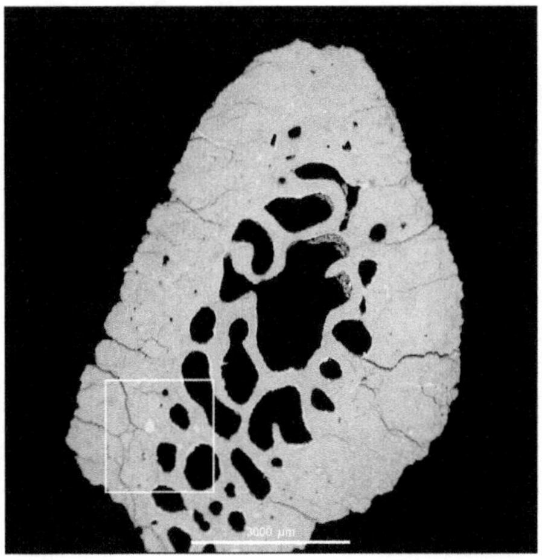

a

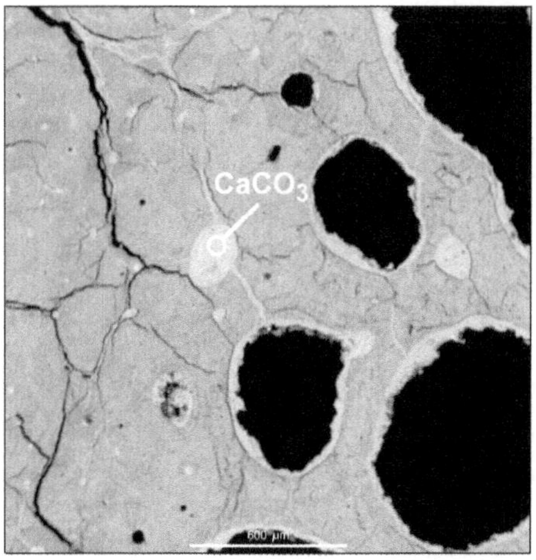

b

Fig. 1a. Mladeč 20c: Rib fragment (overview): Scanning-electron microscopic image (BSE-mode) shows exogenous sinter encrustations and carbonate inclusions; **(b)** Mladeč 20c: Rib fragment (detail): Qualitative element analysis in SEM verifying carbonate inclusions

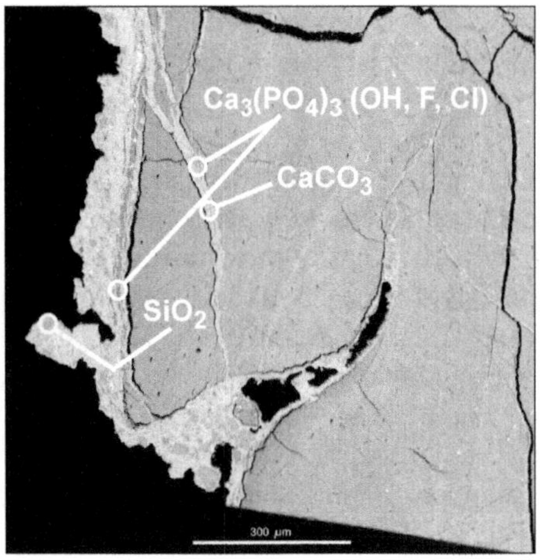

a

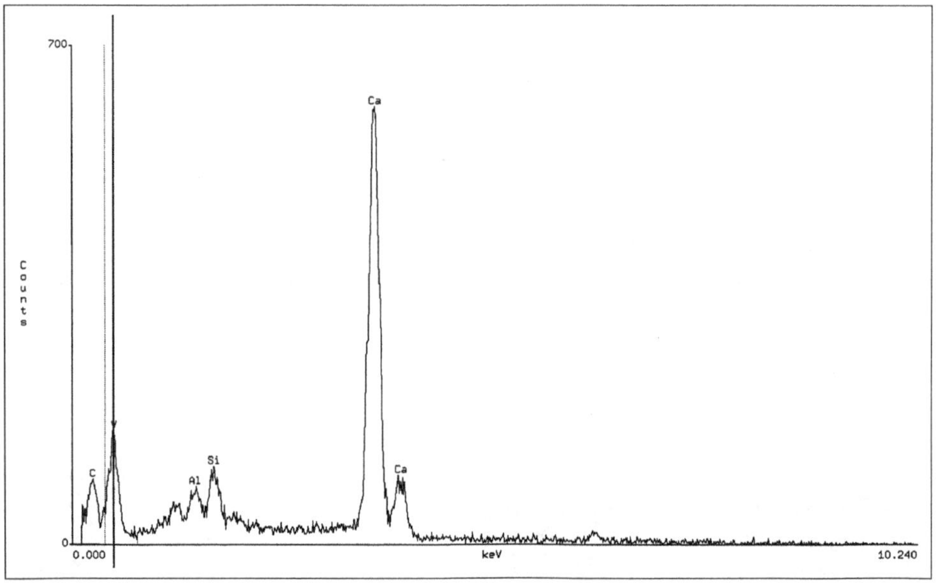

b

Fig. 2a. Mladeč 25a: Rigth proximal radius: Scanning-electron microscopic image (BSE-mode) shows exogenous sinter encrustations and carbonate inclusions; **(b)** Mladeč 25a: Graph of the qualitative element analysis in SEM verifying carbonate inclusions

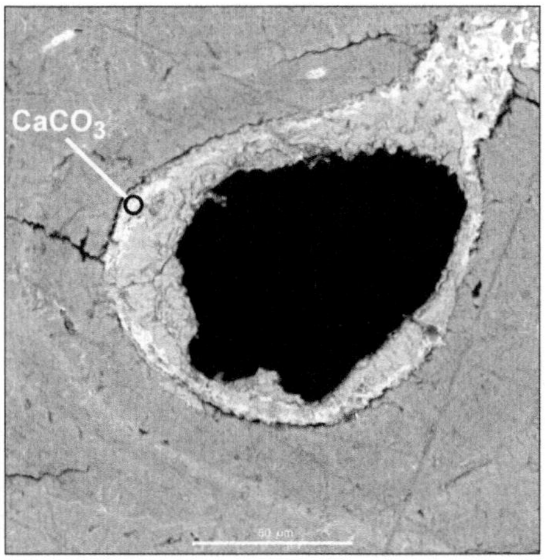

a

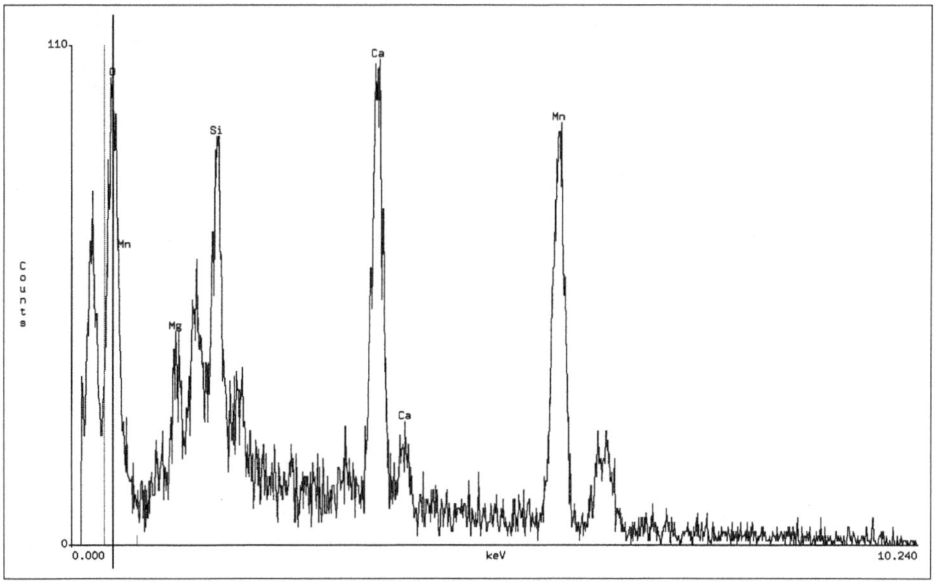

b

Fig. 3a. Mladeč 25a: Rigth proximal radius: Scanning-electron microscopic image (BSE-mode) shows exogenous sinter encrustations and carbonate inclusions; **(b)** Mladeč 25a: Graph of the qualitative element analysis in SEM verifying carbonate inclusions, siliciclastics and manganese

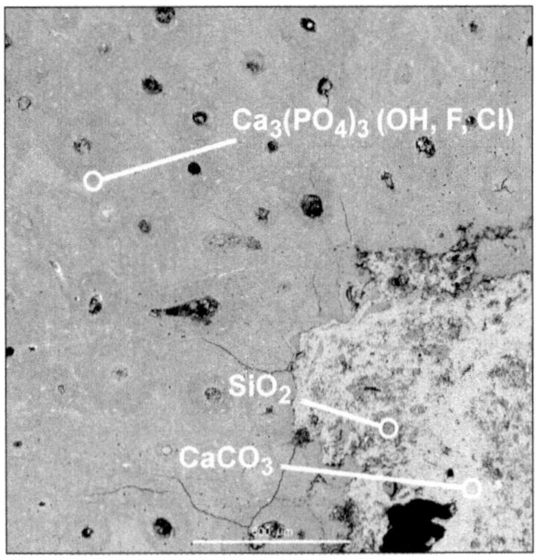

Fig 4. Mladeč 28: Left proximal femur: Cancellous bone areas showing SiO_2 and $CaCO_3$ inclusions

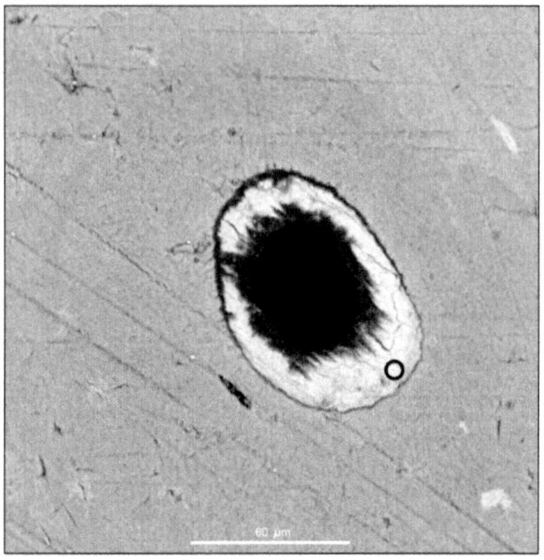

a

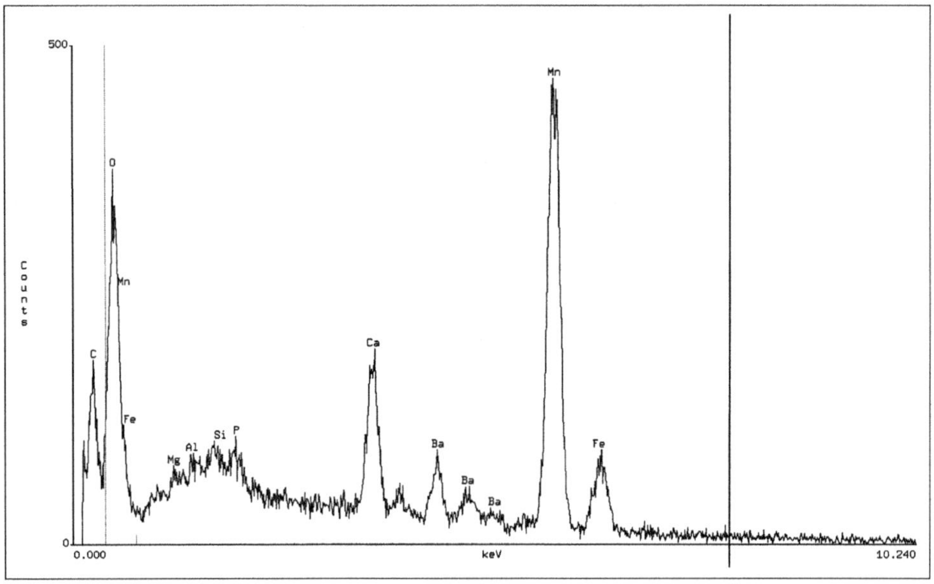

b

Fig. 5a. Mladeč 28: Left proximal femur: Cancellous bone with encrustations in a Haversian channel; **(b)** Same as **(a)**: Graph of the qualitative element analysis of the encrustations in SEM verifying manganese and iron

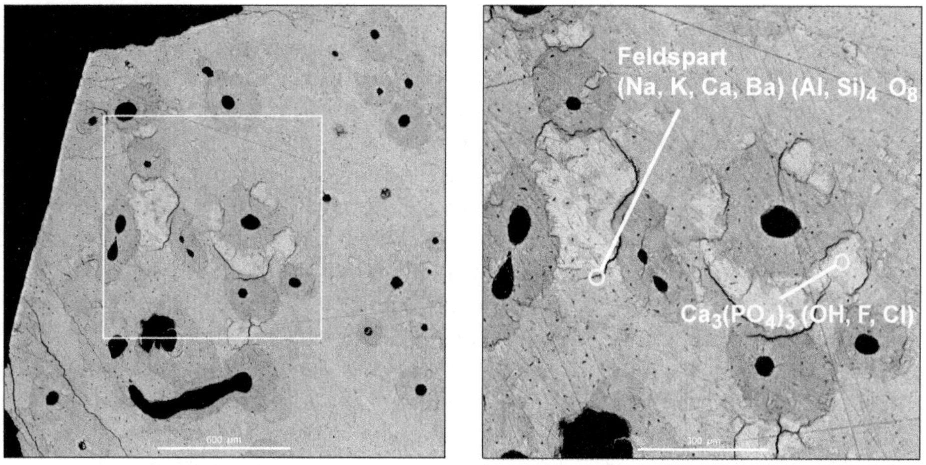

a b

Fig. 6a. Mladeč 102: Proximal femur fragment: Cancellous bone with osteonal structures and different mineralization; **(b)** Mladeč 102: Proximal femur fragment: Cancellous bone with osteonal structures and qualitative element analysis at different spots

On several occasions, Szombathy reported unspecific "black inclusions", "coalish" layers, and "charcoal" finds, which he ascribed to a "fire place"[14] (Szombathy, 1925). But no charcoal fragments were achieved and our element screening of some of the human remains indicates a different origin: Interestingly, we could verify conspicuous amounts of manganese in some preparates, specifically in Mladeč 25a and Mladeč 28 (Figs. 3b and 5b). So, in our view, the dark-brown to black-brown color observed at some of the human remains (diagnosable specifically in the isolated canine's root Mladeč 9a, see Wild et al., this volume, chap. 7) does not underline the assumption of fire use within the Mladeč Cave, e.g., as integrative part of a burial ceremony, similar to that observed at Lagar Velho (Duarte, 2002, 199; Queiroz, 2002, 152) or Sunghir (Binant, 1991); it is more likely that the color reflects an input of exogenous soil components, in particular manganese, as it has been reported already by Jelínek (1987). According to Schultz (1997), such a phenomenon, i.e., an element input of manganese (even iron), could act – among others – as an agent to protect skeletal remains from further destruction (Schultz, 2003, 79).

Conclusions

The present investigation aimed to gain insight to decomposition and taphonomic history of the human remains recovered at the early Upper Paleolithic Mladeč site. Following Szombathy's descriptions, the skeletal elements (N = 37) excavated during his field-campaigns in 1881 and 1882 were dispersed within the cave; most of them were found in the central area of Chamber D, the "Dome of

14 "Mit der etwas tiefer, bei 35 cm Tiefe gelegenen Brandstelle kann der Schädel nicht in nahen Zusammenhang gebracht werden, weil er von ihr durch die in 30 cm Tiefe gelagerte dünne Sinterschicht [...] getrennt war" (Szombathy, 1925, 6); ... „in einer Tiefe von 2.5 m eine unregelmäßige, stark verwitterte Kalksteinplatte von etwa 60 cm Länge und Breite, über der eine 1 cm starke kohlige Schicht lag. In 2.6 m gab es eine ähnliche Kohlenschicht. [...] es waren zwar keine besonderen Funde bei diesen schwarzen Einlagerungen, doch glaubte ich sie ebenso wie die in der Tiefe von 35 cm gelegene Kohlenschicht als alte Feuerstelle betrachten zu müssen." (Szombathy, 1925, 5).

the Dead". The minimum number of individuals investigated here is five to six, taking the NHM collection as basis. All human bones were fragmented (similar to the recovered animal bones) and several of them - although often heavily encrusted by matrix - were undoubtedly broken when the bones were fresh and collagen rich. Szombathy purportedly attributed them to human interference (and, as a consequence, concluded anthropophagy); we do not agree with his diagnosis, since most of the features he identified and discussed, must be undoubtedly attributed to animal chewing. The carnivore damage marks are localized at the articular ends, no spiral fractures on diaphyses of long bones are visible. From these alterations it can be concluded that scavenging took place in the cave (but the remains could also be inserted via the open chimneys into the cave; Svoboda, 2000; Pacher, this volume, chap. 6). Unfortunately, most of the postcranial remains are covered by sinter concretions as mentioned above, ruling out a reliable diagnosis of superficially located bone alterations. Surprisingly, in two specimens we observed fine modifications, which must be interpreted as cut-marks or being similar to cut-marks; but these findings have to be treated with caution and need further detailed investigation. Summing up, the taphonomic pattern observed in the Mladeč human fossil record indicates that carnivore activities and/or human manipulations (to a lesser degree) played a major role for the observed distribution of the remains.

Gathering the results of the chemical analysis, we assume that the purported "coal" layers, which have been used as one of the arguments for human occupation of the cave, represent most probably just packages or concretions of manganese enriched soil or loam. Nevertheless, the nature of the Mladeč human assemblage is also coincident with an accumulation or inclusion by humans.

Acknowledgements

The author wish to thank Ronald Mühl for preparing the ground sections and the block preparations, Dr. Franz Brandstätter, Department of Mineralogy, NHM Vienna, for taking the SEM-photos and the chemical analysis, Wolfgang Reichmann for providing the photographic documentation, Bettina Voglsinger for her assistance during manuscript preparation and Dr. Martina Pacher, Institute of Paleontology, University Vienna, and Dr. Walpurga Antl-Weiser, Department of Prehistory, NHM Vienna, for critical comments on an earlier version of the manuscript.

References

Andrews, P. and Cook, J. (1985) Natural modifications to bones in a temperate setting. *Man* (N.S.) 20, 675–691

Andrews, P. and Fernandez-Jalvo, Y. (1997) Surface modifications of the Sima de los Huesos fossil humans. *Journal of Human Evolution* 33, 191–217

Arsuaga, J. L., Carretero, J. M., Grazia, A. and Martínez, I. (1990) Taphonomical analysis of the human sample from the Sima de los Huesos Middle Pleistocene site (Atapuerca/Ibeas, Spain). *Human Evolution* 5, 505–513

Antl-Weiser, W. (2006) Szombathy's excavations in the Mladeč Cave and the first presentation of the results. In (M. Teschler-Nicola, Ed.) *Early modern humans at the Moravian Gate: The Mladeč Caves and their remains,* this issue, pp. 1–16

Bayer, J. (1922) Das Aurignac-Alter der Artefakte und menschlichen Skelettreste aus der „Fürst Johanns-Höhle" bei Lautsch in Mähren. *Mitteilungen der Anthropologischen Gesellschaft Wien* 52, 173–185

Bayer, J. (1925) *Die ältere Steinzeit in den Sudentenländern.* Anstalt für Sudentendeutsche Heimatforschung (Vorgeschichtliche Abteilung), Heft 2, pp. 73–77

Binant, P. (1991) *Les sépultures du Paléolithique.* Paris: Éditions Errance

Blumenschine, R. (1995) Percussion marks, tooth marks and experimental determinations of the timing of hominid and carnivore access to long bones at FLK Zinjanthropus, Olduvai Gorge, Tanzania. *Journal of Human Evolution* 29, 21–51

Buikstra, J. and Ubelaker, D. (1994) *Standards for Data Collection from Human Skeletal Remains.* Arkansas Archeological Survey Research Series 44. Fayetteville: Arkansas Archeological Survey

Duarte, C. (2002) The burial taphonomy and ritual. In (J. Zilhão and E. Trinkaus, Eds.) *Portrait of the artist as a child. The Gravettian human skeleton from the Abrigo do Lagar Velho and its archeological context* (Trabalhos de arqueologia 22). Lisboa: Instituto Português de Arqueologia, pp. 187–201

Frayer, D. W., Jelínek, J., Oliva, M. and Wolpoff, M. H. (2006) Aurignacian males from the Mladeč Caves, Moravia, Czech Republic. In (M. Teschler-Nicola, Ed.) *Early modern humans at the Moravian Gate: The Mladeč Caves and their remains,* this issue, pp. 185–272

Haglund, W. D., Reay, D. T. and Swindler, D. R. (1988) Tooth mark artefacts and survival of bones in animal scavenged human skeletons. *Journal of Forensic Sciences* 33, 985–997

Haglung, W. D. (1997) Dogs and coyotes: Postmortem involvement with human remains. In (W. D. Haglund and M. H. Sorg, Eds.) *Forensic taphonomy. The post-mortem fate of human remains.* Boca Raton: CRC Press, pp. 367–381

Haynes, G. (1980) Evidence of carnivore gnawing on pleistocene and recent mammalian bones. *Paleobiology* 6, 341–351

Heinrich, A. (2003) Josef Szombathy (1853–1943). *Mitteilungen der Anthropologischen Gesellschaft* 133, 1–45

Hochstetter, F. v. (1882) Fünfter Bericht der prähistorischen Commission der mathemat.-naturwiss. Classe der kaiserlichen Academie der Wissenschaften über die Arbeiten im Jahr 1881. *Sitzungsberichte der kaiserlichen Akademie der Wissenschaften math.-nat. Classe* 85, [90]–[107]

Hochstetter, F. v. (1882) Fünfter Bericht über die Resultate der [...] im Jahre 1881 in den mährischen Höhlen vorgenommenen Untersuchungen. *Anzeiger der kaiserlichen Akademie der Wissenschaften. Mathematisch-naturwissenschaftliche Classe* 19, 39–40

Hochstetter, F. v. (1883) Sechster Bericht der Prähistorischen Commission der mathematisch-naturwissenschaftlichen Classe über die Arbeiten im Jahre 1882. *Anzeiger der kaiserlichen Akademie der Wissenschaften. Mathematisch-naturwissenschaftliche Classe* 20, 46–47

Hochstetter, F. v. (1884) Siebenter Bericht der prähistorischen Commission über die Arbeiten derselben im Jahre 1883. *Anzeiger der kaiserlichen Akademie der Wissenschaften. Mathematisch-naturwissenschaftliche Classe* 21, 47–48

Holland, T. D., Anderson, B. E. and Mann, R. W. (1997) Human variables in the postmortem alteration of human bone: Examples from the U.S. War Casualties. In (W. D. Haglund and M. H. Sorg, Eds.) *Forensic taphonomy. The post-mortem fate of human remains.* Boca Raton: CRC Press, pp. 263–274

Jackes, M., Sherburne, R., Lubell, D., Barker, C. and Wayman, M. (2001) Destruction of microstructure in archaeological bone: a case study from Portugal. *International Journal of Osteoarchaeology* 11, 415–432

Jelínek, J. (1983) The Mladeč finds and its evolutionary importance. *Anthropologie* 21, 57–64

Jelínek, J. (1987) Historie, indentifikace a význam mladečských antropologických nálezů z počátku mladého paleolitu. *In 25 let pavilonu Anthropos 1961–1986.* Brno: Moravské muzeum, pp. 51–70

Milner, G. R. and Smith, V. G. (1989) Carnivore alteration of human bone from a late prehistoric site in Illinois. *American Journal of Physical Anthropology* 79, 43–49

Merbs, C. F. (1997) Eskimo skeleton taphonomy with identification of possible polar bear victims. In (W. D. Haglund and M. H. Sorg, Eds.) *Forensic taphonomy. The post-mortem fate of human remains.* Boca Raton: CRC Press, pp. 249–262

Minugh-Purvis, N., Bence Viola, T. and Teschler-Nicola, M. (2006) The Mladeč 3 infant. In (M. Teschler-Nicola, Ed.) *Early modern humans at the Moravian Gate: The Mladeč Caves and their remains,* this issue, pp. 357–384

Oliva, M. (1989) Mladopaleolitické nálezy z Mladečsých jeskyní. *Časopis Moravského muzea zemského* 74, 35–54

Oliva, M. (1993) Le contexte archéologique des restes humains dans la grotte de Mladeč. In (J. Pavúk, Ed.) *UISPP, XIIe Congres,* Volume 2. Bratislava: Institut d'Archeologie, pp. 207–216

Oliva, M. (2006) The Upper Paleolithic finds from the Mladeč Cave. In (M. Teschler-Nicola, Ed.) *Early modern humans at the Moravian Gate: The Mladeč Caves and their remains,* this issue, pp. 41–74

Pacher, M. (2006) Large mammal remains from the Mladeč Caves and their contribution to site formation processes. In (M. Teschler-Nicola, Ed.) *Early modern humans at the Moravian Gate: The Mladeč Caves and their remains,* this issue, pp. 99–148

Prohaska, T., Teschler-Nicola, M., Galler, P., Přichystal, A., Stingeder, G., Jelenc, M. and Klötzli, U. (2006) Non destructive determination of $^{87}Sr/^{86}Sr$ isotope ratios in early Upper Paleolithic human teeth from the Mladeč

Caves - preliminary results. In (M. Teschler-Nicola, Ed.) *Early modern humans at the Moravian Gate: The Mladeč Caves and their remains*, this issue, pp. 505–514

Prossinger, H. and Teschler-Nicola, M. (2006) Electronic segmentation methods reveal the preservation status and otherwise unobservable features of the Mladeč 1 cranium. In (M. Teschler-Nicola, Ed.) *Early modern humans at the Moravian Gate: The Mladeč Caves and their remains*, this issue, pp. 341–356

Queiroz, P. F. (2002) The Anthracology of the burial. In (J. Zilhão and E. Trinkaus, Eds.) *Portrait of the artist as a child. The Gravettian human skeleton from the Abrigo do Lagar Velho and its archeological context* (Trabalhos de arqueologia 22). Lisboa: Instituto Português de Arqueologia, pp. 152–153

Schultz, M. (1988) Paläopathologische Diagnostik. In (R. Knussmann, Ed.) *Anthropologie. Handbuch der vergleichenden Biologie des Menschen*, Vol. 1. Stuttgart: Fischer, pp. 480–496

Schultz, M. (1997) Microscopic investigation of excavated skeletal remains: a contribution to paleopathology and forensic medicine. In (W. D. Haglund and M. H. Sorg, Eds.) *Forensic taphonomy. The post-mortem fate of human remains.* Boca Raton: CRC Press, pp. 201–222

Schultz, M. (2001) Paleohistopathology of bone: A new approach to the study of ancient diseases. *Yearbook of Physical Anthropology* 44, 106–147

Schultz, M. (2003) Light microscopic analysis in skeletal paleopathology. In (D. J. Ortner, Ed.) *Identification of pathological conditions in human skeletal remains.* San Diego: Academic Press, pp. 73–107

Serre, D., Langaney, A., Chech, M., Teschler-Nicola. M., Paunovic, M., Mennecier, Ph., Hofreiter, M., Possnert, G. and Pääbo, S. (2004) No evidence of Neandertal mDNA contribution to early modern humans. *PLOS Biology* 2, 1–5

Sorg, M. H. and Haglund, W. D. (2002) Advancing forensic taphonomy: purpose, theory, and process. In (W. D. Haglund and M. H. Sorg, Eds.) *Advances in forensic taphonomy. Method, theory, and archaeological perspectives.* Boca Raton: CRC Press, pp. 3–29

Svoboda, J. (2000) The depositional context of the Early Upper Paleolithic human fossils from the Koněprusy (Zlatý Kůň) and Mladeč caves, Czech Republic. *Journal of Human Evolution* 38, 523–536

Svoboda, J. (2006) The structure of the cave, stratigraphy, and depositional context. In (M. Teschler-Nicola, Ed.) *Early modern humans at the Moravian Gate: The Mladeč Caves and their remains*, this issue, pp. 27–40

Szombathy, J. (1900) Un crâne de la race de Cro-Magnon, trouvé en Moravie. Comptes-Rendus du Congrès International d'Anthropologie et d'Archéologie préhistoriques, XII. Session, Paris 1900, pp. 133–140

Szombathy, J. (1904) Neue diluviale Funde von Lautsch in Mähren. *Jahrbuch der k k. Zentralkommission für Kunst- und historische Denkmäler* 2, 9–16

Szombathy, J. (1925) Die diluvialen Menschenreste aus der Fürst-Johanns-Höhle bei Lautsch in Mähren. *Die Eiszeit* 2, 1–34, 73–95

Teschler-Nicola, M., Gerold, F., Bujatti-Narbeshuber, M., Prohaska, T., Latkoczy, Ch., Stingeder, G. and Watkins, M. (1999) Evidence of genocide 7000 BP - Neolithic paradigm and geo-climatic reality. *Collegium Antropologicum* 23, 437–450

Trinkaus, E., Svoboda, J., West, D. L., Sládek, V., Hillson, S. W., Drozdová, E. and Fišáková, M. (2000) Human remains from the Moravian Gravettian: Morphology and taphonomy of isolated elements from the Dolní Věstonice II site. *Journal of Archaeological Science* 27, 1115–1132

Trinkaus, E., Smith, F. H., Stockton, C. and Shackelford, L. L. (2006) The human postcranial remains from Mladeč. In (M. Teschler-Nicola, Ed.) *Early modern humans at the Moravian Gate: The Mladeč Caves and their remains*, this issue, pp. 385–446

Wolpoff, M. H., Frayer, D. W., Trinkaus, E. and Teschler-Nicola, M. (2006) Inventory and photo-documentation of the Mladeč hominid remains. In (M. Teschler-Nicola, Ed.) *Early modern humans at the Moravian Gate: The Mladeč Caves and their remains*, this issue, pp. 159–184

Wolpoff, M. H., Frayer, D. W. and Jelínek, J. (2006) Aurignacian female crania and teeth from the Mladeč Caves, Moravia, Czech Republic. In (M. Teschler-Nicola, Ed.) *Early modern humans at the Moravian Gate: The Mladeč Caves and their remains*, this issue, pp. 273–340

Plates

Plate I.

Mladeč 27: Right femur diaphysis with perimortem breakage lines and sinter encrustations
(a) proximal shaft end, anterior view
(b) posterior view
(c) distal shaft end, dorsal view
(d) lateral view

Plate II.

(a) Mladeč 25c: Right proximal ulna with animal gnaw-marks (circular depressions at the olecranon process)
From left to right: lateral, posterior and medial view
(b) Mladeč 26: Left radius diaphysis with perimortem fracture, distal portion, anterior view
(c) Mladeč 28: Left proximal femur with animal chewing and a cut mark, proximal portion, posterior view

Plate III.

(a) Mladeč 23: Right proximal humerus with sinter along the perimortem fracture
From left to right: anterior, anterio-medial, posterior view
(b) Mladeč 24: Right humerus diaphysis with unclear, probably postmortem alterations
From left to right: medial and lateral view of the proximal shaft end; anterior view of the distal portion

Plate IV.

(a) Mladeč 20b: Rib fragment, caudal view (cut-marks?)
(b) Mladeč 28: Left proximal femur with a cut mark at the inferior margin of the obturator externus groove

Plate V.

(a) Mladeč 20c: Rib fragment, undecalcified thin-ground section (80µm), viewed through the microscope in transmitted light. Magnification x 40
(b) Mladeč 20c: Rib fragment, viewed through the microscope in polarized light using a Hilfsobject red first order (quartz) as compensator. Magnification x 40
(c) Mladeč 25a: Rigth proximal radius, undecalcified thin-ground section (80µm), viewed through the microscope in transmitted light. Magnification x 100
(d) Mladeč 25a: Rigth proximal radius, viewed through the microscope in polarized light. Magnification x 100
(e) Mladeč 25c: Right proximal ulna, undecalcified thin-ground section (80µm), viewed through the microscope in transmitted light. Magnification x 100
(f) Mladeč 25c: Right proximal ulna, viewed through the microscope in polarized light. Magnification x 100

Plate VI.

(a) Mladeč 28: Left proximal femur, undecalcified thin-ground section (80µm), viewed through the microscope in transmitted light. Magnification x 40
(b) Mladeč 28: Left proximal femur, viewed through the microscope in polarized light using a Hilfsobject red first order (quartz) as compensator. Magnification x 40

(c) Mladeč 28: Left proximal femur, viewed through the microscope in transmitted light. Magnification x 100
(d) Mladeč 28: Left proximal femur, viewed through the microscope in transmitted light. Magnification x 100
(e) Mladeč 28: Left proximal femur, viewed through the microscope in polarized light. Magnification x 100
(f) Mladeč 28: Left proximal femur, viewed through the microscope in polarized light using a Hilfs-object red first order (quartz) as compensator. Magnification x 100

PLATE I

PLATE II

a

b

c

a

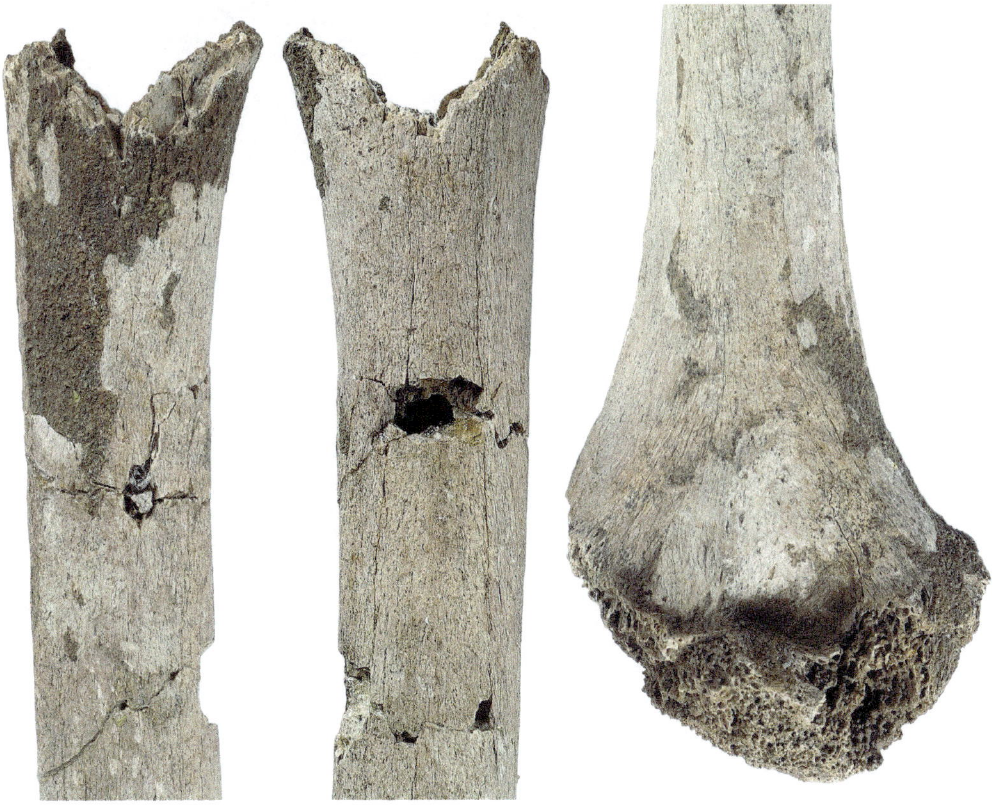

b

PLATE IV

a

b

PLATE V

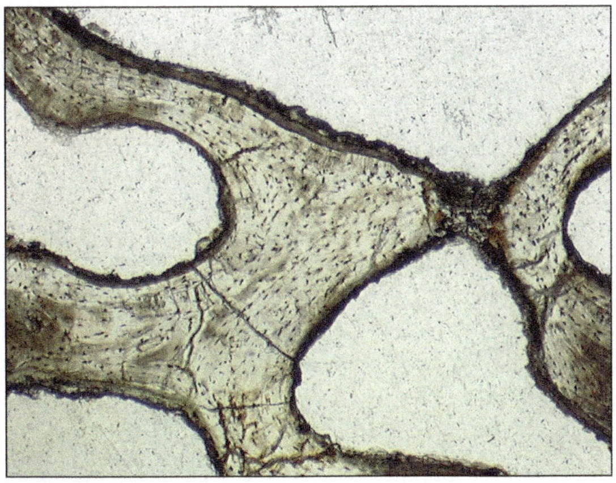

a

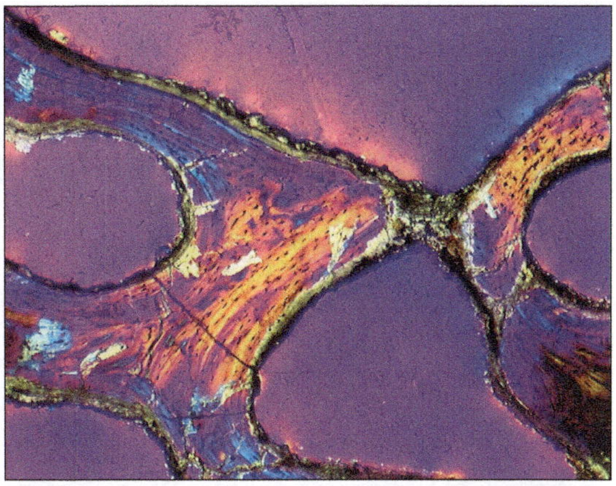

b

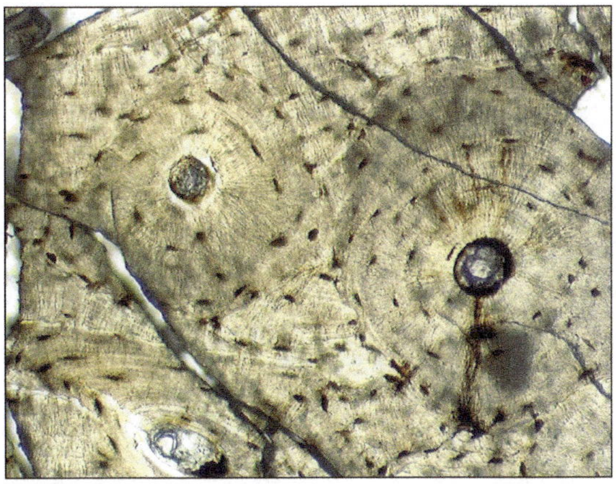

c

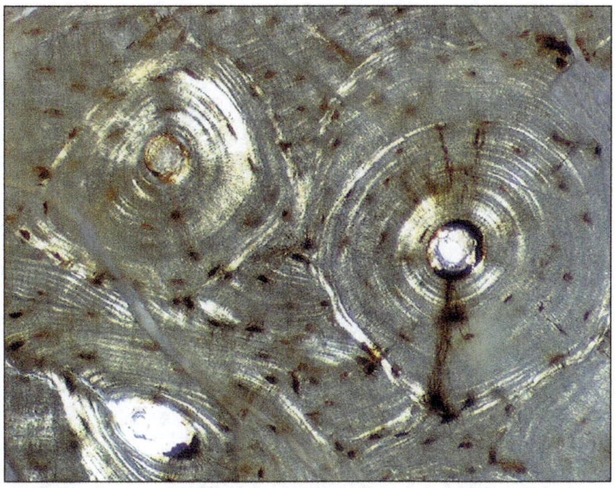

d

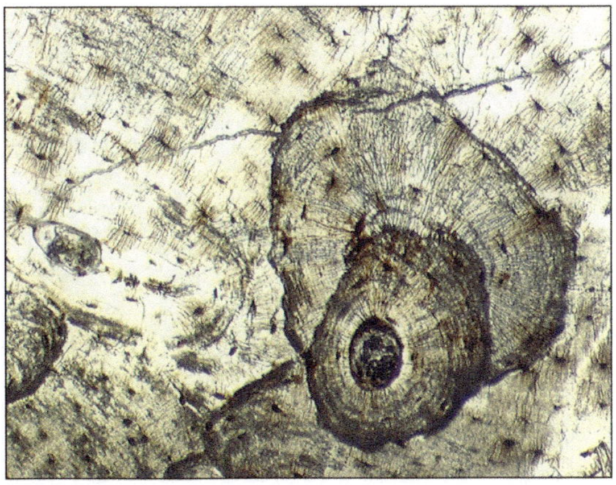

e

f

PLATE VI

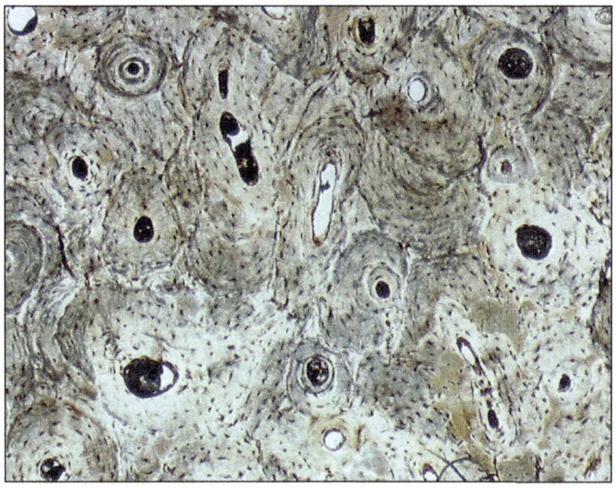

a

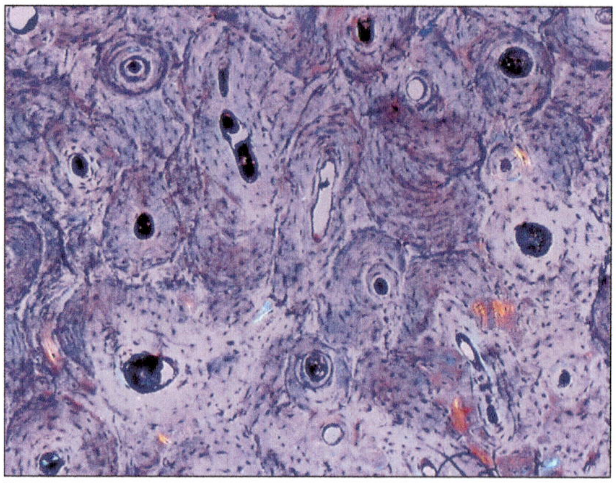

b

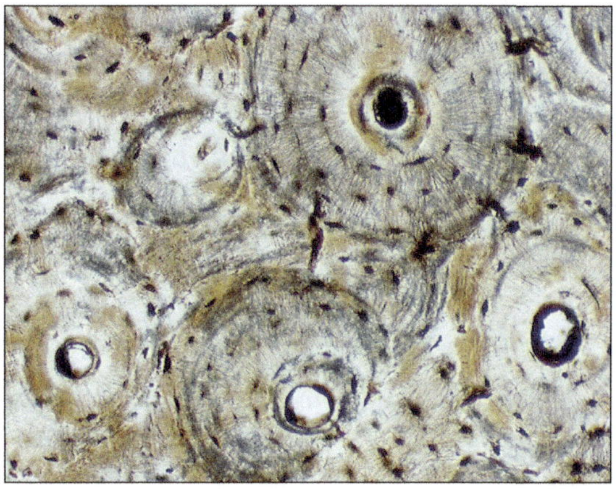

c

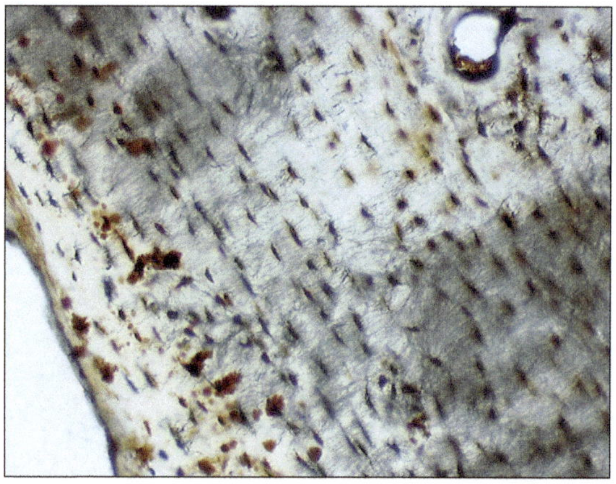

d

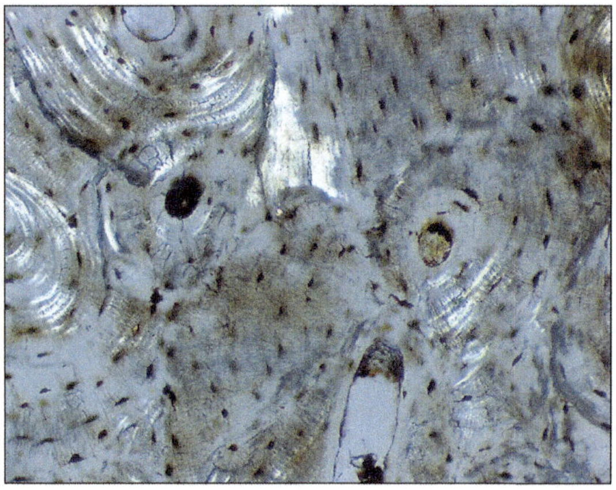

e

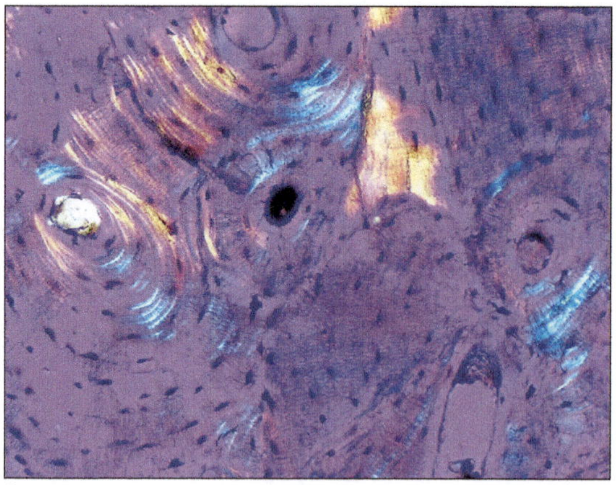

f

Chapter 6

LARGE MAMMAL REMAINS FROM THE MLADEČ CAVES AND THEIR CONTRIBUTION TO SITE FORMATION PROCESSES

Martina Pacher

Introduction

The Mladeč Caves lie west of the village Mladeč near Olomouc in Moravia. The cave system developed in the Devonian limestone of the Třesín Hill, and was discovered in the course of limestone quarrying.

Following Svoboda (this volume), four sites are distinguished inside and above the hill. Site I is situated in a large cave consisting of several halls, collapsed chimneys, and corridors. Older names of this site include "Fürst-Johanns-Höhle" and "Bočkova díra". After its discovery in 1826 or 1829, finds of a "giant" and of animal bones were reported from Hall A behind the entrance (Szombathy, 1925, 4; Maška, 1886). Josef Szombathy, from the Naturhistorisches Museum in Vienna, carried out the first scientific excavations in 1881 and 1882 in Hall D (Szombathy, 1882; Hochstetter, 1883). In 1902, additional parts of the cave were opened and finds of animal bones were determined by Knies (Szombathy, 1904). The exact locality of this survey is not known.

At about the same time, a quarrel over the claim of ownership occurred. In April 1902, the owner of the "Plavatisko", the area of the Třesín Hill directly above the cave, closed the entrance to the Mladeč Cave and opened up a new access directly from his ground (after Smyčka in Szombathy, 1904). The year 1902 is inscribed on the rock wall in the middle of the chimney leading into Hall A (Svoboda, 2000, 527). Thus, it seems likely that the chimney into A was re-opened during the dispute.

From 1903 onwards, Knies examined different localities in Halls D and E, probably around the large debris cone that he first observed (Svoboda, 2000, 530; Szombathy, 1925, 9). At that time, the first arrangements to accommodate public visits inside the cave had already begun (Svoboda, 2000, 530). In 1911, the Krajínski musejní spolecnost v Litovli (Museum Society in Litovel) became the owner of the total area of the cave (Oliva, 1989, 53). In order to better adapt the cave for public visits, large-scale earth removals took place. Some parts of the cave were levelled of up to 3 or 4 meters and as a result, deeper parts of the cave were discovered (Szombathy, 1925, 1, 9). A few juvenile bear remains were mentioned only from the Netopiře jeskyně (bat cave) southeast of Hall E. The corridor "s" served as a second entrance after removal of the filled in sediment. Szombathy (1925, 9) assumed here an original horizontal entrance to Mladeč Cave I, but during the Late Pleistocene this corridor was already filled by earlier, perhaps Middle Pleistocene sediments (Svoboda, 2000, 534). Fürst and Smyčka carried out additional important excavations in 1922. The exact position of their survey is not known but it is assumed to be near findspot "e" (Oliva, 1989, 53). Szombathy (1925, 10) speaks of two separate "fireplaces" with animal bones, human remains, and bone artifacts in the surrounding area. Spot "II" on the ground map of Jelínek (1983) might indicate one of these localities (see Fig. 3b). Northwest of the "fireplaces", mainly bovid remains are mentioned. This locality is the continuation of Szombathy's findspot "d". More recent investigations in Hall D were organised by Jelínek from 1958 to 1963, and concerned mainly Middle Pleistocene layers (Svoboda, 2000, 532; Jelínek, 1983, 1987).

Mladeč Cave II is located about 50 steps west of the present entrance to site I. It was discovered and subsequently destroyed in 1904 during quarrying operations. A week later, Knies examined the site and collected various finds (Szombathy, 1904; Svoboda, 2000, 531).

A third site, "Podkova" cave, lies on the northern slope of the hill, and is separated from the main cave system. Directly above the karstic system lies site IV, called "Plavatisko". Here a Gravettian open-air site was located, but the Třesín Hill was settled throughout prehistory (Jelínek, 1983). For a more detailed description of the sites and the history of investigation at the Mladeč Caves, see Svoboda (this volume) and Svoboda (2000).

Sites I and II are famous mainly for their rich anthropological material. More than 100 specimens of modern humans are reported (see Svoboda et al., 2002, 957). Based on skulls and maxillae, the preserved material represents seven or eight individuals, including one child (see list in Jelínek, 1983). Immediately after the discovery of the human bones, their Pleistocene age was questioned (Maška, 1886). The alleged contemporaneousness of reindeer and human remains as emphasized by Hochstetter (1883), probably favoured a Magdalenian age of the assemblage but this was already doubted by Hoernes (1903). Finally, Bayer (1922) placed the remains in the Aurignacian horizon. He summarized various arguments to prove his assumption. Among them was evidence of cave bears, one perforated bear tooth, and above all, bone points with a massive base. Bayer (1922) regarded these artifacts as diagnostic of the Early Upper Paleolithic, and introduced the Mladeč Caves as an eponymous site for bone points with a massive base.

The various findspots in the Mladeč Caves have produced about 40 bone points but only a few stone artifacts. According to Svoboda (2000, 531), none of the lithic artifacts are diagnostic. Following Oliva (1989, 54), only one of the preserved specimens found at today's entrance area can be ascribed to an Aurignacian tradition. In addition, Valoch (1995, 73) and Jelínek (1983) report an artifact found in corridor "c", which could be of Middle Pleistocene age (Svoboda, 2000, 531).

The rich faunal material consists of large mammals and small vertebrate remains. Various authors have published different faunal lists (see summary in Musil, 2002), but the material has not been examined in detail until today.

Material

Faunal remains analyzed in this study include material recovered during the various excavations at sites I and II. Animal bones from the excavations by Szombathy are housed in the Naturhistorisches Museum Wien. The collection of Knies is stored at the Moravské zemské muzeum in Brno. The material from excavations by Fürst and Smyčka is normally kept at the museum in Olomouc but was available for this study at the museum in Brno. The material from excavations carried out by Jelínek is also housed at the Moravske muzej.

The following study mainly focuses on bones of large mammals, but the collections of Knies and Jelínek also contain a large amount of small vertebrate remains. During the excavation of Szombathy, micro-vertebrates were also found. He documented the „untere Nagetierschicht" in his profile at findspot "a" at 2.6 m below the surface (Bayer, 1922, 182, Abb. 4). Bayer (1922, 183) ascribed the layer to a Mousterian horizon of the glacial accumulation point of the Riss. Woldřich studied the remains but apparently did not publish the results (Woldřich, 1897, 399; Bayer, 1922, 183). The material was not available at the Naturhistorisches Museum. According to the depth of recovery, these small vertebrate remains belong to the Middle Pleistocene horizon of the Mladeč Caves.

The total assemblage from the Mladeč Caves has suffered severe losses during the course of time. Likewise, the large mammal assemblage that was still available for this study, is biased due to various reasons. Unauthorized excavations and collections of bones occurred soon after the discovery of cave I in Hall A. Visitors collected animal bones and pieces of flow stone (after Wolny in Szombathy, 1882,

100; Maška, 1886). In Hall D, near findspots "b" and "d", the excavation areas of Szombathy became disturbed between 1881 and 1882 (see Antl-Weiser, this volume), and in corridor "c", Szombathy assumed the sediments had already been disturbed (Szombathy, 1882, 101). In order to adapt the Mladeč Cave for tourist visits, large-scale earth removals took place several times that led to a levelling of the cave floor in Halls D and E of up to 3 or 4 meters (Szombathy, 1925, 9; Smyčka, 1925 in Oliva, this volume).

An unknown portion of the finds from Mladeč Cave II was taken away by private collectors (Szombathy, 1904, 14). Other finds could not be collected due to their poor state of preservation, such as the bulk of bovid remains at findspot "d" (Szombathy, 1882, 106), or they became destroyed at Mikulov castle at the end of the Second World War (Svoboda, 2000, 531). A certain bias towards the loss of small and fragile elements due to recovery techniques can also be assumed.

Methods

The study of animal bones from the Mladeč Caves consists of a taphonomic part followed by a metric part. In the first part, the material is examined in order to contribute to the knowledge of site formation processes at the Mladeč Caves. In the second part, the remains are characterized by metric analysis. The taphonomic analysis is undertaken by adopting methods used at other cave sites (Pacher, 2000; Pacher, 2004) but is restricted by the loss of bones and the rather poorly documented fieldworks inside the cave.

At first, the remains are determined to skeletal element and species. Then, the total material is examined for bone modifications, and the state of preservation of single specimens is observed. The spatial distribution of the remains is reconstructed by using information given in previous publications.

In the second part of the paper single specimens are measured following v. d. Driesch (1976). Values obtained on the Mladeč bone material are compared to data from faunal remains from other Late Pleistocene localities, mainly from caves and open-air sites from Lower Austria and Moravia.

Taphonomic analysis

Species composition and skeletal element distribution

A total of 632 bones of large mammals from the Mladeč Caves still preserved in the collections of Szombathy, Knies, and Smyčka (Table 1) could be ascribed to 22 species, respectively.

An additional eight specimens are left undetermined to species or genus. Ten specimens are of uncertain provenience and therefore, marked with a question mark in Table 1. The remains were found in a box of the collection of Smyčka containing a conglomeration of extant and fossil remains that were used for a previous reconstruction of a fireplace inside Mladeč Cave I. The origin of the fossil remains could not be traced back with certainty; an antler from *Capreolus capreolus* and an upper molar from *Coleodonta antiquitatis* are such uncertain finds. Roe deer was not mentioned in any previous faunal list (see Musil, 2002). Szombathy (1925, 12) determined woolly rhino for Mladeč Cave II, but it was not mentioned in Smyčka's report about this site (in Szombathy, 1904). Thus, roe deer and woolly rhino remain uncertain finds at the Mladeč Caves, and are not included in Fig. 1. In addition, a mandible from a cave lion, three molar fragments from woolly mammoth, and three antler fragments from reindeer all belong to the uncertain remains. Another crucial find is an antler fragment from a red deer with the brow tine cut

Table 1. NISP (Number of identified specimens) of examined large mammal remains from the Mladeč Caves according to the collections; MNI (Minimum number of individuals; adult/juvenile)

Species	Szombathy NISP	Knies NISP	Smyčka NISP	Total NISP	Total MNI
Canis lupus	–	34	4	38	1/1
Alopex lagopus	–	1	6	7	4/–
Vulpes vulpes	4	20	1	25	3/1
Mustela sp.	–	8	3	11	2/1
Panthera (leo) spelaea	6	–	1+1?[1]	7+1?	1
Crocuta crocuta	–	1	3	4	1/2
Ursus spelaeus	7	6	–	13	2/1
U. deningeri / U. spelaeus	–	–	139	139	3/4
Felis sylvestris	–	–	4	4	1
Rangifer tarandus	11	16	62+3?	89+3?	3/1
Cervus elaphus	–	1	11+1?	12+1?	1/1
Megaloceros giganteus	–	–	2	2	1
Capreolus capreolus	–	–	1?	1?	1?
Alces alces	–	1	–	1	1
Cervidae indet.	–	–	2	2	–
Bos primigenius	–	–	12+4?	12+4?	3
Bison priscus	19	11	128	158	6/1
Bos primigenius f. taurus	–	–	2	2	2
Bovidae indet.	–	–	31	31	
Capra/Ovis	–	–	1	1	1
Equus sp.	6	5	55	66	3/1
Mammuthus primigenius	–	–	8+3?	8+3?	1/1
Coelodonta antquitatis	–	–	1?	1?	1?
Total	**53**	**104**	**475+10?**	**632+10?**	**40/15+1?**
indet.			8	8	

[1] uncertain evidences

off by a metal tool. The specimen was found in the collection of Smyčka, and may represent a younger intrusion into the Mladeč Caves.

Megaloceros giganteus, Crocuta crocuta, and *Capra ibex* are not mentioned in previous faunal lists (see Musil, 2002; Szombathy, 1925). Material from *Meles meles* and *Ursus arctos* is not yet confirmed in the new study, and remains of Mustelidae are ascribed to genera only. Szombathy (1925, 4) assumed remains of *Capra ibex* at the time of the discovery of the Mladeč Cave at Hall A, because of enormous horn-cores mentioned by Sommer in his report from 1829. In the studied material, only one tooth could be ascribed to *Capra/Ovis* in general, due to its poor state of preservation. *Crocuta* is only mentioned from "new parts" recovered in 1902 (Szombathy, 1904, 12). The exact locality of these parts could not be reconstructed.

The majority of remains from the Mladeč Caves come from bovids (33%), ursids (25%), reindeer (15%), horse (11%) and wolf (6%), respectively. All other species are evident by a few bones only. In Ursidae, *Ursus spelaeus* is determined in the collections of Szombathy and Knies whilst in the collection of Smyčka, *Ursus deningeri* surprisingly prevails. At least the bulk of the remains (122 NISP) can be ascribed to the latter species, but no detailed study has yet been undertaken on the remains from bears. In addition, 89 bones from *Ursus deningeri* are preserved in the collection of Jelínek, which leads to a predominance of bear remains with 39.7% of the determined material.

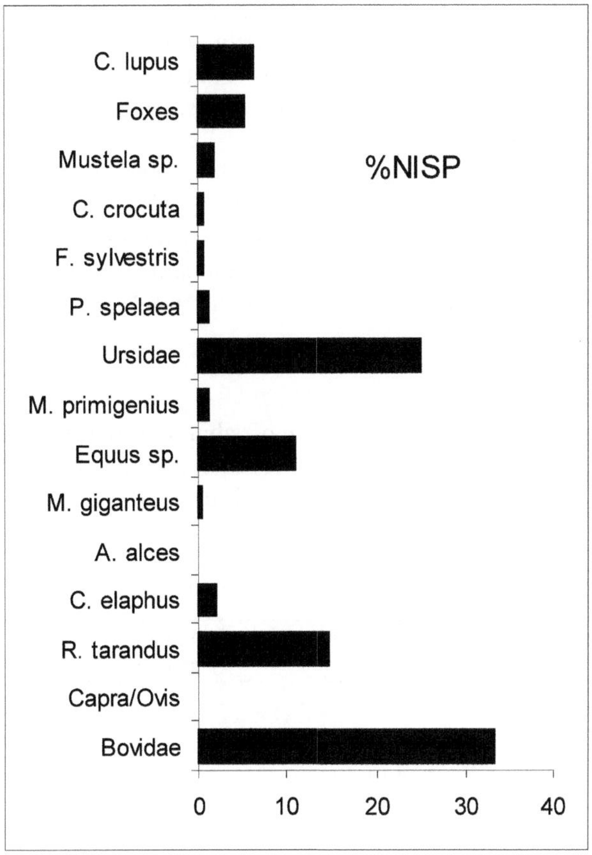

Fig. 1. Percentage of total NISP for large mammals from the Mladeč caves, except uncertain finds (Ursidae without coll. Jelínek)

By counting MNI (minimal number of individuals), steppe bison is represented by six individuals, followed by arctic fox with four. Several species are represented by a MNI of three individuals, and all others by even less individuals. Juveniles are clearly underrepresented in the assemblage, except in ursids.

Besides the skeletal elements, certain species are evident due to artifacts. The collection of Szombathy contains 22 perforated teeth from five species. A revision of the pendants at the Naturhistorisches Museum Wien (Pacher, 2005) confirms one perforated canine each from wolf and bear, nine incisors from beaver, eight incisors from moose, and one incisor from horse. The latter shows no perforation. Slight traces on the root could be interpreted as an attempt to drill the tooth. On some of the teeth from beaver, the assumed perforated part of the tooth is broken. The teeth from *Alces alces* were originally ascribed to reindeer (Bayer, 1922, 178; Hochstetter, 1883, 169). In another old publication, even wild boar was assumed (see Maška, 1886, 55). Oliva (1993, 212) first referred the teeth to moose.

Alces alces is further represented by one vestigial metapodial bone in the collection from Knies, and Smyčka found a bulk of worked metapodials near findspot "e" previously ascribed to a large Cervidae (Oliva, 1993, 212). The size of the specimens as well as the preservation of the distal articular surface, again clearly points to *Alces alces*. Contrary to *Megaloceros giganteus* and *Cervus elaphus*, which are Plesiometacarpalia, *Alces alces* and the smaller *Rangifer tarandus* are Telemetacarpalia, thus reducing the proximal part of the 2nd and 5th metapodial bone (Königswald, 2002, 71). The metapodial bones form unusual artifacts, partly with drilled and carved articular surfaces (Oliva, 1989, Plate IV; Oliva, 1993). Although not all specimens show modifications like the one from the collection of Knies, it seems the selective evidence of metapodials from *Alces alces* at the Mladeč Caves is related to the raw material use of humans.

Several authors refer to the large, flat bone point with a massive base and assume a mammoth rib (Hochstetter, 1883; Bayer, 1922, 173; Szombathy, 1925, 13) as raw material. One point confirms the use of mammoth ivory (Oliva, 1993, Fig. 2), whilst other pieces are made of antler and bone.

One metapodial from horse from the collection of Szombathy was regarded as an artifact, but obvious modifications are missing. The tip is broken (Szombathy, 1925, Tafel 2, Abb. 5).

Skeletal element distribution

Skeletal element distribution was analyzed for the most abundant species (Table 2). Different patterns are evident. Nearly all parts of the skeleton are preserved in Bovidae (Fig. 2b). On the contrary, horse (Fig. 2c) is mainly represented by isolated teeth (44%) and bones of the extremities among them,

Table 2. Skeletal element distribution of the most abundant large mammal species at the Mladeč Caves

	Canis lupus	Equus sp.	Rangifer tarandus	Bison / Bos[1]	Ursidae / coll. Jelínek
Antler/horn			40		
Skull	2	1	2	4 (2 lost)	9 / 6
Mandible	(1 lost)	1	7	4	15 / 9
Isol. Teeth	9	29		23	48 / 17
Atlas	1			2	- / 1
Axis	1			2	
Vertebrae	9		9	28	2 / 4
Ribs	2	2	1	11	- / 15
Scapula				2 / 3	
Humerus	2	1	4	9/ 2	7 / 6
Radius	1	3	3	2	5
Ulna				3 / 1	8 / 2
Carpals			1	1	2
Pelvis	1		1	7	2 / 3
Femur	1	1	3	5	13 / 5
Tibia	2	4	3	5 / 2	10 / 8
Fibula					3 / 1
Metapodials	4	11	10	12 / 4	19 / 10
Astragalus		1		5	3
Calcaneus	1	1	2	6	1
Other tarsals				4	
Ph.1	2	6	2	11	3 / 2
Ph.2		3	2	6	1
Ph.3		2		4	1
Patella					
Hyoid				2	
Sternum				1	
Total	**38**	**66**	**90**	**159 / 12**	**152 / 89**

[1] without *Bos primigenius* f. taurus

metapodials prevail with 17%. Antler fragments are clearly most abundant in reindeer (Fig. 2a) with 45%, followed by several bones of the axial skeleton and the extremities. Twenty-five specimens are shed antlers, 14 are beam fragments, and one specimen is still attached to a small preserved part of the cranium.

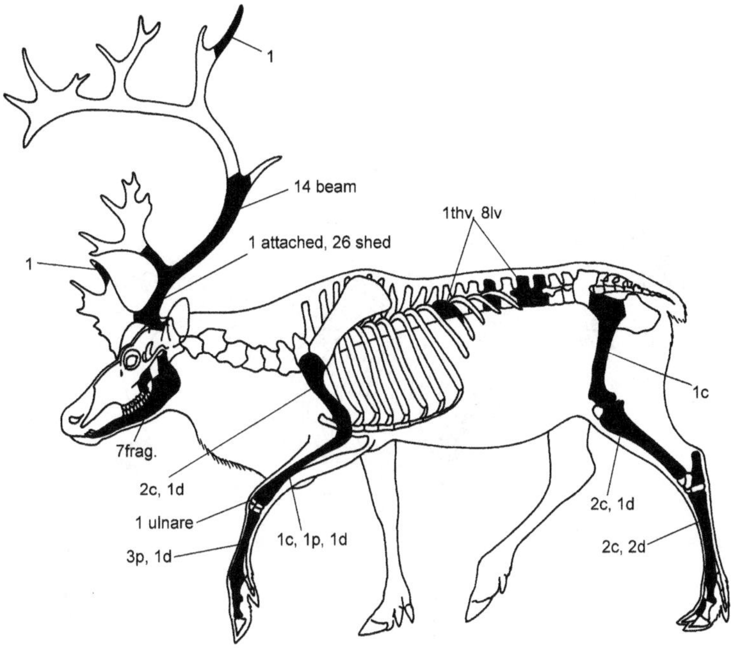

Fig. 2a. Skeletal element distribution of *Rangifer tarandus* (c...complete, d...distal, p...proximal, s...shaft fragment); Fig. 2a–d N. Frotzler, modified

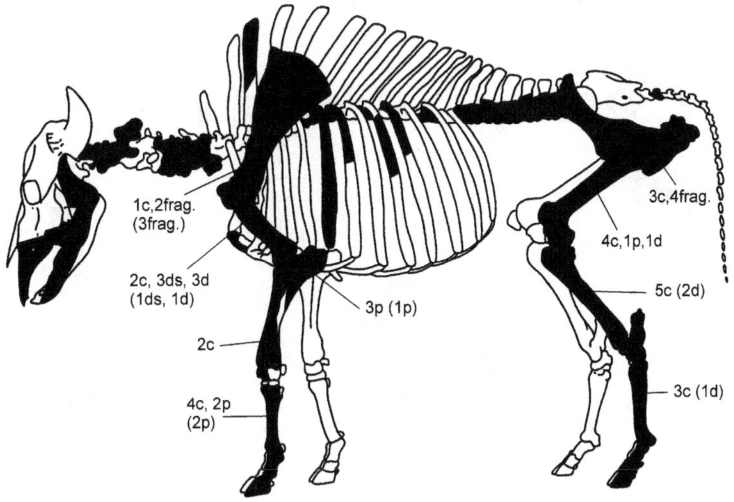

Fig. 2b. Skeletal element distribution of *Bison priscus* (c...complete, d...distal, p...proximal, s...shaft fragment)

Among carnivores, *Canis lupus* (Fig. 2e) is represented by various parts of one single skeleton. Ursids (Fig. 2d) produced an element distribution known from various bear caves. A bias towards isolated teeth and denser elements is obvious in a gross comparison of several deningeri and spelaean sites (Fosse et al., 2002, 84). At the Mladeč Caves, teeth outweigh with 31.6 %, followed by metapodial bones with 12.5 %. Including the 89 bear remains from the collection of Jelínek results in 27.5 % dental remains, and 12.3 % metapodial bones. More fragile and smaller elements are underrepresented or missing. Additional identified species are represented by a few specimens only.

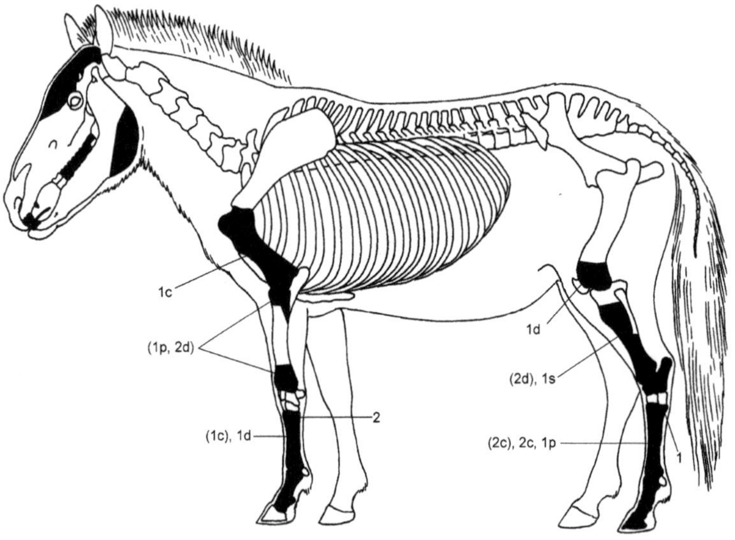

Fig. 2c. Skeletal element distribution of *Equus* sp. (c...complete, d...distal, p...proximal, s...shaft fragment)

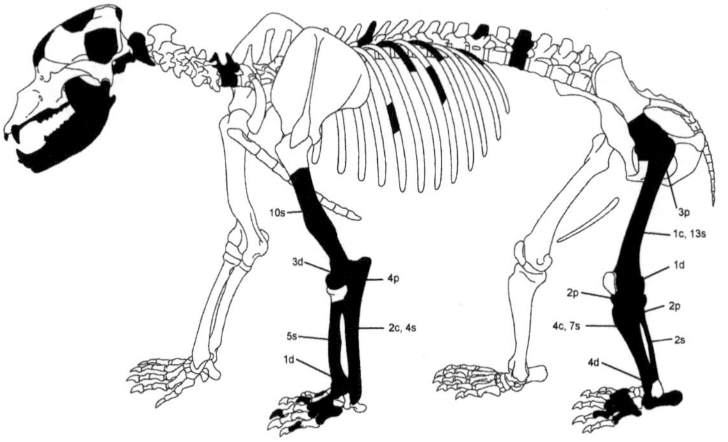

Fig. 2d. Skeletal element distribution of Ursidae, including material from the collection of Jelínek (c...complete, d...distal, p...proximal, s...shaft fragment)

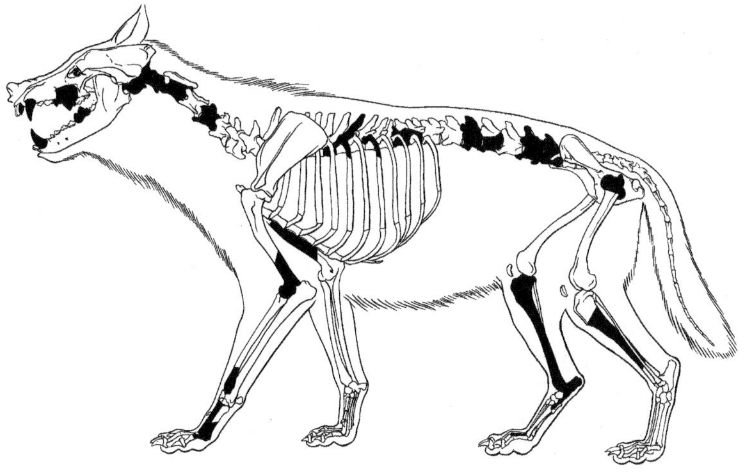

Fig. 2e. Skeletal element distribution of *Canis lupus;* Pales and Garcia 1981, modified

Bone modification and surface preservation

In general, patterns of bone modifications are relatively scarce on the material from the Mladeč Caves. Proposed human influence on the remains could not be confirmed. The crushed femora of a bovid described by Szombathy (1925, 11, Tafel 2, Abb. 4) shows clear traces of carnivore gnawing with furrows in the spongy part of the opened distal shaft, as does a metapodial revealing furrows on the shaft. Splinters on the canine of the cave bear mandible in the collection of Szombathy (1925, 13, Tafel 3, Abb. 7), can be ascribed to natural breakage. The proposed chopping marks (Szombathy, 1925, 11, Tafel 2, Abb. 3) on the occipital condyle of the published skull from *Bison* could not be examined, since the skull is missing. Judging from the foto, the types of traces suggest younger chopping marks that probably occurred during excavation. Nonetheless, two specimens from the collection of Smyčka do reveal cut marks with eye determination. The first one is a distal metapodial bone from reindeer with various vertical cut marks on the shaft. An examination under a microscope reveals the whole shaft area is full of marks made with metal tools, which probably originated in the course of cleaning the bone from the thin calcareous crust. Exfoliation of the uppermost bone surface next to some cut marks indicates their modern origin. On the epicondyles, old scratches are visible. In addition, the bone shows a recent fracture and no traces of marrow cracking (Plate III, Figs. 2a–c). The second bone is a heavily modified proximal fragment of a metacarpal from a bovid. The specimen shows spiral fractures, cones, and gnaw marks. A bundle of vertical scratches on the proximal shaft (Plate III, Figs. 1a–c) resemble cut marks. More scratches are visible under a microscope. Their origin is unclear, probably metal tools are responsible for the scratches.

In addition, the described bone is one of the few specimens revealing a distinct preservation pattern in the material from the Mladeč Caves, distinguished by the degree of bone modification and the overall appearance of the remains. This type of preservation (type 1) comprises mainly fragments of yellow color, often revealing traces of intense carnivore activity as indicated by spiral fractures, cracks, fissures, and cones (Plate II, Figs. 1–3; Plate IV, Fig. 1a). The edges are sometimes rounded and the bone surface appears smooth and polished. Only bones of bovids and *Equus* could clearly be ascribed to this type of preservation. In *Equus*, eight out of 36 bones are concerned. All of them come from the collection of Smyčka. Six of these eight bones show intense gnaw marks (Table 3; Plate II, Fig. 2 and 3). Only two metapodials are completely preserved.

Table 3. Evidence of gnaw marks according to preservation types

	Type 1	Type 2	Type 3	others	$\%NISP_{bone}$
Bovidae	13	5	–	6	15.9
Equus	6	–	–	2	21.6
Rangifer	–	–	–	6	6.7
Ursidae	–	–	22	–	10.4

Small fragments of two humeri, one metacarpal, and one metatarsal could not be unambiguously ascribed to type 1 or to any of the determined bovid species and therefore, are classified as "Bovidae indet.", as are 27 remains of teeth.

Among bovids, twelve specimens, mainly bones of the extremities and the shoulder blade, are ascribed to type 1. Eleven bones show intense gnaw marks. Although mainly fragmented, morphological criteria allowed determination of the bovid remains of type 1 as *Bos primigenius.* On bovid remains from the collection of Smyčka, a second type of preservation (type 2) can be observed. The remains are mainly of bright white color. They are partly covered with a thin, brown, calcareous crust and only a few specimens show moderate traces of carnivore activity (Plate IV, Figs. 2a–b; Plate V, Figs. 1a–b). These bright colored remains could have come from one of the findspots in the northwest of the "fireplaces" near "e". Szombathy (1925, 10) interpreted this locality as a continuation of his findspot "d", since at both localities bovid remains of a more recent impression occurred than from other findspots. Five specimens of type 2 show slight traces of carnivore activity.

Additional remains of *Equus* show different colors from brown to reddish-brown, gray and light-yellow, and only two of these bones show slight traces of carnivore activity. In total, 21.6%, that is eight bones from *Equus* out of 37, are gnawed. Two bones from *Equus,* a cranial fragment and a femur, are of light brown color and appear more recent than other bones. They could also be younger intrusions among the fossil remains. The same is true for two metapodial bones ascribed to *Bos primigenius* f. taurus.

Gnaw marks are also encountered in *Rangifer tarandus;* 6.7% of the bones from reindeer (6 out of 89) show traces of slight carnivore activity (Table 3).

The majority of ursid remains from the collection of Smyčka resemble Middle Pleistocene bear bones from the excavation of Jelínek. In general, Middle Pleistocene remains from the Mladeč Caves are of bright yellow or reddish brown color, and often show traces of manganese-oxide and iron-oxide. One hundred and twenty-two out of 139 bear bones from the collection of Smyčka definitely belong to this type of preservation (type 3), and are ascribed to *Ursus deningeri* (Plate I, Figs. 4 and 5). Gnaw marks are evident on juvenile and adult bones (15:7). The few other bones from the collection of Smyčka could not be definitely ascribed to any of the two bear species.

Radiometric dates

Several radiometric dates have been obtained from bones of both animal and human remains (Wild et al., 2005 and Suppl. Table 1) from the Mladeč Caves. The samples mainly come from the collection of Szombathy. In course of the faunal study, two samples from the collections of Jelínek and Smyčka were also dated. A previous attempt at dating an animal bone from the Olomouc museum and a human rib fragment from findspot "d" failed (Svoboda et al., 2002, 958). Unfortunately, the samples of cave bear and cave lion did not contain enough collagen either (Table 4).

The results of radiometric age determination indicate several phases of input (Table 4). The dates for *Bison priscus* from the collection of Szombathy fall into the Pleniglacial (c. 24–14 kyr BP).

Table 4. Radiometric dates of animal bones from the Mladeč Caves

	Species	Element	Collection	Years BP	Cal. BC
VERA[1]-2217	*Bison priscus*	atlas	Szombathy	16,950+/-60	18,310 - 17,950
VERA-2218	*Bison priscus*	maxilla	Szombathy	21,340+/-100	
VERA-2219	*Bison priscus*	metap.	Szombathy	25,880+/-180	
VERA-2962	*Castor fiber*	tooth	Jelínek	8,515+/-35	7,595 - 7,525
VERA-2963	*Equus sp.*	metap.	Smyčka	42,500+1,400/-1,200	
	Ursus spelaeus	metap.	Szombathy	failed	
	Panthera spelaea	tibia	Szombathy	failed	
	Panthera spelaea	mt3	Szombathy	failed	

[1] Vienna Environmental Research Accelerator

Around the Late Glacial Maximum, tundra-steppe environments around the Mladeč Caves can be assumed. Environmental conditions also favour the determination of steppe bison instead of *Bos primigenius*. One date is slightly older but the metapodial bone contained very little collagen. The obtained age of 25,880±180 years BP has to be treated with caution because of low collagen yield (Wild et al., 2005).

The metapodial bone from *Equus* from the collection of Smyčka represents a Middle Würmian input of remains. The bone is ascribed to material of preservation type 1, revealing intense gnawing damage probably produced by hyenas. The date coincides with the age of various hyena dens in the wider region (e.g., Nagel et al., in press; Hofreiter et al., 2004).

It is interesting to note that the dated human bones indicate a different phase of input than the animals thus far dated. Four dates obtained from human remains from the Mladeč Cave range around 31,000 years BP (Wild et al., 2005). One sample provided a slightly younger age and is regarded as contaminated. The dates confirm the previously assumed age of the human fossils from the Mladeč Caves.

Besides remains of bears, no other fossils from the excavation of Jelínek are considered in this study, with the exception of one single incisor from *Castor fiber* (Plate I, Fig. 1). The tooth comes from Jelínek's sondage III (Fig. 3b). Unfortunately, the tooth is broken at the root, but since it is an isolated single find, a correlation to the perforated beaver teeth from the collection of Szombathy was discussed. Surprisingly, radiometric dates place the tooth in the Boreal interstadial that is characterised by Mesolithic traditions.

Spatial distribution of remains

The large mammal remains were found at different localities inside the cave system, sometimes in association with human remains and artifacts (Table 5). From the excavation of Szombathy, quite detailed descriptions of the sites and their contents have been published. Large mammal remains were not reported from findspot "a" in the first publications (Szombathy, 1882; Hochstetter, 1883), whereas in Szombathy (1925, 5), a rib fragment from reindeer attached by calcareous crust to the human cranium in "a" as well as bones of reindeer and bovids in the near vicinity were mentioned. The bulk of faunal material came from "b" and "d". Five species, *Canis lupus, Vulpes vulpes, Ursus spelaeus, Rangifer tarandus,* and bovids are mentioned from findspot "b". Remains of one single individual are represented by reindeer. Findspot "d" yielded only bovid remains. Szombathy (1882, 106) reported complete skulls, anatomically connected vertebrae and ribs, complete long bones and so on, near the surface. Unfortunately, their poor state of preservation allowed collection of only a few specimens.

In corridor "c", a few bones of *Ursus spelaeus* were found in sediments that were probably already disturbed (Szombathy, 1882, 101). In 1882, in the central part of Hall D, remains of reindeer and bovids were again mentioned (Szombathy, 1925, 6). Two stone tools, ten bone implements, and 22 perforated animal teeth found scattered over an area of 20 m^2 in Hall D belong to the collection of Szombathy (1925, 12). In addition, a vestigial metapodial from horse was included among the bone artifacts (Szombathy, 1925, 11, Tafel 2, Abb. 5) but shows no traces of use or modification.

Several bones collected from findspot "b" and listed in Szombathy (1882, 105) are lost. Among them, a left mandible of *Canis lupus* and the two cranial fragments of bovids published in Bayer (1922, 184, Abb. 6) and Szombathy (1925, 11, Tafel 2, Abb. 3). Morphological characteristics of the occipital region confirm the depicted specimen as *Bison priscus*.

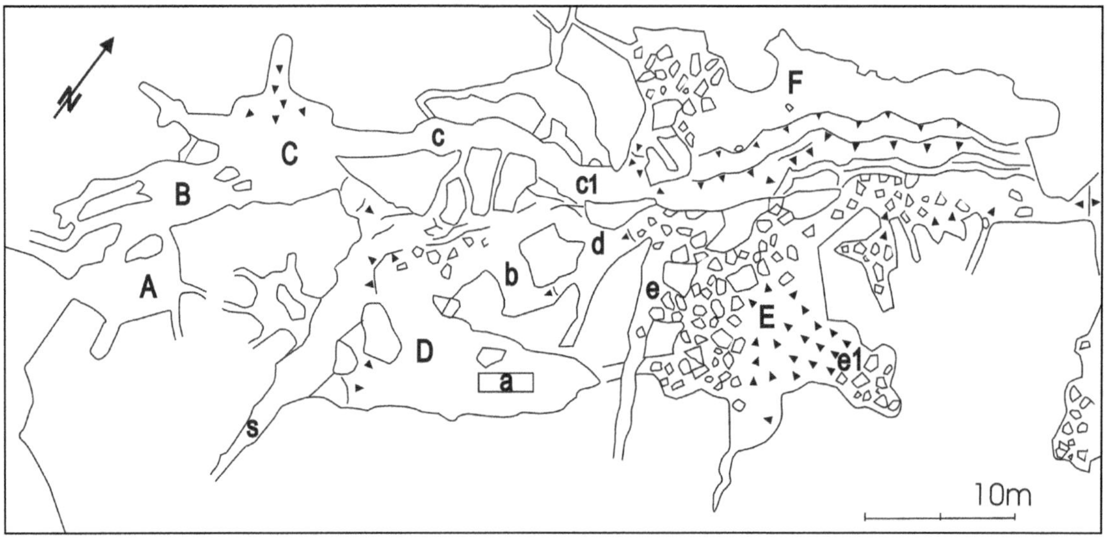

Fig. 3a. Ground map of Mladeč Cave I at the time of Szombathy's excavations (after Jelínek, 1987, 54)

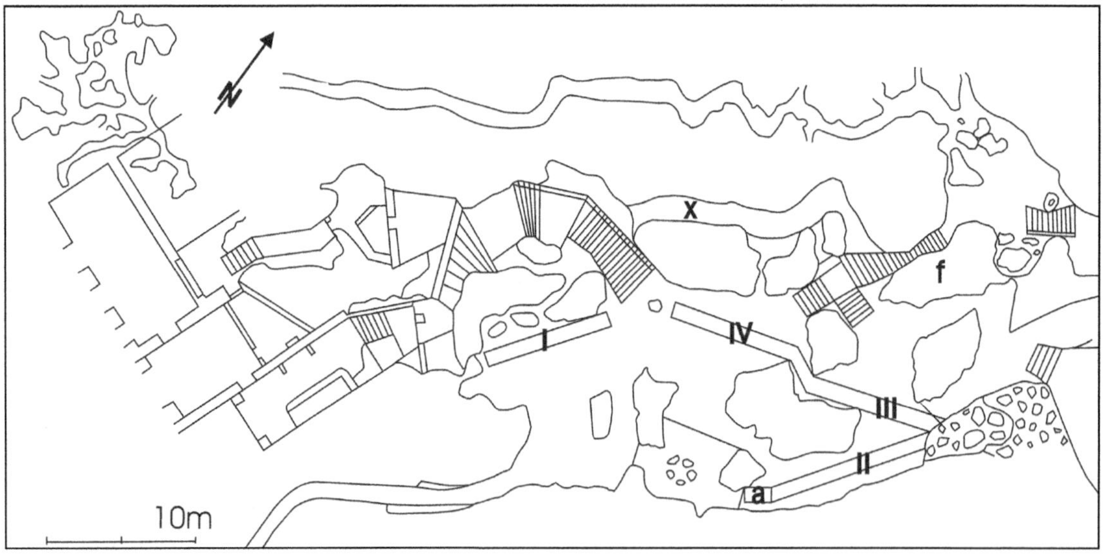

Fig. 3b. Ground map of Mladeč Cave I at the time of Jelínek's excavations (after Jelínek, 1987, 61)

Already at the time of the discovery of the Mladeč Cave, the first fossil finds were reported from today's entrance part, Hall A. According to the description of the bones, Szombathy (1925, 3–4) assumed at least remains of woolly mammoth, ibex, reindeer, and a small mammal the size of a hare. It is unclear whether human remains were also found. The remains of a "giant" were probably bones from animals (see discussion in Szombathy, 1925, 4). In the course of accommodating the cave for visits, additional bone remains were found in Hall A, among them an atlas from a mammoth, two teeth from *Equus*, and an antler fragment (Szombathy, 1925, 7). Following Svoboda (2000, 529), in 1905 Knies mentioned a carved bone point associated with the "giant". Later, in 1937 Skutil found two stone artifacts in secondary loess on the right side of the entrance (Oliva, 1989, 53).

The new parts of the cave discovered in 1902 produced besides remains of wolf, fox, reindeer and bears, also bones of hyena and lion (Knies in Szombathy, 1904, 12). The exact position of the findspot could not be reconstructed.

Knies probably excavated an area around the debris cone between Halls D and E that he first mentioned (Svoboda, 2000, 530). Large mammal bones found by Knies are listed in Table 5. Szombathy's finds from "a" and "b" are of gray color and covered by a thin brownish calcareous crust (Szombathy, 1882). A few specimens from the collection of Knies show an identical preservation pattern. They are probably derived from the same localities in Hall D.

It remains unclear, whether Knies also found bone artifacts and human remains. Five bone points from the collection of Knies in the Moravské zemské muzeum are of unknown provenience and perhaps came from the local people (Oliva, 1989, 53). Human remains were assumed by Szom-

Table 5. Distribution of mammalian species and other finds according to findspots (Net. jes = Netopiře jeskyně)

	Hall A	Hall D (Szombathy)				D, E (Knies)	Hall E (Smyčka)			corr. c	Net. jes.	new parts	Site II
		a	b	d	central 20 m²		spot 1*	spot 2	spot 3			1902	
Bone artifact					x	x?		x					x
Stone artifact					x		x?			x			x
Pendants					x			x					
Homo	?	x	x		x	x?	x	x					x
Micro-vert.		x				x	x						x
C. lupus			x			x	x					x	x
C. l. familiaris													x?
Mustelidae													x
Foxes			x			x	x					x	x
M. meles						x	x						
P. spelaea						x	x					x	
C. crocuta												x	
Ursidae			x			x	x			x	x	x	
R. tarandus	x	(x)				x	x					x	x
C. elaphus						x	x						x
Cervidae	?							x					
A. alces						x	x						x
Bovidae	?	(x)	x	x		x	x		x				x
C. ibex	?												
Equidae	x					x	x						x
M. primigenius	x												
C. antiquitatis													x

* animals from spot 1 and 2

bathy (1925, 9), but Knies himself did not mention such finds from Mladeč Cave I (see Svoboda, 2000, 530).

The exact position of the findspots from Smyčka are not known, but Oliva (1989, 53) assumed them to be near Szombathy's position "e" (see Fig. 3a). According to Szombathy (1925, 10), three findspots are distinguished. In the vicinity of one "fireplace", animal bones, human remains, and stone artifacts were found. This spot probably corresponds to the find of human remains (spot II in Fig. 3b) in the ground map from Jelínek (1983, 58). About two meters to the northeast, animal bones and bone artifacts made from metapodial bones from *Alces alces* became uncovered. Szombathy (1925, 10) mentioned fragmented human remains, but their provenience is unclear. A second "fireplace" was assumed to the south of the first one. Northwest of the two "fireplaces", mainly bovid remains were uncovered. These finds are related to Szombathy's findspot "d".

The large mammal species composition (Table 5) of the findspots near "e" is the same as that of the excavations by Knies (Szombathy, 1925, 12). No details for the three findspots are given except for the continuation of Szombathy's findspot "d", that is characterized by an accumulation of bovid remains.

Mladeč Cave II produced faunal remains of several species. Maška (in Szombathy, 1904, 12) mentioned reindeer, deer, wolf, red fox, bear, a small carnivore, hare, and a dog, which according to Bayer (1922, 183), was probably a misidentification. Szombathy (1925, 12) determined also *Coelodonta antiquitatis, Equus caballus, Bos primigenius,* and probably *Bison priscus* but he did not mention bear remains. *Alopex lagopus, Putorius putorius,* and *Alces alces* were added by Knies and Smyčka to the large mammal list from Mladeč II (Szombathy, 1925, 12).

It was not possible to distinguish between remains from Mladeč I and II, although both the museums in Brno and Olomouc may contain specimens from Mladeč II (M. Oliva, pers. comm.). Knies and Smyčka collected remains from both sites.

Discussion of site formation processes

Various authors gave different interpretations of the finds from the Mladeč Caves mainly based on the artifacts and human remains. Bayer (1922) assumed a human occupation and burial site, and Szombathy (1925, 8; 1882, 106) even assumed cannibalism since human bones had been found scattered and fragmented, in contrast to a completely preserved skeleton from reindeer. Some modified animal bones were regarded as evidence for humans hunting prey and the total assemblage was regarded as contemporaneous.

Knies reconstructed the taphonomic history of a complete reindeer skeleton found in a fissure of the large debris cone. He assumed (in Svoboda, 2000, 530) the reindeer had fallen into the cave and that parts of the skeleton had been re-deposited at two basic directions. Jelínek (1983) favoured the idea of an input of anthropological and archaeological remains through the large chimney in D during the Late Pleistocene. According to Jelínek (1983, 58), the chimney was closed for good during the first Interstadial of the last glacial period.

The problem remains unresolved whether the Mladeč Cave system was a human occupation and burial site, a natural trap, or even a place of ritual human input (see Oliva, this volume; Svoboda, 2000; 2002). The discussion on the depositional history of the Mladeč Caves is mainly based on both, the question of the existence of a natural horizontal entrance and the arguments for a visit of man inside the cave.

The present study allows the inclusion of evidence from the faunal material. The large mammal remains from the Mladeč Caves reveal certain new and important aspects of the total assemblage. Incorporating evidence from faunal material as well as other information provided by previous authors, the following reconstruction of site formation processes at Mladeč Cave I is attempted:

The oldest remains from the Mladeč Caves come from the Middle Pleistocene or even earlier (Svoboda, 2000, 531). The assemblage is characterized by *Ursus deningeri,* small mammal fauna, birds, snakes, and molluscan remains of Middle Pleistocene and even Biharian age (Jelínek, 1983; 1987).

Szombathy first encountered micro-mammals of the Middle Pleistocene at 2.6 m below the surface as shown in the profile from locality "a" (Bayer, 1922, 182, Abb. 4). Surprisingly, the collection of Smyčka also contains a large amount of remains from *Ursus deningeri.* The type of preservation and the morphology of the remains clearly resemble bear remains from the collection of Jelínek (Plate I, Fig. 4 and 5). Svoboda (2000, 531) also reports micro-fauna and snails found by Smyčka; the latter are one of the characteristic finds of the Middle Pleistocene at the Mladeč Caves (Jelínek, 1983, 58). Thus, besides rich Late Pleistocene material, Smyčka must have encountered Middle Pleistocene deposits as well. Unfortunately, not much is known about his excavation campaigns. The localities of the three findspots were near "e" (Oliva, 1989, 53), that is, in Hall E east of a large accumulation of blocks related to the chimney. The Museum Society of Litovel undertook large-scale earth removals leading to a levelling of various parts of the cave of 3–4 meters. By then, Halls D and E became one large cavity (Szombathy, 1925, 9). Even large blocks northwest of E must have been removed (see Fig. 3a). The Middle Pleistocene horizon at about 2.6 m below the surface at the time of Szombathy was reached in the course of the large-scale earth removals in Halls D and E, and Smyčka must have discovered these older deposits in one of his findspots, since the bulk of bear remains from the collection of Smyčka is ascribed to *Ursus deningeri.* Svoboda (2000, 531) also assumed the presence of re-deposited sediments and rubbish from previous excavations at Smyčka's excavation areas.

From 1958 to 1962, Jelínek (1983; 1987) examined the Middle Pleistocene deposits in D and E (Fig. 3b). Late Pleistocene sediment had already been missing due to preceding excavations and tourist accommodation of the cave. Middle Pleistocene sediment built the cave floor and became excavated in four trenches in Halls D and E (Fig. 3b). Thus, the map of Jelínek from today's outline of the cave, gives us an impression of the morphology of the cave during the Middle Pleistocene and prior to the deposition of the debris cone in the "Dome of the Death". Halls D and E form one large cavity, interrupted by various large pillars reaching from the cave floor up to the ceiling.

The predominance of bears *(Ursus deningeri)* in the Middle Pleistocene assemblage suggests a former horizontal entrance into the cave. Taken in mind the morphology of the cave, with its cavities and corridors, the former existence of a horizontal entrance would not be surprising but is not yet proved. Szombathy assumed corridor "s" (Fig. 3a) as the original entrance during the Late Pleistocene, but it was already closed with earlier, probably Middle Pleistocene sediments (Jelínek, 1987).

The next period proved is at 42,000 years BP by a radiometric date obtained from a metapodial bone of *Equus* from the collection of Smyčka. The specimen is ascribed to bones of preservation type 1 that is clearly distinct from the preservation of other bone material at the Mladeč Caves. Type 1 includes only bones of bovids that are ascribed to *Bos primigenius* and *Equus,* and characterized by yellow-green color, intense gnaw marks, and spiral fractures. The high intensity of gnaw marks points to hyena, which is known to break up bones of middle-sized ungulates such as bovids and horses, as regularly indicated by the abundance of these species in hyena dens (e.g., Nagel et al., in press; Musil, 1962). The radiometric date obtained from the type 1 specimen corresponds to ages obtained from hyena bones in the region, such as Teufelslucke (Nagel et al., in press) and other sites (Hofreiter et al., 2004), but there is no clear evidence for a hyena den at the Mladeč Caves. Hyena is not mentioned in any previous faunal list from the Mladeč Caves (see Musil, 2002) and is confirmed in the assemblage by only four specimens. In addition, a hyena den would require an horizontal entrance. Spiral fractures and cones (Plate III, Fig. 1b) are also typical traces resulting from processing of bones for marrow by Paleolithic man, but there is no clear evidence of human interference with the remains.

Unfortunately, the provenience of this particular material could not be reconstructed. The idea that the remains might have come from Mladeč Cave II could not be confirmed. Different faunal lists are given (Szombathy, 1904, 9; Svoboda, 2000, 532) and only in Szombathy (1925, 11) both *Equus* and bovid remains are mentioned. In addition, it remains unclear whether Mladeč II was a small cave with a horizontal entrance (Szombathy, 1904, 14; 1925, 11) or a vertical chimney (see Svoboda, 2000, 532). The extensive modifications point to an input of remains by man and/or carnivores and thus, a locality with a horizontal entrance at 42,000 years BP is required.

The next period is indicated by radiometric data obtained from the human remains from the Mladeč Caves. Four dates obtained range around 31,000 years BP (Wild et al., 2005). All samples taken come from the collection of Szombathy, and comprise findspots "a" and "b", and finds from the central part of Hall D. Additional human remains are confirmed at Smyčka's findspots near "e", and Mladeč II.

The first attempt to further delimit the age of the human fossils was made in 1994. Svoboda (2000, 533) assumed an original solid calcite above the fossil deposits at findspot "a". Thus, samples were taken from remaining calcite layers at the wall of the cave in Hall D. Two sinter data obtained resulted in a conv. ^{14}C age of 34,000–35,000 years BP (Svoboda et al., 2002, 958). The authors conclude the data give the minimum age of the fossils.

By regarding Szombathy's description of Hall D in detail, no evidence for a solid top calcite cover in "a" can be found. On the contrary, the southeastern part of Hall D is described as an even area with a loamy surface but with an elevation of two to three meters (Szombathy, 1882, 101). It was in this loamy area, where Szombathy started his investigations. Only the lower southwestern part of D was repeatedly covered with sinter on the roof as well as on the surface.

The profile recorded at "a" also shows no top layer of sinter (Bayer, 1922, 182; Szombathy, 1925, 11, Tafel 2, Abb. 2) but about 60 cm of malleable, loamy sediment with sand interspersed only by a thin, continuous layer of sinter at a depth of 30 cm. According to Bayer (1922, 182), it was mainly the brown, sandy loam that contained the fossils.

Later, Szombathy mentioned isolated slabs of sinter in or above the loamy sediment (1925, 3, 5). Also, the cranium 1 was attached to a plate-sized piece of sinter, horizontally embedded into the loamy sediment (Szombathy, 1925, 5). Small remnants of a white, solid sinter are still partly visible at the base of the cranium.

The description of Hall D given by Szombathy (1925, 6) proves no continuous layer of sinter but fossiliferous loamy deposits interspersed with fragments of sinter structures. At the central part of Hall D, limestone debris was also mentioned. Only at the southwestern part of Hall D was the sinter floor still intact and preserved at the time of recovery. Thus, the sinter data presented by Svoboda et al. (2002) can give no minimum age of the fossils as proved by direct dating of the cranium 1 specimen. The sinter still attached to this specimen must have developed at about 31,000 years BP or younger, and disintegration of the skeletons must have already taken place prior to the development of the sinter.

Human remains from "b" and the central part of Hall D show only a slight cover with a thin, brown calcareous crust, as do animal bones from these findspots. At findspot "e", human fossils were assumed beneath a thick layer of sinter (Svoboda, 2000, 531) but Szombathy (1925, 10) reports human and animal remains only in the vicinity of a sinter structure. Thus, only cranium 1 is directly related to sinter development after the deposition of the skull. If we assume an original intact sinter floor, it must have reached from the southwestern part of Hall D towards its central part but did not affect the fossil remains, except at "a". It also remains unclear if the sinter structure observed at "e" was part of the same generation of sinter development in the Mladeč Cave.

On the other hand, formation of a solid, white sinter after the deposition of the skull at "a", suggests a stable condition in the area at that time. Perhaps the rather narrow passage from the chimney towards "a" was already sealed up by a large block at that time, and thus prevented younger sediments and bones to reach findspot "a". The closed passage would also explain the very low number of finds at findspot "a" (see Szombathy, 1882, 1925, 5). The younger bovid remains are related to findspots "d" and "e", and were partly encountered on the surface (Hochstetter, 1883; Szombathy, 1925, 6).

Bovid remains are most abundant at the Mladeč Caves. The type 1 assemblage is characterised by *Bos primigenius,* while the younger deposits are dominated by *Bison priscus.* Radiometric dates place three samples of *Bison* into a time span from probably 25,880 years BP until 16,950 years BP. Skeletal element distribution of steppe bison remain reveals a completely different preservation pattern than at other sites with a higher portion of bovid remains, like the hyena den Teufelslucke (Nagel et al., in press), the open-air sites Wallertheim (Gaudzinski, 1995, 311, Table 15), and the Mousterian site Mezmaiskaya cave (Baryshnikov et al., 1996).

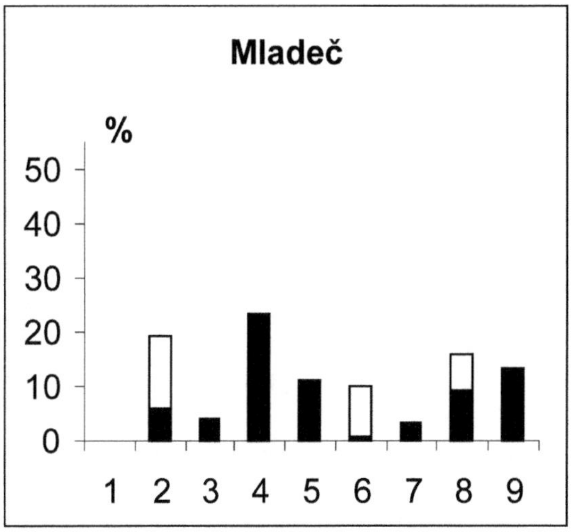

Fig. 4a. Skeletal element distribution of *Bison priscus* from the Mladeč Caves (white bars: teeth, mc, mt)

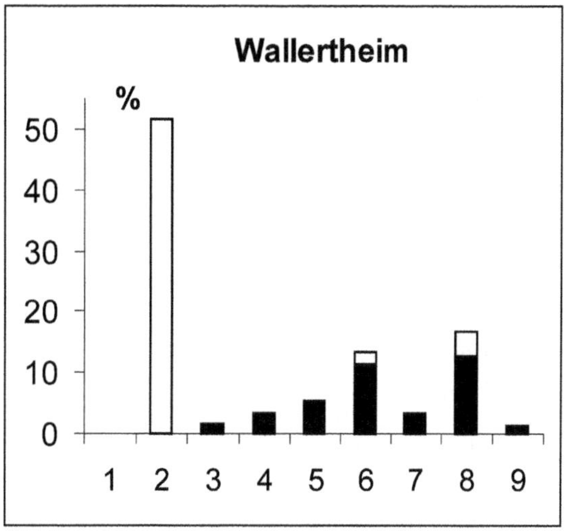

Fig. 4b. Skeletal element distribution of *Bison priscus* from Teufelslucke

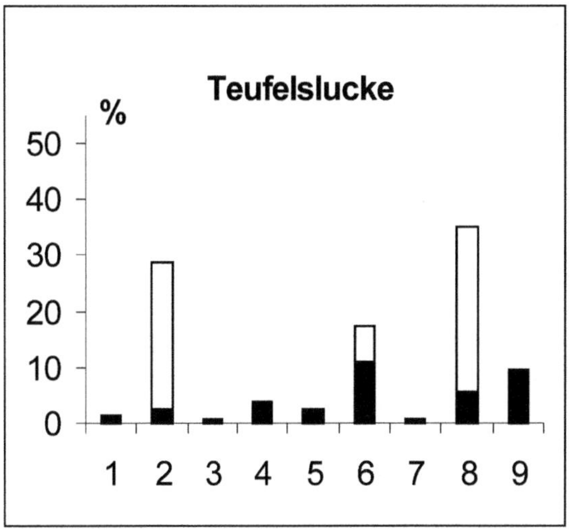

Fig. 4c. Skeletal element distribution of *Bison priscus* from Wallertheim

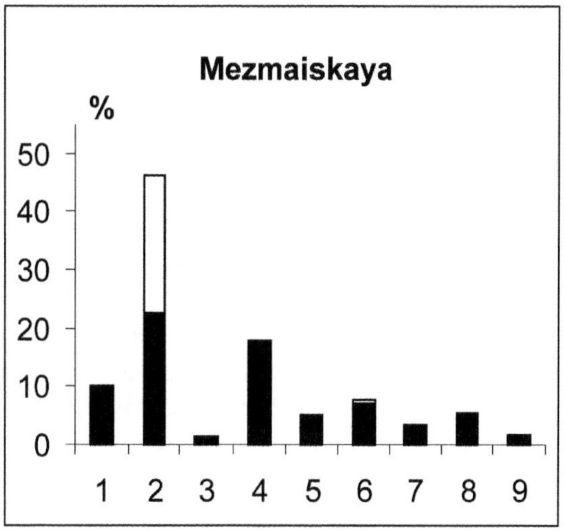

Fig. 4d. Skeletal element distribution of *Bison priscus* from Mezmaiskaya

None of the sites are contemporaneous with the Mladeč Caves. Bovidae are rarely encountered in Upper Paleolithic sites in the region, and thus apparently did not belong to the main hunting game of humans. Skeletal element distribution in Fig. 4a–d was calculated following Stiner (1991). The rough comparison confirms the different character of the *Bison* assemblage at the Mladeč Caves. Compared to the mentioned sites, in which a selective preservation of particular elements due to different reasons is evident, the Bovidae from the Mladeč Caves show a nearly even distribution of all skeletal elements. The picture is slightly blurred, and it was not possible to find conjoining specimens in Bovidae; both is probably caused by the loss of material at the Mladeč Caves. Nonetheless, the description of bovid accumulations containing bones of all parts of the skeleton (Szombathy, 1882, 106) confirms the assumption of an input of originally intact carcasses or complete animals of *Bison priscus*.

Evidence of complete skeletons is further reported for reindeer by Knies (Szombathy, 1925, 9; Svoboda, 2000, 530) from a fissure filling accompanying the large debris cone in D, and by Szombathy (1882, 106; 1925, 6) from findspot "b". Today, the complete skeletons are only partly preserved, as is proved by various long bones from both collections which belong to one individual, each. In addition, 17 bones belonging to one individual of *Canis lupus* are preserved in the collection of Knies. In his diary, Szombathy (see Antl-Weiser, this volume) also refers to various bones of one single fox from Mladeč II.

Bone modification analysis reveals a low degree of carnivore activity, which can be explained by carnivores that survived the fall and gnawed on the bones. Similar situations are observed in natural traps and in other chimney caves (e.g., Galik, 1997).

The scarce and isolated finds of additional species (e.g., mammoth, giant deer, red deer) may have become washed in through the chimney from the surface above the cave. This may also hold true for the abundant but isolated shed antlers from reindeer.

Descriptions and ground maps from Szombathy and Jelínek confirm the idea of a rather restricted distribution of both human and animal bones around the chimney in D and E.

Svoboda (2000, 530) published a reconstruction of the surface at Hall D based on the remains of the top calcite at the wall and the elevations of the cone remains. Following the given outlines of the debris fan, the horizontal distance from findspot "a" to the central part of the cone was 12 m or even less if the slope of the talus is considered. Species and skeletal element distribution at the Mladeč Caves confirms the existence of a debris cone, but in course of the reconstruction of site formation processes, three major differences to previous interpretations became evident.

First, the majority of fossil remains are younger than the development of sinter layers in Hall D. This assumption is confirmed by radiometric dates and also, by descriptions from Szombathy (1882, 101) prior to excavation. The loamy surface of the southeastern part had a higher elevation than the sinter covered southwestern part. Horizontally embedded fragments of sinter within the loamy sediment indicate a former continuation of the sinter floor towards Hall E. The sinter floor below the chimney probably became destroyed in course of the development of the debris cone caused by the later rock fall and the weight of the overlying deposits. Finds were restricted up to a depth of 60–70 cm in Hall D (Szombathy, 1925, 6) or lay on the surface of the cave floor (Svoboda, 2000, 534).

Second, input of fossil remains through the chimney would lead to the development of a debris cone. As a consequence, the large debris did not rest upon a "normal layer" of Pleistocene finds (Szombathy, 1925, 12). The fossils were part of the cone, leading to a rather patchy distribution of the remains. Pillars in Halls D and E as well as different times of input explain an accumulation of bones at different findspots. The impression of the patchy distribution may have been intensified by recovery techniques.

Third, the chimney was open longer than previously assumed (e.g., Jelínek, 1983, 57) as indicated by the reconstruction of site formation processes and radiometric dates. The enormous rock fall between D and E must have been at least partly younger than the development of the fossil accumulation. Large boulders in Hall E lie upside down or on the side as indicated by stalactites still attached on the blocks (Szombathy, 1882, 101–102). The deepest point of Hall E was the central part, whereas towards the east, elevation increased rapidly forming the highest point at "e1" where again smaller blocks are visible. The eastern part of the chamber was not covered with blocks (see Fig. 3a). The rock fall did not reach the total area of findspots "a–d", and therefore, complete bones are preserved. Some specimens from the collection of Szombathy were crushed, like a skull fragment from *Bison*. Fragments were kept in place until today by a thin calcareous crust. At Hall E, Paleolithic finds and fossils were encountered beneath large blocks (Szombathy, 1925, 29).

The large rock fall occurred after the input of the bovid remains and partly closed the chimney. The reindeer found by Knies in a fissure accompanying the talus cone (Szombathy, 1925, 9) was probably one of the latest large mammal inputs into the Mladeč Cave. Later, the opening of the chimney became smaller but allowed accumulation of a rich micro-fauna in a fissure filling recovered by Knies (Szombathy, 1925, 12). A second, smaller rock fall at "e1" closed the chimney for good at the end of the last glacial period or even during the Holocene, since Szombathy (1925, 12) also mentioned modern species in the small mammal assemblage.

The reconstruction of the beginning of the deposition of the talus cone faces some uncertainties. Oliva (this volume) assumes a Middle Pleistocene age of the talus cone. Svoboda (2000, 533) assumed that the central part of the cone is still preserved today, which he ascribed to the Middle Pleistocene, while at the surface of the cone, the Late Pleistocene "layer" became deposited.

The interpretation of the Middle Pleistocene age of the talus cone is based on the very recent investigations at the Mladeč Caves from 1958–1962 (Jelínek, 1983; 1987). At that time, Late Pleistocene deposits had already been missing due to preceding excavations and accommodation of the cave for tourists. Thus, Middle Pleistocene sediments built today's cave floor and became excavated in four trenches in Halls D and E. The fauna recovered was dominated by *Ursus deningeri* and micro-faunal remains were characterised by an abundance of snake vertebrae (Jelínek, 1983, 58). The subsequent micro-faunal analysis by Horaček and Ložek presents a faunal assemblage (see Musil, 2000; Svoboda, this volume) of Middle Pleistocene or even Biharian age (Jelínek, 1987).

Species composition clearly differs from the micro-faunal assemblage of the fissure filling recovered by Knies. In the latter, species of a tundra-steppe environment restricted to a cold and dry phase of the last glacial period occur such as, *Lagopus albus* and *L. alpinus*, *Allactaga jaculus* and *Dicrostonyx torquatus*. Szombathy (1925, 12) ascribed these elements to the "obere Nagetierschicht". Only a few isolated teeth are evident for large mammals, like *Canis* and *Ursus*. Next to the steppe elements, extant species are also identified (Szombathy, 1925, 12).

There is no clear evidence for a correlation of the micro-vertebrate assemblage of the basal sediments and the fauna encountered in the fissure at the debris cone. The only exception could be the evidence of molluscs and numerous snake vertebrae mentioned from the fissure filling (Szombathy, 1925, 12). Snakes are regarded as typical faunal elements of Middle Pleistocene deposits at the Mladeč Cave (Jelínek, 1983, 58). Nonetheless, even if part of the remains was of Middle Pleistocene age, its occurrence together with a Late Pleistocene faunal assemblage would rather indicate a mixing of deposits than an alteration in the interpretation of site formation processes in particular.

The fissure filling accompanying the talus cone comprises a completely different period of time than the micro-vertebrates from the basal sediments. In fact, it represents the last period of input through the chimney after the large rock fall. In general, either owls or small carnivores are responsible for the introduction of micro-fauna into caves. Certain small-sized species, like marmot, foxes and others, are also known to build their burrows in caves, as indicated by the presence of complete skeletons at the Mladeč Cave (e.g., Szombathy, 1925, 12). After the large rock fall, probably a slope became deposited from the entrance of the chimney towards the large blocks. Particular mammalian species known to use caves might have entered through the chimney and used the upper slope for their burrows. Other species became introduced into the Mladeč Cave by avian predators or smaller carnivores.

Reconstruction of site formation and radiometric dates show that the cone development below the chimney in Halls D and E did not start as early as the Middle Pleistocene. The next period clearly evident at Mladeč I comprise the human remains. Their interpretation still poses some problems.

As the animal bones, the human remains are related to the various findspots around the chimney, suggesting a rather restricted area of Late Pleistocene find-distribution (Svoboda, 2000, 534). Given the 20th century loss of some of the human remains, only a small portion of the recovered material is preserved. The list of skeletal elements from the humans published in Szombathy (1925, 6, 10) and Jelínek (1983, 59), suggests an input of originally complete bodies or at least partly intact skeletons. At the time of recovery, the human remains were scattered and dispersed.

The studied animal bone material shows no traces of human exploitation. Although we do not exactly know which species are contemporaneous with the human remains, the total faunal assemblage from the Mladeč Cave differs significantly from human occupation sites in terms of species composition, skeletal element distribution, and bone modification. The only exception is the material ascribed to preservation type 1 that is clearly older than the human remains and of unknown provenience.

The proposed "fireplaces" within the Middle Pleistocene deposits and at Mladeč II later became refuted. Black colored areas were produced by concentrations of manganese-oxides (Szombathy, 1904, 14; 1925, 5; Jelínek, 1983, 58; Svoboda, 2000, 530). The nature of presumed charcoal concentrations at higher levels is more problematic to evaluate, since Late Pleistocene deposits are removed from the Mladeč Caves. Following Szombathy (1925, 6), the charcoal structure in "a" lay at a depth of 35 cm and cannot be related to cranium 1 since a thin layer of sinter at a depth of 30 cm separated them both.

Also, the existence of a "fireplace" recovered by Smyčka within the loamy sediment that was covered by a 70 cm thick sinter with a large sinter structure on top (Szombathy, 1925, 10) seems rather doubtful, given the time that would be necessary for a large sinter structure to develop. No details from Smyčka's second "fireplace" more to the south and at lower depth are available. In general, site formation processes from Smyčka's excavation areas near "e" are more critical to reconstruct. Not much contextual data is available. Faunal analysis clearly shows a mixing of Middle and Late Pleistocene deposits, a levelling of sediments of 3–4 m, and probably a deposition of rubbish and debris from previous excavations in this area (Svoboda, 2000, 531). In addition, the western part of Hall E was more severely affected by the large rock fall. Nonetheless, the whole area is clearly related to the chimney and reveals similar depositional characteristics as findspots in Hall D.

A chimney cave may serve as a natural trap as indicated by at least the younger faunal remains from the Mladeč Cave. Human remains represent the first evident period of input through the chimney. It is not possible to decide at the moment, whether a distinct burial tradition, an input in course of a distinct religious rite, or a washing in of the remains occurred (Svoboda, 2000, 534). Given the rather low number of individuals, even occassional accidents may serve as explanation. In addition, it needs to be tested if the human and artifact remains stand in some connection to the Paleolithic site on top of the Třesín hill directly above the cave.

The artifacts found at the Mladeč Cave also pose some problems. Artifacts and especially pendants found at the same site as human remains would normally be regarded as associated with man. This holds true for the bone points of type Mladeč, which are diagnostic of an early Late Paleolithic.

One the other hand, perforated teeth of beaver and moose are rather unusual implements during the Late Paleolithic (see Pacher, 2005). Radiometric dating of the beaver tooth from the collection of Jelínek, which might have been used as a pendant, even points to a Mesolithic age. Beaver is not a common faunal element of the Late Paleolithic either. Direct radiometric dating confirms its presence on the southeastern border of the Alps at 41,000 years BP (Fladerer, 2000). Moose seemed to be scarce during the Late Paleolithic but also became more abundant in course of the climatic amelioration after the Late Glacial Maximum. Both species are regarded as one of the first indicators of the reforestation (Königswald, 2002, 128). Moose then attained greater importance for human-subsistence during the Alleröd and Early Holocene period, as proved at various sites in Germany (Terberger, 2002, 60).

The other pendants from the Mladeč Cave are teeth from wolf and bear. Previously, cave bear was assumed (e.g., Bayer, 1922, 178) but the rather small size of the tooth does not allow distinction of the two bear species. The incisor from horse shows no clear traces of an attempt at drilling. It is not clear if remains of one large necklace or single pendants have been found (Bayer, 1922, 178; Szombathy, 1925, 12), but if so, the necklace could also be interpreted as Mesolithic or even Neolithic in age not only because of the radiometric date, but also because of species composition. Even wild horses may have survived until Neolithic times in the region (see Bauer, 2001). In any case, the use of moose teeth and metapodials as pendants and bone tools is a rather unique feature in the Mladeč Cave. Additional evidence of drilled moose teeth of uncertain stratigraphic position comes from the Mammoth Cave near Krakow, Poland (Hoernes, 1903, 109).

The few lithic artifacts from various findspots in the Mladeč Caves, show a variety of raw material use (Oliva, 1989, 54; 1993, 213), and are not diagnostic (Svoboda, 2000, 531). Only one of the two artifacts found in 1937 on the right side of the cave entrance is regarded as belonging to an Aurignacian tradition (Oliva, 1989, 53). Here, a chimney into Hall A opens and the two implements were found in secondary loess, which is a clear evidence of allochthonuos remains in a cave. An input of the artifact from the loess deposits above the cave seems likely, perhaps even in course of the re-opening of the chimney into Hall A.

Later on, a quartz polyhedron of unknown age (Valoch, 1995, 73) was found in corridor "c" (see Fig. 3b). A Middle Pleistocene age is assumed (Svoboda, 2000, 531). In addition, artifacts were found at Mladeč II (Oliva, 1993, 213). Maška (in Szombathy, 1904, 11) even mentioned a large "axe" made of silex.

The rather diverse artifact assemblage suggests an input at different periods but up to now, this assumption cannot be proved. In the absence of clear traces of a site frequented by humans, the few artifacts may either be associated with the human skeletal remains or may have been washed into the cave through chimneys or smaller fissures as suggested by the remains in Hall A.

Summary of the taphonomic analysis

The comprehensive study of faunal material from the Mladeč Caves allowed a rather detailed reconstruction of site formation processes, especially for Halls D and E, but also confirms the need for ongoing studies.

The oldest deposits at the Mladeč Cave are of Middle Pleistocene age or even older. They are characterised by an abundance of *Ursus deningeri,* a rich micro-fauna, and molluscan remains. A horizontal entrance to the Mladeč Cave at that time is assumed.

A small but distinct assemblage of animal bones represented by *Bos primigenius* and *Equus* sp. shows intense gnaw marks attributed to hyena activity. Evidence of human activity is uncertain. Unfortunately, the origin of the material could not be reconstructed, but it points to a locality accessible for hyenas or even humans around 42,000 years BP.

The human remains from the Mladeč Cave probably represent a first period of input through the large chimney in D and E that must have opened at least slightly before 31,000 years BP. Later on, mainly bovids, reindeer, and horses fell into the natural trap. Large mammal analysis and spatial distribution of the remains does not correspond to a site frequented by humans, as further indicated by a total absence of traces of human exploitation on the animal bones. Marginal presence of gnaw marks can be explained by carnivores, probably wolves, that survived the fall into the cave and fed on bones from human and animal remains. The isolated specimens of additional animal species might have been washed in through the chimney from the surface above.

A period of sinter development at findspot "a" and probably also at the central part of Hall D, is proved after the deposition of cranium 1. It is unclear whether the sinter structure described by Szombathy (1925, 10) from Hall E represents the same generation of sinter development, or not. In any case, human and animal remains from "b" to "e" were not directly affected by sinter structures. Sinter development prior to the accumulation of the Late Pleistocene talus cone is confirmed by previous dates obtained on sinter samples from the wall of the cave (Svoboda et al., 2002). These dates were concluded to give a minimum age of the fossils. As could be shown, the majority of faunal material accumulated after the development of sinter in the area around the chimney. Bovid remains of the Pleniglacial, but also other large mammal species like reindeer and wolf, are proved to have fallen into the natural trap and accumulated at different findspots around the chimney. Large pillars separate different areas in Halls D and E and hence, favoured the development of local accumulations of fossils. The impression of findspots may have been enforced by recovery techniques, but they are also a typical structure of debris cones. More or less horizontal layers do not occur but fossils accumulate along the slope of the cone.

The large rock fall that separated Halls D and E stopped the input of large mammalian species at least into "a"–"d". Probably the reindeer recovered by Knies in a fissure was the last large mammalian victim of the trap. Raptor birds and small carnivores were responsible for the input of a variety of micro-mammals found in a fissure filling that accompanied the large debris cone. Species composition indicates typical elements of a dry and cold tundra-steppe environment as well as modern species, and indicates deposition of remains from the last glacial period up to the Holocene. After the rock fall and in course of the accumulation of a sedimental slope, the chimney allowed smaller mammals to enter the cave and build their burrows along the upper part of the slope. Probably during the early Holocene the chimney was closed for good by another smaller rock fall in "e1".

Site formation processes presented in this study for the Mladeč Cave are rather different than previously suggested. The reconstruction is based on a comprehensive faunal analysis and includes information about the spatial distribution and characteristics of the remains. Still, in course of this study, new questions arose while others remained unresolved. Some of them could be subject for ongoing studies and may contribute to a discussion of the proposed taphonomic history at the Mladeč Caves.

First of all, an input of humans through the chimney could either be related to a distinct burial tradition, a religious rite, or may be caused by natural reasons. In any case, it should be tested if there is a correlation between artifacts and human remains from the Mladeč Caves with the Paleolithic and prehistoric settlements at the Třesín hill directly above the cave.

Second, the species composition of artifacts and pendants is rather unique for the Late Pleistocene. The distribution of these artifacts should be studied in order to find out if raw material use of moose teeth and metapodial bones is a typical feature of either Late Pleistocene or younger traditions in the region.

Another task not directly related to the human remains, is to resolve the origin of the type 1 preservation material. There are certain evidences of mixing of deposits in the collection of Smyčka, but also indicated by the dated beaver tooth found within the Middle Pleistocene material of the collection of Jelínek. Regarding the large-scale earth removals in the cave, it might not be possible to better reconstruct the depositional context of these materials but additional carefully chosen radiometric samples might help to do so. Radiometric dating in combination with a detailed study of the small mammal assemblage from the fissure filling might also help to confirm the proposed sequence of site formation processes at the Mladeč Caves or may result in a slightly different picture of the taphonomic history of the site.

Metric analyses of the material

The second part of the study comprises a detailed metric analysis mainly of the Late Pleistocene faunal material. The remains available for this study are listed according to museum collections:

NHMW...Naturhistorisches Museum Wien, collection of Szombathy
MZM...Moravské zemské muzeum, collection of Knies
OM...Olomouc muzeum, collection of Smyčka

Canis lupus LINNAEUS, 1758
Material:
(NHMW - 1 mandible lost, 1 drilled canine)
MZM - mcV dist. sin., mcIII sin., mcIV sin., metapodium dist. indet., 2 phalanges proximal, cinf. dext., I3sup. dext., P3sup., caninus, M1sup. dext., m1inf sin., calcaneus dext. juv.,
1 cervical vertebra, 3 lumbal vertebrae,
one individual - occipital fragment, maxilla fragment + P4sup., M1sup. sin., M2sup. sin., atlas, axis fragment, 4 vertebra fragments, humerus dist. sin., humerus shaft fragment, tibia dext., tibia shaft sin., radius dist. sin., femur prox. sin., pelvis acetabulum fragment,
OM - thoracic vertebra fragment, 2 rib fragments, cinf. sin.,

Thirty-eight remains from the wolf *(Canis lupus)* are recorded from the Mladeč Caves. The collection of Knies contains a box with 17 bones definitely ascribed to one individual (bold letters). The surface of the bones is brown and partly covered by a thin, brown, calcareous crust. A second, juvenile

Table 6. Comparison of tooth measurements from *Canis lupus* from the Mladeč Caves

	Cinf.Br	P^3Br	P^3L	P^4L	M^1Br	M^1L	M^2Br	M^2L
MZM	5.3							
MZM		17	7					
MZM					19.4	16.2		
MZM					19.5	16		
MZM							9.3	8.5
MZM				26.2				
OM	11.1							
Willendorf[1]	9.4			23.5–25.5	18.8–21.5	15–17.5	12–14	8.2–10.2
Předmostí[1]	8.5–10			23–25.8	19–22.8	15–17	11–14.2	9–11.3
Pavlov[2]		17.6		26.9	21.8	16.4	14.4	10.2
					23.0	17.0	14.8	10.2
Pod Hradem[3]		16.3	7.2	~27.8	20.6	15.5		
Sveduv stul[4]				25.4	20.1	16.5		

[1] after Thenius (1959, 140), [2] after Musil (1997, 458), [3] after Musil (1965, 27), [4] after Musil (1962, 122)

Table 7. Comparison of metapodial bone measurements of *Canis lupus* from the Mladeč Cave

		TL	Bp	Dp	Bd	SD
MZM	mcV		~10	8.7		
MK 962[1]			14.5	14.2		9.0
Pavlov[2]		62–92	9.5–14.9		8–13	8–11
MZM	mcIII	86.35	12.30	16.50	13.30	9.70
Pavlov[2]		73–85	8.5–11		10–12	8–10
MZM	mcIV	51.8	8.3	11.2	10.5	6.2
MK 959[1]			10.8	14.5	11.5	7.8
Pavlov[2]		75–99	7.4–12		9.5–11	7–9

[1] Wachtberg after Fladerer (2001, 23), [2] Pavlov after Musil (1955, 286)

individual is represented by one calcaneus. Thus, the material comprises at least two individuals. The lower canine (Table 6) and the 4th metapodial (Table 7) are very small specimens and should rather be classified as "Canidae indet."

Only a few measurements are available from bones of the one individual. The size of the left humerus (bd ~44 mm, Td 31.6 mm), the right tibia (bd 29.9 mm), the left radius (bd 33.5 mm), and the left femur (bp 54 mm, SD 18.3 mm) corresponds to specimens from Willendorf and Předmostí (both Thenius, 1959, 140) and smaller specimens from Pavlov (Musil, 1997, 445, 458). The value obtained on the radius is on the lower range of values given for specimens from Barova (Musil, 1960, 30), Předmostí, and Krems-Hundssteig (both in Fladerer, 2001, 23).

Larger specimens from Pavlov (humerus bd 48.4 mm) and Barová cave (humerus bd 47.7 mm; femur bp 71.0 mm) clearly exceed the size of the Mladeč individual. The breadth of the upper M2 is very small.

Alopex lagopus Linnaeus, 1758 and *Vulpes vulpes* Linnaeus, 1758

Material:

NHMW – 72.205 mandible fragment dext. juv., 72.204 humerus sin. juv., 72.202 tibia dext. juv., 72.203 femur juv.

MZM – 2 mandibles dext., 2 mandibles sin., 2 radii dist. dext., radius prox., 3 tibiae dist. dext., 3 tibiae dist. sin., tibia shaft sin., tibia sin. juv., tibia prox. dext. juv., humerus dist. dext., ulna prox. dext., femur dext. juv., pelvis fragment dext., scapula fragment dext., 1 cervical vertebra

OM – 4 mandible fragments sin., cinf. dext., radius dist. dext., mcII dext.

A total of 33 bones of foxes were recorded from all three collections. Distinguishing arctic fox and red fox was possible for the mandibles by comparing size (Fig. 5) and morphological traits (see sum-

Table 8. Measurements of mandibles and teeth from *Alopex lagopus* (A) and *Vulpes vulpes* (V) from Mladeč

Museum	Inv. no.	side		TL	p_1–m_3	m_1–m_3	p_1–p_4	p_2–p_4	Cinf.Br	m_1L	m_1Br	m_1H	p_2H
NHMW	72.214	dext.	V							16.9	6.6		
MZM		dext.	V							16.8	6.6		
		dext.	V							~15.5			
		sin.	A	100.9	51	24				13.9	5.5	12.5	
OM		sin.	A				30.9	26.4		13.5			12.2
		sin.	A				30	25.7	4.2				12
		sin.	A				29.2	25.1	3.8				11.4
		sin.	V				35.3	30.2		16.6			13.2

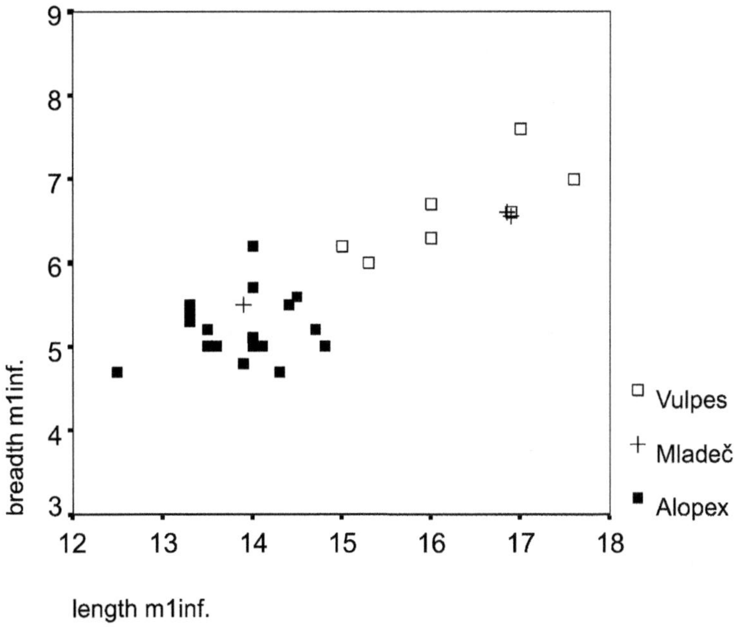

Fig. 5. Comparison of breadth and length of m1inf. from *Vulpes vulpes* and *Alopex lagopus* (data from Musil, 1959a, 75; 1959b, 93; 1955, 291; 1962, 126; 1997, 445; Fladerer, 2001, 28; Table 8)

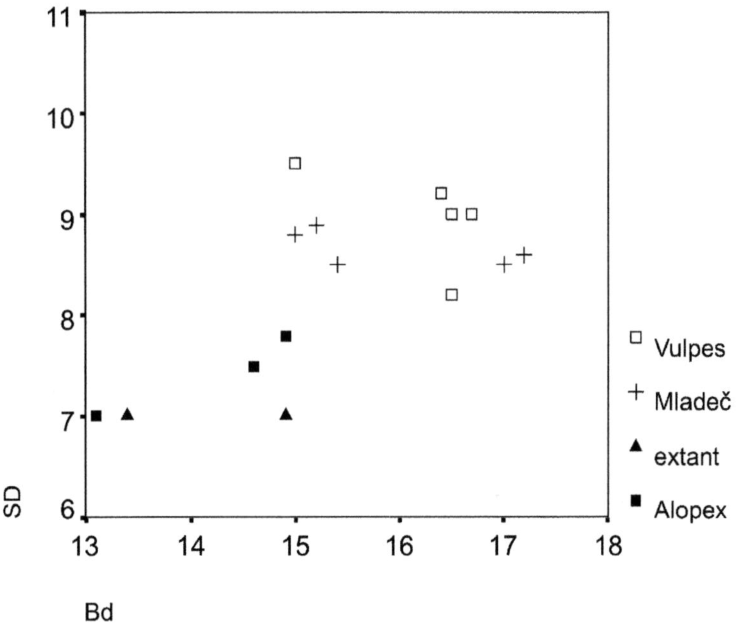

Fig. 6. Tibiae from *Vulpes vulpes* and *Alopex lagopus*, smallest breadth of diaphysis and distal breadth (data from Beneš, 1975, 191; Musil, 1997, 445; Table 9)

Table 9. Comparison of long bone measurements from foxes (A...*Alopex lagopus*, V...*Vulpes vulpes*, MK...Museum Krems)

Museum	Inv. no. / coll.			GL	Bd	Dd	BT	SD	Dpa	SDO	
Humerus:											
NHMW	72.204	sin.	V		20.3	15.5					juv.
MZM	Knies	dext.	V		22.7		9.4				
MK 936	Wachtberg[1]		A	102	16.6	12.7		6.2			
	Pod hradem[2]		A		16.4	12.2					
			V		19.5	16					
			V		21.2	16.4					
	Pavlov[3]	n=5	V		17.7–22	12.4–17					
Tibia:											
MZM		dext.	V		17.2	11.9		8.6			
		sin.	V		15.4	11.5		8.5			
		dext.	V		15.2	10.5		8.9			
		sin.	V		15.0	10.0		8.8			
		sin.	V		17.0	11.9		8.5			
	Pavlov[4]		V		16.5	12					
	Pavlov[5]		A		12–14.5	7–10.5					
	Šipka[6]		V		16.5	11		8.2			
			V		16.4	11.7		9.2			
			V		16.7	11.7		9.0			
	Pekarna[7]		V		15.8	11					
			A		14.9	11		7.8			
			A		14.6	10.7		7.5			
	Griffen	sin.	V	180	16.5			9			
		sin.	A	110.5	13.1			7			
Radius:											
MZM		dext.	V		15.4						
		dext.	V		15.3			7.8			
OM		dext.	A		10.8						
	Svéduv[8]		V		17						
			A		12.2						
MK 939	Wachtberg[1]		A		13.6	7.7		6.1			
	Šipka[6]		A		12			6.6			
	Pavlov[5]		A		9.3–11.2	5.5–7.2					
			V?		11.9–13.8	7.7–9.4					
	Pavlov[4]		V	121.9	16.4	8.4					
			A		14.5	7.6					
Ulna:											
MZM		dext.	V						16.2	12.8	
	Gr.Grotte[9]		V						15	12.4	
	Griffen		V						16.1	13.6	

[1] Fladerer (2001, 30), [2] Musil (1965, 28), [3] Musil (1997, 445), [4] Musil (1997, 458), [5] Musil (1959b, 92–93), [6] Beneš (1975, 194, 192), [7] Beneš (1975, 171), [8] Musil (1962, 124), [9] Weinstock (1999, 7)

mary in Kunst, 1992). Elements of the post-cranial skeleton are separated by the relative slenderness and smaller size of *Alopex* bones (Fig. 6).

The four specimens from the Naturhistorisches Museum appear to be from *Vulpes vulpes*. The bones show the typical type of preservation from the excavation of Szombathy. Measurements of the m1inf. (Table 8) and the juvenile humeri (Table 9) as well as the general size of the other two bones confirm remains of the red fox. The collections of Knies and Smyčka comprise remains of both species. Three measurable mandibles and nine long bones of the collection of Knies are identified to *Vulpes vulpes*. One mandible is identified as arctic fox. The remaining eight fragments could not be unambiguously ascribed to any of the two species. Regarding their general appearance, the majority of them are probably also from red fox.

Three mandibles of the collection of Smyčka (Table 8), one radius (Table 9), and one mcII are attributed to *Alopex lagopus*. Measurements of the mcII (GL 37.3 mm, Bd 5.8 mm) correspond to values given for arctic fox by Schütt (1969) and Poplin (1976). One mandible is attributed to the red fox. One lower canine with a breadth of 4.7 mm was probably from artic fox.

Mustela putorius LINNAEUS, 1758 and *Mustela eversmanni* LESSON, 1827
Material:
MZM – maxilla fragment sin., small maxilla fragment, Csup. sin. + dext., cinf., mandible fragment sin. + dext., p3inf.
OM – mandible dext., humerus dext. juv., femur dist.

Eleven specimens from small mustelids are preserved in the collections of Knies and Smyčka. The remains are identified as *Mustela* but could not be unambiguously ascribed to any of the two species, *M. putorius* or *M. eversmanni*. The material from the collection of Smyčka resembles the Middle Pleistocene material in terms of bone surface preservation. The mandible has a total length of 32.13 mm and a mean height of the ascending ramus of 10.2 mm (measurement 13 after v.d. Driesch, 1976).

Crocuta crocuta spelaea GOLDFUSS, 1823
Material:
MZM – humerus dist. dext.
OM – cinf. dext., 2 pd3 sup. sin.

One specimen from the collection of Knies and three teeth from the collection of Smyčka are identified to *Crocuta crocuta*. At least one adult and two juvenile individuals are counted. The humerus gives a distal breadth of 55.2 mm and lies at the lower range of the distribution of hyena at Teufelslucke, Lower Austria, and the cave from Winden (Table 10).

Table 10. Measurements of humeri from *Crocuta crocuta spelaea*

Collection	Site		Bd	Dd	BT
Knies	Mladeč		55.2	46.4	51.5
IPUW[1]	Windener cave		61.7	47.9	48.6
	Teufelslucke[2]	n=8	55–63		

[1] Institute of Palaeontology, University of Vienna, [2] Ehrenberg and Kernerknecht (1940)

Felis sylvestris Schreber, 1777
Material:
OM – femur dext., tibia sin., ulna dext., humerus sin.

Four intact preserved long bones from the collection of Smyčka are identified to *Felis sylvestris*. The remains probably belong to one large adult male individual. The length measurements listed in Table 11 clearly exceed the maximum values given for male wild cats by Kratochvil (1976) of more than 15 mm and even more than 20 mm in the ulna.

Felis sylvestris is a typical faunal element of warmer climates and an unusual find in Late Pleistocene assemblages. Thus, the remains might be older or younger than Late Pleistocene. Large specimens of wild cats are known from the Middle Pleistocene as well as from the Late Würm and early Postglacial (Kurtén, 1965). The yellow color of the bone surface suggests Middle Pleistocene material, but color is not reliable enough to prove the older age of the fossils.

Table 11. Measurements of long bones from *Felis sylvestris* from Mladeč

	Side	TL	Bp	Dc	Dp	Bd	Kd	Dd	SD	SDO
Femur	dext.	160.7	31	9		28	12			
Tibia	sin.	167.5	30			20.9	11.2			
Ulna	dext.	164.7				11			10.7	14
Humerus	sin.	145.5	24.7		30.9	27.2	10.8	19		

Panthera (leo) spelaea Goldfuss, 1810
Material:
NHMW – 72.190 femur dext., 72.198 tibia sin., 72.191 humerus dist. dext., 72.192 mtIII, 72.192 2 mtIV
(OM – 1 UK-molar fragment sin., 1 mandible fragment)

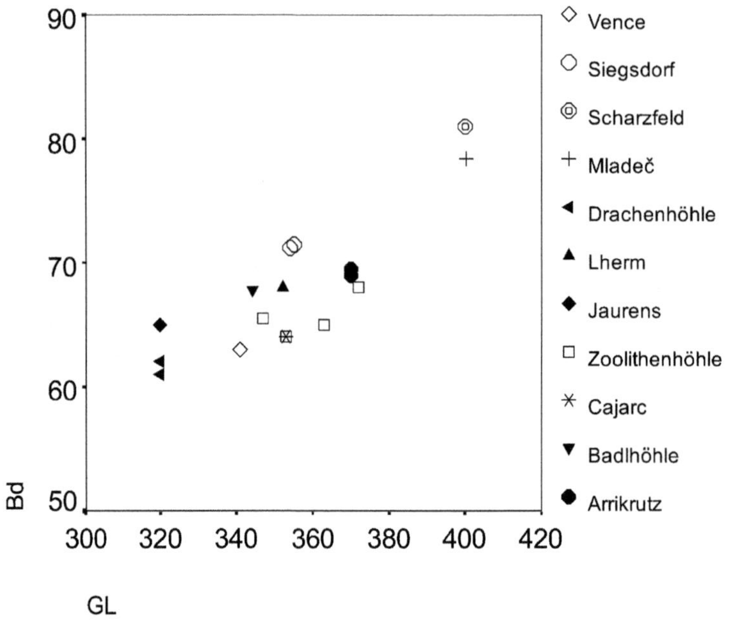

Fig. 7. Tibia from *Panthera (leo) spelaea*, greatest length (GL) and greatest distal breadth (Bd) (data from Altuna, 1981, 42; Ballesio, 1980, 73; Dietrich, 1968, 356; Sickenberg, 1931, 750; Gross, 1992, 125; Hemmer and Schütt, 1970, 143; Table 12)

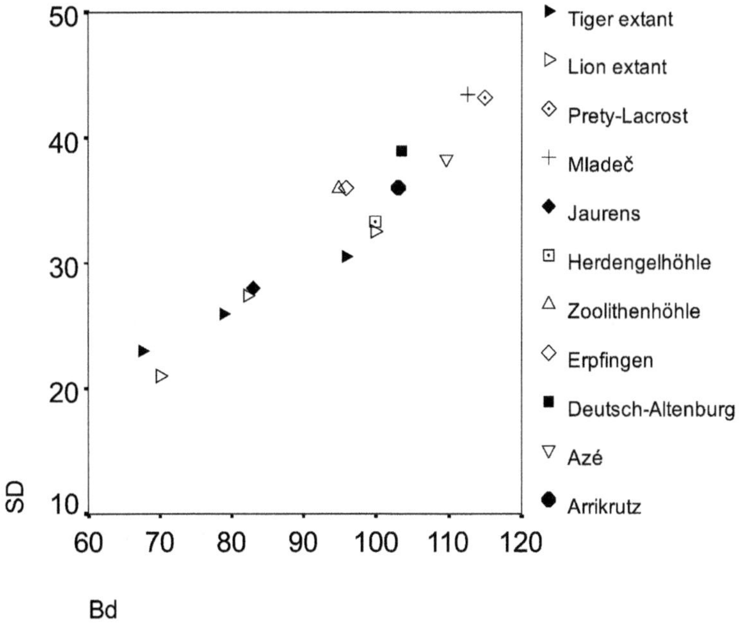

Fig. 8. Humerus from *Panthera (leo) spelaea*, smallest diaphysis breadth (SD) and greatest distal breadth (Bd) (data from Altuna, 1981, 39; Argant, 1991, 214; Ballesio, 1980, 72; Dietrich, 1968, 351; Gross, 1992, 120; Nagel, 1997, 236; Table 13)

The collection of Szombathy comprises five remains from cave lion. The taxonomic position of the Late Pleistocene cave lion is still under debate. Hemmer (2003) recently favours a discrimiantion of the cave lion on sub-species level (*Panthera leo spelaea*).

All specimens are from large individuals (Table 12). The values for long bones even considerably exceed values given for the large cave lion from Arrikrutz (Altuna, 1981). The tibia (Fig. 7) is as large as one specimen from Scharzfeld (Hemmer and Schütt, 1970, 143) and corresponds to length values of 350–405 mm given for the Middle Pleistocene remains from Repolust-cave (Fladerer and Frank, 1997, 286). The size of the humerus (Fig. 8) corresponds to a specimen from Prety-Lacrost (Argant, 1991, 214). There is no evidence for Middle Pleistocene large mammal remains in the collection of Szombathy and the preservation of the material does not resemble Middle Pleistocene material from the Mladeč Caves, either. The long bones probably represent a very large cave lion.

The two metapodials (Table 13) are slightly smaller than the corresponding skeletal elements from Arrikrutz but are still from rather large animals, similar to the specimens from Tilde (Riedel, 1982).

Table 12. Comparison of long bone measurements from *Panthera (leo) spelaea* from Mladeč

			GL	Bp	Bd	Dd	SD	Dc	BT
Mladeč	72.190	Fe	470	111.6	99.2		44.1	54	
Mladeč	72.191	Hu			112.7		43.4		77.8
Prety-Lacrost[1]					115	73	43.2		
Erpfingen[3]	GPIT		360	80.8	96		36		
Mladeč	72.198	Ti	400.3	107.7	78.4	52.3	41.6		
Scharzfeld[2]			400		81				
Badlhöhle[3]	IPUW 80		344.1	93.7	67.8	45.9	36.1		

[1] after Argant (1991, 214), [2] after Hemmer and Schütt (1970, 143), [3] own data

Table 13. Comparison of metapodials from *Panthera (leo) spelaea* from Mladeč

			GL	Bp	Bd	Dp	SD	Dd
Mladeč	72.192	mtIII	149.4	26	27	40.8	19.4	22.5
Arrikrutz[1]			156	30.5			23.5	
Arrikrutz			154	30.5			22	
Tilde[2]			145.3		27.5			24.8
Mladeč	72.192	mtIV	156.5	26.2	25.6	36.5	20.2	24.5
Mladeč	72.192		154.8	26.8		36.5	20.1	
Arrikrutz[1]			159		29		20	
Tilde[2]	1260		152		26.2			24.6

[1] after Altuna (1981, 43), [2] after Riedel (1982, 27)

The collection of Smyčka contains one fragment of a mandible and of a conjoining lower molar. The mandible, too, is from a large individual but it is unknown if the remains came from the Mladeč Caves. The specimens were found in a conglomeration of extant material and fossil remains that came with the collection of Smyčka. This material was used for a previous reconstruction of a "fireplace" inside the Mladeč Caves.

Ursus spelaeus Rosenmüller and Heinroth, 1793
Material:
NHMW – 72.187 mandible fragment sin., 72.197 humerus dist. dext., 72.198 tibia sin., 72.199 femur dext., 72.200 mcIV dext., 72.200 2 mtIV dext., 72.201 mandible fragment sin. juv.,
MZM – I1sup., cinf. dext., I1 + 2sup., metapodium indet. juv., 2 ph.1.

The collections of Szombathy and Knies comprise only 14 bones from *Ursus spelaeus*. The Mladeč Cave is not a typical cave bear site, because of the few remains. For the fragment of a mandible even *U. arctos* cannot be excluded. The length (65.3 mm) and breadth (19.7 mm) of the condyle points either to a subadult female or a brown bear. The humerus (Table 14) is from a very large specimen.

Table 14. Measurements of remains from *Ursus spelaeus* from Mladeč

			GL	Bp	Dp	Bd	Dc	SD	Dd
NHMW	72.197	hu				128		44	
NHMW	72.199	fe	385.8	101.6		84.1	48.2	36.3	
NHMW	72.198	ti	306	96		76.4		34.2	42.2
NHMW	72.200	mcIV	90	20.4	29.1	20.4		18.4	
NHMW	72.200	mtIV	78	20.7		20.3		15.9	
NHMW	72.200	mtIV	90.1	20.4	27.2			17	

Ursus deningeri Reichenau, 1904

Surprisingly, the majority of bear remains from the collection of Smyčka can be ascribed to *Ursus deningeri*. Morphological traits, the yellow color of the bones, and traces of manganese-oxides resemble the 89 bear specimens from the collection of Jelínek (Plate 1, Figs. 4 and 5). A detailed study of the Middle Pleistocene bear remains from the Mladeč Caves will be presented elsewhere.

Mammuthus primigenius Blumenbach, 1799
Material:
OM – 2885 scapula fragment, 3082 tibia prox., 3076 tibia prox. + shaft juv., 3036 two rib fragments, 3039 acetabulum fragment sin. :
without number – a small ivory fragment, femur shaft fragment, juv., (3 lamellae of molars)

Mammoth is evident by eight specimens from the collection of Smyčka at the Mladeč Caves. Three lamellae of molars were found in the conglomeration of extant and fossil remains and therefore, are of uncertain provenience. A rib fragment with a breadth of 39 mm is most likely from mammoth.

Coelodonta antiquitatis Blumenbach, 1807
Material:
(OM – 1 upper molar)

The evidence of whoolly rhino is unclear in the examined faunal material from the Mladeč Caves, although mentioned in previous faunal lists (see Musil, 2002). The isolated upper molar was found in the box with mixed extant and fossil material from the collection of Smyčka.

Equus sp.
Material:
NHMW – 72.193 humerus sin., 72.194 mtIII sin., 72.195 mcIII sin., 72.196 ph.1 post. dext., ph.2 ant. dext., ph.3 dext.
MZM – calcaneus sin., 2 mcII-mcV fragments, 2 rib fragments
OM – 2886 radius dist. dext., 2884 radius dist. dext., 2901+2899 mtIII dext., 2903 ph.1 post. dext., 2904 mcIII dext., 2912 mtIII prox. sin., 2996 ph.2 post. dext., 3006 cranial fragment, 3007 tibia shaft sin., 3008 mtIII sin., 3009 mtIII juv?, 3013 radius prox. dext., 3014 ph.1 ant. dext., 3019 mcIII dist., 3029 astragalus sin., 3030 and 3032–3035 4 P3-M2 sup, 3037 ph.1 dext. juv., 3040 ph.1 ant. sin., 3059 mtIII sin., 3081 tibia dist. sin.,
OM without inventory number – femur dist. dext., tibia dist. sin., ph.2 ant. dext., ph.1 ant. sin., ph.3 dext., 1 small mandible fragment, 6 upper cheek teeth, 5 incisors, 1 caninus, 12 lower cheek teeth

In *Equus* two different preservation types are distinguished. Specimens attributed to type 1 are indicated by bold letters in the inventory list and in Tables 15–18. The remains are characterized by a yellow-green color and intense carnivore activity. A metapodial bone (Inv. no. 2899) from horse of type 1 has been dated to 42,000 years BP. The few specimens fall within the size range of the other horse remains at the Mladeč Caves. Values obtained correspond to middle-sized, heavy horses known from other early Late Pleistocene sites in the region (Figs. 9 and 10). Horses of similar size and time range from sites in Moravia and Lower Austria have been differently classified as *Equus* cfr. *chosaricus* (Thenius, 1966), *Equus ferus* cf. "*solutreensis* Nobis, 1971" (Rabeder, 1996), *Equus germanicus*

Table 15. Long bone measurements of *Equus* from Mladeč

	Inv. no	Bone	GLC	Bd	Dd	BT	BFd	SD	UD
OM	3081	tibia		~76	51.3			49.8	13.9
	no n.			78	50				
Pavlov[1]		n=4		76.1–81.2	49.2–54.1				
NHMW	72.193	humerus	296.38	91.26		82.8		43.91	
Pavlov[1]				81.8					
OM	2886	radius		86.35			72		
Pavlov[1]		n=2					76.5–79		

[1] after Musil (1997, 460)

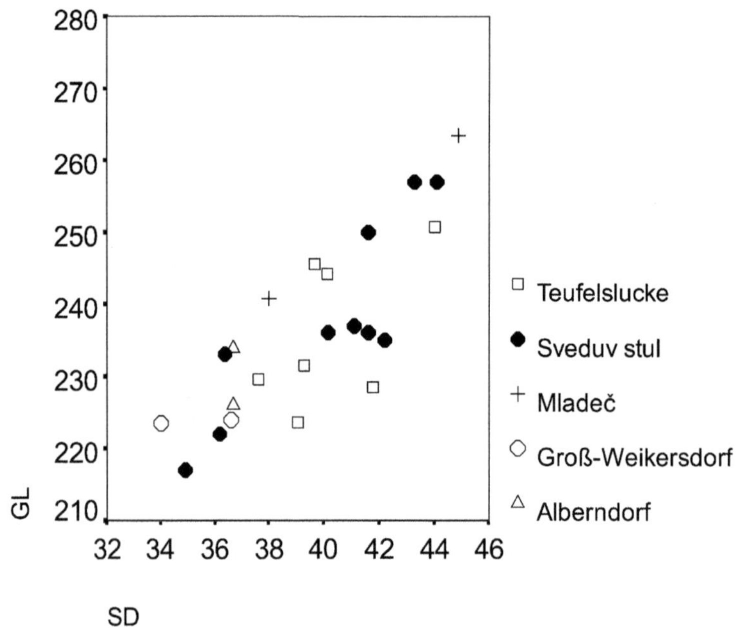

Fig. 9. Metacarpal III from *Equus*, greatest length (GL) and smallest diaphysis breadth (SD) (data from Fladerer, 1996, 254; Musil, 1962, 189–190; Rabeder, 1996, 89; Table 16, and unpublished data from Teufelslucke)

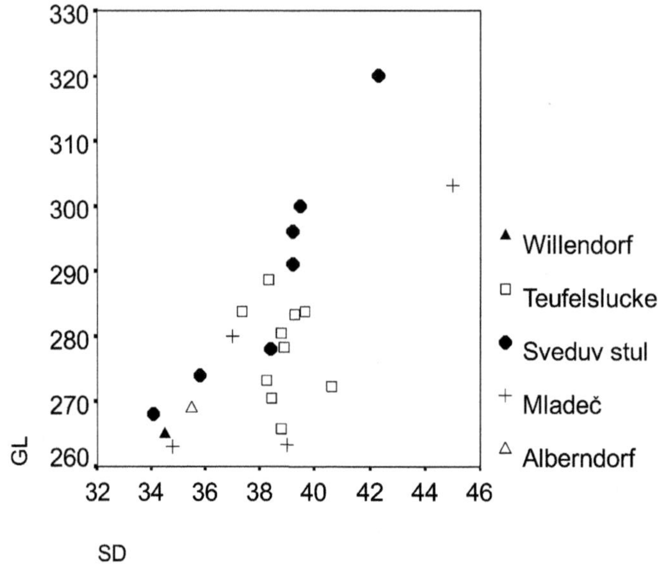

Figure 10. Metatarsal III from *Equus*, greatest length (GL) and smallest diaphysis breadth (SD) (data from Fladerer 1996, 254; Musil, 1962, 189–190; Thenius, 1959, 160; Table 16, and unpublished data from Teufelslucke)

(Musil, 1997), and even as *Equus taubachensis* (Musil, 1990). Given the problems in the taxonomy of horses, I refer to the remains from the Mladeč Caves as *Equus* sp.

Table 16. Measurements of metapodial bones from *Equus* of Mladeč

	Inv. no	GL	GLl	Ll	Bp	Dp	SD	Bd
		mc III						
NHMW	72.195	263.4	259.8	255	62.8	43.2	44.9	61.5
OM	2904	240.3	240.7	233.3	58.4	38.8	38	55
	3019							58.4
		mt III						
NHMW	72.194	303.2	297.2	297.7	53.7	43.2	45	56.1
OM	3008	~263					34.8	52
	3059	263.3	256.6	255.6	54.4	44	39	46.6
	3009	~280			~52.3		37	~55
	2901+2899				52	47.3		53.2
	2912				56	45.3		

Table 17. Measurements from proximal and middle phalanges from *Equus* from Mladeč

	Inv. no	Ph.I	GL	Bp	BFp	Dp	SD	Bd	BFd
NHMW	72.196	post.	80	62.3			38.7	50.8	49.9
OM	2903	post.	94.2	57.2	52.3	38.5	36.4	49.8	49
	3014	ant.		65.9					
	3040	ant.	88	60.6	56.3	45.5	42.5	53.4	50.5
	-	ant.	82.8	59.9	53	~39	37.3	49	47.7
		Ph.II							
NHMW	72.196	ant.	51.6	51.6			48.3	49.3	49.3
OM	-	ant.	53.8	65.2			59.3	62.9	62.9
	2996	post.	56.8	64			55	60.8	60.8

Table 18. Additional measurements of bones from *Equus* from Mladeč

		Inv. no	LF	BF	Ld	HP	GH	GB	BFd	Lmt	GH
OM	PhIII	-	28	47.7	45	27.8					
OM	Astragalus	3029					65.7	69.5	60.5	66	65.7

Megaloceros giganteus BLUMENBACH, 1803
Material:
OM - 1 antler fragment, 1 shed antler fragment

Two small antler fragments from the collection of Smyčka can be ascribed to *Megaceros giganteus,* because of the typical structure of the surface and its morphology. One specimen is probably a fragment above the base (Plate I, Fig. 3). The other one is a shed antler showing the base for the brow tine and the segregation of the beam into two branches.

Alces alces LINNAEUS, 1758
Material:
MZM - mcII-V sin.

Alces alces is represented at the Mladeč Caves by vestigial metapodials and teeth used as raw material for personal implements and tools. One additional metacarpal was found in the collection of Knies (Plate I, Fig. 2). Although not all of the moose metapodials are definitely worked, it is most likely that all remains of *Alces alces* in the Mladeč Caves can be ascribed to raw material use by humans.

Capreolus capreolus LINNAEUS, 1758
Material:
(OM – 1 shed antler fragment)

The evidence of *Capreolus capreolus* is unclear in the examined faunal material from the Mladeč Caves. Roe deer is not mentioned in any previous faunal lists (see Musil, 2002, 73–76), and, in general, does not belong to the typical faunal assemblage of the Late Pleistocene in the region. The shed antler fragment of a male was found in the box with mixed extant and fossil material from the collection of Smyčka.

Cervus elaphus LINNAEUS, 1758
Material:
MZM – mcIII-IV dist.
OM – 1 antler fragment still attached to the parietal bone, 3 antler tip fragments, 3 incisors, 1 m3inf. dext., 1 upper molar, 1 upper milk molar, 1 molar, (1 antler fragment)

The collection of Knies contains a distal metacarpal fragment from *Cervus elaphus* with a distal breadth of 61.9 mm. Four antler remains and seven teeth are preserved in the collection of Smyčka. One antler fragment reveals chopping marks from a metal tool at the base of the brow tine. This specimen is of uncertain provenience.

Rangifer tarandus LINNAEUS, 1758
Material:
NHMW – 72.207 humerus dext., 72.208 femur dext., 72.209 radius sin., 72.210 tibia sin + tibia prox. dext., 72.211 mtIII-IV sin. + dext., 72.213 mandible fragment dext., 72.214 mandible sin.
without inventory number – ulnare sin., antler beam fragment
MZM – shed antler fragment, occipital fragment, maxilla + M3sup. sin., mandible fragment dext., calcaneus dext., pelvis (acetabulum) fragment sin., humerus dist. sin., mcIII-IV prox. sin., mcIII-IV dist., mtIII-IV dist. sin., ph.2 juv., ph1 juv., ph2 sin., ph1 sin., mtIII-IV sin., mtIII-IV prox. sin.
OM – 2894 radius prox. sin., 2923 lumbar vertebra, 2927 lumbar vertebra, 2945 lumbar vertebra, 2965 mandibula fragment, 2967 lumbar vertebra, 2968 lumbar vertebra, 2971 humerus sin., ?2974 femur prox. dext., 2975 mtIII+IV dist. dext., 2976 radius dist. dext., 2978 ph.1 post. sin., 2993 lumbar vertebra, 2996 cervical vertebra, 2998 lumbar vertebra juv., 2999 mcIII + IV prox. sin., 2999 femur dist. sin., 3000 lumbar vertebra, 3005 calcaneus dext., 3018 humerus dist.+shaft sin., 3107–3109 3 antler beam fragments, 3110 shed antler fragment, 3112 antler fragment, 3113 antler beam fragment, 3119 antler beam fragment
Without inventory number – 2 mandible fragments sin., mandible dext., tibia dist + shaft dext., mcIII-IV prox. sin., 1 rib fragment, 1 cast antler, 24 shed antler fragments, 7 antler beam fragments

Eighty-nine animal remains from the Mladeč Caves are from *Rangifer tarandus*. Antler fragments prevail with 40 specimens. Only the beam fragment from the collection of Szombathy, two shed antlers, and the specimen attached to the cranium from the collection of Smyčka are from male individuals, due to their size. The other specimens are either from female or juvenile individuals. A clear segregation is not possible (Sturdy, 1975, 56).

The calcaneus from the collection of Knies has a greatest length of 97.6 mm and the one from the collection of Smyčka reveals 92.7 mm. The greatest breadth is 30.5 mm. Only few measurements for

Table 19. Long bone measurements of *Rangifer tarandus* from Mladeč

	Inv. no.	GL	GLC	Bp	Dp	Bd	Dd	SD	DU	BT	Dc	Td	
Femur													
NHMW	72.208	268.9	259	66.3		57.6		33.4	7.5		29.6		
OM	2971		257	62.8		57.5		21			29.7		
Tibia													
NHMW	72.210	290		61.8		39.2		24.4	7			32.5	
	72.210			60.4									
Pavlov[1]						37.2	32.2						
						37.2	30.2						
Humerus													
NHMW	72.207					47.4				44.9			juv.
MZM						47.3		20.3		46			
OM	2971	225.8		53	62.9	44		19.3		42			
Pavlov[1]						43.1	35.7						
Radius													
NHMW		244.8		45.9		43.2		22.6					
OM	2976					40							
Pavlov[1]						43.6	25.5						
						43.7	25						
				40.6	25								

[1] after Musil (1997, 448)

Table 20. Metapodial bone measurements from *Rangifer tarandus* from Mladeč

	Inv. no.	GL	Bp	BPq	Bd	Gelbr	TD	Td
mcIII								
MZM			32.6	23				
			39.9				15.3	21
mtIII								
NHMW	72.211	262.2	30.1	32	41.6	41.5	16.4	23.3
	72.211	262.9	30.7	32.3	40.8	40.3	17.4	23.3
MZM			41.5				15.3	21.8
		281	33	35.2	42.4		18.8	
		256.3			39.5		18.4	

Table 21. Measurements of mandibles from *Rangifer tarandus* from Mladeč

	Inv. no.	p_1-p_4	Hp_2	m_1-m_3	Hm_1	v3	v12	v13	Hm_3	p_1-m_3	DL
NHMW	72.214	46.1	28.2		39.9	60	84.5	82.2	38.9		
NHMW	72.213	46.0	28.3		30.2						64.5
OM		48	26								
		41	31.9		38						
	2965	40.9	28	53.5	30				38.2	95.8	

v3, v12, v13...measurements 3, 12 and 13 after v. d. Driesch (1976) comparison are available from sites in the region, since remains of reindeer are often heavily fragmented. They belong to the main hunting prey of humans at many sites (e.g., Thenius, 1959; Fladerer, 1996; Logan, 1990; Musil, 1994).

Bos primigenius BOJANUS, 1827 and *Bison priscus* BOJANUS, 1838

Material:

Bison priscus:
NMHW - 72.215 humerus sin. + humerus sin. juv., 72.216 femur dext., 72.217 thoracic vertebra juv. + thoracic vertebra, 72.218 radius dext., 72.219 tibia sin. + dext., 72.220 mcIII–IV sin. + ph.1 sin. + dext. + ph.2 + ph.3, 72.221 axis, 72.222 atlas, 72.223 cranium fragment juv., 72.224 mandibula fragment sin. + mandibula fragment sin. juv., 72.225 scapula sin., (two fragments of the occipital region are lost)
MZM - ulna prox. sin., ph.3 ant. dext., 3 rib fragment, 1 sternum, 1 thoracic vertebra juv., 2 isolated vertebral extremitas, 1 caudal vertebra, 1 thoracic vertebra
OM - 3024 scapula fragment sin., 3071 small scapula fragment, 3055 tibia dext., 3057 tibia dext., 3060 tibia sin., 3041 humerus dist. sin., 3080 humerus dist. + shaft dext. juv., 3079 humerus dist. + shaft sin. juv., 3010 humerus dist., 3081 humerus dist. + shaft sin. juv., 3076 femur sin. juv., 3064 femur prox. dext., 3061 femur dext., 3063 femur sin., 2958 astragalus dext., 3054 astragalus dext., 2970 astragalus sin., 2953 astragalus dext., 1955 astragalus sin., 3068 radius sin., 3052 ulna prox. dext., 3086 ulna prox. dext., 3075 calcaneus sin., 3062 calcaneus dext., 3074 calcaneus dext., 3037 calcaneus sin., 3072 calcaneus dext., 3067 calcaneus juv., 2916 mc dext., 2915 mc dext., 2898 mc prox. sin., 2914 mc sin., 2926 centrotarsale sin., 2961 centrotarsale sin., 2960 centrotarsale sin., 2997 centrotarsale dext., 3067 radiale sin., 3068 mt dext., 3078 mt sin., 3063 mt dext., 2918 ph.1 ant. sin., 2941 ph.1 ant. dext., 2995 ph.1 post. dext., 2977 ph.1 post. dext., 2963 ph.1 post. dext., 2925 ph.1 post. sin., 2917 ph.1 post. dext., 2937 ph.1 post. sin., 2939 ph.1 post. sin., 2956 ph.1 post. dext., 2927 ph.2 post. sin., 2920 ph.2 post. sin., 2959 ph.2 post. dext., 2924 ph.2 ant. dext., 2962 ph.2 ant. sin., 2943 ph.3 post. sin., 2932 ph.3 post. dext., 2933 ph.3 post. sin., 3077 pelvis acetabulum fragment sin., 3094 pelvis dext., 3093 pelvis acetabulum fragment dext., 3085 pelvis sin., 3020 pelvis sin. + dext., 3087 pelvis illium fragment dext., 2992 pelvis ischium fragment sin., 3054 cranium fragment, 3047 cranium maxilla fragment dext., 3053 cranium maxilla fragment sin., 3082 mandibula ramus ascendens fragment sin., 2895 mandibula ramus ascendens fragment sin., 3073 hyalia, 3002 rib fragment prox., 3051 rib fragment prox., 3016 rib fragment prox., 2943 rib fragment prox., 3050 rib fragment prox., 2994 rib fragment, 2953 rib fragment, 2980 rib fragment, 2950 atlas, 2964a axis fragment, 2929 cervical vertebra juv., 2931 cervical vertebra juv., 2934 cervical vertebra, 2938 cervical vertebra, 2939 thoracal vertebra fragment, 2948 thoracal vertebra, 2945 thoracal vertebra, 2944 thoracal vertebra, 2919 thoracal vertebra, 3022 lumbar vertebra?, 2943 lumbar vertebra, 2969 thoracal vertebra, 2982 lumbal vertebra juv., 2940 lumbal vertebra juv., 2942 lumbal vertebra, 2942a lumbal vertebra juv., 2946 lumbal vertebra juv., 2949 lumbal vertebra, 2947 lumbal vertebra, 2964 lumbal vertebra, without inventory number a caudal vertebra, 2 M1sup. sin., 1 M1sup. dext., 3 M2sup. sin., 1 M3sup. sin., 2 P4sup. sin., 1 P4sup. dext., 2 P3sup. sin, 1 P3sup. dext., 2 milk molars sup., 3024 m1–m2inf., 3026 m1–m2inf., 2 m1–m2inf., m3inf. dext., deciduous tooth, p2inf. dext., without number a small scapula fragment, calcaneus dext., hyalia

Bos primigenius:
OM - 2883 scapula fragment sin., 2882 scapula fragment dext., 2891 scapula fragment, 2892 tibia dist. dext., 2887 tibia dist. sin., 3056 humerus dist. + shaft dext., 2889 humerus dist. dext., 2890 ulna prox. sin., 2896 mc prox. sin., 2888 mc prox. sin., 2800 mc prox. sin., 2897 mc dist. sin.

Bovidae indet.:
OM - 2893 mt dist. sin., 3021 humerus dist. sin., 2951 humerus dist. dext., 2951 humerus dist. dext

Bos primigenius f. taurus:
OM - 3124 mt prox. dext., 3012 mc sin.

Table 22. Measurements of metapodial bones from Bovidae from Mladeč

		Inv. no.	GL	Bp	Dp	Bd	SD	Dd	TD
McIII+IV									
NHMW	*Bison*	72.220	236.4	76.6		78.1	43.1	40.1	28.5
OM		2916	243.8	84.0	49.0	86.0	49.3	44.2	
		2915	236.5	79.4	46.4	85.0	48.6	43.3	
		2914	244.2	85.4	48.9	85.2	49.9	46.0	
	Bos	2888		84.6	50.5				
		2896		~77					
	Bos p. f. taurus	3012	198.7	54.9	35.1		31.5	31.3	
MtIII+IV									
	Bison priscus	3068	302.7	66.1		79.0	40.5	43.2	
		3078	294.0	66.2		72.7	40.7	44.0	
		3063	304.4	68.0		79.5	39.8	43.8	
	?Bos prim.	2893				81.0			
	Bos p. f. taurus	3124		~42					

Table 23. Long bone measurements from Bovidae from Mladeč

	Inv. no.	GL	Ll	GLC	Bp	Bd	Dd	SD	UD	BT	Dc	BTd	BTp
Femur													
Bison	3076												
	3064										67.9		
	3061			470	160			48.5			64	46	50.5
	3063			467		127.5		48.3			62.4	46	50
	2999					92.8							
Tibia													
Bos	2892					73.6	53.3						
	2887					88.6	56.2						
Bison	3057	467	413			86.5	61.4	56.4	16				
	3055	465	420					58	17				
	3060	460	405		127.3			55	16				
Humerus													
?Bos	3021					115				100.7			
	2951									110.8			
Bison	72.215	387				99.5		47.3		98			
	72.215	407				105.5		51.2		99.7			
	3080					109.2				101.9			
	3079					106.4		51.7		102.6			
Bos	3056							58.3					
	2889					~81				73.2			
Radius													
Bison	72.218	380.3			111.2	92.5		55.4					
	3068	370			108.5	97.3		58.5	16.5				
Ulna													
		LO	**Dpa**	**SDO**									
Bison	Knies	140.2	101	75.2									

Bovid remains from the Mladeč Caves can be identified to three species. Two metapodials (3012, 3124) from the collection of Smyčka are clearly determined as *Bos primigenius* f. taurus by their preservation and size (Table 22). The domestic animals probably come from the prehistoric settlements at the Třesín Hill and became mixed with the fossil remains.

The bovid remains of the collection of Szombathy are well preserved. A photo from one of the missing cranial fragments from the collection of Szombathy (1925, 12, Tafel 3, Abb. 3) and several post-cranial elements allowed identification of these remains as *Bison priscus*. The few specimens of the collection of Knies are also regarded as *Bison priscus,* mainly because of their similar state of preservation and their provenience from the same localities inside the cave.

The bulk of bovid remains come from the collection of Smyčka and like in *Equus,* two types of preservation are evident. Twelve specimens are ascribed to preservation type 1 by their color and intensity of gnawing traces (Plate II, Figs. 1, 3; Plate IV, Figs. 1a–b; Plate V, Fig. 2a–b). Although highly fragmented, morphological characteristics of some specimens allowed identification of these remains as *Bos primigenius*. Four additional, highly fragmented specimens might belong to this assemblage

Table 24. Measurements of scapula and pelvis from Bovidae from Mladeč

	Inv. no.	LG	BG	GLP	SLC	HS	LA	LFO	SH	SU	SB
Scapula											
Bos	2883	81.0	60.2	96.5	~86.0						
	2882	82.2	60.9	91.2	76.0						
Bison	72.225	79.7	63	97.2	76	524.3					
	3024	81.0	66.8	99.5	83.5						
Pelvis											
Bison	3077						94.3				
	3094						89.5	105			
	3093						94.2				
	3085								57.0	145	30.0
	3020						90.2	116	58	147	31.6

Table 25. Additional measurements of *Bison priscus* bones from Mladeč

	Inv. no.	Bd	GLm	GLl	Dm	Dl	GL	GB
Astragalus								
	2955		109.4	~59	58.3			
	2970	59.4	85.2	87.3	50.3	49.7		
	2953	61.4	87.3	92.2	51.2	50.5		
	2958	59.3	84.2	89.0	48.3	50.0		
	3054	55.1	78.5	84.0	46.3	46.7		
Calcaneus								
	3075						183	61.5
	3062						176	60.5
	3074						181.2	62.8
	3037						117.2	60.5
Centrotarsale								
	2926							79.8
	2961							73.3
	2960							78.0
	2997							80.3

Table 26. Measurements from phalanges of *Bison priscus* from Mladeč

	Inv. no.	GL	Bp	Dp	SD	BdF	Bd	Ld	MBS	DLS
Ph.1 ant.										
	2918	80.0	42.5	43.0	42.0	45.0				
	2941	80.5	43.6	42.0	41.8	45.0				
Ph.1 post										
	72.220	75.5	39.7							
	72.220	76.3	42.3				41.8			
	2995	86.9	40.0	45.4	36.9	38.6				
	2977	84.4	39.9	42.9	34.0	37.0				
	2963	84.4	39.5	44.2	38.2	42.0				
	2925	83.5	38.5	43.9	37.5	38.7				
	2917	84.5	40.5	44.2	38.5	40.5				
	2937	83.0	40.9	45.5	39.2	43.0				
	2939	81.4	39.0	38.2	36.0	38.5				
	2956	77.0	33.0	36.0	29.0	45.0				
Ph.2 ant.										
	2924	55.6	55.4		34.0		38.4			
	2962	56.2	45.0		37.4		39.3			
Ph.2 post										
	72.220	52.1	41.5		30.6					
	2927	55.8	41.9		32.4		33.0			
	2920	58.2	41.9		31.2		35.2			
	2959	58.5	42.3		33.4		36.4			
Ph.3 ant.										
	Knies							70.5	84.0	
Ph.3 post										
	72.220							70	32.8	~85
	2943							72.0	34.0	87.7
	2932							76.3	35.4	93.2
	2933							74.4	32.4	93.6

Table 27. Measurements of additional elements from Bovidae from Mladeč

	Inv. no.	GL	GB	GLF	BFcr	BFcd	H	LCDe	
Atlas	2950	110	187.2	102.4	117.4	130.5			
	72.221	~120		105.3		117.1	103.2		
Epistroph.					108				
	72.221						151.7	132.8	
Thorac. vert.	72.217						390		
		p_2-m_3	m_1-m_3	p_2-p_4	DL	Hm_3	Hm_1	Hp_2	M^1-M^3
Mandible	72.224	171.3	109.4	61.9	128.8	78.3	60.5	52.2	
Maxilla	3053								95.7

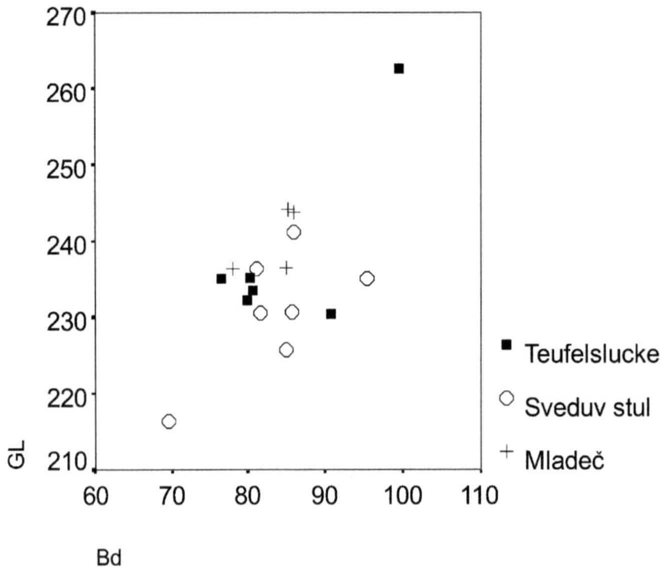

Fig. 11. Greatest length (GL) and greatest distal breadth (Bd) of metacarpals from Bovidae (data after Musil, 1962, 208; Table 22, and unpublished data from Teufelslucke)

but could not be clearly determined and are summarized as "Bovidae indet.", as are 27 isolated teeth (see Table 22). The large material of post-cranial bones and the few cranial remains of type 2 preservation are ascribed to *Bison priscus*. Sixteen remains are from immature individuals. Based on humeri, three juvenile individuals are proved.

Discrimination between *Bos* and *Bison* in post-cranial bones faces many difficulties due to the high variability of certain morphological criteria in both genera. Various authors have dealt with this task, and partly different morphological features are stressed. The study of bovid remains from the Mladeč Caves is based on criteria given by McCuaig Balkwil and Cumbaa (1992), Martin (1987), Sala (1986), and Stampfli (1963). Given the variability of criteria not only single traits but instead, the overall appearance of an assemblage needs to be considered. Plates IV and V show some of the most evident morphological differences between *Bos* and *Bison* in the Mladeč assemblage based on humeri and metapodial bones. The different types of preservation are also shown. In addition, the articular surface and the development of the glenoid process of the scapula show also clear differences in *Bos* and *Bison* from the Mladeč Caves.

Bovid remains are rare at Late Pleistocene sites in the region, with the exception of hyena dens. Measurements correspond to values from *Bison* from Teufelslsucke (own data) and Sveduv stul (Musil, 1962). Only one specimen from Teufelslucke considerably exceeds other values and one specimen from Sveduv stul is smaller than the bulk of remains (Fig. 11). In general, more detailed studies are needed in order to characterise bovid remains of the Late Pleistocene.

Capra/Ovis indet.
Material:
OM – lower molar

Only one lower molar can be ascribed to *Capra/Ovis* in general. The specimen shows traces of dissolution by gastric juices.

Conclusion

The present analysis includes a taphonomic and metric study of the faunal remains from the Mladeč Caves. Quantitatively, Bovidae are the main component of the Late Pleistocene assemblage, followed by reindeer, horse, and wolf. The remains represent individuals that fell into the natural trap in Halls D and E around the late glacial maximum. Additionally, a few specimens of various faunal elements became washed in through the chimney. A first phase of input at 31,000 years BP is represented by the human remains that are also related to the debris cone. The animal bones show no traces of human interference. Gnaw marks of carnivores are scarce with the exception of a distinct small assemblage represented by *Bos primigenius* and *Equus* sp. showing clear traces of hyena activity at 42,000 years BP. The provenience of this material is not known. As far as possible, site formation at the Mladeč Cave is reconstructed and confirms a Middle Pleistocene bear site, followed by a natural trap accumulation lasting at least from about 31,000 years BP until the Holocene. The results obtained emphasize the need of detailed analyses in order to reconstruct site formation processes. New questions raised may be subject for ongoing studies. Faunal composition and element distribution is rather different to Late Pleistocene human occupation sites in the region. Numerous metric data of various species are presented.

Acknowledgements

I would like to thank Dr. Teschler-Nicola and Dr. Florian Fladerer who invited me to join the project on the Mladeč Caves. I am also especially indebted to Dr. Oliva and Dr. Antl-Weiser, who gave me the opportunity to study the collections from Mladeč housed at the Moravské zemské muzeum in Brno and the Naturhistorisches Museum in Vienna. In addition, I would like to thank Dr. Teschler-Nicola, Dr. Antl-Weiser, Dr. Oliva, and Dr. Svoboda for helpful information, a visit to the site as well as fruitful discussions on site formation processes. The study was granted by the NHMW.

References

Altuna, J. (1981) Fund eines Skeletts des Höhlenlöwen (*Panthera spelaea* Goldfuss) in Arrikrutz, Baskenland. *Bonner Zoologische Beiträge* 32/1–4, 31–46

Antl-Weiser, W. (2006) Szombathy's excavations in the Mladeč Cave and the first presentation of the results. In (M. Teschler-Nicola, Ed.) *Early modern humans at the Moravian Gate: The Mladeč Caves and their remains,* this issue, pp. 1–16

Argant, A. (1991) *Carnivores quaternaires de Bourgogne.* Documents des Laboratoires de Géologie Lyon 115, Dép. des Sciences de la Terre. Univ. Lyon: Claude-Bernard, p. 300

Ballesio, R. (1980) Le gisement pléistocène supérieur de la grotte de Jaurens à Nespouls, Corrèze, France: Les Carnivores (Mammalia, Carnivora, 2. Felidae). *Nouvelles Archives de Museum d'Histoire naturelle de Lyon* 18, 61–102

Baryshnikov, G., Hoffecker, J. F. and Burgess, R. L. (1996) Palaeontology and Zooarchaeology of Mezmaiskaya Cave (Northwestern Caucasus, Russia). *Journal of Archaeological Science* 23, 313–335

Bauer, K. (2001) Wildpferd *Equus ferus* Boddeart, 1785. In (F. Spitzenberger, Ed.) *Die Säugetierfauna Österreichs,* Grüne Reihe des Bundesministeriums für Land- und Forstwirtschaft, Umwelt und Wasserwirtschaft 13, Graz: austria medien service GmbH, pp. 773–775

Bayer, J. (1922) Das Aurignac-Alter der Artefakte und menschlicher Skelettreste aus der „Fürst Johanns-Höhle" bei Lautsch in Mähren. *Mitteilungen der Anthropologischen Gesellschaft in Wien* 52, 173–185

Beneš, J. (1975) The Wurmian foxes of Bohemian and Moravian karst. *Acta Musei Nationalis Pragae (Serie B)* 31, 149–209

Dietrich, W. O. (1968) Fossile Löwen im europäischen und afrikanischen Pleistozän. *Paläontologische Abhandlungen, A, Paläozoologie* 3, 323–366

Driesch, v. d. A. (1976) *Das Vermessen von Tierknochen aus vor- und frühgeschichtlicher Zeit*. München: Springer, p. 114

Ehrenberg, K. and Kernerknecht, S. (1940) Die Fuchs- oder Teufelslucken bei Eggenburg Niederdonau, I. Teil. *Abhandlungen der Zoologischen-Botanischen Gesellschaft in Wien* 17, 1–301

Fladerer, F. (1996) Die Tierreste von Alberndorf in Niederösterreich. In (J. Svoboda, Ed.) *Paleolithic in the Middle Danube Region*. Spisy Archeologockého Ústavu av ČR v Brné 5, pp. 247–272

Fladerer, F. (2000) Late Quatenary vertebrate taphocoenoses from cave deposits in southeastern Austria: responses in a periglacial setting. In (M. B. Hart, Ed.) *Climates: Past and Present*. Special Publications, 181, London: Geological Society London, pp. 197–211

Fladerer, F. (2001) Die Faunareste vom jungpaläolithischen Lagerplatz Krems-Wachtberg, Ausgrabung 1930. *Mitteilungen der Prähistorischen Kommission* 39, 1–95

Fladerer, F. and Frank, C. (1997) Große Badlhöhle. In (D. Döppes and G. Rabeder, Eds.) Pliozäne und pleistozäne Faunen Österreichs. *Mitteilungen der Kommission für Quartärforschung der österreichischen Akademie der Wissenschaften* 10, 283–290

Fosse, Ph., Morel, Ph. and Brugal, J.-Ph. (2002) Taphonomie et Etologie des Ursides Pleistocenes. In (Th. Tillet and L. R. Binford, Eds.) *„L'Ours et l'Homme"*. Actes du Colloque d'Auberives-en-Royans, 1997. ERAUL 100, pp. 79–102

Galik, A. (1997) Zur Grabungsmethodik und Fundsituation in der Durrezza-Schachthöhle. *Neues aus Alt-Villach* 34, 17–29

Gaudzinski, S. (1995) Wisentjäger im Wallertheim: Zur Taphonomie einer mittelpaläolithischen Freilandfundstelle in Rheinhessen. *Jahrbuch Römisch-Germanisches Zentralmuseum Mainz* 39, 245–423

Gross, C. (1992) Das Skelett des Höhlenlöwen (*Panthera leo spelaea* Goldfuss, 1810) aus Siegsdorf/Ldkr. Traunstein im Vergleich mit den anderen Funden aus Deutschland und den Niederlanden. München: Diss. Tierärztliche Fakultät Univ. München, p. 130

Hemmer, H. (2003) Pleistozäne Katzen Europas. Cranium 20, 6–22

Hemmer, H. and Schütt, G. (1970) Körpergrösse und Extremitätenmasse Ältest- und Altpleistozäner Europäischer Pantherkatzen (Genus *Panthera*). *Mainzer naturwissenschaftliches Archiv* 9, 132–146

Hochstetter, F. v. (1883) Sechster Bericht der prähistorischen Commission der mathematisch-naturwissenschaftlichen Classe der kaiserlichen Akademie der Wissenschaften über die Arbeiten im Jahre 1882. *Sitzungsberichte der kaiserlichen Akademie der Wissenschaften math.-nat. Classe* 87, 168–170

Hoernes, M. (1903) Der *diluviale Mensch in Europa*. Braunschweig: Friedrich Vieweg und Sohn, p. 227

Hofreiter, M., Serre, D., Rohland, N., Rabeder, G., Nagel, D., Conard, N., Münzel, S. and Pääbo S. (2004) Lack of phylogeography in European mammals before the last glaciation. *Proceedings of the National Academy of Sciences USA* 101, 12963–12968

Jelínek, J. (1983) The Mladeč finds and their evolutionary importance. *Anthropologie* 21, 57–64

Jelínek, J. (1987) Historie, identifikace a význam Mldecských anthropologických nálezu z pacátku mladého Paleolitu. In *25 let pavilonu Anthropos 1961–1986*. Brno: Moravské muzeum, pp. 51–70

Königswald, W. v. (2002) *Lebendige Eiszeit. Klima und Tierwelt im Wandel*. Stuttgart: Konrad Theiss Verlag, p. 190

Kratochvil, Z. (1976) Das Postcranialskelett der Wild- und Hauskatze (*Felis sylvestris* und *Felis lybica* f. catus). *Acta Scientiarium Naturalium Academiae Scientiarium Bohemoslavacae Brno* 10, 1–43

Kunst, K. G. (1992) Großsäugerreste aus dem Nixloch bei Losenstein-Ternberg, O.Ö. In (D. N. Nagel and G. Rabeder, Eds.) Das Nixloch bei Losenstein-Ternberg. *Mitteilungen der Kommission für Quartärforschung der Österreichischen Akademie der Wissenschaften* 8, 83–127

Kurtén, B. (1965) On the Evolution of the European Wild cat, *Felis sylvestris* Schreber. *Acta Zoologica Fennica* 111, 3–29

Logan, B. (1990) The hunted at Grubgraben: An analysis of faunal remains. In (A. Montet-White, Ed.) The Epigravettian site of Grubgraben, Lower Austria: the 1986 and 1987 excavation. *ERAUL* 40, pp. 65–167

Martin, T. (1987) Artunterschiede an Langknochen großer Artiodactyla aus dem Jungpleistozän Mitteleuropas. *Courier Forschungsinstitut Senckenberg* 96, 1–121

Maška, K. J. (1886) *Der diluviale Mensch in Mähren. Ein Beitrag zur Urgeschichte Mährens*. Programm der mährischen Landes-Oberrealschule in Neutitschein für das Schuljahr 1885/86. Neutitschein: Selbstverlag des Verfassers, pp. 50–56

McCuaig Balkwil, D. and Cumbaa St. L. (1992) A guide for the identification of postcranial bones of *Bos taurus* and *Bison bison*. *Syllogeus* 71, 1–277

Musil, R. (1955) Osteologický material z paleolitického sídliště v Pavlově. *Acta Academiae Scientiarum Čechoslovenicae basis brunensis* 27/6, 279–320

Musil, R. (1959a) Poznámky k paleontologickému materiálu z Dolních Věstonic. *Anthropozoikum* 8, 73–82

Musil, R. (1959b) Osteologický material z paleolitckého sídliště v Pavlově. *Anthropozoikum* 8, 83–101

Musil, R. (1960) Die pleistozäne Fauna der Barová-Höhle. *Anthropos* 10 (n.s. 3), 1–37

Musil, R. (1962) Die Höhle „Svéduv stul", ein typischer Höhlenhyänenhorst. *Anthropos* 13 (n.s. 5), 97–260

Musil, R. (1965) Die Bärenhöhle Pod hradem. Die Entwicklung der Höhlenbären im letzten Glazial. *Anthropos* 18 (n.s. 10), 7–92

Musil, R. (1990) Pferdefunde *(Equus taubachensis* und *Equus skythicus)* aus der Kůlna-Höhle in Mähren. *Weimarer Monographien zur Ur- und Frühgeschichte* 26, 1–34

Musil, R. (1994) Hunting game of the culture layer of Pavlov. In (J. Svoboda, Ed.) Pavlov I, Excavations 1952–53. *The Dolní Věstonice Studies* 2 (*ERAUL,* 66), pp. 183–209

Musil, R. (1997) Hunting game analysis. In (J. Svoboda, Ed.) Pavlov I Northwest. The Upper Palaeolithic burial and its settlement context. *The Dolní Věstonice Studies* 4, pp. 443–468

Musil, R. (2002) Fauna moravských jeskyní s paleolitckými nálezy. In (J. Svoboda, Ed.) Prehistorické jeskyne. Katalogy, dokumenty, studie. Prehistoric caves. Catalogues, Documents, Studies. *The Dolní Věstonice Studies* 7, pp. 53–101

Nagel, D. (1997) Revision der mittelpleistozänen Großsäugerfauna aus Deutsch-Altenburg 1. *Wissenschaftliche Mitteilungen aus dem Niederösterreichischen Landesmuseum* 10, 231–249

Nagel, D., Pacher, M. and Morlo, M. (in press). The Late Pleistocene cave hyena (*Crocuta crocuta* spelaea) from the Teufelslucke cave/Austria. In (L. K. Horwitz and Ph. Fosse, Eds.) Hyenids in the Circum-Meditteranean: A taphonomic perspective. *Prehistoire et Anthropologie Méditteréennes*, special issue

Oliva, M. (1989) Mladopaleolitické nálezy z Mladečských jeskyní. *Acta Musei Moraviae, Scientiae sociales* 74, 35–54

Oliva, M. (1993) Le contexte archéologique des restes humains dans la Grotte de Mladeč. *UISPP XIIe Congr. Bratislava* 2, 207–216

Oliva, M. (2006) The Upper Paleolithic finds from the Mladeč Cave. In (M. Teschler-Nicola, Ed.) *Early modern humans at the Moravian Gate: The Mladeč Caves and their remains,* this issue, pp. 41–74

Pacher, M. (2000) Taphonomische Untersuchungen der Höhlenbärenfundstellen in der Schwabenreith-Höhle bei Lunz am See (Niederösterreich). *Beiträge zur Paläontologie* 25, 11–85

Pacher, M. (2004) Taphonomic analyses of cave bear remains from Potočka zijalka (Slovenia): Further analyses and conclusion. In (M. Pacher, Pohar, V. and G. Rabeder, Eds.) Potočka zijalka – palaeontological and archaeological results of the excavation. *Mitteilungen der Kommission für Quartärforschung der Österreichischen Akademie der Wissenschaften* 13, 97–113

Pacher, M. (2005) Bärenzähne als Schmuck im Paläolithikum. In (D. Nagel, Ed.) Festschrift für Prof. Gernot Rabeder. *Mitteilungen der Kommission für Quartärforschung der Österreichischen Akademie der Wissenschaften,* 14, 135–152

Pales, L. and Garcia, A. (1981) *Atlas ostéologiques.* Paris: Ed. CNRS

Poplin, F. (1976) Les Grands Vertébrés de Gönnersdorf Fouilles 1968. In (G. Bosinski, Ed.) *Der Magdalénien-Fundplatz Gönnersdorf* 2. Wiesbaden: Franz Steiner Verlag

Rabeder, G. (1996) Die Säugetier-Reste des frühen Aurignacien von Groß-Weikersdorf C (Niederösterreich). *Beiträge zur Paläontologie* 21, 85–91

Riedel, A. (1982) Il grande Felino della Grotta Tilde. The Lion of the Tilde cave. *Atti del Museo Civico di Storia Naturale – Trieste* 34, 10–42

Sala, B. (1986) *Bison schoetensacki* Freud. from Isernia la Pineta (early Mid-Pleistocene – Italy) and revision of the european species of bison. *Palaeontographia Italica* 74, 113–170

Schütt, G. (1969) Die jungpleistozäne Fauna der Höhlen bei Rübeland im Harz. *Quartär* 20, 79–125

Sickenberg, O. (1931) Die Großsäugetiere der Begleitfauna. In (O. Abel and G. Kyrle, Eds.) Die Drachenhöhle bei Mixnitz. *Speläologische Monographien* 7/8, pp. 747–762

Stampfli, H. R. (1963) Wisent, *Bison bonasus* (Linne, 1758), Ur, *Bos primigenius* Bojanus, 1827, und Hausrind, *Bos taurus* Linne, 1758. In (J. Boessneck, J.-P. Jéquier and H. R. Stampfli, Eds.) Seeberg, Burgäschisee-Süd. Die Tierreste. *Acta Bernensia* II, pp. 117–196

Stiner, M. C. (1991) Food procurement and transport by human and non-human predators. *Journal of Archaeological Science* 18, 455–432

Sturdy, D. A. (1975) Some reindeer economies in prehistoric Europe, In (E. S. Higgs, Ed.) *Palaeoeconomy.* Cambridge: Cambridge University Press, pp. 55–95

Svoboda, J. (2000) The depositional context of the early Upper Palaeolithic human fossils from the Koneprusy (Zlatý kůň) and Mladeč Caves, Czech Republic. *Journal of Human Evolution* 38, 523–536

Svoboda, J. (2006) The structure of the cave, stratigraphy, and depositional context. In (M. Teschler-Nicola, Ed.) *Early modern humans at the Moravian Gate: The Mladeč Caves and their remains,* this issue, pp. 27–40

Svoboda, J. A., v. d. Plicht, J. and Kuželka, V. (2002) Upper Paleolithic and Mesolithic human fossils from Moravia and Bohemia (Czech Republic): some new ^{14}C dates. *Antiquity* 76, 957–962

Szombathy, J. (1882) Über Ausgrabungen in den mährischen Höhlen im Jahre 1881. In (F. v. Hochstetter) Fünfter Bericht der prähistorischen Commission der math.-nat. Classe der kaiserlichen Akademie der Wissenschaften über die Arbeiten im Jahre 1881. *Sitzungsbericht der kaiserlichen Akademie der Wissenschaften* 85, [90]–[107]

Szombathy, J. (1904) Neue diluviale Funde von Lautsch in Mähren. *Jahrbuch der k. k. Zentralkommission für Kunst- und historische Denkmäler* N. F. 2, 9–16

Szombathy, J. (1925) Die diluvialen Menschenreste aus der Fürst-Johanns-Höhle bei Lautsch in Mähren. *Die Eiszeit* 2, 1–34, 73–95

Terberger, T. (2002) Paläolithikum und Mesolithikum. In (U. v. Freeden and S. v. Schnurbein, Eds.) *Spuren der Jahrtausende. Archäologie und Geschichte in Deutschland.* Stuttgart: Konrad Theiss Verlag, pp. 60–107

Thenius, E. (1959) Die jungpleistozäne Wirbeltierfauna von Willendorf i. d. Wachau, N.Ö. In (F. Felgenhauer, Ed.) Willendorf in der Wachau. Monographie der Paläolith-Fundstellen I-VII. *Mitteilungen der Prähistorischen Kommission der Österreichischen Akademie der Wissenschaften* 8-9, 133–170

Thenius, E. (1966) V. Die Cervidae und Perissodactyla (Equidae, Rhinocerotidae). In (K. Ehrenberg, Ed.) Die Teufels- oder Fuchsenlucke bei Eggenburg (NÖ). *Österreichische Akademie der Wissenschaften, math.-nat. Klasse, Denkschriften* 112, 61–82

Valoch, K. (1995) The earliest occupation of Europe: Eastern Central and Southeastern Europe. In (W. Roebroeks and T. v. Kolfschoten, Eds.) *The earliest occupation of Europe.* Leiden: University of Leiden, pp. 67–84

Weinstock, J. (1999) The Upper Pleistocene mammalian fauna from the Große Grotte near Blaubeuren (southwestern Germany). *Stuttgarter Beiträge zur Naturkunde* 277, 1–49

Wild, E. M., Teschler-Nicola, M., Kutschera, W., Steier, P., Trinkaus, E. and Wanek, W. (2005) First dating of Early Upper Palaeolithic human remains from Mladeč. *Nature* 435, 332–335

Woldřich, J. N. (1897) Übersicht der Wirbeltierfauna des „Böhmischen Massivs" während der anthropozoischen Epoche. *Jahrbuch der k. k. geologischen Reichsanstalt* 47, 393–427

Plates

Plate I.

(1) The radiometric dated tooth from *Castor fiber* from the excavation Jelínek
(2) The vestigial metapodial from *Alces alces* from the collection Knies
(3) A small fragment of antler from *Megaloceros giganteus* from the collection Smyčka
(4) One mandible fragment from *Ursus deningeri* from the collection Smyčka
(5) One ulna from a juvenile bear *(Ursus deningeri)* from the collection Smyčka

Plate II.

(1) 2889 – humerus fragment with gnawing marks from *Bos primigenius*, preservation type 1, from the collection Smyčka
(2) 2886 – gnawed radius fragment distal from *Equus* sp., preservation type 1, from the collection Smyčka
(3) 2890 – heavily gnawed ulna fragment from B*os primgenius*, preservation type 1, from the collection Smyčka

Plate III.

(1a) 2973 – proximal metacarpus from *Bos primigenius*, view from volar with cone
(1b) view from dorsal
(1c) details of scratches on the dorsal surface, probably due to preparation
(2a) 2999 – metacarpus from *Rangifer tarandus*, view from dorsal
(2b) details of modern cut-marks and surface exfoliation due to cleaning
(2c) modern cut-marks at the middle of the shaft

Plate IV.

(1a) 2888 – metacarpus fragment from *Bos primigenius*, view from dorsal
(1b) view of the proximal surface
(2a) 2916 – metacarpus from *Bison priscus*, view from dorsal
(2b) view of the proximal surface

Plate V.

(1a) 3080 – humerus from *Bison priscus*, view from cranial
(1b) distal view from caudal
(2a) 3056 – humerus *Bos primigenius*, view from cranial
(2b) distal view from caudal

PLATE I

PLATE II

1

2

3

PLATE III

1b

1c

1a

2a

2b

2c

PLATE IV

1a

1b

2a

2b

PLATE V

1a

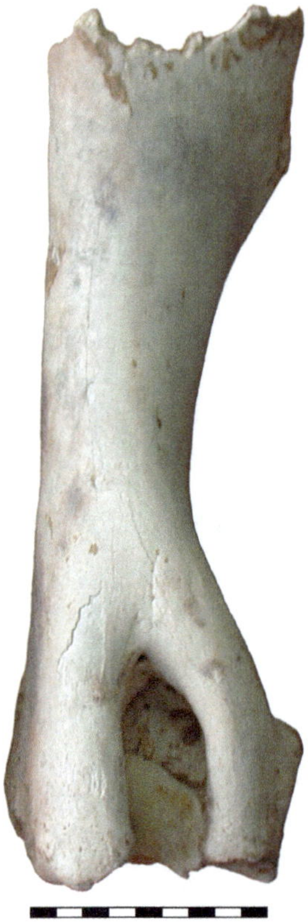

1b

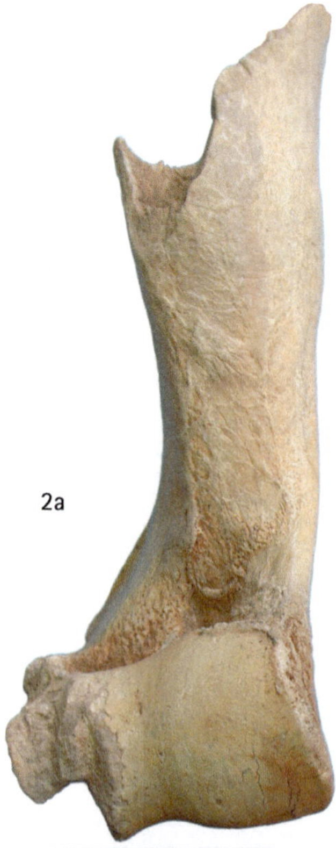

2a

2b

Chapter 7

^{14}C DATING OF EARLY UPPER PALAEOLITHIC HUMAN AND FAUNAL REMAINS FROM MLADEČ

Eva M. Wild, Maria Teschler-Nicola, Walter Kutschera, Peter Steier and Wolfgang Wanek

Introduction

The question of the age of the famous fossil human remains from the Mladeč Caves is of considerable importance for debates concerning our understanding of biological and cultural processes at the beginning of the Upper Paleolithic in Europe.

The first systematic test excavations in the cave of Mladeč, Middle Moravia, Czech Republic, were conducted in 1881 and 1882 by Josef Szombathy, collaborator of the Naturhistorisches Museum Wien (Szombathy 1882; 1925). Initial findings included human and animal skeletal remains, among them human cranial and post-cranial elements of at least five individuals. Judging from the degree of bone fossilization, as well as animal bones in a similar state of preservation, Szombathy assumed that the entire material was of pleistocene origin. Though initially challenged, this assumption has subsequently been corroborated by the artifacts discovered in 1882 (Hochstetter, 1883; Maška, 1886, 55), which Bayer (1922) and other scholars eventually attributed to the early Upper Paleolithic, i.e., Aurignacian culture (Oliva, 1993; Svoboda, 2000). What remained problematic to the entire ensemble was the lack of precise documentation on these finds and of records on the sequence of sedimentary deposits and exact coordinates of the artifacts and associated fossils. In spite of the difficulties arising from incomplete or insufficient documentation, the ensemble of these finds has played - and still plays - an important and essential role to discussions concerning the transition period when both Neandertals and early modern humans inhabited Europe.

The skeletal remains are widely accepted as those of early modern humans since the analysis of Szombathy (1925). However, there is an ongoing discussion as to whether they exhibit distinctive archaic features, indicative of some degree of regional Neandertal ancestry, or are morphologically solely aligned with recent humans and therefore document only a dispersial of modern humans into Europe. The purportedly archaic (or Neandertal) features include a number of qualitative characteristics, such as the sagittal cranial profile and robust supraorbital regions in the Mladeč 5 and 6 males, distinctive occipital bunning in the subadult individual Mladeč 3 as well as Mladeč 5 and 6, large palatal and dental dimensions of Mladeč 8, large crowns of the Mladeč 9a, 10 and 51 canines, and articular hyperthrophy of some of the postcrania. Moreover, although they are robust compared to recent females, the Mladeč 1 and 2 crania exhibit a few of these features (Jelínek, 1983; Frayer, 1986; 1992; Wolpoff et al., 2001; Trinkaus et al., this volume, Wild et al., 2005). This issue is associated with the still unknown mode of emergence of early modern humans in Europe and the fate of the Neandertals. A central point in the entire debate is the age of the Mladeč remains; the question is whether they are indeed of early Upper Paleolithic age and therefore originate from a time period which overlaps with the youngest Neandertals in Europe known so far (Smith et al., 1999; Hublin et al., 1996).

Moreover, Mladeč may also contribute significantly to the actual discussion of the development and spreading of Upper Palaeolithic cultural traits. It is beyond a doubt that the artifacts discovered on this site can be attributed to Aurignacian technology. A traditional concept focussed on the association between early modern humans as manufacturer of the Aurignacian artifacts, but this assumption has been challenged according to recently obtained ^{14}C dates: The AMS dating of a Lottorina shell associated with the Cro-Magnon human remains discovered 1868 near les Eyzies-de-Tayac, Dordogne, and routinely attributed to the Aurignacian culture, could demonstrate that they date to 28 ^{14}C kyr BP and herewith to the early Gravettian (Henry-Gambier, 2002). Direct ^{14}C dating of human fossils from the southwestern Germany site Stetten (Vogelherd cave) showed that these remains originated from the Neolithic (between 3,900 and 5,000 ^{14}C yr BP, Conard et al., 2004). Other purportedly Aurignacian-age modern humans, e.g. from Velika Pećina (Smith et al., 1999), Hahnöfersand (Terberger et al., 2001), Koněprusy (Svoboda, 2000) and La Rochette (Orschiedt, 2002), also turned out younger than initially believed. For the few direct ^{14}C dated human finds with definite early Upper Paleolithic age, such as Peştera cu Oase (Romania, ~35 ^{14}C kyr BP, Trinkaus et al., 2003), Kent's Cavern (U.K., ~31 ^{14}C kyr BP, Stringer, 1990), Peştera Muierii (Romania, ~30 ^{14}C kyr BP, Păunescu, 2001) and Peştera Cioclovina (Romania, ~29 ^{14}C kyr BP, Păunescu, 2001), a lack of archaeological objects at the sites prevents us from proving that early modern humans were the manufacturers of Aurignacian culture.

At the Mladeč site both, human remains and archaeological objects with typical characteristics of the Aurignacian were found. Though not well documented in the excavation protocols from the end of the 19th century an association between the human finds and the cultural remains can be assumed (Szombathy, 1925; Bayer, 1922; Oliva, 1993; Svoboda, 2000). Therefore the finds from the Mladeč Cave gained not only an important role in debates concerning our understanding of biological processes at the transitional period of Middle to Upper Paleolithic but also for the discussion of who were the founders of the Aurignacian culture (e.g., Mellars, 2004).

Radiocarbon dating

Due to the importance of the Mladeč fossils in the various archaeological and anthropological issues many efforts have been made to ^{14}C date the human fossils directly or indirectly, but all of them failed until recently. All these attempts were based on the dating of the organic bone fraction (collagen) present in fossil animal or human bones from this site. Unfortunately the preservation state of the collagen in the bone samples appeared as very poor, which made successful application of the radiocarbon method impossible.

Indirect ^{14}C dating of the human fossils

In 2002, a first reliable estimate of the age of the fossils was derived via the ^{14}C age determination of carbonates originating from different layers of a sinter crust which might have covered the sediment horizon in the so-called "Dome of the Dead", which contained the human remains and artifacts (Svoboda et al., 2002). The uncalibrated ^{14}C ages of the two carbonate samples, which were 5 cm apart from each other in the sinter layer, were determined as 34,160 +520/–490 ^{14}C years BP (GRN-26333) and 34,930 +520/-490 ^{14}C years BP (GRN 26334). Due to these almost identical results it was concluded that the sinter exhibited a relatively rapid formation rate. Since traces of a sinter crust can also be detected on the surface of the human bones it was further concluded that the age difference between the bones and the carbonate layer might be only small. Therefore the sinter dates were assessed as a good estimate for a minimum age of 34–35 ^{14}C kyr BP for the human fossils. In the same study again an unsuccessful attempt to date an animal bone from this site was reported, which failed due to the low collagen content.

Another attempt to date the human fossils was performed at the Vienna Environmental Research Accelerator (VERA) Laboratory in Vienna, and was recently published (Wild et al., 2005). In a first

Table 1 Radiocarbon ages determined for the faunal remains from the Mladeč site

Laboratory number	Sample name, provenience and inventory number	Sample material	Gelatine yield [mg/g sample]	^{14}C-age[a]) [years BP]	calibrated date[b])
VERA-2217	*Bison priscus,* NHM/PA[c]) 72222	bone, vertebra	~8.7	16,940 ± 60	18,310 BC – 17,950 BC
VERA-2218	*Bison priscus,* NHM/PA[c]) 72223	teeth from the upper jaw	~7	21,340 ± 100	***
VERA-2219	*Bison priscus,* NHM/PA[c]) 72220	bone, metapodium	0.3	25,880 ± 180	***
VERA-2962	*Castor fiber* MMB; n.a.[d])	tooth	4.3	8,515 ± 35	7,595 BC – 7,525 BC
VERA-2963	*Equus* MMB; n.a.[d])	bone, metapodium	~10	42,500 +1400/–1200	***

[a]) errors are 1 sigma uncertainties

[b]) determined with the calibration program OxCal (Bronk Ramsey, 2001 and 1985) and the calibration curve INTCAL04 (Reimer et al., 1998); data correspond to the 2-sigma confidence level (95.4%). *** indicate that the uncalibrated ^{14}C age lies outside the time range of the INTCAL04 curve

[c]) Naturhistorisches Museum Wien / inventory number of the Department of Prehistory

[d]) Moravské zemské muzeum, Brno; inventory number not available

The preservation state of the faunal samples can be deduced from the gelatine/collagen yields after the respective sample pretreatment. (EA-IRMS measurements were not performed for the faunal samples and the human test tooth). Considering losses during the handling of the material and the fact that in the case of the teeth an unknown amount of tooth enamel was present in the untreated samples, the majority of the yields fall in the range where reliable ^{14}C dating can be assumed. The collagen yield of VERA-2219 is far below the 5% level of collagen in recent bones (see text). Therefore the age of this sample has to be considered with caution.

onset we tried to avoid sampling the valuable human remains, instead selecting animal bones stored at the Naturhistorisches Museum Wien and the Moravské zemské muzeum in Brno as dating material. We hoped that dating the animal remains might allow to narrow down the time period when the early modern humans occupied the Mladeč Cave, i.e., enable an indirect dating of the human fossils. Five animal bones from different species were ^{14}C dated with the AMS technique at VERA (Vienna Environmental Research Accelerator). Three other samples which also originated from animal finds could not be dated due to insufficient collagen content. The determined ages of the successfully dated animal bones are given in Table 1. It is evident that the time period spanned by these ages is too long (~8.5 ^{14}C kyr BP to ~42.5 ^{14}C kyr BP) to deduce a sufficiently accurate date of the early modern human finds.

Direct ^{14}C dating of the human fossils

After the animal dating campaign it was clear that the only way to achieve a reliable age estimate of the human fossils would be to date the human remains themselves. Although curatorial aspects strictly limit the amount of material consumable for invasive scientific investigations, a sample from a proximal ulna fragment, Mladeč 25c, was acquired for ^{14}C dating. The age determined for this sample is ~26 ^{14}C kyr BP. This age puts the human ulna into the Gravettian cultural period, but must be considered with caution since the preservation state of the collagen was – though above a recommended limit for reliable dating – relatively low (see description of methods and discussion below). Interestingly, the carbonate crust and the color of this specimen differ from the other human finds and it may thus originate from a different find spot (see Teschler-Nicola, this volume, chap. 5). Due to the insecure dating result of the ulna further dates of the Mladeč fossils were demanded to verify this first direct age. Again curatorial considerations made it impossible to sample gram quantities of bone material of the Mladeč specimen. For further investigations we decided to concentrate on the fossil teeth, which were preserved together with the crania and a maxilla of the fossil assemblage. The idea was that the collagen in dentine from the inner part of a tooth might be better protected from degradation and contamination in cases where intact enamel and intact alveolar bone are present. In order to check whether enough dating material would be available by sampling the teeth crowns, two test samples from known age human teeth were taken. One sample originated from a male human skull and is dated to the 19th century. The second sample was taken from a sub-adult individual from an early Bronze Age grave yard. We applied the same sampling and chemical pretreatment method to these samples as we intended to use for the Mladeč teeth (see below). Before sampling, the surface of both samples had been abraded for cleaning, which also caused a considerable reduction of the enamel fraction, which is low in organic material. The tooth samples prepared in this way (~400 mg and ~300 mg) were further processed chemically as described below. Both samples, which in essence consisted of dentine with unknown amounts of enamel, yielded enough collagen (~15 mg) to perform a ^{14}C age determination and the calibrated ages of both samples were in the expected time range. The results of the test measurements showed that the selected sampling method together with the selected chemical method (described below) promised to yield enough dating material and enabled reliable dating.

After successful application of the ^{14}C method to the test samples, in the next step the samples from the Mladeč fossils were taken. The most prominent specimens kept at the Naturhistorisches Museum Wien, i.e., Mladeč 1, Mladeč 2 and Mladeč 8 (samples from Mladeč 2, Mladeč 8 and the ulna, Mladeč 25c, were also used in the DNA study by Serre et al., 2004), were selected for radiocarbon dating. All of them were excavated from the Main Cave. In addition, an isolated canine (Mladeč 9a), which showed an excellent general preservation state, was selected for dating as well. Approximately one half of each crown (Mladeč 1 and Mladeč 2) or part of the roots (Mladeč 8 and Mladeč 9a) was taken for the radiocarbon determinations (see Fig. 1). In order to preserve the entire information about the shape of the teeth, casts were made before sampling.

As already mentioned, one crucial prerequisite for a reliable ^{14}C age determination of bones is the preservation state of the collagen in the bones. According to Hedges and van Klinken (1992), samples with more than 5% of the original collagen still present should be capable to yield reliable ages, whereas samples with a more advanced degradation of the organic bone fraction are assessed as problematic. They suggested determination of nitrogen and carbon content and C/N ratio of the unprocessed bones or some collagen extracts thereof as a test for the suitability of ancient bones for dating. The nitrogen content of fresh, defatted and dried compact bones is in the range between 4% and 5% dw (dry weight) (Petchey, 2005). The nitrogen content of fossil bones therefore provides an estimate for the amount of preserved collagen in a sample, whereas an abnormal C/N ratio $>> 4$ may indicate the presence of large amounts of exogenous carbon or an advanced deamination as a consequence of collagen degradation. Although amino acid analyses of some Mladeč bones performed in the course of a DNA study (Serre et al., 2004), showed that the preservation state of the collagen from Mladeč 2 and Mladeč 25c meets the criteria for DNA analysis – whereas Mladeč 8 (a maxillary bone sample) did not – we tested the material selected for dating using the quality check proposed by Hedges and van Klinken (1992).

The carbon and nitrogen content, and thereof the C/N ratio of powdered dentine from ~10 mg of tooth samples – obtained in addition to the ^{14}C samples – were determined with combined elemental analysis/stable isotope ratio mass spectrometry (EA-IRMS). The measurements were perfor-

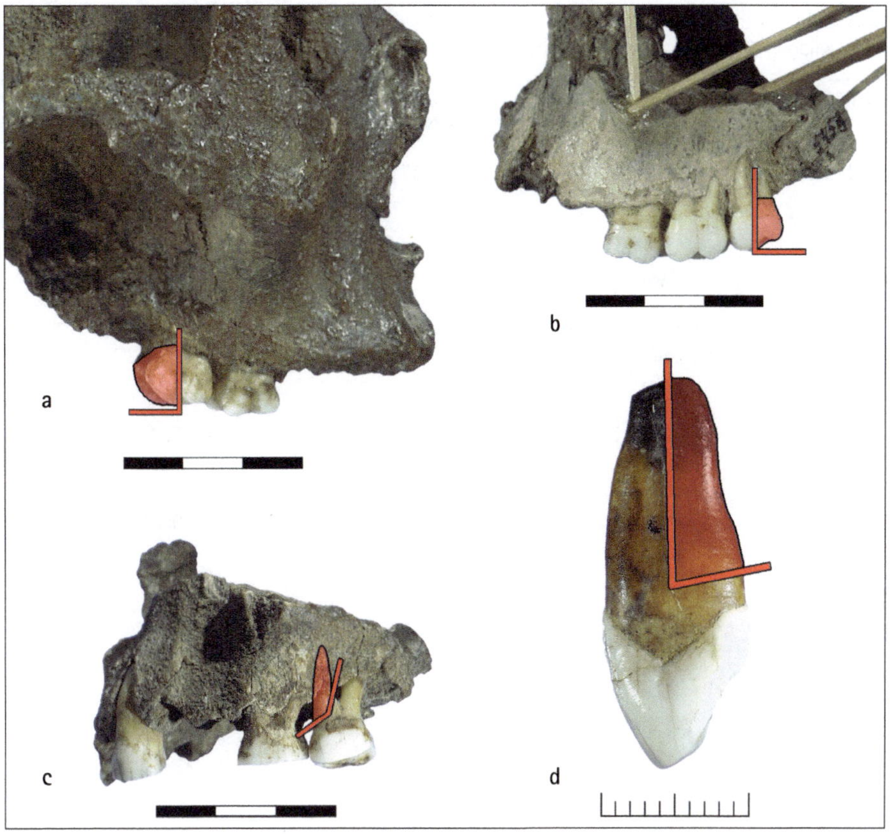

Fig. 1. The photographs show the dated human remains with the sampled parts indicated in red: **(a)** Mladeč 1, lateral view from right, **(b)** Mladeč 2, lateral view from left, **(c)** Mladeč 8, lateral view from left, **(d)** Mladeč 9a, right maxillary canine, mesial view. A centimeter scale is displayed for the pictures a to c, whereas for figure d the minor graduation marks are correspond to 1 mm (Wild et al., 2005). Copyright for the photograph: Naturhistorisches Museum (2004), Anthropologische Abteilung, Burgring 7, 1010 Vienna, Austria

med with an elemental analyzer (EA 1110, CE Instruments) coupled to a gas isotope ratio mass spectrometer (DeltaPLUS, Finnigan MAT) operating in the continuous flow mode. The complete data of this test are given in Table 2. According to the determined parameters of the untreated teeth, a good preservation state of the collagen in the teeth can be assumed and no large quantities of carbon contamination were detected in the samples. Although mainly applied for collagen extracts and appraised as being not very sensitive for the detection of contaminants, the $\delta^{13}C$ and $\delta^{15}N$ values of the untreated dentine support the good preservation of the collagen and the absence of large amounts of exogenous carbon. For the unprocessed ulna this test gave less encouraging results: a low N content (~0.5% dry weight) shows that – although the collagen content is still above the 5% level of the collagen content of recent bones (see above) – the degradation of the collagen is already advanced, and the unusually high C/N ratio of ~14 may indicate the presence of exogenous carbon.

All bone samples dated in this study, i.e., animal bones and the human ulna, were processed according to the routine pretreatment method for fossil bones at VERA. This method is in essence an acid-base-acid method, which leads to an enrichment of the organic material present in the sample by dissolution of the inorganic bone fraction, and removes carbonates (authentic and secondary) and humic acids. Secondary carbonates and humic acids might have been introduced into the samples

Table 2. Results of the EA-IRMS sample quality tests of the untreated human samples plus the EA-IRMS results determined for the ^{14}C dated collagen of Mladeč 8 (bold) together with the ^{14}C ages determined for the individual specimen

Laboratory number	sample name	sample material	C-content¹) [% dw]	N-content¹) [% dw]	C/N ratio¹)	$\delta^{13}C$¹) [‰]	$\delta^{15}N$¹) [‰]	^{14}C-age²) [years BP]
VERA-2736	Mladeč 25c	ulna	6.4 ± 0.2³)	0.47 ± 0.06³)	13.7 ± 1.7³)	-24.6 ± 0.2³)	10.0 ± 0.5³)	26,330 ± 170
VERA-3073	Mladeč 1	right M2 distal half of the crown	11.8⁴)	3.2⁴)	3.7⁴)	-19.1⁴)	10.6⁴)	31,190 +400/-390
VERA-3074	Mladeč 2	left M3 distal half of the crown	6.4⁴)	1.4⁴)	4.7⁴)	-20. 6⁴)	10.3⁴)	31,320 +410/-390
VERA-3075	Mladeč 8	left M2 mesial-buccal root	10.7 ± 0.1⁵)	2.3 ± 0.2⁵)	4.7 ± 0.4⁵)	-21.4 ± 0.3⁵)	11.7 ± 0.4⁵)	30,680 +380/-360
		pretreated collagen	**44.3 ± 0.3⁵)**	**16.1 ± 0.7⁵)**	**2.7 ± 0.1⁵)**	**-20.1 ± 0.4⁵)**	**10.9 ± 0.7⁵)**	
VERA-3076A (white colored collagen)	Mladeč 9a right maxillary canine	lingual half of the root	9.6 ± 0.6⁵)	2.4 ± 0.4⁵)	4.0 ± 0.3⁵)	-19.7 ± 0.2⁵)	9.6 ± 0.6⁵)	31,500 +420/-400
VERA-3076B (brown colored collagen)	Mladeč 9a right maxillary canine	lingual half of the root	-	-	-	-	-	27,370 ± 230

¹) determined by EA-IRMS (elemental analyzer stable isotope ratio mass spectrometry)
²) determined by AMS, errors are 1 sigma uncertainties
³) mean value of 3 EA-IRMS measurements and 1 standard deviation (SD) of the mean
⁴) only a single EA-IRMS measurement was performed for this samples
⁵) mean value of 2 EA-IRMS measurements and standard deviation of the mean

The $\delta^{13}C$- and $\delta^{15}N$-values are defined as the relative deviation (in ‰) of the $^{13}C/^{12}C$ and $^{15}N/^{14}N$ ratio of a sample from the $^{13}C/^{12}C$ of the V-PDB (Vienna - Pee Dee Belemnite) standard and the $^{15}N/^{14}N$ of the atmospheric N_2 standard (at-air), respectively. The standard deviations of the mean values given in the table for multiple measured samples include uncertainties due to sample inhomogeneities. The reproducibility of repeated measurements of a laboratory standard was 0.10‰ (SD) for $\delta^{13}C$ and 0.15‰ (SD) for $\delta^{15}N$. Except for the pretreated collagen of Mladeč 8, the EA-IRMS measurements were performed on the untreated samples and $\delta^{13}C$ reflects the isotopic composition of the total carbon in the sample, which may in the case of the untreated samples originate from the organic and the inorganic sample fraction as well as from exogenous carbon. Accordingly the $\delta^{15}N$ values of these samples reflect the isotopic composition of the total nitrogen.

during burial times. For further clean-up of the sample, gelatine is produced from the remaining residue (collagen). This method is frequently used for the processing of bone samples for ^{14}C dating in several laboratories (Petchey, 2005). A different approach was used for the chemical pretreatment of the Mladeč teeth. For these samples, with amounts between 350 mg and 200 mg, we used a similar approach as Schmitz et al. (2002) who extracted collagen from fossil bones by dissolving the inorganic bone fraction in dilute HCl. We accomplished this procedure for the Mladeč teeth with an alkaline treatment with dilute NaOH solution followed by a treatment with HCl after the demineralization step. After each step the collagen was washed with bi-distilled water. The last step of our routine bone pre-treatment method, the gelatine production was omitted in the case of the tooth sample in order to avoid unnecessary losses of the valuable sample material which come along with each clean-up step. As mentioned above, for collagen originating from "protected" dentine a smaller risk of contamination can be assumed. Even the NaOH treatment of the collagen for the removal of humic acids seems to have been unnecessary as the collagen extracted from the teeth appeared as a white colored substance (except in one case, see below) after demineralization and the NaOH solution stayed uncolored during this treatment, which indicates that no significant amounts of humic acids were present in these samples.

The ^{14}C sample (~300 mg) from the Mladeč 8 tooth yielded approx. 25 mg of collagen. This "large" amount enabled splitting of the extracted collagen; a sub-sample identical to the ^{14}C dated material was measured by EA-IRMS and a C/N weight % ratio of ~2.7 was determined (see Table 2). This value, when transformed into an "atomic" C/N ratio of 3.2, compares to "atomic" C/N ranges from 3.5–2.9 (DeNiro, 1985; Ambrose, 1990) and 3.4–2.6 (Schoeninger et al., 1989) for gelatinized collagen from well-preserved bones given in the literature. The values of the C content, the N content and the $\delta^{13}C$ and $\delta^{15}N$ values (parameters used by some ^{14}C laboratories for assessment of the quality of extracted collagen) determined for this sample also indicate that the collagen extracted from Mladeč 8 can be considered well preserved. Furthermore, these data show that the applied chemical procedure yielded reliable dating material and prove the cleanup of the tooth samples from contaminants.

The isolated canine exhibited a dark brown colored apex and part of the collagen extracted from the canine sample was brownish as well. Therefore the collagen was divided into two samples; VERA-3076A, the white colored fraction, and VERA-3076B, the brownish colored collagen.

The extracted collagen from the human teeth and the gelatine produced from the ulna and the animal bones were subjected to the routine sample preparation and measurement procedure used for ^{14}C dating of archaeological samples at VERA (Wild et al., 1998; Steier et al., 2004).

Discussion

Although an accurate indirect dating of the human fossils from the Mladeč Cave - which was the aim of the present study - was not possible via ^{14}C dating of the faunal remains, the animal data themselves provide important information concerning the inclusion of the faunal remains in the cave (see Pacher, this volume, chap. 6).

The ^{14}C ages of all human Mladeč samples directly dated in this study are listed in Table 2. Whereas all uncalibrated ages of the teeth agree at ~31 ^{14}C kyr BP within uncertainties (except for sample VERA-3076B), the ^{14}C age of ~26 ^{14}C kyr BP of the ulna is significantly younger. As already mentioned above, the C-N data of the untreated ulna indicate that this sample might have been affected by the presence of a considerable amount of exogenous carbon. It is not clear whether the applied chemical method was capable to remove the total amount of this contamination. Therefore the ^{14}C age of this sample has to be treated with caution.

The younger age determined for the brown collagen fraction VERA-3076B of the isolated canine (see Table 2) supports our hypothesis that the color of the apex resulted from a contamination, which could not be removed completely by the chemical procedure. The agreement of the other teeth data confirms the applicability of the chemical cleaning method used for these samples. Incomplete removal of a modern carbon contamination from the samples would hardly result in an agreement of the contamination affected ^{14}C ages.

The ages determined for the human Mladeč samples all lie within a time period for which a generally agreed calibration curve for the transformation of uncalibrated ^{14}C ages older than ~21 ^{14}C kyr BP into calendar time ranges is not yet available. According to the existing, albeit divergent, ^{14}C records for this period determined in different archives, a shift of the "true ages" by several thousand years towards higher ages might be possible (Bard et al., 2004). Attempts to construct calibration curves for the relevant time period based on these data sets have been made (e.g., van Andel et al., 2004). According to van der Plicht et al. (2004), the unresolved discrepancies between the data sets suggest that these "calibration curves" should best be termed "radiocarbon comparison curves". Moreover, the authors warn of erroneous conclusions drawn from calibrated ^{14}C data derived from such curves.

The AMS ^{14}C dating of the Mladeč human remains confirms that they derive from the time period of the middle to late Aurignacian of Central Europe. Given the presence of multiple individuals - male and female, adult and immature - with cranial, dental and postcranial elements, the Mladeč assemblage becomes the oldest directly dated substantial assemblage of modern human remains in Europe. Only the ~35 ^{14}C kyr BP Peştera cu Oase mandible and cranium, from two individuals, are definitely older, but they currently lack postcranial remains and an archaeological association. Moreover, the Mladeč dates on both robust "males" (Mladeč 8 and 9a) and less robust "females" (Mladeč 1 and 2) fall into the same time period, reinforcing the fact that the variability within the assemblage reflects both the original population variability and probably its level of sexual dimorphism.

Therefore, the important human fossil assemblage from the Mladeč Cave joins the less complete early modern human remains from Peştera cu Oase, Kent's Cavern, Peştera Muierii and Peştera Cioclovina in being directly AMS ^{14}C dated to the period of the Aurignacian in Europe. As such, they will be increasingly important for our understanding of the biology of the earliest modern humans in Europe and for the transitional period when both Neandertals and early modern humans occupied Europe, which is still inadequately understood.

Acknowledgement

We thank Wolfgang Reichmann for providing the photographic documentation and Stefan Lehr for his help in sample preparation.

References

van Andel, T. H., Davies, W., Weninger, B. and Jöris, B. (2004) Archaeological dates as proxies for the spatial and temporal human presence in Europe: a discourse on the method. In (T. H. van Andel and W. Davies, Eds.) *Neanderthals and modern humans in the European landscape during the last glaciation: archaeological results of the Stage 3 Project*. Oxford: Oxbow, pp. 21–31

Ambrose, S. H. (1990) Preparation and characterization of bone and tooth for isotopic analysis. *Journal of Archaeological Science* 17, 431–451

Bard E., Rostek F. and Ménot-Combes, G. (2004) A better radiocarbon clock. *Science* 303, 178–179

Bayer, J. (1922) Das Aurignac-Alter der Artefakte und menschlichen Skelettreste aus der „Fürst Johanns-Höhle" bei Lautsch in Mähren. *Miteilungen der Anthropologischen Gesellschaft Wien* 52, 173–185

Conard, N. J., Grootes, P. M. and Smith, F. H. (2004) Unexpectedly recent dates for human remains from Vogelherd. *Nature* 430, 198–200

DeNiro, M. J. (1985) Postmortem preservation and alteration of in vivo bone collagen isotope ratios in relation to palaeodietary reconstruction. *Nature* 317, 806–809

Frayer, D. W. (1986) Cranial variation at Mladeč and the relationship between Mousterian and Upper Palaeolithic hominids. *Anthropos* (Brno) 23, 243–256

Frayer, D. W. (1992) Evolution at the European edge: Neanderthal and Upper Paleolithic relationships. *Préhistoire Européenne* 2, 9–69

Hedges, R. E. M. and van Klinken, G. J. (1992) A review of current approaches in the pre-treatment of bone for radiocarbon dating by AMS. *Radiocarbon* 34, 279–291

Hochstetter, F. v. (1883) Sechster Bericht der Prähistorischen Commission der math.-nat. Classe der kaiserlichen Akademie der Wissenschaften über die Arbeiten im Jahre 1882. *Anzeiger der kaiserlichen Akademie der Wissenschaften, math.-nat. Classe* 20, 46–47

Henry-Gambier, D. (2002) Les fossils de Cro-Magnon (Les Eyzies-de-Tayac, Dordogne): nouvelles données sur leur position chronologique et leur attribution culturelle. *Bulletin et Mémoire de la Société d'Anthropologie de Paris* 14, 89–112

Hublin, J. J., Spoor F., Braun, M., Zonneveld F. and Condemi S. (1996) A late Neanderthal associated with Upper Palaeolithic artifacts. *Nature* 381, 224–226

Jelínek, J. (1983) The Mladeč finds and their evolutionary importance. *Anthropologie* 21, 57–64

Maška, K. J. (1886) *Der diluviale Mensch in Mähren. Ein Beitrag zur Urgeschichte Mährens.* Programm der mährischen Landes-Oberrealschule in Neutitschein für das Schuljahr 1885/86. Neutitschein, pp. 50–56

Mellars P. (2004) Neanderthals and the modern human colonization of Europe. *Nature* 432, 461–465

Oliva, M. (1993) Le contexte archéologique des restes humains dans la grotte de Mladeč . In *UISPP, XIIe Congres* (Vol. 2). Bratislava: Institut d'Archéologie, pp. 207–216

Orschiedt, J. (2002) Datation d'un vestige humain provenant de La Rochette (Saint Léon-sur-Vézère, Dordogne) par la méthode du carbone 14 en spectrométrie de masse. *Paléo* 14, 239–240

Pacher, M. (2006) Large mammal remains from the Mladeč Caves and their contribution to site formation processes. In (M. Teschler-Nicola, Ed.) *Early modern humans at the Moravian gate: The Mladeč Caves and their remains,* this issue, pp. 99–148

Păunescu, A. (2001) *Paleoliticul şi mezoliticul din spaţiul transilvan.* Bucureşti: Editura AGIR

Petchey, F. In (T. Higham) Chemical pre-treatment methods. Radiocarbon web-info: /www.c14dating.com/pret.html, 2005

van der Plicht, J., Beck, J. W., Bard, E., Baillie, M. G. L., Blackwell, P. G., Buck, C. E., Friedrich, M., Guilderson, T. P., Hughen, K. A., Kromer, B., McCormac, F. G., Bronk Ramsey, C., Reimer, P. J., Reimer, R. W., Remmele, S., Richards, D. A., Southon, J. R., Stuiver, M. and Weyhenmeyer, C. E. (2004) NOTCAL04-Comparison/calibration ^{14}C records 26–50 cal kyr BP. *Radiocarbon* 46, 1–14

Reimer P. J., Baillie M. G. L., Bard E., Bayliss A., Beck J. W., Bertrand C., Blackwell P. G., Buck C. E., Burr G., Cutler K. B., Damon P. E., Edwards R. L., Fairbanks R. G., Friedrich M., Guilderson T. P., Hughen K. A., Kromer B., McCormac F. G., Manning S., Bronk Ramsey C., Reimer R. W., Remmele S., Southon J. R., Stuiver M., Talamo S., Taylor F. W., van der Plicht J. and Weyhenmeyer C. E. (2004) IntCal04 Terrestrial Radiocarbon Age Calibration, 0–26 cal kyr BP, *Radiocarbon* 46, 1029–1058

Schoeninger, M. J., Moore, K. M., Murray, M. L. and Kingston, J. D. (1989) Detection of bone preservation in archaeological and fossil samples. *Applied Geochemistry* 4, 281–292

Serre, D., Langaney A., Chech, M., Teschler-Nicola, M., Paunovic, M., Mennecier, Ph., Hofreiter, M., Possnert, G. and Pääbo, S. (2004) No evidence of Neanderthal mtDNA contribution to early modern humans. *Public Library of Science (Biology)* 2, 313–317

Schmitz, R. W., Serre, D., Bonani, G., Feine, S., Hillgruber, F., Krainitzki, H., Pääbo, S. and Smith, F. H. (2002) The Neanderthal type site revisited: Interdisciplinary investigations of skeletal remains from the Neander Valley, Germany. *Proceedings of the National Academy of Sciences USA* 99, 13342–13347

Smith F. H., Trinkaus, E., Pettitt, P. B., Karavanić, I. and Paunović, M. (1999) Direct radiocarbon dates for Vindija G_1 and Velika Pećina Late Pleistocene hominid remains. *Proceedings of the National Academy of Sciences USA* 96, 12281–12286

Steier, P., Dellinger, F., Kutschera, W., Priller, A., Rom, W. and Wild, E. M. (2004) Pushing the precision limit of ^{14}C AMS. *Radiocarbon* 46, 5–16

Stringer, C. B. (1990) British Isles. In (R. Orban, Ed.) *Hominid remains: An update. British Isles and Eastern Germany.* Brussels: Université Libre de Bruxelles, pp. 1–40

Szombathy, J. (1882) Über Ausgrabungen in den mährischen Höhlen im Jahre 1881. *Sitzungsberichte der kaiserlichen Akademie der Wissenschaften* 85, [90]–[107]

Szombathy, J. (1925) Die diluvialen Menschenreste aus der Fürst-Johanns-Höhle bei Lautsch in Mähren. *Die Eiszeit* 2, 1–34, 73–95

Svoboda, J. (2000) The depositional context of the Early Upper Paleolithic human fossils from the Koněprusy (Zlatý kůň) and Mladeč Caves, Czech Republic. *Journal of Human Evolution* 38, 523–536

Svoboda, J. A., van der Plicht, J. and Kuzelka, V. (2002) Upper Palaeolithic and Mesolithic human fossils from Moravia and Bohemia (Czech Republic): some new ^{14}C dates. *Antiquity* 76, 957–962

Svoboda, J. A., van der Plicht, J., Vlček, E. and Kuzelka, V. (2004) New radiocarbon datings of human fossils from caves and rockshelters in Bohemia (Czech Republic). *Anthropologie* 42, 161–166

Terberger, T., Street, M. and Bräuer, G. (2001) Der menschliche Schädelrest aus der Elbe bei Hahnöfersand und seine Bedeutung für die Steinzeit Norddeutschlands. *Archäologisches Korrespondenzblatt* 31, 521–526

Teschler-Nicola, M. (2006) Taphonomic aspects of the human remains from the Mladeč Caves. In (M. Teschler-Nicola, Ed.) *Early modern humans at the Moravian gate: The Mladeč Caves and their remains*, this issue, pp. 75–98

Trinkaus, E., Moldovan, O., Milota, Ş., Bîlgăr, A., Sarcina, L., Athreya, S., Bailey, S. E., Rodrigo, R., Gherase, M., Higham, T., Bronk Ramsey, C. and van der Plicht, J. (2003) An early modern human from the Peştera cu Oase, Romania. *Proceedings of the National Academy of Sciences USA* 100, 11231–11236

Trinkaus, E., Smith, F. H., Stockton, C. and Shackelford, L. L. (2006) The human postcranial remains from Mladeč. In (M. Teschler-Nicola, Ed.) *Early modern humans at the Moravian gate: The Mladeč Caves and their remains*, this issue, pp. 385–446

Wild, E. M., Golser, R., Hille, P., Kutschera, W., Priller, A., Puchegger, S., Rom, W., Steier, P. and Vycudilik, W. (1998) First ^{14}C results from archaeological and forensic studies at the Vienna Environmental Research Accelerator. *Radiocarbon* 40, 273–281

Wild, E. M., Teschler-Nicola, M., Kutschera, W., Steier, P., Trinkaus, E. and Wanek, W. (2005) First dating of Early Upper Palaeolithic human remains from Mladeč. *Nature* 435, 332–335

Wolpoff, M. H., Hawks, J., Fayer, D. W. and Hunley, K. (2001) Modern human ancestry at the peripheries: A test of the replacement theory. *Science* 291, 293–297

Chapter 8

INVENTORY AND PHOTO-DOCUMENTATION OF THE MLADEČ HOMINID REMAINS

Milford H. Wolpoff, David W. Frayer, Erik Trinkaus and Maria Teschler-Nicola

Table 1. Inventory of the Mladeč hominid remains. This table is the full and complete inventory of the remains found at the site and supplants and replaces other inventories such as Smith (1997), Szombathy (1925) and Vlček (1971). Specimens documented photographically (Plates I–XVIII) are in boldface (NHM = Naturhistorisches Museum Wien, MMB = Moravské zemské muzeum)

Number	Identification	Estimated age	Location in cave	Discovered	Repository
1	**Female cranium**	17	Chamber D, Locus a	1881	NHM
2	**Female cranium**	18	Chamber D, Locus b	1881	NHM
3	**Cranial remains**	2–3	Chamber D, Locus b	1881	NHM
4	Frontoparietal fragment	Adult	Chamber E, Locus e	1922	Destroyed, 1945
5	**Male calotte**	Adult	Quarry Cave	1904	MMB
6	**Male calotte**	Adult	Quarry Cave	1904	Destroyed, 1945
7	**Maxilla**[1]				
8	**Male maxilla: right I^2, left C, M^{1-2}**	Adult	Chamber D, Locus d	1882	NHM
9	**Right maxillary C & P^3**	Adult	Chamber D	1882	NHM
10	**Right maxillary M^3**	Adult	Chamber D, Middle	1882	NHM
11	**Cervical vertebra (C3, C4 or C5)**	Adult	Chamber D, Middle	1882	NHM
12	**Left rib 1**	Adult	Chamber D, Locus b or d	1881 or 2	NHM
13	**Right fragmentary clavicle?**	Adult/ immature	Chamber D, Locus b or d	1881 or 2	NHM
14	**Left fragmentary rib 2 or 3**	Adult	Chamber D, Locus b or d	1881 or 2	NHM
15	**Left fragmentary rib 4, 5 or 6**	Adult	Chamber D, Locus b or d	1881 or 2	NHM
16	**Right fragmentary rib 6, 7, 8 or 9**	Adult	Chamber D, Locus b or d	1881 or 2	NHM
17	**Right fragmentary rib 9 or 10**	Adult	Chamber D, Locus b or d	1881 or 2	NHM

Number	Identification	Estimated age	Location in cave	Discovered	Repository
18	**Right fragmentary rib 11**	Adult	Chamber D, Locus b or d	1881 or 2	NHM
19	**Right fragmentary rib 11 or 12**	Adult	Chamber D, Locus b or d	1881 or 2	NHM
20	**5 Rib fragments**	Adult	Chamber D, Locus b or d	1881 or 2	NHM
21	**Left ilium & ischium fragment**	Adult	Chamber D, Locus d	1882	NHM
22	**Right ilium & ischium fragment**	14–15	Chamber D, Locus d	1882	NHM
23	**Right proximal humerus**	Adult	Chamber D, ?Locus b	1882	NHM
24	**Right humerus diaphysis**	Adult	Chamber D, ?Locus b	1882	NHM
25a	**Right proximal radius**	14–15?	Chamber D, Locus d	1882	NHM
25b	**Right radius diaphysis**	14–15?	Chamber D, Locus d	1882	NHM
25c	**Right proximal ulna**	14–15?	Chamber D, Locus d	1882	NHM
26	**Left radius diaphysis**	Adult?	Chamber D, ?Locus b	1882	NHM
27	**Right femur diaphysis**	Adult	Chamber D, Locus a	1881	NHM
28	**Left proximal femur**	Adult	Chamber D, Locus b or d	1882	NHM
29	**Right tibia distal epiphysis**	< 18	Chamber D, ?Locus b	1881 or 2	NHM
30	**Left talus**	Adult	Chamber D, Locus d	1882	NHM
31	**Right metacarpal 3**	14–21	Chamber D, ?Locus b	1881 or 2	NHM
32	**Left metatarsal 3**	14–21	Chamber D, ?Locus b	1881 or 2	NHM
33	Pelvis (animal bone, excluded)		Chamber D, Locus d	1882	NHM[4]
34	**Cervical vertebra (human?)**		Chamber D, Locus d	1882	NHM
35	"Finger bone"	14–21	Chamber D, ?Locus b	1881 or 2	NHM[2]
36	Metatarsal 5	14–21	Chamber D, ?Locus b	1881 or 2	NHM[2]
37	"Cranial rear"	Young Adult	Chamber E, Locus e	1922	Destroyed, 1945
38	Frontal fragment	Adult	Chamber E, Locus e	1922	Destroyed, 1945
39	**Right parietal fragment**	Adult	Chamber E	1903–1911	MMB
40	**Left occipital/parietal fragment**	Subadult or Young Adult	Chamber E	1903–1911	MMB

Number	Identification	Estimated age	Location in cave	Discovered	Repository
41	**Left occipitotemporal fragment**	Adult	Chamber E	1903–1911	MMB
42	5 Parietal fragments	Adult	Chamber E, Locus e	1922	Destroyed, 1945
43	4 Occipital fragments	Adult	Chamber E, Locus e	1922	Destroyed, 1945
44	Parietal fragments	Subadult?	Chamber E, Locus e	1922	Destroyed, 1945
45	Parietal fragments	Subadult?	Chamber E, Locus e	1922	Destroyed, 1945
46	12 Cranial fragments, mandible, ulna, humerus, radius, and two fibula fragments	Child	Quarry Cave	1904	Possibly destroyed, 1945
47	Left maxilla I^2, P^4–M^2	Subadult	Chamber E, Locus e	1922	Destroyed, 1945
48	Left maxilla M^{1-3} (same as #49)	Adult	Chamber E, Locus e	1922	Destroyed, 1945
49	Right maxilla P^3, M^{1-3}	Adult	Chamber E, Locus e	1922	Destroyed, 1945
50	Right maxillary fragment	Adult	Quarry Cave	1904	Destroyed, 1945
51	Left maxillary fragment	Adult	Quarry Cave	1904	Destroyed, 1945
52	Mandible, left I_2–M_3, right P_4–M_2	Adult	Chamber E, Locus e	1922	Destroyed, 1945
53	Right mandible		Chamber E, Locus e	1922	Destroyed, 1945
54	Mandible, left P_4–M_2, right I_2–M_2	Adult	Quarry Cave	1904	Destroyed, 1945
55	Mandible, I_2–M_1	Adult	Quarry Cave	1904	Destroyed, 1945
56	Canine		Chamber E, Locus e	1922	Destroyed, 1945
57	Premolar		Chamber E, Locus e	1922	Destroyed, 1945
58	Premolar		Chamber E, Locus e	1922	Destroyed, 1945
59	Molar		Chamber E, Locus e	1922	Destroyed, 1945
60	Lumbar vertebra		Chamber E, Locus e	1922	Destroyed, 1945
61	Ilium fragment	Adult	Quarry Cave	1904	Destroyed, 1945
62	Clavicle diaphysis		Chamber E, Locus e	1922	Destroyed, 1945
63	Clavicle fragment		Chamber E, Locus e^3	1922	Destroyed, 1945
64	Clavicle fragment		Chamber E, Locus e^3	1922	Destroyed, 1945
65	Clavicle	Adult	Quarry Cave	1904	Destroyed, 1945

Number	Identification	Estimated age	Location in cave	Discovered	Repository
66	Fragmentary scapula	Adult	Quarry Cave	1904	Destroyed, 1945
67	12 Rib fragments	Adult	Quarry Cave	1904	Destroyed, 1945
68	Humerus fragment		Chamber E, Locus e	1922	Destroyed, 1945
69	Humerus distal fragment		Chamber E, Locus e	1922	Destroyed, 1945
70	Humerus diaphysis	Adult	Quarry Cave	1904	Destroyed, 1945
71	Humerus diaphysis	Adult	Quarry Cave	1904	Destroyed, 1945
72	Radius fragment	Adult	Quarry Cave	1904	Destroyed, 1945
73	Ulna proximal fragment	Adult	Quarry Cave	1904	Destroyed, 1945
74	Right femur diaphysis		Chamber E, Locus e	1922	Destroyed, 1945
75	Right femur diaphysis		Chamber E, Locus e	1922	Destroyed, 1945
76	3 Femur diaphysis fragments		Chamber E, Locus e	1922	Destroyed, 1945
77	Left femur distal fragment		Chamber E, Locus e	1922	Destroyed, 1945
78	Most of left femur and right diaphysis fragment	Adult	Stone Quarry	1904	Destroyed, 1945
79	Right tibia proximal fragment		Chamber E, Locus e	1922	Destroyed, 1945
80	Tibia proximal epiphysis		Chamber E, Locus e	1922	Destroyed, 1945
81	Left tibia diaphysis		Chamber E, Locus e	1922	Destroyed, 1945
82	Tibia fragments		Chamber E, Locus e	1922	Destroyed, 1945
83	Tibia fragments		Chamber E, Locus e	1922	Destroyed, 1945
84	Right fragmentary tibia	Adult	Quarry Cave	1904	Destroyed, 1945
85	3 Fibula fragments		Chamber E, Locus e	1922	Destroyed, 1945
86	Fibula fragments	Adult	Quarry Cave	1904	Destroyed, 1945
87	12 Metacarpals	Adult	Quarry Cave	1904	Destroyed, 1945
88	**Left hand proximal phalanx 2**	Adult	Chamber E	1903–1911	MMB
89	**Left hand proximal phalanx 3**	Adult	Chamber E	1903–1911	MMB
90	**Right metacarpal 3**	Adult	Chamber E	1903–1911	MMB

Number	Identification	Estimated age	Location in cave	Discovered	Repository
91	**Right metacarpal 4**	Adult?	Chamber E	1903–1911	MMB
92	Phalanges (damaged)		Chamber E, Locus e	1922	Destroyed, 1945
93	Hand phalanges	Adult	Quarry Cave	1904	Destroyed, 1945
94	Right zygomatic	Adult	Quarry Cave	1904	Destroyed, 1945
95	Femur fragment	Adult	Quarry Cave	1904	Destroyed, 1945
96	Tibia fragment	Adult	Quarry Cave	1904	Destroyed, 1945
97	Fibula fragments	Adult	Quarry Cave	1904	Destroyed, 1945
98	Phalanges	Adult	Quarry Cave	1904	Destroyed, 1945
99	Calcaneus	Adult	Quarry Cave	1904	Destroyed, 1945
100	Rib fragments	Adult	Quarry Cave	1904	Destroyed, 1945
101	Scapula	Adult	Quarry Cave	1904	Destroyed, 1945
102	**Left femur diaphysis**	Subadult	Chamber D	1881/82	NHM

[1] Joined to Mladeč 2
[2] Not found in the 1961 inventory, presumably lost
[3] 2 meters northeast of Locus e
[4] Specimen is non-hominid

References

Smith, F. H. (1997) Mladeč. In (F. Spencer, Ed.) *History of physical anthropology.* An encyclopedia. New York: Garland, pp. 659–660

Szombathy, J. (1925) Die diluvialen Menschenreste aus der Fürst-Johanns-Höhle bei Lautsch in Mähren. *Die Eiszeit* 2, 1–34, 73–95

Vlček, E. (1971). Czechoslovakia. In (K. P. Oakley, B. G. Campbell and T. I. Molleson, Eds.) *Catalogue of fossil hominids, part II: Europe.* London: British Museum (Natural History), pp. 47–64

Plates I–XVIII (scales are all in centimeters)

Plate I.
Mladeč 1: Cranium
Top left: frontal view; top right: occipital view; center left: right lateral view; center right: left lateral view; bottom left: superior view; bottom right: basal view

Plate II.
Mladeč 2: Cranium
Top left: frontal view; top right: occipital view; center left: right lateral view; center right: left lateral view; bottom left: superior view; bottom right: basal view

Plate III.
Mladeč 3: Cranial fragments
(a) frontal bone; left: right lateral view; center: frontal view; right: left lateral view; **(b)** parietal fragments; **(c)** occipital and right temporal bone; center: occipital view; left: right lateral view; right: left lateral view

Plate IV.
Mladeč 5: Calotte
Top left: frontal view; top right: occipital view; center left: right lateral view; center right: left lateral view; bottom left: superior view; bottom right: basal view

Plate V.
Mladeč 6: Calotte (this photos are of a cast)
Top left: frontal view; top right: occipital view; center left: right lateral view; center right: left lateral view; bottom left: superior view; bottom right: basal view

Plate VI.
Mladeč 8: Maxilla
Top left: frontal view; top right: posterior view; center left: right lateral view; center right: left lateral view; bottom left: superior view; bottom right: palatal view

Plate VII.
(a) Mladeč 9a: Right maxillary C
Left to right: buccal, lingual, distal, mesial and occlusal view
(b) Mladeč 9b: Right maxillary P^3
Left to right: buccal, lingual, distal, mesial and occlusal view
(c) Mladeč 10: Right maxillary M^3
Left to right: buccal, lingual, distal, mesial and occlusal view

Plate VIII.
(a) Mladeč 11: Cervical vertebra (C3, C4 or C5)
Top left: cranial view, top right: caudal view; center left: ventral view, center right: dorsal view; bottom left: right lateral view; bottom right: left lateral view
(b) Mladeč 34: Vertebral body
Top left: cranial view, top right: caudal view; center left: ventral view, center right: dorsal view; bottom left: right lateral view; bottom right: left lateral view

Plate IX.
(a) Mladeč 12: Left rib
Top: cranial view; bottom: caudal view
(b) Mladeč 13: Right clavicle
Top: cranial view; center: caudal view; bottom: ventral view
(c) Mladeč 14: Left fragmentary rib 2 or 3
Top: ventral view; center: cranial view; bottom: dorsal view
(d) Mladeč 15: Left fragmentary rib 4, 5 or 6
Top: ventral view; center: cranial view; bottom: dorsal view
(e) Mladeč 16: Right fragmentary rib 6, 7, 8 or 9
Top: dorsal view; center: cranial view; bottom: caudal view

Plate X.
(a) Mladeč 17: Right fragmentary rib 9 or 10
Top: external view; center: cranial view; bottom: caudal view
(b) Mladeč 18: Right fragmentary rib 11
Top: external view; center: cranial view; bottom: caudal view

Plate XI.
(a) Mladeč 19: Right fragmentary rib 11 or 12
Top: external view; center: cranial view; bottom: ventral view
(b) Mladeč 20a: Rib fragment
Top: external view; center: cranial view; bottom: ventral view
(c) Mladeč 20b: Rib fragment
Top: external view; center: cranial view; bottom: ventral view
(d) Mladeč 20c: Rib fragment
Top: external view; center: cranial view; bottom: ventral view
(e) Mladeč 20d: Rib fragment
Top: external view; center: cranial view; bottom: ventral view
(f) Mladeč 20e: Rib fragment
Top: external view; center: cranial view; bottom: ventral view

Plate XII.
(a) Mladeč 21: Left ilium and ischium fragment
Left: ventral view; right: dorsal view
(b) Mladeč 22: Right ilium and ischium fragment
Left: ventral view; center: dorsal view; right: lateral view

Plate XIII.
(a) Mladeč 23: Right proximal humerus
From left to right: anterior, lateral, posterior and medial view
(b) Mladeč 24: Right humerus diaphysis
From left to right: anterior, lateral, posterior and medial view

Plate XIV.
(a) Mladeč 25a: Right proximal radius
From left to right: anterior, lateral, posterior and medial view
(b) Mladeč 25b: Right radius diaphysis
From left to right: anterior, lateral, posterior and medial view

(c) Mladeč 25c: Right proximal ulna
From left to right: anterior, lateral, posterior and medial view
(d) Mladeč 26: Left radius diaphysis
From left to right: anterior, lateral, posterior and medial view
(e) Mladeč 28: Left proximal femur
From left to right: anterior, lateral, posterior and medial view

Plate XV.
Mladeč 27: Right femur diaphysis
From left to right: anterior, lateral, posterior and medial view

Plate XVI.
(a) Mladeč 29: Right tibia distal epiphysis
Top left: anterior view; top right: posterior view
Center left: lateral view; center right: medial view
Bottom left: superior view; bottom right: caudal view
(b) Mladeč 30: Left talus
Top left: anterior view; top right: posterior view
Center left: lateral view; center right: medial view
Bottom left: superior view; bottom right: caudal view
(c) Mladeč 31: Right metacarpal 3
From left to right: dorsal, lateral, volar and medial view
(d) Mladeč 32: Left metatarsal 3
From left to right: dorsal, lateral, plantar and medial view

Plate XVII.
(a) Mladeč 39: Right parietal fragment
Left: ectocranial view; right: endocranial view
(b) Mladeč 40: Left occipital/parietal fragment
Left: ectocranial view; right: endocranial view
(c) Mladeč 41(?): Left occipitotemporal fragment
Left: ectocranial view; right: endocranial view

Plate XVIII.

(a) Mladeč 88: Left hand proximal phalanx 2
Left: dorsal view; right volar view
(b) Mladeč 89: Left hand proximal phalanx 3
Left: dorsal view; right volar view
(c) Mladeč 90: Right metacarpal 3
Left: dorsal view; right plantar view
(d) Mladeč 91: Right metacarpal 4
Left: dorsal view; right plantar view
(e) Mladeč 102: Left femur fragment
From left to right: anterior, lateral, posterior and medial view

PLATE I

PLATE II

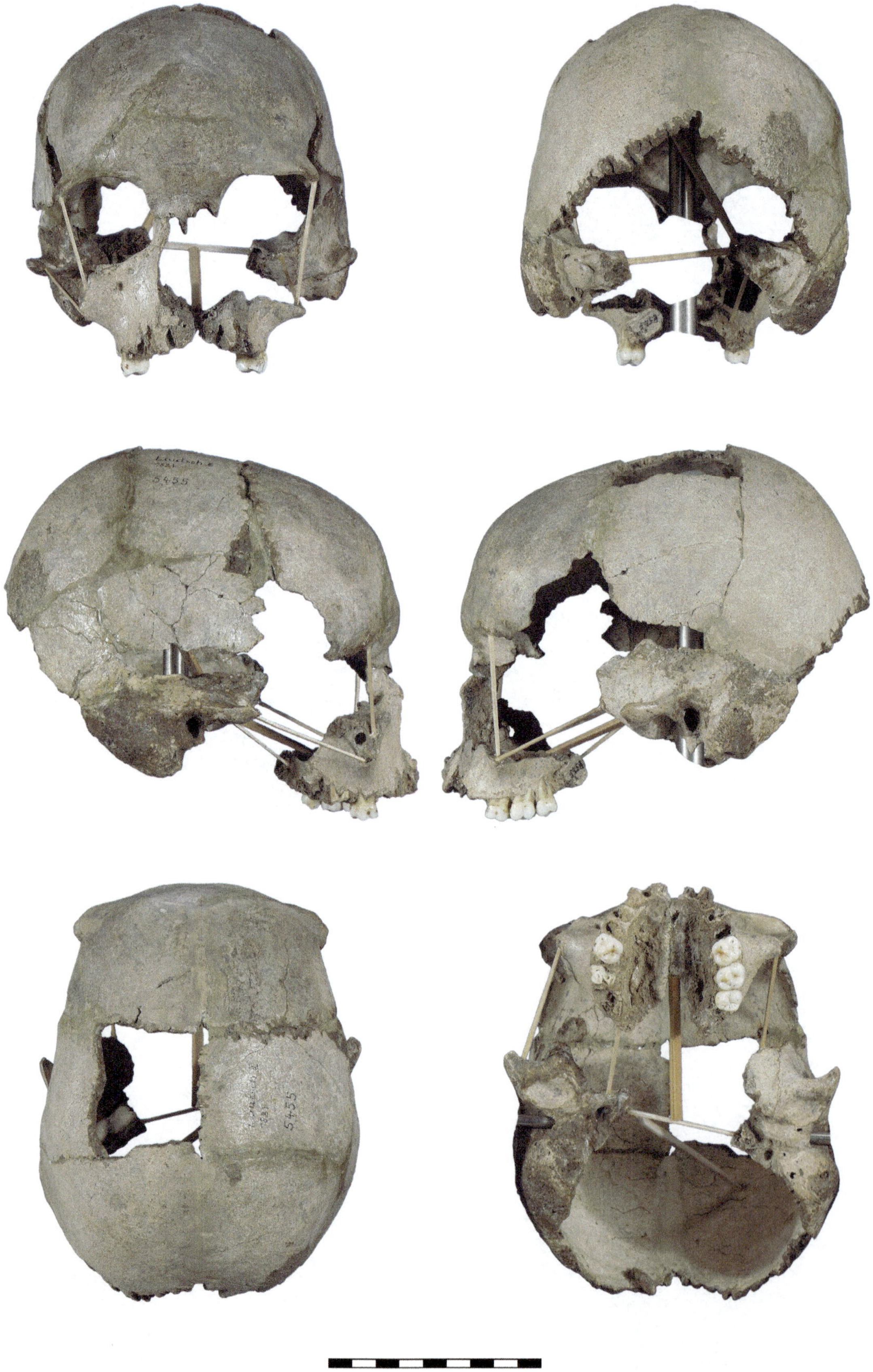

PLATE III

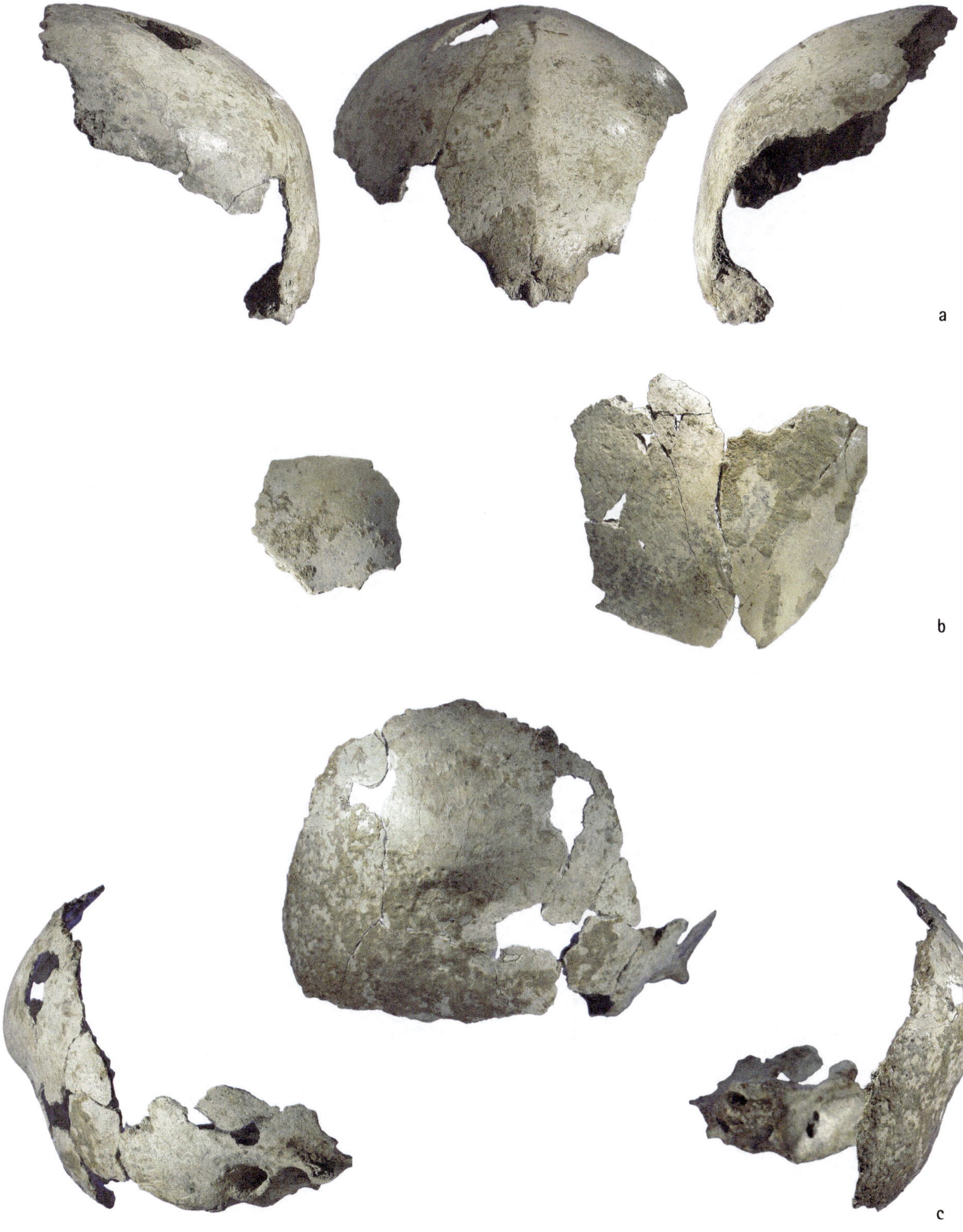

PLATE IV

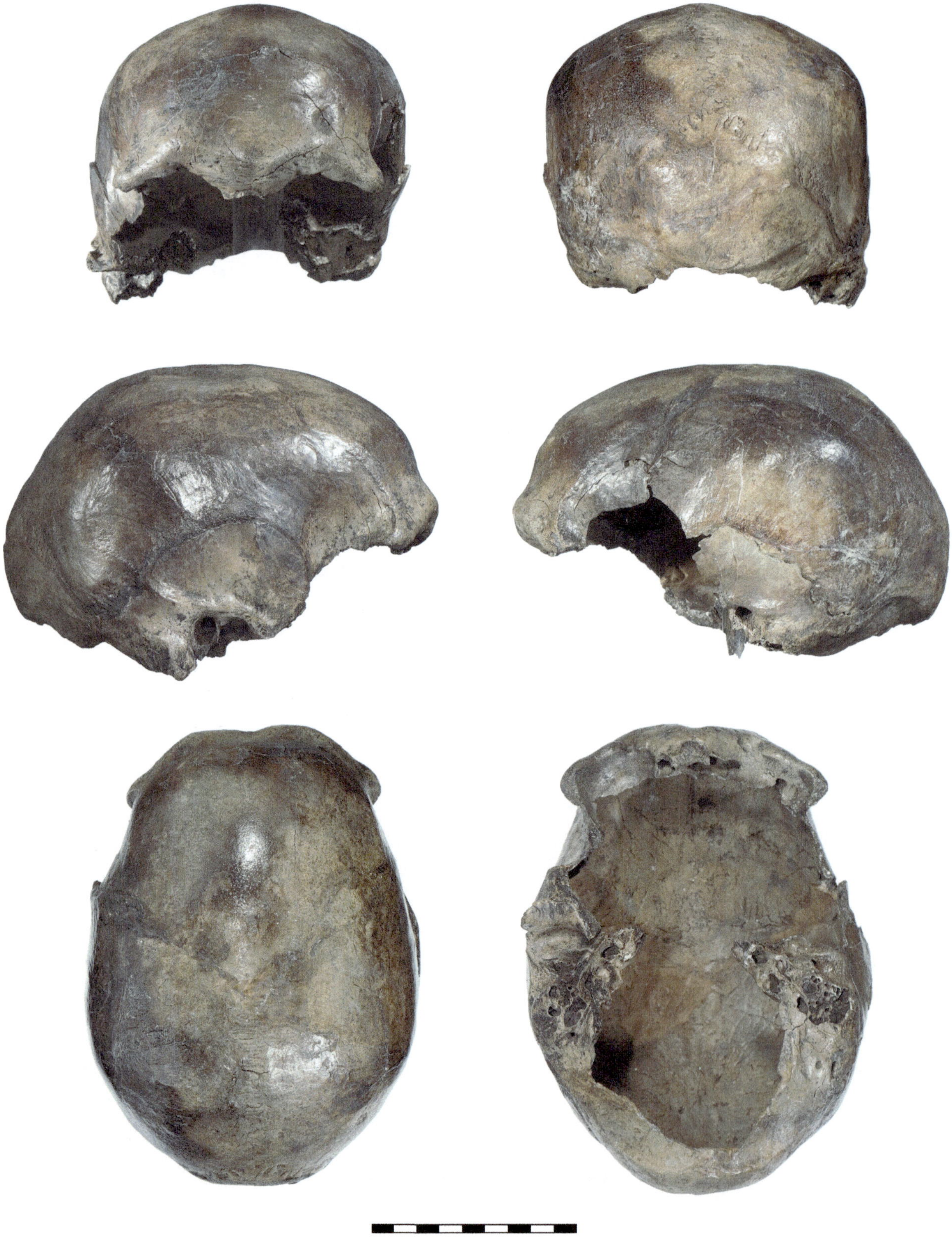

PLATE V

PLATE VI

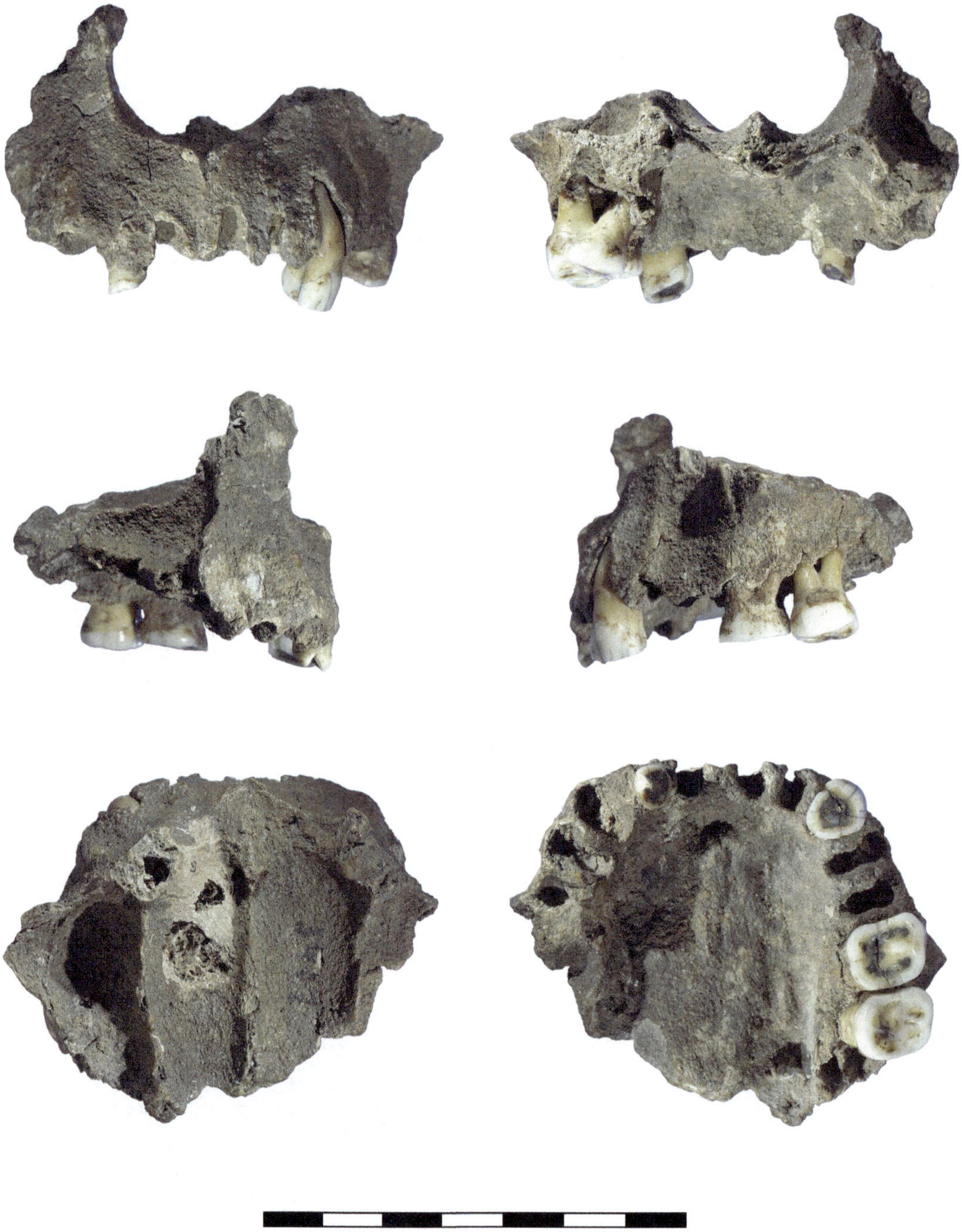

PLATE VII

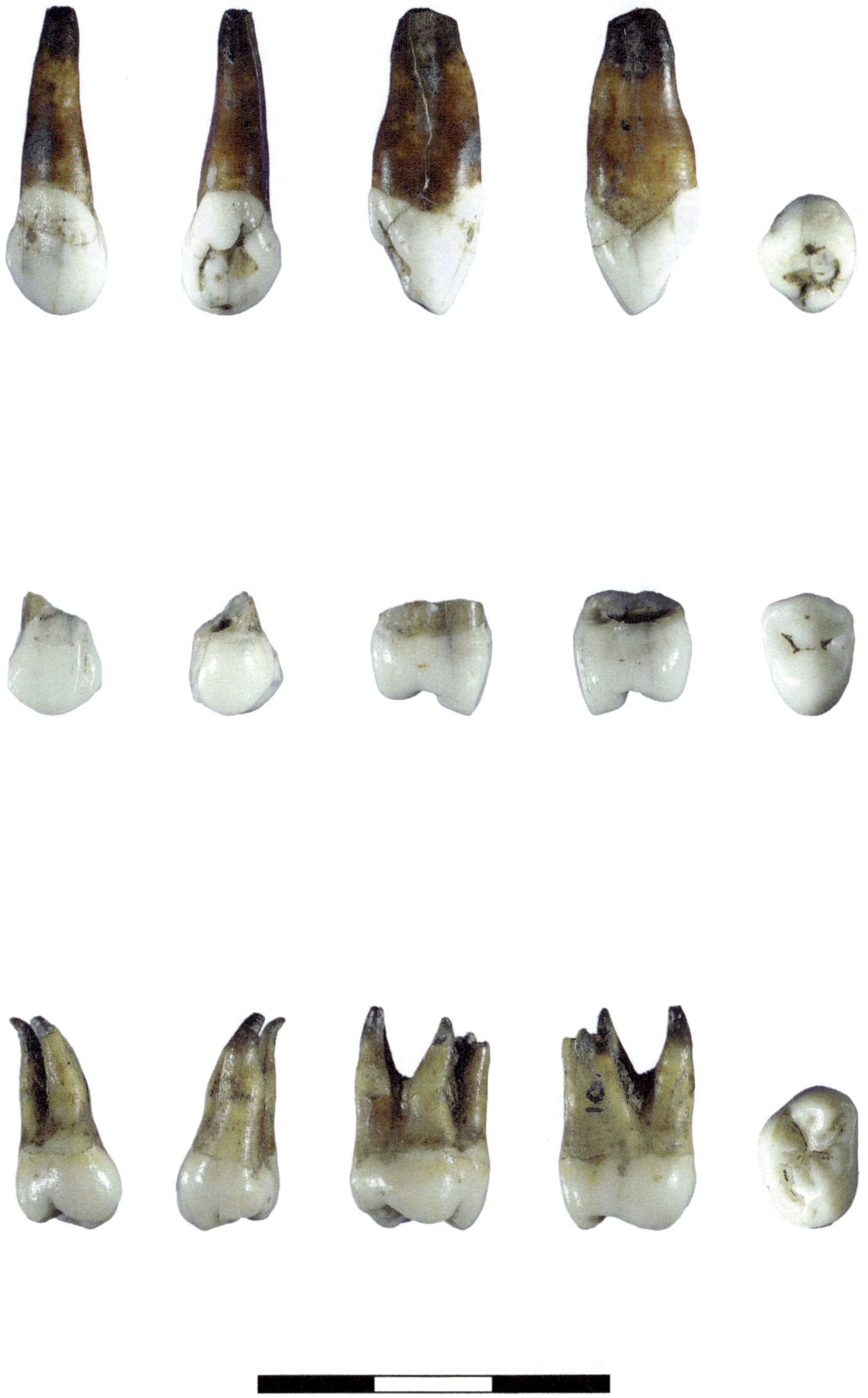

a

b

c

PLATE VIII

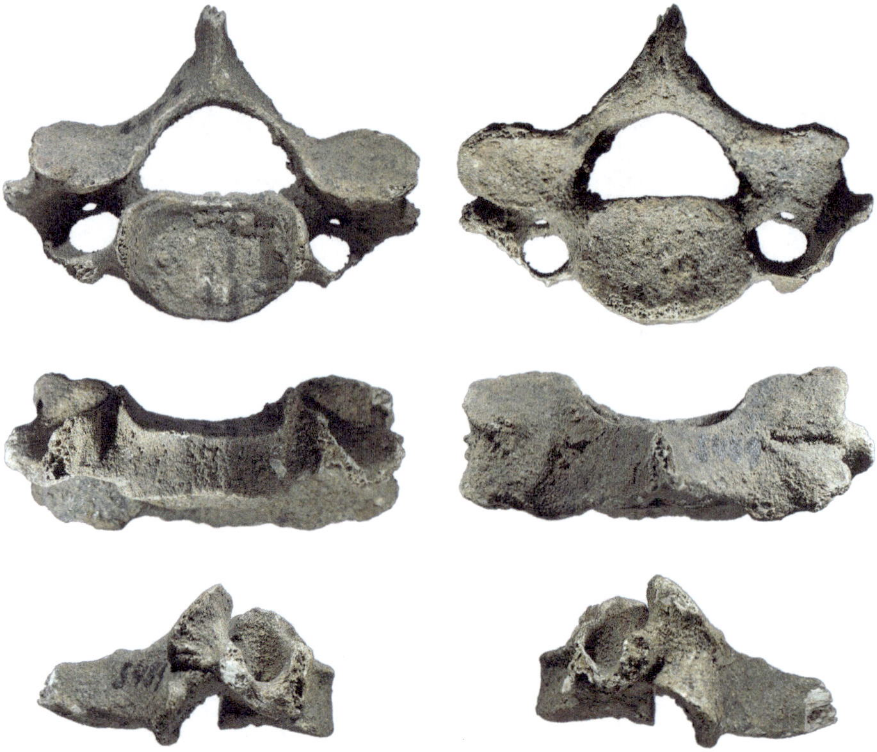

a

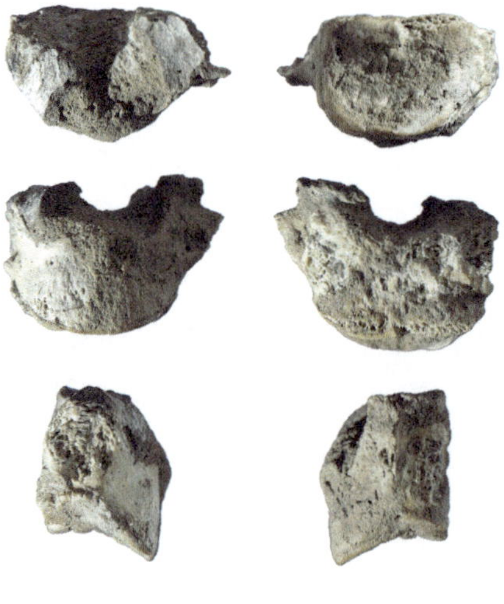

b

PLATE IX

a

b

c

d

e

PLATE X

a

b

PLATE XI

PLATE XII

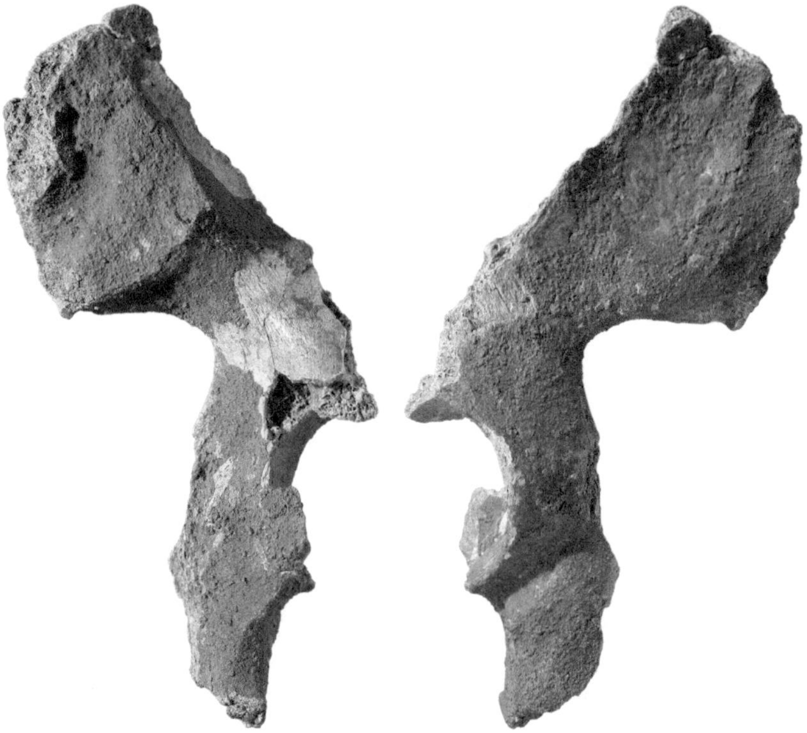

a

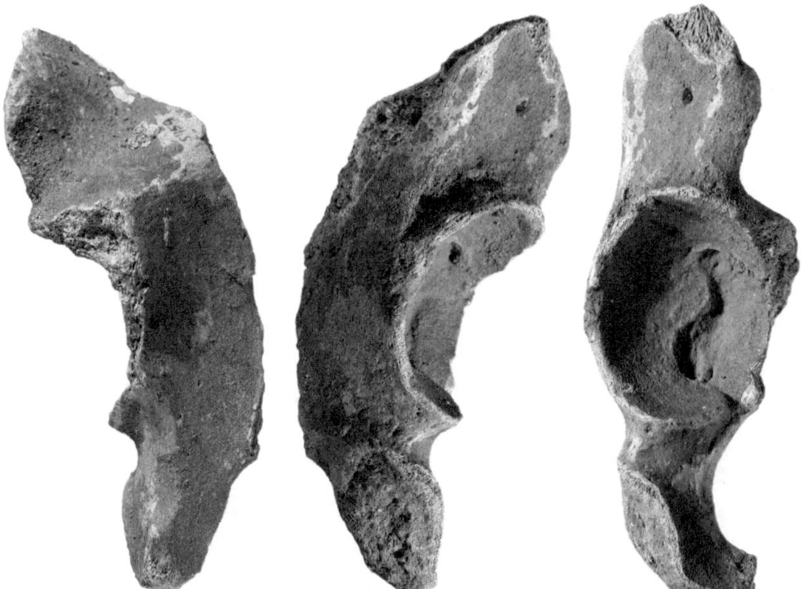

b

PLATE XIII

a

b

PLATE XIV

a

b

c

d

e

PLATE XV

PLATE XVI

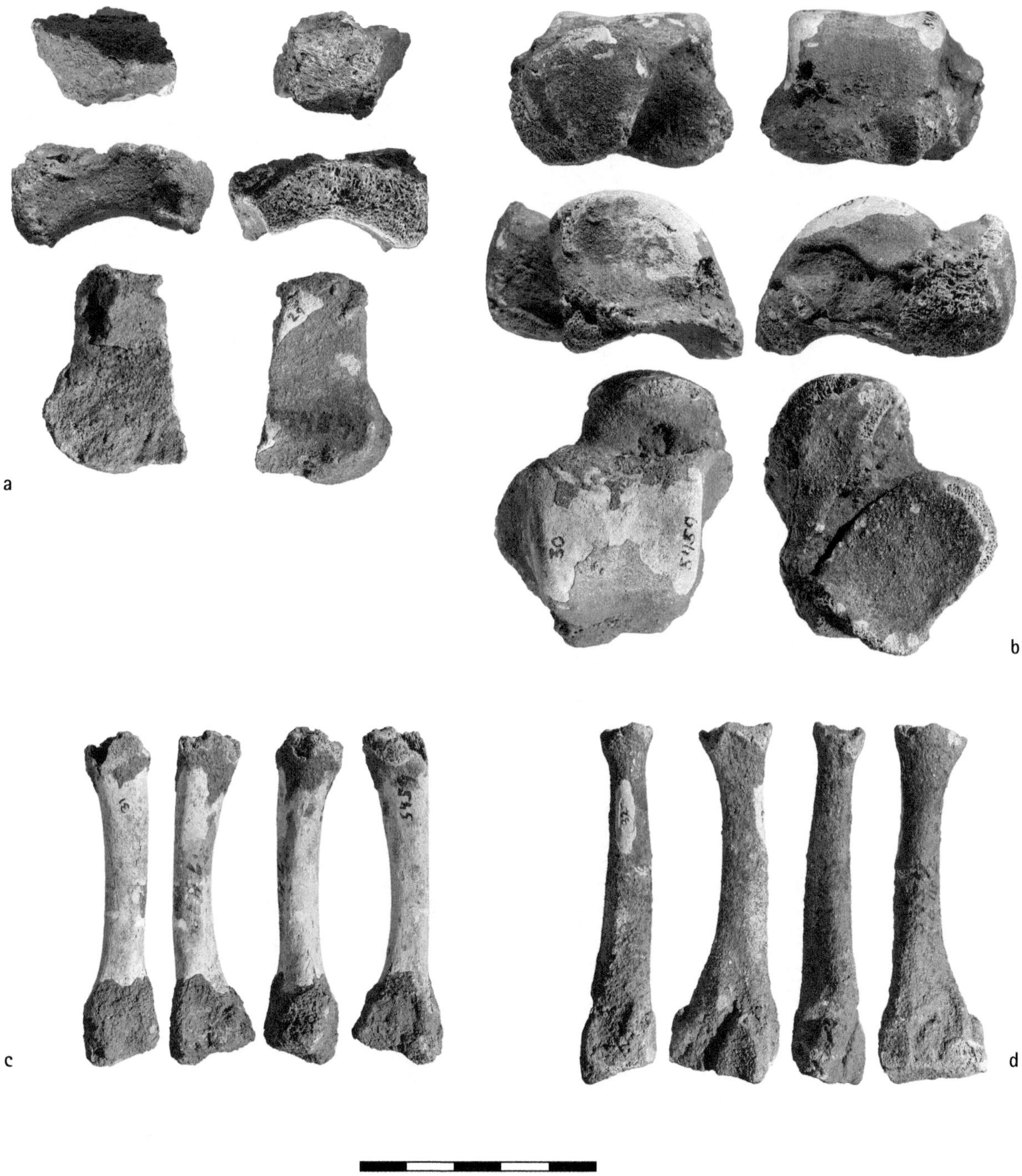

PLATE XVII

a

b

c

PLATE XVIII

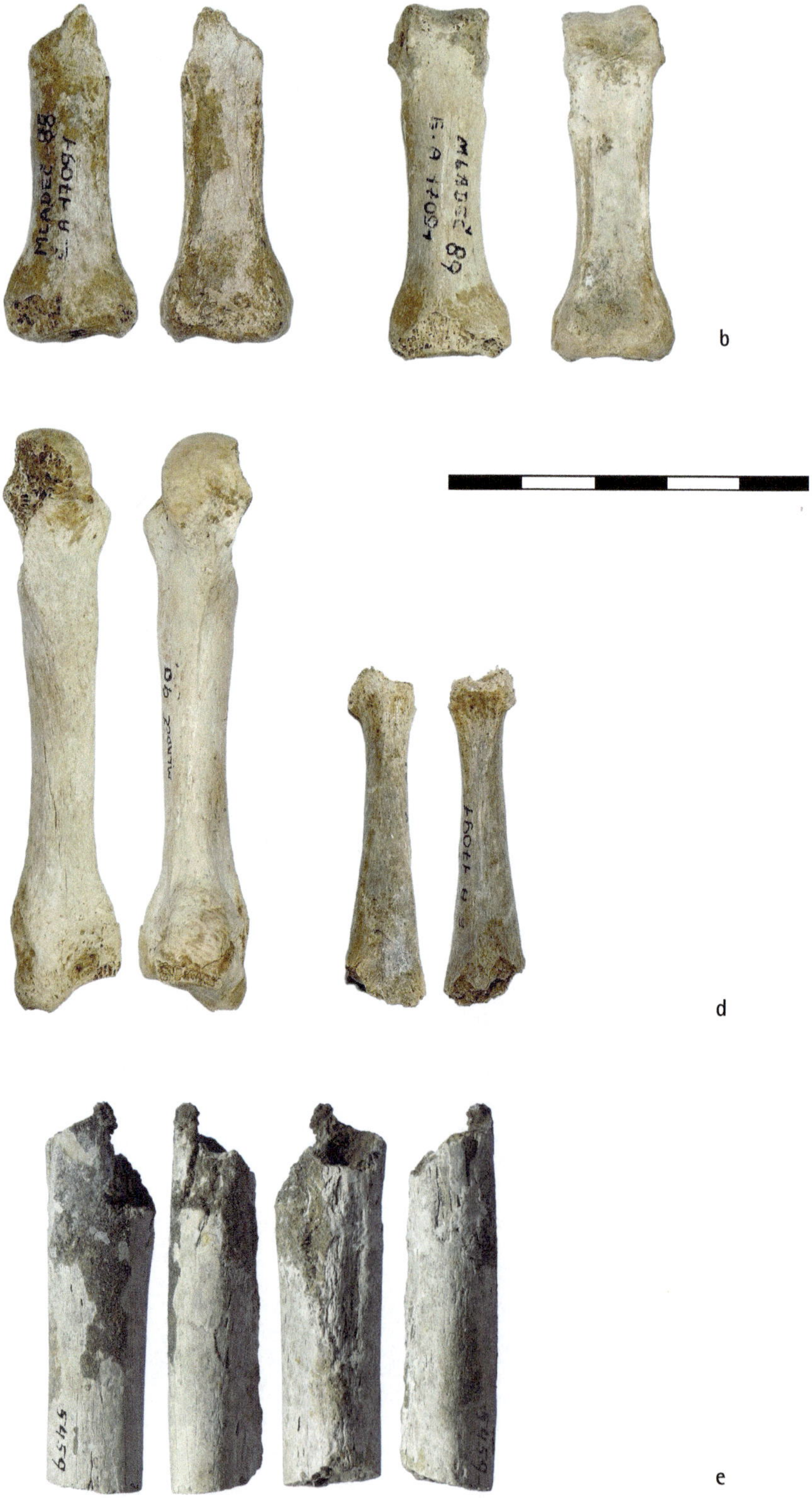

a b c d e

Chapter 9

AURIGNACIAN MALE CRANIA, JAWS AND TEETH FROM THE MLADEČ CAVES, MORAVIA, CZECH REPUBLIC

David W. Frayer, Jan Jelínek, Martin Oliva and Milford H. Wolpoff

Introduction

Fig. 1. Josef Szombathy, at about the time he first visited the Mladeč Caves

In 1925 Josef Szombathy (1853–1943) published a full, and for its time, relatively complete account of the excavations, geology, paleontology, archaeology, and anthropology of the Mladeč Caves[1]. It is unclear how Szombathy learned of Mladeč but in 1881 and 1882 he was commissioned by the Vienna Academy of Sciences to conduct exploratory research in the caves (Fig. 1.). The property was then owned by Prince Johann von and zu Liechtenstein, who as Szombathy commented, provided some "meager" financial support to run the excavations. The days Szombathy spent there were devoted to mapping the Main Cave and putting in test excavations, primarily in an area Szombathy called the "Dome of the Dead." As luck would have it, his excavations, though intended to be preliminary, produced major collections of human remains and prehistoric artifacts. Szombathy identified the locus of some discoveries, made a sketch of the vertical stratigraphy, and saved a great deal of the excavated material. After completing his work at the Main Cave in the late 19th century, all of the human remains and archaeological materials, and all of the faunal materials were brought to the Naturhistorisches Museum Wien, where they still reside.

Szombathy returned to Mladeč twice more, once in 1904 to study some of the new discoveries by Knies and Smyčka in the Quarry Cave and again in 1925 to examine new specimens excavated from the Main Cave. On the last trip, it seems he was expecting to have this material transferred to him in Vienna, but he was only allowed to study them in Litovel. He wrote[2] "it was impossible for me, however, to undertake intensive investigation [...] because I was equipped only with my traveling tool kit" (1925, 73).

1 In local records and also in the scientific literature we find four names in connection with these caves. First, Szombathy calls the caves "Fürst Johanns Höhle", as the entrance of the caves lay on the property of Prince Johann of Liechtenstein who supported Szombathy's excavations. Second, the German name for Mladeč, Lautsch, was also introduced by Szombathy and has become the name of this site in some publications. Third, the local Czech people called the caves "Bočkova díra" (Boček's Hole), after a legendary highwayman who once lived somewhere in the Třesín forests. The final and generally accepted name is Mladeč Caves, after Mladeč, the village in which the caves are located. "Mladeč" is pronounced "Mladitch".

2 This, and all other citations from Szombathy, have been translated from the original German.

Some 40 years after Szombathy began his work at Mladeč he published the geological, archaeological, faunal and anthropological analysis in Bayer's new journal, *Die Eiszeit* (Szombathy, 1925). Besides his own work, he drew upon the results of the excavations in the early 1900's. Until now, this has remained the only comprehensive treatment of the Mladeč Cave and its human remains, although various details have been published by Jelínek (1976; 1983), Frayer (1986; 1997; Frayer et al., 1993), Smith (1983; 1984; 1985; 1997), Caspari (1991), and Wolpoff (1982; 1989a; 1989b; 1999).

In this paper, we provide the first full description and comparative analysis of the identifiably adult male[3] remains from Mladeč. This is the first of two papers on Mladeč, the second focused on the adult female crania (Wolpoff et al., this volume). Our choice of adult males for the first analysis is dictated by the phylogenetic hypotheses described below. We contend that comparisons must be made holding sex constant, especially because the sex ratios differ substantially in Mladeč and the two potentially ancestral samples, European Neandertals and Skhul/Qafzeh. The latter has an inadequate number of female crania complete enough for reliable comparisons to both samples (Wolpoff, 1999). Thus, a similar analysis of the female remains will not be able to clearly address issues of ancestry.

We present certain details about the history of discoveries, the provenience and numbering of all the specimens (many of which no longer exist), the associated industry, and the age of the site, in this first publication. Our information about the cave, its excavation, and the specimens that no longer exist comes from a number of sources as detailed below. For a variety of reasons we rely most heavily on Szombathy's work (1925; 1926). These include the completeness of Szombathy's writings, the fact that he alone witnessed and recorded many of the critical events surrounding the excavations in the Main Cave which has now been stripped of its Upper Paleolithic layers to facilitate tourism, and the tragic destruction of many of the human remains (Table 1) during the closing days

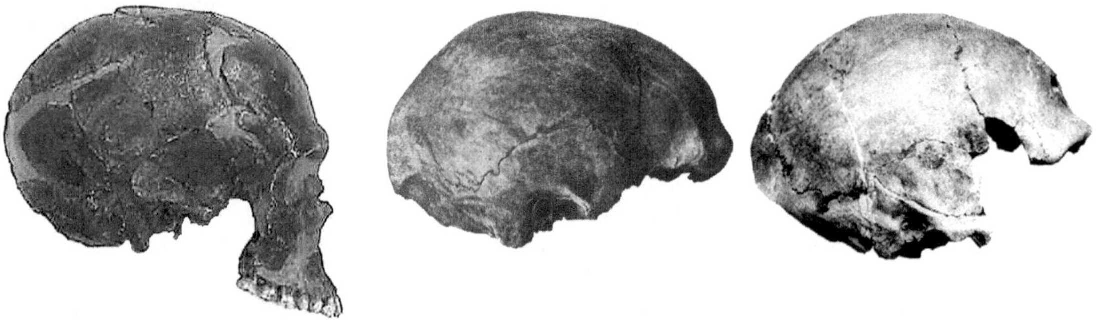

Fig. 2. Mladeč 5 (center) compared with Qafzeh 9 (cast, left) and Spy 2, in lateral view. The three specimens shown here are males, and Mladeč is unquestionably the youngest of the three. In this paper we examine hypotheses of ancestry for the Mladeč remains. Were populations represented by both of these earlier specimens ancestral to Mladeč, or does its ancestry lie uniquely in one of them? We question whether one of these earlier samples can be excluded from ancestry.

3 Our identification of male individuals is based on size and robustness, and the presence of any features diagnostic for sex. It is bound to be conservative for the crania because the age of many fragmentary specimens is unknown, and juvenile males often resemble females for these features. Of the 11 incomplete adult Mladeč crania that can be diagnosed for sex, we believe that five are most probably male. We have been somewhat less conservative in diagnosing the postcranial specimens. Here we accepted any evidence of larger size or greater robustness as suggesting a male status. On the whole, with the exception of the largest and most robust remains, we believe our postcranial designations are less certain than they are for the cranial materials.

of the Second World War. Despite these problems, Mladeč has played prominently in the modern human origins debate because the specimens address critical details about the fate of the earlier European Neandertal populations.

Given its age, Aurignacian affiliations, and Central European locale, we entertain three hypotheses about the ancestry of the Mladeč males (Fig. 2), which by inference are about the ancestry of all the Mladeč folk:

- European Neandertal populations are uniquely ancestral to Mladeč.
- "Early modern" populations penecontemporary with the Neandertals, in particular Skhul and Qafzeh, are uniquely ancestral to Mladeč.
- *Both* European Neandertals *and* the penecontemporary Skhul and Qafzeh population are ancestors of Mladeč.

Neandertals and Skhul / Qafzeh[4] are the two large samples that are immediately earlier than Mladeč and geographically most proximate. Hypothesizing ancestries involving one, the other, or both are not only the usual suggestions made about the evolutionary issue, but also correspond to the most common hypotheses about the origin of the European Upper Paleolithic. Sole descent from Neandertals is a polygenic statement that implies the independent evolution of humans in different regions. Descent from Skhul / Qafzeh is a prediction of the Eve theory, though its valid demonstration is not necessarily proof of this theory because it is also compatible with multiregional evolution. Descent from both is uniquely compatible with multiregional evolution and its demonstration would be a refutation of the Eve theory.

4 In fact, Skhul / Qafzeh is the *only* large sample of putative "early moderns" who are earlier. Our comparisons with this sample are as much dictated by necessity as by the logic of testing the Eve theory. However all other explanations of European origins must examine the same crania and postcranial remains, because these are all there are. Further, while we here, and others, treat this sample as representing Africans (Klein, 2001), notes that for much of this time the Levant can be considered an ecological part of Africa), the anatomical relationship to Africans is not compelling for more than one or two specimens, others resemble Asians, or have no specific resemblances. Therefore, the hypothesis tested with the Skhul / Qafzeh remains is compatible with the Eve theory but is not a hypothesis of unique African origins for Europeans.

Background

Setting

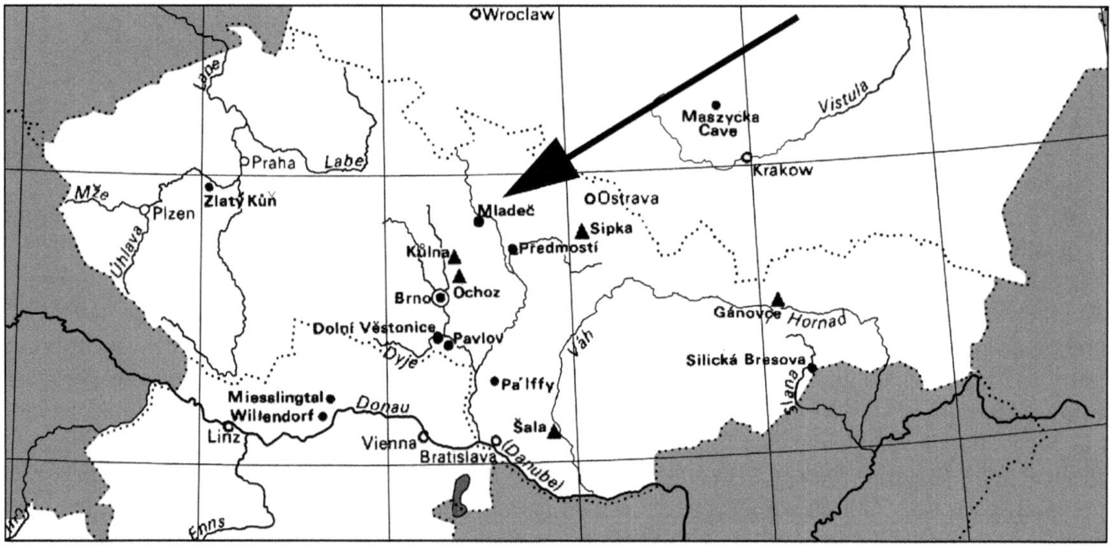

Fig. 3. Map of Central European Late Pleistocene sites, modified after Oakley et al. (1971, Map 1)

The Mladeč Caves are located in a hilly region of Moravia near the northern rim of the Pannonian basin (Fig. 3). The caves themselves are in the Třesín Hill, a flat limestone knoll 4 km northwest of Litovel on the outskirts of the village of Mladeč in central Moravia, Czech Republic. The knoll overlooks the flood plain of the Morava River to the northeast, the main artery connecting north and south Moravia. Geologically Třesín Hill and the Mladeč Caves form part of the isolated central Moravian Devonian limestone rock island. On the south, the subterranean stream Hradečka, a tributary of the Morava River, borders the Třesín Hill.

The present entrance to Mladeč Caves is on the southern slope of Třesín Hill in a landscape of cliffs protruding from Upper Pleistocene loess drifts. The cave proper is formed from a complicated labyrinth of underground spaces and corridors. Intense karstification of the local limestone resulted in a large number of chambers, most of them filled with sediments (Horacek and Lozek, 1984). Svoboda (2000) provides key details of the depositional history and a reconstruction of these caves, as they were when the first specimens were found.

From the southeast, the Třesín Hill can be seen from a great distance as a projecting, partially wooded promontory. Bare cliffs originally formed its southern and southeastern sides but a thick layer of Late Paleolithic loess now covers most of them. Numerous archaeological finds, ranging from the Paleolithic to the present, document the attractiveness of the highly visible Třesín Hill. In the Paleolithic, hunters had an excellent overlook of the broad Morava River valley. In later prehistoric periods the strategic importance of the locality continued as it controlled the passage along the Morava River to the north. The top of the hill, partly under cultivation, is covered with loess. At the beginning of the century Upper Paleolithic stone tools (Gravettian) were found in these sediments. The Podkova (Horseshoe) Cave, which yielded some fauna and Magdalenian stone tools, is also in the Třesín hill, on the northern side just above the flood plain.

There are several other well-known late Middle Paleolithic sites in the region nearby. The Šipka Cave, an important Middle Paleolithic locality noted for the early discovery of a human mandibular fragment, is situated 48 km to the east of Mladeč. The Middle Paleolithic site, Kůlna Cave, is on the northern margin of the Moravian karst, 30 km to southwest. This site yielded a partial maxilla and parietal fragment. There are also several nearby Upper Paleolithic localities; the most important of them, Předmostí with its mass grave and numerous individuals, is about 30 km to the southeast.

History of excavations

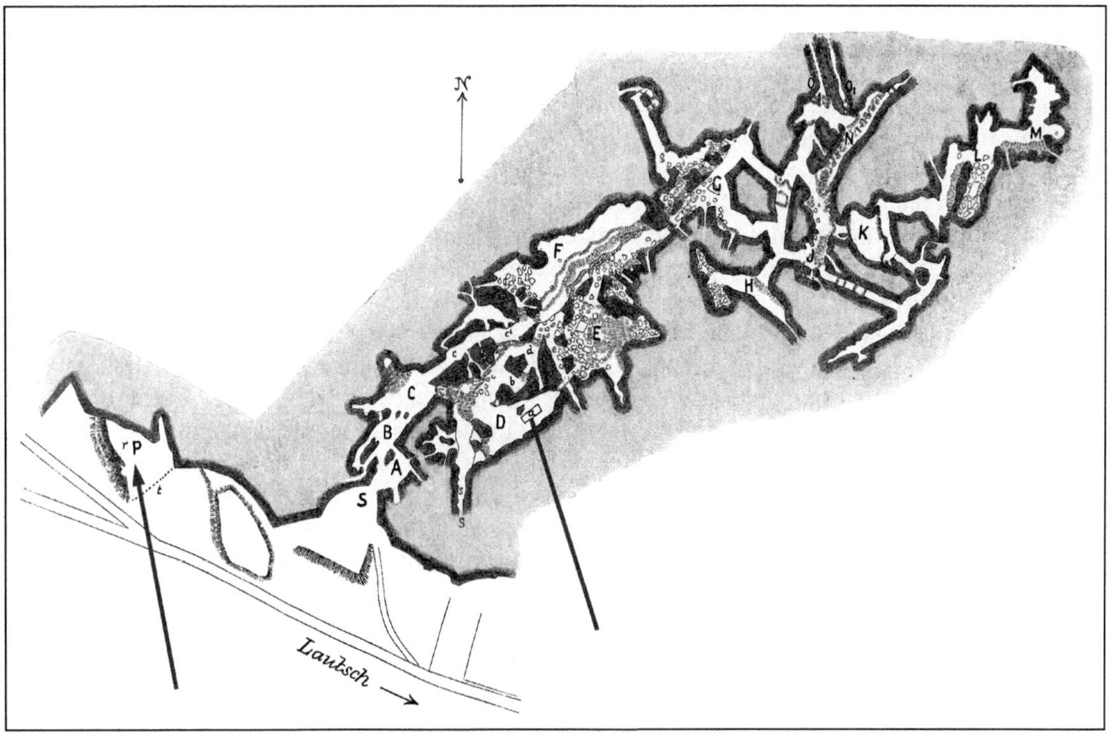

Fig. 4. Plan of the Mladeč Cave floor, after Szombathy (1925), with the direction of the nearby village indicated on the road. The main chambers are indicated by capital letters, and the findspots of the various human remains are designated by letters a, b, e, and r. The positions of the first specimen (a, Mladeč 1) and of the triple burial in the stone Quarry Cave (r, in Chamber P) are marked.

In June 1881, Szombathy excavated a trench in the Main Cave at Mladeč, marked in his plan as "a." (Fig. 4, keyed to Szombathy's locality designations as reported in this paper). Today there is an area in the cave called "The site of Szombathy's find", but it does not correspond to the real findspot, which was a few meters more to the east, near a large rock pillar. In the 1925 monograph, Szombathy wrote that he was digging in undisturbed layers, and he provided a sketch of the stratigraphic situation. Below the surface in a layer with some fragmentary fossil animal bones, mainly of *Bos primigenius,* he found a human skull (Mladeč 1) at the depth of 20–30 cm. In his description of the excavation he wrote (1925, 6):

> *"Attached by limestone deposits to the zygomatic of skull was a piece of reindeer rib 10 cm long and a small piece of bone of unknown origin. Directly next to the skull in undisturbed cave loam were fragments of reindeer and bovine bones. No other human skeletal remains*

were found at this location except for the diaphysis of a slender femur [Mladeč 27][5] which had its ends removed in ancient times since the broken edges are just as covered with limestone deposits as the rest of it. The skull cannot be closely related to the somewhat deeper (35 cm) hearth because of a layer of limestone at 30 cm depth, which separates the two, and which must have taken some time to form.
In the test areas "b" and "d" also dug in the first year (1881) diluviale mammal bones were found from the surface down to a depth of 60 cm. From the area "b" I could identify the following animals: Rangifer tarandus, Bos primigenius, Ursus spelaeus, Lupus spelaeus, and Vulpus vulgaris fossilis. Amongst them were various remains of human skeletons [...]. The human skeletal remains include large fragments of the vault of two different crania [Mladeč 2 and 3], two halves of the upper jaw [Mladeč 7, now known to be part of Mladeč 2], and a few pieces of ribs [part of the set we cataloged as Mladeč 12–20] and extremities [probably some or all of Mladeč 23, 24, 26, 29, 31, 32, 35, and 36]".

The human remains from the 1882 excavation were from the same layer in the Main Cave and include a robust maxilla (Mladeč 8), a maxillary canine and premolar (Mladeč 9) and a molar (Mladeč 10), and a number of fragmentary postcranial remains (see chap. 8, Table 1). Besides these anthropological finds, locus "b" yielded also some archaeological finds, namely bone tools, stone artifacts, and a collection of perforated animal teeth (a total of 22 pieces, and 2 unidentifiable fragments), very probably forming part of one or more necklaces. In these teeth, the single bear canine is relatively small, belonging to *Ursus arctos*. The collection also includes a canine of a wolf, a horse incisor, ten reindeer incisors, nine beaver incisors and two other unidentifiable tooth fragments. The collection of bone tools includes a thin slightly arched point made of a *Bos or Bison* rib, with its tip broken off, two base fragments of similar bone points, and seven other fragments of points of this type. Szombathy reports only two chert artifacts from here, a non-retouched bladelet and a burin.

The entrance to the Mladeč Caves was situated on the property of the Prince of Liechtenstein, while the caves themselves were under the fields of A. Nevrlý, a Mladeč villager. The caves under his field were ceded to him, according to Szombathy, and in 1902 he erected a wall to separate the entrance on the Liechtenstein property from the caves, and opened a new entrance from his fields. Jan Knies, a teacher and amateur archaeologist, started excavations in the Main Cave with the permission and assistance of Nevrlý. According to Knies' records, obtained by one of us (J. J.), it is clear that he was digging in Chamber E of Szombathy's plan. He was working in the elevated part of the Main Cave, excavating the site adjoining the 6 m tall debris cone that is under the big chimney. In the surface layer (beneath the debris from the chimney) Knies found few fragments of human bones (Mladeč 39–41 and 88–91). Like the human remains found in Chamber D, these bones were similar in color, contained dendrites and were covered by a thin limestone layer. He only mentioned them briefly in his notes, and no subsequent publications appeared on the fragments; they remained unknown, even to specialists. After the death of Knies they were deposited at the Anthropos Institute of the Moravské zemské muzeum in Brno, together with field diaries and with other materials acquired by Knies in the Mladeč Caves.

There are eight fragments altogether, four cranial and four podial remains. The four cranial fragments from what we believe are three different individuals, two adults (Mladeč 39 and 41) and a juvenile cranium of two pieces (Mladeč 40). Knies also recovered five complete bone tools and three fragments of similar tools. Three of them are flat Aurignacian bone points of Mladeč type, the fourth

5 Szombathy only numbered a few of the crania; we designated numbers for all of the specimens from the caves, whether or not they survived the war. We tabulate them in chap. 8, Table 1, and for purposes of clarification we refer to specimens by these numbers, even in citations when the original author did not.

is a flat bone tool pointed at both ends, and the fifth point was rounded at the base. He collected the skeletal remains of the macrofauna that included reindeer, wild cattle, elk, horse, and beaver. There is no doubt that these finds come from the surface layer, but their exact horizontal position prior to the excavation is not known.

On March 22nd, 1904, during quarrying operations, the entrance to another Small Cave (the "Quarry Cave," P of Fig. 4) was found some 50 meters west of the site of the 1902 entrance to the Main Cave complex. Knies visited the site on April 1st, 1904 (Knies, 1905). Soon after the discovery of the Quarry Cave (April 20), Dr. Jan Smyčka (physician and mayor of the larger nearby town of Litau) visited the site (Smyčka, 1907). Maška came on August 12th and Szombathy arrived August 25th. These visits illustrate the interest in the discovery of new fossil humans that was prevalent at the turn of the century (see also Radovčić, 1988). Szombathy (1904) mentions that the cave was an isolated triangular chamber with a caved-in ceiling 20 meters long and 6–8 meters broad, oriented in the southeast to northwest direction. Knies obtained permission to excavate the site. On clearing the broken stones it was evident that the caved-in ceiling fully covered the sediments including the skeletal remains they contained. We believe these to belong to three individuals, based on their crania: two adult males [Mladeč 5 and 6] and a child [Mladeč 46]. Numerous additional adult postcranial remains were found together with the two damaged adult crania in the same loam-filled area [Mladeč 61, 65–67, 70–73, 78, 84, 86–87, 93–101]. Most of these were listed by Szombathy, who studied them in 1904 and in 1925 in Litovel where they were deposited in 1923 after first being sent to the Fürst Liechtenstein Museum in Úsov Castle. The remains included two maxillae (Mladeč 50 and 51), two mandibles (54 and 55), rib fragments (67), a clavicle (65), a scapula (66), a fragmentary ilium (61), two humeri (70 and 71), a radius fragment (72), two "shin bones" [one of these is a femur (78) and the other a misidentified ulna fragment (73)], a tibia (84), a fibula (86) and 12 metacarpals and phalanges (87 and 93). The remainder of these adult remains were later collected by Knies and given to the Moravské zemské muzeum in Brno. These were listed by Knies (1905) but never studied by Szombathy, and include a right zygomatic (Mladeč 94), a calcaneus (99), phalanges (98), a scapula (101) and fragments of femur (95), tibia (96), fibula (97), and ribs (100).

In view of the very large number of postcranial remains associated with these three crania and the limited (2 by 3 meter) area of loam the specimens were found in, evidently sealed by a rockfall, we believe it is very likely that the Quarry Cave skeletons were intentionally buried, possibly at the same time. If so, the remains are unlikely to represent a nuclear family, there are two adult males and a child, and the absence of any body decoration is of some interest (especially in view of the remains of one or more necklaces found in the Main Cave). In that they do not represent a family group and might be three males (certainly the two adults are male); this trio resembles the triple burial from Dolní Věstonice (Vlček, 1991).

In 1911, the Mladeč Caves became property of the Litovel Museum Association (Krajinska musejni spolecnost v Litovli), which started an extensive project of clearing the caves and to make them accessible to visitors. The excavations started in 1912 under Smyčka's supervision. These activities profoundly changed the Main Cave, completely clearing out the cave chambers and making it easier to move about in the cave. Szombathy (1925, 73) reports that some human remains were recovered during this process:

> *"Twelve mostly smaller fragments of the top of the skull, that originate from two different individuals [Mladeč 42 and 43] and cannot be pieced together, one fragment of an upper jaw [Mladeč 47] and two clavicle fragments [Mladeč 62] were gray, passed through by dendrites, and covered by thin sinter-loam crusts. These human remains have been selected from a collection of animal bones. This is discernible from the fact that the faunal remains have been cemented together with two of the human fragments (in one case one of the human fragments is connected with two incisors of a reindeer and in another case a part of a reindeer's molar is connected to a human fragment)."*

Part of the Main Cave situated to the northwest of the site and excavated by Szombathy in 1882 was dug again in May 1922. Here, Fürst discovered numerous human and animal bones and some bone tools. Smyčka (1922; 1925) and Fürst (1923–24) published brief reports on this enterprise. Fürst writes that he found fragments of five skulls, which included a male [Mladeč 4], a female [Mladeč 38] and three children [Mladeč 37, 44, and 45], along with a number of other human bones. Szombathy discusses these as part of a section in his publication dealing with "the anthropological finds in the cave after 1903." The list covers many of Knies' finds, all of the Quarry Cave finds, as well as the finds of the Litovel Museum Association from the period after 1911.

The finds collected by Szombathy, which include hominids, their tools, and the associated fauna, are in the Naturhistorisches Museum Wien. The 1904 finds were mostly deposited at the Úsov Museum and later were transferred to the Litovel Museum. Some material from the Quarry Cave collected by Knies was transferred to the Moravské zemské muzeum in Brno. A child's skeleton from this collection was sent to be studied by Matiegka, but it is not clear whether this research took place in Brno or if the specimen (Mladeč 46) was transferred to Prague. Some of the anthropological finds resulting from Knies' excavations in the Main Cave after 1903, three cranial fragments and four hand bones, together with 5 bone tools, formed part of Knies' private collection. One of the authors (J. J.) visited the surviving members of the Knies family in 1959 and received the collection for the Anthropos Institute of the Moravské zemské muzeum. The rich finds discovered in 1922 were added to the collections of the Litovel Museum.

In the 1930's nearly all of the Mladeč finds remaining in Czechoslovakia were consolidated into one collection at the Moravské zemské muzeum, Brno. While it is possible that one child, Mladeč 46, remained in Prague, the only other Litovel materials not transferred to Brno were some of bone tools and the flints Knies collected, and a few of the large faunal remains. Subsequently during World War II the anthropological collections of the Moravské zemské muzeum in Brno were moved to a "safer" place, the Mikulov Castle located on the present Czech/Austrian border. This castle proved to be anything but safe since it was intentionally torched and gutted by fire by the retreating troops and local members of the Nazi party at the end of the war. Of the sixty plus hominid specimens from Mladeč stored there, only cranium 5 could be recovered from the ashes! While the specimen is still reasonably intact, it did suffer some damage, which we discuss in a following section. It is worth noting that most of the important finds from Předmostí and those Dolní Věstonice specimens known at the time were also incinerated in this fire.

The history of excavations at Mladeč and the history of Europe over the time that the excavations took place resulted in the fact that the Mladeč remains have never been together, at a single place, like some of the other large key Upper Pleistocene collections important for their variation (Skhul and Qafzeh, for instance). The 102 specimens listed in chap. 8, Table 1 (this volume) can all be associated with specific excavators and excavations. While the precise positions in the caves are known for only some, knowledge of where the various excavations took place allows us to stipulate provenience for virtually all the specimens with a fair degree of accuracy.

Neandertals at Mladeč?

Szombathy's 1925 monograph was a substantive history of the cave and the excavations that took place there, along with presenting some amount of anthropological data about most of the human remains. However, there seemed to be one major omission from Szombathy's 1925 monograph, concerning the nature of the 1922 finds.

Smyčka mentioned in an article (1925) that the remains discovered by a group of amateurs in 1922 were of "Neanderthal type." Perhaps for this reason, he believed these to be the earliest of all Mladeč finds. We can no longer be certain which specimens this refers to because, with the exception of Mladeč 4, the remains were never pictured or described in any detail. It has been since supposed that Szombathy did not mention them for this reason: his paper was about *human* remains, not *Neandertal* remains.

It was assumed therefore that some Neandertal remains were destroyed in the Mikulov Castle fire of 1945 before they had been described, and this was reported occasionally by careful readers of the Central European literature (e.g., Leakey and Goodall, 1969). Nobody surmised that these were in reality the specimens described by Szombathy as "found after 1903."

Szombathy's visit to Litovel and to Mladeč in 1925 was evidently very brief. In his report (1925) he thanks the Litovel authorities for their kind help, but his description and the study of the material "after 1903" was superficial, as noted above. Besides this, some of his contemporaries repeated Fürst's assertion (1923–24) that bones were found of at least five individuals, of a male, of a female and of three children. This did not agree with the description given by Szombathy and, therefore, it was generally accepted that Szombathy described only some of the specimens discovered after 1903.

Scholars who consulted the 1925 monograph questioned why he did not describe the human remains found in 1922, when it was evident that he must have seen them in Litovel. We now recognize that indeed he did see them, and that he described them as well, emphasizing that the most important discovery was cranium 4 and the specimens associated with it. This is clearly mentioned by Blekta (1932), and so it is completely clear now that there was only one period of major discoveries in the Main Cave after 1903. This consisted the 1922 finds, discovered by amateurs under the chimney in the place marked "e". The chambers to the side of the main passageway leading to the cave's rear were cleaned out last and only incompletely since a much later excavation conducted by the Moravian Museum was able to examine the sediments remaining there. What was most misleading was the title of the section in which he described these remains: *"Die menschlichen Skelettreste der Fürst-Johanns-Höhle von den Grabungen seit 1903"*. "Since 1903," for the most part, meant "in 1922."

This conclusion helps us identify the actual findspot in the cave, which is on the northeastern side of the stone pillar. Those visiting the cave today will see that on the southwestern side of this pillar there is a plaque mistakenly marking the place of the find. In fact, there existed on the southwestern side little space, only a few square meters. Thus the 1922 material, its provenience, and even the circumstances of discovery can be identified. Smyčka was wrong to assert (1922) that these finds are the earliest of all are more primitive and are of Neandertal type. The morphology we can discern from the published photograph (Fig. 7) does show a prominent supraorbital region, but while its supraorbital morphology may be stronger than in the Mladeč 5 and 6 males, this structure was the only basis for Smyčka's description of Mladeč 4 as a Neandertal, and we do not believe it is sufficient. Moreover, it was evidently not earlier than the other finds. The 1922 remains were associated with Aurignacian tools, and the human bone fragments are in exactly the same state of fossilization as the other human skeletal material (e.g., Mladeč 1) and the Upper Pleistocene faunal remains.

Archaeological associations

The archaeological remains are important in addressing key issues. Most importantly these include the provenience of the specimens in the Main Cave, the relation of the Quarry Cave specimens to those in the Main Cave, and the place of Mladeč in the Central European archaeological sequence (in spite of our efforts, direct dating of the human specimens has not been accomplished as of this writing).

Circumstances of the finds

The first group of reliably documented archaeological materials comes from Szombathy's second field season in 1882. At the center of "The Dome of the Dead," in the upper part of the sediments, there were a number of archaeological finds, as described above. These were mixed with bones of reindeer and *Bos* or *Bison*, and with the human skeletal material. Based on Szombathy's account of his work, it seems that no artifacts accompanied the human specimens discovered in 1881; at least, none were recognized by Szombathy during the excavations. However, it is worth noting that a great deal of sediment was removed in a short period of time, so it is possible that some material was overlooked and discarded. Yet, Szombathy gives the impression that no material was ever lost, except for a few specimens damaged by workers.

There was also a large amount of sediment removed in 1882 and it is not always clear from which locality some specimens derived. Following Szombathy's work in the Main Cave, Knies collected five, almost complete bone points which he donated to the Moravské zemské muzeum. Since Knies concentrated on studying the area adjoining the debris cone between Chambers C and D (Knies 1905 and his unpublished research diary IV), we assume these points came from this area. Yet, there is no mention of the finds of bone points in his records.

When the Quarry site was opened in 1904 (site P in Szombathy's plan, Fig. 4), sediments and other materials from the opening of the small abyss-like cave were initially disposed of, without informing the archaeologists (Knies, 1905; Maška, 1905; Smyčka, 1907). In subsequent visits to the Quarry Cave the museum in Litovel obtained three flat bone points, one 17.8 cm long, another is a fragment of a similar but bigger point, and the third is a smaller fragment, probably not a tool. Knies found another flat bone point consisting of five fragments, totaling about 14 cm in length. He also found two flint artifacts (6.5 cm long) and two unworked chert pebbles. In the Litovel Museum there was also a large chert flake, a chert core and a piece of a *grauwacke* from the Quarry Cave. The association of the Quarry Cave remains with those in the Main Cave is unequivocally established through a comparison of these lithic remains. Yet, important anthropological and paleontological specimens along with some bone and stone artifacts, no doubt, were destroyed. Knies writes (1905, 12) that there were broken and scattered bones along the road leading to the top of the Třešín Hill. The same author mentions that for some time the phosphate-rich loam was also exploited at the place (Knies, 1928) and we assume other finds were destroyed during these quarrying operations.

When the area became property of the Museum Society in Litovel in 1911, a real turning point in the history of the research of the Mladeč Caves occurred. As mentioned above, the Society cleared the sediments from all areas now forming part of the route marked-out for tourists. Up to 1922 we do not have much detail about these clearing operations. In that year, however, a group of amateur researchers (Fürst, Smékal, Charvat, Rohm, Novotný, Smyčka, and others) dug their way down to the surface of the cave deposits beneath the large debris cone, finding a large concentration of animal and human bones. From the incomplete and fragmentary records made by the leaders of these excavations (Fürst, 1923–24; Smyčka, 1922; 1925) it is not possible to locate exactly the place where the human remains of 1922 were found. However, one of the participants of these events, Rohm, marked the site as "Point 3" in his sketch of the cave system. This is consistent with the report by Szombathy who located the findspot at locus e, and with our own conclusions as reported above. Szombathy reported the discovery of two fireplaces, one of which had a thin layer of charcoal and was outlined with a circle of stones. Its discoverers describe a "stone hearth" (Fürst, 1923–24; Smyčka, 1925).

No matter when the excavations occurred, there is no doubt at all that a certain amount of bones and artifacts were pillaged, lost or discarded. The simple fact that the excavations were conducted deep in the Main Cave under artificial light must have resulted in the incomplete collection of materials. In addition, numerous items from the collection of the Litovel Museum were also lost. The ultimate fate of the 51 fragments of "bones awls" from the old exposition near the caves (compare Skutil, 1938, footnote 76) is also unclear. Thus, given the history of the excavations, the fact that much of this work was done by amateurs, and the loss of the material subsequent to the final work at Mladeč, the surviving collection is but a small portion of the material that was originally in the Main and Quarry Caves.

Following 1922 few important discoveries occurred in the caves; only Skutil (1938, footnote 77) mentions that Novotný found in the loam below the chimney a flint artifact 75 mm long. In 1937 Skutil himself discovered in the secondary loess right of the entrance two Aurignacian stone tools (Fig. 60 of Skutil, 1938, 32). The later excavations organized by the Moravské zemské muzeum in 1958 and 1962 did not result in the discovery of any further tools dated to the Upper Paleolithic. The most recent discovery of possible human activities in the cave was by one of the authors (M.O.) in 1981, who found red ochre dots on the walls of the "Dome of Dead" (D), and sporadically elsewhere (Oliva, 1987).

Aurignacian character of the archaeological finds

In agreement with publications beginning with Bayer (1922), and including Szombathy (1925), Jelínek (1969; 1983; 1987), Oliva (1987), Smith and Trinkaus (1991), and most recently Svoboda (2000), we show here that the Mladeč archaeological materials are from an early Central European variant of the Aurignacian and that this industry is the only one found at the site. The personal items found in association with specimens are fully compatible with this interpretation. Evidence of body ornamentation is an important aspect of the Mladeč archaeological inventory. It is quite likely that they were introduced into the Main Cave at the same time, the same way. Svoboda (2000, 534) described the situation as follows:

> *"Placing the Szombathy's finds 'a–e' into this plan, all obviously lay on the surface at the foot of the northeastern cone, showing a pattern of regular dispersal of several meters away from the chimney. Following the verbal description by Knies (1905), his finds were associated with the same cone. [...] Most probably the artifacts and the skeletal elements were thrown or washed into the area below the chimney."*

Therefore, we may consider the different sources of archaeological evidence together.

This Mladeč industry is part of a Moravian complex that forms the largest concentration of Aurignacian finds east of the Rhine (Oliva, 1987; 1993). This Central European Aurignacian complex documents not only the extraordinary quantity of localities (about 100 sites in the area west of the Morava River), but also their chronological span reaching from Late Pleniglacial A (Vedrovice II, Valoch et al., 1985) to Pleniglacial B (Oliva, 1986). The industry shows that the human skeletal remains from both the Main Cave and the Quarry Cave at Mladeč can be placed into a temporally limited early or middle phase of the Central European Aurignacian. This is supported by faunal analysis of Aurignacian sites (Hahn, 1977), which places the Mladeč fauna in a pre-Pleniglacial position.

Bone tools are the most diagnostic of the Mladeč artifacts, and the most important component of the bone tools are found in both the Main and Quarry Caves. These are points with broad bases, or Mladeč type points (Fig. 5). We can find numerous analogies in various Central European assemblages (Albrecht et al., 1972; Brodar and Brodar, 1983). These points occur in Central European assemblages in association with thick end-scrapers, Aurignacian blades, foliated points and backed instruments. The relatively variable assemblage of points from Mladeč does not include any examples of a type with rhomboid shape such as those known from Willendorf. Nor are there any with elongated, narrow distal parts, as are typical of Mamutowa Cave. But most diagnostically, there is not a single point with a split-base at Mladeč. This is surprising since the thin cross sections that are typical at Mladeč are otherwise generally found with split-bases (Albrecht et al., 1972, 60). Thus, the archaeological remains from the Quarry Cave are clearly and

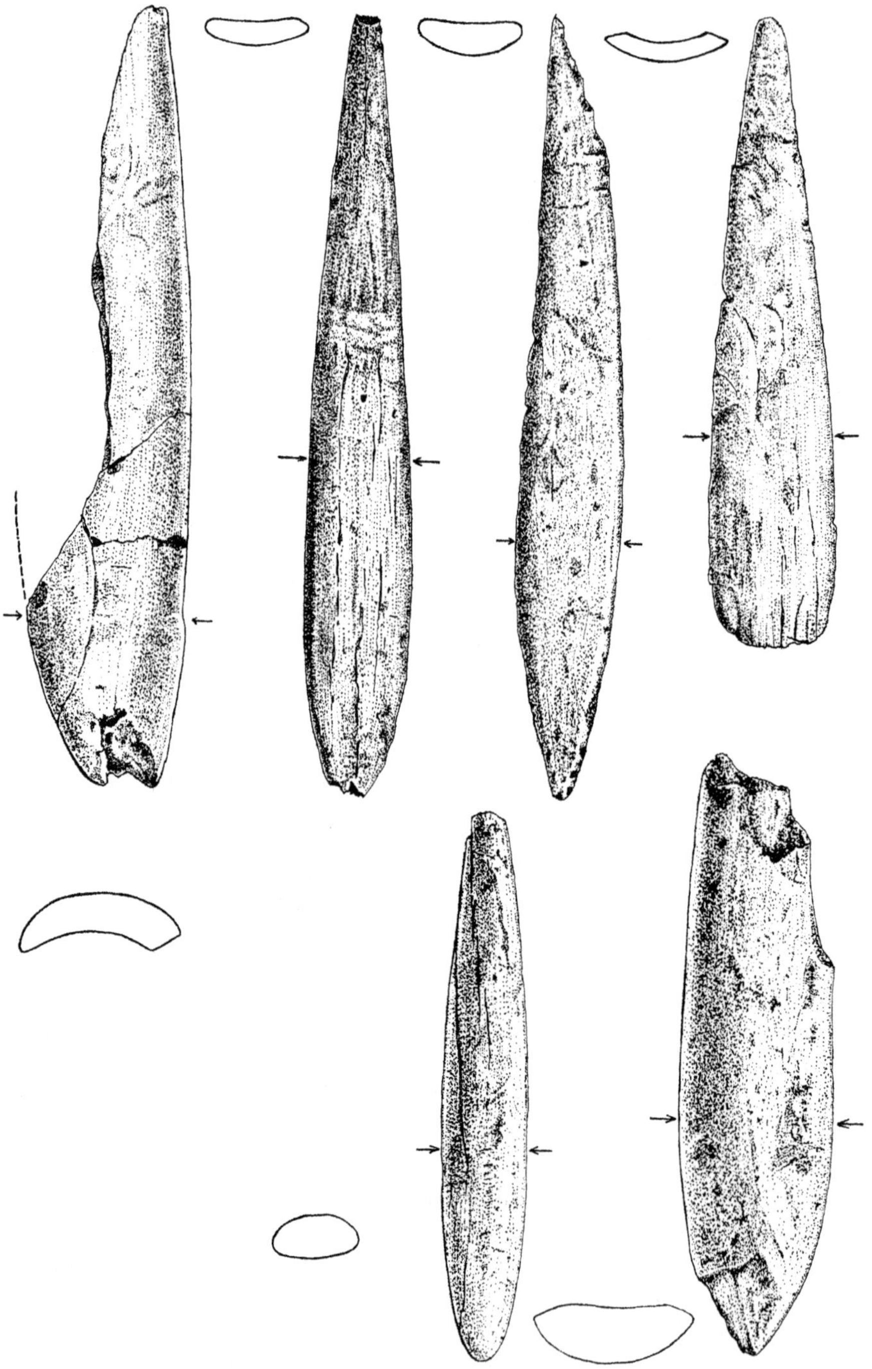

Fig. 5. Massive base bone points from the Mladeč Caves (after Jelínek, 1983)

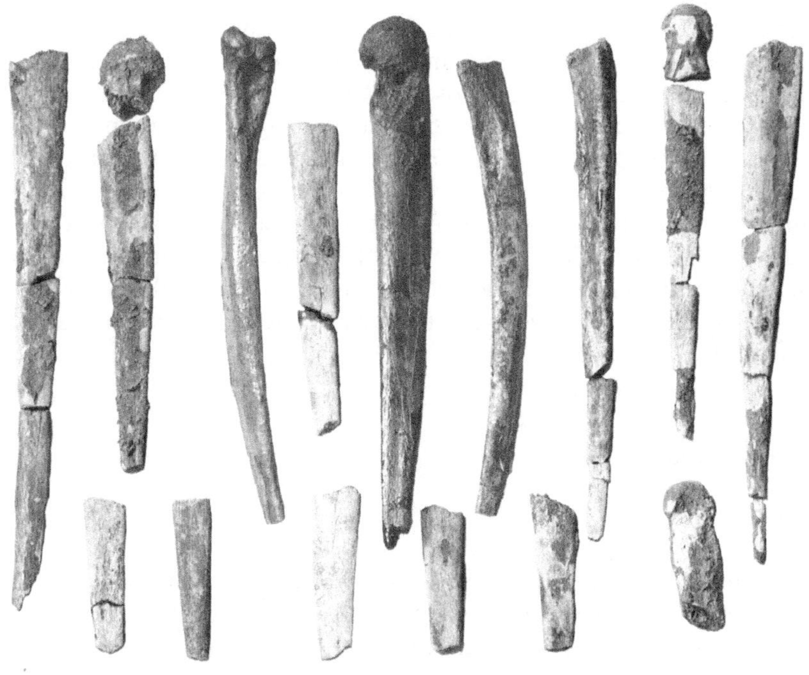

Fig. 6. Mladeč bone awls and awl tips (from Szombathy, 1925, Abb. 9)

unambiguously related to those of the Main Cave through the common presence of Mladeč points.

Mladeč preserves a high frequency of awls (Fig. 6), while at other comparative sites these are rare. As mentioned above, many of the deer metacarpals were also considered to be awls in the older literature. These seem to be non-utilitarian artifacts, but in fact their functional attributions are unclear since in most cases the distal tip is missing. They show an unusual form of processing where the heads were narrowed, and many are ornamented with bilateral incisions. Bilateral incisions appear quite frequently in the Aurignacian, either on the points (namely from Potočka zijalka; Brodar and Brodar, 1983, Figs. 7–12, 14, 16, 21–22) or on flat bone stylets from Vogelherd (Hahn, 1977, Figs. 34, 36, 58–59), Sirgenstein (Hahn, 1977, Fig. 67, 9), Peskö (Hahn, 1977, Fig. 149, 2), and Velika Pećina (Malez, 1967, Fig. VI, 2). At Potočka zijalka (Brodar and Brodar, 1983, Fig. 22, 21) only a single engraved artifact has the undisputed shape of an awl (Brodar and Brodar, 1983, Fig. 21, 16).

The pendants of animal teeth are typical Aurignacian ornaments (Hahn, 1972; White, 1982), but the large numbers that occur in the Mladeč Caves is atypical. Regular rows of incisions on bones are not unusual in the Aurignacian (Marshack, 1972; Hahn, 1972), but the refinement and accomplishment with which they were realized at Mladeč is extraordinary on some specimens.

The question is whether the artifacts found at Mladeč can be seriated in any broad regional scheme? In Western and Central Europe, split-base points are in most cases stratigraphically lower than those with a massive base. This phenomenon probably does not have chronological significance because of (1) the mixed occurrence of the two types (albeit in different proportions) at single sites, and (2) the juxtaposition in various attributed dates for the two point types. Thus, at both Vogelherd and Istállóskö, above layers where split-based bone points predominate are layers with both Mladeč points and split-based points mixed together.

The archaeological finds left on the premises of the Moravian Museum in Brno survived the war intact, as did the materials in the Litovel and Olomouc museums and the anthropological and archaeological materials in the Naturhistorisches Museum Wien. Although these archaeological remains are now in several different places, their study provides a clear answer to the question of archaeological provenience for the human remains. Unlike so many other sites dug more than one hundred years ago, *the Mladeč Caves have only a single cultural type* and it can be readily identified. This is fortunate, since if there were other layers that dated to the Mousterian or to the post-Aurignacian Upper Paleolithic, it might have been difficult to be certain of the relationship between the cultural and the human remains. But as a single component site with some burials, it is not possible to maintain that the Mladeč Caves, and the humans interred in them, are anything other than Aurignacian (Svoboda, 2000, 533).

Materials and methods

Comparative male samples

Our comparisons are all with crania that we believe are male, and are based on our observations and measurements of the original specimens, except when they no longer exist. Determinations of sex for the crania of the comparative samples, like the Mladeč determinations, are not controversial. Our assessments agree with most other workers who have studied large portions of the material, and we note and discuss below those few cases that may be considered uncertain. Sex determinations for teeth are almost always based on associated osteological materials.

The specimens included in the sample of European Neandertals are from the Würm glaciation, as indicated by stratigraphic, archaeological, and in a limited number of cases radiometric dates. The Neandertal cranial sample is composed of the following males: Arcy-sur-Cure 8; St. Césaire 1; La Chapelle; Feldhofer Cave; La Ferrassie 1; Hortus 49; Kůlna 1[6]; the cranial rear from Marillac C10-342; Monsempron 1; Guattari (Monte Circeo); Le Moustier[7]; La Quina 2, 7, 13 and 25; Saccopastore 2; Spy 2; Teshik Tash[8]; Vindija 258, 261, 293, 305. The comparative Neandertal dental sample consists of Arcy-sur-Cure 8, 9; La Chapelle-aux-Saints 1; Le Cotte de St. Brelade (Jersey) 1; La Ferrassie 1; Genay 1; Guattari 2, 3; Hortus, 3, 14; Kůlna 1; Le Moustier 1; Ochoz 1; Le Petit-Puymoyen 2; La Quina 9, 20; Regourdou 1; Saint Césaire 1, 2; Spy 2; Sakajia 1 and Teshik-Tash.

The Skhul and Qafzeh remains date to the later part of the Middle Paleolithic, and by every estimate they are earlier than the Mladeč remains and penecontemporary with European Neandertals. While the Skhul and Qafzeh ages are surely not identical, we believe that the sites sample similar and probably related populations and therefore can be validly combined for com-

6 Although published as "subadult" we regard the specimen as adult because the comparable teeth show more wear than on other young Neandertal adults, such as the Krapina E maxilla which has a posterior interproximal facet on the M^2.

7 This 13 year old is included for cranial capacity and those calvaria measurements not involving the cranial superstructures that would be expected to develop with adult status.

8 Cranial capacity only for this 8–9 year old.

parisons. The Skhul/Qafzeh cranial sample we have used consists of Qafzeh 6, 7[9], and 9[10], and Skhul 4, 5, 6, and 9[11]. The dental specimens included are Skhul 4, 5, 6, and 9A and Qafzeh 4, 6, 7, 8, 9, and 10.

We also make limited, non-systematic comparisons with the small sample of other males from the earliest European Upper Paleolithic, based on our detailed studies of the original remains (or for Předmostí 3 a cast). These are: Brno 2; Dolní Věstonice 11/12, 13–15[12], and 16; Pavlov; Předmostí 1, 3, 7[13], 9, 14 and 18. We believe it is very unlikely that this sample includes specimens significantly

9 Vandermeersch does not ascertain sex for Qafzeh 7. Our comparison of the postcrania (ulna, fibula), and of gnathic and dental size, and morphology, with other specimens of the Skhul/Qafzeh sample indicate that it is a small male. Generally, postcranial and mandibular size and robustness is very similar to Qafzeh 9, and when these can be compared with other skeletal females (Qafzeh 3, Skhul 2 and 7), Qafzeh 7 is markedly larger. This diagnosis, however, is a difficult one because there are few female remains for comparison, and because the specimen lies on the small and gracile end of the male range.

10 Qafzeh 9 has often been regarded as a female. For instance, Vandermeersch (1981, 62), who calls it a "probable female" because of the gracility of the cranium and postcranial skeleton, and from some details of the pelvis; the size of the dentition and jaws does not support his female interpretation. In the postcranial skeleton, there is no question that Qafzeh 9 is a small and rather gracile specimen. Vandermeersch particularly focuses on the small size of the humerus head to support his diagnosis. Its mediolateral diameter (43.3 mm) is smaller than the 50 mm Skhul 4 male (the only other humerus head in the Skhul/Qafzeh sample). However, we are far from certain of the significance of humeral head size. For instance in a different comparison, Předmostí 4 is a *female* with a humerus of similar length and a larger head diameter, while Předmostí 14 is a *male* with a humerus of similar length and a head of the same diameter. Vandermeersch discusses several pelvic features that he regards as supporting the female diagnosis for Qafzeh 9. From his estimates of pubic and ischial lengths, he describes the pubis as being relatively and absolutely long, as is expected in females. However the 75 mm pubic length he estimates is considerably *less* than the other pubes in the Levant sample - the male pubis of Skhul 4, which is 82 mm in length, and Skhul 9, which is 88 mm. Moreover, Skhul 4 has the same ischiopubic index as Qafzeh 9 (approximately 100). The length of the Qafzeh 9 sciatic notch opening was reconstructed at approximately 50 mm (the unreconstructed measurement was 38 mm). Vandermeersch regards this as a female character because it corresponds to the large size of the notch in modern European females. However, the length of the sciatic notch opening in Skhul 4 is 59.8 mm (l) and 56 mm (r), and is between 50 mm and 65 mm in three Předmostí males. According to Vandermeersch, the Qafzeh 9 cotylosciatic breadth index is within both modern male and female ranges. While none of Vandermeersch's arguments supporting a female diagnosis are compelling, analysis of the pubis according to Phenice's (1969) criteria strongly supports a male diagnosis. There is no ventral arc, the form of the subpubic dorsal ramus as seen from the internal view is straight, and the subpubic neck is flat and broad (also see Rosenberg, 1988).

11 Regarding Skhul 9, sex determination again centers on the pelvis. In their diagnosis of the specimen, which they regarded as male, McCown and Keith (1939, 89) state: "In the pelvis of Skhul IX we have met with a mixture of sexual characters, some - such as the length of the pubic ramus and the capacity of the true pelvis - which favor the female sex, while others, such as the area on the sub-pubic arch for the attachment of the external genital organ, favor a male diagnosis." In their subsequent discussion of the sub-pubic region, McCown and Keith further note (p. 85): "The character of the sub-pubic arch is the most striking of the various feminine features [...] the anterior area of the pubic area of the attachment in Skhul 9 is beveled and thin, as in a female [...] the anterior beveled edge of the pubic arch is [...] sharp and everted." Many of the features compatible with a female diagnosis are associated with pubic lengthening and thinning, and are also Neandertal resemblances. On the whole, the male features of the pelvis, the size of the cranium, and other postcranial details support McCown and Keith's male diagnosis.

12 These three are the so-called "triple burial," teen-age boys who died at the same time. DV 15 is pathological. His condition, the chronological age of the three, and unreliable reconstruction in the three, severely limits the measurements we were able to use.

13 This is a very large 12 years old - the largest of all males but with newly erupted M2s and an unerupted P^4. It is only included in a few calvarial comparisons, where it is often the maximum.

younger than 26,000 years and if we have erred it is on the side of caution. These specimens and their provenience are reviewed in Churchill and Smith (2000).

Data collected

Specific sources for the Mladeč data are detailed within the discussions of the specimens below (Table 1).

For the surviving specimens, the observations and measurements reported here are based on studies of the original specimens by the authors conducted between 1974 and 2001. The primary cast of Mladeč 6 was studied in Brno. Our measurements were compared with those few provided by Szombathy and the cast proved to be quite accurate. Other data come from Szombathy's monograph.

As we noted, virtually all[14] of the Předmostí specimens were destroyed with the Mladeč remains, leaving only a few charred limbs uncovered after the end of the hostilities. Unlike the Zhoukoudian hominids, for which every scrap except the isolated worn teeth was cast (Mann, 1981), most of these Moravian specimens are gone forever. For the Předmostí data used in some comparisons, we have relied on primary casts of specimens 3 and 4 located at the Moravské zemské muzeum in Brno (which proved to be very accurate in scale and detail), the publications of Matiegka (1934) and additional details mentioned by Kříž (1903); Maška (1895); Wankel (1884); Skutil (1940); Absolon (1929); Hrdlička (1930); Szombathy (1925) and Morant (1930).

As much as possible, the morphological features we discuss are named following Weidenreich (1951) and Hauser and De Stefano (1989). We almost always used standard measuring points, as defined by Martin (1928). In those cases where we found it necessary to define a position for measurement, it is discussed in the text. Our abbreviations for these and other landmarks and directions are given in Table 2. We calculated indices, angles at the parietal corners, and various projections into the sagittal plane at the midline.

The dental remains poised special problems. Compared to the cranial and postcranial remains, there are only a few teeth recovered from Mladeč and much more than half of these have been lost or destroyed. According to the inventory of the dental material published by Szombathy, a total of 54 teeth were originally discovered. Unfortunately, he did not publish measurements or descriptions for all these teeth, so at present there are metric data for only 32 of the original 54 teeth. Today, in the Naturhistorisches Museum Wien a total of 15 teeth are preserved, all from the maxilla. For specimens that no longer exist, we have used Szombathy's original descriptions and his (limited) photographs. Dental measurements for the Mladeč males come from our study of the teeth in Vienna (Mladeč 8 and 9) and from Szombathy's monograph (1925) for the Mladeč mandibles that were destroyed at Mikulov. Our measurements follow the technique described in Frayer (1978) and Wolpoff (1979) for taking mesiodistal lengths and buccolingual breadths. Szombathy does not provide details on how he measured lengths and breadths for the teeth he published, but considering the 14 teeth (not all of which belong to males) that still exist in Vienna, our measurements differ from Szombathy's in mesiodistal lengths and buccolingual breadths from -9% to +11.5% (mean length difference = 3.8%; mean breadth difference = 1.7%). While the average differences are higher than we like, unfortunately Szombathy (1925) represents the only source for measurements on Mladeč specimens 50, 51, and 54.

Our nonmetric observations are based on the standard descriptions in the literature (especially Kraus et al., 1969). For occlusal morphology we have relied on Dahlberg (1951), for the development of taurodontism on Hillson (1991), and for dental wear, the standardization developed by B. H. Smith

14 A mandible found by Wankel, and a femur, are all that remain.

Table 1. Mladeč specimens discussed in the text (NHM = Naturhistorisches Museum Wien, MMB = Moravské zemské muzeum)

Number	Identification	Estimated age	Location in cave	Discovered	Repository
4	Frontoparietal fragment	Adult	Chamber E, Locus e	1922	Destroyed, 1945
5	Male calotte	Adult	Quarry Cave	1904	MMB
6	Male calotte	Adult	Quarry Cave	1904	Destroyed, 1945
8	Male maxilla: right I^2, left C, M^{1-2}	Adult	Chamber D, Locus d	1882	NHM
9	Right maxillary C & P^3	Adult	Chamber D	1882	NHM
10	Right maxillary M^3	Adult	Chamber D, Middle	1882	NHM
39	Right parietal fragment	Adult	Chamber E	1903–1911	MMB
40	Left occipital/parietal fragment	Subadult or young adult	Chamber E	1903–1911	MMB
43	Occipital fragments (4)	Adult	Chamber E, Locus e	1922	Destroyed, 1945
47	Left maxilla I^2, P^4–M^2	Subadult	Chamber E, Locus e	1922	Destroyed, 1945
48	Left maxilla M^{1-3} (same as #49)	Adult	Chamber E, Locus e	1922	Destroyed, 1945
49	Right maxilla P^3, M^{1-3}	Adult	Chamber E, Locus e	1922	Destroyed, 1945
50	Right maxillary fragment	Adult	Quarry Cave	1904	Destroyed, 1945
51	Left maxillary fragment	Adult	Quarry Cave	1904	Destroyed, 1945
52	Mandible, left I_2–M_3, right P_4–M_2	Adult	Chamber E, Locus e	1922	Destroyed, 1945
53	Right mandible		Chamber E, Locus e	1922	Destroyed, 1945
54	Mandible, left P_4–M_2, right I_2–M_2	Adult	Quarry Cave	1904	Destroyed, 1945
55	Mandible, I_2–M_1	Adult	Quarry Cave	1904	Destroyed, 1945
56	Canine		Chamber E, Locus e	1922	Destroyed, 1945
57	Premolar		Chamber E, Locus e	1922	Destroyed, 1945
58	Premolar		Chamber E, Locus e	1922	Destroyed, 1945
59	Molar		Chamber E, Locus e	1922	Destroyed, 1945

Table 2. Definitions of abbreviations found in the tables. The descriptions of the terms defined are in Martin (1928)

Abbreviation	Definition	Abbreviation	Definition
ant	anterior	k	krotaphion
ast	asterion	L	length
au	auricular point	l	lambda
ba	basion	mf	maxillofrontale
Br	breadth	mm	millimeters
br	bregma	ms	mastoidale
btwn	between	na	nasion
C	canine	o	opisthion
cc	cubic centimeters	occ	occipital
co	coronale	op	opistocranion
FH	Frankfort Horizontal	pr	prosthion
fmo	frontomalarorbitale	proj	projection
fmt	frontomalartemporale	pt	point
gl	glabella	st	stephanion
ht	height	sup	superior
i	inion	zm	zygomaxillare
iob	innerorbital breadth	zpm	zygomatic process of the maxilla

(1984, Fig. 3). This graded series of wear stages is similar to Brothwell's chart (1981), although it includes more categories for each tooth and extends the wear codes to the anterior teeth and premolars.

Cranial remains

Mladeč 5

Mladeč 5 is the long, low calvarium of a robust adult male (Figs. 7, 8, 9, 10, 11, and 13), and may include more associated remains, that comprise one of the adults in the triple burial found in the Side Cave. It was discovered in 1904 by workmen and was studied by Szombathy in 1904 (1926), who suggests that the Mladeč 50 and 51 (maxilla halves) and mandible 54 might belong to the same individual. In fact, some or possibly all of the adult postcranial remains from the Quarry Cave could also belong to Mladeč 5. These are detailed in chap. 8, Table 1. As discussed above, the remains were taken to the Mikulov Castle in 1945. Fortunately, the calvarium was left in a box in a hallway of the castle where it was protected by collapsing walls, and the fire damage was relatively minor. Our study is based on the original specimen housed at the Moravské muzeum, Brno, and on the pattern (i.e., primary cast of the original specimen) made of the cranium before 1945. This pattern has served as the basis of all Mladeč 5 molds that have been made to date.

Our assessment of sex is based on the large size of the vault, the prominence of the muscle attachments, and the marked development of the cranial superstructures, especially the supraorbital tori. This specimen has invariably been regarded as a male.

An estimate of age, based on endocranial suture closure, must take into account the complete obliteration of the coronal and sagittal sutures for their entire preserved extent and the obliteration of all but the superior portion of the lambdoidal suture (this part is closed, but a distinct groove marks its course). Judging from the high degree of internal suture obliteration at various places in the cranial vault, and the number and depth of the pacchionian pits, this individual was older than a young adult at the time of his death.

Distortion and deformation

Comparing the cranium as it exists today with the pattern made of it before 1945 reveals a number of consequences of the unfortunate history of this vault. Below we will describe the condition of the specimen as it was before the Mikulov fire and add relevant changes due to this fire that characterize the skull at present.

The right parietal contour was undistorted, and remains so even after the fire. However, the anterior portion of the right temporal squama, the anterolateral corner of the parietal and the posterior aspect of the greater wing of the sphenoid were lost because of the fire. This section, some 50 mm in height from the base of the temporal squama and about 24 mm in anteroposterior breadth, has been restored in plaster and painted to look like the original. The contour of the section appears correct, but the anterior border of the temporal squama differs from its original condition. On the left, most of the temporal squama and the articulating portion of the greater wing of the sphenoid were separated from the parietal and somewhat twisted away from the midline, bringing the left part of the cranial base medially. Associated with this twisting, a large crack extends from the left periodic region posterior to the coronal suture, tapering toward the midline and ending approximately 25 mm from the sagittal suture, 10 mm posterior to the coronal suture. The posterior margin of the parietal anterior to this crack is slightly flanged along it. As a result of the twisting of the left side of the

Fig. 7. The three best-preserved Mladeč males, in lateral view. From above these are crania 6, 4, and 5 (from Szombathy, 1925)

vault, this side appears to have a distinct inferomedial angulation when seen from the rear. Abbildung 3b of Tafel 8 in Szombathy (1925) clearly shows the inward cant of the left side of the cranium, demonstrating that this distortion cannot be totally a consequence of the fire. However, the fire has clearly accentuated the distortion of this region. The parietal's anterior border has now twisted outward along this crack; the resulting gap has been filled with plaster and painted to look like the original bone. This does not hide the fact that the edges no longer meet along this gap and the parietal's anterior edge is as much as 3 mm more lateral than the posterior edge of the frontal. The anterolateral edge of the parietal has been lost, as has some additional anterior portion of the temporal squama. As a whole the temporal is even more twisted than it was, its base is more medial while the top of the squama is more lateral and separated from the inferior border of the parietal by a 3.6 mm gap. The full extent of this twisting can be easily observed in inferior view by comparing the angulation of the left and right petrosals. Also, associated with this new damage, about 20 mm of the most inferior part of the left frontal is missing.

There are other areas of deformation and/or distortion. One of these is a large, shallow depression, lateral and posterior to bregma on the right parietal, approximately 40 mm in length and 30 mm in breadth. The medial margin almost touches the sagittal suture, and the anterior margin is just posterior to the coronal suture. This depression existed before the 1945 fire, however at some point it was broken and repaired with plaster; this is obvious internally, but less obvious externally because the plaster was painted to match the bone. Szombathy, who considered this an "old injury", reports that the depression "penetrates the bone, creating a 3.5 cm long fracture on the inside, oriented in the sagittal direction" (1925, 75). However, on the left side we can palpate a weaker depression in the same position. Weidenreich (1951) described similar parasagittal depressions in the Solo crania where they are associated with and accentuated by a sagittal torus. The Mladeč 5 vault also possesses such a torus, on the frontal and much more strongly on the anterior half of the parietal bones. Thus, there is some reason to believe that the right side depression may not be postmortem, but this leaves unexplained its much deeper expression on the right than on the left. Another area of postmortem damage is about 42 mm posterior to the supraorbital on the left side of the frontal, extending 12 mm from the midline with an 8 mm anteroposterior dimension. The area is depressed, due to external table collapse and crushing into the diploë. On the left side of the occiput, just superior to the nuchal line, there was a small triangular-shaped portion of bone broken along the lambdoidal suture. No distortion is associated with this break, which Szombathy described as having extended completely through the vault. After the fire, the area was evidently further damaged and was repaired by being filled with plaster painted to resemble the bone.

Both mastoid processes show breakage. On the left side, the process is broken at its base, level with the digastric groove. This break extends through the petrous portion of the temporal, medial to the mandibular fossa, to a point parallel with the fossa roof. The right mastoid is much more complete, although also broken. The missing portion is limited to the area medial and anterior to the most inferior point on the process (which we believe is preserved) and cancellous air cells are exposed up to the position of the tympanic ridge. A small damaged section of the most medial portion of the pyramidal process is also preserved. It was further damaged by the fire; more cancellous air cells were exposed as even more of the very thin bone in this region was broken away.

A number of other observations seem to be consequences of the fire. Some additional portions of the cranial base are missing. On the nuchal plane, 20 mm of bone that once existed anterior to the lowest inion position are no longer present. This is particularly important in understanding the morphology in the vicinity of the right occipital temporal suture. On the original pattern it is clear that this suture extended from asterion through the center of the paramastoid crest, while today virtually all the occipital portion (including the occipital half of the paramastoid crest) has been broken away and posterior to the paramastoid crest replaced by painted plaster making the course of the suture impossible to follow on the cranium and underestimating the size of the paramastoid crest.

A small nick is missing from the temporal at the most lateral extent of the glenoid fossa. Moreover, there are numerous new cracks in the bone surface. Some of these are quite large, for instance a transverse opening across the occipital plane that has been filled with clay, and as elsewhere painted to look like the original bone. These cracks do not follow the stress lines in the bone as described by Tappan for weathering cracks (1973). A final consequence of the fire is a color change from ash-gray passed through by dendrites as described by Szombathy, to an uneven dark to very dark brown.

In our assessment of the metric characteristics we took both the original distortion and the additional changes due to the fire into account. We relied primarily on the right side of the original specimen. Transverse breadths across the whole skull were not measured directly, but were calculated from projections of the right side to the sagittal plane. When necessary, the cast was used to help ascertain the position of landmarks.

Pathology and trauma

Two healed wounds occur on the Mladeč 5 vault and appear to be blunt trauma limited to the external table and superficial diploë. The larger of these is situated directly over the middle of the left orbit, approximately 21 mm above the supraorbital arch (Fig. 8). This deep depression measures 4.3 mm in sagittal length and 8 mm transversely. There are no signs of infection or periosteal bone growth, indicating that the wound occurred long before death. The second wound is located on the right parietal about 30 mm lateral and anterior to lambda. This very small wound is oval shaped, with 6 x 9 mm diameter. Like the larger wound there is no evidence of infection, suggesting a complete healing of an old injury (see chap. 16, this issue). We could observe no cutmarks.

Fig. 8. Frontal views of Mladeč 5 (left, showing the cranial wound) and Mladeč 6. Both preserve the supraorbital region, Mladeč 5 being the more complete. Mladeč 6 preserves the superciliary arch with some lateral torus structure, a supraorbital notch, and the superior portion of the nose (both are casts).

Total vault

The calvarium of Mladeč 5 extends on the midline from the frontomaxillary suture (with about 5 mm of the most superior portion of the frontonasal suture preserved), posteriorly to a point slightly anterior to the approximate position of the inferior nuchal line (but see above), some 25 mm anteroinferior to the nuchal torus. The frontal extends laterally to the frontozygomatic sutures and is largely complete except for missing portions of the posterior orbital plates. Most of the greater wing of the sphenoid is preserved on the right side, while only the most superior portion is present on the left. The join of the sphenoid with the parietal results in a pterionic region of the "H" form. Of the temporal bones, the right is more complete, originally missing only portions of the zygomatic root and the mastoid (as described above). The superior border of the temporal squama has been broken away from the parietal along the thinnest portion of the bevel between them. Thus, while the superior surface of the temporal squama appears to be evenly rounded, it was in fact quite angular with its most superior extent at least 16 mm above the existing temporal. The break on the zygomatic root extends to a position parallel to the roof of the mandibular fossa. On the left side, the anterior breakage is more extensive so that only the posterior portion of the temporal is preserved (i.e., beginning at the position of the external auditory meatus). The parietal bones are completely preserved, as is the occipital plane and the superior part of the nuchal plane of the occiput. Of the cranial base only the petrosals are preserved. As described above, damage from the fire has resulted in the loss of some additional areas around the periphery of the vault.

The general appearance of the specimen is long, low, broad, and very robust, with prominent supraorbitals, a well-developed occipital bun, and a distinct nuchal ridge. Because of this configuration, Mladeč 5 has been alternatively described as Neandertal-like (Jelínek, 1969, 1976, 1983; Smith, 1982; Wolpoff, 1982, 1999; and see Fig. 2), as of modern, but very robust form (Howells, 1982; Stringer, 1982, 1989), as resembling Jebel Irhoud 1 or as "archaic compared to their Western European counterparts" (Stringer et al., 1984, 117). The issue raised by these publications is whether the cranium appears to be archaic because it is robust, because it resembles a "primitive" form of modern human that is distinct from Neandertals, or because it resembles Neandertals.

The maximum cranial length of 205.6 mm is very great (Table 3), actually out of the Skhul/Qafzeh range. It is within the Neandertal range and close to the mean (although larger than it). Glabella-lambda length patterns similarly. Again, the Skhul/Qafzeh mean length is markedly less. As measured from nasion, however, the cranium does not appear to be as long. The two length measures from this point are less than the Neandertal mean, although exceeding the Skhul/Qafzeh range. Clearly, the projecting supraorbitals contribute significantly to the marked expression of maximum cranial length, in spite of the furrow at the Mladeč 5 glabella position that depresses it relative to the supraorbital arch. Compared with the Neandertals, Mladeč 5 is similar in length, while it is invariably much longer than the Skhul/Qafzeh males.

We examined several indirect measures of cranial height because the cranial base is incomplete. The bregma-inion distance is large, exceeding the Neandertal maximum and right at the top of the Skhul/Qafzeh range. We calculated the vertical height from the auricular point to bregma, as projected into the sagittal plane. This distance, 111.2 mm (Table 4) is between the Neandertal and the Skhul/Qafzeh mean values (Fig. 24). It is at the Neandertal maximum, and just below the Skhul/Qafzeh minimum. The height/length index determined for the auricular height and the nasion-opisthion length (distance from nasion rather than glabella to avoid pneumatization) is 56.7% (Table 3), or very low vaulted. The index is close to the Neandertal average, and well below the Skhul/Qafzeh range. A related index was calculated for the nasion-lambda length. This distributes in a similar manner. The long Mladeč 5 vault is absolutely low, and relatively even lower as measured from the auricular point. Height measures are generally closer to the Neandertals than to the Skhul/Qafzeh specimens.

Table 3. Cranial vault dimensions for males

	Mladeč 5	Mladeč 6	Neandertal		Skhul/Qafzeh	
			Mean (n)	Range	Mean (n)	Range
Capacity (cc.)	1650		1531 (8)	1300–1641	1552 (5)	1518–1587
Lengths (mm)						
gl-op (Cranial L)	205.6	200.5	203.8 (5)	199.4–209.0	197.4 (4)	193.0–202.5
gl-l	194.0	186.3	190.2 (6)	184.5–197.0	184.6 (3)	184.0–185.6
na-op	196.0	197.3	200.5 (5)	194.5–206.5	186.5 (3)	183.5–189.0
na-i	186.5	188.5	191.7 (5)	188.2–196.5	182.0 (2)	179.0–185.0
na-l	188.2	181.2	189.8 (5)	183.5–196.5	178.3 (3)	176.5–180.8
br-i	160.2	165.0	147.7 (6)	142.5–153.5	158.0 (4)	155.4–161.0
Breadths (mm)						
Maximum cranial Br	156.0	166.5	154.6 (5)	147.0–158.5	144.4 (5)	139.6–152.0
Biparietal	154.0	153.5	152.7 (7)	147.0–158.3	142.6 (5)	139.6–144.5
Biauricular	150.0		139.2 (6)	127.6–146.6	131.1 (4)	123.8–135.0
Bimastoid	136.0	137.0	110.6 (5)	101.4–117.0	115.2 (3)	106.5–128.0
Arcs (mm)						
na-op	318.0	321.0	286.9 (5)	260.0–293.5	299.5 (2)	296.0–303.0
na-i	342.5	340.0	315.8 (4)	306.0–326.0	322.0 (2)	318.0–326.0
gl-l	256.0	258.5	239.5 (6)	230.0–247.0	243.0 (2)	240.0–246.0
br-i	202.5	200.0	184.4 (5)	171.0–194.0	200.0 (3)	183.0–217.0
au-br	159.0		151.7 (5)	150.0–153.0	149.3 (2)	149.0–149.5
*Indices (*100)*						
Cranial Br/L	75.9	83.0	75.9 (5)	73.7–77.7	72.2 (4)	69.6–75.1
Biparietal Br/Cranial L	74.9	76.6	75.4 (5)	73.7–76.6	72.1 (4)	69.6–74.6
Auricular ht/na-l	59.1		57.1 (4)	55.2–60.3	63.2 (2)	63.0–63.4
Auricular ht/na-op	56.7		54.2 (4)	52.3–56.9	60.9 (2)	60.7–61.1
gl-l arc/chord	132.0	138.8	126.0 (6)	121.3–129.2	131.4 (2)	129.3–133.5
br-i arc/chord	126.4	121.2	125.4 (5)	119.6–132.6	127.5 (3)	117.8–139.1
au-br arc/chord	118.6		117.0 (4)	114.3–118.4	114.9 (2)	113.9–115.8

Table 4. Distances from the auricular point to midline landmarks, in sagittal projection (mm) for male crania. The Mladeč 5 values are like those of other Early Upper Paleolithic European males. Neandertal males generally deviate far less from these Europeans than the Skhul/Qafzeh males do (Fig. 24). This similarity is reflected in a comparison of the measurements

Distance to:	Mladeč 5	Neandertal		Skhul/Qafzeh	
		Mean (n)	Range	Mean (n)	Range
nasion	103.5	106.9 (5)	103.2–112.9	95.9 (2)	92.7–99.1
glabella	114.5	111.3 (4)	105.8–117.0	106.3 (2)	103.2–109.3
bregma	111.2	109.4 (5)	105.4–111.0	113.2 (3)	112.1–114.7
lambda	103.0	105.6 (5)	101.2–108.5	108.0 (4)	104.5–114.1
opistocranion	99.2	102.5 (5)	99.7–106.3	103.5 (3)	102.6–104.6
inion	84.8	87.5 (5)	82.4–91.8	98.5 (4)	84.8–96.4

While a reduction in basal length of the cranium is said to characterize the early modern populations of Europe (Howell, 1951), we have found that this is not the case for the male comparisons reported here. If we use the nasion-inion diameter as a measure of the length of the cranial base, mean for all the earlier European Upper Paleolithic males is 191.7 mm, virtually identical to the Neandertal mean of 192.1 mm while Mladeč 5 value is below these means, it would appear that there is no significant difference in the European samples. However the Skhul / Qafzeh sample is markedly smaller, its maximum value is smaller than Mladeč 5 in the length of the cranial base (Table 3).

The maximum cranial breadth, measured across the supramastoid crests on the temporal bones, is 156 mm; this is very close to the Neandertal mean, and is greater than the Skhul / Qafzeh mean and maximum. Clearly the vault is very broad. The biparietal breadth of Mladeč 5 is almost as great as the maximum breadth, and breadths across the cranial base are above the maxima for both the Neandertal and the Skhul / Qafzeh samples.

In sum, we view the braincase as generally matching the Neandertals in overall form and dimensions (see Fig. 2). Only cranial height measures lie between these samples. However, certain Mladeč 5 basal breadths are quite large even compared with the Neandertals; the auricular and bimastoid dimensions exceed the maxima of both comparative samples. These basal breadths are unusual for other European early Upper Paleolithic males.

The cranial capacity Szombathy reports for Mladeč 5 is 1500 cc. (1926), a determination we regard as much too small. We have calculated two regressions to determine cranial volume. Formulae were determined from crania with actual endocast determinations. These are Dolní Věstonice 3 (1322 cc from Jelínek, 1954), Pavlov I (1472 cc, from Vlček, 1991), and Předmostí adult crania 3 (1608 cc), 4 (1518 cc), 9 (1555 cc) and 10 (1452 cc, all from Matiegka, 1934). These 6 crania formed the basis for developing two regressions for cranial capacity estimation, a least mean squares linear determination, and a power curve based on a least mean squares fit of logs. In both cases we used a volume estimation for the independent variable. The volume was estimated two different ways, using measurements that avoided including cranial superstructures. Biparietal breadth avoids the basal pneumatization, and the vertical height from the auricular point to bregma is the only comparable height measure for Mladeč 5. The two formulae are based on the following variables:

$$V_1 = \text{(nasion-opistocranion)*(biparietal breadth)*(auricular height)*}10^{-4}$$

$$V_2 = \text{(nasion-lambda)*(biparietal breadth)*(auricular height)*}10^{-4}$$

In both cases the volume estimate variables were calculated from the products of the measurements in millimeters. Linear and power curve regressions were determined for both volume variables from the sample of endocast capacities.

$cc = 3.21 * V_1 + 562.36$ (average error of 35 cc) $cc = 36.29 * V_1^{0.656}$ (average error of 34 cc)

$cc = 3.51 * V_2 + 510.57$ (average error of 33 cc) $cc = 31.49 * V_2^{0.685}$ (average error of 33 cc)

Four cranial capacities could be estimated from these formulae. The multiple estimates were averaged and the resulting capacity rounded to the nearest 5 cc., and presented in Table 3. Our Mladeč 5 determination is quite close to Billy's (1972). Our estimate of approximately 1650 cc. is larger than any male in the comparative samples. This undermines the popular conception that Neandertals have larger vaults than the Europeans who follow them in time.

The total vault has a "teardrop" shape as seen from above (Fig. 9), and like many Neandertals a flattened cranial rear and lateral angulation of the lateral superior orbital border can be seen. The Mladeč 5 breadth / length index is identical to the Neandertal mean, while relatively broader than any

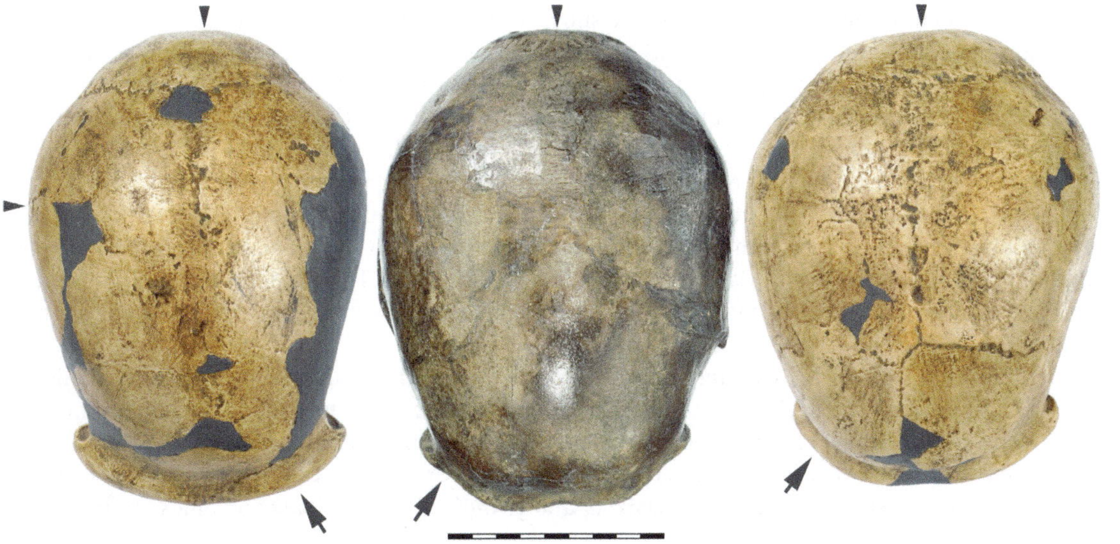

Fig. 9. Superior view of Mladeč 5 (center) compared with Spy 1 (left) and Spy 2 (right). Some regionally common similarities of the Mladeč specimen to these crania that can be seen in this view include the flattened cranial rear, the lateral angulation of the lateral superior orbital border. The position of maximum cranial breadth is anterior as it is in Spy 1, not the more common Neandertal posterior position seen in Spy 2. However, the general cranial dimensions and many specifics such as the postorbital constriction more closely resemble Spy 2 (Spy 1 and 2 are casts).

Skhul/Qafzeh male. Mladeč 5 is characterized by a maximum breadth that is low on the parietal bones, and posterior to the position of the external auditory meatus. The greatest breadth of the brain case is on the supramastoid crest of the posterior temporal squama; this is also the case for the La Ferrassie 1 and Spy 2 Neandertals, as well as for Skhul 5 and Předmostí 3. The parietal bosses are pronounced and are located approximately over the mastoids, which is slightly anterior to the position of the bosses in most Neandertal crania and in Skhul 5, although similar in position to Spy 1 (Fig. 9). At the anterior of the cranium, the supraorbital region projects markedly enough for the structure to have a superior surface, but it is not continuous over glabella where there is a depression separating the superciliary arches into two distinct sides.

The undistorted right cranial contour, as seen from the rear (Fig. 10), shows a fairly straight parietal side, slightly expanded at the base of the vault and inward leaning toward the top. It is most similar in form to the (the markedly higher) Skhul 5 and Předmostí 3 vaults and also resembles Spy 2. At the top of the vault, the parietal bones are also straight, angling upward toward the peak at the sagittal suture to form a low keel (a configuration of angle and curvature unmatched in the comparative sample, though most closely approached by Spy 2 and less so by Předmostí 3).

As seen from the side, the supraorbitals project moderately forward and are separated from the low curved frontal by a shallow although distinctly expressed supratoral sulcus, a form not dissimilar to many of the higher-vaulted Neandertals (Spy 2, La Chapelle). However, the supraorbital projection and supratoral sulcus expression are not extreme for the European Upper Paleolithic; they both are exceeded by the Pavlov and Předmostí 3 males.

Relative to the position of the auricular point, the frontal (and generally the upper face) extends quite anteriorly (Table 4). The distances from the auricular point to nasion and glabella, projected into the sagittal plane, are large. The nasion projection is slightly below the Neandertal mean but much greater than the Skhul/Qafzeh sample maximum. The glabella projection is even more dramatic, exceeding the Neandertal mean in spite of the furrow at glabella. These relationships are not an artifact of an unusual position for the auricular point position.

Total vault curvature in the sagittal plane shows only a moderate rise (Table 3). For instance, the arc/chord index for glabella-lambda is 132, about the same as the early Central European Upper Paleolithic male mean and average for the two males from Skhul/Qafzeh. This is above the range of the much flatter vaulted Neandertal males. The curvature of the vault's posterior, as measured by the bregma-inion arc/chord index, is much more flattened in Mladeč 5, as well as in the early Central European Upper Paleolithic and Neandertal samples. This contrasts with the rounded posteriors of the Skhul/Qafzeh males. The transverse arc/chord index as defined along the auricular point-bregma distance, shows a pattern of curvature relations similar to that of the vault posterior in the sagittal plane in that the Skhul/Qafzeh condition is most different. The Mladeč 5 and Neandertal values are more curved, while the males of the Skhul/Qafzeh sample show more flattened paracoronal contours.

Behind the middle of the parietal bones, the vault shows strong angulation with only moderate posterior projection. The sagittal distances from the auricular point to lambda, opistocranion, and inion respectively (Table 4) are each below the means of the comparative samples. The Skhul/Qafzeh sample means are always larger than Neandertals, and for inion the difference is considerable. Mladeč 5 is always smaller than Skhul/Qafzeh, and two of the three Mladeč 5 values are below the minima for the Levant males.

Distinct lambdoidal flattening begins some 55 mm anterior to lambda in the sagittal plane, and extends about 15 mm onto the occiput below lambda. The condition is very similar to that in Neandertal males, as well as in Předmostí 3. It is unlike the evenly rounded form of this region in most of the Skhul/Qafzeh male crania, and is also dissimilar to certain early Upper Paleolithic European specimens such as Brno 2 and Pavlov where the parietal bones above lambda are flattened, but the occipital below the lambdoidal suture is convexly rounded and projects markedly posteriorly.

Independent of this lambdoidal flattening, but beginning at the same position anterior to lambda, a groove runs along the sagittal suture to lambda and then bifurcates to extend along both sides of the lambdoidal suture for some 27 mm the breadth of this very shallow groove is about 11 mm in most places. This is a much weaker expression of a similar depression as found in Mladeč 6. It is not found in any of the other early Central European Upper Paleolithic males, nor in any of the male Neandertals.

Frontal

The frontal is rather steep from above glabella to metopion, while long and fairly flat from metopion to bregma. Both of these sections are slightly curved in the sagittal plane. The total curvature resulting from these two portions of the frontal is marked, exceeding the maximum curvatures for the comparative samples, as measured by the glabella-bregma arc/chord index (Table 5). The steepness of the forehead above glabella is due to a well-developed and centrally located frontal boss. Above this boss there is a very weakly developed frontal keel. The form of this region is most similar to Spy 2 and is approached by La Chapelle.

The inferior temporal line forms a strong distinct ridge for the full length of the frontal. At the inner border of the temporal fossa, the temporal line diverges into an inferior and a superior line. This superior line arches strongly upwards – a condition common in the earlier European Upper Paleolithic males as well as the males from Skhul/Qafzeh. The Neandertal males differ considerably from this condition, with temporal lines that generally travel in a straight line posteriorly from the temporal notch, with little upward extension. While the Neandertals show variation in this morphology, only Spy 2 shows the arching characteristic of the later specimens.

The vertical inner wall of the Mladeč 5 temporal fossa is somewhat convex. Its anterior surface encroaches up onto the wing of the frontal, resulting in an almost horizontal surface just behind the

Table 5. Dimensions (mm) and indices of the male frontal bones

	Mladeč 5	Mladeč 6	Neandertal		Skhul/Qafzeh	
			Mean (n)	Range	Mean (n)	Range
Breadths						
Minimum frontal	106.2		109.0 (8)	102.2–111.0	105.1 (5)	96.0–110.2
Maximum frontal (co-co)	126.0	125.0	123.5 (6)	120.4–127.0	118.1 (4)	112.0–127.0
Minimum/maximum index	84.3		89.1 (6)	86.6-91.2	88.1 (4)	80.7-92.8
Outer orbital Br (fmt-fmt)	120.9		120.4 (7)	116.0–124.3	117.1 (4)	109.0–124.1
Inner biorbital (fmo-fmo)	109.0		113.1 (7)	107.5–119.7	108.9 (4)	96.0–118.8
bi-stephanion	113.6	93.0	118.5 (5)	112.4–122.0	113.2 (4)	109.0–121.0
bi-st/maximum breadth index	90.2	74.4	95.5 (5)	91.9–98.1	95.9 (4)	92.4–98.5
ant orbit angle	39.0	37.0	33.9 (3)	32.7–34.7	30.9 (2)	30.2–31.5
ant interorbital (mf-mf)	25.2	26.2	23.3 (3)	22.0–25.5	19.9 (2)	19.6–20.2
fmo-fmt	9.9		11.4 (6)	10.0–12.8	11.1 (5)	9.0–12.0
Lengths						
gl-b	113.6	117.6	113.1 (5)	105.5–118.8	105.3 (4)	98.6–110.5
na-b	116.0	120.5	113.7 (5)	106.8–121.9	109.0 (3)	107.8–110.5
Arcs						
gl-b	127.0	127.5	119.3 (6)	110.0–126.0	116.0 (2)	113.0–119.0
na-b	140.0	140.0	127.2 (5)	119.0–135.0	130.5 (2)	126.0–135.0
gl-b arc/chord Index	111.8	108.4	107.2 (5)	105.2–109.8	107.5 (2)	107.2–107.7
Supraorbital Tori						
na-fmo	56.0		61.3 (5)	59.4-63.4	54.8 (3)	49.7–59.2
na-fmt	63.9		68.1 (5)	66.1–71.7	61.1 (3)	54.9–65.7
Medial projection from internal	17.5		22.8 (4)	19.0–24.5	18.8 (2)	17.5–20.0
Projection at orbit center [1]	14.5	17.4	22.4 (9)	18.5–20.5	19.7 (4)	15.5–22.5
Lateral projection	23.0		23.8 (5)	22.0–26.0	23.4 (4)	22.5–24.1
External length from sulcus	19.0	19.3	20.6 (7)	17.5–25.1	20.3 (6)	16.5–23.7
Medial height [2]	20.5	19.9	17.9 (7)	14.8–20.5	16.8 (4)	15.0–19.1
Height at the highest point [2]	18.0	17.7	13.6 (7)	8.8–17.2	13.0 (4)	11.8–16.0
Height at the orbit center [1,2]	16.2	16.2	13.6 (9)	10.4–15.9	12.3 (5)	9.2–16.5
Lateral height [2]	8.5		10.9 (7)	9.1–12.3	11.6 (4)	10.2–13.8
Center/Medial ht index [3]	79.0	81.4	76.4 (7)	50.7–87.8	74.7 (2)	62.9–86.4
Lateral/Medial ht index [3]	41.0		64.3 (6)	51.1–79.1	81.3 (2)	72.3–90.2

[1] This is thickness as defined in Smith and Ranyard (1980) or Wolpoff et al. (1981).
[2] The orbital center is not the midorbit position of Smith and Ranyard (1980), or Wolpoff et al. (1981).
[3] Does not include Qafzeh 9.

coronally oriented aspect of the anterior temporal crest. Contributing to this morphology is the fact that this posterior face of the zygofrontal suture, which forms the lateral edge of the temporal fossa surface, is anteroposteriorly oriented and short. This total configuration is unusual in the west. With the exception of Spy 2 (and to a lesser extent Skhul 4), the region is quite unlike the other comparative European and Near Eastern crania.

The Mladeč 5 temporal lines converge strongly toward the midline. A sharp temporal ridge is developed along the inferior line, which is stronger than the superior line on the right side, although not on the left. The angulation between the sides and top of the frontal squama is marked by the superior line which forms a lower, more rounded ridge (a more torus-like structure) for the entire length of the bone, separated from the more medial squama by a shallow groove that can be palpated along most of its length. The bistephanion breadth is small, similar to the Neandertal mean but below that of Skhul / Qafzeh. This is not an artifact of the breadth of the frontal. Relative to the maximum frontal breadth this distance is also small.

In fact, the frontal bone is large. Mean sagittal length of the frontal as measured from nasion is greater than the Neandertal mean and much larger than the Skhul / Qafzeh measurement. Measured from glabella the Mladeč 5 length is almost identical to Neandertal sample, but once again, the Skhul / Qafzeh mean is much shorter. The same can be said for the maximum frontal breadth. Mladeč 5 is just above the mean Neandertal value, while the Skhul / Qafzeh mean is less. Even the minimum frontal breadth, while less than the Neandertal mean, exceeds the Skhul / Qafzeh average. Thus, for most gross frontal dimensions the Skhul / Qafzeh sample remains the smallest (but it is not the smallest in cranial capacity).

In the relation of minimum to maximum frontal breadths, Mladeč 5 has a small relative breadth compared with the earlier samples. Mladeč 5 mainly differs from these males in having a narrower postorbital region. Laterally, the Mladeč 5 outer orbital breadth *(fmt-fmt)* is similar to the Neandertal and the Skhul / Qafzeh means. This contrasts with other, narrower European earlier Upper Paleolithic male values. The inner biorbital breadth *(fmo-fmo)* for Mladeč 5 is relatively smaller, below the Neandertal mean but still large for an earlier Central European Upper Paleolithic male. The difference between these two Mladeč 5 biorbital breadths reflects the flatter upper face of Mladeč 5, with its more forward facing orbital pillars. The change in orbital pillar orientation from more sagittal to more forward facing (i.e., coronal) results in an outer biorbital breadth dimension that is large relative to the inner biorbital breadth. Thus, in this case the dimensional similarity of Mladeč 5 and the Neandertal sample does not reflect a similarity in morphology.

To further examine the upper facial flatness variation more precisely, we determined a value estimating the projection of nasion anterior to the outer orbital rims, by calculating the sagittal distance of nasion anterior to the bi-*fmt* line (Table 11, and see Wolpoff et al., 1981). This estimate approximates the transverse angulation of the superior portion of the face; the greater the projection of nasion, the smaller the angle between the two superior orbital margins. The calculated values clearly separate the Neandertals, with their marked lateral facial angulation (large nasion projection), from the Skhul / Qafzeh specimens with their *very* flat upper faces. Mladeč 5 is well below the Neandertal range, but at the top of the Skhul / Qafzeh range, and the Mladeč specimen could be described as having an angled upper face compared with Levantine males. In part, this upper facial flatness determination results from the depressed nasal root discussed above, and the actual superior orbital contour is more angled than this projection seems to indicate (Fig. 9)[15]. The flatter face of Mladeč 5 is primarily a consequence of its greater lateral facial length (*au-fmo*, see Table 11) compared with Neandertals, a dimension that exceeds the maxima for the two comparative samples.

The Skhul / Qafzeh male nasion projection mean is the flattest of all the sample means. The Mladeč 5 facial flatness value barely lies within the top (i.e. most angled portion) of the Skhul / Qafzeh range. Thus, while the Mladeč 5 upper face is relatively flat in comparison with the Neandertal males, the degree of flatness is quite different from the normal Skhul / Qafzeh condition. Moreover, the Mladeč 5 face has rather different dimensions. This is because the Skhul / Qafzeh sample is much smaller in absolute facial dimensions than any of the Europeans. In fact, all of the male earlier Central European Upper Paleolithic faces are flat for the same reason, because these specimens add an elongated lateral facial length *(au-fmo)* to an already projecting upper facial region such as that found in the Neandertals. In contrast, the Skhul / Qafzeh faces are flat because both the upper nasal projection and the lat-

15 In superior view the Mladeč females appear to have even flatter upper faces than Mladeč 5, yet their nasion projection estimations are virtually the same. The difference lies in the fact that the nasal root is not as depressed in these females as it is in the male.

Table 6. Cranial thickness (mm) in adult males

Position	Mladeč 5	Mladeč 6	Mladeč 39	Mladeč 40	Mladeč 43	Mladeč 4	Neandertal Mean (n)	Neandertal Range	Skhul/Qafzeh Mean (n)	Skhul/Qafzeh Range
Frontal										
lateral eminence	8.5	9.5				8.0[1]	7.8 (9)	6.5–9.5	6.7 (5)	4.1–8.5
Parietal										
vertex		4.2					7.2 (5)	5.5–8.5	9.0 (1)	
lambda	6.0						7.4 (7)	6.0–8.0	9.6 (3)	8.5–11.5
asterion	8.5	8.4					6.8 (9)	5.5–9.2	11.5 (3)	7.3–15.0
mastoid notch	7.0						8.3 (7)	4.0–13.4	11.8 (2)	11.0–12.5
pterion	5.0						6.8 (3)	6.5–7.5	5.1 (2)	4.8–5.5
anterior eminence	6.5	7.1					7.0 (8)	5.0–9.1	6.8 (4)	4.0–12.7
middle eminence	6.0	8.0	6.0				7.9 (10)	6.2–10.3	7.0 (6)	4.5–11.0
posterior eminence	6.5	6.4	7.0	5.0			8.1 (8)	6.0–9.6	6.3 (3)	4.5–9.0
Occipital										
lambda	7.5			7.7			7.7 (5)	6.0–9.0	8.9 (3)	8.5–9.0
endinion	11.0				15.0		12.4 (5)	9.9–14.3		
occ mastoid suture	8.5						7.8 (5)	7.0–8.6	14.0 (2)	12.0–16.0

[1] This number was reported by Szombathy "from the middle of the bone" and is not exactly equivalent to the lateral eminence position compared here, but the values are often quite close.

eral facial length *(au-fmo)* have *reduced* (the nasal region more so than the lateral facial region) relative to the Near Eastern Neandertal (Shanidar, Trinkaus, 1983; Amud, Tabun) condition.

Frontal squama thickness was measured at the lateral eminence, the approximate center of the lateral side of the frontal boss (Table 6). The Mladeč 5 bone is thick, 8.5 mm, which is only approached by the Neandertal mean and at the very top of the range of the Skhul/Qafzeh males. It is considerably thicker than other early Upper Paleolithic males (4.4 mm, n=5).

The remaining frontal feature of Mladeč 5 is its supraorbital structure. There is a continuous bar of bone over the orbits for the entire length of the anterior frontal margin, with a very shallow depression in the structure just above glabella (the supraglabellar fossa), which shows very faint remnants of the metopic suture. The anterior-facing exterior surface of this structure is moderately pitted with at least 10 large foramina and in small detail provides an excellent example of the vermiculate pattern. The shape of the superior-facing surface is more complex. Over the highest point of the orbit, where the supraorbital arch attains its maximum vertical height, a thick ridge connects the top of the supraorbital to the frontal squama behind it, at its lateral anterior face (Fig. 8). Lateral to this ridge, up to the margin formed by the superior temporal line, the forehead is very flat, a condition which Szombathy (1925, 75) referred to as a "flattened trigonum supraorbitale". Medially, between these ridges, a very shallow supratoral sulcus separates the supraorbitals from the vertical aspect of the frontal squama. In the Neandertal males the supratoral sulcus is continuous between the temporal lines. For at least two specimens, La Chapelle and Spy 2, a connecting ridge can be palpated but it is weakly expressed and does not interrupt the sulcus. Of these males, Spy 2 most closely resembles the Mladeč 5 condition. Brno 2, Předmostí 3, and Pavlov lack these ridges and have a continuously developed supratoral sulcus that extends between the temporal lines. The Mladeč 6 condition is somewhat different from these (see below).

The supraorbital arch is not divided from the lateral torus structure by a supraorbital groove (Fig. 8). Its vertical height varies, being greatest just lateral to glabella and thinning slightly to the highest point of the orbit, which is the position of the supraorbital notch. Lateral to this, the feature bends sharply inferiorly and thins to about half this height, then thickens slightly at the lateral frontal trigone. Thus, the thinnest part of the supraorbital is at the lateral third of the undivided structure, on the lateral torus. This structure is one of the most striking features of the cranium. There has been some discussion as to whether this is a true supraorbital torus and the extent to which it does or does not resemble the supraorbital torus in Neandertals.

According to Smith and Ranyard (1980, 589), the Neandertal supraorbital torus can be characterized as: "... basically an osseous bar, extending continuously across the inferior margin of the frontal bone. The torus forms an arch over the superior margin of each orbit and appears depressed superiorly in the midline by the presence of a supraglabellar fossa." This description fits some, though not all, of the Skhul / Qafzeh specimens as well. The question of interest is to what extent the Mladeč 5 supraorbital region resembles this condition. Again according to Smith and Ranyard (1980, 600): "The supraorbital region of the European representatives of *Homo sapiens sapiens* typically exhibit a considerably different form from that of European Neandertals." Several of the early Central European Upper Paleolithic males, particularly Předmostí 3 and Pavlov appear as scaled down versions of the Neandertal condition as exhibited by certain specimens such as La Ferrassie 1, if the supraorbital notches are ignored. In particular, they share three traits with the Neandertals; a continuous supraorbital structure that is thickest in the superciliary region, a fairly even development of supraorbital height[16], and an even, roughly horizontal upper orbital margin. Mladeč 5 differs from this condition and is less like some of the Neandertals in three regards: (1) the contrast between the medial and lateral aspects of the supraorbital region is greater; (2) the supratoral sulcus as mentioned above is less well excavated; (3) the superior orbital margin is not horizontal but rather exhibits a double arched condition with the most superior point on the arch much higher than the lowest point on the superior orbital margin. The highest point on the superior orbital margin is located at the supraorbital notch; because the supraorbital is thickest in this region, the double arched appearance over the orbits is accentuated. In terms of the supraorbital region Mladeč 5 is not at all unique within the early Central European Upper Paleolithic sample. For instance, Brno 2 exhibits an even more extreme double arched form with both the shape of the superior orbital margin and the maximum height of the superciliary arches directly over the medially located highest point of the orbits contributing to the expression. Mladeč 6, as discussed below, has an even shallower supratoral sulcus.

The above comparisons are, of course, individual. The general pattern is more easily discerned by comparing the samples metrically (Table 5). Mladeč 5 is characterized by a very thick supraorbital structure, thinning most markedly just medial to its most lateral aspect, and then thickening slightly to the point where lateral height in the table was measured. This thinning is also reflected in the narrow orbital pillar dimension *(fmt-fmo)*. The male Neandertal average is thinner than Mladeč 5 in the most medial portion, at the highest point of the orbit (invariably measuring the height of the superciliary arch), and at the center of the orbit. But the Neandertals are taller only in the lateral portion, at the frontal trigone. The Skhul / Qafzeh males follow the Neandertals in this sequence, although height is slightly less, medially and at the high point of the orbit. Examining the index of central to medial height, the Neandertal mean exceeds Skhul / Qafzeh while Mladeč 5 *exceeds* the Neandertal mean. This is unusual for an early Central European Upper Paleolithic male. However, the ratio of lateral to medial height (41) shows Mladeč 5 to be more like the early Central European Upper Paleolithic males, in that the lateral aspect of the supraorbital is relatively small (mean = 54, n = 3). The Neandertals have somewhat thicker lateral aspects, but both of these European samples contrast with the Skhul / Qafzeh males in which the lateral ratio of 81.3 mm is very high and the structure is of a much more uniform height.

---

16 The vertical dimension of the supraorbital structure, see Weidenreich (1943).

Of course, individual Neandertals vary considerably in both the height of the torus and the evenness of its development. Thus, specimens such as La Ferrassie 1 and Spy 2 thin markedly over the lateral third of the structure. Moreover, Spy 2 thickens slightly again at the lateral frontal trigone so that there is the same sequence of relative thickness and thinness over the length of the torus. The absolute dimensions differ from Mladeč 5, in that Spy 2 is thicker laterally. The late dated Vindija 261 specimen shows the most pronounced relative thinning in the center of the lateral third of the torus, and very closely approaches the Mladeč 5 condition (Wolpoff et al., 1981, Fig. 10 and discussion).

In terms of its center-orbital projection from the internal surface of the frontal, the Mladeč 5 supraorbital arch is quite small relative to the comparative samples, in fact under their minimum values. However when projection is measured from the supratoral sulcus[17], on the outside of the bone, the Mladeč 5 dimensions are similar to both comparative sample means. These are two alternative ways of measuring the projection of the supraorbital region, and their different patterns of variation reflect the differences in shape and orientation. Lateral projection from the internal surface is also quite similar between the Mladeč 5, Neandertal, and Skhul / Qafzeh values.

This morphological complex, reflecting the differences between Mladeč 5 and the Neandertals, is quite common in several of the other European early Upper Paleolithic male crania, for instance Brno 2. However, as noted above other specimens such as Předmostí 3 and Pavlov are more Neandertal-like in the evenness of height across the lateral third of the structure and the horizontality of the orbit's superior border. The Skhul 4 and 5 crania, as well as Qafzeh 6, are similar to Předmostí 3 in this regard. However, the Skhul / Qafzeh male sample is far from uniform. For instance, Qafzeh 7 has a very projecting supraorbital structure with a deep supratoral sulcus. Its form is strongly everted, and most of the supraorbital surface faces in an anterosuperior direction. If height were measured along this axis, the figure would be substantial. A true vertical measure of height would make Qafzeh 7 specimen somewhat thinner, with a distinct centrolateral thinning similar to that in the Vindija supraorbitals. Seen against this complex background, the Mladeč 5 morphology does not reflect a simple case of overall decrease in robustness from a Neandertal or Skhul / Qafzeh condition.

On the inferior face of the supraorbital, the expression of the orbital notches is asymmetric. On the right side, which we believe is the normal side; a notch is centered about 30 mm from the midline, 4.5 mm in width, and about 2 mm in depth. This notch is located at the highest point of the orbit and there is a shallow groove medial to the notch. On the left side, the notch is divided into two elements that are more distinct. The lateral notch is in the same position as on the right, about 30 mm from the midline. Medial to this notch there is a groove, much deeper and accentuated by a ridge of bone between it and the notch. This ridge extends superiorly to the top of the supraorbital arch. Unlike the left side, the area above the supraorbital notch is convex in shape. The distinctive morphology of the left supraorbital notch seems to be a consequence of the healed injury occurring directly above the supraorbital. It appears that the area on the brows, including the supraorbital notch, was remodeled with a deep vascular channel medial to the supraorbital notch. From the preserved bone, we can find no evidence for infection of the orbital region.

The orbital plate, forming the roof of the orbit, arches upward, posterior to the superior orbital border. The most superior point on the orbital roof is well above and posterior to the superior orbital border. The Skhul / Qafzeh males are similar in this region, but for the Neandertal males, the orbital plate is markedly flatter and rises only slightly above the superior orbital border.

On the part of the frontal that lies between the orbits, the interorbital dimensions are extremely broad. The distance between the internal orbital angles exceeds the maxima for the comparative samples, while the anterior interorbital breadth is at the very top of the Neandertal range. The most superior breadth dimension across the nasal bones (Table 10), 12.7 mm, contrasts in being ab-

17 Weidenreich's (1943) "torus length".

solutely narrow, and of course relatively narrow compared with the anterior interorbital breadth. The nasal root is markedly depressed, as we noted above. Unfortunately, the angle of the nasal bones cannot be ascertained because the nasals themselves are not actually preserved on the original specimen, and while the casts shows the remnants of nasal bones, these are only what appear to be some unidentifiable internal fragments of bone stuck along the frontonasal suture. What remains on the original specimen is only the anterior surface of the frontal's contribution to the frontonasal suture.

The superior border of the nasal bones extends about 9 mm above the frontomaxillary suture, and thus nasion approaches the position of glabella, lying at the very base of the supraorbital arch. The frontonasal suture forms an inverted "U". This form probably corresponds to the interdigitation of the frontal bone with a nasal pillar such as is also found in the Mladeč 6 male and the Mladeč 1 female. For further discussion of this form see the description of the Mladeč 6 morphology.

Internally, breaks on the inferior frontal margin suggest the presence of a relatively large frontal sinus (Skinner and Sperber, 1982; such a sinus is common in Pleistocene Europeans according to Szilvássy et al., 1987). The internal aspect of the coronal suture is closed and totally obliterated. Anterior to it, along the left side of the *crista frontalis* are two circular pacchionian depressions, about 8.5 mm in diameter.

Parietal bones

The Mladeč 5 right parietal is complete, and portions of the left are preserved as described above. The temporal lines are visible for their entire extent on the bone, forming a weak ridge. There is, however, no angular torus. The superior temporal lines are high on the vault, for instance passing just above the parietal bosses. The minimum distance between them lies below the means of the comparative samples (Table 5), in fact it is below their ranges. Given the greater size of the Mladeč 5 parietal bones relative to the Neandertals (see the parietal radius data in Table 7), this position high on the vault indicates an increased temporalis muscle attachment area and suggests a more powerful temporalis muscle than is the norm for the earlier Neandertals or for Skhul / Qafzeh. The difference between Mladeč 5 and the Neandertals is most evident for the posterior portion of the muscle. This region of attachment is both higher and more posteriorly expanded in the Mladeč 5 vault. The most posterior extension of the superior temporal line is virtually at the lambdoidal suture, a position similar to that in most of the other earlier Central European Upper Paleolithic crania. The superior temporal line reaches the lambdoidal suture about midway between lambda and asterion, and travels inferiorly along the suture to asterion where it turns anteriorly, contributing to the mastoid crest. In this regard, the Skhul / Qafzeh sample more closely resembles the European Neandertals, where the superior temporal line meets the lambdoidal suture in a much lower position, just above asterion, and turns anteriorly to form the mastoid crest almost immediately.

The inferior temporal lines are more difficult to discern, especially in the anterior location. Posteriorly they can be traced from the supramastoid crest from which they evenly curve posterosuperiorly to a point as close as 24 mm from the lambdoidal suture. This is some 36 mm superior to asterion. From here they continue in their curvature and as they course anteriorly they reach the most prominent part of the parietal boss.

The Mladeč 5 parietal bones are larger than the Neandertal mean values in all linear dimensions. In fact, with the exception of the transverse chord from bregma to asterion the dimensions are near or even above the Neandertal maxima. The height measures of the bone are large even for a Central European early Upper Paleolithic male. The four borders are most similar to the Skhul / Qafzeh means, although the anterior edge is somewhat longer. Whether measured this way, or by the parietal radius, it is clear that this is a fairly large bone.

Table 7. Parietal dimensions (mm) and indices

	Mladeč 5	Mladeč 6	Neandertal		Skhul/Qafzeh	
			Mean (n)	Range	Mean (n)	Range
Chords						
br-l	119.1	118.9	110.8 (9)	102.9–120.2	118.6 (4)	111.8–129.8
k-ast	87.5		83.2 (5)	76.8–87.0	88.3 (4)	79.9–91.3
l-ast	94.8	84.0	90.8 (5)	84.4–98.3	93.8 (5)	89.2–96.8
k-br	106.3		99.1 (6)	92.4–105.8	96.7 (4)	94.0–100.3
k-l	139.0		133.9 (6)	126.7–140.0	138.2 (5)	132.8–146.5
br-ast	148.0	148.2	135.9 (5)	130.1–139.3	142.0 (4)	140.4–144.4
br-st	64.4	54.5	67.6 (5)	62.5–71.0	60.2 (4)	51.3–68.8
Min btwn temporal lines	107.5	56.0	116.9 (5)	111.6–122.0	113.2 (4)	109.0–121.0
Parietal radius	72.6		69.0 (5)	68.2–70.2	71.2 (3)	70.5–72.1
Arcs						
br-l	129.0	131.0	119.8 (8)	107.0–128.0	132.7 (3)	121.0–144.0
l-ast	103.0	92.0	106.1 (5)	101.0–120.0	100.9 (4)	96.5–110.0
k-br	121.0		116.1 (6)	107.0–122.5	111.0 (3)	108.0–113.0
k-l	162.5		168.9 (6)	156.5–178.0	168.0 (4)	162.5–175.0
br-ast	179.0	170.0	166.5 (5)	164.0–168.5	173.0 (3)	165.0–182.0
br-st	66.0	71.0	71.0 (5)	66.0–76.0	63.0 (3)	52.0–75.0
Parietal angles (degrees)						
br-l	110.2		110.4 (5)	104.6–118.4	116.2 (3)	105.1–126.8
k-ast	74.1		74.3 (5)	66.7–79.4	75.3 (3)	67.3–80.5
l-ast	84.5		81.8 (5)	76.3–89.5	83.9 (3)	81.0–86.7
k-br	94.1		93.8 (5)	85.4–97.8	84.2 (3)	82.0–87.8
Arc/Chord indices						
br-l	108.3	110.2	108.9 (8)	104.0–115.6	109.7 (3)	108.2–110.9
l-ast	108.6	109.5	116.8 (5)	112.3–122.1	108.7 (4)	103.3–114.9
k-br	113.8		118.9 (6)	115.8–126.7	115.4 (3)	112.7–118.4
k-l	116.9		127.0 (6)	121.1–132.5	121.4 (4)	118.5–126.9
br-ast	120.9	114.8	122.8 (5)	120.6–126.1	121.0 (3)	116.3–129.1

The Mladeč 5 parietal bones are squarer than rectangular in form. The sagittal border is longer than the temporal border, and the coronal border is longer than the lambdoidal border - a relative sequence of sizes shared by the comparative sample means, and most individual specimens. A general method for examining parietal shape was suggested by Delattre and Fénart (1958), and was used by Piveteau (1967) in discussing the very rectangular parietal from Lazaret. The method is based on circumscribing a circle around the four points defined by the parietal corners (using asterion as the inferoposterior point), and then calculating the angles formed by the intersection of the transverse chords (bregma-asterion and lambda-krotaphion). These angles can be used to describe parietal shape *independent of size,* while the radius of the circumscribed circle is a measure of size. Data for these angles and for the radius given in Table 7 show the Mladeč 5 parietal to be shaped *very similarly* to the Neandertal mean, except for its larger inferior and smaller superior angles. The superior and anterior angles are quite unlike the Skhul/Qafzeh sample.

Thus, the Mladeč 5 parietal is nearly identical to the European Neandertals in shape, and to the Skhul/Qafzeh samples in size. We believe that the most significant aspect of the Mladeč 5 parietal form is probably found in its large posterior angle, corresponding to its expanded occipital.

The Mladeč 5 anterior and posterior parietal border arc/chord indices are low relative to the Neandertals, and most closely approximate the indices for the small Skhul/Qafzeh sample. The difference between the comparative samples and the Mladeč specimen is minimal for the superior border. We believe this difference in anterior and posterior border curvature reflects the fact that Mladeč 5 has parietal bones transversely more flattened and angular (i.e., less evenly rounded) than most European Neandertals. The flat nature of the Mladeč 5 parietal bones is directly reflected in transverse indices. Thus, the index for the krotaphion-lambda distance, angling across the entire bone, is very low – below the ranges of the two comparative samples.

The earlier Central European Upper Paleolithic sample generally shows either virtually no difference or some reduction in parietal thickness, compared with the Neandertals (Table 6), where sample sizes exceed 1. In this context the Mladeč 5 parietal bones are thin, and contrast with the thick frontal bone. Except at asterion they are below the mean values for the Neandertals, and they are generally also below the early Central European Upper Paleolithic means. Mladeč 5 is very much thinner than the Skhul/Qafzeh mean values around the parietal edges, but of equivalent thickness or even thicker centrally.

Occipital

The Mladeč 5 occiput is quite large, especially in its occipital squama (Table 8). It is flattened superiorly (as part of the lambdoidal flattening described above). Below the region affected by lambdoidal flattening most of the remaining occipital plane is also flattened, resulting in a virtually vertical posterior surface (in the approximate Frankfort Horizontal orientation) for the more inferior aspect of the occipital plane. There is a clear line of demarcation between these two flat surfaces, about 20 mm inferior to lambda. The vertical flat surface is about 45 mm in breadth. Lateral to it, the sides of the occipital plane angle sharply (about 45°) toward asterion.

Compared to the Neandertals, the Mladeč 5 occipital plane is quite long, as measured to the most superior extent of the nuchal line. It exceeds the Neandertal range and is quite large for an early

Table 8. Occipital bone and nuchal region dimensions (mm) and indices

	Mladeč		Neandertal		Skhul/Qafzeh	
	5	6				
			Mean (n)	Range	Mean (n)	Range
Breadths						
Biasterionic	120.0	126.5	125.4 (5)	123.0–129.9	117.6 (6)	108.5–129.0
Nuchal attachment area	132.0	143.0	101.6 (5)	87.0–113.0	90.8 (3)	83.0–97.0
Nuchal/bi-ast index	110.0	113.0	81.1 (5)	69.7–91.9	79.1 (3)	74.8–82.3
Lengths						
l-i	66.9	62.7	58.0 (6)	51.0–62.8	67.9 (6)	60.9–76.0
l-lowest i	69.0	70.5	61.1 (5)	58.0–64.5	65.2 (4)	60.9–69.6
l-ast	94.8	84.0	90.8 (5)	84.4–98.3	93.8 (5)	89.2–96.8
i-ast	68.7	72.7	72.0 (4)	67.0–76.0	67.1 (5)	59.8–80.0
Arcs						
l-i	73.5	69.0	64.3 (6)	58.0–70.0	68.3 (4)	64.5–71.0
l-ast	103.0	92.0	106.1 (5)	101.0–120.0	97.3 (4)	95.7–100.0
l-i arc/chord index	109.9	110.0	111.0 (6)	108.3–113.7	106.9 (4)	100.5–110.8
l-ast arc/chord index	108.6	109.5	116.8 (5)	112.3–122.1	108.7 (4)	103.3–114.9
i-biasterionic projection	33.5	35.8	36.1 (4)	27.8–44.3	33.1 (5)	25.2–47.3

Upper Paleolithic European. However, this length is slightly less than the mean for the Skhul/Qafzeh sample. An alternative position for the top of the nuchal plane, which we call "lowest inion", is located at the most inferior point on the tip of the inion prominence at the midline. The position is more inferior than is the inion position defined by Martin (1928, 615)[18] that is used above. The length of the occipital plane as measured to it exceeds the means of both comparative samples. By either measure the Mladeč 5 occipital plane is quite long. The occipital plane is flat, as indicated by the arc/chord index of the lambda-inion dimension. This flattening is common in all of the samples. The Mladeč 5 value is well within the Neandertal range (it is flatter than the Spy occiput, see Fig. 2), but barely within the range of the Skhul/Qafzeh males, whose occipital bones are even flatter by this measure.

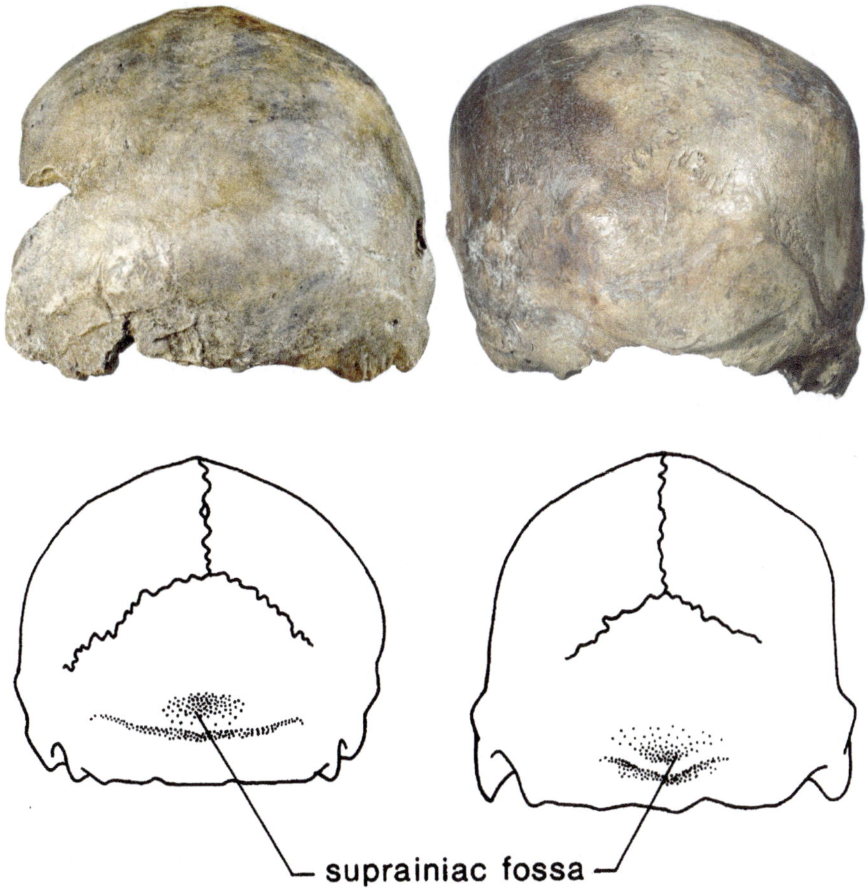

Fig. 10a, b. Mladeč 5 (right) and a cast of 6 in occipital view. Below them is a diagram showing the suprainiac fossa differences, comparing a Neandertal (left) and the usual early Upper Paleolithic character state. Mladeč 6 has a weakly expressed suprainiac fossa similar to the Neandertal condition, while Mladeč 5 lacks a suprainiac fossa. Caspari argues that the Neandertal fossa is not truly homologous with the common fossa form in the European Upper Paleolithic, which makes it all the more interesting that the only suprainiac fossa at Mladeč resembles Neandertals

18 The Martin definition for inion is the line connecting the highest points on the superior nuchal line, where it crosses the median sagittal plane.

Many particulars of the nuchal torus region and nuchal plane are preserved (see Caspari, 1991 for further details, our description replies heavily on Caspari's work). The lower border of the vertically flat occipital plane is bounded by the superior nuchal line, which forms a ridge that extends across the entire occiput. There is a weak supreme nuchal line that can be discerned above this ridge. Centrally, the supreme nuchal line dips inferiorly to form a distinct inion prominence (external occipital protuberance) that is situated slightly above the position of the superior nuchal ridge. This prominence is 14 mm broad and about 3 mm in sagittal height. Lateral to it, the supreme line rises to a maximum height of 12.5 mm above the superior nuchal ridge. Lateral to this position, the supreme line parallels the superior nuchal ridge, approaching but not meeting it. This line can be traced to asterion, where it meets the mastoid crest.

The superior nuchal ridge attains its greatest vertical thickness at the midline, where it forms a downward pointing triangle, inferior to the inion prominence described above. Laterally, it thins progressively as it passes in a superolateral arc. The arc reaches its highest extent 42 mm lateral to the midline (15 mm from the lambdoidal suture). From this point, the weak ridge bends inferiorly and anteriorly, paralleling the lambdoidal suture, and meets the mastoid crest at asterion.

Inferior to this ridge the nuchal surface is sharply angled, at about 30° to the vertically flat posterior part of the occipital plane. Between the two most superior points on the superior nuchal ridge, the most posterior portion of the nuchal plane is deeply excavated. This results in a markedly overhanging nuchal ridge for the central portion of the bone. The portion of the nuchal surface remaining prior to the 1945 damage (extending to the approximate position of the inferior nuchal line) is roughly developed and slightly convex. Combined with the lambdoidal flattening, the angulated nuchal plane sets off the back of the occiput in a well-developed occipital bun (chignon)[19]. The chignon of Mladeč 5 only differs from the Neandertal condition in two ways, it is transversely narrower than any male Neandertal bun and it lacks a suprainiac fossa. None of the Skhul / Qafzeh remains have lambdoidal flattening or occipital bunning (McCown and Keith, 1939; Vandermeersch, 1981).

The reduction in occipital (biasterionic) breadth between the Neandertal and the earlier Central European Upper Paleolithic samples is dramatic. Mladeč 5 is intermediate; although much closer to the Neandertal mean than other Central European Upper Paleolithic males, its value is below the Neandertal range (Table 8). Biasterionic breadth most closely approximates the average for the Skhul / Qafzeh sample. The Mladeč 5 nuchal muscle attachment area breadth exceeds the occipital breadth, a relation that generally is not seen within any of the comparative samples (including the early Central European Upper Paleolithic one) as is indicated by the index between these two breadths. In fact, the direct maximum transverse measurement of the nuchal muscle attachment area exceeds every specimen in the comparative samples. The relatively and absolutely great breadth of the Mladeč 5 nuchal muscle attachment area reflects what we perceive as a marked development of the nuchal musculature in this specimen. The index comparing breadth of the nuchal muscle attachment area to occipital bone breadth shows that these muscles take up the full breadth of the occiput, and indeed extend beyond it, which is unlike any specimen in the comparative samples.

The tendency to expand the nuchal muscle attachment area in breadth is characteristic of all of the earlier Central European Upper Paleolithic males (Caspari, 1991), although none of these attain the Mladeč 5 extreme. This is just the opposite of the biasterionic breadth comparisons where the European early Upper Paleolithic male mean is small. Thus, in this context the Mladeč condition is an extreme case of the tendency to broaden the nuchal muscle attachment area that prevails throughout the entire earlier European Upper Paleolithic sample.

---

19 A "hemi-bun" is distinguished from a bun by the fact that the lambdoidal flattening begins more posteriorly, only involving the occiput (Smith, 1982). By this definition, the Mladeč 5 and 6 morphologies, while different, are those of a true bun.

Morphologically, the expansion of the nuchal muscle attachment area is accomplished by a shift in the anterior course of the superior nuchal line. In Neandertal males, such as La Ferrassie 1 and Spy 2 (and probably the others as well), the superior nuchal line curves evenly from a point lateral to the suprainiac fossa, extending onto the paramastoid crest *medial to the mastoid process.* In Mladeč, the superior nuchal line begins its curve about the same distance from the midline. It swings markedly anteriorly and is parasagittally oriented at the approximate position of the retromastoid process. At this point, the line swings laterally following the contour of the supramastoid surface and extends on to the medial posterior face of the mastoid where it forms the mastoid crest. Thus, the whole posterior of the mastoid is incorporated into the nuchal muscle attachment area. It is possible that the contribution of the nuchal line to the paramastoid process in Neandertals helps account for its marked development.

The exact details of the extent to which the total area devoted to nuchal musculature is consequently expanded in Mladeč 5 remain unclear. This is because the central, most inferior portion of the nuchal plane is not preserved. We estimated the sagittal length of the nuchal plane by calculating the sagittal distance of inion posterior to the biasterionic line. This distance, 33.5 mm, is similar to the comparative samples. Considering the small magnitude of differences in this projection, compared with the great amount of expansion in breadth of the nuchal muscle attachment area, the total amount of nuchal musculature is clearly greater in Mladeč 5.

At the most anterior lateral portion preserved on the right side of the bone, a prominent paramastoid crest remains (none of the bone remains medial to this). The crest, which is actually on both the temporal and occipital since the suture runs along its ridge, is more strongly developed than that of La Chapelle, although weaker than the La Ferrassie 1 crest. On the right side the crest is elongated, extending to the rear of the mastoid process. On the left side the entire length of the crest is broken away, only its most anterior and posterior most bases can be seen.

On the left side there is a well-developed occipitomastoid crest that begins 5 mm anterior and 16.5 mm medial to the base of the digastric sulcus, and extends about 20 mm posteriorly. The crest projects about 2.5 mm inferior to the bone surface and, thereby, is somewhat lower than the paramastoid crest described above. In comparison to the early Central European Upper Paleolithic sample of both sexes, the occipitomastoid crest shows the strongest development in Mladeč 5. In comparison to the Neandertal sample, the occipitomastoid crest in Mladeč 5 is about the same size as in La Chapelle and considerably larger than in La Ferrassie.

Unlike the seemingly thin parietal bones, thickness of the occipital squama (measured at lambda, see Table 6) nearly equals the Neandertal mean, although it is much smaller than the means for the Skhul / Qafzeh sample. Thickness at the occipital mastoid suture exceeds Neandertals but is markedly smaller than the Skhul / Qafzeh mean and range. Endinion thickness is somewhat below the Neandertal mean, although well within the range. However, it is half the value of Brno 2, the other earlier Central European Upper Paleolithic male allowing this measurement.

Temporal bones

Compared to both samples, the temporal squama is long (Table 9). It had a superior surface that was roughly triangular in shape before its thin beveled edge was broken off. The posterior portion of the superior surface was almost a straight line, extending some 60 mm anterosuperiorly from the parietal notch. At its highest point the superior edge of the squama is 65 mm posterosuperiorly to the most inferior point on the squama's anterior border. Again this anterior portion of the superior surface is a straight line, extending from the anterosuperior surface of the squama.

The squama is oriented vertically and lacks virtually any curvature. On its most posterior extension there are two well-developed crests: superiorly the supramastoid crest *(crista supramastoidea),* an extension of the inferior temporal line; and below this the mastoid crest *(crista mas-*

Table 9. Temporal and mastoid process dimensions (mm)

	Mladeč 5	Neandertal		Skhul / Qafzeh	
		Mean (n)	Range	Mean (n)	Range
Squama					
Maximum length	73.9	64.2 (6)	59.5–62.4	58.9 (2)	56.7–61.0
Supraglenoid gutter length	28.9	27.5 (5)	24.1–31.8	23.3 (4)	21.6–26.7
Postglenoid process					
Height	6.1	5.6 (6)	0.1–8.2	7.2 (5)	3.7–9.8
Basal thickness	8.6	4.0 (6)	0.1–6.5	5.3 (5)	3.5–6.6
Mastoid process					
au-ms	34.7	35.9 (5)	33.0–38.0	37.3 (5)	33.4–43.5
Auricular-mastoid ht	32.1	36.2 (5)	35.3–37.8	37.1 (5)	31.9–43.0
Height from digastric	8.5	6.8 (5)	4.0–9.1	9.4 (5)	7.1–11.5
Basal length	43.8	37.8 (5)	34.1–42.7	39.3 (4)	37.1–40.8
Glenoid fossa					
Length	25.4	20.5 (5)	18.7–21.5	19.4 (4)	16.5–21.6
Articular length	15.6	12.8 (5)	11.6–14.6	11.3 (4)	9.8–12.4
Breadth	20.9	25.8 (5)	21.7–28.0	27.3 (3)	26.4–28.9
Depth	6.4	5.8 (5)	3.0–9.2	6.5 (4)	5.1–8.4

toideus), an extension of the confluence of the superior temporal line and the nuchal line. Separating these crests is a 16 mm wide, deep supramastoid groove *sulcus supramastoidea)* on the right side. On the left side the crests approach each other much more closely, and the supramastoid groove is only 7 mm in breadth. The supramastoid crest continues anteriorly to form the root of the zygomatic process of the temporal. Unfortunately, even on the better-preserved right side, the lateral surface of this process is broken along the mandibular fossa (beginning 11 mm anterior to porion), and at its most medial extent the zygomatic root just includes the articular surface of the glenoid fossa.

The superior surface of the zygomatic root in the position just above the glenoid fossa forms a gutter for the posteroinferior-most fibers of the temporalis muscle. Dimensions of this gutter for Mladeč 5 are within the Neandertal range. The gutter is small for an early Central European Upper Paleolithic specimen. In this regard the Skhul / Qafzeh males are unlike both the Neandertals and the post Neandertal Europeans, in that their supraglenoid gutters are very small. In Mladeč 5 the supraglenoid gutter is angled at about 30° to the Frankfort Horizontal. However, this angle is almost parallel to the Frankfort Horizontal in Skhul 5. In contrast, the Neandertal males have a much higher angulation, perhaps in concert with their less well-developed posterior temporal bones, and Qafzeh 9 resembles the Neandertal condition. La Chapelle, the least angled of the Neandertal males, is most similar to Mladeč 5.

The glenoid fossa, preserved only on the right side, is deep and very long. Its sagittal length exceeds the comparative sample ranges. The horizontal anterior extension of the articular surface (articular length) is long, much longer than the Neandertal and Skhul / Qafzeh means and actually above their maxima. The anterior wall of the fossa is sharply angled to both the horizontal surface and to the roof of the fossa. Posterior to the roof of the fossa, there is a sagittally thick (exceeding the comparative samples maxima) but only moderately tall post-glenoid process, which separates the back wall of the fossa from the anterior ring of the external auditory meatus.

Thus, the mandibular fossa combines a deeply set, well rounded roof with a well delineated top, that is posteriorly bounded by a post-glenoid process and anteriorly bounded by a distinctly angled anterior face (the "vertical" portion of the auricular surface). The horizontally oriented anterior portion of the articular surface, approximately 9 mm in length, is in turn strongly angled to the fossa's anterior face.

This fossa form can be matched in some of the specimens of the European earlier Upper Paleolithic and Skhul / Qafzeh samples. However, excepting Spy 1 and La Quina 5 (both regarded as females in this study), the mandibular fossa in most of the Neandertals systematically differs from this condition. For most of the sample, the fossa is usually shallower, even discounting the effects of age (Hinton, 1981*a*) and the arthritic condition on several of the specimens. The main difference lies in the form of the articular surface. Instead of distinctly angled, more or less vertical posterior and horizontal anterior portions, in most Neandertals the articular surface comprises the anterior wall of the fossa in its entirety, forming one long flat face with an orientation varying between but not reaching either the horizontality or the verticality of the Mladeč 5 fossa. Boule (1911–1913) noted this form in La Chapelle and La Ferrassie, but mistakenly regarded it as a symplesiomorphy shared with living apes. Only Spy 1 and La Quina 5 (females, in our assessment) approach the Mladeč 5 condition. In both of these specimens the articular surface is also divided into two parts at a strong angle to each other. These differ from the Mladeč specimen in that the anterior (horizontal) portion of the surface is sagittally much shorter.

On the medial wall of the glenoid fossa there is a deep narrow groove, the Glasserian fissure, dividing the medial glenoid wall from the petrous pyramid. This fissure is at the very posterior of the glenoid roof, actually an extension of the groove separating the glenoid roof from the anterior wall of the external auditory meatus. The morphology of this region is common (but not unique) in Neandertal and Skhul / Qafzeh males.

On the remaining petrous pyramid there is no unusual morphology. On the left side the base of the broken ossified styloid is preserved. The tympanic ridges appear to have been sharp and fairly projecting, although much of the structure is broken away. Breaks in the bone expose the carotid canal, the jugular fossa, and a number of air cells. On the right side the sphenotemporal suture remains, but none of the sphenoid is preserved on the cranial base.

The external auditory meatus is elliptical in shape, the long axis almost vertical, leaning slightly forward at its superior point. The form is close to that in Spy 1 and 2, although in most of the Neandertals the elliptical meatus leans distinctly backwards. The long axis of the opening is 10.0 mm while the short axis is 8.5 mm transversely.

There is some damage to the mastoid processes on both sides. Moreover, asymmetry of the regions preserved posterior and superior to the processes affects the appearance of features somewhat differently. The right process is much better preserved, lacking only the thin cortical surface medial to its tip, and thus exposing the pneumatized cells along its inferomedial aspect. In our view, the most inferior part of the process remains, forming what evidently was a broad, flat surface similar. At its base, the process has large sagittal (26 mm) and transverse (26.5 mm) dimensions. Mastoid basal length, measured as the distance from the posterior point on the digastric sulcus to the *ms* point, exceeds the means of the two earlier samples.

In contrast, however, the vertical projection of the process is quite small, whether measured from the Frankfort Horizontal (auricular-mastoid height) or from the digastric groove (digastric height). Projection beneath the Frankfort Horizontal is less than the means of the comparative samples. The Mladeč 5 auricular mastoid height is very small, even below the Neandertal range. Moreover, considering the projection of the mastoid as measured below the digastric sulcus, Mladeč 5 lies between the Neandertal and the Skhul / Qafzeh means. Mladeč 5 is within the ranges of the comparative samples. In sum, these data show that the projection of the mastoid is quite small.

On the left side the base of the process is broken more superiorly and pneumatized cells are exposed much deeper into the bone. Thus, at first glance the base of the process appears to have been much larger and the region appears to have been asymmetric, lacking a digastric groove. However, close inspection shows that virtually the entire extent of the digastric groove and the paramastoid crest lateral to it have also been broken away exposing air cells continuous with those of the mastoid itself. As described above, the most anterior and posterior edges of the mastoid can be observed, and laterally the distinct edge of the digastric groove can also be seen. These features provide the opportunity to compare the gross dimensions at the base of the left and right mastoid processes, and these are virtually identical. Appearances can be deceiving, and we believe that when the breaks are taken into account, the two sides of the vault are actually symmetric in this region.

Again on the right, the mastoid crest divides at the anterior portion of the lateral process face, with a superior branch wrapping to the front of the process and a well marked vertical branch running to the anteroinferior surface where it terminates at its tip. Posterior to this vertical branch, the posterolateral surface of the process is divided by a shallow groove, deepest at its most inferior aspect. The orientation of this surface aligns it with the nuchal plane, and we view this surface as an anterolateral extension of the nuchal muscle attachment area.

Just posterior to the mastoid process are doubled, asymmetrical expressed mastoid foramina. The main foramen on both sides is located just behind the lateral border of the mastoid, just below the mastoid crest. On the right the foramen is circular, while the left foramen is chevron-shaped. About 6.8 mm to it on the left there is an accessory foramen. On the left side, the foramen is smaller and located 13.5 mm from the main foramen.

The Mladeč 5 mastoid, and the region surrounding it, addresses the question of whether the Mladeč features can simply be explained by robustness. We believe the answer is no, because Mladeč 5 is unlike any of the robust earlier Central European Upper Paleolithic specimens from other sites in its combination of features. Similarity often lies in the incorporation of the mastoid's posterolateral face into the nuchal muscle attachment area, a morphology also seen in Předmostí 3 and 4, Cioclovina, and Pavlov. On the other hand, virtually every other mastoid in specimens of the early Central European Upper Paleolithic sample is larger than the Mladeč 5 mastoid in basal dimensions and projection of the mastoid tip. Simply put, so small a mastoid in an otherwise very robust European specimen is incongruous.

The size and degree of projection of the mastoid process in the Neandertals is much more similar to Mladeč 5, and this specimen clearly falls within the Neandertal range. Most of the Neandertal mastoids however lack a posterolateral face that is oriented with the nuchal plane. Indeed, generally the Neandertals do not seem to be as robust in this region, insofar as the posterior cranial base is not as heavily pneumatized and the muscle attachments not as well marked. For instance, the mastoid crest is often not as prominent or as well-developed as in Mladeč 5, and instead in many cases this crest can be better described as a tubercle just posterior to the external auditory meatus. Thus, we consider the mastoid tubercle, described as a unique feature of the Neandertals (Santa Luca, 1978; Hublin, 1998), as a structure homologous with the mastoid crest that is expressed very weakly because in the Neandertals the superior nuchal line does not extend onto the mastoid process and therefore does not contribute to the expression of this feature. Some direct evidence supporting this hypothesis of homology derives from our examination of a cast of the Cioclovina female. In this specimen the prominent superior nuchal line extends onto the temporal to a position superior to the rear of the mastoid's face and then turns sharply inferiorly so that it extends down the posterior surface of the mastoid. Thus, even though the line is very well-developed, it does not reach the mastoid's lateral face, and this specimen has a mastoid tubercle.

Sphenoid

The superior portion of the greater wing was originally preserved, from the most anterior portion of the temporal to the sphenofrontal suture at the top of the temporal suture and superior to *infratemporale*. At its superior border there is a sphenoparietal articulation, some 20 mm in length. As detailed above, some damage occurred on the more posterior aspect of the greater wing in the 1945 fire.

On the right side the greater wing of the sphenoid was originally preserved. It has since been broken away, but fortunately the bone fragment remains on the cast, displaced slightly laterally. The long sphenoparietal articulation appears to be common in Europe, for both the male Neandertal and earlier Central European Upper Paleolithic samples. The feature varies, but the Neandertal range encompasses the earlier European Upper Paleolithic range, and the earlier Central European Upper Paleolithic average is above the Neandertal average. In contrast, the Levant males have a very short sphenoparietal articulation, smaller than any of the European specimens.

Endocranial surface

Endocranially, the major part of the cranial base has not been preserved (only the pyramids of the temporal bones and part of the left temporal bone remain), and we can examine only part of the endocranial surface of the occipital bone; most of it is damaged or missing. The study of the endocranial side of both temporal bones is hindered by damage, especially heavy on the left side.

The well-developed frontal crest is 5 mm high. For a 43 mm length it separates the two frontal lobes. At the endometopion position the crest is replaced by the *sinus longitudinalis superior.* Between endometopion and endobregma there are two Pacchionian granulations along the median sagittal line on the right side and one on the left. The coronal suture is paralleled on the parietal side by Breschet's sinus, more visible on the left parietal bone where it starts from a deep Pacchionian granulation. The details of the impressions of the *Arteria meningica media* on both endocranial sides differ, but both are richly branched and have numerous anastomoses. Near endobregma there is a *"lac sanguine"* on the left parietal bone. A second one with a smaller Pacchionian granulation occurs at the endobelion position. Another small granulation is on the right parietal bone. The course of the lambdoidal suture in the region of lambda is well defined. Finally, the impressions of the cerebral gyrification are prominently displayed. As is the normal condition, they are more pronounced in the bottom part of both parietal bones than in the top.

The pyramids of the temporal bones have been preserved and the *pars squamosa* has been better preserved on the right side of the skull. Also on the right there is a deeply impressed *sinus petrosquamosus,* regarded by Saban (1984) as a plesiomorphic character. The impressions of the anterior, middle, and posterior branches of the meningeal artery, and the impressions of the temporal gyri are prominent. These extend onto the parietal region where the three distinct branches of the middle meningeal artery are easily traced. Each of these has an independent stem extending to the inferior border of the parietal on the left side, where the entire morphology of the internal surface can be observed. There is one pacchionian pit on the left parietal, just anterior to lambda.

On the occipital squama's endocranial surface there are asymmetric cerebral fossae, deeper on the left side and shallower on the right. Neither extends superiorly to lambda. They appear similar in size, although the left is slightly larger, an asymmetry related to the asymmetry of the cerebral hemispheres. The cerebellar fossae have been partially preserved. The sagittal sinus is bro-

Fig. 11. Cranial base of Mladeč 5. Note the superior extent of the nasal bones, the glabellar depression, and the deep glenoid fossa

ken superior to the endinion position. The transverse sinus is preserved mainly on the right and is doubled with the ridges approximately 10 mm apart.

Since most of the cranial base is missing, our observations are limited. On the right side of the frontal bone there is an arch corresponding to Broca's Cap. On the right temporal squama in front of the pyramid there is a marked depression corresponding to the impression of the third temporal gyrus, as well as the impression of *Arteria meningica media*. Only part of the left cerebellar fossa has been preserved on the squamous portion of the occipital bone.

Mladeč 6

The Mladeč 6 vault is the second adult of the triple burial from the small Quarry Cave (Figs. 7, 8, 10 and 13). We studied the primary cast at the Moravské zemské muzeum, with supplementary information from Szombathy's description and illustration. The vault is an adult male, similar in many respects to Mladeč 5, but less complete. Szombathy reconstructed it out of a number of fragments. He suggested that the Mladeč 55 mandible might belong to the same specimen. In fact, some or all of the adult postcranial remains from the Quarry Cave could belong to Mladeč 6. While we believe it very likely that all of the adult gnathic, dental and postcranial remains from the Quarry Cave belong either to Mladeč 5 or 6, there is not and never a way to divide the sample between the two specimens. With the exception of the crania, all of the material was destroyed and never molded and, excepting the mandibles, never photographed.

According to Szombathy (1925, 76) "the sutures display an advanced grade of obliteration" which would indicate that the specimen is older than a young adult. Exocranially all sutures are fused, but to varying degrees are visible.

The specimen has been variously said to be more or less primitive than Mladeč 5, but neither description seems justified to us. The fact is that like Mladeč 5, the calvarium is extraordinarily robust and unlike modern European males from any living population. The specimen was destroyed during the fire at Mikulov Castle. Our description is based on a primary cast at the Moravské zemské muzeum, Brno, and on the notes and photographs published by Szombathy.

The vault is comprised of a frontal with adhering nasal remnants, most of the parietal bones, the superior portion of the occiput, and the posterior portions of the temporal bones. Sagittally, the calvarium is preserved from a point on the superior nasal bones, 12 mm below the nasofrontal suture, to a point on the occiput in the approximate position of the inferior nuchal line. About half the orbital margin remains on the right side, somewhat less on the left. Breaks occur superomedially to the preserved margin on both sides, although they do not reach the midline. From a point about 44 mm posterior to nasion, the breaks on both sides travel posterolaterally until they reach the coronal suture. On the right side the parietal is preserved along its inferior margin (including parts of the beveled suture for the temporal squama) for all but its anterior most corner. However, because of the condition of the cast, it is impossible to ascertain the original shape of the temporal squama's superior edge. On the left, somewhat more of the inferior portion is broken away and generally only half of the bone remains. Only the very back of the left temporal squama is preserved, while on the right the preservation is somewhat better, including the posterior portion of the mastoid and the temporal both medial and superior to it. The occiput is partially preserved, with the remaining aspect approximately superior to the inferior nuchal line. Based on the surface of the cast, we believe that some of the right occipital squama along the meeting of the upper and lower scales might have been reconstructed. The affected area begins about 20 mm from the midline and extends somewhat superiorly, onto the occipital plane. An alternative explanation of the observations is that there was a poorly cleaned casting line crossing the bone here, although this is less likely since the cast seems to have been made from a one-piece mold with the vault set on a pedestal. The Szombathy text is of no help in this problem. The suspected area continues laterally almost to asterion. It does not involve any of the sutures, or any of the morphology of the right temporal posterior. Our description of the bone, and its morphology, assumes that the details for this region on the left side are correct.

Some amount of deformation, without any obvious associated cracking or missing bone areas, can be seen when the skull is placed in superior view. The front portion appears to be slightly twisted in a counterclockwise direction, so that the left supraorbital is somewhat more posterior than the right. What appears to be equivalent asymmetry along the back of the skull is actually a consequence of reconstruction on the right side of the occipital squama, described above.

Mladeč 6 shows at least one healed cranial wound. This depressed region is located on the left frontal, centered 22 mm lateral to the frontal torus and 32 mm anterior to the coronal suture. The wound is oblong with the long axis (13 mm) running 45° to the sagittal plane. The breadth of the wound is about 6 mm and the depth about 1 mm surrounding the oblong wound is a circular area of bone that appears to have been in the remodeling process as the bone healed. Like the wound on Mladeč 5, this represents a healed wound that occurred long before death. Closer to the midline 50 mm above glabella is shallow depression that may also be a healed wound. The depressed area is oriented in approximately the same direction as the healed wound described above; however, due to the condition of the cast, it is impossible to be certain that this depressed area is the result of a premortem injury.

The vault as a whole

While long, low, and broad like Mladeč 5 (Table 3), the Mladeč 6 calvarium is somewhat broader and higher, and distinctly more domed (in Szombathy's words "roof-like") in sagittal view. Thus, for instance, the arc/chord index for glabella-lambda of 138.8 is much higher than for Mladeč 5 and in fact exceeds the ranges for the comparative samples. The nasal root is markedly depressed. The supraorbital region is somewhat less projecting, and the forehead lacks a true supratoral sulcus (in part because of the marked frontal keel that extends down the entire length of the bone to a position just above glabella). The very top of the vault is flat from bregma to vertex (about 55 mm posterior) and the difference in cranial height between bregma and vertex appears to be substantial.

Beginning 19 millimeters posterior to vertex, lambdoidal flattening extends the full 55 mm to lambda, and, 25 mm additionally onto the occiput, where the flattened surface merges into the occipital prominence. Along the parietal portion of this flattened region, the sagittal suture and the region surrounding it are distinctly depressed. The depressed area is triangular, ranging from about 12 mm in breadth behind vertex to 40 mm in breadth at lambda. At its most inferior extent the depression is bounded medially by the superior temporal line. As mentioned above, we regard this as a much stronger expression of a similar morphology found in Mladeč 5.

An occipital bun, according to Trinkaus and Le May (1982, 27), "[...] may be described as a posterior projection of the occipital squama, which is evenly rounded in *norma lateralis* and slightly compressed in a craniocaudal direction [...] occipital buns are associated with a depression, or flattening at lambda, which involves the adjacent portions of the occipital and parietal bones."

By these definitions, Mladeč 6 has an occipital bun. However, it is not as prominent as that of Mladeč 5. The posterior vertical face of the bun is shorter, and the craniocaudal flattening is less extensive. This is because the lambdoidal flattening is at a higher (i.e., more open) angle relative to the nasion-inion line in this higher vault, and also because the nuchal plane is convex rather than concave in the region just below the nuchal line. Moreover, the bun has less posterior projection; the posterior face is angled inferoposteriorly relative to the nasion-inion line rather than being vertical. While the lateral aspects of the occipital projection (bun) in Mladeč 5 are transversely compressed in the region of the lambdoidal suture, in Mladeč 6 the occipital projection is evenly arched transversely without any clear angulation or transverse compression at the lambdoidal suture. These factors combine to result in the appearance of a more poorly delineated, less projecting bun.

In sum, despite their general similarities there are a number of contrasts in sagittal form between cranium 5 and 6. While slightly shorter in maximum length, the contour of the vault is higher and more evenly rounded to vertex in cranium 6, but posterior to this the plane of lambdoidal flattening is both steeper and longer. The bun is not as projecting or as steep on its posterior face, as discussed above, while the nuchal plane is markedly more vertical in orientation.

In posterior view, the Mladeč 6 parietal sides are more superomedially sloping then even the right (undistorted) side of Mladeč 5 and the sides of Předmostí 3, the most sloping of the Upper Paleolithic males. The angle between the sides and the top portion of the vault is distinct, but much less acute than in Mladeč 5 both because of the greater slope of the sides and because of the marked angulation of the top of the Mladeč 6 parietal bones with the horizontal. The superior surface of the Mladeč 6 vault is highly domed in this view, with the flattened horizontal area along the sagittal suture only 20 mm wide at its very center (which is the top of the vault in posterior view).

The vault is broader than Mladeč 5; its greatest breadth, 166.5 mm (projected to the midline from the right side), is on the supramastoid crest, a structure that is much more pronounced than in Mladeč 5. The breadth at the mastoid crest is almost as great, and this maximum cranial breadth exceeds that of every other specimen in the comparative sample. The maximum biparietal breadth is not as great, the value of 153.5 mm is close to that of Mladeč 5, and to the Neandertal mean of 152.7 mm these are markedly broader than the Skhul/Qafzeh males. The great cranial breadth is also reflected in the cranial indices. Whether calculated from the maximum cranial breadth that is above the ranges of the comparative samples, or from the biparietal breadth that is at the top of the Neandertal range and above the Skhul/Qafzeh range, this vault is extraordinarily broad. In contrast, the other earlier Central European Upper Paleolithic males have relatively and absolutely narrower crania, thereby resembling the Skhul/Qafzeh specimens in a way that the Mladeč crania do not. Given the obvious relation of the Mladeč males to the Central European early Upper Paleolithic males, this could suggest that the vault narrowing of the Central European early Upper Paleolithic sample is not homologous to that of the Skhul/Qafzeh sample and therefore the similarity is not a marker of relationship.

The difference between the Mladeč 6 total cranial breadth on the one hand, and the biparietal and bimastoid breadths on the other, is entirely due to the development of the supramastoid crest. Apart from general muscularity, we believe that the total cranial breadth reflects the dramatic pneumatization of the specimen's cranial base, resembling but exceeding the condition in Mladeč 5. Yet, contrasting with the superstructures and muscle attachments discussed above, as seen in posterior view the nuchal lines are weak in the central portion of the bone and the area of attachment for the nuchal muscles is not as rugose as most other early Central European Upper Paleolithic males. Superiorly, the parietal boss is less prominent and more posteriorly located than in Mladeč 5.

While Mladeč 6 is a broader skull, like cranium 5 it narrows significantly anteriorly. The exact details of the narrowing cannot be established because the lateral portion of the frontal is missing on both sides from the approximate centers of the orbits. Glabella is prominent and the superior orbital borders slope significantly posterior laterally away from it. The equivalent region is more angled than Mladeč 5, although in the more lateral portion of the Mladeč 5 superior orbital margins (not preserved in Mladeč 6) there is a significant posterolateral angulation. Also, unlike Mladeč 5 there is no tendency for the supraorbitals to divide at glabella.

The superior temporal line can be seen most anteriorly where it is preserved on the right side as a 30 mm long ridge developing into a line some 25 mm anterior to stephanion. Just past this point on the parietal it arches superiorly. The line continues in its posteromedial sweep coming as close as 26.5 mm to the sagittal suture about midway between bregma and the vertex position. Here the superior temporal lines are 56 mm apart. On the left side, where the bone seems better preserved, the line is in a perfectly symmetric position, but here is expressed as a weak ridge. At vertex the superior temporal lines are 69 mm apart. The line (ridge) continues superior to the parietal boss.

The frontal view reveals massively developed supraorbitals and the frontal keel (Fig. 13), which dominates the sagittal profile producing an even more peaked appearance at the top of the vault than is seen in posterior view.

We did not estimate a cranial capacity for Mladeč 6, mainly because we felt we could not determine a reliable cranial height. Maximum height above the glabella (or nasion) to inion line may only provide a poor cranial height estimate because inion appears rather high on the vault's posterior (see Table 8). We have not used the 1450 cc. estimate Frayer (1986) reports, based on the Poissonnet, Olivier and Tissier (1978) regression, because the great similarity of the preserved measurements with those of Mladeč 5 suggest the value is too small. Our unpublished research shows this formula to be unreliable on specimens outside the sample used to determine it.

Nasal bones

While only 12 mm of the nasal bones remains, there is sufficient bone left to show that the superior nasal width is marked, it is about 60% of the anterior interorbital width (Table 10). The 16.4 mm superior breadth of the nasal bones exceeds Mladeč 5. The Mladeč 6 interorbital widths themselves (*mf-mf* and distance at the anterior orbital angle) are quite large, exceeding the maxima for the two comparative samples (Table 5).

The nasal bones are strongly keeled, their sides facing at a greater than 50° angle to the coronal plane and meeting along a rounded midline. For the 11 mm that are preserved, the nasal bones are straight in the sagittal plane. Above them there is a distinct depression of the nasal root. The nasal angle is moderately high. The angulation begins at the top of the nasals and in this regard is similar to the Neandertal condition as well as later crania from the European Upper Paleolithic (Fig. 12). However, the nasal angle does not seem to be as high as in many of these specimens.

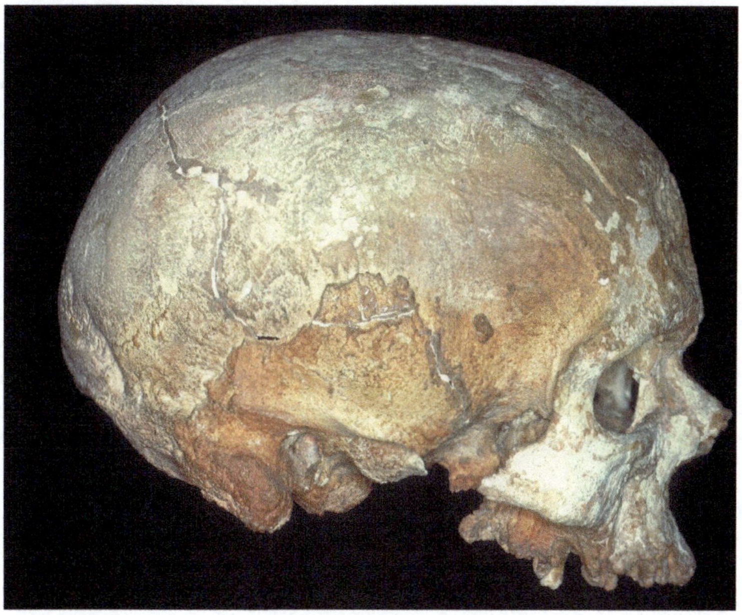

Fig. 12. Cro-Magnon 1 shows a high nasal angle, typical of later Upper Paleolithic specimens

Table 10. Nasal bone dimensions

Dimensions across both nasal bones	Mladeč 5		Neandertal 6		Skhul/Qafzeh	
			Mean (n)	Range	Mean (n)	Range
Minimum breadth		14.0	13.8 (2)	11.3–16.3	6.5 (1)[1]	
Superior breadth	12.7	16.4	15.4 (2)	12.7–18.0		
Superior br/anterior iob index	50.2	62.6	64.2 (2)	57.7–70.6		

[1] Qafzeh 6

Like Mladeč 5, the frontonasal suture is superior to the frontomaxillary suture (a consequence of the midline keel). The bones are both elevated above the level of the nasal processes of the maxilla, forming a nasal pillar with sides that face laterally relative to the parasagittal plane. The morphology of the Mladeč 5 nasofrontal suture suggests the same nasal configuration. This combination of a nasal pillar and midline keel is common in European faces and is forensically useful in identifying them (Gill and Gilbert, 1990). It is also common among the early Central European Upper Paleolithic males, and can be observed in the Předmostí male crania 3 and 9. Few Neandertal males preserve the superior nasal bones, but the best preserved of these (Guattari) has the same nasal configuration, forming a nasal pillar.

In the Skhul / Qafzeh males the form of the nasal bones appears to be mixed. The very top of the nasal bones preserved in Skhul 5 are quite flat, but the nasal bones in Qafzeh 6 project above the maxilla and form a pillar similar to Mladeč 6 and Guattari.

Frontal

The frontal of Mladeč 6 has a moderately developed supraorbital arch (Figs. 8, 13). This is evident on the better-preserved right side, which clearly shows the lack of a separation into medial and lateral supraorbital elements. Like Mladeč 5 the superciliary portion is much more strongly developed, but unlike this specimen the structure is projecting rather than depressed just above glabella. The supraorbital notch is broader and located more medially, virtually at the superomedial corner of the orbit. The notch morphology can only be observed on the right, since this region on the left was broken open and filled with plaster. In absolute dimensions the supraorbital structure of Mladeč 6 closely resembles that of Mladeč 5. However, visually it appears to thin more than Mladeč 5, from its medial to its central aspect. From its central portion, laterally, the height of the structure is more evenly developed. What little remains of its most lateral portion show the structure to be thinner than Mladeč 5.

Central supraorbital height[15] is 81.4% of the medial height (Table 5), thus relatively larger than Neandertals and Skhul / Qafzeh, which are similar in this proportion. The inferior border of the supraorbital arch does not appear as highly arched as Mladeč 5; in this, it is similar to most Neandertals as well as certain other earlier Central European Upper Paleolithic males such as Pavlov and Předmostí 3.

Where comparable, the Mladeč 6 projection is similar to or greater than that of Mladeč 5. Projection at the center-orbit from the internal surface is much greater, although still below the Neandertal and Skhul / Qafzeh means. Projection[17] measured to the anterior face of the frontal squama is the same as Mladeč 5 and close to the Skhul / Qafzeh and Neandertal means.

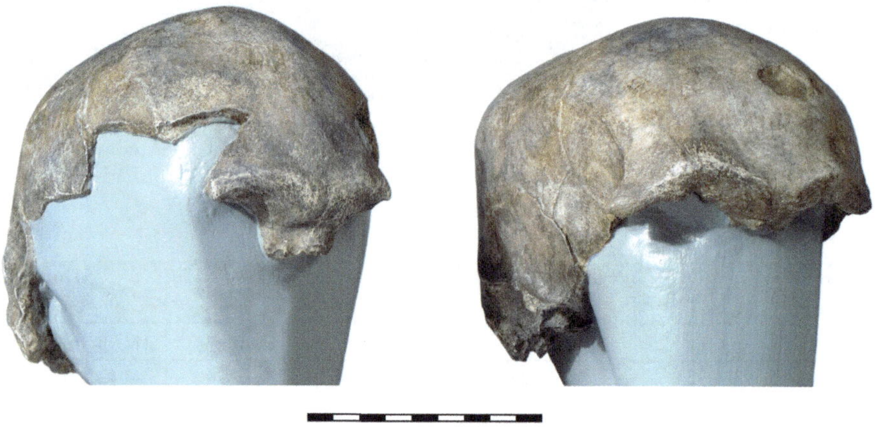

Fig. 13. Comparison of Mladeč 5 (right) and 6 (left) in 3/4 view. The form of the arched supraorbital regions, and the frontal keel in Mladeč 6 can be seen in this view (both are casts).

While a true supratoral sulcus does not exist, lateral to the midline there is a shallow sulcus between the supraorbital structure and the frontal squama. The lack of a frontal boss and the presence of a frontal keel along the sagittal midline, instead, result in a much more angular appearance of the lateral aspects of the squama as seen in a frontal view. At its broad top, the frontal squama is quite flat on each side, actually slightly depressed lateral to the frontal keel. There is no marked angulation between the sides and the sloping top of the bone.

The frontal is longer, although not broader, than the frontal of Mladeč 5. The length dimensions (measured from glabella and from nasion) lie at the top of or above the Neandertal and the (much smaller) Skhul/Qafzeh means. In contrast, the bistephanic breadth is very small (especially relative to the maximum frontal breadth), reflecting the high temporal lines. The temporal lines are very unusual, substantially higher than Mladeč 5, or any specimen in the comparative samples.

In terms of sagittal curvature, the Mladeč 6 frontal is flatter than that of Mladeč 5. Its flatness approximates the Skhul/Qafzeh sample, but it is not as flat as the Neandertal mean values. The difference in curvature between the two Mladeč males is a consequence of the strongly developed, transversely expanded frontal boss in Mladeč 5, contrasting with the straight sagittally elongated frontal keel of Mladeč 6.

Finally, frontal squama thickness (Table 6), measured at the approximate position used for Mladeč 5, shows the squama to be even thicker. Its thickness exceeds the Neandertal means (it is at the maximum) and is above the Skhul/Qafzeh maximum.

Parietal bones

The right parietal of Mladeč 6 is more complete than the left, but the bone surface preservation on the left is somewhat better so both will be used in the description that follows. The sagittal suture is keeled from bregma to the approximate position of vertex. It is here that the keel separates to form two distinct ridges that form the sides of the flattened triangle at the cranial rear discussed above. These ridges form a border between the rounded sides and the flattened back (i.e., lambdoidal corner) of the bone. As mentioned above, the superior temporal line on the left forms a low ridge for its entire course on the bone. The superior temporal line, in passing superior to the parietal boss, meets the diverging branch of the ridge emanating from the sagittal keel, some 33 mm anterior to the lambdoidal suture. From this point virtually to the suture itself the ridges parallel each other, coming no closer together than 6 mm, and there is a low flat elevated surface between them for this distance. Just at the lambdoidal suture, the superior temporal line courses inferiorly to parallel the suture. This is some 35 mm along the lambdoidal suture from lambda and at a much higher position compared to Mladeč 5. Therefore, the posteromedial hypertrophy of the posterior temporalis contrasts even more greatly with the much lower position for the muscle as marked by the most posterior extent of the superior temporal lines in the Neandertal and the Skhul/Qafzeh samples.

Where comparable, parietal size (Table 7) is generally close to that of Mladeč 5, although somewhat shorter along the lambdoidal border. Curvatures (as measured by the arc/chord indices) along this and the superior border also closely resemble Mladeč 5. Sagittal curvature does not clearly distinguish any of the samples, or the Mladeč crania. However, the posterior border curvature in Mladeč 6 is low compared with Neandertals, but not Skhul/Qafzeh. In this regard, it is much like Mladeč 5. A marked difference lies in the Mladeč 6 index for the transverse chord from bregma to asterion. It is much less curved than Mladeč 5 but similar to the Neandertal and Skhul/Qafzeh males. The difference in diagonal curvature reflects the less pronounced parietal bossing of Mladeč 6.

Parietal thicknesses (Table 6) are approximately equal to those for Mladeč 5, excepting the greater thickness at the center of the bone (anterior and especially middle positions). In these regions the thicknesses are closest to the Neandertal means, and are well above the Skhul/Qafzeh ranges. We believe that the marked asterionic thickness of both Mladeč specimens, compared with the Neandertals, reflects the general expansion of robustness in the region of the nuchal muscle attachment area.

Occipital

The occiput is distinctly angled at the superior nuchal line, forming the boundary between the occipital and the nuchal planes. Seen on the left, this line is weakly expressed as it sweeps from the very low inferior inion position to its highest position on the bone, some 21 mm superiorly. At this highest position (about 28 mm lateral to the midline), it parallels the supreme nuchal line, which is about 23 mm above. These two lines travel together as more or less parallel arcs, outlining a distinct nuchal torus for virtually the entire breadth of the bone, until the supreme line can no longer be discerned, 40 mm from the midline.

At the midline the supreme line forms the base of a shallow, weakly expressed, horizontally elliptical suprainiac fossa, resembling the Neandertal condition but much weaker (Fig. 10). This fossa has the most Neandertal-like form of any we have observed in the European early Upper Paleolithic sample since there is no external occipital protuberance below it, and the fossa lies directly above the torus. Another similarity is in its elliptical shape, 31 mm long and 10 mm broad at the midline. According to Caspari (1991, 153), the transverse occipital torus lies below the fossa and is quite unlike the arrangement in other Upper Paleolithic males:

> *"The torus is high and is bilaterally developed; at its highest it is close to 30 mm high. It arches across the occipital, with its superior border forming an apex and descending, this descent occurring along the lambdoidal suture. The torus, however, does not protrude much posteriorly, and in this respect is not well developed. The torus does serve to provide a more vertical surface to the back of the occiput, which would be acutely angled without it. In its height, width, and lack of relationship with the superior nuchal line, this torus more closely resembles those of Neandertals than other Upper Paleolithic crania. [...] The suprainiac fossa [...] differs from these resorptive surfaces when they are found on most Upper Paleolithic specimens, in that it has no relationship to pronounced nuchal lines."*

The supreme line distinctly borders the suprainiac fossa, along its base. Superiorly the border is better defined on the right than on the left. Lateral to the fossa the supreme line continues its arch as described above. However the nuchal torus, which is defined by the superior and supreme lines below the suprainiac fossa, thins dramatically to just 6 mm at a position 30 mm from the midline. Along this torus, 48 mm from the midline, there is a prominent retromastoid process both vertically broader and markedly more prominent than the nuchal torus. While other males in the early Central European Upper Paleolithic have prominent nuchal tori, such as Pavlov, none extend as laterally or show a retromastoid process. In Neandertal males a nuchal torus far from the midline is generally present, but only Spy 2 has a distinct retromastoid process. None of the Skhul/Qafzeh males have this process. In Mladeč 6 both the process and the associated torus totally disappear 8.5 mm short of the occipital temporal suture. This obliteration of the torus seems to be a consequence of preservation; since on the right side the equivalent ridge clearly carries on to the temporal where it merges with the mastoid crest. Where the ridge crosses the occipital temporal suture, it is some 15 mm inferior to asterion.

When the supreme nuchal line diverges from the superior line medially, it swings strongly upward, rising about 25 mm above this position, 30 mm lateral to the midline. Just at the midline, where it is at the approximate vertical center of the occipital bun's flattened surface, the supreme line dips slightly. This contrasts with the Mladeč 5 condition where the supreme nuchal line never diverges widely from the superior line and at the midline forms an extra inion prominence that actually touches the superior nuchal line. Between these two lines, the bone surface bulges moderately outward. Superior to these lines, much of the occipital plane is involved in the bun described above.

The occipital plane is as curved as in Mladeč 5, although the length of the occipital plane as measured to inion is less (Table 8). However, the vertical height of the low, downward pointing inion triangle is large, and therefore the occipital plane length measured to the lower inion position is larger than in Mladeč 5. As measured to inion, the more comparable of the occipital plane definitions, the short Mladeč 6 plane is markedly below the Skhul / Qafzeh mean value (although not as low as the Neandertal mean). The length to lowest inion, however, exceeds the comparative sample means.
The nuchal plane is convex below the superior nuchal line, and the muscle attachment area is not especially rugose except for a ridge that can be seen on the remaining right side. This ridge is about 13 mm posterior to the back of the mastoid and extends from the occipitomastoid suture, more or less paracoronally, for 24 mm until the broken edge of the bone is reached. We believe this is an inferior nuchal line with an unusually lateral extension. In two of the Neandertal males (La Chapelle and La Ferrassie 1) a homologous, but lower and broader ridge can be found in this position (Caspari, 1991). In these specimens, it is clearly a lateral extension of the inferior nuchal line. No other early Central European Upper Paleolithic males show this.

The nuchal muscle attachment area is even broader than in Mladeč 5. In fact, this is the broadest of all the nuchal muscle attachment areas studied. Like Mladeč 5, its breadth exceeds the biasterionic breadth. This relationship between breadths only characterizes the Mladeč males. In general, other males of the Central European Upper Paleolithic also have relatively broad nuchal planes, certainly much broader than the earlier comparative samples (for instance, fully above the Skhul / Qafzeh range), but the nuchal plane breadth is never broader than the biasterionic breadth. This is true in spite of the fact that the absolute value of the biasterionic dimension of Mladeč 6 also exceeds the means of all the comparative samples, as well as exceeding Mladeč 5. The Mladeč 6 biasterionic breadth is most similar to the Neandertal mean, as is the distance from inion to asterion. The approximated length estimate for the nuchal muscle attachment area (measured as the sagittal projection behind the biasterionic line) is larger than in Mladeč 5. This sagittal length estimate is close to the means for the comparative samples. With roughly equal nuchal plane length estimates and much greater breadths, we believe that these nuchal dimensions reflect a more extensive development of the nuchal musculature in Mladeč 6, as compared with Mladeč 5, and indeed in contrast with all the comparative specimens.

Temporal bones

Only the most posterior aspect (22 mm) of the left temporal is preserved. On the right, the bone is preserved as far as 52 mm anterior to asterion, and our morphological description will be limited to this side. The back of the temporal squama has a steep posterior border, unlike Mladeč 5. On the squama, a very prominent mastoid crest is separated from a well-expressed supramastoid crest by a shallow sulcus, 19 mm in breadth. Much of posterolateral face of the mastoid process is preserved, to what appears to be a position that is well superior and posterior to the (unpreserved) most inferior point on the process tip. While some of the anterior face is missing, the preserved digastric groove just medial to the mastoid shows that it was extremely small in the transverse basal dimension. In this regard, the mastoid and adjacent expanded paramastoid processes are a good match for many Neandertal crania, and contrast markedly with earlier Central European Upper Paleolithic crania from other sites, and with the Skhul / Qafzeh sample. Because of its preservation, the mastoid of Mladeč 6 does not allow standardized measurements. The posterior face of the mastoid process, with its mastoid crest configuration, is oriented along the nuchal plane, and as in Mladeč 5 it appears to act as an anterior extension for the nuchal muscle attachment area.

At the most posteromedial extent of the bone, on the right side, the posterior root of a paramastoid crest can be seen, actually on both the occipital and the temporal since like Mladeč 5, the suture runs along the ridge of the crest. However, in Mladeč 5 the crest forms the medial wall of the digastric sulcus, while in cranium 6 there are several low ridges on the temporal separating (at least the most posterior portion of) the digastric sulcus from the paramastoid crest. Just posterior to these, some 7 mm inferior to the confluence of the nuchal torus with the mastoid crest, and 10 mm anterior to the occipitotemporal suture, there is a large mastoid foramen.

Mladeč 4

The Mladeč 4 vault, unlike Mladeč 5 and 6, was discovered in the Main Cave at Mladeč, during Fürst's excavations of Chamber E in May 1922. The discovery was one of five individuals, also including Mladeč 38 (a female frontal) and the child Mladeč 37, and the possibly adult fragments Mladeč 44 and 45, as well as some very fragmentary postcranial remains. All of these were destroyed in the Mikulov Castle fire of 1945. According to Szombathy (1925), the specimen was put together out of four main pieces, comprising a frontoparietal fragment. The frontal was preserved mainly on the right side from the orbital margin, although missing both glabella and nasion. Much of the parietal was not recovered, but the remaining portion extends about 80 mm behind the coronal suture. The vault was slightly compressed in the vertical direction, affecting the bone from its vertex to its base and resulting in cracking along the lateral aspects of the specimen.

Szombathy's description and his single figure (1925, Abb. 8, 2, reproduced here as Fig. 14) is all that remains of the specimen. For this reason, we quote him at length (1925, 73):

> *"Because the skull is somewhat crushed, the shape in norma verticalis cannot be described. The frontal bone is broad with flattened frontal protuberances [bosses]. It rises above the supraorbitals smoothly, in an even curve that steadily increases in its arc. The superciliary arch on the right does not project much but is moderately massive and broad. The thickening of the brow ridge stretches from the mid point of the upper orbital margin. The transition between the superciliary arch and the flat, clearly defined trigonum supraorbitais is smooth. The temporal line on the frontal is strongly developed and can also be observed on the anterior of the parietal bone, at a distance of 60 mm from the sagittal suture. The suture elements around bregma are still open, while those lateral portions of the coronal suture are closed. The inner side of the vault is without detail. The frontal sinus is not visible from the internal surface."*

The illustration Szombathy provided (Fig. 7) shows a lateral view of what was evidently a male. Seen in this view, the sagittal contour and the curvature of the vault appear quite similar to Mladeč 5, especially in supraorbital projection, lowness of the frontal angle, the development of the frontal boss, and the expression of the distinct supratoral sulcus. Also like this specimen the temporal line appears to have formed a ridge across the entire length of the frontal squama. The marked expression of these features is of some interest, given the relatively young age of the Mladeč 4 adult. Szombathy (1925, 73) reported a frontal thickness of 8 mm at the middle of the bone. This is not directly comparable to the thickness at the lateral eminence reported in (Table 6), but our experience is that thickness at these two places is usually similar. If we make this assumption, we can conclude that Mladeč 4 is similar to the other Mladeč males, as thick or thicker than the Neandertal males and thicker than is normal in the Skhul / Qafzeh sample.

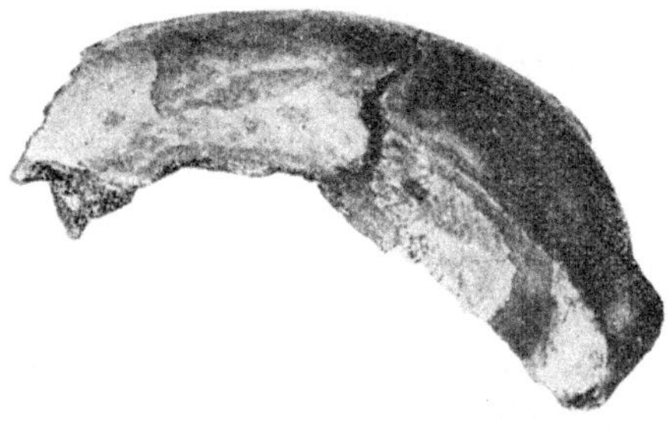

Fig. 14. Lateral view of the Mladeč 4 cranial fragment. This is the only illustration of the specimen

This incomplete specimen is quite important, because it is central to the issue of whether the differences between the two crania from the Main Cave that we regard as female and the two males from the Quarry Cave described above, could possibly be attributed to a (heretofore unrecognized) difference in the time of deposition. Mladeč 4 is from the Main Cave and is archaeologically, faunally, and stratigraphically linked with the two female vaults, but shares the morphology of the males from the Quarry Cave. This strongly supports our contention that the differences we can detail between the male crania 5 and 6 and the female crania 1 and 2 (Frayer, 1986; Jelínek, 1983) do indeed reflect sexual dimorphism in the population and not the presence of two anatomically different populations.

Mladeč 39

Sometime between 1903 and 1911, Knies discovered a large piece of central right parietal. This is now housed at the Moravské zemské muzeum where we studied the original. Found with it were Mladeč 40, 41, and 88–91. The piece includes about 30 mm of sagittal suture that we judge to include the approximate center of the bone. The anterior and posterior surfaces of the bone are roughly parallel, angling posterior laterally from the suture at about 55°. The fragment's length along the posterior surface is 65 mm (see chap. 8, Plate XVII).

We believe the specimen is a young adult. On the one hand, the sagittal suture is unfused and we can find no traces of a temporal line. Also, the vault is very thin (Table 6) at the approximate vertex position on the sagittal suture, only slightly exceeding Mladeč 2. These suggest a younger age, but in contrast, the central portion of the parietal is considerably thicker, for instance at the approximate middle of the bone the thickness is 6 mm, the same as in the Mladeč 5 adult male. At the posterior center of the bone the thickness of 7 mm is greater than the five other specimens from Mladeč for which this dimension is known. These central parietal thicknesses suggest an adult status.

The parietal thicknesses show considerable reduction in the Central European early Upper Paleolithic sample, as compared with the Neandertal males. Mladeč 39 is within the Neandertal range for both thicknesses, but below the Neandertal means. The combination of thinness along the sagittal suture with marked thickening at the center of the bone is unusual. Mladeč 2 is one of the few specimens that even comes close to Mladeč 39 in this combination.

While it is impossible to accurately ascertain the shape of the top of the vault from such a small fragment, we did attempt to position the fragment on a number of vaults. We found that it is an astonishingly exact match for Mladeč 6 at the vertex position, with its rear border just anterior to the flattened back of the cranium. It does not closely match any of the other specimens from Mladeč or from the comparative early Central European Upper Paleolithic sample.

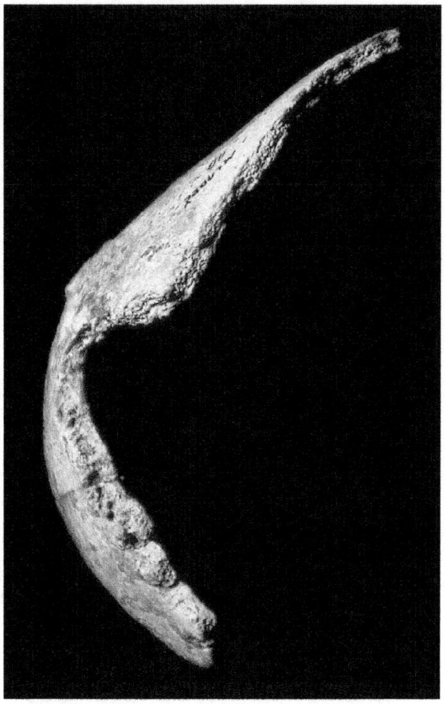

Fig. 15. Mladeč 40 cranial rear in lateral view

Mladeč 40

Preservation

Mladeč 40 is a substantial portion of a subadult or young adult posterior cranium, comprised mostly of a large parietoccipital fragment (Fig. 15). Preserved is part of two parietal posteriors, contacting a *planum occipitale* along a long segment of lambdoidal suture that extends across the midline. The left parietal includes a number of extrasutural bones in the lambdoidal suture, and extends maximally 82 mm anterior to this suture, which is 71 mm in length on the left side of the fragment. The right parietal is preserved only in a 25.5 mm length of extrasutural bone adhering to the occipital border of the lambdoidal suture, almost reaching lambda at the midline. The occipital consists of a 71.5 mm wide portion of the occipital plane, extending maximally 37.5 mm below lambda, a position that is clearly well above the position of the (missing) biasterionic line. The greatest parasagittal length of the specimen is about 100 mm, and the greatest preserved coronal breadth in approximate anatomical position is 105 mm (see chap. 8, Plate XVII).

Sex and age

Examination and comparison of cranial thicknesses (Table 6) suggest an adolescent or young adult age for this specimen. Two other facts are important, but not at all conclusive, in ascertaining the age at death for Mladeč 40. There is no evidence of temporal lines, which may be a consequence of cranial surface preservation but does seem to indicate a younger age. At the same time, the marked development of the extrasutural bones would seem to indicate just the opposite; mainly, an older age at death. Cranial thicknesses also give mixed information – the posterior eminence thickness is below the adult range at Mladeč (and less than all other early Upper Paleolithic specimens) while the occipital thickness at lambda is greater than that of the robust Mladeč 5 male. It is possible that Mladeč 40 is a female. However, in spite its size, we do not regard this specimen as a gracile female due to the male-like shape of the posterior vault and the marked thickness of the occiput at lambda. Therefore, we contend that Mladeč 40 is most reasonably interpreted as an adolescent male. Given the thinness of the vault in the posterior of the parietal, this specimen cannot be the same individual as Mladeč 39.

Description

Only part of the posterior portion of the left parietal is preserved, an irregular quadrangular fragment with longer sides measure 60 mm in length. The lambdoidal suture forms the third side. It is 49 mm long, and its superior most 29.2 mm adjoins the occiput. Finally the fourth side is 37 mm in length. It is situated 35 mm from the sagittal suture and roughly parallels it. The inferior part of the parietal, approaching the asterion position, is 17 mm long. This portion is not connected with the occipital. It is slightly thicker approaching the asterion position, 4.8 mm at the most inferior point preserved on the suture and 5.3 mm on the squama just anterior to this point. In the central part of the lambdoidal suture the bone thickness is 4.0 mm, and diagonally at the apex of the angle defined by its superior and anterior edges fragment is 5.0 mm thick. At its most inferior extent the lambdoidal suture has a characteristically wider configuration, confirming that the break is in the neighborhood of the asterion. For the superior part of the lambdoidal suture 29.2 mm of the parietal bone is connected with the occipital bone; for half of this length there is what may be one or several well-articulated Wormian bone(s). This extrasutural bone is formed to the detriment of the parietal bone and does not reach to the *planum occipitale* of the occipital scale.

Careful examination of the surface of the parietal bone fragment reveals no evidence of a temporal line. This surface is slightly bent, or saddle shaped toward its rear, due to the somewhat protruding occiput of the skull. Three factors are important in the assessment of this observation: (1) the a bulging scale of the occipital bone is quite common in Neandertal crania (especially from western Europe; (2) the interpretation of this morphology in Mladeč 40 is supported by the presence of Wormian bones in the lambdoidal suture; and (3) the youth of the specimen may influence the expression of this feature.

On the endocranial surface of the parietal there are preserved the impressions of a part of the posterior (occipital) branch of *Arteria meningica media,* and also at least some evidence of its secondary branches. The posterior branch has almost horizontal course. No anastomoses have been formed. The gyrification is well reproduced. This is also described in the study of the endocast reported below.

On the right side, along virtually the entire preserved length of the sagittal suture (24 mm), there are three continuous extrasutural bones. The inferior two of these extend as much as 21 mm anterior to the suture in the parasagittal direction. Combined with the left side, the entire length of the superior aspect of the lambdoidal suture provides evidence of numerous extrasutural bones with a similar unusual anteroposterior length.

The preserved portion of the occiput includes most of the *planum occipitale* on the left side and a part of the right side of the scale of the occipital bone. It is limited by the lambdoidal suture, with which at a distance of 32 mm left of lambda there are what remain of the above-mentioned Wormian bones. Other Wormian bones have been preserved on the right side of the lambdoidal suture. As mentioned above, they are approximately the same size and extend to as close as 3 mm from the lambda position. The entire preserved exocranial surface of the *planum occipitale* is evenly vaulted (rounded). It is 71 mm in transverse breadth and a maximum of 37.5 mm in sagittal length.

The endocranial surface features a medium-sized occipital endocranial crest preserved at the length of 25 mm from lambda. From this view it could possibly be suggested from the course of this crest that the Wormian bone in the right part of the lambdoidal suture has been formed to the detriment of *planum occipitale*. However it is more likely, and more commensurate with the morphology of the specimen that the Wormian bones were originally bilaterally symmetric to either side of the lambda position and it is the endocranial crest and the associated sagittal sinus that show asymmetry. It would seem that the sagittal sinus is deflected to the right, superiorly, which would indicate a much larger left cerebral fossa. To the left of the endocranial crest part of a medium-deep left cerebral fossa is preserved. Its configuration is described with the endocranial cast.

The lambdoidal suture is preserved for 71 mm on the left parietal, 17mm of this is on the parietal, while the superior most 32 mm is preserved only on the occiput (to lambda), and the middle 30 mm connects the parietal with the occipital bone (23 mm of which is formed by the Wormian bone or bones). To the right of lambda, along a length of 25 mm, a small bony fragment adheres to the occipital. It is difficult to ascertain whether this is a small fragment of the parietal bone, or part of a Wormian bone, but it is more likely that in the lambdoidal suture there were at least three asymmetrically situated Wormian bones. Although these Wormian bones might have contributed to the bulginess of the occiput, the saddle-shaped bend of the fragment of the left parietal bone indicates, that it has been caused mainly by the shape of the parietal bones and of the upper scale of the occipital bone. The lambdoidal suture, with the exception of the Wormian bones, is of a simple shape.

The thickest place of the preserved fragment of left occipital bone is in its central part, reaching 7.7 mm at lambda (Table 6). At the point where the posterior branch of *Arteria meningica media* is found on the endocranial side of the bone, the left parietal thickness is 5.9 mm. In its vicinity, on both sides of this arterial impression, are minimum bone thickness values as little as 4.4 mm.

The endocranial cast is not actually preserved on the midline, as is defined by the right and left portions of the lambdoidal suture segments described above, except at the lambda position itself. Instead, it deflects to the right inferomedially, reflecting a larger cerebral fossa. On the cast of the endocranial side of the parietal bone we can see the simple typical pattern of the posterior branch of *Arteria meningica media*. Most of it runs in *sulcus temporalis superior*. Between the parietal and occipital parts of the cast there is well visible the lambdoidal suture and the Wormian bones in it. The cast of *planum occipitale* clearly shows *sulcus lunatus* on the left side and the occipital endocranial crest. To the right of it we can see the impression of the right part of the lambdoidal suture and the right Wormian bone.

Mladeč 40, in sum, had a bulging occiput, documented by a saddle-shaped fragment of the parietoccipital. The simplicity of the lambdoidal suture (in spite of the presence of Wormian bones), and the rounded vaulting of the bone surface of the *planum occipitale* are also unusual. On the endocranial cast there is a simple pattern of the posterior branch of *Arteria meningica media* and *sulcus lu-*

natus. This also can be considered as a retention of characters found much more commonly in the Neandertal populations of the region.

Mladeč 43 (destroyed)

Four occipital fragments attributed to a single bone were recovered from Fürst's 1922 excavations in the Main Cave, along with Mladeč 4, 37, and 38. Szombathy studied them in the museum at Litovel, and provides a 15 mm measure of occipital thickness at inion (Table 6). This region is thick in the Central European early Upper Paleolithic male sample, Mladeč 43 is exactly at the Central European early Upper Paleolithic male mean and at the maximum of the female range. For this reason we regard the specimen as male, and note that Mladeč 43 is above the Neandertal male range (although only just above the Skhul/Qafzeh female mean[20]).

20 There are no Skhul/Qafzeh male data for this, the female mean (not reported in Table 6) occipital thickness at inion is 13.8 mm, n=2.

Jaws and teeth

There are only a small number of jaws and teeth, compared with the plentitude of cranial and postcranial, and most of the teeth are associated with jaws. The teeth are from a mixture of adolescent and adult individuals. The youngest individuals are Mladeč 9 (male), 47, and females 1 and 2 (Mladeč 2 is only slightly older). The youngest is judged on the basis of tooth wear, and the older because of the unerupted or just erupted third molars. Older specimens are mostly male, such as Mladeč 8 (based on heavy incisor, canine and first molar wear) and Mladeč 50/51/54, which Szombathy described as exhibiting substantial dentin exposure on most teeth. As for dental remains of infants or children, there are neither deciduous teeth nor unerupted permanent teeth. Thus, unlike the cranial and postcranial material that includes infants and children, the dental remains are comprised solely of older subadult and adult individuals.

Mladeč 8

Mladeč 8 was discovered by Szombathy in 1882 in Chamber D, locus "d" of the Main Cave. In the process of excavation a piece of the frontal process of the maxilla on the left side and the tip of the anterior nasal spine were broken away. The specimen was shipped to the Naturhistorisches Museum Wien and our observations are based on studying the original there.

Preservation

The specimen consists primarily of the alveolar portion of the maxilla, the base of the left zygomatic process of the maxilla, the right lateral border of the piriform aperture (extending superiorly 16 mm from the lower nasal border and 17 mm of bone lateral to it), portions of the left and right maxillary sinus, and the complete nasal floor. Parts of the outer surface are covered by a thin layer of matrix, as can be seen in Figure 16. The left alveolar border is more complete than the right and contains the canine and first and second molars. Sockets for the incisors, premolars, and third molar are present, their internal walls filled with matrix. On the right, the alveolus is broken away through the middle of the M^1 socket. A heavily worn I^2 is still located in the right maxilla, while empty sockets exist for the incisors, canine, and premolars. In the anteroposterior plane the specimen is complete from prosthion to virtually the posterior nasal spine; the tip of the spine is broken away, a fact Szombathy used to suggest the specimen was trampled before it was fossilized.

Age and sex

Mladeč 8 shows a high degree of wear on the I^2, C', and M^1. In fact, of all the currently preserved dental remains from Mladeč, this specimen shows the greatest amount of wear. Based on this, we note that it is considerably older than Mladeč 1 and 2 which we age at more-or-less 16 years; the second molar, for instance, being much more worn than the first molar of Mladeč 1 and the incisor is worn below the cementoenamel junction on its lingual side. Tooth wear also exceeds Předmostí 3, especially on the M^2 and relating to the remaining crown height. This is a specimen we age at approximately 35. Therefore, Mladeč 8 must have been at least in its mid-thirties at death, probably older.

The canine has the largest crown area in the European Upper Paleolithic (Frayer, 1978) and, where comparable, all of the dental and most of the palatal dimensions are larger than Mladeč 1 or Mladeč 2. Thus, the specimen seems very likely to be male.

Description

This maxilla is quite unlike the Mladeč 1 and 2 females. Besides being generally larger in virtually all measures, it has less subnasal prognathism and its appearance is massive and robust. The palate is broad, but vertically and anteroposteriorly short. The nasal aperture is narrower (Table 11) than the comparative samples, and lies below their minima. Internal and external nasal lines define the lower border of the nasal aperture (Fig. 16). The external lines are continuous with the lateral nasal margin and merge together at the midline to form the anterior nasal spine. They form a sharp demarcation between the floor of the nose and the subnasal portion of the maxilla. The internal, or posterior line, is visible on the internal surface of the lower nasal margin. It extends as a straight line coronally along the inferior nasal margin posterior to the anterior line by about 5 mm at the lateral nasal margins and by 10 mm at the midline. Between these lines a shallow gutter is delineated. The anatomy of this region is quite similar to the Gibraltar female and Guattari 2. In these specimens, the interior line also extends to the lateral wall of the aperture, but it is closer to parallel and the distance between the lies here is not much different than near the midline. The marginal lines are about the same distance apart in Guattari, but are further apart in Gibraltar.

Eleven millimeters above the lower border, on the right internal wall that is better preserved, there is the broken base of what appears to have been a projection extending medially into the nasal chamber. We believe this is the base of a medial projection, as described by Schwartz and Tattersall (1996) and Franciscus (1999). This broken base extends superiorly for 6 mm, and is 11 mm long in the parasagittal plane. Its inferior border is not smooth to the internal wall of the aperture, but forms a distinctly raised rim along its entire length, and therefore interrupts the wall in an area where there are normally no thick projections. This lower border is at about 30° to the base of the nasal floor, with the posterior part more superior. It could not represent an abrupt inward turn of the internal nasal wall, because this would require that the nasal aperture have an abnormally low height and very unusual shape. The interpretation that it is the base of a medial projection is more reasonable, but if so the projection extends to the lateral nasal border, while in Gibraltar the projections begins about 9 mm posterior to that border.

Below the nose, the maxilla is paracoronally flattened between the pronounced canine buttresses; a line through the most anterior aspects of the canine roots passes through the center of the central incisor roots. The canine pilasters extend vertically to the level of the nasal floor, and are about 7 mm lateral to the lateral nasal margin. They are medially bordered by a sulcus. Szombathy reports that the

Table 11. Dimensions of male facial bones (mm)

	Mladeč		Neandertal		Skhul/Qafzeh	
	5	8	Mean (n)	Range	Mean (n)	Range
au-fmo	88.0		82.2 (6)	79.8–84.7	76.1 (3)	72.3–79.5
nasion proj bi-fmt line	20.7		31.2 (5)	28.0–35.8	15.3 (2)	9.0–21.7
biangular breadth		69.2	78.3 (3)	72.2–86.1	72.9 (2)	71.5–74.3
alveolar ht		17.5	28.7 (4)	26.5–30.4	25.8 (3)	23.2–28.2
nasal breadth		29.1	33.2 (5)	30.0–35.1	31.8 (3)	30.8–32.5
zpm base-alveolar margin		12.8	12.7 (3)	9.5–15.7	12.2 (4)	10.0–15.6
zpm base-prosthion		45.3	56.1 (3)	54.2–59.9	51.1 (4)	49.7–53.6
zpm base-post M^3		27.5	19.7 (2)	19.5–19.8	25.5 (4)	24.8–27.0

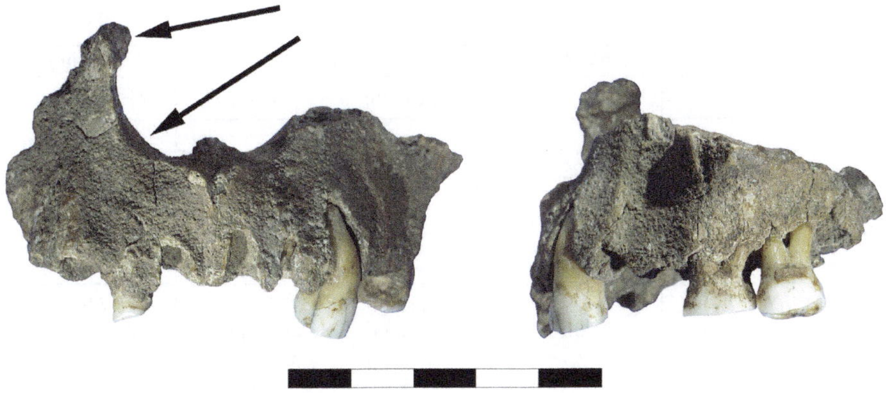

Fig. 16. Mladeč 8 in frontal and lateral views. Note the prominent nasal spine, vertical face of the subnasal region, and the doubled inferior border of the nasal aperture that borders the shallow gutter traveling along its anterior edge, marked with the lower arrow. The higher arrow points to the broken base of the medial projection on the internal wall. According to Schwartz and Tattersall (1999) this is a Neandertal autapomorphy

anterior nasal spine was sharp and projecting, before it was broken in the process of excavation. In the sagittal plane the maxillary contour just below it is convex. The contour becomes straighter laterally, but some curvature remains and the root of the left canine arevery markedly curved, with the apex of the curve some 5.4 mm superior to the cementoenamel junction.

Lateral to the canine buttresses the face of the maxilla curves markedly. Its external face is virtually a straight line from the center of the P^3 root to the most posteriorly preserved point that is behind the M^2. Actually, this line is slightly bowed outwards, with the anterior of the M^1 root at its apex. The anterior surface of the zygomatic process of the maxilla is at the position of the middle of the P^4 root, and the center of the base of this process is at the middle of the M^1 root. This position is more anterior than the process in Mladeč 1 and 2.

The zygomatic process of the maxilla is high. Its base is approximately 12.8 mm superior to the alveolar margin, which is very much like the comparative samples. What little remains of the base of this process appear as an angle between two straight surfaces and does not form a true malar notch. The anterior face of the process swings laterally, and for its preserved extent it appears to be exactly in the paracoronal plane, and thereby perhaps reflects a cheek that is flat and laterally oriented, at least along its base. Just anterior to the zygomatic process of the maxilla is the bottom portion of a broad canine fossa.

From the top, the maxillary sinuses can be seen located over the P^3's at their most anterior point and extend posterior to behind the M^3 socket (on the left). The most inferior portion of the maxillary sinus extends only slightly below the roof of the palate laterally and does not extend below the maxillary torus.

From the side, the impression of the specimen exhibits little alveolar projection anterior to the zygomatic process of the maxilla, for instance greater than Předmostí 3 but less than Mladeč 1. The minimal anterior projection of the lower face is reflected in the fact that from the base of the zygomatic process of the maxilla to prosthion the distance is less than for the diminutive Předmostí 3, and below the minima for the comparative samples. In the archaic samples this distance is greater while the distance from the base of this process to the back of the third molar is less. The ratio of these (front to back) is 1.7 times greater in the Neandertal males, showing them to have much more of the snout anterior to the cheek position (or, put another way, a more posterior cheek position). The ratio for the Skhul/Qafzeh sample is intermediate. The alveolar prognathism, as reflected by the angle of the subnasal region to the tooth row, is only moderate.

Table 12. Palate and tooth row dimensions (mm) for males

	Mladeč 8	Neandertal		Skhul/Qafzeh	
		Mean (n)	Range	Mean (n)	Range
Palate lengths					
Alveolar	55.5	60.2 (2)	57.3–63.0	65.6 (2)	64.1–67.0
Palate	46.5	59.6 (1)[1]		56.3 (2)	52.8–59.7
Anterior palate	34.5	45.0 (2)	43.1–46.9	44.7 (3)	44.5–44.9
Tooth row lengths					
pr-postcanine	24.6	25.2 (4)	23.9–26.2	25.7 (5)	24.5–27.2
pr-M^1	46.1	46.4 (4)	45.2–50.0	48.3 (5)	47.2–49.0
pr-post M^3	57.2	64.0 (3)	61.1–67.7	64.5 (5)	63.7–66.5
Tooth row lengths between roots					
I^1-C	23.0	23.4 (3)	22.8–24.0	24.2 (3)	23.3–25.8
C-M^2	43.9	41.1 (4)	37.6–44.4	42.5 (4)	41.9–43.2
P^3-M^2	35.6	33.8 (4)	30.3–37.1	34.8 (4)	34.5–35.1
Palate depths at					
P^4/M^1	11.5	18.0 (2)	13.0–23.0	18.3 (3)	17.0–20.0
M^1/M^2	11.6	18.0 (2)	15.0–21.0	19.2 (3)	17.0–20.5
Breadth between roots of					
I^2	31.0	31.6 (4)	28.6–36.0	32.2 (4)	30.6–34.8
C	48.3	46.0 (3)	45.2–47.3	46.8 (4)	44.7–49.9
P^4	61.0	60.2 (3)	58.1–61.9	60.3 (3)	59.8–60.9
Internal at P^3	37.4	37.9 (3)	36.8–39.0	37.3 (4)	35.3–38.9

[1] Saccopastore 2

As Szombathy discusses, the palatine bone surface is not quite complete to the position of the posterior nasal spine. The greater palatine foramina are large and the grooves anterior to them are deep and well excavated. This small, fairly shallow palate with its small alveolar height (17.5 mm, even smaller than the Mladeč 1 female) contrasts with the robust dentition.

Dentition

Four teeth (left C', M^2, M^3, and right I^2) are preserved in this male maxilla. Sockets for all other teeth occur on the left side, while on the right only the I^1, C', P^3, P^4, and the mesial half of the M^1 alveoli exist, with the area behind M^1 broken away (Fig. 17). Each of the sockets is covered (or filled) with reddish-brown matrix indicating that all missing teeth were lost postmortem, some time after the specimen was deposited. The matrix has been cleaned from the teeth and roots, except for a thin coating that covers some of the M^2 occlusal surface.

Occlusal wear on the four preserved teeth indicates Mladeč 8 is an older adult, as we noted above. There is a significant differential in wear between the anterior and posterior teeth in that both the I^2 and C exhibit greater wear than the M^1 and substantially more wear

Fig. 17. Occlusal view of Mladeč 8. The dentition shows considerable wear, and enamel is missing on the lingual surface of the lateral incisor

than the M^2. On the right I^2, the crown is worn to nearly the cervical border, representing stage 8 level of wear. For this tooth, little of the crown remains other than an enamel rim on all but the lingual border, surrounding the secondary dentin, which is perforated by the root canal. Just at the position of the lingual enamel rim, the transverse breadth measurement is 7.4 mm we have not included this in the tables because it is *not* the crown breadth of this incisor. Wear on this tooth angled in both the mesiodistal and labiolingual planes. From the labial aspect, the remaining enamel on the mesial corner has a height of 0.7 mm, while at the opposite (distal) corner the height is 2.0 mm differences between these two heights account for a 15° angle across the occlusal surface. In the labiolingual plane the crown is also angled (about 20°), the result of heavier wear on the lingual aspect of the incisor. The distolabial edge is the highest and the mesiolingual edge the lowest. The occlusal surface is slightly convex and sharp-edged. Wear producing this type of crown angulation is unlikely to result from normal occlusion and in our view is probably related to oral manipulation.

The canine shows flat occlusal wear with complete dentin exposure covering the surface which is encircled by an enamel rim, representing a stage 7 level of wear. The enamel rim is slightly elevated in comparison to the occlusal surface and small enamel chips have spalled off around the edges. Unlike the incisor, there is no mesiodistal angulation to the wear plane. There appears to be a distinct labiolingual angulation (the labial side higher), but we contend that this is a consequence of the displacement of the crown out of its socket either postmortem or as the result of continued eruption (if the opposing tooth was lost). The extreme wear has removed all occlusal morphology, but from what remains of the labial surface, it is clear that there are no hypoplastic lines. Large interproximal contact facets exist on the mesial and distal surfaces for the I^2 and P^3, respectively.

The complete root of this tooth is visible since the canine is displaced out of its alveolus and extends below the occlusal plane defined by the left M^1. Also, the labial face of the alveolar socket is broken away, exposing the entire labial root surface. Both conditions result from postmortem changes. While the crown is quite large, the root is relatively short (15.1 mm) measured from the root tip to the cementoenamel junction. However, its length (7.1 mm) and breadth (10.2 mm) dimensions at the cervix are large, and the combination of the short root length and broad circumference dimensions produces a stocky root. The root is also bowed in the inferosuperior direction, so that the mesial face is concave.

Of the two remaining molars, M^1 shows considerably more occlusal wear. On this tooth, the only remaining occlusal enamel is a small ridge near the middle of the buccal region. Otherwise, no surface detail survives, but rather the crown is comprised of dentin surrounded by an enamel rim (= Stage 7). There is moderate cupping of this surface, since the perimeter is slightly raised in comparison to the middle of the occlusal surface. Besides the heavy occlusal wear, there is a large interproximal facet for the (missing) P^4 and a smaller one for the M^2. Thus, along with occlusal attrition, there has been considerable tooth loss due to interproximal wear. This has resulted in a roughly square crown shape. Due to wear, the cusp pattern cannot be determined, although enough remains of the lingual face to observe that there was no Carabelli's cusp. X-rays of the pulp chamber indicate it was normal in shape, although small. There is no indication of taurodontism.

The left M^2 shows the least amount of wear. The tooth is slightly displaced out of its socket in an inferior direction, so that its present occlusal surface is below that of the associated M^1. The mesiolingual cusp exhibits the most wear with a large dentin patch, while a pinpoint dentin exposure occurs on the distolingual cusp. The buccal cusps are blunted, but show no dentin exposure. This state of attrition corresponds to a Stage 3 wear category. On the mesial face a large, concave interproximal facet exists and a smaller distal facet is present, indicating the M^3 was in occlusion for some years before death. Given the size of both of these facets, some mesiodistal length has been lost to due interproximal attrition.

Occlusal morphology shows a 4+ pattern with a very large distolingual cusp. It is also clear that the cusp surfaces are not wrinkled. As in the M^1, the pulp chamber is small and does not extend into the roots. The roots are bifurcated and show little tendency for fusion. Finally, there are small portions of dental calculus adhering to the buccal and distal enamel faces, but stains on the enamel of the other surfaces indicate there were heavy deposits of calculus which have been subsequently lost or cleaned away.

Using lengths along the tooth row to examine some of the characteristics of the dentition as a whole in the males (Table 12), the comparisons reveal only relatively small difference between the small comparative samples. The anterior tooth row is a little smaller than the Neandertal mean and the posterior tooth row is a little larger, but the differences are small and the dimensions lie within the Neandertal range. The Skhul/Qafzeh comparison is somewhat different in that the Mladeč anterior length is smaller than any male from this sample, while the posterior length is greater.

The Mladeč 8 dentition is large, often exceeding the Neandertal and Skhul/Qafzeh means. This is especially true for the canine, which is on average unusually large in Neandertals. The Mladeč 8 canine breadth is at the top of the Neandertal range, and above the range for the Skhul/Qafzeh sample (Table 13). In fact this is a relatively large tooth whether length, breadth or area are compared. This canine is one of the largest in the entire Upper Paleolithic sample. The two molars are also large, especially in their breadths which again at the maxima (or exceed) the ranges in both archaic samples. The length dimensions are smaller, most likely a function of the advanced interproximal wear. Coupled with the robust palate dimensions and differential tooth wear, Mladeč 8, found in another context, might be identified as a Neandertal.

Table 13. Dimensions for male maxillary teeth. Length and breadth are in millimeters, areas in square millimeters

	Mladeč				Neandertal		Skhul/Qafzeh	
	8	9	50/51	47	Mean (n)	Range	Mean (n)	Range
C length	9.1	8.7			8.6 (10)	8.2–9.8	8.8 (6)	8.1–9.8
breadth	10.7	10.0			10.0 (9)	9.2–10.7	9.5 (6)	8.6–10.3
P^3 length		8.1			7.5 (11)	7.0–8.4	7.7 (7)	7.0–8.6
breadth		10.6			10.4 (13)	9.2–11.3	10.6 (7)	10.2–11.1
P^4 length				6.7	7.3 (8)	6.3–8.4	7.3 (6)	6.7–7.9
breadth				9.8	10.4 (10)	9.2–11.1	10.3 (6)	9.2–10.8
M^1 length	10.8		11.0	11.1	11.4 (10)	9.9–13.1	11.8 (10)	10.9–12.5
breadth	13.6		12.3	13.0	12.1 (10)	11.1–13.0	12.8 (10)	11.9–13.5
M^2 length	10.7			10.0	11.0 (10)	9.9–12.2	11.3 (6)	10.0–12.6
breadth	13.7			13.0	12.7 (9)	11.4–13.8	12.6 (6)	12.1–13.0
C area	97.4	87.0			85.9 (9)	75.0–101.9	83.9 (6)	72.0–100.7
P^3 area		85.9			78.2 (11)	64.4–92.4	81.9 (7)	71.6–94.7
P^4 area				65.7	75.9 (8)	58.0–92.7	75.4 (6)	65.6–85.8
M^1 area	146.9		135.3	144.3	138.5 (10)	110.4–165.1	151.3 (10)	139.2–164.7
M^2 area	146.6			130.0	139.4 (9)	111.8–158.1	142.5 (6)	123.6–158.1

Mladeč 9

Mladeč 9 consists of an isolated right maxillary permanent canine and third premolar (Fig. 18). The two teeth show a similar degree of wear and the interproximal contact facets fit together, making it likely that these two teeth come from the same individual. Moreover, based on size (Table 13), we consider the two teeth as belonging to a male (see below).

The maxillary canine is intact, except for the root tip that has broken away. Length of the root along the labial surface from the cementoenamel junction to the broken root tip is 14.6 mm, with a reconstructed total length of 18.0 mm. Just above the crown the root measures 6.1 mm long and 10.0 mm broad. Thus, although the Mladeč 9 canine has a longer root than Mladeč 8, it is more constricted in its basal dimensions and has a less stocky appearance. Lengths and breadths of the crown are similar between Mladeč 8 and Mladeč 9 with the latter slightly smaller. While we do not normally determine sex for isolated teeth, given its large size, we feel Mladeč 9 is almost certainly male. Crown dimensions for this tooth are at or above the Neandertal and Skhul / Qafzeh means and the tooth represents the fourth or fifth largest canine in the Upper Paleolithic sample (Frayer, 1978). Wear on the canine is minor, restricted to a small dentin patch on the tip and additional polishing on the labial surface. Small interproximal facets occur on the mesial and distal surfaces. These are in their normal positions (indicating the I^2 and P^3 were not crowded) and are shallow, attesting to the young age of the individual. Along the labial surface of the crown there are some very faint hypoplastic lines which course across the mesial to the distal surface. These are minor in expression and do not represent substantial developmental stress.

The most interesting aspect of this canine is the lingual surface of the crown (Fig.19), which possesses clear evidence of shoveling along the mesial and distal edges and a lingual tubercle *(tuberculum dentale)*. Szombathy noted both of these features. The shoveling is more pronounced along the distal edge, where a thickened pillar runs from the base of the canine to nearly the occlusal margin. This pillar is raised about 7 mm from the central fovea and in its greatest dimension is 3.0 mm thick in the mesiodistal direction. The mesial pillar is slightly less distinct. The greatest mesiodistal thickness of 2.4 mm is where a large median ridge runs just distal to the shoveled mesial border. This secondary ridge fills the central fovea and is separated from the mesial pillar by a shallow groove. Along the base of the canine is a well-developed lingual tubercle. This raised area occurs mesial to the midline of the canine and projects distinctly in the lingual direction. A shallow fissure separates the lingual tubercle from the distal pillar, while along the mesial edge the lingual tubercle is continuous with the shoveled mesial border, although it bulges distinctly away from it (see chap. 8, Plate VII).

In many respects the morphology of the Mladeč 9 canine is similar to some Neandertals that have canine shoveling coupled with a distinct lingual tubercle (Patte, 1962). For example, in some of the isolated teeth from Krapina (such as tooth 36, see Wolpoff, 1979), the mesial and distal edges are strongly shoveled and a well-developed median ridge parallels the mesial edge. According to Patte's terminology the tooth is less "premolarized" than Krapina 36. The Krapina E (Krapina 49) maxillary canine even more closely resembles the Mladeč tooth, combining the same elements of marginal pillars separated from the bulging lingual tubercle distally but not mesially, and a well-developed central lingual ridge. Krapina 139 and 147 are also similar, although Mladeč 9 is more "premolarized" in that the lingual tubercle is larger and more discontinuous on its distal border. Other Krapina teeth conform to varying parts of this pattern on the lingual surface, and it is clear that Mladeč 9 falls within the Krapina range of morphological variation.

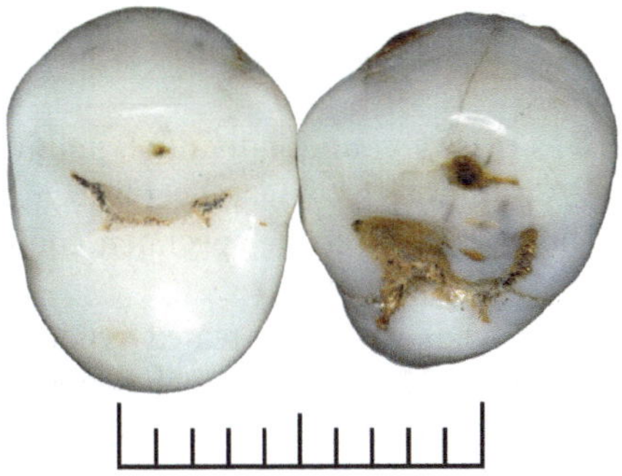

Fig. 18. Mladeč 9 canine and premolar in occlusal view

Fig. 19. Buccal and lingual view of the Mladeč 9 canine

Later Neandertals can differ considerably from this condition. At the extreme the lingual surface of the Le Moustier canine is quite smooth and concave from its base to the tip. Hortus III preserves a condition much like that of some Zhoukoudian canines, combining strongly developed marginal pillars that are continuous with multiple basal tubercles superiorly bonded by a concave lingual face. The Monsempron canines are more similar to the Mladeč 9 condition. The Aurignacian canine from Vindija G_1, Vi 287, also combines a basal tubercle that is continuous with a weak marginal pillar, and marked with a groove along its distal surface. It has a weak distal pillar, and a moderately developed central ridge mesial to the midline.

In the Skhul/Qafzeh sample, no specimen duplicates the Mladeč 9 morphology. The most similar is the Skhul 10 canine, which preserves only the unworn tip. On it are preserved two distinct marginal pillars and a central ridge. Nothing is known of the base of this tooth, but no preserved canine base in the Skhul/Qafzeh sample resembles the Mladeč 9 canine. For example, Qafzeh 9 is a barely worn tooth and its entire lingual surface is smoothly bulged, lacking any development of marginal ridges, a central ridge or a basal tubercle. The morphology preserved on the Qafzeh 6 and 7 maxillary canines are similar and do not resemble the Mladeč 9 condition.

The associated maxillary third premolar consists only of the crown, which was broken away from the root at the cementoenamel junction (Szombathy believed recently). The dimensions of the crown are 8.1 mm x 10.6 mm on the crown surface pinpoint dentin exposure occurs on the buccal cusp. The pulp cavity is small, showing no signs of taurodontism. The lingual and buccal cusps are large making the mesiodistal fissure appear deeply buried in the occlusal face. A small extra cuspule occurs on the right mesial margin, just below the canine interproximal facet.

The Mladeč 9 canine is metrically identical to the Neandertal mean and just slightly smaller in breadth and area from the Skhul/Qafzeh mean, although this tooth is not fundamentally different from either sample. However, the same is not true for the P^3. While within the range of both the Neandertal and Skhul/Qafzeh samples, it has much greater length (8.1 mm compared to the Neandertal mean of 7.5 mm or the Skhul/Qafzeh mean of 7.7 mm). P^3 breadth is not markedly different from either the archaic samples, but the occlusal area is larger than either the Neandertal or Skhul/Qafzeh mean, a product of the large length dimension.

Mladeč 10

This specimen is represented by a single upper right M^3. It derives from a young individual, judging from the occlusal wear, which consists of only minor polishing. A faint interproximal facet is located on the mesial surface and the roots are completely formed. Thus, we suspect the specimen comes from a 18–19 year old (see chap. 8, Plate VII).

The occlusal surface is basically three-cusped, although there is a small, well-demarcated hypocone that occurs as a distal cingulum. Compared to Mladeč 2, the hypocone on Mladeč 10 is much better developed. The face of the crown is highly wrinkled, in that small cuspules occur on the occlusal surface.

The roots show very little convergence, with the buccal roots well separated from each other and the lingual root strongly divergent from the base. Internally, the pulp chamber is large and extends more than half way up the lingual root. This tooth, then, shows a moderate degree of taurodontism.

Dental and gnathic remains that have been destroyed

Mladeč 50, 51, and 54

Discovered in the Quarry Cave, Mladeč 50 and 51 are small associated fragments of maxilla. They are only described as retaining heavily worn first molars. Few other details are provided for the maxillary M^1's except that they were 11 mm in length and between 12.0 mm – 12.5 mm in breadth. Szombathy associated these maxillary fragments and the Mladeč 54 mandible on the basis of heavy tooth wear and their overall massiveness, and speculated that they "most probably belonged" to the Mladeč 5 calvarium. While the association is not certain, size and robustness verify the supposition that these specimens are male.

The Mladeč 54 mandible is comprised of two pieces. The larger of these is an almost complete right corpus and part of the left side that Szombathy studied at the Litovel Museum. It is figured in his Tafel 8, Abb. 6 (reproduced here in Figs. 20, 21 and 22). The specimen is broken at the left canine position and posteriorly at the very back of the right ramus, but the remaining portion does not include the gonial angle. The second piece does not fit onto the first. This left corpus fragment holds the P_4–M_2. According to Szombathy all of the mandibular teeth "are worn down to the dentin," except the M^2's. No other surface details are mentioned. On the right side Szombathy noted that the M^3 socket was "empty and quite shallow" and suggested that this tooth "had been lost during the individual's lifetime."

Table 14. Dimensions for male mandibular teeth. Length and breadth are in millimeters, areas in square millimeters

	Mladeč		Neandertal		Skhul/Qafzeh	
	54	52	Mean (n)	Range	Mean (n)	Range
I_2 length	6.2		6.3 (7)	6.0–6.9	6.6 (6)	5.8–7.9
breadth	7.5		7.7 (10)	7.2–8.3	7.5 (6)	6.8–9.0
C length	8.0		7.9 (10)	7.4–8.4	8.1 (5)	7.7–8.8
breadth	10.0		9.4 (11)	8.8–10.1	8.7 (5)	8.2–9.9
P_3 length	7.4		7.8 (11)	6.7–8.5	8.0 (6)	7.2–8.5
breadth	9.0		9.5 (11)	8.8–10.2	9.1 (6)	8.4–9.6
P_4 length	7.4		7.2 (10)	6.4–8.4	7.8 (6)	7.4–8.2
breadth	8.9		9.3 (9)	8.5–10.2	9.2 (6)	8.4–9.8
M_1 length	10.5*	11.1	11.5 (12)	10.7–12.6	12.2 (7)	11.4–13.1
breadth	11.5	10.5	11.1 (12)	10.2–11.7	11.9 (7)	11.4–12.7
M_2 length	12.0	10.0	11.8 (10)	11.1–12.9	11.7 (5)	10.7–12.2
breadth	12.0	10.0	11.4 (10)	10.3–12.1	11.5 (6)	10.3–12.2
M_3 length		10.0	11.7 (9)	10.8–12.5	12.3 (6)	10.6–13.2
breadth		10.0	11.5 (9)	10.5–13.4	10.9 (6)	9.9–12.0
C area	80.0		75.0 (10)	65.2–82.3	70.8 (5)	64.3–86.7
P_3 area	66.6		74.3 (11)	61.8–83.3	73.2 (6)	60.1–81.6
P_4 area	65.9		66.9 (9)	54.7–84.8	72.0 (6)	61.7–80.0
M_1 area	120.8	116.0	127.9 (12)	108.6–142.4	144.5 (7)	128.8–161.0
M_2 area	144.0	100.0	134.8 (10)	114.3–150.0	134.4 (5)	109.2–148.8
M_3 area		100.0	135.5 (9)	112.9–166.9	134.7 (6)	109.4–158.4

* dimension reduced from heavy occlusal wear

The mental protuberance is triangular, small, and strongly projecting inferior to the lower rim of the mandible, making the chin appear narrow and pointed. The mental spine, according to Szombathy, is a small multi-notched prominence some 15 mm above the base. Inferior to it the digastric fossae have lengths of about 30 mm and breadths of 8 mm. Just at the symphysis the inferior border of the corpus is narrow, but more laterally this border thickens, becoming a flat but narrow bone plate that extends to the first molar position. The mylohyoid line is distinct for its entire course. However, under the molars it is particularly strongly molded so that the alveolar portion of the corpus is some 20 mm thick beneath the third molar, while in this position below the mylohyoid line the bone thins to 4–6 mm.

A small mental foramen is located under the P_4, about midway between the corpus base and the alveolar margin. The root of the ramus begins between the first and second molars, and its anterior edge appears to cross the alveolar margin at the middle of the M_3 socket. The symphysis is 36.5 mm high (Table 15), and between the first and second molars corpus height is 36 mm. The symphysis height is a small value, within the range of the European Neandertals but smaller than all of the Skhul / Qafzeh males. The more posterior corpus height is greater, at the top of or above the comparative sample maxima. The 20 mm breadth at the M_3 position is also quite large, at or exceeding the comparative sample maxima. Obviously, Mladeč 54 is quite large. The size is particularly significant because of the marked trends for mandibular *size and robustness* reduction that characterized this period. In general the corpus has more or less even heights throughout the tooth row; the alveolar and basal margins are approximately parallel to each other. This characterizes several other male Central European mandibles from the early Upper Paleolithic, but is rare in earlier specimens.

Dental dimensions of Mladeč 54 are given in Table 14. The teeth are large and approximate the size of the Neandertal and Skhul / Qafzeh means, despite the advanced wear (Fig. 23). The mandible no longer exists, but Szombathy's description of the wear ("worn down to the dentin") suggests that the length measurements of the anterior teeth (and possibly others as well, most certainly M_1 length) are not really comparable to those in our data set since we generally do not include the lengths of teeth with marked interproximal attrition. Doubtlessly, the teeth are substantially reduced from their original size; yet, even with the wear, the incisor dimensions are not markedly different from either the Neandertal or Skhul / Qafzeh samples and canine breadth and area are also large, near the maxima of the ranges for both comparative samples. European Neandertal mandibles are characterized by their anterior tooth expansion, and the few Mladeč anterior teeth we have resemble them in this regard.

The two premolars are generally smaller than the Neandertal and Skhul / Qafzeh means in both dimensions, but lie within the ranges of both. Given their purportedly extreme wear, unworn length

Table 15. Mandibular dimensions (mm) for males

	Mladeč 54	Mladeč 52	Neandertal		Skhul / Qafzeh	
			Mean (n)	Range	Mean (n)	Range
Symphysis height	36.5	36.0	37.5 (9)	35.0–39.8	39.9 (4)	36.6–44.1
Symphysis breadth		16.0	16.0 (9)	13.2–17.4	15.1 (4)	13.7–16.1
Symphysis Index		44.4	42.5 (9)	35.2–49.6	38.4 (4)	35.1–44.0
Height at M_1/M_2	36.0	32.0	33.1 (7)	30.6–35.6	34.6 (3)	32.8–36.3
Breadth at M_1/M_2		16.0	15.7 (9)	13.0–18.0	16.7 (5)	12.9–18.1
M_1/M_2 Index		50.0	48.2 (7)	45.1–55.2	44.8 (3)	37.2–52.4
Breadth at M_3	20.2		16.3 (6)	14.7–17.8	17.6 (2)	15.2–20.0

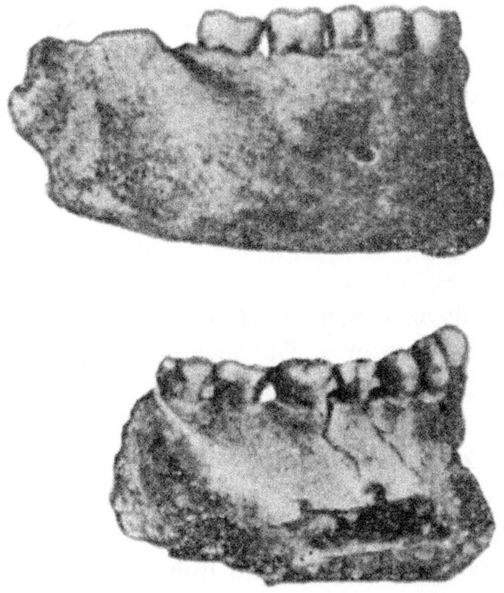

Fig. 20. Mladeč mandibles 54 (above) and 52 (reversed). The male specimen is Mladeč 54; we compare it here with Mladeč 52, the only other reasonably complete mandible from the site. We did not feel confident in assigning sex to Mladeč 52

Fig. 21. Mandibular bases for the male specimen Mladeč 54 (below), and Mladeč 52

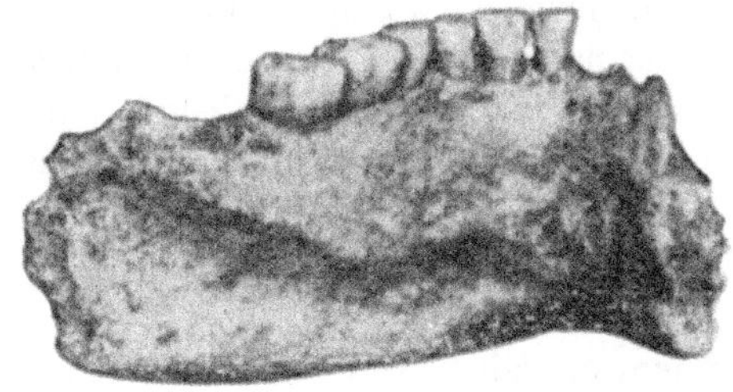

Fig. 22. Internal view of the Mladeč mandible 54 (reversed), showing the strong mylohyoid line

dimensions must have been considerably larger, and in fact the molar breadths above the means of the comparative samples, and at the top of the ranges. A different relation holds for the areas, and the M_1 is particularly small, we believe because of the shortened mesiodistal length. We conclude that in contrast to the premolars, the Mladeč 54 molars are not reduced, whether compared to the Skhul / Qafzeh sample or to the Neandertals.

Finally, while this specimen is from the Main Cave, we think there is no possibility that this mandible is associated with Mladeč 8, which has a canine area 13% larger than the Neanderthal and 16% larger than the Skhul / Qafzeh mean. Even the smaller canine of Mladeč 9 is larger than both archaic sample means, so it is apparent from this admittedly small sample that these three Mladeč males are characterized by having large upper and lower canines.

Mladeč 55

Also from the Quarry Cave, this mandible "probably belonged to Mladeč 6" according to Szombathy, especially if Mladeč 54 was associated with cranium 5. The specimen is comprised of the right half of a mandibular corpus with I_2–M_1, lacking the gonial angle and the ascending ramus. Szombathy only mentions the teeth are "worn down to the dentin." Unfortunately, he did not include dental measurements and provided only a sketchy description of the mandible. The chin is not completely intact, but according to Szombathy it was well-developed. From the tuberculum to the third molar position, the base of the corpus thickens, developing into a lateral posterior torus some 15 mm thick and as much as 13 mm in height. The mylohyoid line is short, ending at the fourth premolar position. The sub-maxillary fovea is shallow and attains a thickness of only 12 mm.

There is a basal tuberosity and flange-like torus projecting laterally from it that is similar to the anatomy found on the Vindija 250 Neandertal. Here, the basal margin is also quite thick and supports a distinct, projecting torus that is as much as 8 mm in height posteriorly. The lateral projection is stronger anteriorly, where it actually creates a lateral flange at the M_1-position, its most anterior extent.

Jaws of uncertain sex

Szombathy briefly described (or mentioned) several fragmentary jaws, some with teeth, that we believe could be male but have no way to further examine this possiblity because they all were subsequently destroyed. We present what is known of these specimens below, and include them in our male dentognathic comparisons. In considering these comparisons, our uncertainty should be kept in mind.

Mladeč 47

According to Szombathy this left maxillary fragment contained I^2, P^4, M^1 and M^2. Measurements for the posterior teeth are given in Table 13. Unfortunately, Szombathy did not publish dimensions for the I^2, although he did describe it as being large. Only a small portion of the socket for the M^3 was preserved, and it is shaped "as if the wisdom tooth had just been in the process of being developed, but had not erupted." Also the lateral incisor showed little wear, although Szombathy was not certain it belonged to the maxilla since it did "not totally fit into the alveolus."[21] He also notes that wear on the M^1 was slight and the P^4 and M^2 were unworn. Thus, based on his brief description, this specimen appears to be a subadult, possibly as little as 11–12 years old at death.

21 It is possible there was some matrix or sand in the alveolus.

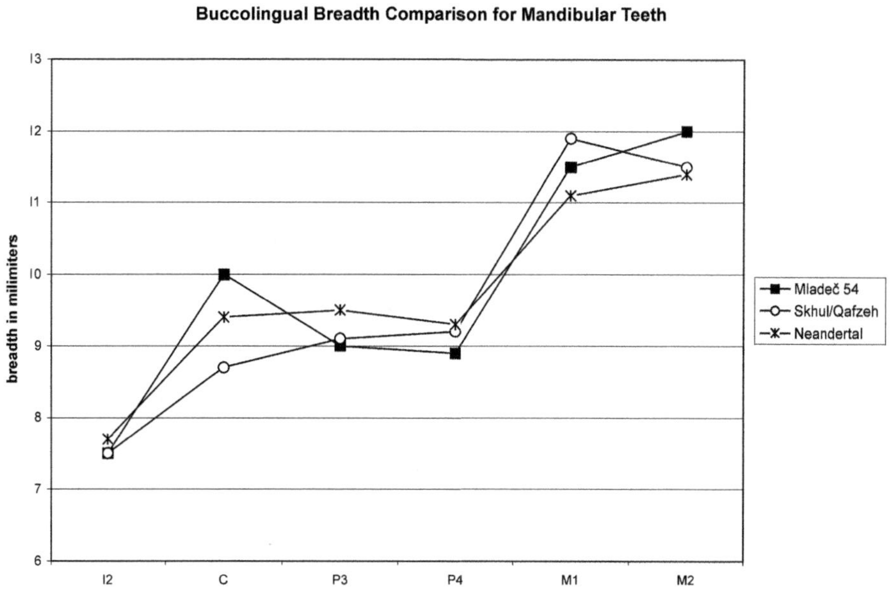

Fig. 23. Comparison of buccolingual breadth dimensions in Mladeč 54 and the means for male Neandertals and Skhul / Qafzeh. The breadth dimension was used in this comparison because it is not affected by interproximal attrition, and even occlusal wear only affects breadth when it reaches the base of the crown

Morphologically, Szombathy mentions the M^1 had "four rounded cusps" and that the mesiolingual cusp was "especially strong." On the M^2, the distolingual cusp was small and separated from the trigone by an "arch-like furrow." From this description it seems clear that the M^2 showed a reduced hypocone and, in fact, Szombathy refers to this tooth as "three-cusped."

The Mladeč 47 P^4 is moderate in size, both dimensions being just below the average for other earliest European Upper Paleolithic males and far below the Neandertal and Skhul/ Qafzeh means, although within their ranges. The breadth dimensions of the two molars are above the means of the two comparative samples. Thus making the same comparisons, the molars (large) and premolars (small) have the same relation as in the Mladeč 54 mandible, described above. However, in this maxilla the lengths are more reliable as the teeth are barely worn. The molar lengths are below the means of the comparative samples, and this contributes to the fact that the occlusal area comparisons give mixed results, the M^1 is relatively large and the other two posterior teeth relatively small. Virtually every dimension is smaller than the corresponding teeth in the Mladeč 8 male. The differences are not enough to dismiss the possibility that Mladeč 47 is male.

Perhaps the most interesting possibility is the reported large size of the lateral incisor; this may be a Neandertal characteristic, but we shall never know.

Mladeč 48/49/52

Szombathy argued these left and right maxillary fragments (Mladeč 48 and 49) "probably belonged" to the mandible we have numbered Mladeč 52 (Tafel 8, Abb. 5 in Szombathy). In both jaws tooth wear is slight and the M^3's had apparently just erupted. Consequently, Mladeč 48/49/52 appears to be a young adult, between Mladeč 47 and 8 (or 54) in age. Szombathy mentions that the upper M^2's showed reduced hypocones and that the M^3 was two-cusped. Of the maxillary M^3's from Mladeč, this specimen appears to have the greatest reduction in the hypocone.

The Mladeč 52 mandible (Fig. 20) was comprised of two lower jaw portions, broken on an inferodistal angle at the position of the right canine. The pieces do not be fit together and thus the shape of the dental arch cannot be determined with certainty. On the left the outer table of the basilar portion of the corpus is broken away from the symphysis to the first molar position (Fig. 21). Reconstruction of this region is evident in Szombathy's figure, and we are not confident of the symphyseal profile. Szombathy claims, "the chin is strongly shaped *as far as is recognizable with the loss of substance of the outer wall*" (our italics). Also on this side, the gonial angle and the posterior portion of the ascending ramus have been broken away. The larger left fragment is shown in Szombathy's Tafel 8, Abb. 5, in lateral and inferior views.

The mental eminence seems to have been broad, and at its inferior tip it extends strongly in an inferoposterior direction. On the internal surface the digastric fossa is shallow and the mental spine is small. A large mental foramen is positioned under the P_4, at about the lower third of the corpus. The root of the ascending ramus is at the second molar position, and the anterior edge of the ramus crosses the alveolar order just posterior to the back of the third molar.

Table 15 shows all of the measurements of the mandible that Szombathy provided for posterity. These suggest the specimen was generally smaller than Mladeč 54. At the symphysis the Neandertal and Skhul/Qafzeh samples are taller but narrower than the Neandertals. Mladeč 52 is shorter than the means of both (actually below the Skhul/Qafzeh range) but with a thickness that matches the Neandertal mean. Thus the breadth/height (or "robustness") index is great, above the Neandertal mean and above the Skhul/Qafzeh range.

At the first molar position the corpus height is below the means and outside of the Skhul/Qafzeh range, while the breadth, once again, is relatively larger so the robustness index is greater than the means for the other samples. Where we can compare, Mladeč 52 is smaller than 54, especially in the molar row. If we could generalize and think of these differences as the result of a process of reduction, it appears than the corpus heights have reduced more than the breadths.

The view of the left side shown by Szombathy reveals a complete dentition. Of the 10 teeth preserved, only the dimensions of the molars were reported. He gives no details of crown surface morphology for any teeth. However, he describes a "right canine and one atrophied incisor that is barely 4 mm thick and wart-shaped" which are present in the lower jaw (Mladeč 52). As no photos of the right side of this specimen exist it is not completely clear if Mladeč 52 had an impacted left canine and a retained dI_2, but this seems a reasonable interpretation of the Szombathy's account.

Szombathy's measurements of the lower molars are given in Table 14. These teeth, especially the M_2, are notably smaller than the teeth of Mladeč 54.

Mladeč 53

Mladeč 53 is a right mandibular fragment found in the Main Cave. Apparently no teeth were preserved in the jaw since Szombathy describes the sockets as empty and provides no measurements for isolated teeth that could have been attributed to it. This mandible is represented by the right half an edentulous corpus and partial ramus, lacking the base from a point anterior to the gonial angle, and missing the condyle. The break involves the more anterior portion of the corpus as well, so that the socket for the left I_2 remains while the symphysis is cut through at the midline. The mental eminence is preserved. It is very rounded and moderately projecting in form. While the *tuberculum mentale* is not particularly prominent, the mental spine is strongly developed. The tooth sockets are not matrix filled, suggesting that the teeth may have fallen out of the sockets after interment in the cave.

Mladeč 56, 57, 58, 59

These specimens, a canine, two premolars and molar, are only described as "loose teeth" and Szombathy does not even identify them as mandibular or maxillary. He gives no dimensions for the four teeth nor does he provide information on their morphology.

Functional analysis of the male crania

Assuming, then, that there is a significant Neandertal ancestry for Mladeč, comparison of the Mladeč males with the male Neandertals is valid for examining evolutionary change, and reveals several differences that we believe reflect changes in masticatory function. One of these is the most posterior extension of the temporal lines. Although the Mladeč crania are no shorter than the Neandertals, their temporal lines reach a higher position on the vault both at the coronal suture and posterior to it, and extend to reach the lambdoidal suture at a significantly higher position. The posterior border of the temporalis attachment follows the lambdoidal suture for at least half of its length. This indicates a more massive development of the temporal musculature, particularly lengthening and increasing the number of the most posteriorly oriented fibers. These differences result in a greater magnitude of the muscle's horizontal component of force (Hylander and Johnson, 1985), and improve the fibers' elastic resistance to tension (Champan, 1985). Such a difference could reflect increases in anterior loading and/or increased involvement of the muscle in the lateral motions in jaw movement (Lieberman, 1997).

A related variation involves the form of the mandibular fossa. Possibly confusing the interpretation of this region is the fact that both age (Hinton, 1981a) and dental function (Moffett et al., 1964; Hinton, 1979) are known to affect the very features that distinguish Mladeč 5 from most of the Neandertals. The mandibular fossa of this male is deeper than that of most Neandertal males. Its articular surface is divided into two aspects, a vertical wall that is the posterior part of the articular surface and the anterior face of the mandibular fossa, and a horizontal surface anterior and inferior to this wall. In the usual Neandertal male condition there is but a single articular surface extending from the roof of the mandibular fossa to the posterior edge of the temporal fossa, angled to the horizontal. Hinton (1981a) shows that the depth of the mandibular fossa and the slope of its anterior face (i.e., the posterior portion of the articular eminence) both increases with increasing tooth wear in younger Eskimos (approaching the Mladeč 5 condition). In contrast the fossa depth and the slope of the posterior part of the eminence both decrease with increasing tooth wear in Native Americans from the Southwest (approaching the Neandertal condition).

Clearly, the age/function relationship is complex. Hinton argues that the pattern of increasing fossa depth and slope of the posterior eminence face is a consequence of constant loading of the horizontal anterior face of the articular eminence by the joint reaction force of the corresponding mandibular condyle. The latter pattern, approaching the Neandertal condition, results from loads placed on the more vertically oriented posterior face of the articular eminence. Assuming that the dental age of Mladeč 5 (which, of course, is unknown) is comparable to the range of ages represented by the Neandertal sample (i.e., that neither age or the degree of tooth wear is a relevant variable in the comparison), Hinton (1981a, 452) suggests a contrast that might be applicable in the comparison of Mladeč 5 and the Neandertal sample:

> *"long-term increases in fossa depth and slope are most predominant in those human groups that regularly perform strenuous loading of their molars and anterior teeth, both of which activities would tend to occur with the condyle out on the eminence crest (especially when masticating or otherwise processing a very large, tough bolus). In contrast, decreases in fossa depth and slope*

observed in groups in which posterior tooth loss is extensive would imply that the condyle is habitually situated back up the posterior slope of the articular eminence while under load."

In explanation, Hinton (1981b) has argued that molar loss is frequently accompanied by increased incisor utilization. For this reason, we contend that if incisor utilization in the Neandertals was more often and/or of greater magnitude that that ever observed in living humans (as one might suppose from the large size and differential wear on Neandertal incisors), posterior tooth loss may be a less important influence on the remodeling of the Neandertal mandibular fossae than is the powerful incisor biting itself[22]. The potential importance of anterior tooth loading unaccompanied by posterior tooth loss in creating a shallow, flattened mandibular fossa may be supported by the very flattened mandibular fossae that characterize *Australopithecus afarensis,* a Pliocene hominid species with even larger anterior teeth.

This would suggest that certain masticatory-related features had changed in the Aurignacian males. In spite of their large canines, and other evidence of expanded anterior teeth, Mladeč contrasts with Neandertals in showing a regular loading pattern emphasizing *both* anterior and posterior regions of the dentition. The relatively greater emphasis on anterior loading that characterizes Neandertals is not specifically found in this sample. The increases in temporalis attachment area, and the shape of the glenoid fossa, support this interpretation.

A second area of difference involves the cranial rear, its shape and the musculature of the nuchal region. One difference in the region varies in the two adult Mladeč males. At the cranial rear, Mladeč 6 has a short vertical occipital face, similar to Neandertals, while the face of the occipital plane is tall in Mladeč 5. Caspari (1991) argues that the vertical posterior is oriented to resist stress on the occipital plane generated by the nuchal muscles. She contends that the larger size and longer vertical surface of the cranial rear in Mladeč 5 creates a different distribution of strain that is usual in Neandertals and Mladeč 6, with a short and narrow vertical occipital face. This is why, she contends, Mladeč 6 shares another feature with Neandertals that Mladeč 5 lacks, an elliptical suprainiac fossa.

The actual breadth of the nuchal musculature on the nuchal plane of the occiput is quite large in both Mladeč specimens, exceeding the Neandertal mean even though the breadth of the bone between the asterion points is similar to the Neandertals. Evidently, reduction in occipital breadth is somewhat independent of the requirements of the nuchal musculature. The expanding attachment area for these muscles may be a simple measure of increased muscular robusticity in these early (so-called) modern Europeans, but it is important to understand the details. Caspari (1991) suggests several muscular complexes that differ in the Mladeč male nuchal planes, compared to the usual condition in Neandertals. She notes the prominent width of major extensor insertions in the Neandertal sample, and attributes this to significant extensor function. In contrast, the Mladeč occipitals have narrower *m. semispinalis* insertions, although with relatively more pronounced markings for other muscles that are less important extensors. This difference is attributed to variation in facial size. Moreover, Mladeč occipitals have more strongly developed superior nuchal lines. Caspari argues that these reflect the increased habitual loading of the sternocleidomastoid and trapezius muscles, both of which are active in upper limb activities, sometimes acting as synergists to each other. The high vertical faces of the male Mladeč temporal and posterior parietal regions reflect the same, according to Caspari. This is particularly evident above the mastoid region, which contributes to the verticality of the cranial sides of these specimens, and thereby differs from many Neandertals. The vertical temporal squamae relate to prominent,

22 We would add, however, that posterior tooth loss is frequent in the older Neandertals, and quite possibly plays a role in the development of extremely flat fossae such as in the La Ferrassie and La Chapelle males.

posteriorly oriented mastoids, reflecting a response to significant stress on the temporal created by nuchal forces.

While differing from Neandertals, Caspari (1991, 247–8) does not find the Mladeč remains are exactly like other early Upper Paleolithic males:

> *"They exhibit features characteristic of the later early Upper Paleolithic sample, but these features are not so well developed. Mladeč 6 is the most Neandertal-like of the Upper Paleolithic sample, having a broad occipital and very wide semispinalis insertions. This morphology and its associated toral components thus resemble Neandertals. Additionally, the superior nuchal line of Mladeč 6 is the least well developed of the Upper Paleolithic males. Laterally, this specimen conforms to more "typical" early Upper Paleolithic morphology. Other Mladeč remains, while lacking the "Neandertal-like" second-layer morphology, also exhibit weaker development of the medial superior nuchal line than their Pavlovian successors."*

She therefore notes that the behavioral changes influencing the cranial rear do not occur all at once. These changes are masticatory, in part, but cannot be described as a simple reduction following from less powerful loading of the anterior teeth because many of the changes involve greater size and/or robustness in posterior cranial structures. Caspari postulates significant increases in carrying and hauling from the shoulders, perhaps with tumplines, in these Aurignacian (and later) Europeans.

The place of Mladeč in human evolution

In a recent paper, Tattersall and Schwartz (1999) give three alternative theories for the fate of the European Neandertals: replacement by conflict or competition, swamping by the genes of incoming populations, or rapid *in situ* evolution into moderns. These alternatives would imply that either Neandertals did not contribute to the ancestry of the Mladeč Aurignacians, that they contributed very little, or that they were the unique ancestors of the Mladeč folk. Logically, however, there are more possibilities and historically the focus has been on them, and not the more restricted set that Tattersall and Schwartz propose.

Through considerations of their ancestry, the Mladeč males address this issue, a point that has not escaped any of the scientists who worked on them. Around the time of the Mladeč discoveries, Schwalbe (1906) noted that certain Central European fossil crania such as Brux and Podbaba[23] lie between the Neandertals and "recent *Homo*". He proposed using the term "var. *Homo fossilis*" to refer to them. Szombathy (1925) suggested that Mladeč 5 would appear to also belong to this group. But for him the term itself was unacceptable, and Szombathy considered this fossil one of the Pleistocene races of *Homo sapiens* – the Brux race. In fact, for Szombathy, the issue of the place of Mladeč in human evolution was an issue of which race the specimens from the site belong to (1925, 79–85). In this he reached the surprising conclusion that three races are represented among the four most complete Mladeč crania, what he called the Brux, Cro-Magnon, and Chancelade race. Yet, he realized that this is not the result of different races being represented. Instead, he wrote (1925, 89) that the "form elements" of these races are present "in different intensities and variable distribution." These conclusions are quite similar to those reached in Weidenreich's analysis of the Shandingdong (Upper Cave) remains from Zhoukoudian. It was Weidenreich's (1939, 172) contention that the three Shandingdong crania show certain common features of an Asian sort, but typify "three different racial elements, best classified as primitive Mongoloid, Melanesoid and Eskimoid". He wrote "[...] re-

23 As events have transpired, neither of these crania have acceptable provenience, so this turns out to have been a failed exercise in morphological dating.

cent North Chinese may be considered as more advanced types, but traceable to ancestors like those represented by the Upper Cave man". Weidenreich's analysis addressed the general question of racial origins, whether there once were pure races that subsequently mixed to form today's "hybridized" populations[24]. He took the Shandingdong specimens as proof that there never were. Weidenreich reasoned that if races had been "pure" in the past, and mixed with each other more and more over time until achieving their present state in which no pure races are left, we would expect that variation of a past sample should be less than today's because there was less intermixture in the past. But Shandingdong showed, if anything, *more* variation. Szombathy did not draw conclusions like this, but rather wrote in the concluding sentence of his monograph (1925, 89) "This correlative existence of the three late Paleolithic human forms in a fairly small area and in the same time [...] lend the Fürst-Johanns-Höhle and its small neighbor cave near Lautsch a highly peculiar impression."

Subsequently, Jelínek (1983) also recognized the great variability in the population from Mladeč, but his focus was more on the phylogenetic questions the variability addressed. He especially stressed the presence of Neandertal-like characteristics, which he summarized by noting, "[t]he special significance of the Mladeč finds [...] lies in the fact that they clearly indicate the Neandertal heritage" (1983, 64). In addition, he stressed and provided details about the role of sexual dimorphism in explaining variation at the site. Similar conclusions were reached by Frayer (1986), Smith (1982), and Wolpoff (1989*b*). These conclusions are reiterated here.

Several analyses have reached somewhat different interpretations of the place of Mladeč. Howell (1984) questioned the Aurignacian provenience, mistakenly asserting (p. xxi) that "the enclosing sediments [surrounding the skeletal remains] are derived from a talus cone from an aven, and any presumed age attribution (for example to the Podhradem interstadial) is purely inferential." We hope to have dispelled this with the details presented here, and restate the fact that both culturally and in terms of faunal associations Mladeč is a single component site; but truth be known, Szombathy already addressed the issue in 1925 (p. 7):

> *"I was the only expert to have seen the finds "in situ" in 1881 and I have no doubt about the identical ages of all human and animal remains found at that time. My report [Szombathy, 1883] states confidently "that we are dealing with a site in which the remains of animals and humans were buried at the same time." [...] The repeated examination of the stratigraphic relationships did not turn up one detail that would speak against the common age of all the finds."*

Other later studies reconstructing the cave and ascertaining its archaeology and provenience of the specimens in it have reached the same conclusions (Jelínek, 1969; 1983; 1987; Oliva, 1987; Smith and Trinkaus, 1991; Svoboda, 2000).

Bräuer and Broeg (1998) conducted a non-metric analysis of the Mladeč remains and reached conclusions quite opposite from us. It is appropriate to discuss why. Their paper disputes the evidence for regional continuity in Central Europe. However, the definition for regional continuity they employed is incorrect[25], and this provides the basis for one reason why we find their conclusions in-

---

24 Today's races would be considered hybridized, as opposed to just mixed, if their ancestors were distinct types or breeds; Weidenreich's role, in this regard, was to show that races are the results of mixed ancestry but are not hybrids between "once pure races."

25 Bräuer and Broeg (1998) equate regional continuity with gradualism. Moreover, they wrote (1998, 106) "during the long period of coexistence, mixing and gene flow between Neandertals and the dispersing modern populations might have occurred to varying degrees in different regions [...] such gene flow could have mimicked some degree of continuity." In fact however, Multiregional evolution is not "mimicked" by gene flow, but is based on it (Wolpoff et al., 2000)

Deviation (mm) of Mladeč 5 Auricular point Distances from Male Means

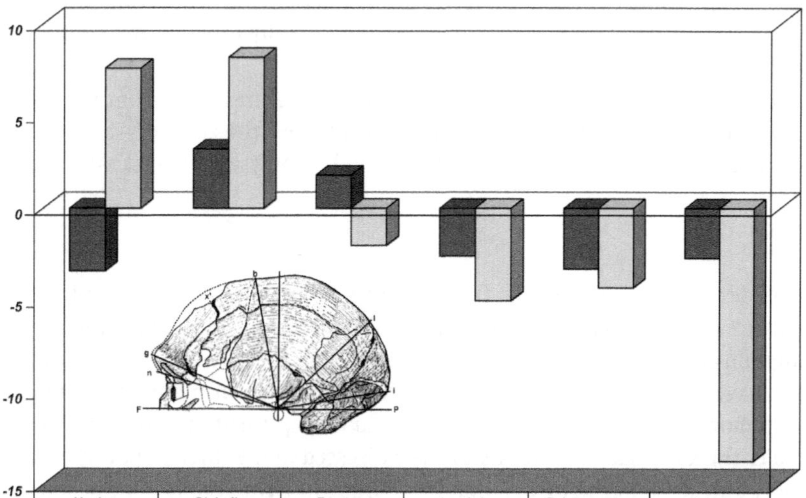

	Nasion	Glabella	Bregma	Lambda	Opistocranion	Inion
■ Neandertal	-3.4	3.2	1.8	-2.6	-3.3	-2.7
□ Skhul/Qafzeh	7.6	8.2	-2	-5	-4.3	-13.7

Fig. 24. Comparison of sagittal auricular distances for Mladeč 5 and male cranial sample means, shown as the deviation (in millimeters) of the Mladeč male from the potentially ancestral Neandertal (n=5) and Skhul/Qafzeh (n ranges from 2 to 4) samples. These distances are shown on the drawing of the Skhul 9 vault, modified from McCown and Keith (1939). For an example of reading this graph, the sagittal projection of the nasion-auricular distance in Mladeč 5 is 3.4 mm less than the male Neandertal mean, and 7.6 mm greater than the male Skhul/Qafzeh mean. In every comparison Mladeč 5 is closer to the Neandertal mean

valid. This is because what these authors looked for was some kind of evidence that the post-Neandertal populations of the region were actually Neandertal populations. In actuality, the evidence for regional continuity in Central Europe is based on a decreased, but substantial frequency of Neandertal *features* in these later folk (Frayer, 1993). We recognize it is not reasonable to calculate such frequencies for the two Mladeč male crania, but we find it significant that numerous features common or ubiquitous in Neandertals (Bräuer and Broeg's "regional continuity traits", and see Hublin, 1998; Schwartz and Tattersall, 1996; 1999) are even preserved in this small sample. For instance, Mladeč 6 has a suprainiac fossa of elliptical form (contra Bräuer and Broeg, see Caspari, 1991), extensive lambdoidal flattening, and a short posterior face on the occipital. Mladeč 5 has a very Neandertal-like sagittal contour (Fig. 2)[26], a well-developed occipitomastoid crest, minimal mastoid projection, and evidence of midfacial prognathism (insofar as the marked anterior projection of the upper face predicts this). Mladeč 8 has a groove along the inferior nasal margin and the remnant of a medial projection on the internal wall of the nasal aperture.

In fact, the Mladeč males have sagittal dimensions and profiles that deviate far less from the Neandertals than they deviate from the Skhul/Qafzeh males (exemplified by Mladeč 5, see Fig. 24). In this regard they are like other early Upper Paleolithic European males (Wolpoff, 1989b). In general

26 Bräuer and Broeg (1998, Fig. 6) show a contour for this specimen that does not match our knowledge of the specimen, our photographs, or any other published contours (i.e., Frayer, 1986).

the resemblances of the Mladeč males to Neandertals are based on metric and non-metric variation, which reveals significant evidence of Neandertal ancestry for the Mladeč folk, which is what we mean by local continuity. Bräuer and Broeg's (1998) conclusions are not confirmed in our systematic analysis of the adult male remains.

But, of course, many other cranial and postcranial features show that the Mladeč folk are different from Neandertals. The phylogenetic question of how the Mladeč sample relates to the earlier Neandertals and other potential ancestors from the Levant is systematically assessed below.

Relationships

The Mladeč males are not Neandertals. Our interest is in how these males might be related to Neandertals through ancestry. We assume the phylogeny of the Mladeč adult male crania reflects the phylogeny of the population from which they came. This phylogeny may be easier to examine in the adult males because they have a more dramatic or pronounced expression of many characteristics than females or juveniles, and moreover, there is an exceptionally poor representation of females in the Skhul / Qafzeh cranial sample. The key issue of relationships is expressed in the three hypotheses raised in the introduction. We reject out-of-hand the theory that the European Neandertals are the sole ancestors of Mladeč. This is a polygenic interpretation that would mean the Mladeč population evolved its features independently from a Neandertal ancestry, in parallel with similar changes that were taking place in other regions. If Neandertals were a different species than modern humans, such a theory would require also that the Mladeč population was part of this species. On the other hand, if Neandertal variation was at the level of subspecies, this theory would require the evolution of a very unlikely number of homoplasies.

Let us, then, consider the other two hypotheses about the descent of the Mladeč male crania:

(1) They are descendants of Skhul and Qafzeh populations[4] and not of Neandertals
(2) They are descendants of *both* European Neandertal populations and the penecontemporary Skhul and Qafzeh.

According to the Out of Africa theory of total replacement the first hypothesis is the expected one. Adherents to this theory have been fairly strong in their contentions about the ancestry issue, especially those who regard the Neandertals as a different species[27]. For instance, in writing about diagnostic features in the Neandertal nasal cavity, the inferior marginal groove and medial projection on the lateral wall of the nasal aperture, Schwartz and Tattersall (1999, 304–305) claim:

> *"The Neanderthal nasal cavity is uniquely derived relative not only just to Homo sapiens but to anthropoids in general [...] the nasal region [...] cannot be dismissed as being merely a vari-*

27 Not everybody with an "Out of Africa" position thinks Neandertals are a different species, and in fact some such as Bräuer (1992) assert that Neandertals could have interbred with Upper Paleolithic Europeans but just didn't do it very often. Bräuer, nonetheless, considers himself an adherent to the "Out of Africa" position (Bräuer and Stringer, 1997), but other observers such as Relethford (1999; 2001) note that this position is actually a variant of Multiregional evolution. Models including population mixing are reticular and cannot be considered a variation of an "Out of Africa" explanation (Templeton, 1997). One comment we received about the present paper developed this theme, noting that many "Out of Africa" models "appear to be indistinguishable from the Multiregional Model *as now stated* [our italics]." Multiregional evolution, however, has never been stated differently from this by any of its originators or supporters, and it has not changed over the key role of genic exchanges in the model of evolution it proposes (contra Smith et al.,1989, see Wolpoff et al., 2000), which from the beginning (Wolpoff et al., 1984) was based on clinal balances of features, reticulate evolution of populations, and the dispersal of successful genes by gene flow and population movements.

ant of Homo sapiens craniofacial anatomy [...] it is at best obscurantist at this point in time to continue to try to cram the morphology of this hominid into that of Homo sapiens. [...] If Neanderthals were any kind of vertebrate other than hominid, no systematist would hesitate for a moment in recognizing them as a distinct species.

Following from claims such as these, the replacement theory has the clearest predictions about the anatomy of the Mladeč descendants - it should manifestly *not* resemble Neandertals - and therefore replacement is the most straightforward theory to attempt to refute.

There are, then, two questions we may ask:

(1) Which of the two source groups is the more probable ancestral population?
(2) Is there any evidence of intermixture in the ancestry?

For the Out of Africa replacement theory to be correct, "Skhul / Qafzeh" must be the sole answer for the first, and the answer to the second is "no." Unique descent from Skhul / Qafzeh requires that, at the least, the greatest resemblances of the Mladeč males will be with the Skhul / Qafzeh males. The first is a *minimum* requirement, because any evidence of mixture, as proposed in question 2, in the ancestry of Mladeč would be a refutation of the first hypothesis.

There are several ways to approach the questions. The great differences in morphology between the overall shape between the Mladeč males and some specimens in the Skhul / Qafzeh sample are readily observable. The Mladeč crania can be forensically identified as European, whereas using standard forensic techniques (e.g. Gill and Gilbert, 1990), the crania from Skhul and Qafzeh are not identified as Europeans: some of these specimens resemble Africans, some East Asians, and others cannot be unambiguously identified. From a forensic perspective, the one group these samples are *not* is European.

Some authors have argued that the crania are not the best place to look for evidence of geographic origin, and that the postcranial remains are where the African origins of the Earliest Europeans are reflected (Holliday, 1997; 1999; 2000; Holliday and Falsetti, 1995; Pearson, 2000a; 2000b)[28]. These arguments are based on postcranial proportions that are not preserved at Mladeč. In fact, and to the contrary, the anatomy of individual Mladeč postcranial bones *does not* point towards an African origin. Instead, insofar as the bones show anything specific about possible geographic origins, there are decidedly *non* African aspects of Mladeč postcranial anatomy (e.g. Gilbert, 1976; Kidder, 1999; Walensky, 1965), including the relatively and absolutely large joint sizes (especially Mladeč 22 and 25), the significant femoral curvature (c.f. Mladeč 75 and 78), and the proximolateral femoral flange (Mladeč 28). Turning to the crania, which we expect to be more diagnostic, Mladeč 5 closely approximates several European Neandertals in overall lateral profile (for example, see Fig. 2), small mastoid process, and other features detailed above (and see Figs. 9

28 This is a surprising approach, in that every forensic publication that addresses the issue urges researchers to avoid using postcranial remains as much as possible. For instance, Krogman and Işcan conclude (1986, 297–300): "there is so much variation in long bone morphology that they can only be used in a very limited, corroborates way." Further, in discussing the various intermembral indices as a source of this information they note (1986, 291): "because the ranges were wide with a great overlap, individual identification as to race via these bones should be considered extremely hazardous." White observes (2000, 375): "the skull is the only part of the skeleton that is widely used in estimating geographic ancestry." Even more curious, some authors supporting the Eve theory argue that the lower limb proportions reflect climate of ancestry in the Upper Paleolithic Europeans because their ancestry is African and that there is not enough time in the Upper Paleolithic for selection to change the heat-adapted proportions into cold-adapted ones, but then reverse themselves and explain the limb proportions in the 24,500 year old Lagar Velho child as the result of cold adaptation and not ancestry.

Table 16. Distribution of metric data, summarizing how often measurements of the Mladeč crania lie closer to the Neandertal or Skhul/Qafzeh means. The question is asked for all instances, and for only those instances when the Mladeč data lie between the means of the comparative samples.

	Mladeč values lying between means		Mladeč values anywhere (between the means, or above/below them)	
Closer to →	Neandertal	Skhul/Qafzeh	Neandertal	Skhul/Qafzeh
Mladeč 5	13	20	62	50
Mladeč 6	9	9	30	29

and 10). Certain features such as the elliptical suprainiac fossa on Mladeč 6 exemplify anatomy said to be unique in the Neandertals (Hublin, 1998). Mladeč 8 preserves two nasal features said to be uniquely characteristic of Neandertals (Schwartz and Tattersall, 1996; but see Franciscus, 1999): the shallow groove along the inferior nasal margin, and the remnant of a medial projection on the internal wall of the nasal aperture.

However, to transform these and similar comparisons into statements with statistical meaning requires overcoming some formidable obstacles. There is only the partial maxilla and the 2 adult male Mladeč crania complete enough to analyze. The comparative samples are extremely small. They are not systematically constituted, in that different comparisons can have different sample sizes and some observations compare different specimens than others. Finally, while we know that many of the metric comparisons involve observations that are not independent of each other, we cannot specify the pattern of dependence, or for that matter the underlying variance/covariance matrix of the samples. For these reasons, we proceed with caution, and focus on some simple tactics that make as few assumptions as possible.

We systematically summarize the many metric comparisons among the Mladeč male crania and the comparative sample crania we made in Tables 3-11. Our results are reviewed in Table 16, where we tallied the cases when Mladeč 5 and 6 more closely approached the Neanderthal or the Skhul/Qafzeh mean[29]. These data are rough approximations of the phenetic affinity either of the Mladeč skulls makes to the Neandertal or to the Skhul/Qafzeh sample means.

Of course, unique ancestry in one of these samples does not mean we would expect all the resemblances of the Mladeč crania be with that sample because we assume that sampling error and other sources of variation would insure this not be the case. If under the replacement model we do not necessarily expect all the resemblances to be in a single direction, it is difficult to specify exactly how many resemblances to one potentially ancestral population would be enough to refute the hypothesis of complete replacement by the other. A hypothesis of replacement with limited mixture is even more difficult to specify. On the other hand, since the Multiregional model does not limit the possible patterns of multiple ancestries, its refutation must be accomplished with a demonstration that the replacement model is correct. This is a more straightforward hypothesis to refute.

We think it best to examine a different null hypothesis. To simplify the analysis we use a Binomial test to examine a null hypothesis of equal ancestry from both comparative samples. Assuming that the equal ancestry hypothesis and replacement hypotheses have predictions about metric similarity that do not overlap[30], it follows that:

29 We only counted cases when means of the comparative samples differed by at least 0.5 mm.

30 If the predictions cannot be distinguished, it would mean that the Neandertal and Skhul/Qafzeh samples are so similar that their phylogeny cannot be resolved.

- if the equal ancestry hypothesis cannot be disproved, it follows that the replacement theory must be incorrect;
- the replacement theory is also rejected if analysis reveals a preponderance of resemblances between Mladeč and the Neandertals;
- the replacement theory will be supported if analysis disproves the equal ancestry hypothesis and reveals a preponderance of resemblances between Mladeč and Skhul / Qafzeh.

Moreover, the equal ancestry hypothesis addresses what various supporters describe as a modified Eve theory. This is the appearance of modernity in Africa and the spread of modern populations around the world with just enough intermixture to account for limited regional continuity in different places (Bräuer, 1992; Smith et al., 1989; Stringer, 1992; Stringer and Gamble, 1993; but see Relethford, 1999 and Templeton, 1997 for a discussion of whether this can truly be described as the Eve replacement theory). If the equal ancestry hypothesis cannot be disproved, we note that this result implies much more mixture than the modified Eve theory would predict, if that theory is considered distinct from Multiregional evolution. One reason for this is that the results of gene flow between unequal sized populations is the predominance of genes from the larger population, and size differences alone would predict that Skhul / Qafzeh would more greatly affect the Mladeč gene pool than the Neandertals would, if gene flow alone were the cause of change.

As we show in Table 16, the comparisons cannot refute the null hypothesis of equal ancestry for either of the Mladeč males. This means that the Skhul / Qafzeh sample cannot be the unique ancestor. Overall, in less than half (45%) the metric comparisons the Skhul / Qafzeh sample was more similar for Mladeč 5, whereas in Mladeč 6 the total comparisons were about equally divided between the two possible ancestral groups. In the more restricted comparison, when the specific Mladeč cranial metric was constrained to fall between the means of the two older samples, the measurement was closer to the Skhul / Qafzeh mean in 60% of the comparisons for Mladeč 5 and in 50% of the comparisons for Mladeč 6. In either case, we cannot disprove the hypothesis of equal ancestry from these data. In this regard we conclude the data indicate that neither group can be ruled as a potential ancestor, which makes the answer to question 1 above, "both."

The distribution for Mladeč 6 clearly cannot be distinguished from the equal ancestry hypothesis, as it is equally similar to both samples, in either comparison. Mladeč 5 is more often closer to Skhul / Qafzeh for values lying between the means of the comparative samples, but is more often closer to the Neandertals considering all comparisons. This means that the Mladeč 5 values far more often lay beyond the Neandertal average than beyond the Skhul / Qafzeh average, "beyond" being defined by the direction of difference between the two comparative samples. Because we are interested in the probability of one out of two possible outcomes, and since we can specify the underlying distribution as reflecting equal ancestry, we used the normal approximation of the binomial distribution for examining the probability that these numbers, summarized in Table 16, could be found in a distribution where the expected relationship is the same for each comparative sample (Siegel, 1956). In those comparisons where Mladeč 5 is between the means of the comparative samples, the probability of 13 or fewer out of 33 (13+20) traits resembling Neandertals is 0.15. Examining the distribution that includes all comparisons, whether the Mladeč 5 values lie between the means or not, the probability of 62 or more out of 112 traits resembling Neandertals is 0.11. We cannot reject the equal ancestry hypothesis at the 0.05 level.

The second question about intermixture in the Mladeč ancestry in a sense is moot, because of the answer to the first is that neither group is a more probable ancestor. Nevertheless, we examined this question using non-metric data for the Mladeč crania and the most complete adult male crania from each of the comparative samples (4 Neandertals, 5 Skhul / Qafzeh, see the Figure 25 x-axis for details). We examined 30 non-metric traits from all parts of the cranium, as described in the text and listed in Table 17. These non-metric traits were scored as present or absent, so that the differences could be validly combined without weighing one more than another. Three of the nonmetric variables completely separated the Neandertal and Skhul / Qafzeh samples. Of them, the Mladeč crania were like the Neandertals in two, and like Skhul / Qafzeh in one. Seven additional traits almost completely separated the comparative samples[31]. Of these, the Mladeč crania were like the Neandertals in four and like Skhul / Qafzeh in two. For the seventh trait, one Mladeč cranium was like each comparative sample. In spite of the predominance of Neandertal resemblances for this subset of 10 traits, the normal approximation of the binomial distribution shows the equal ancestry hypothesis cannot be rejected at the 0.05 level.

Because this analysis and the metric one were based on characteristics of the groups themselves, for a third analysis we addressed the relationship of individuals in a way that ignored group assignments. We calculated the pairwise differences between each of the two Mladeč crania and the 8 other specimens from the non-metric traits. These are shown in Figures 25 and 26. Pairwise difference analysis is commonly applied to DNA sequence data to derive information about past population demography. It has also been applied to sequence data to investigate the closeness of relationship that a single ancient individual has to samples of living humans from different regions of the world (Kings et al., 1997). In these genetic analyses, the number of nucleotide differences between all possible pairs of individual DNA sequences is counted, and the results are presented as the frequency distribution of the number of differences. The assumptions are that each difference represents a mutation and that individuals who share fewer pairwise differences are more closely related because fewer mutations separate them. An equivalent assumption underlies all phenetic clustering techniques, where similarity is assumed to reflect relationship. Such procedures consider individuals who cluster more closely to be more closely related to each other. They do not necessarily assume a full independence of the traits, just as independence cannot be assumed for nucleotide differences in the non-recombining mtDNA molecule. The required assumption is that traits more closely linked are randomly distributed throughout the data set. The procedure is conservative, in that the absence of data for a specimen is considered the absence of difference. Missing data in our comparative samples are not randomly distributed. The Skhul / Qafzeh crania have more missing data than the Neandertals do. This means that in this specific analysis, the results will be weighed to show more similarities with the Skhul / Qafzeh remains.

We use pairwise analysis here for a similar purpose, to examine the relationship of the Mladeč crania to the individuals in our two comparative samples of potential ancestors. The number of differences between each Mladeč cranium and the others were tallied, and the figures aligned the specimens in order of increasing difference.

The average pairwise difference between Mladeč 5 and the Neandertal sample is 14.8 mm, and between it and the Skhul / Qafzeh sample is 14.0 mm, virtually the same. For Mladeč 6 the corresponding comparisons are 7.8 mm and 11.6 mm differences, so it is closer to the Neandertal sample. A Sample Runs test (Swed and Eisenhart, 1943) was used to examine whether the ordering of Neandertal and Skhul / Qafzeh crania, based on the number of pairwise differences from the Mladeč crania, is random (the null hypothesis). Randomness can be rejected at the p=.05 level when there are 2 or less, or 9 or more runs[32] from the same site, for a sample of this size. There are 5 runs for Mladeč 5 and 3 runs for Mladeč 6 – randomness in the order of pairwise similarities cannot be rejected. Again, these data fail to reject the equal ancestry hypothesis, and thereby disprove the notion that the Mladeč crania are uniquely related to Skhul / Qafzeh.

Our analyses are limited by the small sample sizes and ignorance of the underlying variance/covariance matrices for the data. The significance tests we used above are the ones we believe are valid

31 Meaning that all of one sample were the same for the character state, and only one specimen in the other sample differed from the opposite character state.

32 A run is one or more crania from the same group in a row. For instance, the Mladeč 6 analysis (Fig. 26) has three runs because the order changes categories (Skhul / Qafzeh and European Neandertal) three times.

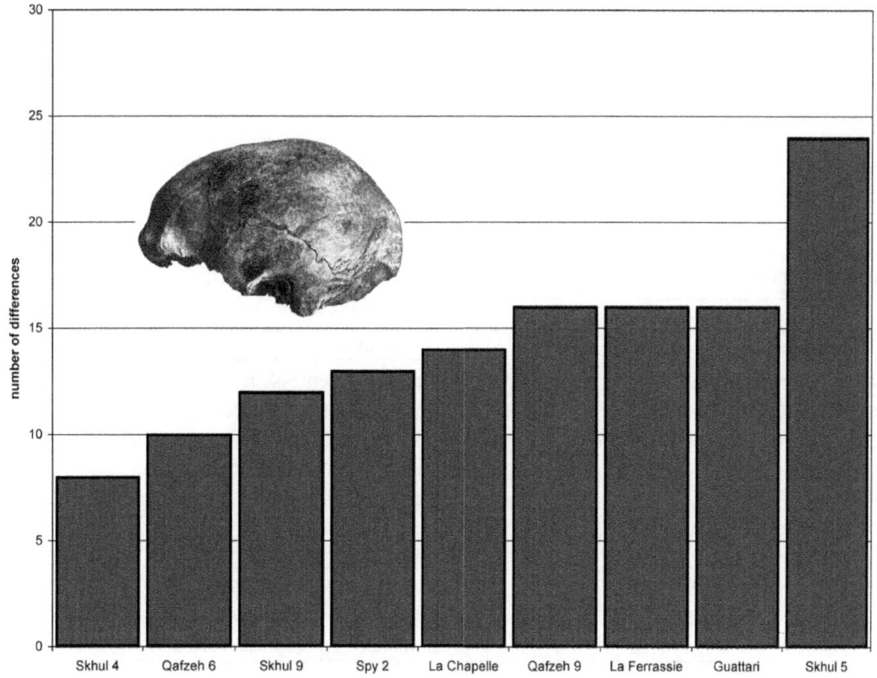

Fig. 25. Pairwise differences between Mladeč 5 and the most complete Neandertal and Skhul/Qafzeh males. 30 nonmetric traits are used in this analysis

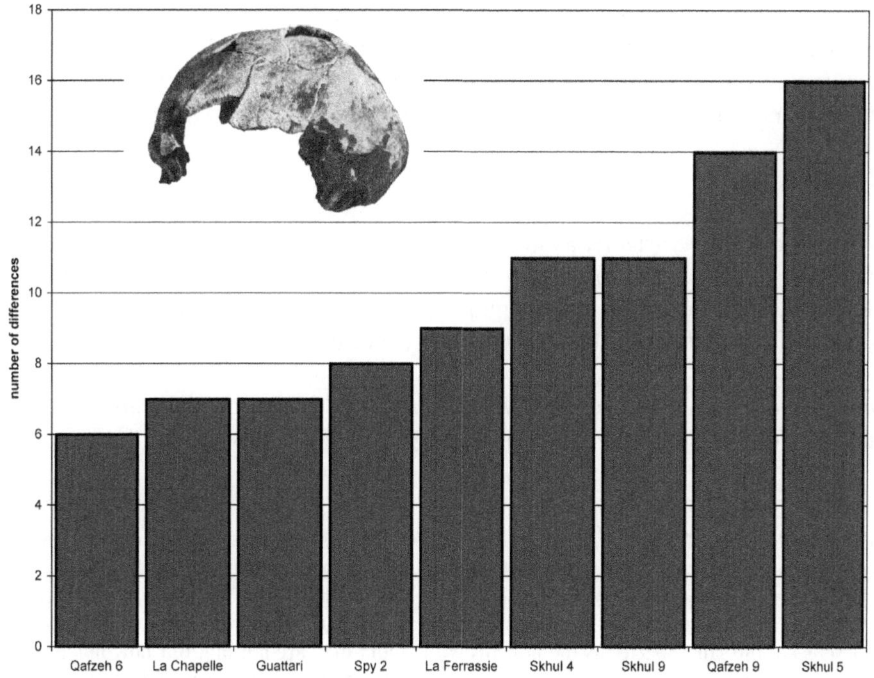

Fig. 26. Pairwise differences between Mladeč 6 and the most complete Neandertal and Skhul/Qafzeh males. 22 nonmetric traits are used in this analysis, less than the number for Mladeč 5 because the vault is less complete

Table 17. Characteristics used in the pairwise difference analyses summarized in Figures 25 and 26. These were developed from the comparisons in the text, with the criteria of representing structures on all parts of the cranium and maximizing the size of the comparative sample. We chose observations we felt could be unambiguously and repetitively scored with accuracy

Whole cranium	*Temporal*
"teardrop" shape (seen from top)	mastoid-supramastoid crests well separated
cranial rear rounded (seen from back)	mastoid process projects minimally*
occipital bun	glenoid articular surface flattened
asterionic parietal thickness (>9 mm)	supraglenoid gutter long*
lambdoidal occipital thickness (>8 mm)*	external auditory meatus leans forward*
Occipital	mastoid tubercle*
vertical occipital face short	*Frontal*
sagittal groove along vault posterior	glabellar depression
occipital plane long (>60 mm)	frontonasal suture arched
suprainiac fossa, elliptical form	supraorbital center dips downward
paramastoid crest prominent	broad frontal (>125 mm)
occipitomastoid crest prominent	central frontal boss
broad occiput (>120 mm)	frontal long (gl-br>113)
retromastoid process prominent	frontal keel
nuchal torus extends across occiput	anterior temporal fossa border angled*
	lateral supraorbital central thinning*
	medial height of supraorbital large (>19 mm)

* not preserved and therefore not used in Mladeč 6 analysis

for the metric and nonmetric comparisons we could make. As we noted, these fail to disprove the hypothesis of equal ancestry for the Mladeč male crania. On this basis, and because of other similarities and comparisons reviewed in the text, we firmly conclude it is very unlikely that the Skhul / Qafzeh remains are the unique ancestors of Mladeč. Replacement hypotheses for the origin of these Europeans, whether complete replacement or mostly complete replacement[33], can be ruled out.

A hybridity issue

The Mladeč males are quite variable, as much so as the male sample from Skhul and this raises the question of whether Mladeč, like some have claimed for Skhul, is a hybrid population. In fact, the interpretation of hybridity does not even require a sample size of more than one. In the analysis of the 24,500 year old Lagar Velho child from Portugal (Duarte et al., 1999, 7608) it was reasoned that:

> *"[...] the morphological mosaic of the Lagar Velho 1 child therefore indicates admixture between early modern humans spreading through Iberia and local Neandertal populations. Such*

33 It has never been clear what "not quite complete replacement" means, and it certainly means different things to different authors. Minimal mixture proposed by replacement theorists responds to the need to posit just enough mixture with Neandertals to account for the similarities of early Upper Paleolithic Europeans to Neandertals after the replacement (Bräuer, 1992). Stringer (1982; 1992) asserts that the amount of mixture required is not sufficient to show that Neandertals are the same species as the humans with whom they are mixing. But how much mixture does this mean? It must be enough to be compatible with Hawks' (1997) estimate that at least 25% of the ancestors of Upper Paleolithic people would have to be Neandertals, to account for the anatomical comparisons Frayer (1993) published, the very observations the mixing is meant to account for. Any reticular model that fits this description is just a restatement of Multiregional evolution (Relethford, 2001).

> *morphological mosaics, with character states distinctive to each parental group (directional dominance), plus intermediate (additive) or divergent (overdominance or underdominance) configurations characterize hybrids."*

Could a similar description of hybridity also apply to the adult Mladeč male remains?

In examining an equal ancestry model[34] for Mladeč, we recognized that the potentially ancestral populations we considered, Neandertals and Skhul / Qafzeh, are not the only possible ancestors. They are the most credible ancestors we know of because of their locality (Neandertals), phylogeny (Skhul / Qafzeh, the earliest *population* of so-called early modern humans), and circumstances (we have no knowledge of nearby penecontemporary populations such as Asians). There must surely have been others. But even if the Neandertals and Skhul Qafzeh were the only ancestors, it is reasonable to ask whether we would be justified from our analysis to regard Mladeč as a "hybrid" population.

The idea of hybridity comes from the supposition that the mode of European evolution when modern European populations originated was a mixing between local Neandertal populations and incoming "moderns" from the south or east. Yet all human populations are admixed populations. Multiregional evolution rests on the contention that human populations have been systematically exchanging genes throughout their evolution and thereby incorporates the ethnogenic tenet that populations eventually either become extinct, split apart, or merge with other populations (Moore, 1994; 1995). Mladeč, indeed all Europeans, are an example of this, mixed populations living at the western periphery of the Eurasian range, at lower number and with more susceptibility to environmental changes than more centrally located human groups. But we have not used "hybridized" to describe this admixture for several reasons. It presumes more difference between the ancestors than we can verify. Furthermore, it implies that the cases of "hybridization" such as Mladeč and Lagar Velho reflect an immediate ancestry in Neandertal and (presumably) modern parents, who presumably would have to be interacting all across the European continent for many thousands of years, to account for this distribution of "hybrid" offspring. In fact, the greatest significance of the Lagar Velho 1 child, in our view, is precisely that: with Mladeč, it shows that the process of mixture spread fully across Europe, and took place over thousands of years. Models of Neandertal evolution based on the fate of natives in colonized regions such as Tasmania cannot be applied to this process, which took place on an isolated island over no more than a handful of generations.

How much mixture? Of late, some of the Eve theorists have taken to arguing that there could have been interbreeding between the two species, Neandertals and moderns, but it was minimal – just enough to account for the unique similarities they show, but not enough to have any significance. As one courageous worker put it (Pearson, 2000b, 589):

> *"[...] [postcranial] slenderness, in comparison with the Neandertals, coupled with differences in body proportions, provides evidence for population replacement (or enough gene flow to be considered a flood) in the Middle-to-Upper Paleolithic transition in western Europe."*

Is this true? Apart from the "just enough gene flow" contention, which in actuality is no longer a variant of the Eve theory but is a restatement of Multiregional evolution, is the amount of Neandertal contribution to the European gene pool as minimal as Pearson suggests? It is important to go beyond assertion to resolve this, and we are now in a position to attempt a numeric estimate. From the

34 That means reticulate evolution – we do not describe this as hybridization as this would imply the earlier presence of subspecies or "pure races" that mixed, and it is our experience and belief that neither of these reflects the true biological description of the relation between the ancestors of the Mladeč remains.

perspective of the earlier Upper Paleolithic we can bracket a minimal range for the estimate. Here (and see Wolpoff et al., 2001) we cannot reject a hypothesis of *equal* ancestry for Neandertals. If 50% is the largest minimum estimate, the smallest must be Hawks (1997) calculation that at least 25% of the ancestors of Upper Paleolithic people would have to be Neandertals, based on his analysis of the anatomical comparisons Frayer (1993) published. This range, 25% to 50% Neandertal ancestry for the earliest post-Neandertal population is the minimum for two reasons. First, more evidence (for instance, an increase in the number of Neandertal traits found in the Upper Paleolithic populations) can only increase the magnitude of the estimate, not decrease it. Second, selection acting on all human populations makes the later Neandertal populations less like the earlier ones and more like subsequent Europeans. The dispersion of new technologies in the Châtelperronian can only have accelerated this process. These factors create the potential for an *underestimate* of Neandertal ancestry because selection is a second reason why later Europeans may differ from earlier Neandertals, besides gene flow from other populations. Selection could make an admixture estimate higher than it actually was. This is what would lead to an underestimate of the amount of Neandertal contribution. It is difficult to reconcile these findings with the description of "a flood."

Even without our analysis, the length and geographic distribution of the changes across Europe demonstrate that a simple swamping of Neandertals genes to oblivion is impossible. In fact, the time scale and spatial extent of the Multiregional process in Europe implies the importance of an evolutionary force that has been ignored in many of the recent models of modern European origins: selection. And once we recognize the potential of changes due to selection, the whole issue of migration and mixture becomes muddied by the fact that the different causes cannot always be clearly distinguished. Why would we expect that the consequences of selection acting on Neandertals in response to their Upper Paleolithic behaviors be much different from the consequences of selection acting on other early Upper Paleolithic populations who might have migrated into Europe during the interstadial?

Conclusions

The exact details of ancestry for Mladeč may never be worked out, but we may address the boundary conditions, the limits within which such details must lie. The evolutionary changes in Europe very often resulted in the same anatomical consequences that genic exchange with other populations could have created. And why shouldn't they? European populations were never isolated from the rest of the world for long periods of time. Both ideas and genes, especially useful ideas and successful genes, were exchanged throughout the Pleistocene and there is little wonder that both the causes of selection and their genetic consequences were broadly similar across large parts of the world. How, in principle, could the effects of gene flow and local selection ever be clearly separated? How might we distinguish anterior dental reduction introduced to Europe with incoming populations from anterior dental reduction taking place *in situ*[35] because of changing selection? Changes such these might make gracile populations appear more similar to each other than their phylogenies would imply. But in this analysis, as much as possible we rely on traits that are not simple reflections of gracility or robustness. In any event, the fact is that we cannot disprove a hypothesis of equal ancestry for Mladeč, and these considerations help make this hypothesis conservative in that the process of change would be expected to produce more similarities between Mladeč and Skhul/Qafzeh.

In broad outline, there is no doubt that some populations entered the very sparsely populated European region during Würm interstadials, and mixed with the indigenous natives. However, we

35 Evident at late Neandertal sites such as Hortus (de Lumley, 1972) and Vindija (Wolpoff et al., 1981).

have no direct evidence for the magnitude of these population movements, and migration is only one of the mechanisms promoting the exchange of genes. There is no reason, biological or cultural, to suppose that the genes of small local populations were swamped out by such a process and that the Neandertal contribution to later Europeans could be described as a drop of cream in a cup of coffee, as some have asserted. The human story, as far as we understand it, is a constant record of population splits, competitions, replacements, and mergers. When people meet, whether for the first time or yearly or seasonally, friendly or not biological and cultural information is exchanged. In the last two European interstadials, peoples were in contact and exchanged ideas and mates. Some groups died out and were replaced by their neighbors, others mixed culturally and biologically, to varying degrees, while retaining some local, regional continuity in both culture and genes (as described by Maryuma and Kimura, 1980; Svoboda, 1993; Whitlock and Barton, 1997). The Mladeč remains may be the earliest post-Neandertal example of this ethnogenic process.

Acknowledgments

We thank M. Teschler-Nicola for her kind invitation to contribute to this volume, and are deeply grateful to the directors, curators, and personnel of the Moravské zemské muzeum in Brno and the Naturhistorisches Museum Wien, for permission to work on the Mladeč specimens, the rooms and facilities that were graciously provided to us, the extraordinary hospitality extended to us, and the great patience shown in waiting for this, the first of a number of Mladeč papers, to appear. We thank H. Kritscher, H. Poxleitner, L. Seitl, J. Szilvássy, and (once again) M. Teschler-Nicola. We also thank the many curators and museum directors with responsibility for the comparative samples we discuss here, for access to the fossil collections in their care and for the hospitality and help extended to us in the course of our research. We are indebted to C. Bauer, who did the full translation of the Szombathy monograph for us. Of our colleagues and co-researchers, the conversations with and contributions of R. Caspari, J. Hawks, N. Minugh-Purvis, F. Smith, O. Soffer, and K. Valoch were exceptional, and we express our great debt to them.

References

Absolon, K. (1929) New finds of fossil human skeletons in Moravia. *Anthropologie* (Prague) 5, 79–107

Albrecht, G., Hahn, J. and Torke, W. G. (1972) Merkmalanalyse von Geschoßspitzen des mittleren Jungpleistozäns in Mittel- und Osteuropa. *Archaeologia Venatoria* 2

Bayer, J. (1922) Das Aurignac-Alter der Artefakte und menschlichen Skelettreste aus der "Fürst-Johanns-Höhle" bei Lautsch in Mähren. *Mitteilungen der Anthropologischen Gesellschaft in Wien* 52, 173–185

Billy, G. (1972) L'évolution humaine au paléolithique supérieur. *Homo* 23, 2–12

Blekta, J. (1932) Kras mezi Konicí a Litovlí. Věstník přírodovědeckého klubu v Prostějově 22, 1–48

Boule, M. (1911–1913) L'homme fossile de la Chapelle-aux-Saints. *Annales de Paléontologie* 6, 111–172; 7, 21–56, 85–192; 8, 1–70

Bräuer, G. (1992) Africa's place in the evolution of *Homo sapiens*. In (G. Bräuer and F. H. Smith, Eds.) *Continuity or replacement? Controversies in Homo sapiens evolution.* Rotterdam: Balkema, pp. 83–98

Bräuer, G. and Broeg, H. (1998) On the degree of Neandertal-modern continuity in the earliest Upper Paleolithic crania from the Czech Republic: evidence from non-metrical features. In (K. Omoto and P. V. Tobias, Eds.) *Origins and past of modern humans: towards reconciliation.* Singapore: World Scientific, pp. 106–125

Brodar, S. and Brodar, M. (1983) *Potočka Zijalka. Visokoalpska Postaja Aurignacienskih Lovcev.* Ljubljana: Dela l razr. SAZU 24/13

Brothwell, D. R. (1981) *Digging up bones.* 3rd edition. London: British Museum (Natural History)

Caspari, R. (1991) *The evolution of the posterior cranial vault in the Central European Upper Pleistocene.* PhD Dissertation, University of Michigan. Ann Arbor: University Microfilms

Champan, A. E. (1985) The mechanical properties of human muscle. *Exercise and Sport Sciences Reviews* 13, 443–501

Churchill, S. E. and Smith, F. H. (2000) Makers of the early Aurignacian of Europe. *Yearbook of Physical Anthropology* 43, 61–115

Dahlberg, A. A. (1951) The dentition of the American Indian. In (W. S. Laughlin, Ed.) *Papers on the Physical Anthropology of the American Indian.* New York: Viking Fund, pp. 138–176

Delattre, A. and Fenart, R. (1958) Essai de systématisation du pariétal son utilisation au cours l'étude de sa croissance. *Bulletins et Mémoires de la Société d'Anthropologie de Paris* (série10) 9, 245–295

Duarte, C., Maurício, J., Pettitt, P. J., Souto, P., Trinkaus, E., van der Plicht, H. and Zilhão, J. (1999) The Early Upper Paleolithic human skeleton from the Abrigo do Lagar Velho (Portugal) and modern human emergence in Iberia. *Proceedings of the National Academy of Sciences USA* 96, 7604–7609

Franciscus, R. G. (1999) Neandertal nasal structures and upper respiratory tract "specialization." *Proceedings of the National Academy of Sciences USA* 96, 1805–1809

Frayer, D. W. (1978) *Evolution of the dentition in Upper Paleolithic and Mesolithic Europe.* University of Kansas Publications in Anthropology 10, 1–201

Frayer, D. W. (1986) Cranial variation at Mladeč and the relationship between Mousterian and Upper Paleolithic hominids. In (V. V. Novotný and A. Mizerová, Eds.) *Fossil man. New facts, new ideas. Papers in honor of Jan Jelínek's Life Anniversary. Anthropologie (Brno)* 23, 243–256

Frayer, D. W. (1993) Evolution at the European edge: Neanderthal and Upper Paleolithic relationships. *Préhistoire Européenne 2,* 9–69

Frayer, D. W. (1997) Perspectives on Neanderthals as ancestors. In (G. A. Clark and C. M. Willermet, Eds.) *Conceptual issues in modern human origins research*). New York: Aldine de Gruyter, pp. 202–234, 437–492 (bibliography)

Frayer, D. W., Wolpoff, M. H., Smith, F. H., Thorne, A. G. and Pope, G. G. (1993) The fossil evidence for modern human origins. *American Anthropologist* 95, 14–50

Fürst, J. (1923–1924) Nálezy v jeskyních Mladečských. *Vlast sborník střední a severní Moravy* 2, 6–7

Gilbert, B. M. (1976) Anterior femoral curvature: its probable basis and utility as a criterion of racial assessment. *American Journal of Physical Anthropology* 45, 601–604

Gill, G. W. and Gilbert, B. M. (1990) Race identification from the midfacial skeleton: American Blacks and Whites. In (G. W. Gill and S. M. Rhine, Eds.) *Skeletal attribution of race: Methods for forensic anthropology. Anthropological Papers of the Maxwell Museum of Anthropology* 4, 47–53

Hahn, J. (1972) Aurignacian signs, pendants and art objects in central and eastern Europe. World Archaeology 3, 252–266

Hahn, J. (1977) Aurignacien: Das ältere Jungpaläolithikum in Mittel und Osteuropa. *Fundamenta* 9 (Series A)

Hauser, G. and De Stefano, G. F. (1989) *Epigenetic traits of the human skull.* Stuttgart: Schweizerbart'sche Verlagsbuchhandlung

Hawks, J. (1997) Have Neandertals left us their genes? In (L. Cavalli-Sforza, Ed.) *Human evolution: Abstracts of papers presented at the 1997 Cold Spring Harbor Symposium on Human Evolution arranged by L. L. Cavalli-Sforza and J. D. Watson.* Cold Spring Harbor: Cold Spring Harbor Laboratory, p. 81

Hillson, S. (1991) *Teeth.* New York: Cambridge University Press

Hinton, R. J. (1979) Form and function in the temporomandibular joint. In (D. S. Carlson, Ed.) *Craniofacial biology.* Monograph 10. Ann Arbor: Center for Human Growth and Development, pp. 37–60

Hinton, R. J. (1981a) Changes in articular eminence morphology with dental function. *American Journal of Physical Anthropology* 54, 439–455

Hinton, R. J. (1981b) Form and patterning of anterior tooth wear among human groups. *American Journal of Physical Anthropology* 54, 555–564

Holliday, T. W. (1997) Body proportions in Late Pleistocene Europe and modern human origins. *Journal of Human Evolution* 32 (5), 423–447

Holliday, T. W. (1999) Brachial and crural indices of European Late Upper Paleolithic and Mesolithic humans. *Journal of Human Evolution* 36 (5), 549–566

Holliday, T. W. (2000) Evolution at the crossroads: Modern human emergence in western Asia. *American Anthropologist* 102 (1), 54–68

Holliday, T. W. and Falsetti, A. B. (1995) Lower limb length of European early modern humans in relation to mobility and climate. *Journal of Human Evolution* 29 (2), 141–153

Horáček, I. and Ložek, V. (1984) Z výzkumu výplně Mladečské jeskyně u Litovle. *Československý kras* 35, 98–100

Howell, F. C. (1951) The place of Neanderthal man in human evolution. *American Journal of Physical Anthropology* 9, 379–416

Howell, F. C. (1984) Introduction. In (F. H. Smith and F. Spencer, Eds.) *The origins of modern humans: A world survey of the fossil evidence*. New York: Alan R. Liss, pp. xiii–xxii

Howells, W. W. (1982) Comment on "Upper Pleistocene evolution in south-central Europe" by F. H. Smith. *Current Anthropology* 23, 688–689

Hrdlička, A. (1930) *The skeletal remains of early man*. Smithsonian Miscellaneous Collections 83. Washington, DC: Smithsonian Institution

Hublin, J.-J. (1998) Climatic changes, paleogeography, and the evolution of the Neandertals. In (T. Akazawa, K. Aoki and O. Bar-Yosef, Eds.) *Neandertals and modern humans in western Asia*. New York: Plenum Press, pp. 295–310

Hylander, W. L. and Johnson, K. R. (1985) Temporalis and masseter muscle function during incision in macaques and humans. *International Journal of Primatology* 6, 289–322

Jelínek, J. (1954) Nález fosilního člověka Dolní Věstonice III. *Anthropozoikum* 3, 37–91

Jelínek, J. (1969) Neanderthal man and *Homo sapiens* in central and eastern Europe. *Current Anthropology* 10, 475–503

Jelínek, J. (1976) A contribution to the origin of *Homo sapiens sapiens*. *Journal of Human Evolution* 5, 497–500

Jelínek, J. (1983) The Mladeč finds and their evolutionary importance. *Anthropologie* (Brno) 21, 57–64

Jelínek, J. (1987) Historie, identifikace a význam mladečských anthropologických nálezů z počátku mladého paleolitu. In (J. Jelínek, Ed.) *25 let pavilonu Anthropos 1961–1986*. Brno: Anthropos Institute, Moravian Museum, pp. 51–67

Kidder, J. H. (1999) Human origins: The fossil record. *American Journal of Physical Anthropology* 109 (2), 275–276

Klein, R. G. (2001) Southern African and modern human origins. *Journal of Anthropological Research* 57 (1), 1–16

Knies, J. (1905) Nový nález diluviálního člověka u Mladče na Moravě. *Věstník klubu přírovědeckého Prostějov* 9, 3–19

Knies, J. (1928) K 100. výročí objevení Mladečských jeskyň. Litovelský kraj 5, 45–49

Kraus, B. S., Jordan, R. E. and Abrams, L. (1969) *Dental anatomy and occlusion*. Baltimore: Williams and Wilkins

Krings, M., Stone, A., Schmitz, R. W., Krainitzid, H., Stoneking, M. and Pääbo, S. (1997) Neandertal DNA sequences and the origin of modern humans. *Cell* 90, 1–20

Krogman, W. M. and Işcan, M. Y. (1986) *The human skeleton in forensic medicine*. 2nd ed. Springfield, IL: Thomas

Kříž, M. (1903) *Beiträge zur Kenntnis der Quartärzeit in Mähren*. Kremsier (Moravia): Steinitz

Leakey, L. S. B. and Goodall, V. M. (1969) *Unveiling Man's origins*. Cambridge: Schenkman

Lieberman, D. E. (1997) Making behavioral and phylogenetic inferences from hominid fossils: Considering the developmental influence of mechanical forces. *Annual Review of Anthropology* 26, 185–210

Lumley, H. de (1972) La grotte moustérienne de l'Hortus. Marseille: Université de Provence.

Malez, M. (1967) Paleolit Velike Pećine na Ravnoj gori u sjeverozapadnoj Hrvatskoj. *Arheološki Radovi i Rasprave* 4/5, 7–68

Mann, A. E. (1981) The significance of the *Sinanthropus* casts, and some paleodemographic notes. In (B. A. Sigmon and J. S. Cybulski, Eds.) *Homo erectus. Papers in honor of Davidson Black*. Toronto: University of Toronto Press, pp. 41–62

Marshack, A. (1972) Cognitive aspects of Upper Paleolithic engraving. *Current Anthropology* 13, 445–477

Martin, R. (1928) *Lehrbuch der Anthropologie: Kraniologie, Osteologie*. Vol. II. Jena: Fischer

Maryuma, T. and Kimura, M. (1980) Genetic variability and effective population size when local extinction and recolonization of subpopulations are frequent. *Proceedings of the National Academy of Sciences USA* 77, 6710–6714

Maška, K. J. (1895) Diluviální člověk v Předmostí. *Čas vlast Muzej Spolku Olomouc* 12, 1–7

Maška, K. J. (1905) Poznámky k diluviálním nálezům v jeskyních mladečských a stopám glaciálním na severovýchodní Moravě. *Časopis Moravského muzea zemského* 5, 1–3

Matiegka, J. (1934) *Homo předmostensis: Fosilní Člověk z Předmostí na Moravě*. 1. Lebky. Prague: Česká Akad Věd Umění

McCown, T. D. and Keith, A. (1939) *The Stone Age of Mount Carmel: The fossil human remains from the Levalloiso-Mousterian*. Vol. 2. Oxford: Clarendon Press

Moffett, B. C., Johnson, L. C., McCrabe, J. B. and Askew, H. C. (1964) Articular remodeling in the adult temporomandibular joint. *American Journal of Anatomy* 115, 119–142

Moore, J. H. (1994) Putting anthropology back together again: the ethnogenetic critique of cladistic theory. *American Anthropologist* 96, 925–948

Moore, J. H. (1995) The end of a paradigm. *Current Anthropology* 36, 530–531

Morant, G. M. (1930) Studies of Paleolithic man IV. A biometric study of the Upper Paleolithic skulls of Europe and their relationships to earlier and later forms. *Annals of Eugenics* 4, 109–214

Oakley, K. P., Campbell, B. G. and Molleson, T. I. (1971) *Catalogue of fossil hominids, Part II: Europe*. London: British Museum (Natural History)

Oliva, M. (1986) Finds from the Pleniglacial B from the territory of Czechoslovakia and the question of the Epi-aurignacian settlement. *The World Archaeological Congress: Pleistocene Perspective*. Vol. 2. Southampton

Oliva, M. (1987) *Aurignacien na Moravě*. Kroměříž: Studi muz Kroměřížska 87

Oliva, M. (1993) *Le contexte archéologique des restes humains de la Grotte de Mladeč*. USIPP, 2nd Congress, 2, 207–216. Bratislava: Institut d'Archaéologie

Patte, E. (1962) *La dentition des Néanderthaliens*. Paris: Masson et Cie

Pearson, O. M. (2000a) Activity, climate, and postcranial robusticity: implications for modern human origins and scenarios of adaptive change. *Current Anthropology* 41 (4), 569–607

Pearson, O. M. (2000b) Postcranial remains and the origin of modern humans. *Evolutionary Anthropology* 9 (6), 229–247

Piveteau, J. (1967) Un pariétal humain de la grotte du Lazaret. *Annales de Paléontologie des Vertébrés* 53, 167–199

Poissonnet, C. M., Olivier, G. and Tissier, H. (1978) Estimation de la capacité craîenne à partir d'un os de la voûte. *Bulletins et Mémoires de la Société d'Anthropologie de Paris* (série 13) 5, 217–221

Radovčić, J. (1988) *Dragutin Gorjanović-Kramberger and Krapina early man: The foundation of modern paleoanthropology*. Zagreb: Školska knjiga and Hrvatski prirodoslovni muzej

Relethford, J. H. (1999) Models, predictions, and the fossil record of modern human origins. *Evolutionary Anthropology* 8, 7–10

Relethford, J. H. (2001) *Genetics and the search for modern human origins*. New York: Wiley-Liss

Rosenberg, K. R. (1988) The functional significance of Neandertal pubic length. *Current Anthropology* 29, 595–617

Saban, R. (1984) Anatomie et Evolution des Veines Meningées chez les Hommes Fossiles. Paris: ENSB-CTMS

Santa Luca, A. P. (1978) A re-examination of presumed Neandertal-like fossils. *Journal of Human Evolution* 7, 619–636

Schwalbe, G. (1906) Studien zur Vorgeschichte des Menschen. *Zeitschrift für Morphologie und Anthropologie* 1, 5–228

Schwartz, J. H. and Tattersall, I. (1996) Significance of some previously unrecognized apomorphies in the nasal region of *Homo neanderthalensis*. *Proceedings of the National Academy of Sciences USA* 93, 10852–10854

Schwartz, J. H. and Tattersall, I. (1999) Toward a definition of *Homo neanderthalensis* and *Homo sapiens* I. The nasal region. In (J. Gibert, F. Sánchez, L. Gibert and F. Ribot, Eds.) *The hominids and their environment during the Lower and Middle Pleistocene of Eurasia. Proceedings of the International Conference of Human Paleontology*. Granada (Spain): Cervantes, Baza, pp. 299–310

Siegel, S. (1956) *Nonparametric statistics for the behavioral sciences*. New York: McGraw-Hill

Skinner, M. F. and Sperber, G. H. (1982) *Atlas of the radiographs of early man*. New York: Alan R. Liss

Skutil, J. (1938) Pravěké nálezy v Mladči u Litovle na Moravě. Litovel: Krajinská musejni společnost

Skutil, J. (1940) Paleolitikum v bývalém Československu. *Obzor Prehistorický* 12, 41–43

Smith, B. H. (1984) Patterns of molar wear in hunter-gatherers and agriculturalists. *American Journal of Physical Anthropology* 63, 39–56

Smith, F. H. (1982) Upper Pleistocene hominid evolution in south-central Europe: A review of the evidence and analysis of trends. *Current Anthropology* 23, 667–703

Smith, F. H. (1983) On hominid evolution in south-central Europe. *Current Anthropology* 24, 236–237

Smith, F. H. (1984) Fossil hominids from the Upper Pleistocene of central Europe and the origin of modern Europeans. In (F. H. Smith and F. Spencer, Eds.) *The origins of modern humans: A world survey of the fossil evidence*. New York: Alan R. Liss, pp. 137–209

Smith, F. H. (1985) Continuity and change in the origin of modern *Homo sapiens. Zeitschrift für Morphologie und Anthropologie* 75, 97–222

Smith, F. H. (1997) Mladeč. In (F. Spencer, Ed.) *History of physical anthropology. An encyclopedia.* New York: Garland, pp. 659–660

Smith, F. H., Falsetti, A. B. and Donnelly, S. M. (1989) Modern human origins. *Yearbook of Physical Anthropology* 32, 35–68

Smith, F. H. and Ranyard, S. M. (1980) Evolution of the supraorbital region in Upper Pleistocene fossil hominids from south-central Europe. *American Journal of Physical Anthropology* 53, 589–609

Smith, F. H. and Trinkaus, E. (1991) Les origines de l'homme moderne en Europe centrale: Un cas de continuité. In (J.-J. Hublin and A. M. Tillier, Eds.) *Aux origines d'Homo sapiens.* Nouvelle Encyclopédie Diderot. Paris: Presses Universitaires de France, pp. 251–290

Smyčka, J. (1907) Litovel a okolí za pravěku. *Pravěk* 3, 140–150

Smyčka, J. (1922) Nálezy diluviálního člověka v Mladči u Litovle na Moravě. *Obzor Prehistorický* I, 111–120

Smyčka, J. (1925) Kdy přišel první člověk do litovelského kraje. Vlast sborn střed sever Moravy 3. *Příl Olomouc Litovel,* 5–7

Stringer, C. B. (1982) Comment on Upper Pleistocene evolution in south-central Europe: A review of the evidence and analysis of trends. *Current Anthropology* 23, 667–703

Stringer, C. B. (1989) The origin of early modern humans: A comparison of the European and non-European evidence. In (P. Mellars and C. B. Stringer, Eds.) *The human revolution: Behavioural and biological perspectives on the origins of modern humans.* Edinburgh: Edinburgh University Press, pp. 232–244

Stringer, C. B. (1992) Neanderthal dates debated. *Nature* 356, 201

Stringer, C. B. and Gamble, C. (1993) *In search of the Neanderthals.* London: Thames and Hudson

Stringer, C. B., Hublin, J. J. and Vandermeersch, B. (1984) The origin of anatomically modern humans in western Europe. In (F. H. Smith and F. Spencer, Eds.) *The origins of modern humans: A world survey of the fossil evidence.* New York: Alan R. Liss, pp. 51–135

Svoboda, J. (1993) The complex origin of the Upper Paleolithic in the Czech and Slovak Republics. In (H. Knecht, A. Pike-Tay and R. White, Eds.) *Before Lascaux: The complex record of the Early Upper Paleolithic.* Boca Raton: CRC Press, pp. 23–36

Svoboda, J. (2000) The depositional context of the Early Upper Paleolithic human fossils from the Koněprusy (Zlatý kůň) and Mladeč Caves, Czech Republic. *Journal of Human Evolution* 38, 523–536

Swed, F. S. and Eisenhart, C. (1943) Tables for testing randomness of grouping in a sequence of alternatives. *Annals of Mathematics and Statistics* 14, 66–87

Szilvássy, J., Kritscher, H. and Vlček, E. (1987) Die Bedeutung röntgenologischer Methoden für anthropologische Untersuchung ur- und frühgeschichtlicher Gräberfelder. *Annalen des Naturhistorischen Museums Wien* 89, 313–352

Szombathy, J. (1883) Über Ausgrabungen in den mährischen Höhlen. *Sechster Bericht der prähistorischen Commission der math.-naturw. Classe der kaiserlichen Akademie der Wissenschaften* 87, 168–170

Szombathy, J. (1904) Neue diluviale Funde von Lautsch in Mähren. *Jahrbuch der k. k. Zentral-Kommission für Kunst- und historische Denkmäler* 2, 9–16

Szombathy, J. (1925) Die diluvialen Menschenreste aus der Fürst-Johanns-Höhle bei Lautsch in Mähren. *Die Eiszeit* 2, 1–34, 73–95

Szombathy, J. (1926) Die Menschenrassen im oberen Paläolithikum, insbesondere die Brüx-Rasse. *Mitteilungen der Anthropologischen Gesellschaft in Wien* 56, 202–219

Tappan, N. (1973) Structure of bone in the skulls of Neanderthal fossils. *American Journal of Physical Anthropology* 38, 93–98

Tattersall, I. and Schwartz, J. H. (1999) Hominids and hybrids: The place of Neanderthals in human evolution. *Proceedings of the National Academy of Sciences USA* 96, 7117–7119

Templeton, A. R. (1997) Testing the out of Africa replacement hypothesis with mitochondrial DNA data. In (G. A. Clark and C. M. Willermet, Eds.) *Conceptual issues in modern human origins research.* New York: Aldine de Gruyter, pp. 329–360, 437–492 (bibliography)

Trinkaus, E. (1983) *The Shanidar Neandertals.* New York: Academic Press

Trinkaus, E. (1989) Issues concerning human emergence in the Later Pleistocene. In (E. Trinkaus, Ed.) *The emer-*

gence of modern humans. Biocultural adaptations in the Later Pleistocene. Cambridge: Cambridge University Press, pp. 1–17, 232–276 (bibliography)

Trinkaus, E. and Le May, M. (1982) Occipital bunning among later Pleistocene hominids. *American Journal of Physical Anthropology* 57, 27–35

Valoch, K., Oliva, M., Havlíček, P., Karásek, J., Pelíšek, J. and Smolíková, L. (1985) Das Frühaurignacien von Vedrovice II und Kupařovice I in Südmähren. *Anthropozoik* 16, 107–203

Vandermeersch, B. (1981) *Les Hommes Fossiles de Qafzeh (Israël)* Paris: Centre National de la Recherche Sciéntifique

Vlček, E. (1991) Die Mammutjäger von Dolní Věstonice. Anthropologische Bearbeitung der Skelette aus Dolní Věstonice und Pavlov. *Archäologie und Museum* (Basel) 22

Walensky, N. (1965) A study of anterior femoral culvature in man. *Anatomical Record* 151, 559–570

Wankel, J. (1884) První stopy lidské na Moravě. *Časopis vlast. muzea společnost Olomouc* 1, 89–96

Weidenreich, F. (1939) On the earliest representatives of modern mankind recovered on the soil of east Asia. *Peking Natural History Bulletin* 13, 161–174

Weidenreich, F. (1943) The skull of *Sinanthropus pekinensis:* A comparative study of a primitive hominid skull. *Palaeontologia Sinica* (New Series D) II, 1–484

Weidenreich, F. (1951) Morphology of Solo man. *Anthropological Papers of the American Museum of Natural History* 43, 205–290

White, R. (1982) Rethinking the Middle/Upper Paleolithic transition. *Current Anthropology* 23, 169–192

White, T. D. (2000) *Human osteology.* 2nd edition. New York: Academic Press

Whitlock M. C. and Barton, N. H. (1997) The effective size of a subdivided population. *Genetics* 146, 427–441

Wolpoff, M. H. (1979) The Krapina dental remains. *American Journal of Physical Anthropology* 50, 67–114

Wolpoff, M. H. (1982) Comment on "Upper Pleistocene hominid evolution in south-central Europe," by F. H. Smith. *Current Anthropology* 23, 693

Wolpoff, M. H. (1989a) Multiregional evolution: the fossil alternative to Eden. In (P. Mellars and C. B. Stringer, Eds.) *The human revolution: Behavioural and biological perspectives on the origins of modern humans.* Edinburgh: Edinburgh University Press, pp. 62–108

Wolpoff, M. H. (1989b) The place of the Neandertals in human evolution. In (E. Trinkaus, Ed.) *The emergence of modern humans. Biocultural adaptations in the Later Pleistocene.* Cambridge: Cambridge University Press, pp. 97–141, 232–276 (bibliography)

Wolpoff, M. H. (1999) *Paleoanthropology.* 2nd edition. New York: McGraw-Hill

Wolpoff, M. H., Frayer, D. W. and Jelínek, J. (2006) Aurignacien female crania and teeth from the Mladeč Caves, Moravia, Czech Republic. In (M. Teschler-Nicola, Ed.) *Early modern humans at the Moravian Gate. The Mladeč Caves and their remains,* this issue, pp. 273–340

Wolpoff, M. H., Hawks, J. D. and Caspari, R. (2000) Multiregional, not multiple origins. *American Journal of Physical Anthropology* 112, 129–136

Wolpoff, M. H., Hawks, J. D., Frayer, D. W. and Hunley, K. (2001) Modern human ancestry at the peripheries: a test of the replacement theory. *Science* 291, 293–297

Wolpoff, M. H., Smith, F. H., Malez, M., Radovčić, J. and Rukavina, D. (1981) Upper Pleistocene human remains from Vindija cave, Croatia, Yugoslavia. *American Journal of Physical Anthropology* 54, 499–545

Wolpoff, M. H., Xinzhi, Wu and Thorne, A. G. (1984) Modern *Homo sapiens* origins: a general theory of hominid evolution involving the fossil evidence from East Asia. In (F. H. Smith and F. Spencer, Eds.) *The origins of modern humans: A world survey of the fossil evidence.* New York: Alan R. Liss, pp. 411–483

Chapter 10

AURIGNACIAN FEMALE CRANIA AND TEETH FROM THE MLADEČ CAVES, MORAVIA, CZECH REPUBLIC

Milford H. Wolpoff, David W. Frayer and Jan Jelínek

Introduction

The two female crania from Mladeč were both found in the Main Cave by Szombathy in 1881. This paper presents their description and comparisons, and the descriptions of several much more fragmentary crania. There is no reason to assume their chronological age differs from the other surviving Mladeč cranial material from the Main Cave or the remains from the Quarry Cave (Svoboda, 2000). Consequently, we do not repeat the geological and archaeological discussion in Frayer et al. (this volume).

Mladeč 1 is the most complete of all the cranial remains from Mladeč. When first discovered it was regarded as male, but with the subsequent recovery of the Quarry Cave specimens (Mladeč 5 and 6), it became apparent that Mladeč 1 was female. While certainly not identical to Mladeč 1, the more incomplete calotte Mladeč 2 and the fragmentary face which articulates with it (Mladeč 7) constitute the second female. Both specimens are young adults based on dental criteria. The other specimens we very briefly describe in this chapter are fragments now destroyed (Mladeč 38 and 42) and the very fragmentary vault piece found by Knies (Mladeč 41) in the Main Cave. The latter, in the Moravské zemské muzeum Brno collections, is a small vault fragment that is unidentifiable (and unsexable). Based on the inventory provided by Szombathy (1925) and our revisions of it, certainly more female remains existed in the Main Cave, but these are the only survivors with useful information preserved. As with the Mladeč males, the tragedy at Mikulov castle robbed us all of a rich collection of early Upper Paleolithic female remains. Compared to the males, Mladeč 1 and 2 are considerably more gracile and differ between each other in various aspects of their preserved anatomy. These two contrast with Neandertal females much more than the Mladeč males contrast with Neandertal males. These two facts pose the main questions beyond the comparative descriptions that we consider here: the nature of sexual dimorphism at Mladeč, and the contrasting patterns of male and female evolution. Mladeč 1 and 2 are currently housed in the Naturhistorisches Museum Wien where they were studied by the authors at various times between 1974 and 2001.

Comparisons

For the most part, the rationale and details of the comparisons we made are discussed in the paper on the male remains from Mladeč by Frayer and colleagues (this volume). The comparisons are made by sex: the Mladeč females are compared with females from other samples, and only compared with the Mladeč males in the discussion of sexual dimorphism. Our comparisons involve means and ranges, as an alternative measure of variation, and we caution the reader to remember that these are observed ranges that may dramatically underestimate the expected ranges of variation for the prehistoric samples.

Adult comparative sample compositions

The European Neandertal sample

The specimens included in the sample of European Neandertals are, unless specifically noted, limited to those representing the period of the Würm glaciation, as indicated by stratigraphic, archaeological, and in a limited number of cases radiometric data. The female Neandertal sample consists of the following crania: Gibraltar (Forbes Quarry) 1; La Ferrassie 2; Marillac C10-41; La Quina 5[1], 8, 10, and 27; Saccopastore 1; Šala; Salzgitter-Lebenstedt; Spy 1[2]; and Vindija 202, 205, 224, 225, 252, 259, 260, 262, 279, and 284. For a variety of reasons, sex determinations for some of the females are somewhat more problematic. The list of females we use here represents our best determinations at this time.

The fact is that there are few female European Neandertals, and to expand our comparisons, in some cases we have also used Krapina C and E maxilla. The similarities of Saccopastore 1 and Gibraltar to the pre-Würm Krapina C (cranium 3) are often noted, and seem to be at least part of the basis for attributing an early date to the later two females (for instance, Stringer et al., 1984). However, we regard these similarities as due to sex and not to geological age, and therefore do not believe they can be used to infer geological age through a morphological date (for instance, as Payá and Walker, 1980; do). Clearly, a full understanding of Neandertal female morphology can only be resolved through the recovery of a well-dated Western European female late Neandertal, but this may be an expectation that is unlikely to be fulfilled because of the burial customs of the Neandertal folk themselves.

In addition to the above, the European female Neandertal dental sample includes Châteauneuf-sur-Charente 2 and Hortus 8.

The Skhul/Qafzeh sample

The Levant sites are clearly from the later part of the Middle Paleolithic, and by every date estimate they precede the Mladeč remains in age. While their ages are surely not identical, we believe that they sample similar and probably related populations and because they are potentially ancestral, they can be validly combined for comparisons, under the assumption that the ancestors of the Mladeč folk arrived in Europe by migration from the southeast. This is an idea that springs to the minds of many who continue to read the persistent descriptions of these folk as "proto Cro-Magnoids". We include this sample in our comparisons to allow an examination of the morphological basis that could provide evidence to support or reject the contention of this migratory origin for the Mladeč population from southwest Asia.

The female Skhul/Qafzeh cranial sample we have used consists of Qafzeh 3 and 5, and Skhul 2 and 7. The bases for considering Qafzeh 9 and Skhul 9 males are presented in Frayer et al. (this volume). The female sample from Skhul/Qafzeh is very small, and all of the specimens comprising it are incomplete. Comparisons with these females are often of limited value because of their fragmentary nature, and in a number of cases we do not make them at all. When our comparisons are limited to a single individual, we identify it.

The European early Upper Paleolithic female sample

The third comparative cranial sample is from the earliest European Upper Paleolithic of Central Europe, the sample of specimens closest to Mladeč in age. Because many were collected so long ago,

---

1 While La Quina 5 is often considered to be female, the dentition is among the largest of the Würm European specimens.

2 Spy 1 is problematic because in some respects it resembles the Feldhofer Cave male calotte.

convincing absolute dates are virtually unknown for this sample and in most cases the dating is by archaeological association. We have been conservative in the determination of this sample, not wishing to confuse the comparisons by including specimens that may actually be much younger. The sample is comprised of individuals that have the highest probability of actually representing European populations during the earliest phases of the Upper Paleolithic[3]: Zlatý Kůň, which may be later than the Aurignacian (Svoboda, 2000); Dolní Věstonice 1-3; Cioclovina (Păunescu, 2001); Předmostí 4, 10, and the older subadult 5. For the most part we do not make systematic comparisons with this sample, but discuss individual anatomical comparisons, or occasionally the comparison of mean values for measurements, when they inform the discussion. The specimens and their stratigraphic and archaeological context are reviewed in Churchill and Smith (2000).

Dates and associations from the literature on this sample are not always authoritative, because of the dates and in some cases the circumstances of discovery for the specimens. Female, or possibly female, specimens that have been thought to date to the earliest Central European Upper Paleolithic, but that either lack convincing confirmatory archaeological and/or geological evidence, or are now known simply to be later in time, include the following: Brno 3, Hahnöfersand, Svitávka; Velika Pećina, and Kelsterbach (Otte, 1979; Smith et al., 1999; Valoch, 1982; Svoboda et al. 2002; Vlček, 1971; Stringer et al., 1984 whose skepticism [pp. 68–69] we share). We believe it is very unlikely that we have included specimens in our sample significantly younger than 26,000 years B. P., and if we have erred it is on the side of caution.

Data collected

In almost all cases the observations and measurements reported here are based on studies of the original specimens in Vienna by the authors. Details of the measurements and measurement techniques and of the data sources are reported in Frayer et al. (this volume). For our Předmostí data, we have relied on the primary cast of cranium 4 located at the Moravské muzeum in Brno, the publications of Matiegka (1934), and additional details mentioned by Kříž (1903), Maška (1895), Wankel (1884), Skutil (1940), Absolon (1929), Hrdlička (1930), Szombathy (1925), and Morant (1930). The other specimens that could only be studied as casts were Cioclovina, and Dolní Věstonice 1 and 2. In all cases when the use of reproductions was necessary, accuracy of the casts was ascertained and if necessary the cast measurements were scaled by comparison with published measurements of the originals.

As much as possible, the morphological features we discuss are named following Weidenreich (1951). We almost always used standard measuring points, as defined by Martin (1928) and White (2000). In those cases where we found it necessary to define a position for measurement, it is discussed in the text. Our abbreviations for these and other landmarks and directions, especially as used in the tables, are given in Table 1.

We calculated indexes, angles at the parietal corners, and various projections into the sagittal plane at the midline. The only other calculated variable was cranial capacity. Direct determinations of the Mladeč hominids cranial capacities were never made. Szombathy ascertained a capacity for cranium 1 by taking the average of calculations made from the Manouvrier, Lee-Wackler, and Froriep formulae and the Welcker tables. This was approximately 1620 cc. Billy (1972) reports a capacity of 1550 cc. for this specimen. Szombathy estimated the cranium 2 capacity by taking proportions of cranial measurements with cranium 1, arriving at a range of 1470–1480 cc. Frayer (1986) reports a 1370 cc. determination, which he calculated from the Poissoinnet et al. (1978) regression.

3 This means the Szeletian, Aurignacian, and early Gravettian.

Table 1. Definitions of abbreviations found in the tables. The descriptions of the terms defined are in Martin (1928)

Abbreviation	Definition	Abbreviation	Definition
alv	alveolare	k	krotaphion
ant	anterior	L	length
ast	asterion	l	lambda
au	auricular point	mf	maxillofrontale
ba	basion	mm	millimeters
Br	breadth	ms	mastoidale
br	bregma	na	nasion
btwn	between	o	opisthion
C	canine	occ	occipital
cc	cubic centimeters	op	opistocranion
co	coronale	pr	prosthion
FH	Frankfort Horizontal	proj	projection
fmo	frontomalarorbitale	pt	point
fmt	frontomalartemporale	st	stephanion
gl	glabella	sup	superior
ht	height	UFH	upper facial height (na-alv)
i	inion	zm	zygomaxillare
iob	innerorbital breadth	zpm	zygomatic process of the maxilla
ju	jugale	zt	zygotemporale

We have calculated two regressions to determine the cranial capacities of these specimens, as well as several others from the Central European early Upper Paleolithic sample, using those individuals with actual endocast determinations. These are Dolní Věstonice 3 (1322 cc from Jelínek, 1954), Pavlov (1472 cc, from Vlček, 1991), and Předmostí adult crania 3 (1608 cc), 4 (1518 cc), 9 (1555 cc), and 10 (1452 cc, all from Matiegka, 1934). There is also an endocast capacity for the Předmostí juvenile cranium 22 (1335 cc), but this was not used in the derivation of the regression formulae.

These 5 crania formed the basis for developing two regressions for cranial capacity estimation, a least mean squares linear determination, and a power curve based on a least mean squares fit of logs. In both cases we used a volume estimation for the independent variable. The volume was estimated two different ways, using measurements that avoided including cranial superstructures. Nasion-opistocranion provides a length measure that does not include the superciliary area, and nasion-lambda is the most complete length possible on Mladeč 2. Biparietal breadth avoids the basal pneumatization, and the vertical height from the auricular point to bregma is the only comparable height measure for Mladeč 1 and 2. The two formulae are based on the following variables:

$$V_1 = \text{(nasion-opistocranion)*(biparietal breadth)*(auricular height)*}10^{-4},$$

$$V_2 = \text{(nasion-lambda)*(biparietal breadth)*(auricular height)*}10^{-4},$$

where in both cases the volume estimate variables were calculated from the products of the measurements in millimeters. Formula determined from the first could be used to estimate cranial capacity for Mladeč 1 and Cioclovina (1475 cc). Formula determined for the second could be used for these, and Mladeč 2. Linear and power curve regressions were determined for both volume variables from the sample of endocast capacities.

$cc = 3.21 * V_1 + 562.36$ (average error of 35 cc), $cc = 36.29 * V_1^{.656}$ (average error of 34 cc),

$cc = 3.51 * V_2 + 510.57$ (average error of 33 cc), $cc = 31.49 * V_2^{.685}$ (average error of 33 cc).

Four cranial capacities could be estimated for Mladeč, only the last two formulae could be used for Mladeč 2. The multiple estimates were averaged and the resulting capacities rounded to the nearest 5 cc. These estimates are given in Table 2. Our Mladeč 1 determinations are quite close to those published by Billy (1972), and the Mladeč 2 determination is close to Frayer's (1986). In all, we believe these capacities are probably as accurate as could be ascertained without direct volume determinations since they are based on regressions developed within the (biological) sample to which they are applied.

The female vaults

Mladeč 1

Mladeč 1, a virtually complete cranium, was found at locus "a" in Chamber D of the Main Cave at Mladeč by J. Szombathy in 1881 (Figs. 1, 2 and 5). Near it was recovered a femur diaphysis Mladeč 27. The vault was reconstructed from a number of pieces. The observations below are based on our study of the original specimen in the Anthropological Division of the Naturhistorisches Museum Wien (see this volume, chap. 8, Plate I).

We believe that an accurate age at death can be ascertained for Mladeč 1. Third molar crypts are present, and open to the alveolar margin on both sides, but neither tooth remains. According to Skinner and Sperber (1982, 298) these crypts are "poorly preserved and indefinite", with the left crypt "preserving in its roof a trace of alveolar radicular crest formation for root bifurcation." Their radiographs show fused roots for the second molars. Apical closure for the second molar roots is at about 14 years in populations that average a 12 year occlusal eruption for this tooth. If an 18-year occlusal eruption is assumed for the third molar, root development for this tooth would begin at about 15 years. Following this assumption, the suggestion of bifurcation in the root of one of the third molars indicates an age-at-death of 16–17 years.

The question, of course, is whether an 18-year-old age for M^3 occlusal eruption can be assumed. Root development of the third molar relative to the crown wear on the M^2 indicates that this may be an overestimate of the age at death, and certainly shows that the third molar eruption is not likely to have been significantly later than 18 years. Skinner and Sperber estimate an age of 14.9 [13.5–17.0 (90% confidence interval range)] for the specimen (assuming it is female, see below), which would indicate that the M^3 was erupting younger than 18 – a condition common for all hominids earlier than Mladeč. For an independent confirmation of the age at death, we examined the tooth wear on the anterior molars. M^2 shows little wear, with most of the ori-

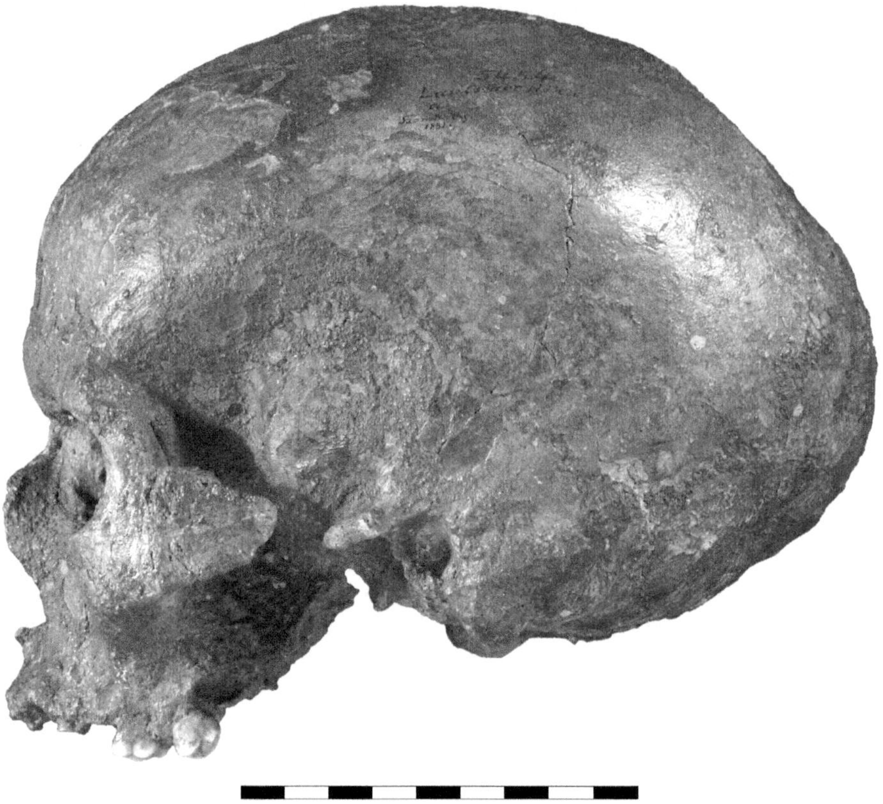

Fig. 1. Lateral view of Mladeč 1

ginal cusp heights remaining. Szombathy reports that when the specimen was discovered the external cranial and facial sutures were open. To verify his observations with an estimate of the dental age, we compared Mladeč 1 to the Krapina E maxilla (aged at 16 years according to criteria developed by Wolpoff (1979), but with erupted third molars) and to several late Upper Paleolithic specimens in the Naturhistorisches Museum Wien with unerupted third molars but some wear on the second molars. The Mladeč 1 molars are equally or slightly more worn than any of these, suggesting that at death the specimen was not far from third molar eruption. A cast of the molar dentition was compared with other Upper Paleolithic specimens aged by Skinner and Sperber (1982). Mladeč 1 shows slightly more molar wear than the Lachaud adolescent (15–16), La Pique 62.1 (14–17), and the Roc de Cave juvenile (14–16). A comparison was also made with the maxilla of Mladeč 2. Here, the first and second molars of Mladeč 1 are slightly less worn, and Mladeč 2, as described below, has an erupted third molar. In sum, our experience with tooth wear in Upper Pleistocene hominids from Europe suggests that 16 and 17 years are reasonable approximations for the ages at death for these two Mladeč specimens. For these reasons, we estimate the age of death of Mladeč 1 to have been 16±1 years. The specimen was clearly an older subadult.

Mladeč 1 has often been regarded as a male (Szombathy, 1901, 1904, 1925; Morant, 1930; Matiegka, 1934; Riquet, 1970; Henke, 1987). We believe this perception comes from the size and robustness of this young specimen (especially compared with modern European females), and its purported similarities to the Cro-Magnon I male (Szombathy, 1901, 1904). On the other hand, other studies of the specimen have suggested the skull is a female (Jelínek, 1983; Frayer, 1978,

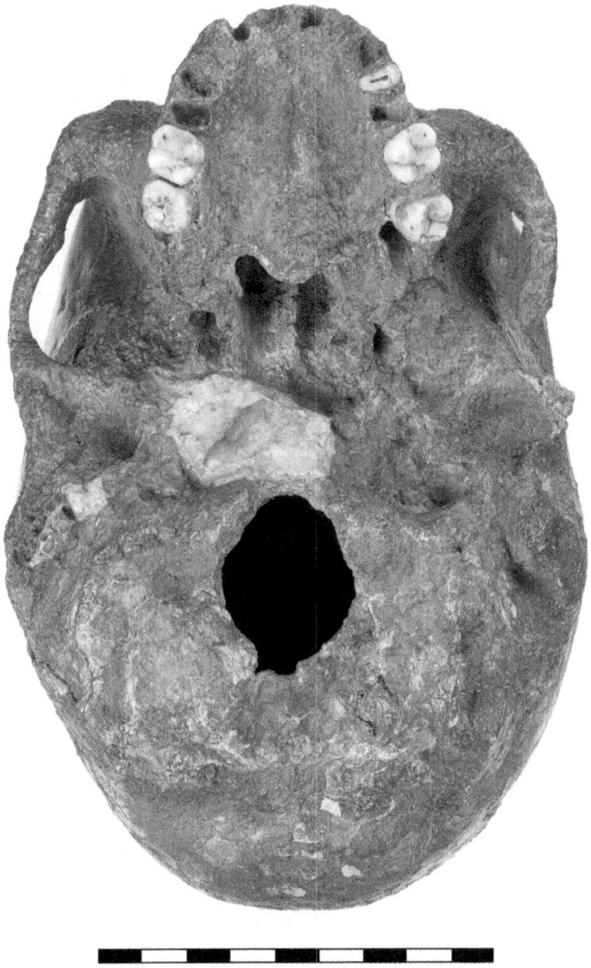

Fig. 2. The Mladeč 1 cranial base

1980, 1986; Wolpoff, 1982, 1999; Smith, 1982, 1984, 1997). The reason for this different interpretation lies in the basis of comparison. The "male" characters seem pronounced when comparisons are made with some of the later males from Western Europe, but the female characteristics are most evident when comparisons are made with the Mladeč males (e.g., Fig. 23 and other comparisons below).

We offer the following considerations to help accurately determine the sex of this specimen. Compared with Mladeč crania 4, 5, and 6 (see Figs. 19 and 23 for the comparison with Mladeč 5), the Mladeč 1 superior orbital margins are sharp and the superciliary arches are vertically tall, but only slightly projecting. Visually, compared with adult women from the early Upper Paleolithic of Europe, the expression of supraorbital projection is much less than Cioclovina, somewhat less than in Předmostí 4 females, but only slightly less than Zlatý Kůň. The frontal squama is high and rounded, and the nuchal region is much more gracile than those of Cioclovina and Předmostí 4. There is an elevated nuchal line at the midline, but no evidence of a nuchal ridge or torus. Although the mastoid process projects more below the cranial base than do the mastoids of the Mladeč males, its overall dimensions are smaller. The mastoid is considerably smaller and less projecting than the mastoids of the later Předmostí females.

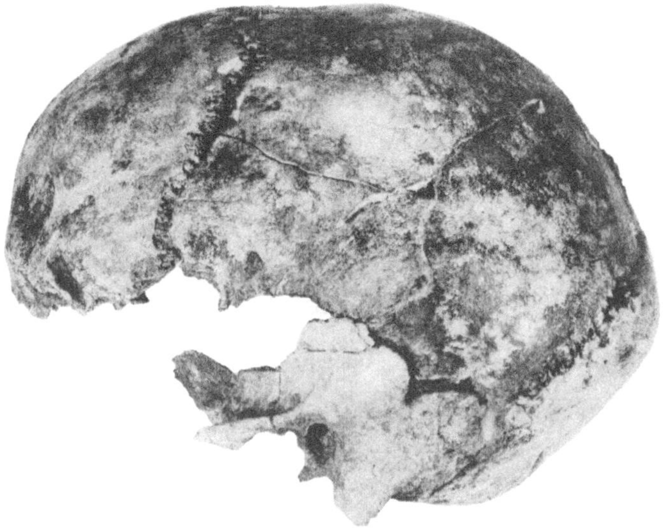

Fig. 3. Lateral view of Předmostí 5, after Matiegka (1934)

Some of these comparisons may be influenced by the fact that the specimen is not a mature adult. While the final size of the palate, face, and vault had been attained, some of the cranial superstructures and muscle attachment areas might have developed further had the specimen lived longer. Moreover, in the Mladeč 4 male, a specimen clearly not very much older than Mladeč 1, but younger than Mladeč 5 and 6 on the basis of suture closure, the cranial superstructures and muscle markings are much better developed than in the Mladeč 1 female. However, as a 16-year-old female, we do not believe that the age related changes could have substantially altered the overall morphological features of the specimen. Therefore, comparison with the Předmostí 5 female is relevant to the problem of the Mladeč 1 age (Figs. 3 and 4). This Předmostí specimen died at a very similar age to Mladeč 1, Matiegka (1934) suggests an age of 15–16 years according to its dentition. We concur that she was slightly younger than Mladeč 1 at death, based on the superior position of the third molar, still in its crypt (in so far as it can be judged from the photographs published by Matiegka (1934)). Předmostí 5 is not unusually small; she has total cranial vault measurements almost invariably between those of the two adult Předmostí females (no. 4 and no. 10). Like Mladeč 1, her superciliary arches are distinct; vertically tall, but with little projection. The mastoids are very well-developed and projecting. There is lambdoidal flattening and the occipital has a very short vertical face. The nuchal plane shows distinct rugosity. This combination of "masculine" and even Neandertal-like features is expressed in a cranium that is diagnosably female. Similarly, the "masculine" supraorbital region of Mladeč 1 is less well developed than the Předmostí 4 supraorbitals, an older aged female (Fig. 8).

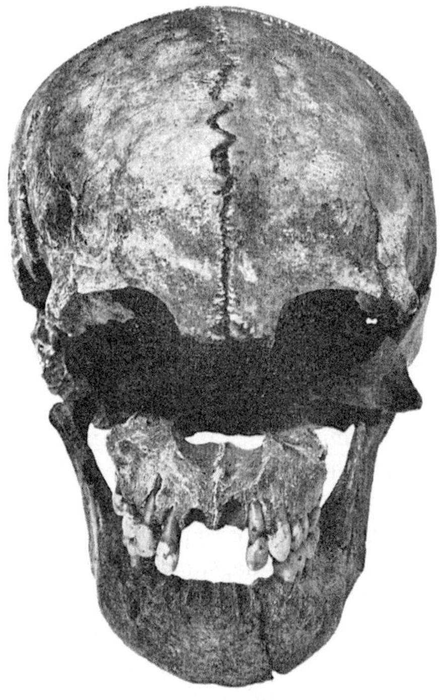

Fig. 4. Facial view of Předmostí 5, after Matiegka (1934)

We recognize that further growth and/or bone remodeling might affect the development of the supraorbital region, the nuchal region, and the expression of muscle markings on the vault. However, we do not believe it possible that these features could change to the extent that they would come to resemble the Mladeč males. Instead, we believe that if they had changed with increasing age, they would more closely resemble the Předmostí 4 (Fig. 8) or Zlatý Kůň (Fig. 24) females.

Thus, while Mladeč 1 might be considered as male if it had been found in a much later context, its association with the Mladeč 4, 5 and 6 crania and its general similarities to the Předmostí and Zlatý Kůň females leads us to sex it as a female. Indeed, its confusion with later males gives some insight into the mode of variation and the direction of evolutionary change in these early Europeans (Frayer, 1980).

Preservation

The skull is virtually complete, although it is missing the following portions: a small posterior section of the left zygomatic arch, the lateral and inferior portions of the right mastoid, a good portion of the lateral vault wall on the right (described below), a large triangular piece from the right portion of the lambdoidal suture laterally and inferiorly (mostly involving the occipital side of the suture), and all of the maxillary dentition except for the first and second molars. The missing portion of the lateral vault wall on the right side extends from about 17 mm posterior to the orbital corner to the mid-parietal, superiorly to 50 mm from the sagittal suture, and inferiorly to the border of the parietal except at its most posterior corner where temporal squama is also missing. This entire area has been reconstructed symmetrically with the left side and we consider this reconstruction to be quite accurate. The missing area around the lambdoidal suture has also been reconstructed.

Matrix has been left over much of the vault's surface, and staligmatic material even further obscures details of the *pars basilaris* and some of the sphenoid anterior to it, especially covering details at the occipitosphenoidal synchrondrosis. Apart from this area, the heaviest concentration is on the face, and covering the nuchal plane. However, except for the parts of the cranial base the matrix

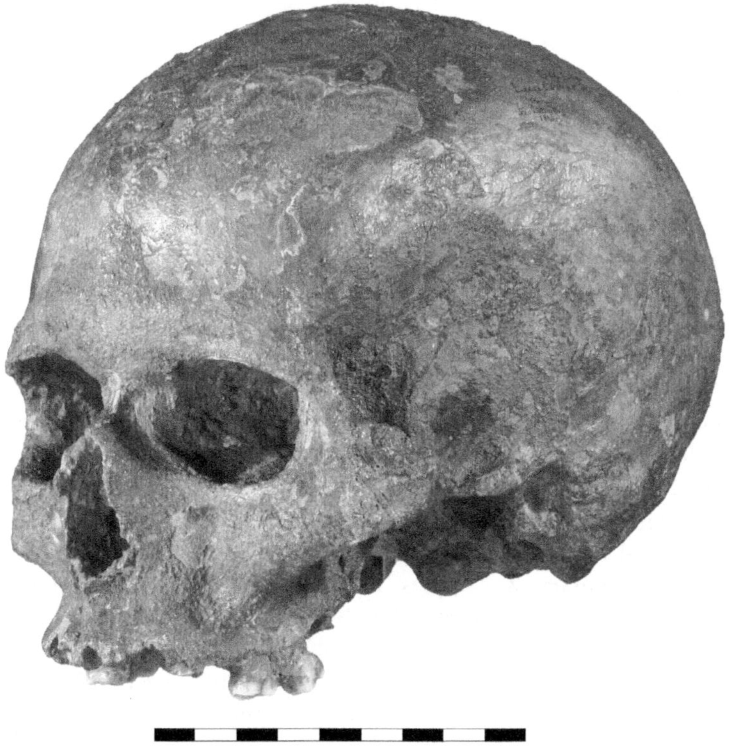

Fig. 5. Mladeč 1 in an angled view

is very thin and most underlying details can be easily discerned. There is also a considerable amount of matrix on the interior part of the skull, making direct measurements of most cranial thicknesses and cranial capacity impossible, which is why we took a regression approach.

Dr. Hermann Prossinger (see Prossinger and Teschler-Nicola, this volume) of the Institute for Anthropology, University of Vienna, has interpreted a CT-scan of the vault for us. He notes the "whole cranium is covered with some sort of shellac or varnish." He writes further that there is a

> *"tell-tale x-ray signature of gypsum (primarily the smoothness of its attenuation) on the right side, where the gypsum is "embedded" in the fossilized bone (most likely so that the curator could keep the gypsum in place during the attempted reconstruction of the parietal). [Moreover, the base of the vault] is encrusted with some material that has a very large attenuation coefficient. It appears inordinately bright in the CT-images. Finding the border between this encrustation and the bases is no simple task. [...] There are many stones and similarly strongly attenuating materials in many places of the endocranium (plugging many foramina, for example). [...] The region of the frontal sinus is filled with a filament-like structure, which I think is some kind of deposited material. There are other sediments inside the cranium, as well as (in patches) all over the exterior surface."*

Despite these limitations, the overall preservation of the skull is good, so that it is possible to determine most standard landmarks and details of the external surface with accuracy. However, some areas of the external table have been flaked away, leaving small regions where the most external portion of the vault is missing. According to Szombathy, much of this flaking occurred during the original excavation of the skull and was associated with removing the adhering staligmatic pieces and reindeer ribs.

The vault as a whole

The cranium is very large, well rounded, and fairly robust in some features, especially considering its age at death. The cranial capacity we have determined for it (Table 2) is the largest of any female in the Central European early Upper Paleolithic. Its maximum length of 198.5 mm is also the largest value for any earlier Central European Upper Paleolithic woman, and exceeds the means of both the Neandertal and the Qafzeh female samples. Other measures of sagittal length reflect this marked size. Comparison of lengths taken from nasion and glabella show that the glabellar prominence of the specimen contributes significantly to this length. Interestingly, even with the larger supraorbitals of the Neandertal and Skhul/Qafzeh females, Mladeč 1 is longer than most of these more archaic specimens. Length measures including the face, such as prosthion-inion, also show Mladeč 1 to be longer than Neandertal females, as do direct measures from basion to glabella and to lambda. The one length measure that does not differ significantly between Mladeč 1 and the Neandertal females is the prosthion-mastoidale distance. The Neandertals are very large, compared with the earlier Central European Upper Paleolithic females, well outside their range in that the earlier Central European Upper Paleolithic and Neandertal female ranges do not even overlap.

The maximum cranial breadth occurs on the parietal bones, as it does on nearly all of the female specimens in the comparative samples. The breadth value, of 141.5 mm is below the Neandertal mean. Qafzeh 3 is considerably larger. Mladeč 1 narrows considerably at its base, although not as much as Neandertals. For instance, biauricular breadth is about 13 mm smaller than the maximum breadth of the vault, slightly less than the mean difference in Neandertal females. The bimandibular fossa breadth, across the outsides of the glenoid cavities (bi-*gp*), exceeds biauricular breadth, unlike every Neandertal female vault. The mastoids are more vertically oriented in the Mladeč 1 vault, and therefore the distance between their tips is toward the top of the Neandertal

Table 2. Female cranial vault dimensions

	Mladeč females 1	Mladeč females 2	Mladeč male 5	Qafzeh 3	Neandertal Mean (n)	Neandertal Range
Capacity (cc)	1540.0	1390.0	1650.0		1297.0 (4)	1245.0–1367.0
Lengths (mm)						
gl-op	198.5		205.6	182.5	193.6 (3)	179.0–201.0
gl-l	185.0	175.0	194.0	175.5	178.0 (3)	166.0–185.5
na-op	193.0		196.0	176.9	188.6 (4)	176.0–199.5
na-i	179.0		186.5	174.0	183.2 (4)	170.5–196.9
na-l	182.0	173.0	188.2	170.5	177.0 (4)	165.5–186.6
br-i	160.5		160.2	151.0	144.5 (5)	139.8–152.3
pr-i	192.0				184.3 (2)	177.0–191.5
pr-ms	129.5	124.0			128.4 (2)	125.9–130.9
ba-gl	113.4				109.0 (1)[2]	
ba-i	86.0				77.3 (2)	74.1–79.4
Breadths (mm)						
Cranial	141.5	141.0	156.0	156.0	142.7 (4)	138.3–146.0
Biparietal	141.5	140.0	154.0	156.0	142.7 (4)	138.3–146.0
Auricular	128.8	132.4	150.0		126.5 (4)	119.9–130.1
Bimastoid	108.0	113.0	136.0		101.3 (4)	96.2–110.4
Bimandibular fossa	129.0	132.3			124.1 (4)	119.2–126.0
Heights (mm)						
ba-br	138.0				113.3 (2)	110.5–116.0
op-br	155.0				133.2 (2)	132.5–133.9
Arcs (mm)						
na-op	306.0		318.0		278.5 (2)	270.0–287.0
na-i	345.0		342.5	317.0	287.2 (3)	282.0–297.5
na-o	395.0			378.0	331.0 (1)[1]	
gl-l	248.0	241.0	256.0	254.0	219.2 (3)	214.0–225.5
br-i	210.0		202.5	194.0	176.6 (4)	170.0–186.5
au-br	154.0	147.0	159.0		147.5 (4)	140.0–152.0
*Indices (*100)*						
MaxB/Max L	71.3		75.9	85.5	74.1 (3)	68.9–81.6
MaxB/na-l	77.8	81.5		91.5	80.8 (4)	76.6–88.2
Parietal B/Max L	71.3		74.9	85.5	74.1 (3)	68.9–81.6
au ht/na-l	61.8	59.9	59.1		59.8 (4)	54.6–66.3
au ht/na-op	58.2		56.7		56.0 (4)	52.5–62.3
gl-l arc/chord	134.1	137.7	132.0	144.7	123.3 (3)	119.3–128.9
br-i arc/chord	130.8		126.4	128.5	122.3 (4)	118.4–127.3
au-br arch/chord	117.1	119.5	118.6		119.6 (4)	116.9–121.2

[1] Gibraltar1

range in spite of the generally broader dimensions of the Neandertal female crania. Like the Mladeč males, the bimastoid distance is large relative to the Neandertal and the earlier Central European Upper Paleolithic means, suggesting that a broad distance across the mastoids characterizes the entire Mladeč sample.

Cranial height as measured from bregma (vertex is coincident with bregma), greatly exceeds the Neandertal maximum. From the auricular point (Table 3), height also exceeds the Neandertal range (although to a lesser extent). The difference in cranial height characterizes the general relation between the Neandertal and the earlier Central European Upper Paleolithic female samples, although in the male comparisons (Frayer and colleagues, this volume), cranial height for Mladeč 5 is almost the same as the Neandertal mean. Expressed as a ratio to nasion-opistocranion or nasion-lambda lengths (Table 2), the relative auricular height is almost the same as in Neandertal females. In fact, of all these specimens, the vault with the *greatest* relative height is Gibraltar 1. The difference in cranial height, then, reflects the larger size of Mladeč 1 and not the appearance of a different cranial proportion.

As seen in lateral view (Fig. 1), the Mladeč 1 face appears quite vertical, except for its alveolar prognathism which results in an index of prognathism (gnathic index) of 104.2 (118.3 from the auricular point, see Table 11). The gnathic index is slightly less than in the Neandertals, while the auricular gnathic index is slightly greater. We take this to mean that the degree of prognathism is essentially the same. The middle and upper facial regions are quite flat transversely, but the region of the maxilla surrounding the nasal aperture is very prominent and the angle of the nasal bones is high. This nasal angulation conforms to the Neandertal and the early Central European Upper Paleolithic pattern in which the angulation begins at nasion, and is distinct from the flat vertical orientation of the superior portion of the nasals preserved in the Qafzeh 3 female.

A significant contrast in lateral view comes with the relative height of the face. Although the Mladeč 1 facial height is the largest of all Central European Upper Paleolithic women, compared with the nasion-lambda dimension,[4] it is a relatively low face in relation to the Neandertal females (Table 11). This ratio (38.5) is much below the Neandertal mean and range.

Table 3. Distances from the auricular point to midline landmarks, in sagittal projection (mm) for female crania

au projection to:	Mladeč females 1	Mladeč females 2	Mladeč male 5	Neandertal Mean (n)	Neandertal Range
prosthion	110.0	110.8		116.3 (2)	112.6–120.0
nasospinale	103.2			107.5 (2)	107.3–107.6
nasion	93.0	87.8	103.5	103.9 (4)	95.0–111.2
glabella	100.1	96.0	114.5	112.4 (3)	107.9–116.0
bregma	112.4	103.7	111.2	105.4 (4)	96.2–109.7
lambda	109.5	98.9	103.0	97.9 (4)	91.7–104.1
opistocranion	105.1		99.2	83.9 (4)	77.9–92.1
inion	86.6		84.8	83.9 (4)	77.9–92.1
opisthion	42.4			41.1 (2)	35.3–46.9

4 So that Mladeč 2 can be compared, the length to opistocranion was not used.

Table 4. The cranial base and the position of basion for female crania

	Mladeč 1	Neandertal	
		Mean (n)	Range
ba-hormion	30.1	24.2 (2)	22.8–25.6
ba-pr	108.9	109.0 (2)	104.0–114.0
ba-na	104.5	103.0 (2)	101.9–104.0
ba-Glenoid pt	68.3	66.1 (2)	65.2–67.1
ba-ms	57.8	53.2 (2)	52.0–54.4
o-ms	60.9	60.8 (2)	58.6–63.9
ba-au projection	24.8	11.4 (2)	11.2–11.6
Front Occ Condyle-pr	104.4	111.1 (1)[1]	
Front Occ Condyle-na	104.5	104.0 (1)[1]	
Front Occ Condyle-i	88.4	81.0 (1)[1]	

[1] Saccopastore 1 only

At first glance, projection of the upper face anterior to the cranial base seems to be dramatic in this specimen. For instance, the nasion-basion diameter (Table 4) exceeds both the Neandertal and the earlier Central European Upper Paleolithic ranges. The basion-glabella distance is also large (Table 2). Thus, as measured from basion the upper face seems to project markedly, as we have observed it does in the Mladeč males (Frayer and colleagues, this volume). However, the corresponding measurements of the upper face from the auricular point are somewhat less (Table 3). In this case the Mladeč 1 values are below the Neandertal ranges. The projection from the auricular point to nasion is about 90% the Neandertal mean. Contrasting the measurements from basion with the measurements from the auricular point, the variation in estimates of facial projection from these two different regions of the cranial base indicate that there is a distinction between the relative positions of basion on the one hand, and of the biglenoid and the biauricular lines on the other.

That such a difference exists is easily observable by inspection (Fig. 6). In Neandertal females such as Gibraltar and Saccopastore 1 the biauricular line passes at or posterior to basion, while in Mladeč 1 and the other earlier Central European Upper Paleolithic females such as Předmostí 4 and Dolní Věstonice 3 this line passes *anterior* to basion.

We examine the possibility that in the more recent sample the auricular position is the same and it is the front of the foramen magnum that is further from it than in the Neandertals. This would account for the contradictory comparisons of facial projections discussed above. In measurements of the distance between basion and hormion (because of the adhering staligmatic material the occipitosphenoidal synchrondrosis cannot be found in Mladeč 1), two Neandertal females average 24.2 mm (Table 4), but the Mladeč 1 value is 30.1 mm. Although Mladeč 1 cannot be directly compared, the length of *pars basilaris* averages 17.9 mm for these two Neandertal females, and 25.7 mm for two earlier Central European Upper Paleolithic females, reflecting the same difference. The males reveal a similar pattern, four Neandertals averaging 21.7 mm and two earlier Central European Upper Paleolithic specimens averaging 24.7 mm for the *pars basilaris* length. Moreover, in the males of the early Central European Upper Paleolithic sample the biauricular line also passes anterior to basion while in the Neandertal males the line is posterior to basion. In sum, whether directly measured from the foot of the vomer or from the sphenoid, the front of the foramen magnum is further from these points in the Central European early Upper Paleolithic sample of both sexes.

The same does not hold for the position of the posterior foramen magnum. In the Neandertal females the biauricular line averages 41.1 mm anterior to opisthion and Mladeč 1 is virtually identical, 42.4 mm. The fact that the foramen magnum length is considerably greater in Mladeč 1 (Table 8) accounts for this difference.

What of the distance from basion to the dentition? A measure from the front of the occipital condyle to the second molar position (Table 13) is much shorter in Mladeč 1 than in Saccopastore 1. The distance from basion to the M^3 posterior border varies similarly, with the mean for two Neandertal females at 65.8 mm while we estimate Mladeč 1 at about 55 mm. The Neandertal dentition is further from the cranial base, and this distance is reduced in Mladeč, presumably as an aspect of facial reduction but this does act to shorten the leverage of molar forces around the fulcrum of the occipital condyles.

A similar effect is the closer positioning of the Mladeč tooth row to the glenoid fossa and the mandibular condyle than is the case for the Neandertal females (Table 4). The more flattened Mladeč 1 face results in essentially the same distances from the auricular point or the glenoid point to *zygomaxillare* (direct measurements) as the Neandertal females have, in spite of the fact that projection of the lower face is reduced in Mladeč. Thus, the Mladeč posterior teeth have moved even further behind *zygomaxillare* and thereby away from the most anterior masseter attachment, and toward the cranial articulation of the mandible. Put another way, more of the posterior tooth row is directly covered by temporalis and masseter musculature, and the anterior teeth have moved closer to the most anterior masseter attachment. It is because of this facial flattening that in spite of the changes described above, the facial length measure *(au-fmo)* of Mladeč 1 is virtually identical to the Neandertal female mean.

In a different analysis of the cranial base, the basion position orthogonally projected onto the nasion-lambda line was determined. The position of basion was calculated as a ratio of the distance between the position of the basion projection to nasion, with the full nasion-lambda distance. This ratio is 49.0 for Saccopastore 1 and 53.1 for Gibraltar. Mladeč 1 is less, 43.3 mm.[5] The Mladeč basion position is more posterior relative to the nasion-lambda line.

We believe there are two differences between Mladeč and the Neandertal sample that in combination resolve these contradictory observations. The evolution of Mladeč, and subsequently of the rest of the earlier Central European Upper Paleolithic sample, has involved a more posterior positioning for *both* the full face and basion, and a deepening of the vault lowering the basion position. As Lieberman (1975) and Laitman et al. (1979) have pointed out, the reduction in anterior positioning of the Neandertal face had consequences that could potentially affect the form of the supralaryngeal vocal tract, trachea, and esophagus in the submandibular region. We believe the more posterior position of the foramen magnum in the earlier European Upper Paleolithic is a response to the parallel reduction in facial size and projection. Together, these posterior shifts in the positions of the face and the cranial base *helped maintain a consistent form and unchanging function* for the structures in the submandibular region, during a period of marked evolutionary changes in the structure of the face. In this respect, our interpretation of the evolutionary changes in all the elements of this region differs from Lieberman (1975) and Laitman et al. (1979). They regard the developments in this region as corresponding to a functional *change* in the supralaryngeal area, while we interpret the differences between the Neandertal and the earlier Central European Upper Paleolithic samples as maintaining the same functional relationship while the anterior projection of the face reduced.

5 The three other females in the earlier Central European Upper Paleolithic sample are just about the same, averaging 45.5 (43–48.3). The same relation holds for the males: four Neandertals average 51.0 (48.4–53.2), while two earlier Central European Upper Paleolithic males average less, 46.9 (43.5–50.3).

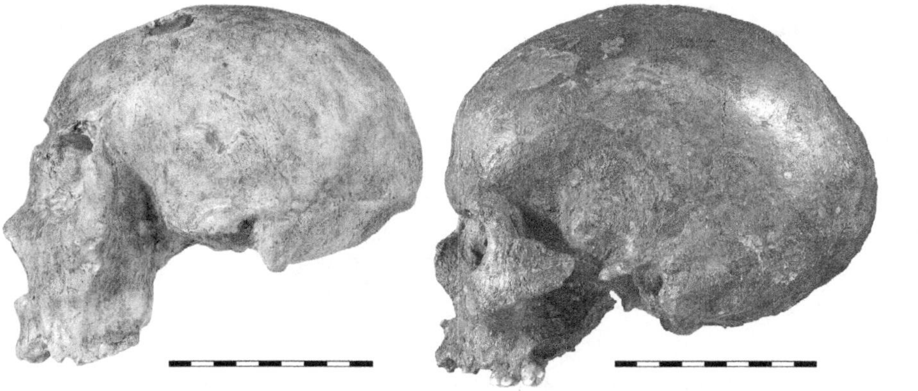

Fig. 6. Comparison of the Neandertal female from Saccopastore (left) and Mladeč 1. This comparison addresses many things, but perhaps the most important is the logic behind comparing specimens of the same sex. A surprising number of contrasts said to distinguish Neandertals are not seen here. Saccopastore does not have a less projecting mastoid process, the vertical face of her occiput is not sorter, she does not have a better expressed occipital bun, and her cranial height as would be seen in life – the cranial height above the top of the orbit – is not markedly less. Of course Saccopastore is a Neandertal and Mladeč 1 is not, but this issue is not about the taxonomic identity of the specimens, it is over how and why they differ. Holding sex constant is an important way to address it

We conclude that the position of the auricular point has been more conservative than the position of the foramen magnum, and that both the face and the front of the foramen magnum change their positions in concert relative to the biauricular line. Consequently, because the region is more stable, the measurements from the auricular point are the more relevant for understanding the changes in basal and facial morphology. For this reason, we regard the reduced projections of the Mladeč 1 face relative to auricular point as best describing the actual morphological change (Table 3).

However, understanding the comparisons of these projections for the Mladeč 1 prognathism index as calculated from the auricular point (Table 11) is limited by the fact that there are no faces allowing this calculation in the Skhul/Qafzeh female sample, and confused by the fact that more specimens contribute to the mean distances to the upper part of Neandertal faces than to the mean distances to the lower part. Comparing mean values, it appears that while projections to the upper part of the face (nasion and glabella) are markedly reduced, the reductions to the lower part are somewhat less reduced. However, this is not actually the case. The two Neandertal females that are complete have the same index that Mladeč 1 has (117 for the Neandertals, 118 for Mladeč 1, and for that matter four Neandertal males have a mean index of 117). The difference between the ratio of the Neandertal means and the mean of the Neandertal ratios comes from the fact that more specimens contribute to the mean values for the upper part of the face than contribute to the mean values for the less often preserved lower part of the face. We contend that the mean of the ratios gives the more accurate comparison.

Other observations of the vault in lateral view show it to be evenly rounded from glabella to a position about 78 mm posterior to bregma. Beginning here the bone is flattened for a distance of 37 mm, to the position of lambda. Below this there is a distinct occipital bun. The anterior aspect is more curved than the Neandertals, the glabella-lambda arc/chord index, for instance, is above the Neandertal range, but this curvature is dramatically less than Qafzeh 3. The more posterior portion contrasts and is less curved. The arc/chord index for bregma-inion, while still above the Neandertal range, is closer to the mean and very similar to the index for Qafzeh 3. We attribute this difference in curvature to the Mladeč 1 bunning, which is very Neandertal-like. In a comparative context, the

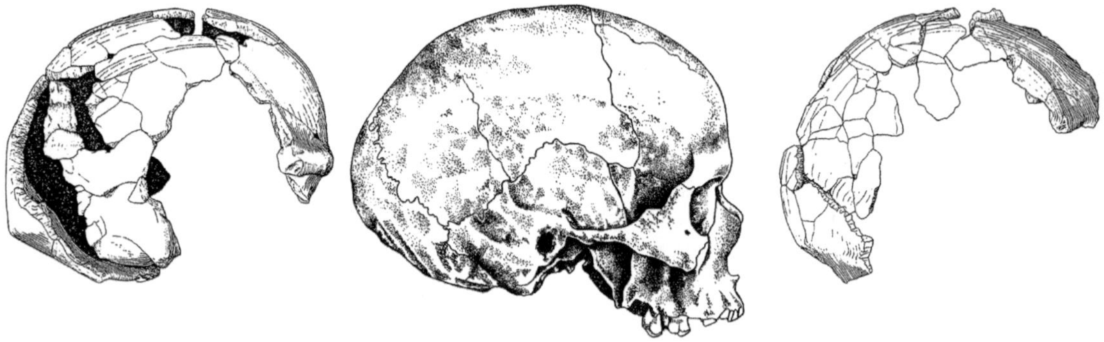

Fig. 7. Mladeč 1 compared with the Qafzeh females 3 (left) and 7 (right), Mladeč 1 drawn by Karen Harvey and the Levant women after Vandermeersch (1981). These Levant females are potential ancestors for Mladeč 1 under both the replacement and Multiregional hypotheses

degree of occipital bunning within the Neandertal female sample varies considerably. In some, such as Spy 1 and Gibraltar the parietal flattening associated with this morphology extends well onto the occiput. In others such as La Quina 5 this flattening involves only the posterior of the parietal bones; the superior surface of the occiput is evenly curved. Bunning in the earlier European Upper Paleolithic females, including Mladeč 1, conforms to the latter pattern. In contrast Qafzeh 3 lacks bunning of any kind (Fig. 7), while Qafzeh 5, has a very elongated flattened area involving the parietal bones and occiput.

The posterior aspect of the vault is quite projecting. Indeed, this is characteristic of the entire earlier Central European Upper Paleolithic sample, but in this regard the Mladeč 1 specimen is at the top of the range or above, so the posterior projection of this specimen is greater than most (Fig. 8).

Our final observations from the lateral view concern the orientation of several aspects of the cranial base. Posterior to the foramen magnum, the nuchal plane of Mladeč 1 is much more angled than the nuchal planes of the Neandertal females, which range from somewhat less angled to nearly horizontal. In contrast, the Skhul/Qafzeh females appear to have an even higher angulation of the nuchal plane.

Anterior to the foramen magnum, the *pars basilaris* is also more angled superiorly in the anterior direction, compared with the Neandertal females. These data combine to suggest that the foramen magnum itself, and the occiput surrounding it, is in a more inferior position in Mladeč 1, and the vertical separation of the auricular point from the cranial base is also greater. This seems to represent an inferior expansion of the posterior portion of the braincase, and is characteristic of the comparative earlier Central European Upper Paleolithic sample, in some cases to an even greater degree. This observation is compatible with the increased Mladeč distances of the front of the foramen magnum from the biauricular line, described above. This increase is not as much *posterior* as it is *inferior.* These comparisons are quite evident in Fig. 6.

In sum, we have described three changes in Mladeč 1 that together account for the increased flexure of her cranial base when compared with the Neandertal females. These are (1) the more inferior position of the foramen magnum reflecting the expansion of the inferioposterior cranial base, and (2) the more posterior positions of the face, so that the same functional relationships in the submandibular region are maintained. It is in these changes, rather than in the changing flexure of the cranial base resulting from them, that the sources of evolutionary change can be found. However, to fully understand the influence of these changes, the expanded braincase must also be taken into account. Mladeč 1, like the other early Upper Paleolithic females of Europe, has significant brain size expansion compared with the Neandertal males. In this change, the females are not similar to the

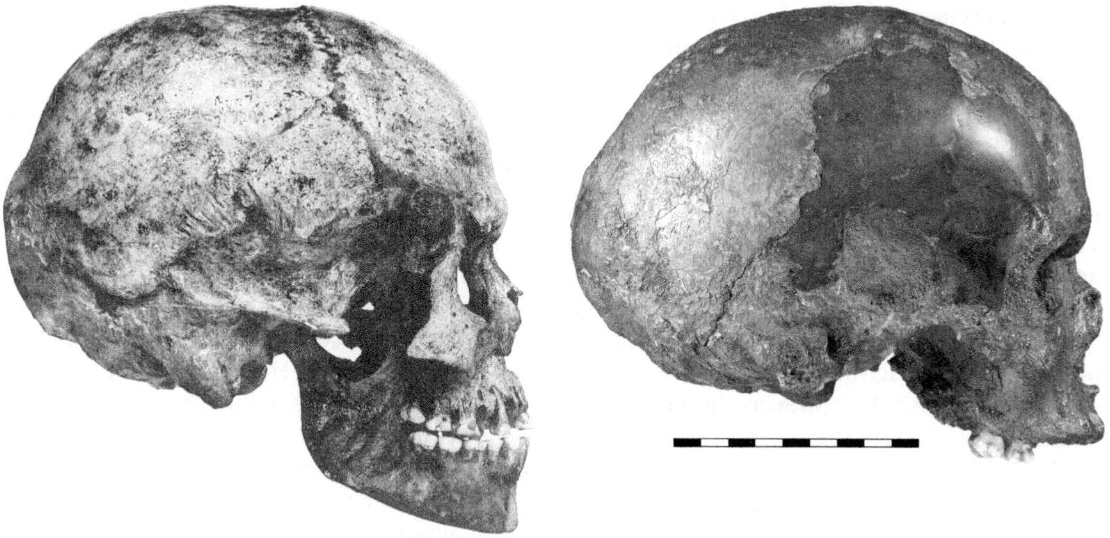

Fig. 8. Mladeč 1 (right) compared with Předmostí 4 (left, after Matiegka, 1934). These early Upper Paleolithic females share many features. In particular, they both have "masculine" details such as supraorbital development. Nasal angulation (including the high position of the base of the nasal angle), the size of facial foramina such as the zygomatic foramen, visible in this view, and the angle and prominence of the lambdoidal flattening are more Neandertal-like in the earlier Mladeč specimen

comparison of males, and the amount of sexual dimorphism in brain size is reduced in Europe (Frayer, 1986).

The increased basal flexure of the earlier Central European Upper Paleolithic females, clearly seen in Mladeč 1, is associated with some reduction in the basal length of the vaults (Weidenreich, 1943; Howell, 1951). Thus, the nasion-inion distance (Table 2) reduces from 183.2 mm in the Neandertals to 179 mm in Mladeč 1, a consequence, we believe, of the vaults' reduced anterior projection. The Skhul/Qafzeh females are even more reduced. However, as discussed in Frayer and colleagues, this volume, the Mladeč males do not show a corresponding reduction in basal length. Since the crania of the earlier Central European Upper Paleolithic males are like the females in that they are more flexed than those of their Neandertal male counterparts, and we assume the Mladeč males are similar to these others, the changes in flexure and in basal length do not appear to be functionally related.

From the vertical aspect, postorbital constriction is slight. The vault bulges strongly behind the anterior border of the temporal fossa, and then continues to broaden slightly to the parietal bosses. The bosses are positioned directly over the mastoids, anterior and superior to the position observed in most of the comparative specimens. Behind the bosses the skull narrows sharply to the sides of the occipital bun, which is thereby pronounced in appearance. This general configuration of the parietal bosses and the area posterior to them is characteristic of some Neandertal females (for instance, La Quina 5). The whole contour is not much like Qafzeh 3, the Skhul/Qafzeh female best preserving these details.

Also from the superior view, there is marked glabellar prominence anterior to the otherwise flat, superior facial region. Glabellar projection does not obscure the even more projecting superior nasal region or the prognathism in the alveolar region. On the right (undistorted) side, the zygomatic arch can be observed. The comparison with Gibraltar is illustrative. Glabellar prominence is similar in this Neandertal female, although the width of the projecting area is narrower. Lateral to it, the superior orbital rim is more strongly angled posterolaterally. The view of the nose is equally projecting in this view, but below it the less prognathic maxilla cannot be seen.

As seen from the rear the parietal walls are vertical, with their greatest breadth at the parietal bosses. Szombathy describes this shape as pentagonal. The contour of the superior surface is slightly domed (Fig. 9). This condition is common for the earlier European Upper Paleolithic females, and not unlike the left contour of the Spy 1 vault, which, however, has lower and more posterior parietal bosses and the parietal wall turns more markedly medially at its base. In the Levant females, the Qafzeh 3 morphology is quite similar to that of Mladeč 1 except for the more posterior position of the boss. The mastoids of Mladeč 1 are vertically aligned, parallel with the parietal sides, differing in this respect from both Neandertal and Skhul / Qafzeh females and some of the other earlier European Upper Paleolithic sample (for instance, Předmostí 4) in which the processes show a definite medial orientation. In other earlier European Upper Paleolithic females such as Předmostí 10 however, the mastoids are more vertical and resemble the orientation of Mladeč 1.

There is a large irregular ossicle at the lambda position (about 14 mm in breadth and 15 mm in height), and a second much smaller one between the large ossicle and the right parietal. Szombathy located lambda within the large ossicle, evidently by extending the sutures to the point where they would have met had there been no ossicle. This ossicle extends along the right portion of the lamb-

Fig. 9. Posterior view of Mladeč 1

doidal suture, almost to the broken area. There are at least four additional very small ossicles along the left portion of the lambdoidal suture.

The basal view of Mladeč 1 (Fig. 2) shows the changes in foramen magnum position discussed above. Besides the features associated with a more posterior foramen magnum position (elongated *pars basilaris*, etc.), a number of other differences may be noted that align the Mladeč specimen with the earlier Central European Upper Paleolithic sample. The orientation of the long axes of the glenoid fossae in the Neandertals is almost exactly transverse to the long axis of the skull, while in Mladeč 1 and other earlier European Upper Paleolithic female specimens the long axes of the fossae have a distinct anteromedial orientation. The temporal fossa is in a more anterior position for all of the earlier Central European Upper Paleolithic females, including Mladeč 1.

The Mladeč 1 temporal fossa length, 44 mm, is almost as large as the 44.5 mm Neandertal mean (Table 9). Breadth of the fossa seems to differ markedly, with the Neandertal fossa much narrower. However, this may be an artifact of small sample size. Area enclosed by the fossa can only be compared for Mladeč 1 and the Gibraltar female. The Neandertal area is only 74% that of the Mladeč female. Perhaps this is attributable to the small sample size, but we suggest it might also indicate that the Neandertal females have smaller temporal musculature (see discussion of most posterior extension of temporal line).

In facial view, the Mladeč 1 face is shorter, flatter, and more strikingly overshadowed by a much higher forehead than any of the Neandertal females (Fig. 10). Some of the most dramatic differences are in the reduced heights of the face (Table 11). In contrast, upper facial breadths (biorbital, bizygomatic, and bijugal) are inconsistently different. Bijugal breadth is above the Neandertal female mean, reflecting the greater facial flatness of Mladeč 1, while the other breadths are reduced compared with the Neandertal females. There are greater differences in facial heights than in breadths, decreases in orbital and nasal heights result in a relatively low nose and low orbits. For instance, the nasal index for Mladeč 1 is much less than the female Neandertal mean (and below the range). Other associated central midfacial reductions include narrowing of the upper maxillary breadth and the bi-infraorbital foramen breadth compared with the Neandertals. However, as noted above the lateral aspects of the midface expand, so that midfacial and bijugal breadths increase. The facial height reductions are most dramatic of these changes, with Mladeč 1 much smaller than and well below the range of the 2 Neandertal females.

Another consequence of these changes is the complete lack of overlap in the orbital indices. Mladeč has significantly lower orbits; the heights are less and the orbital index reflects this. No orbital heights are known for the Skhul/Qafzeh females, but the orbit breadth of Qafzeh 3 is smaller than Mladeč or any Neandertal female. Another difference in orbital shape stems from the orbital margins. The Mladeč 1 orbits are rectangular with orbital margins that are approximately straight and at right angles to each other. Most of the other earlier European Upper Paleolithic females conform to this pattern; the rectangular form of the orbits is expressed equally or even more strongly in some of the other individuals such as Předmostí 10. However, in Mladeč 2 the orbital contours are rounded and thereby more like the Neandertals. In the Neandertal females the four orbital borders are rounded. Moreover, the juncture of the inferior margin with the lateral margins is not angular. Instead both medial and lateral margins slope significantly from the base (inferior margin), to the vertical midpoint of the orbit and this entire surface can be described as a junction. In sum, the Neandertal female orbits are not only absolutely taller, but have an orbital shape that is markedly rounded, when compared with most of the early Upper Paleolithic females from Central Europe.

The last difference in the orbits is observed in their lateral orientation. The outer and inner orbital margins in Mladeč 1 both lie in the same paracoronal plane, and thus the orbital angulation can be described as flat. This configuration is characteristic of the other earlier European Upper Paleolithic females, and to the extend it can be discerned, of Qafzeh 3. In the Neandertal females, the

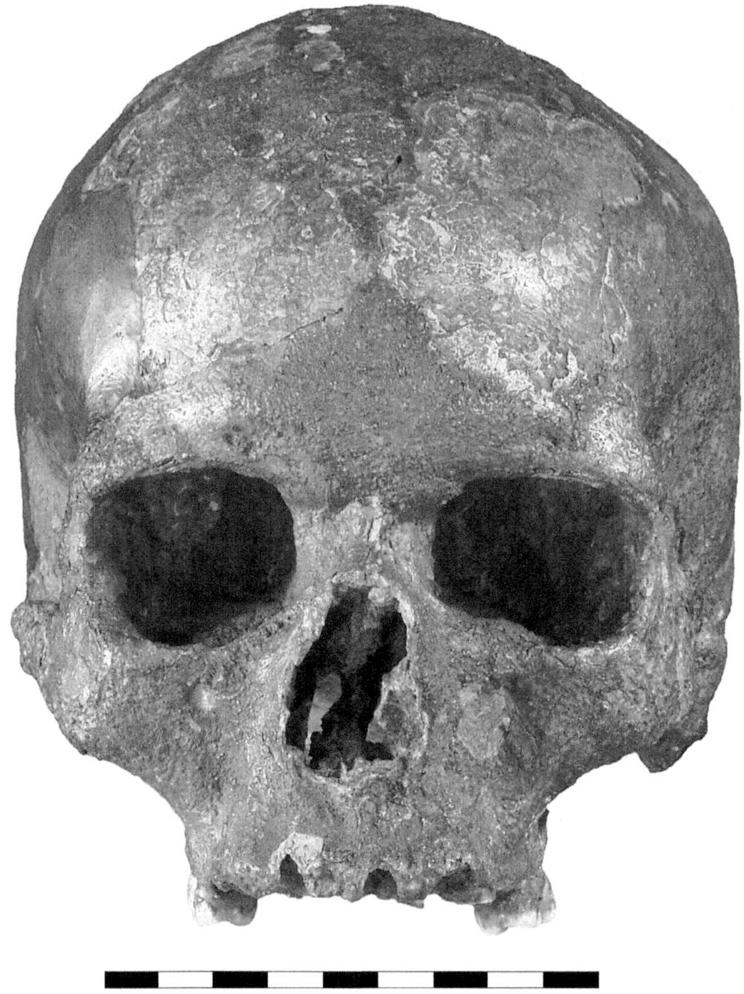

Fig. 10. Mladeč 1 face

orbital pillars are posterior to the nasal borders, and thus the orbits are angled to the paracoronal plane. This distinction reflects the difference in facial orientation, and the fact that the Neandertal females are characterized by both a marked anterior projection of the nose and the maxilla surrounding it, and a posterior displacement and lateral orientation of the orbital pillars.

However, a distinct similarity is found in the development of the frontofacial paranasal sinus system. Europeans, including Neandertals, often have large frontal and large maxillary sinuses, and in this regard differ on average from the normal condition in many other populations (Szilvássy et al., 1987). For instance, Africans have relatively small maxillary sinuses. Mladeč 1 has the European condition of large paranasal sinuses (Fig. 11). Only the frontal sinuses can be observed in the Qafzeh females, these are quite large. The maxillary sinus Szilvássy and colleagues report for a Qafzeh male, cranium 6 are also quite large.

Lower on the face, the Neandertal midface is puffed outwards in the region surrounding the nose, and between the nasal aperture and the orbits, resulting in a weakly expressed canine fossa. Below the nose, however, the anterior of the maxilla is flat and vertical, with no specific alveolar prog-

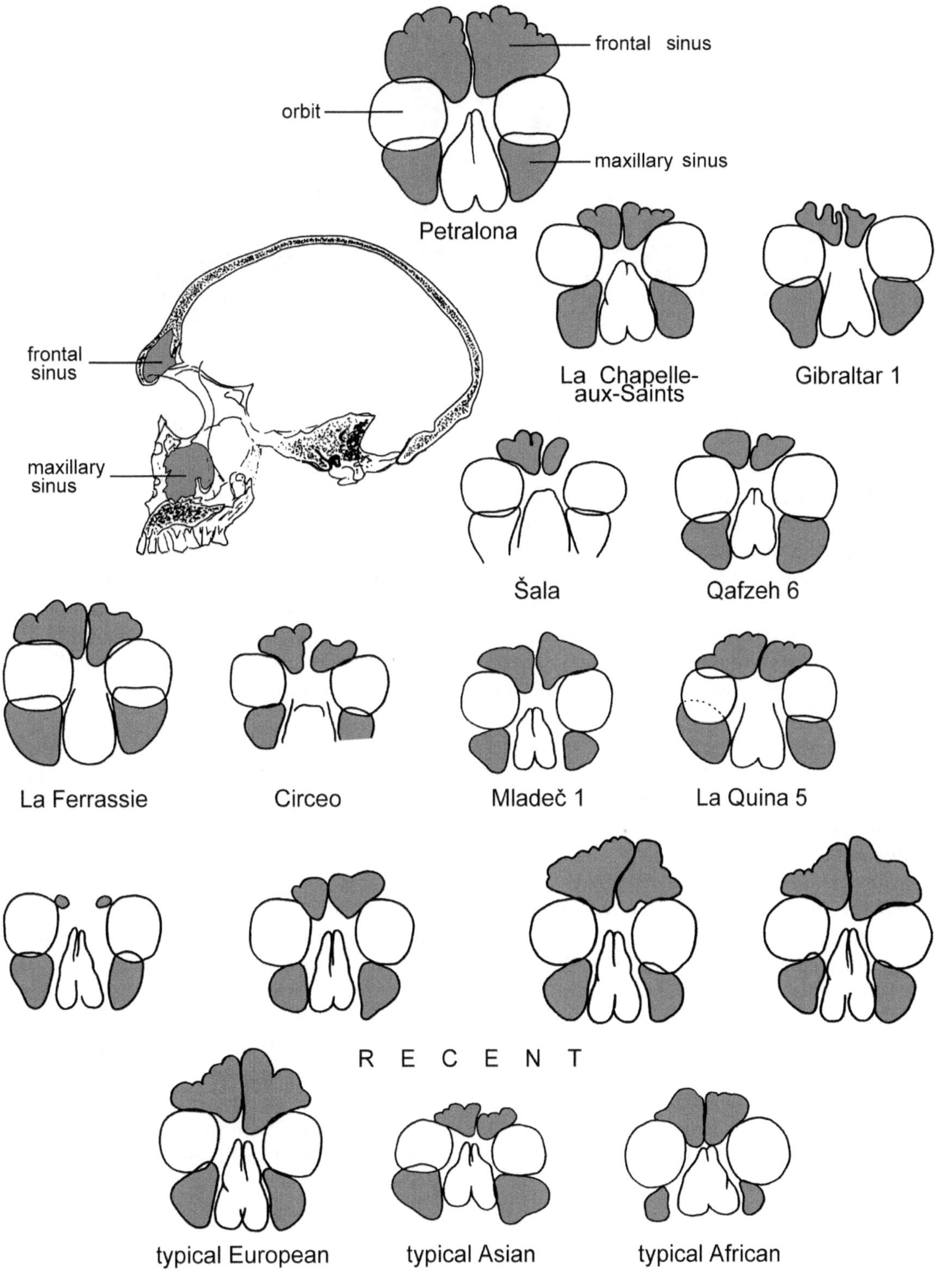

Fig. 11. Geographic distribution of paranasal sinuses (frontal and maxillary) in fossil and recent crania, shown here with a cross-section of La Ferrassie indicating the sinus positions. In the European pattern, these sinuses are about equal in size; frontal sinuses predominate in Africans and maxillary sinuses in Asians according to these authors. From Szilvássy, Kritscher and Vlček (1987), and Heim (1976, Fig. 49)

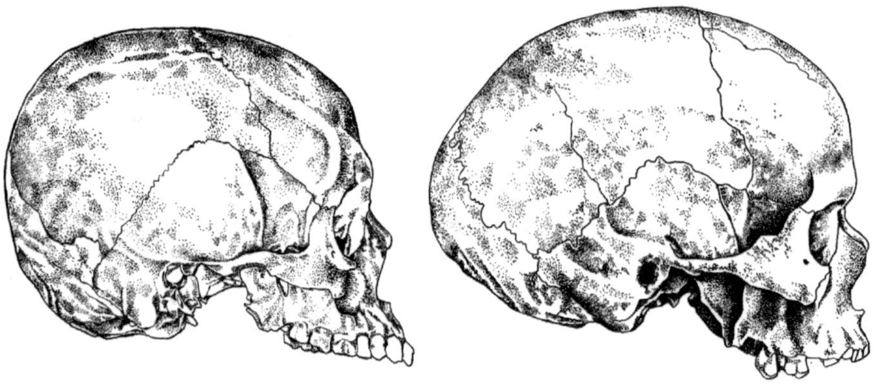

Fig. 12. Comparison of a Masai woman (left) and Mladeč 1, both drawings by Karen Harvey. This comparison contrasts the different morphologies of alveolar prognathism, and fails to reveal any special similarities that might suggest that Mladeč has an especially close relationship with Africans

nathism. The Mladeč 1 condition is quite the opposite. The midface is more deeply excavated lateral to the nose below the orbits, a canine fossa is present, and the inferior maxilla shows distinct alveolar prognathism, although nothing to the extent of prognathism characteristic of Africans (Fig. 12). In the lower border of the zygomatic process of the maxilla, however, the Mladeč 1 specimen is more intermediate. Most of the earlier Central European Upper Paleolithic female faces (Cro-Magnon 2, Fig. 16), Předmostí 10 and 22, for instance) show distinct maxillary notches between the lower border of the process and the external wall of the palate. In the Neandertals, and to a great extent Mladeč 1, the lower border flows evenly onto the wall, producing what Howell (1957) once described as shelving of the malar onto the maxilla.

In sum, we would say that while some similarities exist, the Mladeč 1 cranium is less like the Neandertal females than the Mladeč males are like the Neandertal males. To some extent this is because the Mladeč female preserves the face while the males do not (except for the Mladeč 8 maxilla), but the fact remains that Mladeč 1 evinces only a few features that can be considered transitional.

This contrasts with the comparisons that can be made for the Levant females. These show many fewer specific similarities to the females of the early Upper Paleolithic of Central Europe. Qafzeh 3, for instance, is a rounder and more compact but much more robust vault. It lacks the occipital extension and bunning of Mladeč 1, the nuchal plane is higher and more vertically oriented, and the forehead is higher and more rounded, with a more pronounced frontal boss just above glabella. On the other hand, the supraorbitals are continuous, and much thicker and more projecting on the Qafzeh female (as they are in the other Skhul / Qafzeh females). At nasion, the nasal bones are flat across and extremely broad, in contrast to the narrow and "pinched" appearance of Mladeč 1. The nuchal plane is longer and more robust, and the cranial base is broader. In many respects the Qafzeh female retains many more archaic features. As we concluded with the males (Frayer et al., this volume), we find difficulty in applying the appellation "proto-Cro-Magnoid" to this Levant sample.

Frontal

The Mladeč 1 frontal is large, particularly high, and well rounded. Sagittal lengths (Table 5) are above the Neandertal ranges (by a small amount) even earlier Central European Upper Paleolithic means. Maximum frontal breadth varies similarly, with Mladeč above the Neandertal mean. However, breadths at the anterior of the frontal, the minimum frontal breadth and inner and outer biorbital breadths are smaller than the Neandertal means. The anterior of the Mladeč frontal is relatively narrow, as reflected in the minimal/maximum frontal breadth index. For most of these dimensions and indices the Skhul / Qafzeh sample is intermediate between the Neandertal females and Mladeč 1. The strongest exception is the minimum frontal breadth, which is very large in this sample.

Sagittal rounding of the squama is reflected in the high glabella / bregma arc / chord index, 112.6. This is above the Neandertal range. The squama is also high and evenly rounded transversely, moderately contrasting with the Neandertals in this regard. There is no frontal boss and the frontal lacks a sagittal keel. However, in the area of metopion there is a distinct but very small bump, measuring no more than 30 mm sagittally and 14 mm transversely. There are similar structures in a somewhat higher position on the Předmostí 4 frontal, but none in the other early Central European Upper Paleolithic females.

Squama thickness can be determined at the approximate position of the lateral eminence (Table 6). This is not a particularly diagnostic measurement; the means of the samples do not differ markedly. Mladeč 1 is slightly below the mean values, but within the ranges of all samples.

The temporal line borders the superior surface of the squama where it is preserved on the left. Unlike most early Central European Upper Paleolithic females, the line on Mladeč 1 does not form a distinct ridge. This is almost certainly a consequence of the specimen's young age; for instance, in Předmostí 5 (Fig. 3), the temporal ridge only extends a short distance posterior to the temporal notch. The line is discernable in Mladeč 1, where it is superior to the angulation between the squama's top and its sides. The markedly upward arching of the temporal line at the supraorbital notch is a feature shared with some of the other early Central European Upper Paleolithic females as well as with the Skhul / Qafzeh females. In the Neandertal females, however, the temporal line barely arches, usually traveling in a straight-line posterior to the temporal notch. Spy 1 is most similar to the Mladeč condition. The Mladeč 1 bistephanion breadth of 108 mm, measured between the temporal lines, is considerably less than the maximum frontal breadth, reflecting the high position of the these lines. The distance is in the low portion of the Neandertal range. However, the Neandertal females lack an equivalent angulation of the frontal squama at the temporal line. We note (Frayer and colleagues, this volume) that the same contrasts concerning the temporal line characterize the Mladeč males compared with the Neandertal males, and in a parallel manner, the Spy 2 male comes closest to matching the Mladeč male condition. Unlike Mladeč 5, the anterior face of the temporal fossa in Mladeč 1 is more vertical; a condition shared with the comparative females.

The supraorbital area is separated from the frontal squama by a shallow but distinct supratoral sulcus. This sulcus is positioned superiorly to the superciliary arches, which thereby form more of a bulge on the frontal squama above the orbits than a projecting bar anterior to this squama. The sulcus expression is weaker than in any other early Upper Paleolithic adult female. This could be a consequence of age at death in Mladeč 1. At the extreme, the supratoral sulcus in this early Upper Paleolithic adult female sample can be markedly well expressed (Fig. 8).

The supraorbital region is clearly divided into a superciliary arch and a weakly developed lateral toral structure, which is mainly expressed as a lateral thickening of the superior orbital border. The projection of the superciliary arches is very small. As measured from the anterior face of the frontal squama (following Weidenreich), a figure of 19.0 mm was obtained, but this is misleading since most of the distance spanned is close to parallel with the frontal squama and not orthogonal to it as Weidenreich designed the measurement to express. We do not perceive that the supraorbital structures are particularly projecting, or laterally prominent in this specimen. In contrast, vertical height of the superciliary arch is quite marked. Measured medially and at the highest point on the orbit, the vertical height of the superciliary arch is above both the Neandertal and Skhul / Qafzeh ranges. The ab-

Table 5. Dimensions (mm) and indices of frontal bones in female crania

	Mladeč females 1	Mladeč females 2	Mladeč male 5	Skhul/Qafzeh Mean (n)	Skhul/Qafzeh Range	Neandertal Mean (n)	Neandertal Range
Breadths							
Min Frontal	103.5	101.1	106.2	108.0 (1)[3]		103.8 (5)	101.0–107.0
Max Frontal	126.5	118.0	126.0			117.4 (5)	108.0–127.0
Min/Max Index	81.8	85.7	84.3			88.6 (5)	82.8–93.5
fmt-fmt	109.7	113.4	120.9			114.4 (5)	106.5–121.8
fmo-fmo	103.0	107.0	109.0	104.0 (1)[4]		106.2 (6)	97.5–112.0
st-st	108.0	102.0	113.6			113.6 (5)	104.4–112.8
Bi-st/Max Br index	85.4	86.4	90.2			96.8 (5)	96.5–97.3
Orbit Angle Br	30.0	33.5	39.0	30.7 (2)	30.3–31.0	32.2 (4)	29.0–35.0
mf-mf	25.5	29.6	25.2			24.1 (4)	21.0–26.9
fmt-fmo	8.5	6.0	9.9	12.2 (2)	11.5–12.9	9.8 (11)	8.5–11.9
Lengths							
na-br	114.4	103.0	116.0	109.0 (1)[3]		104.5 (5)	100.6–111.7
gl-br	110.1	100.0	113.6			102.5 (4)	97.3–109.6
Arcs							
na-br	133.0	121.0	140.0	123.0 (1)[3]		116.1 (4)	111.0–121.0
gl-br	124.0	112.0	127.0			107.8 (4)	99.5–117.5
Arc / Chord Indices							
na-br	116.5	117.5		112.8 (1)[3]		110.1 (4)	107.9–112.1
gl-br	112.6	112.0	111.8			105.2 (4)	102.3–107.2
Supraorbitals							
na-fmt	57.7	60.0	56.0	62.0 (1)[3]		63.8 (6)	60.3–66.4
na-fmo	52.5	56.9	63.9	55.2 (2)	53.5–56.9	57.6 (6)	55.0–61.0
Medial projection		11.2	17.5	15.7 (2)	14.3–17.0	19.4 (5)	17.0–21.0
Orbit center projection[1]		7.0	14.5	17.8 (2)	16.5–19.0	19.1 (11)	11.1–27.5
Lateral projection		17.7	23.0	18.0 (1)[3]		20.9 (4)	17.5–24.0
Length from sulcus	19.0		19.0	18.2 (2)	18.0–18.4	20.6 (5)	16.9–24.5
Medial height[2]	21.0	20.5	20.5	17.5 (2)	16.1–18.9	15.8 (6)	13.4–18.0
High pt height[2]	16.8	11.6	18.0	10.8 (2)	9.2–12.3	10.5 (10)	6.5–14.1
Orbit center height1[1, 2]	6.1	4.0	16.2	11.0 (3)	7.1–14.5	10.0 (12)	4.3–14.9
Lateral height[2]	6.7	4.9	8.5	10.8 (2)	10.5–11.1	9.6 (10	6.8–12.3
Central/medial ht Index	29.0	19.5	79.0	78.7 (2)	76.7–80.7	67.5 (6)	32.1–89.8
Lateral/medial ht Index	32.1	23.9	41.0	62.0 (2)	58.7–65.2	59.4 (6)	49.4–78.3

[1] The orbit center is not the same as the midorbit position in Wolpoff et al. (1981), Smith and Ranyard (1980). Measured at the center of the orbit, it is medial to the position where midorbit measurements are taken in these publications.

[2] This is "thickness" in Wolpoff et al. (1981), Smith and Ranyard (1980)

[3] Qafzeh 3

[4] Skhul 2

Table 6. Cranial thickness (mm) in females

	Mladeč females 1	Mladeč females 2	Mladeč male 5	Skhul/Qafzeh		Neandertal	
				Mean (n)	Range	Mean (n)	Range
Frontal							
Lateral eminence	6.0	4.0	8.5	6.4 (4)	5.0–7.5	6.5 (7)	3.8–9.5
bregma		5.5		9.7 (1)[1]		6.2 (4)	5.5–7.2
Parietal							
bregma		4.0		6.0 (1)[2]		5.9 (3)	5.1–7.0
vertex		3.0		9.2 (2)	8.5–9.8	6.0 (2)	4.5–7.0
lambda		5.5	6.0	10.7 (3)	8.7–13.6	9.5 (1)[4]	
asterion	6.0	5.8	8.5	10.1 (3)	7.1–15.1	6.3 (3)	5.5–8.0
Mastoid notch	5.5	4.0	7.0	8.2 (1)[3]		5.5 (3)	5.0–6.0
Anterior eminence	3.0	6.3	6.5	8.8 (3)	7.7–9.5	7.3 (4)	4.5–11.0
Middle Eminence	4.0	4.7	6.0	10.3 (3)	7.8–12.3	8.2 (6)	5.3–11.3
Posterior eminence	6.5	5.5	6.5	8.7 (4)	7.0–9.5	7.2 (3)	5.0–10.0
Occipital							
lambda	5.0		7.5	10.4 (3)	9.2–12.0	7.6 (4)	5.5–11.0
inion	13.0		11.0	13.8 (2)	13.5–14.0	10.8 (4)	6.0–14.0

[1] Qafzeh 3
[2] Qafzeh 5
[3] Skhul 2
[4] Spy 1

sence of a lateral supratoral structure is probably age related, torus is found in both center-orbital and lateral positions. In some cases the expression of this feature is marked (for instance, Zlatý Kůň).

At the most lateral aspects of the superciliary arch, there is a well-developed supraorbital groove, separating the medial and lateral portions of the brows. At the most medial extent of this groove, distinct supraorbital foramina occur bilaterally. Both left and right foramina are positioned within the groove and well above the superior orbital margin (5.5 mm). Both are separated from the superior orbital margin by a distinct bar of bone. There is no frontal foramen. Zlatý Kůň exhibits a similar morphology in this region, although the supraorbital foramen is not as highly placed on the margin and the bar of bone separating the foramen from the superior orbital margin is much thinner. In fact, in general Mladeč 1 most closely resembles Zlatý Kůň, and we believe the similarity would have even been greater had the Mladeč female died at a somewhat older age.

The Mladeč 1 supraorbital region is modern in the European context, although among living European populations it is unlikely that any female could be found that comes close to matching it in robustness. Comparisons with the Neandertal condition are quite different. The Neandertal females generally have a shallower supratoral sulcus than the males (especially Gibraltar and La Quina 5). The region differs grossly from the modern European condition in that the frontal slope is lower and the supraorbital torus forms a continuous bar of bone (of greatly varying thickness) over the orbits. Torus heights at both the middle and the lateral aspects of the orbits are a significant proportion of the medial supraorbital height (see the supraorbital height indices). In the Mladeč specimen these heights are a much smaller proportion of the medial height. A higher frontal slope for the Neandertal females would make the supratoral sulcus much more similar to the Mladeč 1 condition, but the contrast of browridge with superciliary arch would still remain.

With regard to the Levant sample, the contrasts are somewhat different, and these females are even *less* like Mladeč than the Neandertals are. The Skhul/Qafzeh hominids generally have a continuous supraorbital torus and not superciliary arches. The supraorbital height indices show that both center-orbital and lateral heights are a very large proportion of the medial height – possibly the largest contrast. In the Skhul / Qafzeh females, the supraorbitals are positioned anterior to the frontal squama and are set off from it by supratoral sulci that are even more deeply excavated than is generally true for the European Neandertal females. Indeed within the Levant, high frequencies for a deeply excavated supratoral sulcus may be a distinctive feature. Levant Neandertal females such as Tabun and Zuttiyeh also have very deeply excavated supratoral sulci.

The morphological pattern and the size of the Skhul / Qafzeh supraorbitals vary somewhat. For instance, in Skhul 2 it is a thick, evenly developed bar of bone, thinning only at the very lateral edges. In Qafzeh 3 the torus is also thick and continuous, but it thins in the center-orbital region, lateral to the position of its large supraorbital notch. In Qafzeh 5 the torus is considerably thinner, at least in a vertical direction. However, none of these particularly resemble Mladeč 1. There is also marked variation in the slope of the frontal in this sample. Most of the frontal bones appear to be lower than Mladeč 1. The exception is Qafzeh 3, a frontal that easily matches this European female in both frontal slope and curvature. The fact is, however, that none of these females especially resemble the Mladeč 1 condition in the combined details of the supraorbital region.

At the anterior of the frontal, upper facial flatness, as measured by the nasion projection anterior to the bi-*fmt* line, is small (Table 11). The Mladeč 1upper face is considerably flatter than the Neandertals. However, the Levant females are even flatter than any of the Europeans. This is evident from an inspection of Qafzeh 3. In terms of upper facial flatness, the Skhul / Qafzeh hominids are extremely flat while the Neandertal faces are peaked along the midline accentuating the projection of the noses. Mladeč 1 is intermediate between these, although on the whole sharing the somewhat peaked upper facial morphology with the Neandertal women.

Orbital depth in Mladeč 1, 56.4 mm, is quite reduced compared with the Gibraltar female value of 64.1 mm. This is the only comparison that we can make with another female specimen.

The distance between the orbits, measured by the anterior interorbital breadth, barely differs between the samples. As measured between the interior corners of the orbits (Table 5), the distance is less than the Neandertals. For the Skhul / Qafzeh sample, while no distance between the orbits can be measured, the interorbital breadth was likely much greater in Qafzeh 3.

The roof of the Mladeč 1 orbit increases in height behind the inferior margin of the superciliary arch (both sides). It reaches its maximum height in an even arch some 15 mm posterior to this inferior margin. The morphology contrasts with the region in the Neandertal females, where the orbital plate is flat, and rises only slightly behind the inferior margin of the supraorbitals. The Skhul / Qafzeh females resemble the Neandertal sample in this feature.

Overall, then, the frontal of the Mladeč 1 female is fundamentally unlike the frontal bones of both the European Neandertal females and the Levant females from Skhul and Qafzeh, and it is unclear which of these would make a better morphological precursor since they also differ substantially from each other. Moreover, the frontal morphology is quite distinct from the male condition as represented by Mladeč crania 4, 5, and 6. We are hesitant to apply a cladistic approach and determine character states for a feature that is seen to vary between the males and females of the same sample. The more traditional approach of comparative anatomy leads to the conclusion that the Mladeč 1 morphology is distinctly European, and can be matched in any of a number of other early Upper Paleolithic crania from the region although it is far more robust than modern European females.

Parietal bones

Our description will be based on the left side because of the reconstruction involving a good portion of the right parietal. Comparing chords for the four borders of the parietal Mladeč 1 and the Neandertal females (Table 7), the Mladeč 1 parietal is expanded and more rectangular in shape. The ra-

Table 7. Parietal dimensions (mm) and indices

	Mladeč females 1	Mladeč females 2	Mladeč male 5	Skhul/Qafzeh Mean (n)	Skhul/Qafzeh Range	Neandertal Mean (n)	Neandertal Range
Chords							
br-l	117.0	112.5	119.1	117.2 (1)[1]		107.8 (4)	101.0–117.6
l-ast	90.5	80.0	94.8	87.3 (2)	87.0–87.6	85.6 (4)	80.0–89.8
k-ast	91.3	81.0	87.5			81.0 (3)	75.4–87.3
k-br	101.5	97.0	106.3			92.1 (3)	86.7–96.5
k-l	135.2	131.0	139.0			128.8 (3)	125.2–133.0
br-ast	147.0	133.3	148.0	144.3 (1)[1]		133.4 (4)	126.0–139.5
br-st	76.3	54.0	64.4	51.3		70.5 (5)	63.3–76.0
Parietal radius	71.2	66.2	72.6			64.8 (3)	64.6–65.1
Arcs							
br-l	126.0	129.0	129.0	121.0 (1)[1]		115.3 (4)	108.0–126.0
k-br	123.0		121.0			109.2 (3)	103.5–116.0
l-ast	103.0	90.0	103.0	95.0 (1)[2]		95.1 (4)	90.0–103.0
k-l	168.0	162.0	162.5			167.0 (3)	161.0–178.0
br-ast	180.0	163.0	179.0			163.3 (4)	154.0–172.0
br-st	77.0	61.0	66.0			76.2 (5)	68.0–82.0
Parietal angles (degrees)							
br-l	110.4	116.4	110.2			107.7 (3)	101.8–115.1
br-k	90.9	94.3	94.1			90.7 (3)	84.3–96.6
ast-k	79.7	75.5	74.1			77.5 (3)	71.4–84.3
l-ast	78.9	73.8	84.5			84.1 (3)	82.7–85.6
Arc / Chord indices							
br-l	107.7	114.7	108.3			106.9 (4)	104.4–108.9
k-br	121.2		113.8			118.0 (3)	116.1–120.2
l-ast	113.7	112.5	108.6			112.3 (5)	106.5–117.3
br-ast	112.4	112.3	120.9			122.5 (4)	118.6–124.8
k-l	124.3	123.7	116.9			129.7 (3)	126.6–133.8

[1] Qafzeh 5
[2] Qafzeh 3

dius of the circumscribed circle also shows the expansion of the Mladeč 1 parietal size. The parietal angles measure less difference in shape. We conclude that the Mladeč parietal is larger and slightly more rectangular, but fundamentally close to the same shape as parietal bones of the Neandertal sample. In terms of curvature, as measured by the arc/chord index of each parietal border, Mladeč 1 has essentially equally curved anterior and posterior borders. For the superior border, Mladeč 1 is the same as the Neandertal mean. The arc/chord indices for the diagonal measures show Mladeč 1 to be less curved than the Neandertal females.

The temporal lines are high on the parietal bones, considering the sex and age of the specimen. Because of the condition of the bone surface, we cannot identify an inferior line. The superior line follows a markedly medial arch posterior to the coronal suture, passing through the apex of the parietal boss. Posterior to the boss, the inferior line arcs strongly downwards and anteriorly, forming the supramastoid crest on the temporal squama. Behind the boss, the superior line arcs posteriorly to meet and parallel the middle and inferior portion of the lambdoidal suture, and then anteriorly to form the mastoid crest. In none of the Neandertal females does the superior temporal line extend as posteriorly as the lambdoidal suture, although the Spy 1 line almost reaches the suture. However, in this Neandertal the superior temporal line only reaches the lambdoidal suture at its very most inferior point, and does not bring a large portion of the nuchal musculature as far backwards as is the case in the Mladeč female. This is the same contrast that is found in the Mladeč males. As we noted above, the cross-sectional area enclosed by the temporal fossa may be smaller in the Neandertal sample, perhaps accounting in part for this difference in the most posterior extent of the temporal line.

The parietal thicknesses vary considerably from the front to the back of the bone (Table 6). However, they all are small, below the ranges of the Skhul/Qafzeh and Neandertal samples. In all of these thicknesses, and in the thicknesses at the posterior of the bone (asterionic, mastoid notch), Mladeč 1 is reduced relative to the Neandertals, while the Skhul/Qafzeh thicknesses are *greater* than the Neandertals.

Occipital

The Mladeč 1 occiput is almost complete, and undistorted. The bone is quite large, its height as measured from lambda to opisthion is above ranges of all comparative samples (Table 8). Biasterionic breadth, however, is comparatively smaller, below the means of the Neandertal and the (virtually identical) Skhul/Qafzeh samples, and below the Skhul/Qafzeh range. The bone is also reduced in squama thickness (Table 6). For instance at lambda, the thickness value is below the ranges for the comparative samples.

As is the case for the males, the nuchal muscle attachment area takes up a large proportion of the occipital breadth, the ratio distinguishing Mladeč 1 from both the Neandertal and the Skhul/Qafzeh females (Table 8). Comparisons of the absolute values of the nuchal plane breadth are not as distinct, mainly because the more archaic samples have generally broader occipitals and this contributes to their nuchal plane breadth values.

Both along the sagittal plane and the borders, arcs measured on the occipital squama are very large relative to the other European samples. The Levant female Qafzeh 3, however, is even more curved. In part the large arcs are a consequence of bone size, but the arc/chord ratio is also relatively great.

In some respects the occipital of Mladeč 1 is the most Neandertal-like portion of the vault and markedly contrasts with the rounded occipital of Qafzeh 3. Besides the rounded parietal borders, the sagittal contours are rounded because Qafzeh 3 combines very flat nuchal and occipital planes with a marked occipital angulation as seen in the sagittal plane (Table 8). The Levant female is more different from Mladeč 1 than any of the Neandertals in:

- absence of an occipital bun,
- expression of a marked centrally located nuchal torus,
- more vertical orientation of the nuchal plane,
- equality of the nuchal plane's sagittal length to the length of the superior occipital plane.

The occipital bun of Mladeč 1 is better expressed than in Saccopastore 1 (Fig. 6), although not as projecting as in La Quina 5. The rounding for the superior portion of the bun begins at the approximate position of the lambdoidal suture (as in La Quina 5), and the rounding extends evenly down to the position of the external inion. In Neandertals, the inferior border of the bun is almost invariably bounded by a nuchal line, below which the nuchal plane is strongly concave, setting off the bun. In Mladeč 1, the posterior of the bun is evenly rounded, its vertical face is shorter, and at and below the nuchal line, the nuchal plane is convex.

Laterally, a fairly flat surface extending from about 30 mm from the midline to the mastoid region sets off the Mladeč 1 bun. Again, this resembles Neandertals such as La Quina 5 and Spy 1, al-

Table 8. Occipital bone and nuchal region dimensions (mm) and indices

	Mladeč females		Mladeč male	Skhul/Qafzeh		Neandertal	
	1	2	5				
				Mean (n)	Range	Mean (n)	Range
Breadths							
Biasterionic	110.4	120.0	126.5	119.5 (2)	119.0–120.0	115.1 (5)	108.5–122.0
Nuchal attach area	106.5	132.0	143.0	101.0 (1)[1]		93.3 (4)	84.0–105.3
Lengths							
l-ast	90.5	94.8	84.0	87.3 (2)	87.0–87.5	85.6 (4)	80.0–89.8
l-o	106.0			92.9 (1)[1]		88.9 (3)	85.0–92.9
l-i	75.1	66.9	62.7	59.6 (2)	53.0–66.1	57.2 (5)	54.5–60.5
i-o	50.0			56.5 (2)	53.0–60.0	45.6 (3)	42.1–47.6
i-ast	64.0	68.7	72.7	68.5 (2)	62.1–69.5	69.3 (5)	64.0–74.0
Inion proj (asterion)	32.4	33.5	35.8			37.3 (5)	25.7–47.0
Arcs							
l-ast	103.0	103.0	92.0	95.0 (1)[1]		95.1 (4)	90.0–103.0
l-o	134.0			108.0 (1)[1]		110.2 (3)	109.0–112.5
l-i	86.0	73.5	69.0	67.5 (2)	66.0–69.0	63.0 (5)	60.0–69.0
i-o	48.0			58.0 (2)	55.0–61.0	46.8 (3)	43.5–49.0
Condyle							
Length	26.0					18.8 (1)[8]	
Breadth	11.0					13.7 (1)[8]	
Foramen Magnum							
Length	38.1					31.5 (2)	30.0–33.0
Breadth	30.7					29.3 (1)[8]	
Indices							
Nuchal/Bi-ast br	96.5	110.1	113.0	84.9 (1)[1]		82.3 (4)	73.1–91.1
l-i/i-o	150.2			105.1 (2)	100.0–110.1	125.8 (3)	116.0–144.0
l-i arc / chord	114.5	109.9	110.0	114.5 (2)	104.4–124.5	110.1 (5)	107.1–114.0
Foramen Magnum L/Br	124.1					112.6 (1)[8]	

[1] Qafzeh 3
[8] Saccopastore 1 only

though not Saccopastore 1. This area is similar to the preserved portions of Cro-Magnon 2 (the occipital plane and the superior aspect of the nuchal plane), although the Cro-Magnon occiput is less projecting (Fig. 16), the flattened area is more perpendicular, and the nuchal plane just below the nuchal line is more concave. The region in Mladeč 1 is quite different from the Předmostí 4 female, in which the occipital plane is much more vertical (i.e., less posteriorly projecting) below lambda and cannot be said to form a bun (Fig. 8). As in the Cro-Magnon female, the nuchal plane is concave beneath the nuchal line in this Předmostí female.

The superior nuchal line in Mladeč 1 seems very weak. It travels in a broad curve across the back of the vault, with a slight ridge forming a downward angulation at the midline producing a small elevated triangle at inion. The base of this elevated triangle is level with the most superior extent of the superior nuchal line. Laterally the superior nuchal line does not reach the mastoid, but rather forms a crest along the paramastoid process. Just superior to this ridge there is a shallow concave patch about 21 mm in breadth and 8 mm in height. This might be perceived as a small suprainiac fossa. The lower border of this fossa is a distinct ridge, paralleling a superior line above it for some

5.5 mm. We interpret this superior line as a supreme nuchal line. Whether because of preservation, obscuring matrix, or the actual morphology, we cannot trace the line beyond this position.

This region differs from both the Neandertal and the Skhul/Qafzeh females. The normal Neandertal condition is for the suprainiac fossa to border a nuchal line or weak ridge that dips slightly inferiorly, but does not form a triangular eminence. In Qafzeh 3, the preserved details of the nuchal line and central torus resemble the Neandertals. Just superior to the center of the nuchal torus, where it crosses the sagittal suture, the occipital plane is flat and vertical for a rectangular region about 20 mm in height and 30 mm in breadth. Just superior to this, some 34 mm above inion (= opistocranion) and only just below lambda, there is a shallow concave region that is literally a suprainiac fossa but not obviously homologous to the Neandertal form, because of its position. Some of the other earlier European Upper Paleolithic specimens more closely resemble Mladeč 1. For instance, the Předmostí 4, Dolní Věstonice 3, and Cioclovina females also have a triangular eminence, even more projecting than that of Mladeč 1, but unlike it not bounded superiorly by a shallow patch. Předmostí 10 does not appear to have a triangular eminence and is more like the Neandertals in this region.

The vault wall is rather thick at inion (Table 6) as compared with the Neandertal and Skhul/Qafzeh females. The inion prominence is in a rather low position. In both its absolute distance from asterion (Table 8) and its posterior extension behind the biasterionic line is below the Neandertal mean value (although above the Qafzeh 3 value). Along the sagittal plane, the nuchal plane is moderate in length. For instance, the ratio of occipital to nuchal muscle attachment area lengths is about 150%, above the means of the comparative samples and above the Neandertal range. It is dramatically greater than the Qafzeh 3 value (94.0%), because as noted above the nuchal plane is longer than the occipital plane.

As noted above, the nuchal muscle attachment area is relatively and absolutely broad, maximum breadth occurring near the position of asterion. The nuchal region is roughly triangular in form, corresponding to the narrowing at the base of the occipital bun as seen in inferior view. The surface is moderately rugose and convex sagittally and transversely. The inferior nuchal line, about half way between inion and opisthion, is distinct. Between the inferior line and inion what remains of the external occipital crest is sharp (distinguishing it from the low and broad development of the crest in those Neandertals that have it at all). Matrix obscures the nuchal surface between the inferior nuchal line and inion, but the rounding characteristic of the nuchal plane posterior to it continues evenly to the condylar fossae. The rim of the foramen magnum is slightly elevated, and there are true postcondyloid tuberosities on both sides. Lateral to the rim, broad and low occipitomastoid and paramastoid crests are about 8 mm apart. The occipitomastoid sutures ridge the paramastoid crests on both sides. Similarly extensive paramastoid crest development is not unknown in the European Upper Paleolithic. For instance, the Cro-Magnon 2 female also has a paramastoid crest, at least as large as that of the Mladeč female. These crests are positioned posterior to the condylar rear, rather than lateral to the condyles as they are in the Neandertals.

Compared with the Neandertal females, the shape and convexity of the nuchal plane is much like La Quina 5, although dissimilar to Saccopastore 1. The development and configuration of the occipitomastoid and paramastoid crests closely resemble some of the Neandertal females, such as Saccopastore 1, but differ from the very marked expression of the paramastoid crest in Spy 1 and Gibraltar (neither of these preserves the region of the occipitomastoid crest).

Finally, part of the region is preserved in Qafzeh 3. The paramastoid crest is broad and low, similar to the expression of the crest in Mladeč 1 and Saccopastore 1.

The Mladeč 1 foramen magnum is longer than any in the two Neandertals, but there is no significant breadth difference. The shape index measured by the ratio of breadth to length therefore reflects the greater relative length. The occipital condyles are also absolutely and relatively long compared with the Neandertals. The surface of the condyle is not doubled, instead forming a continuous face from front to back. Both condyles have deep condylar fossa. Although as discussed above the paramastoid crest is distinctly developed, lateral to the left condyle we find no evidence of a paramastoid process. On the right, matrix covers this region, but the area bulges inferiorly. It is impossible to determine if this is due to a paramastoid process underlying the matrix or simply matrix thickness.

Sagittal measures reflecting the length of occipital base show that the bone has a marked anteroposterior dimension. The distance from the front of the condyle to inion (Table 4) is larger than the Neandertal mean. Anteriorly, the length of the bone is also expanded. As discussed in the section concerned with the position of basion, the basion-hormion distance is long (the *pars basilaris* cannot be measured directly), reflecting the posterioinferior positioning of the foramen magnum compared with Neandertals.

Temporal bones

The temporal squama is tall and long relative to the comparative samples (Table 9). Both dimensions greatly exceed the Neandertal ranges. The posterior border is virtually a straight line from the most superior point on the squama to the parietal notch; in this respect it is like Předmostí females 5 and 10 and also similar to the Neandertal females. The posterior corner of the squama lacks a marked development of the supramastoid and mastoid crests, as it is preserved. Our observations in this region are rendered difficult by a combination of cortical surface loss and adhering matrix. We can delineate the shallow groove separating these crests (about 12 mm in breadth). On the external face of the mastoid process, the mastoid crest can be delineated on the left side, following a course inferiorly along the center of the process to its tip. Since the superior nuchal line is continuous with the paramastoid crest and does not extend onto the mastoid, there is no compound crest formed on the mastoid process as in Mladeč 5.

At the base of the squama, the digastric groove is deep and moderately broad. The mastoid process projects well below the Frankfort Horizontal. Metrically, it exceeds the Neandertal range (projecting more strongly). However, the process does not project dramatically below the digastric sulcus because the cranial base bulges inferiorly just medial to the mastoid. With this reduced expression, the mastoid projection below the sulcus is within the Neandertal range. The basal dimensions of the process are also not particularly large. In fact, basal length is about the same as the Neandertal mean.

Anterior to the mastoid process, the external auditory meatus (much better cleaned on the right side) is elliptical in shape, with a long axis vertical to the Frankfort Horizontal. The axis orientation is similar to Spy 1, differing from La Quina 5 in which the elliptical axis leans posteriorly and Gibraltar in which the long axis is horizontal. Dimensions of the meatus are approximately 12 mm by 8 mm.

Posterior to the petrosal crest, the tympanic ring surrounding the external auditory meatus is continuous with the anterior face of the mastoid, so that this anterior face forms the posterior border of the meatus. This contrasts with the usual Neandertal condition in which there is a posterior aspect to the tympanic ring resting upon the anterior face of the mastoid and often separated from it by a narrow groove. Předmostí 4 and Cro-Magnon 2 conform to the Mladeč 1 condition, but other earlier European Upper Paleolithic females (for instance Předmostí 22) more closely resemble the Neandertals.

On the anterior face of the meatus, a weakly developed post-glenoid process is contiguous with the meatus wall. Its height is below the female Neandertal mean while its breadth is larger. The process is barely distinct, and can only be clearly differentiated from the meatus wall at its most inferior point.

The glenoid fossa (described on the left because of matrix in the right fossa) is like the Neandertal females in the slope of the fossa's anterior face and the lack of a horizontally oriented surface on

Table 9. Temporal and mastoid process dimensions (mm) in female crania

	Mladeč females 1	Mladeč females 2	Mladeč male 5	Skhul/Qafzeh	Neandertal	
				Mean (n)	Mean (n)	Range
Squama						
Maximum length	72.6	68.5	73.9		58.2 (5)	57.6–59.3
Height above FH	44.0	33.0			36.0 (5)	32.7–40.0
Supraglenoid gutter length	26.6	26.0	28.9	23.2 (1)[2]	24.8 (6)	21.5–29.7
Supraglenoid gutter breadth	15.1	19.1			17.1 (3)	15.6–19.5
Temporal Fossa						
Length	44.0	41.3			44.5 (3)	43.5–46.5
Ant mastoid to ant glenoid	54.9				49.5 (1)[4]	
Breadth	27.3				20.4 (1)[4]	
Postglenoid process						
Height	5.9	6.0	6.1	2.8 (1)[2]	6.3 (6)	1.0–8.4
Base thickness	5.3	6.2	8.6	5.8 (1)[2]	4.1 (7)	1.3–6.3
Mastoid process						
au-ms	37.0	29.7	34.7		31.3 (6)	29.0–33.1
Auricular-mastoid ht	37.6	30.9	32.1		31.1 (6)	27.0–34.1
Digastric height	7.1	3.2	8.5	6.9 (1)[3]	5.0 (6)	4.2–7.4
Basal length[5]	35.0	32.9	43.8		34.7 (7)	30.2–40.5
Glenoid fossa						
Fossa length	18.5	18.9	25.4	22.0 (1)[2]	19.3 (7)	16.0–22.8
Articular length	11.2	11.0	15.6	13.0 (1)[2]	10.8 (7)	8.5–12.8
Fossa breadth	29.8	25.0	20.9		23.6 (5)	21.4–25.7
Fossa depth	6.8	5.0	6.4	2.4 (1)[2]	5.0 (7)	3.0–7.5

[2] Skhul 7
[3] Qafzeh 3
[4] Gibraltar 1

the articular eminence. Most particularly Saccopastore 1, Gibraltar, and La Quina 5 also lack a horizontally oriented surface on the articular eminence. In these Neandertal specimens, however, the fossa is shallower and the anterior face is more sloping. The Spy 1 fossa has a short horizontal surface on its articular eminence, and the slope of the fossa's anterior face is about the same as in Mladeč 1. The entire articular eminence can be said to form the anterior face of the fossa in Mladeč 1 and in three out of the four female Neandertals.

On the other hand, the Mladeč 1 glenoid region differs from the Neandertals in a number of ways. One of these, discussed above, is the orientation of the long axis of the articular surface. In the Neandertal females this is close to being transverse (paracoronal), while in Mladeč 1 there is a marked anteromedial orientation. Another difference is found on the medial wall of the fossa of the Neandertal females, where a fissure, which is the continuation of the fissure separating the post-glenoid process from the tympanic bone, divides the infratubercle process from the *spina angularis*. The fissure is not found in Mladeč 1. The Mladeč 1 glenoid fossa is the broader and deeper than any Neandertal specimen. The single Skhul/Qafzeh specimen, Skhul 7, is similar in length, but much shallower.

Much of the detail on the cranial base is obscured by matrix. In so far as we can determine the angulation of the petrous portion of the temporal and the position of the carotid and jugular foramina do not appear unusual (or different from the comparative specimens). In its remaining details, the cranial base appears to be essentially modern.

Sphenoid

The sphenoid is almost completely preserved, but unfortunately much of its external surface is obscured by matrix. One of the more diagnostic features on the cranial base is the position of the foramen ovale. The foramen ovale of Mladeč 1 is positioned lateral and posterior to the base of the lateral pterygoid plate, a position common in most modern crania. This contrasts with Neandertal females (best seen in La Quina 5) where the foramen is more directly behind the lateral pterygoid plate base and actually posterior to the anterior border of the glenoid fossa (a consequence of the more anterior foramen magnum and associated basal structures of the Neandertals discussed above). In addition, there is no pterygoid-spinous ossification, judging from the intact right pterygoid plate and the foramen ovale is not divided. Matrix obscures other nonmetric observations of the sphenoid's base.

Only the anterior portion of the lesser wing, holding the pterygoid fossa where the plates merge, is preserved. The remaining portions of lamina are encrusted with matrix, much more so on the left than on the right. Fourteen millimeters of lamina remains on the right lateral plate, and somewhat less remains on the medial plate. In height (palatal-basal) this pterygoid process is almost identical to Gibraltar. However, the angulation of the process is at about 45° to the Frankfort Horizontal in Mladeč 1, while in both Gibraltar and Saccopastore 1 the angle is less and thus the pterygoid process is more horizontally inclined.

Lateral to the pterygoid plates, the base of the greater wing of the sphenoid forms a broad, flat surface, bounded by the internal border of the temporal fossa. The width of this surface is roughly double that of the Neandertal females Gibraltar and Saccopastore 1. In the parasagittal plane this surface is distinctly angled in a anterosuperior direction. The angulation, of about 30°, is considerably greater than in the Neandertal females. This reflects the more inferior position of basion in the Mladeč female.

The lateral wall of the greater wing does not differ significantly between Mladeč 1 and the females of the Neandertal comparative sample, in that the sphenoparietal articulation is long. The form of pterion is "H". Within the orbits, the orbital lamina of the sphenoid is obscured by matrix.

Nasal bones

The Mladeč 1 nasal bones are complete from the frontonasal suture inferiorly to a position just above the lower border of the orbit. The most inferior portions of the nasal bones are absent on the right side. On the left the nasal is complete to the lateroinferior corner, although centrally the bone is missing. The most striking feature of the nasals is their narrow "pinched" appearance, which results from a central ridge that is raised well above the nasal processes of the maxilla. The elevation of the nasals involves the whole bone, so that what we refer to as the nasal pillar is comprised of both nasal bones in their entirety, and extends along their full height from nasion to the inferior break. The lateral walls of this pillar are oriented parallel to the parasagittal plane, and thus face in a fully lateral direction.

The closest match to this morphology is found in Předmostí 4, although the nasal bones are considerably broader in this specimen. In the noses of the female Neandertals the internasal suture is also raised above the nasal processes of the maxilla. However, in these specimens (Gibraltar and Saccopastore 1) this suture forms a ridge, and the lateral portions of the nasals are angled to the parasagittal plane and merge insensibly into the nasal processes of the maxilla which continue this angulation to the medial orbital borders. The superior portion of the nasals preserved in Qafzeh 3 is

Table 10. Nasal bone dimensions

	Mladeč females			Mladeč males		Qafzeh	Neandertal	
	1	2	38	5	6	3		
							Mean (n)	Range
Dimensions								
Minimum Breadth	6.5	8.4	9.0		14.0	12.9	13.7 (2)	12.0–15.4
Superior Breadth	10.6	10.1		12.7	16.4	16.7	14.3 (2)	12.0–16.5
Inferior Breadth	18.0						21.6 (2)	19.8–23.4
Nasomaxillary suture length	23.5	23.0					31.4 (2)	27.4–35.3
Indices								
Sup width/anterior iob	41.6	33.8		50.2	62.6		58.2 (3)	47.3–71.7
Min/Max br across bones	36.1						61.2 (2)	56.6–65.8

even less like the Mladeč 1 region and the regions of other early European Upper Paleolithic specimens than the Neandertals are. In this region the face of the bones is very flat, neither ridged as in the Neandertals nor pinched into a nasal pillar.

Relative to glabella the Mladeč 1 nasal root is markedly depressed. In profile the nasal bones are strongly curved and, following the nasal profile of the maxilla, it is clear that this curvature continued in the missing portions of the bones. The extent of the curvature is so great that while the superior portions of the bones are close to perpendicular to Frankfort Horizontal, the most inferior portion must have been almost parallel to this plane. The superior nasals are straight from nasion to about 7 mm inferiorly, and then begin a more gentle curvature. As a result of this curvature, the nasal angle is quite low when the entire length of the nasal bones is considered.

The angulation of the Mladeč 1 nasals to the face, or to the Frankfort Horizontal, resembles the Neandertal females. In Gibraltar the orientation also goes from virtually vertical to parallel relative to the Frankfort Horizontal, although the Saccopastore 1 nasals do not curve quite as much and therefore do not reach equal horizontality. However, the Neandertal nasals begin their angulation in a position further below nasion than do the nasals of Mladeč 1, and their curve is more gentler attaining an equal arc over a longer bone. The superior portion of the nasals preserved in Qafzeh 3 shows absolutely no curvature at all. If the nose of this specimen was like Mladeč 1, enough is preserved to expect to see marked curvature. If the nose was like the Neandertal female noses, the curvature would probably begin below the preserved portion. However, there is no reason to suppose that Qafzeh 3 was like either of these.

Superiorly, the Mladeč 1 nasals are very narrow (Table 10). The superior and minimum nasal breadths are below the Neandertal female sample range. This small superior dimension is reflected in the very small ratios of superior breadth to the interorbital breadth, and the minimum breadth relative to maximum breadth. In both ratios the value is smaller than all comparative specimens. Superiorly, the Neandertals and Qafzeh 3 are considerably broader than Mladeč 1. Moreover, in the Neandertal females the minimum breadth is very similar to the superior breadth, and thus the nasals do not appear as "pinched" as in Mladeč. The Qafzeh 3 nasals are somewhat less pinched. At their inferior border the breadth of the Mladeč 1 nasal bones is smaller than the female Neandertals. The nasomaxillary suture is much shorter than the Neandertals, lying below their range.

Zygomatic bones

Description of the zygomatic bone is based on the left side where there is less matrix and no distortion. A distinct crest follows the course of the zygomaticomaxillary suture from the lower orbital margin (where the suture originates at the lateral border of the infra-orbital foramen) to the base of the bone.

On the anterior surface of the zygomatic crest, just above the base, there is a well-defined zygomatic tubercle. The breadth of the base including the tubercle is 12.1 mm. The zygomaxillary crest marks the angulation between the paracoronally oriented zygomatic process of the maxilla and the lateral face of the zygomatic bone. The angle is approximately 70°. This lateral face of the zygomatic is quite flat. Along the entire lower border the attachment for the masseter is well-developed and results in a thick (6–8 mm) rugosity, terminating in a marked tubercle at the temporal side of the zygotemporal suture.

Comparing Mladeč 1 with the Neandertal females, the lengths (or heights) of the zygomatic bone taken along the zygomaxillary suture or from the bi-*zm* line to the base of the orbit (Table 12), and the heights of the bone from *zm* to *fmt* are almost all the same. However, the Mladeč bone extends more posterolaterally, so that the *zm-ju* distance is greater for Mladeč 1 than for any other specimen, in spite of the fact that the mean bizygomatic breadth for the Neandertal females is greater than the Mladeč 1 value (Table 11). The lateral extension of the zygomatic bone increases from the Neandertal Mladeč 1, even though the bizygomatic breadth decreases. This corresponds to changes in the curvature and angulation of the bone, and a higher position for the superior border of the zygomatic arch in the specimens of the earlier Upper Paleolithic.

Maxilla

The maxillae are complete, although a thin matrix coating covers their surface. Starting at the nasofrontal suture, the maxillary pillars are vertical inferiorly to the lacrimal foramen. At this position they blend evenly into the lower orbital margins, angling at about 45° to the coronal plane. The border of the pillars along the nasal bones is distinguished by a very distinct angle. In contrast, the orbital pillars in the Neandertal females appear to be "pinched" and the medial orbital walls of the two orbits are not parallel to each other. The angulation in the paracoronal plane is about the same, but the nasal bones continue this orientation and thus appear as an extension of the maxillary pillars rather than being separated from them by a sharp angulation. This difference reflects the anterior projection of the Neandertal midface.

Another reflection of this midfacial projection is found in the orientation of the maxilla at the inferomedial corner of the orbit. In the Neandertal females the face of the maxilla in this region approaches horizontality and is continuous with the maxillary border of the lower nasal bones. Thus from the orbital margin at this position to the maxillary border of the nasal aperture the surface of the maxilla is approximately horizontal, and the maxillary surface retains this orientation even more medially, at least to the position of the infraorbital foramen. In Mladeč 1 the maxillary face is also horizontal, but the horizontal region only involves the medial surface of the orbit and extends inferiorly to the position of the most inferior part of the nasal bones, or the inferomedial orbital angle. Lateral to this the Mladeč 1 maxillary face is vertical. The maxillary face is more perpendicular along the sides of the nose, and there is a distinct groove expressing an angulation of the surfaces between the orbital pillars adjacent to the orbital margins and the maxilla adjacent to the nasal aperture. Thus the region lateral to the middle of the nasal aperture is somewhat puffed in Mladeč 1 and is more like the Neandertal females than the other earlier Central European Upper Paleolithic females are. Another expression of this "puffiness" is reflected in the great distance from the nasal aperture at the most inferior point on the nasal bone to the closest point on the inferior orbital margin. This distance is 18 mm in Mladeč 1 but only 12.5 mm in Předmostí 4. In the puffier Neandertal faces the distance is greater, for instance 21 mm in Gibraltar and 23.5 mm in Saccopastore 1.

The nasal aperture is teardrop in shape. The maxillary walls expand laterally inferior to the aperture's top so that the greatest breadth of the piriform aperture is well above the nasal sill. The maxillary borders surrounding the aperture have a strong anterior projection, angling away from the paracoronal plane to such an extent that they almost fully parallel the parasagittal plane at the aperture's border. The inferior border of the nose is moderately guttered lateral to the anterior nasal spine, which is elevated above the most inferior portions of the border. There are two distinct nasal lines at the inferior margin. The more inferior of these begins just lateral to the anterior nasal spine, and arcs evenly superolaterally to merge with the lateral border of the nasal aperture. The more superior of the lines originates at the anterior nasal spine where they merge to form its tip, and parallel the inferior lines laterally to the point where they enter the nasal cavity and merge with the internal aspect of the aperture's walls. These two lines are about 2 mm apart and are separated by a groove. The anterior nasal spine is in the form of a prominent triangle, projecting markedly forward from the lower nasal margin. A sharp vertical crest extends inferiorly from the spine's tip, running for about 5 mm below the inferior margin. The guttering described above extends from the sharp superior line and occupies a space medial to the inferior termination of the lower border.

The Neandertal females, as represented by Saccopastore 1, Gibraltar, and La Quina 5, are characterized by a very broad lower nasal margin with a distinct angulation between the floor of the nasal aperture and the anterior face of the maxilla. Some of the specimens, for instance Gibraltar and to a lesser extent Saccopastore have a sharp line at this margin (the remaining portion is more rounded in Spy 1). The two best-preserved margins are almost flat transversely. Moreover, there is only a single nasal line that extends from the anterior nasal spine to merge with the lateral walls of the nasal aperture in these three Neandertal female specimens. Inferior to the anterior nasal spine there is no medial crest on the anterior maxillary face.

However, the morphological features described above are not unique to all Neandertal females. In particular, two late Neandertal females from Vindija cave (Wolpoff et al., 1981) diverge significantly and instead approach the Mladeč 1 condition. Nasal breadths in these maxillae, 28.5 mm and 26.2 mm respectively for Vi 225 and 259, are only slightly greater than the Mladeč 1 value of 25.9 mm. These maxillae combine the double line form of the inferior nasal margin and the anterior maxillary crest inferior to and continuous with the prominent anterior nasal spine that are also characteristic of the Mladeč 1 female.

As measured from the juncture of the zygomaxillary suture with the lower orbital margin, the upper maxillary breadth is reduced in the earlier Central European Upper Paleolithic females when compared with the Neandertal females. Similarly, the minimum distance between the infraorbital foramina is reduced. In both these dimensions Mladeč 1 lies below the earlier European Upper Paleolithic mean.

More inferiorly on the maxilla, however, the midfacial breadth really does not show a significant change. This contrasts with the breadth of the inferior nasal aperture (i.e. nasal sill), which is markedly smaller in Mladeč 1 (Table 11).

Lateral to the area immediately surrounding the nasal margin the maxilla is vertical and transversely flattened. The angulation between this flat zygomatic process and the projecting nasal margin results in a distinct canine fossa. In the Neandertal females the zygomatic process of the maxilla is convex in the vertical plane, and distinctly posterolaterally angled transversely. Thus, the anterior surface of the maxilla flows more evenly into the portion of the maxilla surrounding the nasal aperture. While these surfaces are usually separated by a shallow groove in the position of the infraorbital foramen, this groove represents only a very weak expression of the canine fossa.

The lower orbital margin is rounded, thick and pillar-like. It projects above the floor of the orbit. Předmostí 4 has a similarly pillar-like margin, and this contrasts with the Neandertal condition. In Gibraltar and Saccopastore 1 the lower orbital margins are much less distinct. The most anteroinferior extent of the orbital margin is continuous with the anterior face of the maxilla.

The size of the infraorbital foramina as measured by its approximate area is about the same in the two European samples. The shape of this opening, however, is somewhat different from specimen to specimen. For instance, in Mladeč 1 the breadth is slightly greater than the height while in the Mladeč 2 maxilla (see below) the height is much greater than the breadth.

Table 11. Facial dimensions (mm) for females

	Mladeč			Skhul/Qafzeh	Neandertal	
	1	2	8[1]			
				Mean (n)	Mean (n)	Range
Breadths						
Biorbital breadth	103.5				105.5 (3)	98.5–111.0
Bizygomatic breadth	137.0				140.0 (2)	138.0–142.0
Bijugal breadth (ju-ju)	114.0				111.6 (3)	102.7–117.0
Midfacial breadth (zm-zm)	103.1				94.0 (3)	85.0–102.4
Bi-infraorbital foramen br	52.6	55.0			60.3 (2)	55.9–64.6
Upper maxillary breadth	62.7	59.0			66.6 (3)	59.0–73.9
Biangular breadth	64.6	67.2	69.2		70.0 (2)	68.6–71.3
Lengths						
Na bi-fmt proj	17.9	19.6			27.8 (6)	26.5–31.6
Superior zt suture-fmt	26.0				33.6 (2)	32.6–34.6
pr-superior zt sut	85.8				88.4 (1)[3]	
pr-fmo	83.9	87.0			99.8 (2)	97.7–101.8
pr-fmt	91.7	91.3			107.7 (2)	104.5–110.8
pr-zm	59.8				69.7 (2)	66.5–72.9
pr-zpm base	45.0	45.8	45.3	41.5 (1)[2]	53.7 (2)	52.0–55.3
zpm base-alveolar margin	9.9	8.7	12.8	9.0 (1)[2]	15.1 (2)	13.0–17.1
Length Indices						
Gnathic	104.2				106.0 (2)	100.0–111.9
Gnathic from auricular	118.3	126.2			117.2 (2)	108.1–126.3
ufh/na-l	38.5	38.2			48.3 (2)	47.7–48.9
Heights						
br-alveolar point	178.1	163.6			165.1 (3)	150.6–173.7
Orbito-alveolar	44.0	42.1			50.7 (2)	48.0–53.6
na-alv (ufh)	70.1	66.0			81.8 (2)	79.0–84.5
na-zm	69.5				73.3 (3)	72.5–74.6
Pommette height	34.0	30.5			34.7 (2)	30.0–39.3
Alveolar height	20.3		17.5		21.5 (7)	16.3–25.3
Nasal aperture						
Nasal height	50.4				59.4 (2)	57.9–60.9
Nasal Breadth	25.9	25.5	29.1		33.5 (2)	32.9–34.1
Breadth/Height	51.4				56.5 (2)	54.0–58.9
Orbits						
Orbit height	31.5	36.1			38.4 (4)	38.0–39.3
Orbit breadth	39.8	40.5		37.3 (1)[4]	40.5 (4)	38.9–42.3
Orbit depth	56.4				64.1 (1)[3]	
Height/Breadth	79.1	88.0			92.3 (2)	89.8–94.7

[1] Male palate
[2] Qafzeh 5
[3] Gibraltar 1
[4] Skhul 2

The infraorbital foramina are close to the orbit. In vertical distance, the separation is only 7.0 mm, even the maxilla of Mladeč 2 greater (10.5 mm). In the two Neandertal female faces allowing this determination, Saccopastore 1 and Gibraltar, the distances are greater than both of these are (15.3 mm and 11.2 mm). This contrasts with the distance from the foramen to the alveolar margin, which is more-or-less, the same. Thus, the infraorbital foramen position only changes relative to the orbital margin, and one might infer that facial reduction had taken place with some shortening of the vertical dimension of the maxillary sinus.

Inferior to the infraorbital foramen a very distinct groove progresses inferiorly and slightly medial toward the alveolar margin, terminating some 13 mm above it. This canine fossa clearly delineates the laterally extending flat cheeks from the markedly rounded nasal and subnasal regions of the face.

We believe that the changes in the Mladeč 1 maxilla can be attributed to two factors, a reduction in the volume of the nasal cavity and a reduction in the amount of vertical force transmitted through the midface as a consequence of anterior tooth loading. The reduction in nasal cavity volume is indicated by several changes, including the breadth of the nasal aperture (and of the surrounding midface), the breadth of the superior nasal bones, and the height of the nose. The position of the infraorbital foramen provides a clue to the region undergoing reduction in vertical dimension. This foramen marks the position of the maxillary sinus since the infraorbital nerve and blood vessels travel above the sinus. That the positions of the infraorbital foramina do not change relative to the alveolar margin indicates that the sinus does not change in vertical height. As we discuss below, there is no significant reduction in postcanine tooth size, and we contend that the lack of change in the posterior dentition accounts for the stability of the lower portion of the midface. However, the midface does reduce, as indicated by the reductions in facial height, nasal height, and orbit height. We propose that these reductions are all a consequence of reduction in the upper portion of the nasal cavity, decreasing its volume. Effects of this reduction include the reduction in orbit height and the changing relation of the infraorbital foramen to the orbital margin. Thus, the reduction in facial height is primarily in the upper portion of the midface, and with the decreasing nasal breadth these reductions are the primary consequences of decrease in the volume of the nasal cavity.

The second factor influencing this region is a response to decreasing force through the midface that results from anterior tooth loading. Vertical forces are transmitted through the maxillary pillars, forming the frontal process of the maxilla (Endo, 1966; Russell, 1985). In the Neandertal female sample these pillars are thickened, and as we describe above in the region surrounding the nasal aperture they project anteriorly so that the medial portion of the region below the orbital rim is oriented horizontally. As we noted, the horizontal orientation extends no more laterally than the position of the infraorbital foramen. In part the configuration of this region reflects the great volume of the nasal cavity in the Neandertals. However, the maxillary pillars themselves are distinctly thickened in this sample, adding to the projection of the bony surface and the horizontality of the region surrounding the nasal aperture. The horizontal orientation is important in resisting the horizontal component of anterior tooth loading that affects bending moments in the central portion of the midface. Thus, apart from the reduction in this region that reflects the decreased volume of the nasal cavity, an additional change in the earlier Central European Upper Paleolithic females including Mladeč 1 is found in the thinning of the bone forming the maxillary pillars. As a whole, the region therefore appears to be subject to less stress due to anterior tooth loading.

In Mladeč 1 the masseter attachment extends onto the base of the zygomatic process of the maxilla by as much as a half-centimeter. In the Gibraltar female the extension onto the maxilla is about double. However, since the base of the zygomatic process of the maxilla itself is in a more posterior position in the Neandertal females, it is not clear whether the masseter attains a more anterior position in the Neandertals. Indeed, the muscle may actually be more posterior relative to the palate (see below).

The earlier European Upper Paleolithic hominids are said to contrast with the Neandertals in the morphology of the inferior border of the zygomatic process of the maxilla (Howell, 1951; Rak, 1986). We note that in all Neandertal females the lower border of the zygomatic process of the maxilla blends evenly into the external walls of the palate. This border forms a straight line angled at about 45° to the horizontal between the zygomaxillary sutures to the palate's external walls, and there is no angle between the two surfaces. The Neandertal condition has been described as a "flying buttress". In Mladeč 1 the form of this region is more angled than in the Neandertals, but not as distinctly angled as in many other specimens. Just medial to the zygomaxillary suture the inferior border of the zygomatic process of the maxilla begins an even curve which arcs gently to merge continuously with the external wall of the palate. There is no distinct angulation marking where these surfaces meet. However, the base of this arch is very close to the alveolar margin, separated from the M^1 root by no more than 9 mm. The arc is shallow, so that its curvature does not deviate significantly from a straight line.

The alveolar region of Mladeč 1 is characterized by a fairly high degree of alveolar prognathism, extending over the full anterior of the maxilla between the canine sockets. The canine root pilasters are weak but visible, extending up to the level of the lower nasal margin. At their most superior extent they are somewhat lateral to the lateral borders of the nasal aperture. In the subnasal region, the prominence of the anterior nasal spine combined with the alveolar prognathism outlines a markedly concave surface in the sagittal plane. This concavity extends laterally across the subnasal region to the positions of the canine root pilasters. In the Neandertals, the western females lack alveolar prognathism, although there is a shallow transverse groove in the subnasal regions of Saccopastore 1, Gibraltar, and La Quina 5. In the Vindija maxillae there is a closer approach to the Mladeč 1 condition. Especially in Vi 259 there is a distinct concavity below the projecting anterior nasal spine and a moderate amount of alveolar prognathism.

As mentioned above, the base of the zygomatic process of the maxilla is low relative to the tooth row. In Mladeč 1 this distance is only two thirds the mean for two Neandertal females, as it is in Qafzeh 5. However, the distance is as great as in the Neandertal females for Dolní Věstonice 3, the only other Central European Upper Paleolithic specimen allowing the measurement. The thickness of the zygomatic process of the Mladeč 1 maxilla at its base is marked, greater than Saccopastore 1 and Gibraltar, and almost double the thickness of Cro-Magnon 2 (no other early Central European Upper Paleolithic specimen preserves the region). Moreover, the zygomatic process of the maxilla of Mladeč 1 remains thick in other positions. At the zygomaxillary suture it is thicker than any of the comparative female specimens and almost double the Neandertal female means (Table 12). The height of the process at this most medial point on the cheek (pommette height) is quite large for an early European Upper Paleolithic female – almost at the Neandertal female mean. Relative to the tooth row, the center of the zygomatic process of the maxilla base is at the first molar position while its anterior face is positioned between P^4 and M^1. In the Neandertal females the zygomatic process of the maxilla is somewhat more posterior relative to the tooth row. The center of

Table 12. Zygomatic dimensions (mm) for females

	Mladeč 1	Neandertal	
		Mean (n)	Range
Length of zm suture	28.3	30.1 (3)	27.2–33.5
Bi-zm line to base of orbit	23.0	23.2 (3)	18.0–27.5
zm-ju	29.5	21.4 (2)	19.5–23.2
zm-fmt	46.5	46.5 (2)	46.0–46.9
Thickness at zm	13.3	6.7 (2)	6.5–6.9

Table 13. Position of the face in female specimens

	Mladeč		Neandertal	
	1	2		
			Mean (n)	Range
au-fmo	80.1	80.5	81.3 (4)	76.6–88.7
au-zm	79.4		76.1 (3)	74.3–79.4
ba-pr	108.9		109.0 (2)	104.0–114.0
ba-na	104.5		103.0 (2)	101.9–104.0
Anterior Occipital Condyle-post M^2	60.5		69.8 (1)[1]	
Glenoid pt-M2	64.3	76.0	79.0 (1)[1]	
Glenoid pt-post C	88.3	98.0	101.0 (1)[1]	
Glenoid pt-zm	56.9		58.9 (3)	55.8–62.0
Glenoid pt-center zpm root	64.1	74.5	70.3 (2)	68.2–72.3
Glenoid pt-pr	113.5	117.7	119.4 (2)	117.8–120.9

[1] Saccopastore 1 only

the process is positioned at the M^2 while the anterior face is at the M^1. Of course, in the Neandertals the tooth row itself is in a more anterior position, which leaves open the question of whether the zygomatic process of the maxilla is posteriorly positioned relative to anything else. We believe that it is. For instance, in lateral view the zygomatic process of the maxilla base in the earlier European Upper Paleolithic female sample is approximately level with the position of the anterior orbital margin. In the Neandertal females the zygomatic process of the maxilla base is clearly positioned well posterior to this margin.

The posterior surface of the zygomatic process of the maxilla, which is the anterior face of the temporal fossa, is vertically oriented – the common orientation in the Central European early Upper Paleolithic European females. In contrast, the Neandertal females appear to have a posterosuperior angulation to the temporal fossa's anterior wall.

The Mladeč 1 palate is elliptical in shape, with its greatest breadth across the M^3 positions. The anterior margin of the palate is highly arched and the incisor roots are far forward of the bicanine line. The length of the premaxilla is approximately 14 mm. In the Neandertal females the anterior margin is very flat and the length of the premaxilla is about half as great. Qafzeh 5 appears to be more arched than is usual in the European Neandertals.

Along the midline the incisive foramen is located directly posterior to the central incisor sockets, well below the roof of the palate on its anterior face. There are low ridges paralleling the midline and separated by a groove at the bimaxillary suture. These begin at the incisive foramen, and extend to the position of the M^1, increasing in elevation. Posterior to this, the palate's roof is smooth; there is no ridging or development of a palatine torus. The Neandertal females have an incisive foramen position that is generally more posterior and on the roof of the palate rather than on its anterior face. None of the females in the comparative samples appear to have either the anterior ridge we described for Mladeč 1, or a palatine torus, but this is not the case for other Mladeč specimens.

The breadth across the tooth roots of the palate, in all positions (Table 13), is smaller than both the Neandertal females and Qafzeh 5. Internal breadths between the tooth socket walls are smaller anteriorly (this reflects the larger anterior teeth in the earlier samples) but toward the palate's rear the Mladeč 1 breadths are more similar to the others. However, this might be a misleading comparison since the internal breadths in the molar region can only be taken on a single specimen, Saccopastore 1.

In terms of the standard measures of palate depth, the Neandertal sample (of two) is extremely variable, and thus we do not regard its mean values with confidence. We propose instead a measure of palate depth determined by taking the height at the internal palate wall because this reflects the same morphology but allows a much greater sample size. The comparison of means for the height of the internal palate wall at the P^4/M^1 and M^1/M^2 positions shows the Mladeč 1 value is above the Neandertal mean. The single Skhul / Qafzeh specimen, Qafzeh 5, is also shorter than Mladeč 1.

Palatine bones

The palatine bones are almost perfectly preserved, but covered with matrix. At the mid-palatine suture, the two bones join in an uneven plane, so that the left projects more anteriorly than the right. The posterior nasal spine is very well-developed. Lateral to this, the palatine bones are well arched as seen in palatal view, forming deep impressions on the posterior border of the palate. Thus, staphylion is located 5.8 mm anterior to the posterior nasal spine. The crest for the insertion of *tensor veli palatini* is extremely high and well arched. The only specimen with a comparable region preserved, Saccopastore 1, shows no corresponding development.

Dentition

The only remaining teeth in the palate are the left and right M^1 and M^2, although sockets for all the other teeth are present (Fig. 13). All these teeth have been lost post-mortem since each of the empty sockets is filled with matrix, except for the left P^3 that contains the root (broken off at the cementoenamel junction). As described earlier, these teeth show little wear. Morphological and metric details of these teeth are described in the dental section below. Sockets for all the remaining teeth are present and filled with matrix, except for the left P^3 that contains the broken root of the tooth. The crown was present when discovered and, according to Szombathy, was subsequently broken off. Because of this matrix it is difficult to observe the orientation of the inner surfaces of the sockets for most teeth, except left I^1. For this case, the socket walls are straight, showing no posterior curvature. The lingual margins of the M^3 sockets are barely elevated above the plane defined by the lingual alveolar borders of M^1 and M^2. Because of this position and the fact that the sockets appear to be

Fig. 13. Palates of Mladeč 1 (left) and 2, after Szombathy (1925). These two female specimens died within a year or so of each other; Mladeč 2 is slightly older

fully open, we believe that the third molars were probably past alveolar emergence and close to their final occlusal position. However, the absence of a distal interproximal facet on the M^2's indicates that the occlusal eruption had not been fully attained. Finally, given the presence of complete alveolar borders for all the anterior teeth and the premolars, there is no indication of any premortem tooth loss, so that all missing teeth were lost postmortem.

The P^3 crown, in fact, must have been lost after discovery since there is no matrix covering the open root surface or imbedded in exposed hollows and cracks. For this tooth the only observation possible is that the pulp chamber is small and does not extend into the root chambers.

On the right side the M^1 and M^2 are in their normal occlusal positions, but on the left, the M^2 has been displaced distally and inferiorly out of its socket. A large mass of breccia fills the interproximal space between M^1 and M^2, separating these teeth by about 5.0 mm. Thus, the occlusal surface of the left M^2 is not in the occlusal plane. Preservation of all teeth is excellent and no asymmetry in wear or morphology occurs. The following morphological description is based on the left side, while metric analysis uses the average of the two sides.

Wear on the M^1's is slight with only pinpoint dentin exposure on the mesiolingual cusp and very minor blunting of the buccal cusps (Stage 2). On the mesiolingual cusp there is a large oval-shaped facet (3.8 mm long, 2.8 mm broad) that extends from the midline of the cusp to the lingual border where it continues onto the lingual face. On the distolingual cusp a round facet (2.5 mm in diameter) occurs in the center of the cusp. Both of these represent normal occlusion with their mandibular opponents. On the mesial face, a large, shallow interproximal facet occurs for the P^4. Lack of depth to this facet indicates little mesiodistal length was lost from interproximal wear.

Seen from the occlusal aspect, the crown is roughly square in shape, although the mesial crown breadth is about 1 mm greater than the distal. The crown is of the 4- type. The distolingual cusp is well separated from the trigone and large, although reduced relative to the other three cusps. The internal faces of the cusps are smooth, with no sign of occlusal wrinkling. Radiographs show a small pulp chamber with a slight enlargement into the mesial root. However, this is limited and does not qualify as taurodontism. The roots are widely divergent and well anchored into the bone.

The M^2's exhibit less attrition than the M^1's with only minor polishing and blunting of (especially) the lingual cusps. This degree of wear corresponds to Stage 1. There are no polished facets on the lingual cusps (as in the M^1's) and the mesial interproximal facet is only lightly etched into the anterior tooth wall. As mentioned above, there is no distal interproximal facet.

The trigone cusps dominate the crown with the hypocone only represented by a small cuspule on the center of the distal border. This arrangement constitutes a 3+ pattern. The pulp chamber resembles the M^1 pattern with a slight expansion into the mesial root, but not to the extent that this tooth can be considered taurodont. The roots are well separated from each other, showing no sign of convergence within the alveolus.

While many of the teeth are missing, something can be said of their size. A number of lengths can be taken along the tooth row, between the socket walls (or roots). The comparisons that can be made (Table 14) involve only females, since they are based on cranial material. They show Mladeč 1 to have an almost Neandertal-like total anterior tooth length. The great distance from prosthion to M^1 and the marked distance between the four incisor sockets verify this contention. The posterior tooth row lengths are even greater, exceeding the small sample of other female specimens.

Actual tooth size is only known for the first two maxillary molars (Table 15). These teeth are large, exceeding the averages of the female Neandertals and Skhul / Qafzeh specimens. However they are smaller than the corresponding teeth in the Mladeč 8 male palate, and the same size or smaller than the teeth in the Mladeč 47 and 50/51 palates described in Frayer and colleagues (this volume).

Mladeč 2

In 1881, Szombathy was excavating a test pit in the part of the cave he designated "b" (see Szombathy, 1925; Svoboda, 2000). He discovered the Mladeč 2 vault within the top 60 cm, as well as two halves of a maxilla we catalogued as Mladeč 7 that were nearby, and the juvenile vault Mladeč 3. Mladeč 7 was subsequently shown to be the same individual as Mladeč 2, and the pieces were attached. A number of postcranial remains were also found in this test pit. Of these, we believe that Mladeč 12–20 (ribs), Mladeč numbers 23 (right proximal humerus), 26 (right radius diaphysis), and metapodials 32, 35, and 36 could be part of the Mladeč 2 individual, although we shall never know.

Preservation

Mladeč 2 consists of the calotte that originally carried that number, and most of a face that we have added to it (see this volume, chapt. 8, Plate II). The calotte includes a nearly complete frontal, left and right parietal bones, and major portions of superior aspects of nasal bones. While in general the bones present are undistorted and complete, there are several missing areas. One of these is the most anteroinferior margin of the frontal, which results in the loss of most of the posterior portions of the orbital roof. There is also a square portion missing from bregma to 47 mm along the sagittal suture, running about 47 mm laterally and inferiorly onto the parietal wall and down the coronal suture. In addition, while both temporal bones are complete in their total lengths, they lack the superior part of the squama. On the right temporal the most medial aspect of the petrosal is broken away, as is the posterior portion of the mastoid tip. Finally, both zygomatic processes of the temporal are broken so that they do not reach the zygotemporal suture.

As Szombathy describes, the skull was found in ten main pieces. In almost all respects his reconstruction appears to be quite accurate, with most of the breaks in the bone well-positioned in their original location. The only area that is incorrectly placed is the left temporal. This piece was indirectly connected to the vault along the occipitoparietal suture, but using plaster between the bone surfaces. We believe that the piece is too low in the sagittal plane, and somewhat rotated in a counterclockwise direction as seen from below, resulting in asymmetry of the glenoid fossae and anterior temporal borders. We did not reposition the left temporal, since the substance used to attach it to the parietal was impossible to dissolve. As much as possible, measurements across the base attempt to compensate for the incorrect position of the left temporal.

The calotte is virtually free of matrix on both the ecto- and endocranial surfaces. A thin coating occurs on the left and right mastoids, extending onto (and into) the external auditory meatus. There is also matrix adhering to the face of the lambdoidal suture for its entire length. The bone is a cream-colored and numerous dark flecks (dendrites) occur, especially, on the endocranial surface. This surface is uneroded. However, externally the bone surface is somewhat less well preserved. The very outer layer of cortex is missing in many areas and some of the surface features such as the temporal lines are difficult to delineate (Fig. 14 and 15).

Where the cranial pieces were glued together a plaster-like substance was used as described above. In preparation the plaster was evidently smeared where these pieces fit together in many cases obscuring morphological detail. This is probably most serious on the frontal where a large triangular piece with its apex at about metopion and its base at the orbital borders was connected with a good deal of plaster on the left side. The plaster was smeared from the frontal squama, onto and over the orbital border, and onto the orbital plate.

Szombathy originally suggested that the maxilla (Mladeč 7 by our cataloging) might belong to the same specimen as the cranium. He believed that the two maxillary halves represented the same specimen as the Mladeč 2 vault because they were lying close together. Like other specimens from Mladeč, the bone is gray or ash-white in color and contains small dendrites. In the Naturhistorisches

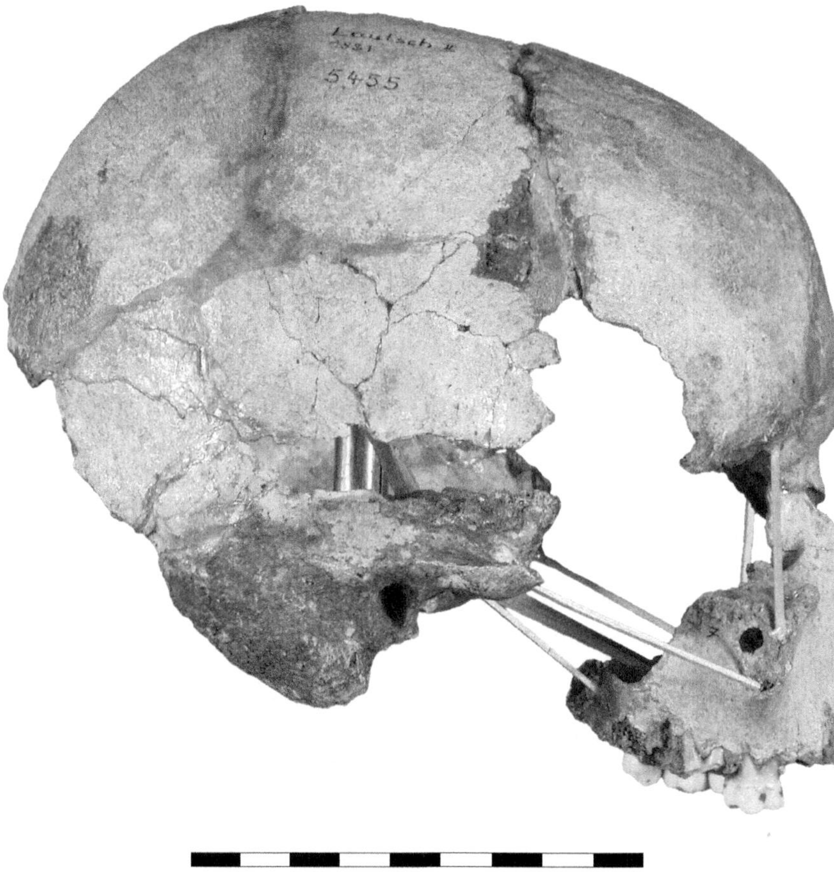

Fig. 14. Lateral view of the reconstructed Mladeč 2

Museum Wien, we had the palate cleaned and some additional cleaning was done on the frontonasal suture of Mladeč 2. From this we discovered a join between a right frontal process of the maxilla (including the lacrimal duct and the portion superior to it) and the inferior lateral segment of frontal process of the maxilla that remained on the Mladeč 2 calotte, about 5 mm inferior to the frontomaxillary suture. We positioned the face on the vault using this join, the continuity of the orbital margin, horizontality between the first molars and symmetry of bilateral points on the anterior of the palate from lambda (since the left temporal was improperly positioned, symmetry from the cranial base could not be determined). The sagittal orientation was the most difficult. We relied primarily on the join itself, the orbital margin, and the angle of the molar row. We will refer to the reconstructed cranium and face as Mladeč 2.

The facial skeleton of Mladeč 2 is undistorted and consists of two halves of the maxilla, originally broken apart along the midpalatal suture. The right half is more complete and includes the inferior portion of the nasomaxillary suture, the lower part of the frontonasal pillar (12 mm superior to the lower orbital margin), the lacrimal foramen, the inferior orbital border to the maxillosphenoidal suture, the zygomatic process of the maxilla which includes the medial most 13 mm of the zygomaxillary suture (and a small piece of zygomatic adhering to it), nearly the entire maxillary sinus, the lateral border of the piriform aperture, the entire alveolar portion of the maxilla, the anterior inferior portion of the palatine bone, and the nasal floor (Fig. 22). But, the very most anteroinferior region of the maxilla is damaged, so that nothing remains of the lower nasal mar-

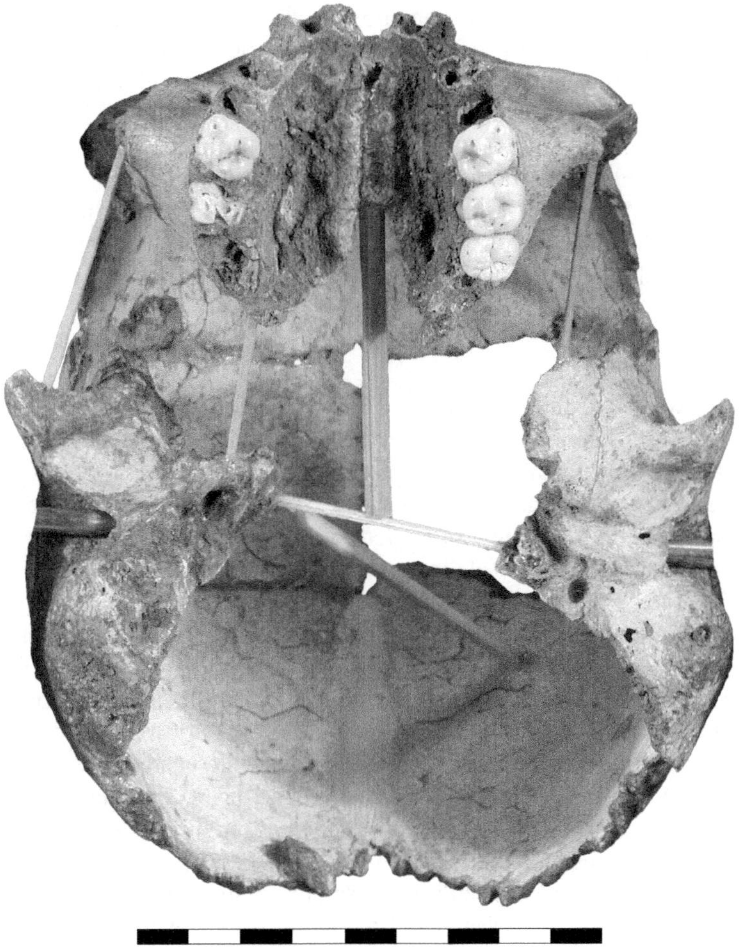

Fig. 15. Basal view of the reconstructed Mladeč 2

gin or anterior nasal spine. Sockets for all permanent teeth are present, while the only tooth crown present is M^1.

The left side is less complete, comprising mainly the entire alveolar portion with sockets for the incisors, canine, and premolars. M^1, M^2, and newly erupted M^3 are still located in the jaw. Besides the alveolar arch, the fragment includes the posterior lateral part of the palatine bone, the inferior most portion of the zygomatic process of the maxilla, the basal portion of the maxillary sinus, and the floor of the piriform aperture. Extensive damage to the subnasal area has destroyed the inferior nasal margin.

During the reconstruction, the two halves of the maxilla were separated for cleaning and then positioned together, but precise contact along the mid-palatine suture did not occur. Perhaps this is because they were originally cemented together with a number of reindeer bones and these covered the mid-palatine suture. According to Szombathy the two halves do not fit together perfectly, due to erosion of the mid-palatine suture on the left side. However, by orienting the pieces with respect to the incisive foramen, transverse palatine suture, and left and right M^1's, it is possible to produce an accurate reconstruction of the dental arcade. In our reconstruction a small amount of filler was used to reconstruct the missing portion of left mid-palatine suture.

Association of the palate with the cranium allows a clear determination of sex and age. In so far as sex is concerned, the gracility of the cranium, especially the supraorbital region, indicates the sex was female. Dimensions of the teeth and palate size, in comparison with Mladeč 1 and Mladeč 8, also suggest that this specimen was probably female. Szombathy reports that before reconstruction all the cranial sutures were open, and the sutures are of simple form, concluding that the specimen was a juvenile. As was discussed in the section on Mladeč 1, the dental age of this specimen is 16±1 years, thus slightly older than Mladeč 1. The third molar of Mladeč 2 had apparently just erupted, since there is only slight polishing wear on its occlusal surface. Based on this criterion Mladeč 2 must be regarded as a young adult, but in fact we regard her as virtually identical in age to Mladeč 1.

Total skull

After reconstruction, we found that the Mladeč 2 face is particularly similar to the face of the Cro-Magnon 2 female (Fig. 16). Therefore, even though this Western European female is not part of our general comparative sample, we will make certain comparisons and discuss relevant aspects of the vault in the text below. These two females share very small flat faces, broad interorbital areas, moderate superciliary height development (in the context of modern European females), minimal projection of the superciliary arches, and relatively large basal dimensions of the mastoids. Of course, the skulls are far from identical. Mladeč 2 is the more prognathic (alveolar prognathism and index of prognathism, Table 11). Her orbits are higher and markedly rounder, the top of the vault is flat (in Cro-Magnon 2 the top is keeled), vertex is 45 mm behind bregma (in Cro-Magnon 2 vertex and bregma are coincident), the malar notch is only weakly developed, the mastoid is less projecting, and the cranial base is much broader.

Mladeč 2 is slightly smaller than Mladeč 1. The cranial capacity we have determined is 90% the Mladeč 1 value (Table 2) and while Mladeč 1 is the largest female vault in the Central European early Upper Paleolithic, this Mladeč 2 capacity is below the female mean. The Mladeč 2 cranial lengths can only be measured to lambda. These are very small, at the bottom of the Central European early Upper Paleolithic female range, for the most part the same as Qafzeh 3, and smaller than the Neandertal female means. Mladeč 2 is a very curved skull in the sagittal plane, resembling Qafzeh 3 in this regard and contrasting with the much flatter Neandertal female vaults – this can be seen in comparing various chord/arc indexes.

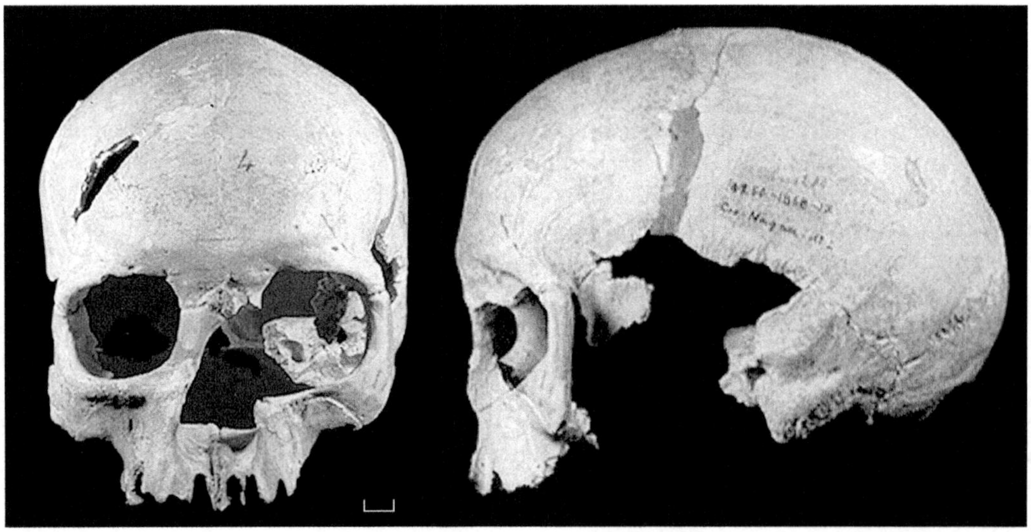

Fig. 16. Cro-Magnon 2, a female with marked facial similarities with Mladeč 2

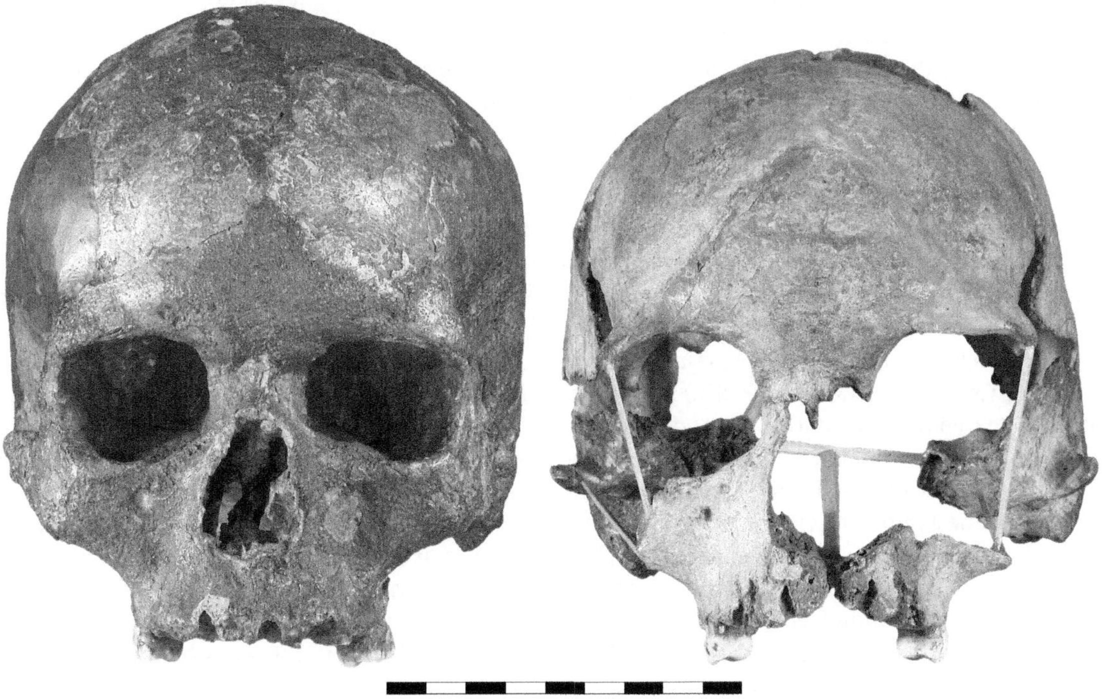

Fig. 17. Comparison of Mladeč 1 and 2 in facial view

The breadths of the vault are quite the opposite, in that this is a very broad vault, with its greatest breadth at the cranial base. The maximum cranial, biparietal and estimated occipital breadths are the same as Mladeč 1, but because of the much shorter vault, relative breadths for Mladeč 2 are quite high. While relative breadths are even greater in Qafzeh 3, they are much less in the elongated Neandertal female crania. The absolute values for the vault breadths are close to the Neandertal female means, while the basal breadths are above the ranges. Biasterionic breadth can be estimated from the parietals and is above the Neandertal female mean.

Cranial height from the auricular point (Table 3) is low, very small, below the Central European early Upper Paleolithic female minimum and below the Neandertal female mean. Relative to nasion-lambda length (Table 2), however, this height is not much different from Mladeč 1 or the female Neandertals.

In sum, the vault as preserved suggests the cranium was short, low, and broad. Except for details at the rear of the preserved cranial base, it is generally a smaller, more gracile, more rounded vault than Mladeč 1. The exceptions are the marked breadth across the cranial base and the rugose development of the mastoid processes, which result in unusually great relative breadth proportions. These comparisons illustrate what most practicing paleoanthropologists know full well – even the smallest samples can encompass very significant variability.

In lateral view, the glabellar area shows a very slight anterior projection and a weak sulcus above it, like Mladeč 1 and Cro-Magnon 2. The nasal bones protrude below. Relative to the auricular point (Table 3) the glabellar and nasion projections are smaller than Mladeč 1 and much smaller than the Neandertal females. The frontal forms a nearly vertical plane to metopion and then curves sharply angling up to the low bregma. The elevation of the cranium continues along this line to a point about 45 mm behind bregma and from here the contour drops sharply along a gentle curve toward lambda (Fig. 14). Interestingly, as in the Mladeč males' vertex also forms the apex of triangle of bone flattened across the rear of the parietal bones. The base of the triangle at the lambdoidal suture is about

50 mm in breadth. Thus, the posterior parietal area is flattened, although this is not evident from a strictly lateral view. This vault contour, with a vertex well separated from bregma, is also seen in Cioclovina. However in Mladeč 1, vertex and bregma are coincident and, consequently, the back of the parietal bones and occiput are not as steeply sloped.

The sagittal and transverse contours are more curved than the Mladeč 1 female, according to the arc/chord indices (Table 2). The curvatures are greater than the Neandertal females, although markedly less than the highly-curved Qafzeh 3. The transverse curvature is less, at the female Neandertal mean (the European samples do not differ in transverse curvature).

The face is prognathic – both the facial angle is high and the alveolar region shows additional prognathism. Measured from the auricular point, what we might call the "auricularognathic" index is higher than Mladeč 1, and identical to the Neandertal maximum. However, the magnitude of the difference between the two Mladeč females (8 index points) is not unusual, even in these small samples. The two Neandertal females differ by 18 index points and the two Skhul/Qafzeh males differ by 17 index points. The marked prognathism in this Mladeč female is matched or exceeded by a Neandertal and a Skhul/ Qafzeh hominid. It is possible that the greater prognathism in the two Mladeč females and the two Skhul/Qafzeh males has significance. However, given the closeness of the mean indices for the combined sex samples (117 for the Neandertals and 120 for two Skhul/Qafzeh males), especially in the context of the marked sample variances, it is difficult to clearly delineate a significant pattern to this variation. The anterior position of the face is also indicated by the marked sagittal distance of prosthion from the auricular point (Table 3). This is 95% the Neandertal female mean value. The auricular-nasion distance only reduces to 85% the Neandertal mean, which accounts for the large *index* of prognathism as calculated from the auricular point.

There is an even greater measurement from the glenoid point to prosthion although this might be misleading. Unlike the measurements from the auricular point, which are projected into the sagittal plane, the measurements from the glenoid point to positions along the tooth row (Table 13) are directly measured between the points. These show a marked anterior positioning of the posterior dentition. But, distances from the glenoid point to the anterior of the palate, at the midline, are more difficult to interpret because they also incorporate great breadth of the vault. While the auricular distances (Table 3) are probably a better indication of the face's midline position since these are projected into the sagittal plane, together these clearly reflect the marked facial prognathism of the specimen.

Unlike Mladeč 1, the lower orbital border is anterior to the upper border in the Frankfort Horizontal (this morphology is like that of Cro-Magnon 2). The anterior face of the zygomatic process of the maxilla is located over the first molar as it is in Mladeč 1. Anterior to it the midnasal projection does not seem to be as great (as described above). Prognathism of the alveolar region is similar, although the area between the lateral nasal border and the canine alveolus does not appear quite as deep or curved as in Mladeč 1.

The discussion of prognathism in the sagittal plane indicates that the dentition is more anterior in Mladeč 2 than in Mladeč 1, while the upper part of the face is not more anteriorly projecting. Indeed, the lateral portion of the upper face is also in a similar position in these two specimens (Table 13). This portion is very anterior; in fact, the lateral facial length is the same as in the much larger female Neandertal faces. This combination shows that both Mladeč faces are transversely flatter than Neandertals. Moreover, measurements that reflect the more medial position of the Mladeč 2 middle (below nasion) and lower face are very similar to the Neandertal female mean values. The *combination* of transverse flattening and lower facial prognathism is unique in this specimen. The Mladeč 2 face *could* be transformed from a Neandertal female condition mainly through vertical shortening and reduced anterior projection for the midline facial positions of nasion and glabella. The lateral portion of the face and the regions above (bregma) and below (prosthion) the upper nose remain in a more stable location. It is as if the same guy who created the Neandertal face by pulling on the nose changed his mind and pounded it back in!

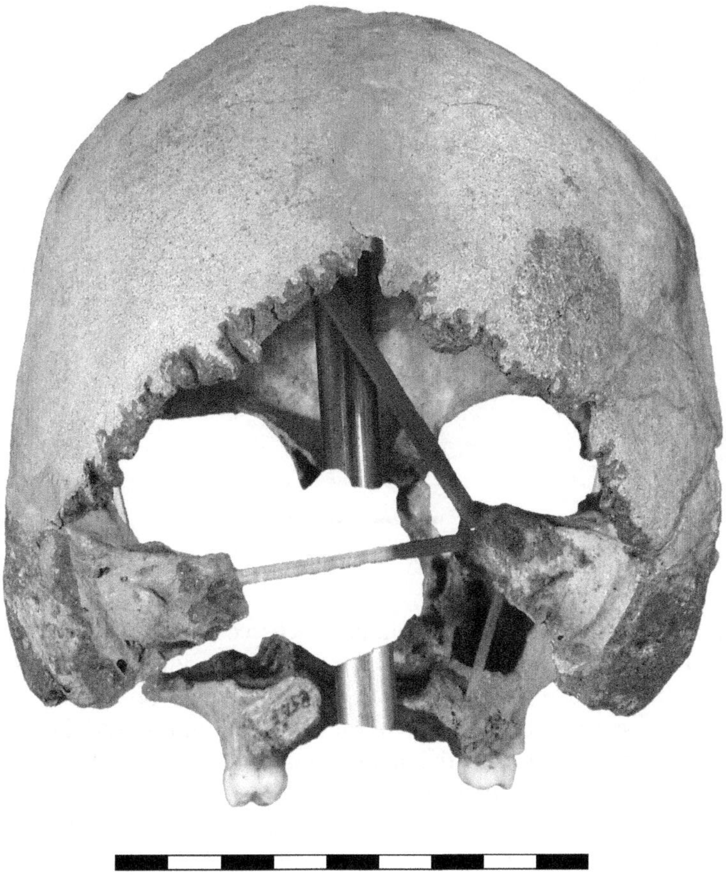

Fig. 18. Mladeč 2 from the rear. The greatest breadth of this cranium is at its base, and there is a broadly depressed sagittal groove

As seen from the front, the coronal contour forms an almost circular arc (Fig. 22). The brows are weakly developed and below them the very wide innerorbital area contrasts with the very narrow superior nasal bone breadth. The index of the superior nasal breadth to the anterior interorbital breadth (33.8, see Table 10) is by far the smallest of any other female discussed here. The brows are very much like Cro-Magnon 2 and more weakly developed than in Mladeč 1. However, the superciliary development falls well within the early Central European Upper Paleolithic, exceeding the three Dolní Věstonice females.

The orbits are high (Table 11) and their contour is rounded. Indeed, it is the great height of the orbits that convince us that no bone is missing from the contact between the maxilla that was designated as Mladeč 7 and the small piece of maxilla adhering to the Mladeč 2 vault. Because of the high orbits, there is a contrast between the low broad face and the high rounded orbits in this specimen as compared with the low, broad face and low, broad orbits of Mladeč 1 (Fig. 17). The orbit breadth does not vary significantly between the samples. Because the orbit breadth resembles the Neandertal condition, and the orbit height is almost within the Neandertal range, the shape index for the orbits of Mladeč 2 more closely approximates the round orbits of the Neandertals than does any other Central European early Upper Paleolithic female. Even still, however, the index is below the Neandertal range.

From the rear, the parietal bones are somewhat less parallel sided than in Mladeč 1, canting laterally inferior to the weakly expressed parietal bosses. Also in contrast to Mladeč 1 the mastoids angle medially to the supramastoid crest. This can be more accurately seen on the right than the left,

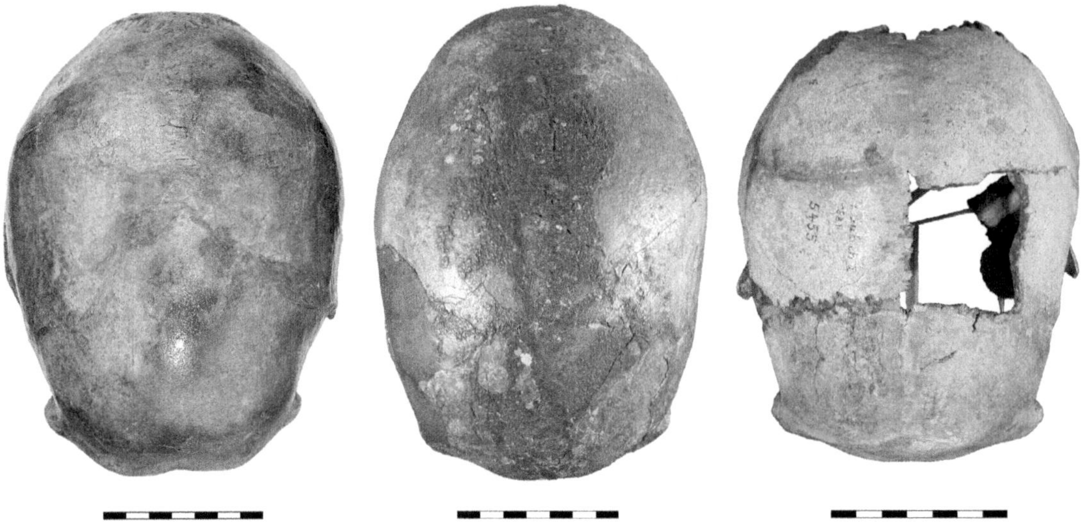

Fig. 19. Superior view of three Mladeč crania. From the left these are crania 5, a male, and the females 1 (center) and 2 (right)

due to the improper position of the left mastoid (as discussed above). In the pneumatization of the cranial base and contour of the parietal walls this otherwise very gracile specimen is rather male-like. Superior to the parietal bosses the paracoronal plane is evenly curved and is not as pentagonal as seen in Mladeč 1. However, if the skull is tipped somewhat anteriorly, the flattening across the back of the parietal bones results in a horizontal contour across the top. Finally, the most posterior aspect of the mastoids shows a flat, somewhat posteriorly oriented surface (Fig. 18).

In superior view, there is a broad glabellar prominence. Lateral to this, the superior orbital borders extend in a straight line, which roughly corresponds to the paracoronal plane. There is only slight postorbital constriction. The temporal fossa is very short (no breadth can be ascertained). From the *ft* point to the parietal boss, little curvature of the frontal and parietal lateral walls occurs. The bosses are evident, but not especially prominent; in fact, in Mladeč 2. Posterior to the bosses, the cranium curves evenly and markedly toward the lambdoidal suture. This contrasts with the flattened surface in this region outlining the occipital bun in Mladeč 1 (Fig. 19). The cranial base is so broad that the superior portions of the mastoids are visible in superior view. The superior surfaces of the mastoids are located just posterior to the bosses when observed vertical to Frankfort Horizontal.

On the maxilla in superior view the lacrimal foramen and inferior orbital plate can be observed and the anterior border of the orbit is preserved to just past the suture with the zygomatic. This border runs in a straight line laterally, as does the maxillary area inferior to it, reflecting the flattened face in Mladeč 2. Since the sphenoid is broken away, the maxillary sinus is open. This large structure occupies a very low position, actually extending well below the roof of the palate, laterally to near the zygomaxillary suture, and posteriorly to well beyond the M^3 socket.

Inferiorly, only the orbital plates and the petrous portions of the temporal bones along with the mastoids are preserved on the cranial base. On the frontal, the base of the frontal sinus is visible bilaterally, divided by the anterior portion of the frontal spine. For the posterior vault, the asymmetry due to the position of the left temporal/sphenoid fragment is most apparent. For instance, from the auricular point to nasion on the left side is 119 mm, while on the right the correct value is 110 mm. The distances from the auricular point to bregma are about the same, but to lambda the inaccurately positioned left side is about 4 mm closer to lambda than the right.

The great breadth of the cranial base discussed above is evident in this view. Had the zygomatic arches been preserved, they would surely have greatly exceeded the bizygomatic breadth of Mladeč 1. For instance, the breadth across the zygomatic processes of the temporal bones (where they are broken just anterior to the glenoid fossa) is already 145 mm and, therefore, exceeds the total bizygomatic breadth of Mladeč 1 (Table 11), which is in a more anterior position.

The long axes of the glenoid fossae appear to angle and turn laterally as they do in Mladeč 1. However, this in an artifact of the improperly positioned left temporal. Actually, the orientation of these long axis is almost exactly paracoronal. The tooth row is much more anterior to the glenoid fossae than it is in Mladeč 1 (Table 13 shows the direct measurements between these). The distances to the tooth row, in fact, approach the Neandertal condition.

Frontal

Compared with Mladeč 1, the frontal of Mladeč 2 is smaller, higher and more arched, generally more gracile, and even thinner. Like Mladeč 1 there is no frontal boss (Fig. 20). Sagittal lengths are particularly short. Frontal chords from both nasion and glabella are actually below the Neandertal means (Table 5). The sagittal lengths are considerably shorter than Qafzeh 3.

The frontal breadths are moderate, not as large as Mladeč 1. Maximum breadth is greater than the Neandertal mean. However, these samples differ little for maximum frontal breadth, and this is almost the case for minimum frontal breadth. The Mladeč and Neandertal females are quite similar; both contrasting with the much greater minimum breadth is Qafzeh 3. The ratio of minimum to maximum breadth is less than the Neandertal mean value in both Mladeč females. Frontal thicknesses are extremely small, near or below the minima for the comparative samples (Table 6).

The temporal lines arch strongly superiorly as they leave the temporal notch, more so than in Předmostí 4, but similar to Zlatý Kůň and Mladeč 1. It is our observation that the temporal line divides into superior and inferior lines some 20 mm posterior to the supraorbital notch. Anterior to this division, the line forms a weak ridge. The inferior line follows a smooth curve to the coronal suture, while the superior line arcs markedly upward from the point where the lines divide. None of the other early Central European Upper Paleolithic female specimens show this division of the temporal lines (into a superior and inferior line) on the frontal. In their more posterior aspect the temporal lines are much more weakly expressed and are barely discernable where they cross the coronal suture. On the coronal suture, the inferior and superior lines are about 10 mm apart. The high position of the temporal lines (also characteristic of the Mladeč males) is reflected in the bistephanion breadth (Table 5). This measurement, taken between the left and right superior lines, is very small (102 mm). It lies be-

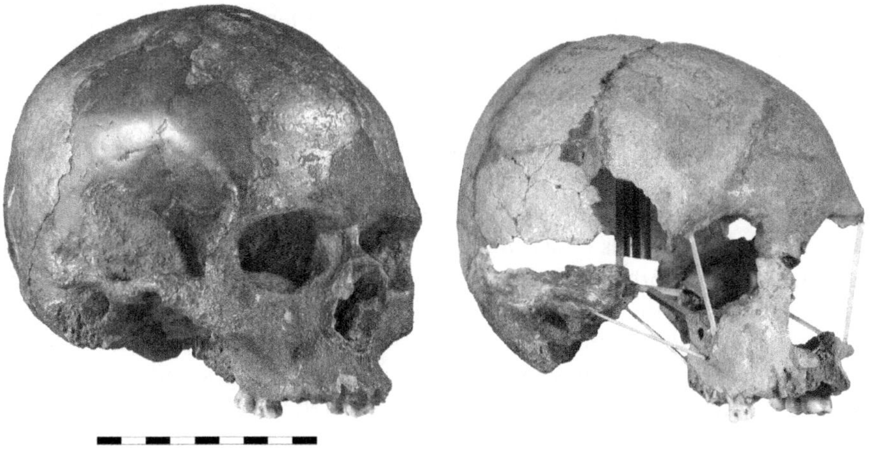

Fig. 20. Comparison of the Mladeč females in angled view

low the minima of the comparative sample. The low value is not an artifact of small frontal size, as is shown by the fact that the index of the bistephanion breadth to the maximum frontal breadth is also below the minima for the comparative samples. Turning to the inferior lines, breadth across the points where these pass over the coronal suture measures 112 mm. In either case, the inferior and superior lines cross the coronal suture above the maximum breadth of the frontal as opposed to being coincident with it.

While in most comparisons this frontal is generally smaller than that of Mladeč 1, the superior facial breadths are greater, and the distances across the orbits approximate the Neandertal female means. Distances between the orbits exceed the Neandertals. These breadths also exceed the Skhul / Qafzeh females (different specimens are involved in different comparisons).

There is a weak supratoral sulcus separating the squama from the supraorbital region. The highly arched squama is neither keeled nor does it show a boss or a small bump similar to that in Mladeč 1. The arc / chord index from glabella is virtually identical to Mladeč 1, above the means of the comparative samples. The vault appears more arched than Mladeč 1 because the glabellar projection is minimal and the squama is more evenly curved in the sagittal plane. The Skhul / Qafzeh females have narrow anterior squamae, with the large central boss offset by a lower flatter region medial to the temporal lines. Moreover, the Levant frontal bones are not as high and have supratoral sulci deeper than those in the European Neandertal females.

Like Mladeč 1 the Mladeč 2 supraorbital region is modern in its form. The region is also similar to that of Mladeč 1 in that the superciliary arch is vertically tall, but projects very little. The glabellar area is continuously thick in its vertical dimension across central supraorbital region and extends to about the mid-orbital position. Mladeč 2 has the most gracile supraorbital region of the early European Upper Paleolithic females, less developed than the region in Mladeč 1. The glabellar prominence is weak, although it is very broad, and the nasal root shows little depression below it. Lateral to the superciliary arches the superior orbital margins are extremely thin and lack any toral development. Even at the temporal notch the thin orbital border continues. The temporal line lies just behind the orbital margin at the zygofrontal suture, separated by only 5 mm. The supraorbital notch is only poorly developed, as in Mladeč 1. It is best expressed on the right side where it can mainly be seen as a weak depression at and just above the superior orbital margin. Mladeč 2 is markedly similar to Cro-Magnon 2 in the supraorbital region. However, the Cro-Magnon female combines a more projecting glabellar region and a prominent frontal boss with a vertical face to outline a shallow supratoral sulcus. Moreover, the central and lateral orbital margins are thicker.

Metrically, the heights of the superciliary arches (medial and high point supraorbital heights, see Table 5) are greater in the early Central European Upper Paleolithic females than in the Neandertals (the Skhul / Qafzeh females are intermediate). The Mladeč 2 dimensions exceed the means of all of these and are at the high ends of the ranges. This contrasts with the lack of toral development over the center-orbit and lateral portions of the Mladeč 2 orbit. Even at its most lateral aspect the Mladeč 2 supraorbital region shows little development. For instance, on the orbital pillar the *fmt-fmo* distance is below the means of all the comparative samples, below the Neandertal range, and so far below the Skhul / Qafzeh range that is close to half the minimum value. Reflecting these variables, the ratio of orbit center to medial height is much less than even the minimum Neandertal and Skhul / Qafzeh values, although it is closer to the Neandertal range than it is to the range of the Levantine females. Lateral supraorbital reduction is reflected in the very low lateral to medial ratio, again very small compared with the Neandertal and Skhul / Qafzeh means and ranges but more like the former than the latter. Both of these ratios in Mladeč 2 are below the Mladeč 1 values.

Unlike Mladeč 1, it is possible to measure the projection of the supraorbital region anterior to the internal face of the frontal squama in this specimen. Projected lengths to the medial and orbit center positions are large in the Neandertals and only slightly less in the Skhul / Qafzeh female. These supraorbital lengths for Mladeč 2 are very small, invariably lower than the minima in the two comparative samples. However, projection of the Mladeč 2 supraorbital region at the lateral position is close to the same as the comparative samples.

The distances along the supraorbital margins, from nasion to *fmt* and to *fmo* are somewhat greater than they are in Mladeč 1. These longer supraorbitals reflect the more angled upper face in the Mladeč 2 female. Thus, the projection of nasion anterior to the bi-*fmt* line (Table 11) is greater. Like Mladeč 1 the upper face is somewhat less angled than the Neandertal females.

Behind the superior orbital margin the orbital roof angles upward. Insofar as the preserved region can be compared with Mladeč 1, these females do not differ. The superior orbital border possesses a supraorbital notch, which is not bridged as in Mladeč 1. The notch is extremely shallow and is located in a medial position, just lateral to the orbital angle.

The interorbital region is extremely broad, both anterior interorbital breadth *(mf-mf)* and the breadth between the superomedial orbital angles (orbital angle breadth) exceeds Mladeč 1 (Table 5). Orbital angle breadth exceeds Neandertal and Skhul/Qafzeh means, while anterior interorbital breadth exceeds all other specimens compared.

Internally, where it is not obscured by plaster or matrix, the surface of the frontal squama is irregular, but shows no distinctive grooves for the meningeal arteries. This surface lacks pacchionian depressions and details of the superior sagittal sinus are either broken away or covered by plaster.

Nasals

Whether considered absolutely or relative to the very broad interorbital area the preserved superior portion of the Mladeč 2 nasal bones are surprisingly narrow (Table 10). The superior nasal breadth is below the ranges of the comparative samples and the ratio of this width to the anterior interorbital breadth is even more dramatically below these ranges. Similarly, the minimum nasal breadth is quite small. The minimum breadth is markedly less than the superior breadth, so like Qafzeh 3 the nasals can be described as having a distinct waist. Similar to the other Mladeč specimens, the nasals form a pillar at the internasal suture. As a whole, the bones project above the frontal processes of the maxilla and are separated from these by a groove. The nasals are set at a moderately high nasal angle. For their preserved length the bones are straight and are angled parallel to the inferior portion of the frontal squama above the supratoral sulcus. They have a higher angulation than the equivalent section of the Mladeč 1 nasals show. While only 12 mm of the bones remain, their continuation on the nasal border of the maxilla's frontal process shows that below this portion the nasal bones must have angled very strongly and, thereby, had they been preserved, would have projected strongly in front of the face. If anything, the angulation of the nasals in Mladeč 2 would have exceeded that in Mladeč 1 had these bones been complete.

The nasal bone lateral length is almost the same as Mladeč 1. Like it, this value is below the Neandertal range.

Parietal bones

The parietal is fairly small, especially along the superior and posterior borders (Table 7). Since the biggest difference in parietal dimensions between the Central European early Upper Paleolithic females and the Neandertal females is in the much *longer* superior border of the Neandertal parietals, the contrast in Mladeč 2 is dramatic. Of the transverse dimensions, the krotaphion-lambda chord does not differ significantly between the Neandertals and Mladeč 2. For the bregma-asterion chord, Mladeč 2 is also like the Neandertal mean in size (Qafzeh 5 is much larger than both). If the parietal radius is taken as a measure of overall size, Mladeč 2 only very slightly exceeds the Neandertals. Mladeč 1 is much larger in all of these dimensions, and we believe that for the most part their magnitudes correspond to the differences in their cranial capacities.

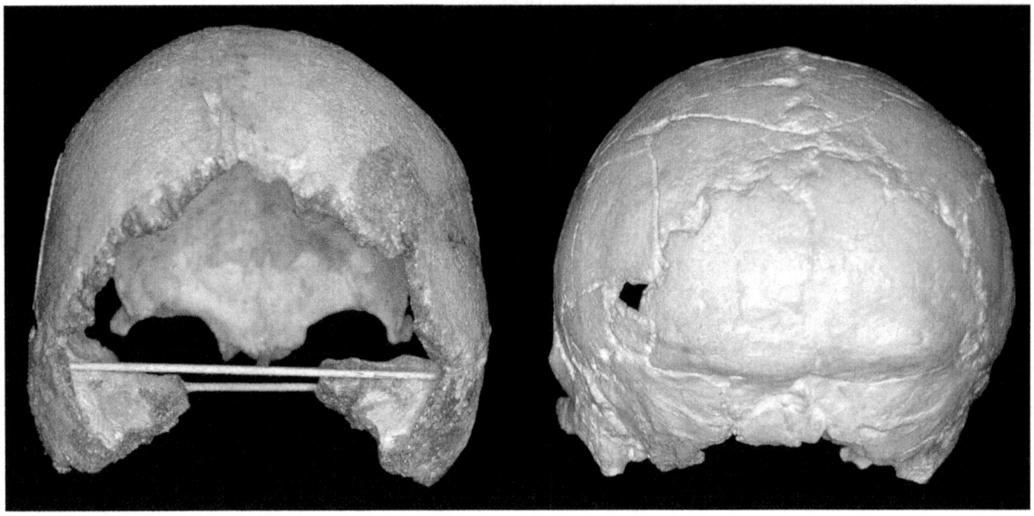

Fig. 21. Comparison of Mladeč 2 and the La Quina 5 female crania in posterior view. Both specimens shown are casts. Differences in cranial contour are evident, including the shape of the sides (rounded in the Neandertal, more slab-sided with a distinct angle at the temporal line in Mladeč), and the position of the greatest breadth (mid-vault in the Neandertal, at the cranial base just over the mastoid in Mladeč)

The parietal angles reveal a pattern similar to Mladeč 1. The top of the bone appears expanded relative to the Neandertals while the rear of the bone is contracted.

Curvature of the parietal as a whole is best indicated by the arc/chord indices for the transverse dimensions. These are very similar to Mladeč 1, only somewhat more curved along the sagittal border. This is also the case in the Neandertal comparison, and the Neandertal parietals are also flatter transversely.

The course of the inferior temporal line is very difficult to discern, but from its high position at the coronal suture (the chord and arc to it from bregma is shorter than those of all the other females) it seems to pass through the apex of the parietal boss and eventually becomes continuous with the supramastoid crest on the temporal. The superior line travels somewhat more medially after it crosses the coronal suture, reaching its highest position some 33 mm behind stephanion. After this the line curves to meet the lambdoidal suture about 39 mm above asterion and for most of the distance that it parallels the lambdoidal suture there is a weak, but distinct angular torus. The line continues onto the mastoid.

Internally, the bone shows good preservation of the surface details. The superior sagittal sinus is weakly expressed and lateral to it are two shallow pacchionian depressions. The anterior and posterior branches of the middle meningeal artery reach the inferior border of the parietal separately and the middle branch is an offshoot of the anterior branch. There is no Breschet's sinus. A distinct double branched form of this sinus is characteristic of European Neandertals (Heim, 1976) and we have found in other early Central European Upper Paleolithic specimens, such as Zlatý Kůň and Vindija 208.

Although none of the occiput was preserved, some of its metric characteristics can be ascertained from the rear of the parietal bones (Fig. 21). For instance, as previously mentioned the biasterionic breadth as estimated from the parietals, 117.1 mm, is very large, between the Neandertal mean and (larger) Skhul/Qafzeh means. On the other hand, the lambda-asterion chord and arc dimensions are quite small. This suggests a short, widely flared occipital plane (commensurate with the small size of the vault) atop a very broad nuchal plane. The arc/chord index is the same as the Neandertal female mean.

Temporal bones

The temporal bones contrast markedly with the frontal of this specimen. While the frontal is gracile and quite modern in appearance, the temporal bones are very robust and in some aspects archaic. The temporal squama is shorter than Mladeč 1 (Table 9), and also appears to be vertically smaller. The squama's length is above the Neandertal range. The height is much smaller and within the Neandertal range, as measured above the Frankfort Horizontal to the highest position on the beveled edge of the parietal's inferior border (not necessarily the highest point on the temporal squama since much of the superior border was broken away). Thus, the proportions of the Mladeč 2 temporal more closely approximate the long, vertically short temporal bones of the Neandertals than do those of Mladeč 1. The squama, like the other vault bones, is quite thin.

The supraglenoid region is of the same length as Mladeč 1 but is somewhat wider. In fact, the gutter breadth is almost at the Neandertal maximum (the Neandertal mean is greater than the Central European early Upper Paleolithic mean). In spite of the small size of the vault, a substantial posterior *temporalis* bundle lay in this gutter. Behind the auricular point position, where the supramastoid crest is continuous with the top of the zygomatic process, a shallow extension of the gutter continues all the way to the rear of the bone. This condition is unlike Mladeč 1, but does resemble Cioclovina and Předmostí 4 to some degree. The superior temporal line and the nuchal line, where they appear as the supramastoid and mastoid crests, come as close together as 7 mm. They are separated by an extremely shallow groove. Both sides show a very similar morphology in this region.

The superior nuchal line extends from asterion to the midline of the mastoid, and then arcs inferiorly to the apex of the process. This line delineates a posterior mastoid surface with a backwards orientation that seems to represent an extension of the nuchal plane similar to that described for Mladeč 5. This configuration also resembles the Předmostí 4 and Cioclovina females, but differs from Dolní Věstonice 3 and Cro-Magnon 2, which have a sharp edge rather than a rearward oriented face at the posterior of the process.

Medial to this posterior mastoid face, a relatively shallow digastric sulcus is bordered by a projecting paramastoid crest. The paramastoid crest projects inferiorly to the approximate level of the mastoid's apex. This differs substantially from the other early European Upper Paleolithic females, which either lack the paramastoid crest (Předmostí 4), or have a projection of the mastoid's apex inferior to it (Mladeč 1, Cioclovina, Dolní Věstonice 3, and Cro-Magnon 2). Lateral to the paramastoid crest the posterior of the temporal is not preserved.

The mastoid (unbroken on the left) is rather small in size, blunted at its apex, and vertically oriented in lateral view. The mastoid process *appears* massive, an impression gained from its shape. This appearance results from the fact that the form is squat and fairly evenly thick to the apex rather than triangular as it is in most other early Central European Upper Paleolithic females. Dimensionally the Mladeč 2 mastoids are small, below the Neandertal female means in the dimensions of the process. Mastoid height as measured by the auricular point-*mastoidale* distance is so low that it is barely within the Neandertal range and the projection of the mastoid's apex inferior to the digastric sulcus is below it.

The difference between the Mladeč 2 mastoid and the Neandertal structures lies in three aspects of its morphology. The Neandertal mastoids contrast with this specimen in their sharp posterior borders and lack of posteriorly oriented surface, a much more triangular form with a narrow apex, and a deeper and more posteriorly extending digastric sulcus which defines a broader basal length dimension. Thus, the Mladeč 2 mastoids present a mix of features, some of which specifically resemble the Neandertals and others do not. The Skhul/Qafzeh females, for the most part, resemble the Neandertals in these features. Qafzeh 3, with the best preserved mastoid region, has a mastoid process that closely resembles the European Neandertals in size and form. Although broken at the paramastoid crest, the remaining crest already projects almost as inferiorly as the mastoid's apex. While the expression of the paramastoid crest resembles that in Mladeč 1 (as described above) and is less pronounced than in Mladeč 2, the combination of mastoid and paramastoid form in this specimen is more Neandertal-like than any of the Mladeč specimens.

The external auditory meatus is oval, with a long axis that slants in an anterosuperior direction as in Mladeč 1. Also like Mladeč 1, the anterior wall of the mastoid forms the posterior wall of the meatus. Posterior to the tympanic ridge, the meatus wall merges with the mastoid's face.

The glenoid fossa is deep and has a distinct vertical anterior face, separated from a horizontal articular surface anterior to it by a sharp angle. This is clearly not an age-related difference. Compensating for the wrongly positioned left side, the orientation of the two fossae is close to paracoronal, and thus the anteromedial angulation of the fossa's long axis is only slight. The glenoid fossa differs substantially in Mladeč 1. Her fossa has a significant anterolateral orientation. In Mladeč 2 the post-glenoid process merges with the anterior wall of the external auditory meatus and the posterior face of the glenoid fossa, including articular surface, extends to the tympanic ridge. The dimensions of the post-glenoid process exceed Mladeč 1 (as well as Skhul 7). The height of the process is close to the Neandertal female mean value while the base is thicker. Length measures of the glenoid fossa are virtually identical to Mladeč 1, while the depth and breadth are somewhat less. The Mladeč 2 dimensions resemble the Neandertals when these earlier females are distinct.

Besides the metric similarities, the paracoronal orientation of two of the three Mladeč fossae (crania 1, 2, and 5) and the merger of the post-glenoid process with the anterior external auditory meatus wall resemble the Neandertal condition. Of these three, the Mladeč 2 fossa is the least like the Neandertals in the perpendicular angulation of its anterior face.

Much of the petrous portion of the temporal bone remains on the right. Just anterior to the supratubercular process a broad groove extends medially from the roof of the glenoid fossa. The angular spine is not present. The region is probably similar in Mladeč 1 although the adhering matrix makes observations of the details difficult. This groove is absent in Předmostí 4. A similar groove in the Neandertal females Gibraltar, Spy 1 and La Quina 5 is narrower, deeper, and more posteriorly positioned. The groove is lacking in Saccopastore 1. The morphology of this region on the medial glenoid fossa surface is variable in the earlier Central European Upper Paleolithic females. The males of this sample, including Mladeč 5, more consistently resemble the Neandertal males (and females).

Medial to the region described above the petrous is broken away on both sizes exposing the cancellous internal structure of the bone but leaving little else for comparison. The vaginal foramen is clearly present on the left side and there is no evidence preserved of an ossified styloid process within it. Finally, dehiscence of the tympanic plate (foramen of Huschke) occurs bilaterally.

Sphenoid

Little of the sphenoid bone remains in this specimen. On the left side the base of the greater wing medial and anterior to the glenoid fossa is preserved. There is nothing particularly diagnostic concerning this area.

Maxilla

Of the portions remaining of the Mladeč 2 face, there are some similarities to various comparable parts of Mladeč 1 and Cro-Magnon 2 (Table 11). These include the widths of the central part of the midface, nasal narrowing, the large medial height of the cheeks (pommette height), the subnasal angulation away from the plane of the face above, and the low base of the zygomatic process of the maxilla. However, there are also a number of contrasts. In particular, the main morphological differences are found in:

- the thinner root of the Mladeč 2 zygomatic process (in the anteroposterior direction),
- the lesser amount of anterior projection around the lateral nasal border (the distance from the medial inferior orbital corner to the lowest point on the nasal maxillary suture is only 10 mm compared with 18 mm in Mladeč 1),
- the shallower palate (lacking a distinct angle between the roof and the premaxillary area, but possessing a distinct palatine torus confined to the mid-palatine suture), the markedly greater gnathic index,
- the much higher rounder orbits (the latter two are also distinctions from Cro-Magnon 2).

There are a number of similarities shared by the two Mladeč females that contrast with the Cro-Magnon 2 condition. The maxilla is less well preserved than that of Mladeč 1, lacking a good deal of bone along the mid-palatine suture and not extending to the base of the zygomaxillary sutures. However, comparing preserved portions, these two faces are similar in some respects. Along the lower border they lack a distinct malar notch and have an only weakly developed canine fossa. The anterior border of the base of the zygomatic process of the maxilla is over the mesial portion of the first molar. Along the orbital margin, both share a tubercle just lateral to the most medial point of the zygomaxillary suture. The tooth row is evenly curved anterior to the bicanine line, and there is a moderate amount of prognathism along the alveolar margin. The alveolar margin is curved following the inverse of the mandibular curve of Spee.

The Mladeč 2 maxilla is low and broad, for instance the orbital alveolar height to M1/M2 is less than Mladeč 1, but we believe the maxilla was broader because the most lateral of the facial breadths

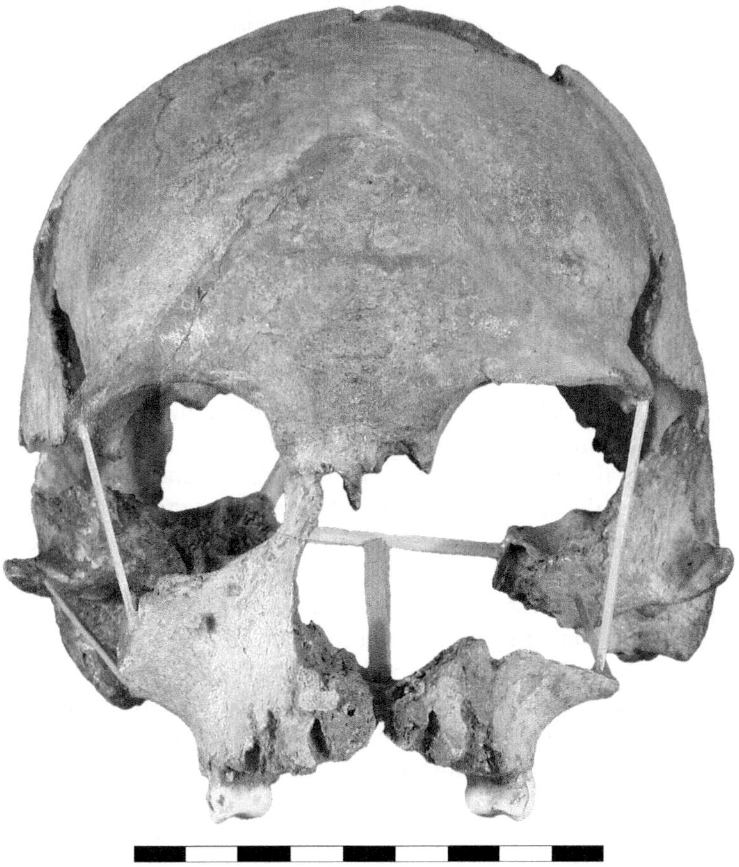

Fig. 22. Mladeč 2 in facial view

preserved (bi-infraorbital foramen and biangular breadths) are greater. Moreover, the greatest distance across the preserved elements of the zygomatic portion of the maxilla (which does *not* extend laterally enough to include the *zygomaxillare* point on either side) is almost as large (95 mm) as the total midfacial breadth of Mladeč 1 (103.1 mm). The zygomatic processes swing quite laterally and a weak canine fossa extends from under the infraorbital foramen. The frontal pillars are not as puffy as in Mladeč 1, so that the sides of the nose anterior to the orbital rim are never horizontal, as they are in the other Mladeč female. The breadth of the nasal aperture is only slightly less than Mladeč 1. It is likely that the nasal index would have been greater, since the height of the Mladeč 2 face is significantly less. The nasal breadth is very small compared with the Neandertals, well below their range. However, if we assume that nasal height is proportional to upper facial height, the expected nasal height of Mladeč 2 would result in a nasal index within the Neandertal range. These data suggest that in relative terms, the Mladeč 2 nose is broad.

Lateral to the lacrimal foramen a suture line connects an accessory infraorbital foramen with the orbital margin. This small accessory infraorbital foramen is located superiorly and medial to the infraorbital foramen, 8.5 mm from it and 4.2 mm below the orbital margin. The main infraorbital foramen is 4.7 mm in breadth, preserved on the left side. This is somewhat less than the foramen breadth in Mladeč 1, and between the two Neandertal female values (3.6 mm and 4.7 mm.). As mentioned above, unlike Mladeč 1 the height of the foramen is greater than the breadth. The center of the foramen is 10.5 mm below the orbital margin, and 28.7 mm above the alveolar margin. In absolute distance this is closer to the orbit than in the Neandertal females, but the Mladeč 2 face is low and in actuality the ratio of orbital distance to distance from the alveolar margin falls within the Neandertal range. The suture line continues posteriorly into the orbit where it appears to join the maxillosphenoidal suture. The base of the zygomatic process of the maxilla is low but seems to be somewhat more angled (relative to the external palate wall) than it is in Mladeč 1. The distance from the alveolar margin to the base of the process is smaller than in any other of the females considered here. However, the height from this base to the lower border of the orbit, the pommette or medial cheek height, is larger. This in fact is just above the Neandertal minimum.

In terms of facial heights, Mladeč 2 is slightly shorter than Mladeč 1 and a good deal shorter than the Neandertal females. As mentioned above the Mladeč 2 face is very prognathic. The index of prognathism based on sagittal projections to nasion and prosthion from the auricular point is very high – virtually identical to the Neandertal maximum. Further reflecting these size relations are several measures that combine all three of these elements (facial height, breadth, and prognathism). These are the measures from prosthion to various lateral elements on the face. The combination of a short and prognathic face makes the Mladeč 2 face appear to be even lower.

From the palatal orientation, the shape of the tooth row is broadly parabolic anterior to the third molars, although lingually these are closer together than the second molars. In the anteroposterior direction the palate is short; the anterior palate length is less than that of any other female discussed (Table 14). However, like the breadths of the face, the breadths of the palate are marked, compared to the breadths of the palates in the other samples. Thus, the breadth across the incisor roots is large, at the Neandertal mean. The breadths across the more posterior teeth are not as great, below the Neandertal means. Mladeč 2 is similar to Qafzeh 5 in these breadth dimensions.

The palatal length and breadth dimensions are combined in measures taken along the tooth row. The more anterior lengths are reflective of palate breadth, and therefore are large enough to be close to or within the Neandertal range. However, because of the short anteroposterior dimension of the palate, tooth row lengths encompassing most or all of the palate, such as the distance from prosthion to the back of the M^3, are comparatively much shorter.

The most distinctive feature of the palate is the moderately developed mid-palatine torus. This structure begins as a low ridge directly behind the incisive foramen and increases in width and height posteriorly. At its widest point (approximately at the transverse palatine suture) it is 15.5 mm. The

Table 14. Palate and tooth row dimensions (mm) for females

	Mladeč females 1	Mladeč females 2	Mladeč male 8[1]	Qafzeh 5	Neandertal Mean (n)	Neandertal Range
Palate Lengths						
Alveolar length	55.2		55.5		62.5 (1)[2]	
Palate length	48.5		46.5		53.1 (1)[2]	
Anterior palate length	36.2	33.4	34.5		41.2 (2)	37.5–44.8
Tooth row lengths						
pr-postcanine	23.9	23.5	24.6	22.4	25.6 (4)	24.3–27.3
pr-post M1	46.7	39.5	46.1	41.3	44.7 (3)	39.5–41.1
pr-post M3		55.0	57.2		59.7 (3)	58.4–62.0
I1-C (roots)	22.4	21.0	23.0	21.7	22.9 (4)	20.0–25.9
C-M2 (roots)	43.6	41.0	43.9		34.0 (6)	29.8–42.0
P3-M2 (roots)	35.0	33.8	35.6		32.7 (4)	31.6–33.7
Palate wall depths						
P4/M1	14.2	9.1		12.2	13.5 (4)	8.4–23.9
M2/M3	15.3	10.7			14.8 (2)	11.1–18.5
Palate depths						
P3/P4	13.6	10.0			16.5 (2)	9.0–24.0
P4/M1	17.5	10.5	11.5		17.5 (2)	11.0–24.0
M1/M2	16.5	10.5	11.6		23.0 (1)[2]	
Palate breadths						
Incisive foramen	4.7			3.9	3.8 (2)	3.3–4.2
External I2 (roots)	28.2	30.2	31.0	29.0	30.2 (2)	29.3–31.9
External C (roots)	42.7		48.3	43.5	44.1 (2)	43.1–45.0
External P4 (roots)	54.0	53.8	61.0	54.1	59.4 (2)	57.5–61.0
External M2 (roots)	65.0	63.2			67.0 (3)	65.5–68.5
Internal P3 (roots)	30.3	29.8	37.4	32.7	33.9 (1)[3]	
Internal M2 (roots)	43.1	41.4			42.7 (1)[2]	

[1] Male palate
[2] Saccopastore 1 only
[3] Gibraltar 1 only

greatest height is about 2.5 mm from the palatal roof. The torus extends onto the palatine bones (where it is not as broad) and terminates at the level of mid-M^3. These heights show the palate to be shallow and fairly even in its depth.

Since the two halves of the palate were separated when we studied the material, an internal view was possible. From this perspective, the mid-palatine torus can also be seen. This thickening produces a maximum inferosuperior diameter of 8.5 mm for the hard palate. The shape of the floor of the hard palate is a feature distinguishing Mladeč 2 from Mladeč 1. The slope of the palate's roof only deepens slightly behind the incisive foramen. There is no true vertically oriented anterior face to the palate, rather only a shallowly sloped superior one anterior to the foramen. The palate slope is not simply a consequence of the palatine torus. Our measurements of the internal palate wall along the alveolar margin also reveal a gradually deepening palate roof, shallower at all points compared to Mladeč 1. Thus, the palate depth measurements were not small as a consequence of the torus development alone. Either measurement set shows Mladeč 2 to have one of the shallowest palates of the European females, and if the single measurement from the Levant (the palate wall height measurement of Qafzeh 5) is indicative Mladeč 2 has a very shallow palate in comparison with these females as well.

Palatine bones

As mentioned above small portions of the palatine bones remain on both halves of the maxilla. On the right side the palatine torus extends across in the mid-palatine suture on to the palatine bones. Since matrix still adheres to most of the palatine bones, it is impossible to determine more precise information concerning their morphology.

Dentition

The maxilla for this specimen is similar in preservation to Mladeč 1 in that only teeth of the posterior tooth row are preserved and that empty sockets for the other teeth are filled with matrix. Only the left M^1, M^2, and M^3 and the right M^1 are preserved. All other teeth were erupted, but lost postmortem, except for the right M^2 which appears to have been broken off during (or since) excavation. For this tooth the root is exposed in the socket, but is not covered with matrix. Besides preservation, Mladeč 1 and Mladeč 2 are also similar in that they both show minor wear on the existing teeth, although wear in Mladeč 2 slightly exceeds Mladeč 1. Consequently, we aged this specimen about 17 years at death.

Left and right M^1's show pinpoint dentin patches, with some minor asymmetry between the two teeth. The left M^1 exhibits dentin exposures of approximately equal size on each of the four cusps, while on the right the mesial exposures are larger than the distal ones. These differences are very minor, representing only idiosyncrasies of the normal attrition process. Relatively large interproximal facets occur on the mesial and distal walls, but these do not cut significantly into the surface and little reduction in mesiodistal length has occurred.

From the occlusal aspect, it is clear that the M^1's form a 4+ pattern, and the distolingual cusp is large relative to the other cusps, representing about a quarter of the total occlusal surface. Compared to Mladeč 1, the hypocone of Mladeč 2 is considerably larger and more bulbous. Like Mladeč 1, there is no surface wrinkling and radiographs show a small, normally configured pulp chamber.

The left M^2 has little wear – there is minor cusp blunting and polishing on all four cusps, but no dentin exposure. Interproximal facets are represented as etchings of the mesial face, with no cavitation of the mesial wall. The cusp pattern is of the 4-type and, compared to Mladeč 1, the hypocone is considerably more developed. Internally, the pulp chamber is small and does not extend into the roots.

The Mladeč 2 third molar has reached the occlusal plane, but based on the extremely minor surface polishing, it appears to have been only newly erupted. Its mesial face rests against the distal wall of M^2, but no interproximal facet has formed. Together, this evidence suggests that the M^3 came into functional occlusion a short time before the individual died.

Unlike the M^1 and M^2, the crown of the M^3 shows a great deal of wrinkling and formation of small cuspules within the inter-cusp surfaces. The hypocone is broken into two small cuspules, located on the distal border and there are fovea and crenulations on all the cusps of the trigone. This type of wrinkling is similar to the isolated maxillary third molar (Mladeč 10) described by Frayer and colleagues (this volume).

Tooth row lengths along the anterior teeth, and the breadth across the four incisor roots, suggest that like Mladeč 1 the missing anterior teeth were probably quite large. The three molars reduce in size from mesial to distal. The M^1 and M^2 are smaller than Mladeč 1 but larger than Neandertal and Skhul/Qafzeh female means. M^3 is reduced, below the means but still within the ranges.

Other adult female cranial remains from the Main Cave

The Mladeč 38 frontal (destroyed)

The specimen is a female frontal discovered by Fürst along with Mladeč 4 (a male) and the Mladeč 37, 44 and 45 children. It was described as an 8.5 cm by 9.5 cm fragment from the anterior portion of the bone, preserving a portion of the nasals and extending on to the frontal squama. Szombathy (1925, 73) describes the forehead profile as flatly arched, and notes the following:

> *"The superior orbital borders are rather flat across, merging centrally to form a weak glabellar prominence and delimiting a strongly expressed trigonum supraorbitale. Nasion, although covered over with matrix, does not appear to be deeply depressed. The nasals are rather narrow (minimum breadth about 9 mm), but are strongly prominent. In contrast to Mladeč 4, this specimen exhibits a voluminous frontal sinus."*

The minimum breadth of the nasal bones, in fact, is slightly larger than the two female nasals (Table 10) but much smaller than the Mladeč 6 value of 14 mm. This measurement is small, whether compared with Qafzeh 3 or the Neandertal female sample, where it is below the range.

Mladeč 41

Discovered by Knies along with Mladeč 39, 40, and 88–91, this small, poorly preserved fragment consists of a portion of unidentifiable cranial bone (this volume, chap. 8, plate XVII, c). The fragment is very roughly triangular with a 43 mm height and a 35 mm breadth. The exterior surface is broken into several pieces, which are slightly displaced from each other, and the interior surface still contains a large piece of adhering matrix. The fragment is quite thin, nowhere exceeding 5 mm in thickness.

Mladeč 42 parietal fragments (destroyed)

Five "very small fragments of adult parietal were found in locus "e" of Chamber E by members of the Litovel Museum Association, along with Mladeč specimens 43, 47, and 62.

Conclusions: Sexual dimorphism and phylogenetic issues

Almost three decades ago when two of the authors (D.W.F. and M.H.W.) first encountered the Mladeč skeletal remains, the specimens in the Naturhistorische Museum Wien were studied before the material in Brno. In many respects the Vienna material conformed to our expectations then about the variability in Upper Paleolithic skeletal remains, which for the most part (especially in the late Upper Paleolithic sample) is represented by males and females with greater size and robusticity, but essentially "modern" morphology. Since the "classic" Upper Paleolithic specimen (Cro-Magnon 1) from France was the most famous, the most accessible and the best known (it was included in most cast collections even at small colleges in the US), our expectations of Aurignacian specimens were influenced by it. Thus in many ways, Mladeč 1 was concordant with what we expected to find in the early Upper Paleolithic. Its high forehead, reduced brows, and small facial dimensions all resembled Cro-Magnon 1. Yet, it possessed a more pronounced bun and distinctive nuchal area that hinted at links to earlier European populations. In addition, the large size and robusticity of the Mladeč 8 palate and Mladeč 30 talus (both males) suggested a morphological pattern atypical of other (generally later) Upper Paleolithic specimens. However, we underestimated and did not appreciate the importance of the great differences between male and female morphology at Mladeč until we later examined Mladeč 5 and a cast of Mladeč 6 in the Moravské Muzeum, Brno. We did not clearly discriminate between what features were robust and what features were Neandertal-like, nor did we fully realize that these were not necessarily the same thing. At our first glance, however, it was apparent that Mladeč 5 was very different from what we expected to find in the Upper Paleolithic. It also was at odds with what W. W. Howells (1982) once observed about Upper Paleolithic crania - that they were instantly recognizable as modern. In general we concurred with Howells' view, but on first sight, the Mladeč material in Brno destroyed this perspective. Mladeč 5 distinctly differed from Cro-Magnon 1, Mladeč 1 and virtually all the early Upper Paleolithic skulls we had seen to that point. It possessed a Neandertal-like frontal, was a low, broad skull and had a well-developed occipital bun and robust nuchal area. These were instantly recognized by us as something different - still modern, but possessing a variety of features which greatly contrasted with the considerably more gracile Mladeč 1 and 2 in Vienna. Clearly, the separation of the Mladeč remains, with the chance allocation to Brno of the primarily the Quarry Cave material which included the two male vaults, and to Vienna primarily the female crania and juvenile from the Main Cave hindered male and female comparisons. Moreover, in those days of Soviet domination and communism in the former Czechoslovakia, it was difficult to go to Brno, so the material there was not known to most American and British paleoanthropologists. In fact, other than Szombathy's description in 1924 (which we subsequently consulted) we can find no photographs or line drawings in the literature of Mladeč 5 or 6 prior to 1978. Finally, since most of the earlier studies included only complete crania in their reviews of the Upper Paleolithic (Riquet, 1970), the Mladeč males were eliminated from consideration and "off our radar screen." Of course, one of us (J.J.) was under no such illusions since he was well-acquainted with the total sample. His publication in 1969 reviewed the Central European material and discussed the great variability at Mladeč, reviewed the distinctive sexual dimorphism at Mladeč, and even suggested evidence of a Neandertal heritage for the early Upper Paleolithic (Jelínek, 1969).

As we recognized the influence of dimorphism on our comparisons, we made special efforts to hold sex constant as we worked on the material and analyzed our results. We now can sum-

Table 15. Dimensions for female maxillary teeth. Length and breadth are in millimeters, areas in square millimeters

	Mladeč 1	Mladeč 2	Mladeč 5	Neandertals Mean (n)	Neandertals Range	Skhūl/Qafzeh Mean (n)	Skhūl/Qafzeh Range
M[1] length	11.2	10.5	10.8	10.9 (5)	10.0–11.5	10.9 (3)	9.9–11.5
breadth	12.2	12.3	13.6	12.0 (6)	11.4–12.8	11.4 (4)	11.1–11.8
M[2] length	11.5	10.4	10.7	10.2 (5)	9.9–10.3	8.8 (1[2])	
breadth	11.9	12.3	13.7	12.3 (5)	11.4–13.4	12.2 (1[2])	
M[3] length		9.0		9.3 (3)	8.7–9.9	9.2 (3)	8.8–9.6
breadth		11.3		12.3 (3)	11.3–13.8	11.1 (3)	10.3–11.7
M[1] area	136.0	129.2	146.9	124.4 (5)	116.0–132.3	124.9 (3)	109.9–133.4
M[2] area	137.4	126.7	146.6	125.7 (5)	112.9–136.5	106.2 (1[2])	
M[3] area		101.3		114.9 (3)	97.6–130.4	101.5 (3)	93.7–112.3

[1] Male palate
[2] Skhul 7

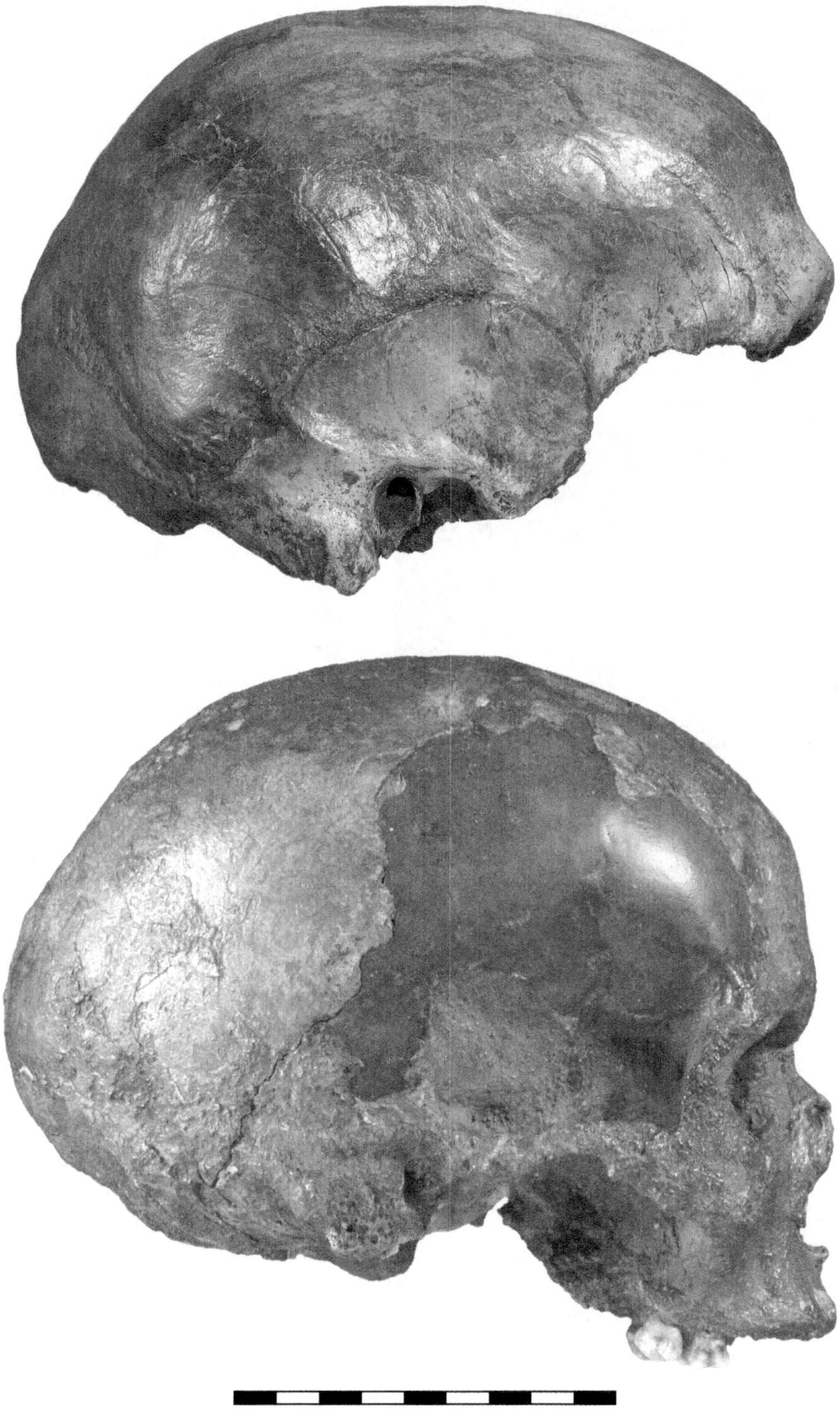

Fig. 23. Mladeč 5 and 1, the best-preserved male and female crania from the site

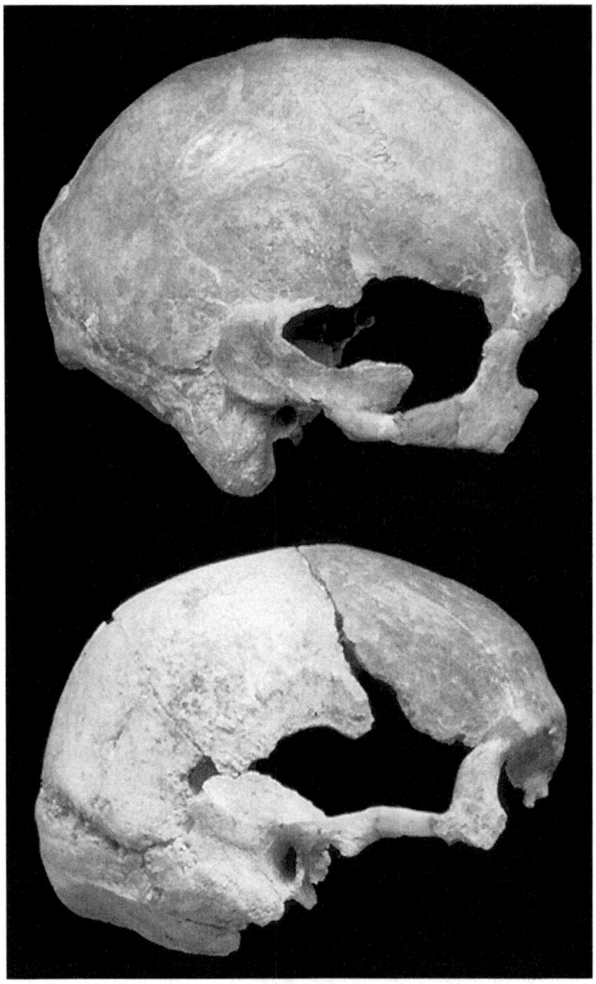

Fig. 24. Comparison of the Zlatý Kůň (below, female) and Pavlov (male) vaults. This variation in Central European crania from the early Upper Paleolithic address sexual dimorphism in the Mladeč sample, as it is similar to the comparison of Mladeč 1 and 5 (Fig. 23)

marize the pattern of dimorphism through the comparison of Mladeč 5 and 1 (Fig. 23, and virtually all of the tables) and also in the corresponding comparison of the Central European male from Pavlov and the Zlatý Kůň female (Fig. 24). The distinctive morphology of the four crania portray textbook size and shape differences used to separate males and females. Male crania are long and low with massive brows and distinctive muscle markings, while females are more gracile in all features. Some of these differences are related to greater robusticity as some have argued, but on the male skulls there is a constellation of features, beyond just size and muscularity, which represent the retention of Neandertal morpology. These are consistently absent in females and along with size differentials make the male/female contrast greater than what is found in subsequent European samples. It is important to note that this is not the only example in the Central European Early Upper Paleolithic. Similar conclusions result from comparing different sex pairs of Předmostí specimens such as 4 and 5 or other males and females samples from Dolní Věstonice. But these are beyond our focus here.

These sex differences are not just size related, since males from Mladeč consistently show features that are diagnostic of European Neandertals as defined by Stringer et al. (1984) and others[6]. In the males these include (but are not limited to) overall lateral profile, an occipital bun with flattening extending onto the parietal, a elliptical suprainiac fossa (in Mladeč 6), a small mastoid process, and the broad lateral incisor and specific nasal features in the Mladeč 8 palate which Schwartz and Tattersall (1996) consider unique to Neandertals. Combined with the generally greater size and robustness in the males, these mainly male features lead to distinctive differences between the males and females from Mladeč. These distinctions are more than simple robusticity differences and, as argued elsewhere (Frayer, 1986), may signal different timings of modernization for early Upper Paleolithic males and females[7]. For example, while there are some

6 The Neandertal features are not necessarily a consequence of the robustness that is associated with being male, which is why we discuss these different components of male morphology separately.

7 Such a difference, of course, does not mean that women became modern in Europe before men did, but relates to our perception of what "modernity" means, and emphasizes that "modernity" is often uncritically used interchangeably with "gracility" (Wolpoff and Caspari, 1997).

similarities in occipital morphology and in nasal angle, it is difficult for us to point to many other morphological similarities between Neandertal and Mladeč females; indeed the women of these two samples show many, fundamental differences (see Figure 6). In contrast, numerous links occur between Neandertal and Mladeč males (Frayer, 1997; Frayer et al., this volume, Fig. 9) that go beyond mere size and we believe signal a significant phylogenetic relationship (Wolpoff et al., 2001). These traits, such as suprainiac fossa, occipital bunning, lambdoidal flattening, mastoid features, expanded lateral maxillary incisor size, and others became uncommon in Upper Paleolithic males after 18,000 years ago (Frayer, 1993). In females, the same traits drop out much earlier, at least judged from the current samples of early Upper Paleolithic females at Mladeč. Obviously, gathering larger samples are important to test these observations, but based on the patterns at Mladeč (and other central European sites not reviewed in this work) it is important to entertain the possibility that different evolutionary forces affected males and females through time.

In this regard it is regrettable that we are unable to perform a systematic ancestry analysis in females as we did in males. This is in part because of the very small sample sizes of European Neandertal females, but is primarily related to the nearly complete absence of adequately preserved Skhul/Qafzeh females. Yet, viewed from the existing samples in Europe and the degree of difference between Neandertal and Mladeč females, a question of "dual ancestry" would be much more difficult to sustain, not because one ancestral source is predominant, but because even visual inspection of the Mladeč females compared to a Neandertal female (Fig. 6), a Qafzeh female (Fig. 7), or most distantly to an African female (Fig. 12) suggests that no special link to any of them is obvious. Females and males of the early Upper Paleolithic certainly co-existed, but the types of evolutionary forces operating on females seem to be different from the males. Whether this has biological meaning or simply is the result of inadequate sampling awaits future work, but it is obvious to us that no case for a special or unique African ancestry can be established at this time. In this regard, the females provide the same information as the males.

Acknowledgments

We thank M. Teschler-Nicola for her kind invitation to contribute to this volume, and are deeply grateful to the directors, curators, and personnel of the Naturhistorisches Museum Wien, for permission to work on the Mladeč specimens, the rooms and facilities that were graciously provided to us, the extraordinary hospitality extended to us, and the great patience shown in waiting for this paper to appear. We thank H. Kritscher, H. Poxleitner, L. Seitl, J. Szilvássy, and (once again) M. Teschler-Nicola. We also thank the many curators and museum directors with responsibility for the comparative samples we discuss here, for access to the fossil collections in their care and for the hospitality and help extended to us in the course of our research. We are indebted to C. Bauer, who did the full translation of the Szombathy monograph for us. Of our colleagues and co-researchers, the conversations with and contributions of R. Caspari, J. Hawks, N. Minugh-Purvis, and F. Smith, were exceptional, and we express our great debt to them.

References

Absolon, K. (1929) New finds of fossil human skeletons in Moravia. *Anthropologie (Prague)* 5, 79–107

Billy, G. (1972) L'évolution humaine au paléolithique supérieur. *Homo* 23, 2–12

Churchill, S. E. and Smith, F. H. (2000) Makers of the Early Aurignacian of Europe. *Yearbook of Physical Anthropology* 43, 61–115

Endo, B. (1966) Experimental studies on the mechanical significance of the form of the human facial skeleton. *Japanese Faculty of Sciences, University of Tokyo* (Section 5) 3, pp. 1–106

Frayer, D. W. (1978) *Evolution of the dentition in Upper Paleolithic and Mesolithic Europe.* Lawrence: University of Kansas Publications in Anthropology (vol. 10)

Frayer, D. W. (1980) Sexual dimorphism and cultural evolution in the Late Pleistocene and Holocene of Europe. *Journal of Human Evolution* 9 (5), 399–415

Frayer, D. W. (1986) Cranial variation at Mladeč and the relationship between Mousterian and Upper Paleolithic hominids. In (V. V Novotný and A. Mizerová, Eds.) *Fossil man. New facts, new ideas. Papers in honor of Jan Jelínek's Life Anniversary. Anthropologie (Brno)* 23, 243–256

Frayer, D. W. (1993) Evolution at the European edge: Neanderthal and Upper Paleolithic relationships. *Préhistoire Européenne* 2, 9–69

Frayer, D. W. (1997) Perspectives on Neanderthals as ancestors. In (G. A. Clark and C. M. Willermet, Eds.) *Conceptual issues in modern human origins research.* New York: de Gruyter, pp. 202–234, 437–492 (bibliography)

Frayer, D. W., Jelínek, J., Oliva, M. and Wolpoff, M. M. (2006) Aurignacian male crania, jaws and teeth from Mladeč Caves, Moravia, Czech Republic. In (M. Teschler-Nicola, Ed.) *Early modern humans at the Moravian Gate. The Mladeč Caves and their remains,* this issue, pp. 185–272

Heim, J.-L. (1976) *Les hommes fossiles de La Ferrassie: Le Gisement. Les Squelettes d'adultes: Crâne et squelette du tronc.* Paris: Masson et Cie.

Henke, W. (1987) The application of multivariate statistics to the problems of Upper Paleolithic and Mesolithic samples. *Human Evolution* 2 (2), 149–167

Howell, F. C. (1951) The place of Neanderthal man in human evolution. *American Journal of Physical Anthropology* 9, 379–416

Howell, F. C. (1957) The evolutionary significance of variation and varieties of "Neanderthal" man. *The Quarterly Review of Biology* 32, 330–347

Howells, W. W. (1982) Comment on "Upper Pleistocene evolution in south-central Europe" by FH Smith. *Current Anthropology* 23, 688–689

Hrdlička, A. (1930) *The skeletal remains of early man* (Smithsonian Miscellaneous Collections 83). Washington DC: Smithsonian Institution

Jelínek, J. (1954) Nález fosilního člověka Dolní Věstonice III. *Anthropozoikum* 3, 37–91

Jelínek, J. (1969) Neanderthal man and *Homo sapiens* in central and eastern Europe. *Current Anthropology* 10, 475–503

Jelínek, J. (1983) The Mladeč finds and their evolutionary importance. *Anthropologie* (Brno) 21, 57–64

Kříž, M. (1903) *Beiträge zur Kenntnis der Quartärzeit in Mähren.* Kremsier (Moravia): Steinitz

Laitman, J. T., Heimbuch, R. C. and Crelin, E. S. (1979) The basicranium of fossil hominids as an indicator of their upper respiratory systems. *American Journal of Physical Anthropology* 51, 15–34

Lieberman, P. (1975) *On the origins of language.* New York: MacMillan

Martin, R. (1928) *Lehrbuch der Anthropologie: Kraniologie, Osteologie.* Vol. II. Jena: Fischer

Maška, K. J. (1895) Diluviální člověk v Předmostí. *Čas vlast Muzej Spolku Olomouc* 12, 1–7

Matiegka, J. (1934) *Homo předmostensis: Fosilní Člověk z Předmostí na Moravě. 1. Lebky.* Prague: Česká Akad Věd Umění

Morant, G. M. (1930) Studies of Paleolithic man IV. A biometric study of the Upper Paleolithic skulls of Europe and their relationships to earlier and later forms. *Annals of Eugenics* 4, 109–214

Otte, M. (1979) *La Paleolithique Supérieur Ancien du Belgique.* Brussels: Musées Royal d'Art et d'Histoire

Payá, A. C. and Walker, M. J. (1980) A possible hominid fossil from Alicante, Spain? *Current Anthropology* 21, 795–800

Păunescu, A. (2001) Paleoliticul şi mezoliticul din spaţiul transilvan. Bucureşti: Editura AGIR

Poissonnet, C. M., Olivier, G. and Tissier, H. (1978) Estimation de la capacité crâienne à partir d'un os de la voûte. *Bulletins et Mémoires de la Société d'Anthropologie de Paris* (série 13) 5, 217–221

Prossinger, M. and Teschler-Nicola, M. (2006) Electronic segmentation methods reveal preservation status and otherwise unobservable features of the Mladeč 1 cranium. In (M. Teschler-Nicola, Ed.) *Early modern humans at the Moravian Gate. The Mladeč Caves and their remains,* this issue, pp. 341–356

Rak, Y. (1986) The Neanderthal: a new look at an old face. *Journal of Human Evolution* 15 (3), 151–164

Riquet, R. (1970) La race Cro-Magnon, abus de language ou réalité objective? In (G. Camps and G. Oliver, Eds,) *L'Homme de Cro-Magnon.* Paris: Arts et Métiers Graphiques, pp. 37–58

Russell, M. D. (1985) The supraorbital torus: "a most remarkable peculiarity". *Current Anthropology* 26 (3), 337–360

Schwartz, J. H. and Tattersall, I. (1996) Significance of some previously unrecognized apomorphies in the nasal region of *Homo neanderthalensis. Proceedings of the National Academy of Sciences USA* 93, 10852–10854

Skinner, M. F. and Sperber, G. H. (1982) *Atlas of the radiographs of early man.* New York: Alan R. Liss

Skutil, J. (1940) Paleolitikum v bývalém Československu. *Obzor Prehistorický* 12, 41–43

Smith, F. H. (1982) Upper Pleistocene hominid evolution in south-central Europe: A review of the evidence and analysis of trends. *Current Anthropology* 23, 667–703

Smith, F. H. (1984) Fossil hominids from the Upper Pleistocene of central Europe and the origin of modern Europeans. In (F. H. Smith and F. Spencer, Eds.) *The origins of modern humans: A world survey of the fossil evidence.* New York: Alan R. Liss, pp. 137–209

Smith, F. H. (1997) Mladeč. In (F. Spencer, Ed.) *History of physical anthropology. An encyclopedia.* New York: Garland, pp. 659–660

Smith, F. H. and Ranyard, G. C. (1980) Evolution of the supraorbital region in Upper Pleistocene fossil hominids from south-central Europe. *American Journal of Physical Anthropology* 53, 589–609

Smith, F. H., Trinkaus, E. Pettitt, P. B., Karavanic, I. and Paunovic, M. (1999) Direct radiocarbon dates for Vindija G1 and Velika Pecina Late Pleistocene hominid remains. *Proceedings of the National Academy of Sciences USA* 96, 12281–12286

Stringer, C. B., Hublin, J.-J. and Vandermeersch, B. (1984) The origin of anatomically modern humans in western Europe. In (F. H. Smith and F. Spencer, Eds.) *The origins of modern humans: A world survey of the fossil evidence.* New York: Alan R. Liss, pp. 51–135

Svoboda, J. (2000) The depositional context of the early Upper Paleolithic human fossils from the Koněprusy (Zlatý kůň) and Mladeč Caves, Czech Republic. *Journal of Human Evolution* 38, 523–536

Svoboda, J. A., van der Plicht, J. and Kuzelka, V. (2002) Upper Paleolithic and Mesolithic human fossils from Moravia and Bohemia (Czech Republic): some new ^{14}C dates. *Antiquity* 76, 957–962

Szilvássy, J., Kritscher, H. and Vlček, E. (1987) Die Bedeutung röntgenologischer Methoden für anthropologische Untersuchung ur- und frühgeschichtlicher Gräberfelder. *Annalen des Naturhistorischen Museums Wien* 89, 313–352

Szombathy, J. (1901) Un crâne de la rasse de Cro-Magnon trouvé en Moravie. Congrès International d'Anthropologie et d'Archéologie préhistoriques, XIIe Session, Paris 1900. *L'Anthropologie* 12, 133–140

Szombathy, J. (1904) Neue diluviale Funde von Lautsch in Mähren. *Jahrbuch der k. k. Zentral-Kommission für Kunst- und historische Denkmäler* II (1), 9–16

Szombathy, J. (1925) Die diluvialen Menschenreste aus der Fürst-Johanns-Höhle bei Lautsch in Mähren. *Die Eiszeit* 2, 1–34; 73–95

Valoch, K. (1982) Neue paläolithische Funde von Brno-Bohunice. Časopis moravského muzeo. Sci. soc. 67, 31–48

Vandermeersch, B. (1981) *Les hommes fossiles de Qafzeh (Israël).* Paris: Centre National de la Recherche Sciéntifique

Vlček E. (1971) Czechoslovakia. In (K. P. Oakley, B. G. Campbell and T. I. Molleson, Eds): *Catalogue of fossil hominids, Part II: Europe.* London: British Museum (Natural History), pp. 47–64

Vlček, E. (1991) Die Mammutjäger von Dolní Věstonice. Anthropologische Bearbeitung der Skelette aus Dolní Věstonice und Pavlov. *Archäologie und Museum* (Basel) 22

Wankel, J. (1884) První stopy lidské na Moravě. *Časopis vlast. muzea společnost Olomouc* 1, 89–96

Weidenreich, F. (1943) The skull of *Sinanthropus pekinensis:* A comparative study of a primitive hominid skull. *Palaeontologia Sinica,* New Series D, Number 10 (whole series No. 127)

Weidenreich, F. (1951) Morphology of Solo man. *Anthropological Papers of the American Museum of Natural History* 43 (3), 205–290

White, T. D. (2000) *Human osteology.* 2nd ed. New York: Academic Press

Wolpoff, M. H. (1979) The Krapina dental remains. *American Journal of Physical Anthropology* 50, 67–114

Wolpoff, M. H. (1982) Comment on "Upper Pleistocene hominid evolution in south-central Europe," by F. H. Smith. *Current Anthropology* 23, 693

Wolpoff, M. H. (1999) *Paleoanthropology.* 2nd ed. New York: McGraw-Hill

Wolpoff, M. H. and Caspari, R. (1997) What does it mean to be modern? In (G. A. Clark and C. M. Willermet, Eds.) *Conceptual issues in modern human origins research.* New York: Aldine de Gruyter, pp. 28–44, 437–492 (bibliography)

Wolpoff, M. H., Hawks, D. J., Frayer D. W. and Hunley, K. (2001) Modern human ancestry at the peripheries: a test of the replacement theory. *Science* 291, 293–297
Wolpoff, M. H., Smith, F. H., Malez, M., Radovčić, J. and Rukavina, D. (1981) Upper Pleistocene human remains from Vindija cave, Croatia, Yugoslavia. *American Journal of Physical Anthropology* 54, 499–545

Chapter 11

ELECTRONIC SEGMENTATION METHODS REVEAL THE PRESERVATION STATUS AND OTHERWISE UNOBSERVABLE FEATURES OF THE MLADEČ 1 CRANIUM

Hermann Prossinger and Maria Teschler-Nicola

Introduction

History

The story of the recovery of the Upper Paleolithic Mladeč/Lautsch site (Moravia, Czech Republic) is reasonably well known (Szombathy, 1925; Svoboda, 2000; Oliva, 1989; Jelínek, 1983). In 1881 and in 1882, Joseph Szombathy, assistant at the *k.u.k. Naturhistorisches Museum,* Vienna, carried out the first systematic excavations in the Johann von und zu Liechtenstein cave, commissioned to do so by the Imperial Academy of Sciences (Szombathy, 1882). All the human fossils collected by Szombathy were handed over to the newly established *Anthropologisch-ethnographische Abteilung* of the *k.u.k. Naturhistorisches Museum,* Vienna (Szombathy, 1925). Included in this trove and still stored in this *Abteilung* was the almost complete cranium of a young adult (16±1 years of age) female (Wolpoff et al., this volume), inventoried as Mladeč 1. This specimen has recently been dated at 31 thousand ^{14}C years BP (Wild et al., 2005).

Many illustrations of the cranium show the left side, hiding the fact that a large part of the right parietal is missing: the specimen had been damaged during the discovery and removal phase in 1881 (Szombathy, 1925). This (false) impression introduces a difficulty when assessing the morphology and is further corroborated by the fact that the missing parietal parts have been reconstructed during Szombathy's directorship of the *Anthropologische Abteilung of the Naturhistorisches Museum,* Vienna (perhaps even by Szombathy himself) using gypsum. (Indeed, in his 1925 publication, Szombathy himself concedes of having been once mislead in one morphological assessment due to the right-hand side having been reconstructed.) After this reconstruction had been completed, the exterior of the cranium had been covered with a curing agent and perhaps painted with shellac, changing the color of the fossilized bone and making the color difference between the gypsum reconstruction and the rest less apparent (see Wolpoff et. al., this volume, chap. 8, Plate I). The same covering had been applied to layers of encrustations (brownish calcite – we henceforth call this sinter – and considerably thick whitish staligmatic material) adhering to the many parts of fossilized cranium, notably at the base (see chap. 8, Plate I, below right) and inside the nasal aperture (see chap. 8, Plate I, top left).

In this paper, we present a suite of electronic segmentation and removal procedures. As these are applied solely to the CT-scan of the specimen, they are reversible, non-invasive and leave the (priceless) original intact and unaltered. The result of the electronic segmentation procedure permits us to present, for the first time, the isolated fossil with all later additions/reconstructions and the encrustations removed. Because the methodologies use a CT-scan, we can also apply them to the interior of the cranium. As a result, we also present the heretofore-unknown endocranial status of this specimen.

The result of this segmentation process and electronic "cleaning" or "preparation" is a 3D data file, which has been used to produce images, with and without the encrustations (Plates I–IV) for comparison and assessment purposes. Because a considerable part of Mladeč 1 has been left encrusted, morphological assessment using the (original) museum specimen is fraught with risks. One outcome of the segmentation process presented here is therefore the opportunity to reappraise the morphological features present in Mladeč 1. Another is that a more detailed record of the restoration work (especially the gypsum reconstructions of the crushed parietals) and the taphonomy has become possible. A third is the possibility of determining the 3D-coordinates of landmarks covered with encrustations in the original specimen.

Methodology

Scanning details

The CT-Scan of the cranium we used in this analysis has a slice thickness of 1 mm; the voxel resolution within a slice is 0.488281 mm × 0.488281 mm (for a total of 512 × 512 voxel^2 per slice). The CT-scan image file was reformatted to produce (cubic) iso-voxels 0.428281 mm on each edge (303 × 436 × 342 voxels[3]). The x-ray density resolution was recalibrated so that the image gray values range from 0 (air) to 4095 (highest mineral density). Background noise due to the scanning process was eliminated using standard medical image software.

Segmentation procedures

All image editing, image algebra and image processing was done using ANALYZE® (Biomedical Imaging Resource, Mayo Foundation). The processed, segmented image files, along with two image files – one consisting of the segmented encrustations (staligmatic minerals and the sinter) and one of the gypsum – were surface rendered in 3DVIEWNIX® (Udupa, 1999). The rendered images were then further processed in PHOTOSHOP® (Adobe Systems Incorporated).

There are four types of foreign material that cover or replace the fossilized bone: (1) gypsum used to reconstruct missing parts, (2) staligmatic mineral, (3) layers of sinter, and (4) curing agent and shellac.

To illustrate this, we show, in Fig. 1a, the CT-image of the transversal slice no. 118. In the occipital region, we observe (along the line AB) gypsum, sinter encrustation and fossilized bone. The x-ray signal characteristics of the gypsum are sufficiently different from the other materials to be easily visually identified in the CT-scan images (Fig. 1b): it has a very smooth appearance because it has a small range of gray values. Unfortunately, the Hounsfield (1973) numbers are very close to the midrange of all other absorption numbers (i.e.,

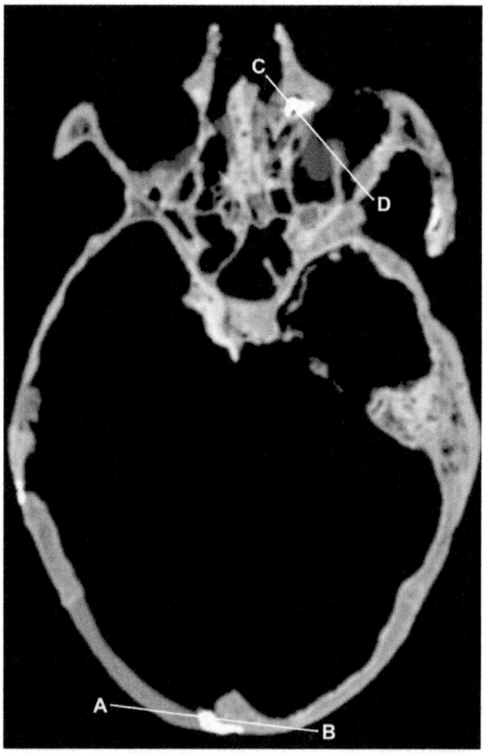

Fig. 1a.

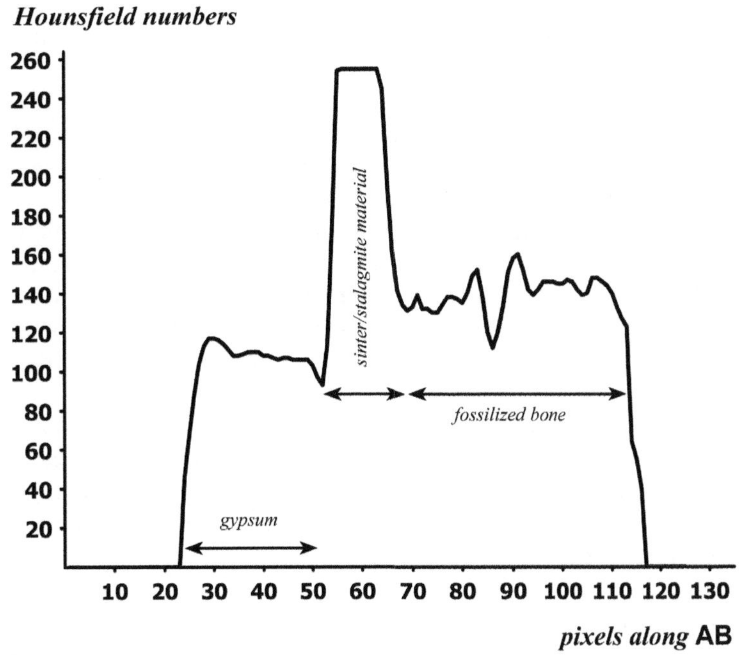

Fig. 1b

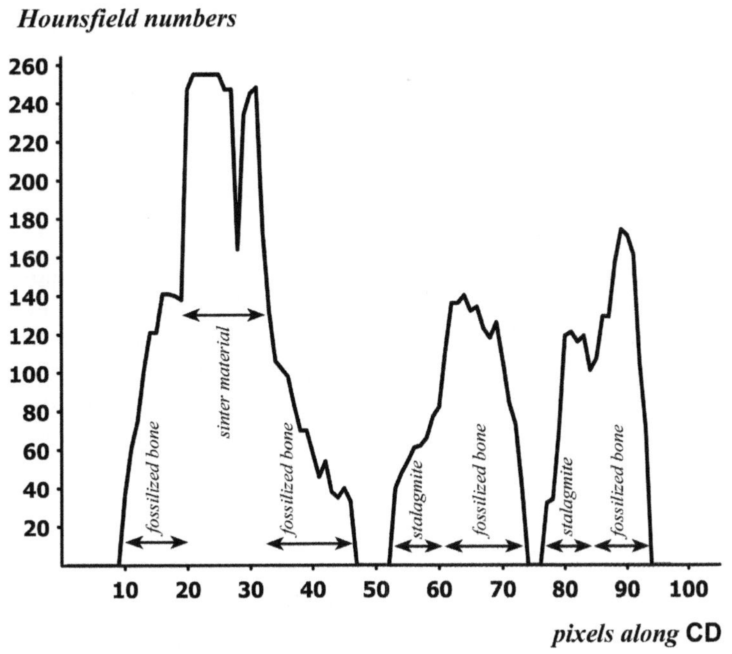

Fig. 1c

Fig. 1. The characteristic Hounsfield number profile of the gypsum reconstructions. **(a)** A CT slice (transverse slice no. 118); **(b)** The profile along the line AB. **(c)** The profile along the line CD those of much of the fossilized bone), so one cannot apply a discrimination algorithm to automatically segment the gypsum reconstructions (a methodology introduced in Prossinger et al., 2000).

The sinter is often thin and covers large regions of the specimen almost uniformly. It usually has higher Hounsfield numbers than does the fossilized bone, but there are plenty of regions with the sinter and staligmatic material having comparable Hounsfield numbers, so one cannot program a gradient segmentation algorithm (Prossinger et al., 2003).

Because of the very high Hounsfield numbers of most of the staligmatic mineral and the sinter (and very few parts of the specimen are equally x-ray dense), setting a threshold allows segmentation of most of the staligmatic mineral and the sinter. Following the line from C to D, as shown in Fig. 1c, we first encounter, at the lateral surface of the nasal cavity, fossilized bone, then sinter, then fossilized bone, then staligmatic material (with very low Hounsfield numbers), then further fossilized bone.

A difficulty is, however, that, in parts, during the fossilization process, bony matter has been replaced with sinter and thus the x-ray signature of some fossilized bone is as high as that of the sinter encrustations. One has to manually image-edit the removed high Hounsfield number parts of the fossilized bone, in order to add back to the specimen parts that had been removed and yet were actually not sinter encrustations.

As a last step, all shellac that had covered removable material (stalagmite, sinter, gypsum) was also image-edited away. In many parts of the specimen where no encrustation material had been removed, the shellac layer has remained. Due to the finite resolution of the CT-scan process, this layer could not always be segmented. Whenever not, the surface in such places is then slightly thicker than it would be in the original, but this difference cannot be detected in the CT-scan. The minimum thickness detectable in the CT-scan is 0.488 mm, so only a shellac layer thicker than this can be detected. However, a thickness bias of ~0.5 mm rarely alters morphological assessments.

Results

Gypsum

There are two gypsum reconstructions (made visible in Plates I–IV): a large region encompassing the posterior part of the right frontal bone and the anterior part of the right parietal and a medium-sized portion of the right parietal directly above the suture with the occipital bone. The large anterior reconstruction region (frontal and parietal) is due to a mishap during the removal of the cranium (Szombathy, 1925). As noted, during removal parts of the parietal were broken and irretrievably destroyed. In Plate II (*left*, top), one can see that there are small encrustations between the reconstructions and the surviving fossilized bone.

Encrustations

There are many places on the cranium that have retained encrustations: within the nasal aperture (Plate I, *left*, top), on the basiooccipital base (Plate I, *right*, top), on the zygomatic arches (Plates I–III), within the orbitae (Plate I, *left*, top) as well as on many endocranial surfaces (Plates II–IV).

Description of ectocranial encrustations

Craniofacial complex (Plate I, left)

The lateral wall of the right orbita has a large surface covered with encrustations, ~4–7 voxels thick. The lacrimal and ethmoid bone on the medial side of both orbitae have thin encrustation layers, ~1–3 voxels thick. On both sides, there are encrustations in the maxillary fossae. Several tooth sockets contain encrustations. Remarkably, these are not completely filled (Plate I, *right,* top); we will address this issue below.

Cranial base (Plates I–IV)

The thickest encrustation covers the region of the sphenooccipital suture. The encrustation is particularly thick on the outer surface of the clivus (Fig. 2, Plates I–IV). Both right and left medial and lateral pterygoid plates are covered with encrustations: even the region between them is filled with encrustation material (Plates I, *right,* top). Electronic segmentation in this region is particularly difficult, due to the extreme thinness of the plates (refer to Conclusion for further elaborations on this issue). The region around both right and left condylar foramina are also covered with encrustations, many of the canals are plugged with them (Plate I, *right,* top).

Zygomatic arch (Plates I–III)

The zygomatic arches are only fragmentarily preserved; even so, they are still in articulation with the other bones of the cranium. There is a large, triangular encrustation at the distal, anterior notch of the left zygomatic arch (Plates I, *left,* top); the right zygomatic arch is more complete and is encrusted in parts (Plates I–II).

Neurocranium (Plate II, left, top)

There are many small regions of encrustation. Noteworthy are the remnants that were not removed when the large gaps were reconstructed with gypsum. In five places, the gypsum contacts encrustations, not fossilized bone.

Description of endocranial encrustations

Anterior and medial cranial fossa (Plates II and IV)

Half the right anterior cranial fossa has been reconstructed; a large region of the left anterior cranial fossa is covered with a thin (~1–5 voxels thick) layer of sinter material. The left middle cranial fossa and the greater wing of the sphenoid bone are also covered with a comparably thin layer. The crista galli is covered with encrustations.

Clivus and anterior occipital bone (Plates II and IV)

The basilar surface of the occipital from the clivus to the foramen magnum has a thick encrustation – several mm at its thickest place (Plates II–IV). The right anterior part of the squama (posterior cranial fossa) of the occipital is encrusted in various places (Plates III and IV), while the left part has very few such encrustations (Plate III, *right,* top). The posterior neurocranium is remarkably free of any encrustations on its inner surface (Plate III).

Maxilla (Plate III, right, top)

Both the right and left maxillary sinus contain encrustations. Surprisingly, these sinuses are not completely filled with them, although the encrustations are thick in parts.

Some quantitative findings

Encrustation thickness at clivus

The Hounsfield number profile can be used to determine the thickness of encrustations, as shown in Fig. 2. A graph of the curve of Hounsfield numbers along a straight line – preferably perpendicular to the Frankfort Horizontal – is made. The boundaries of the transition encrustation/fossilized bone are marked and the pixel values on the abscissa can be used to estimate the thickness. The encrustations on the inner and outer surface of the clivus are particularly thick: 7.8 mm and 6.8 mm, respectively. These values are needed in order to correct the measurements made on the cranium with calipers (when measuring overall height of the cranium, for example).

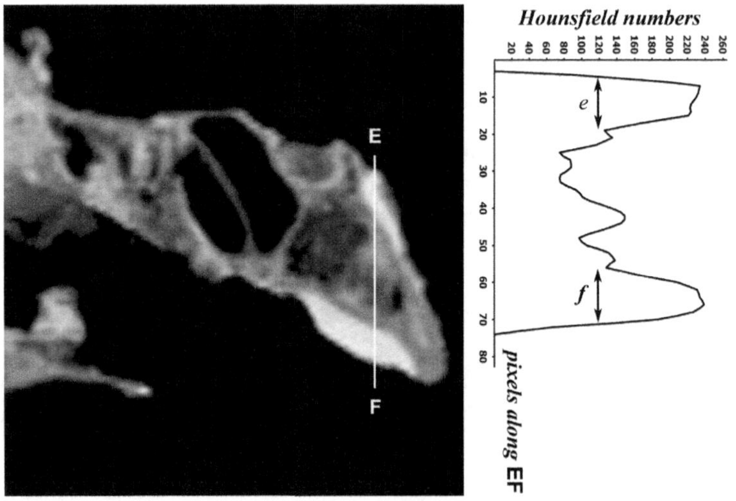

Fig. 2. A methodology used to determine the thickness of the staligmatic material deposited on the outer and inner surface of the clivus. The transverse slice shown here is (close to) perpendicular to the Frankfort Horizontal, so pixels counts directly convert to millimeters. (Foreshortening due to the Pythagorean Theorem is negligible.) The Hounsfield number profile allows the determination of the boundary between staligmatic material and fossilized bone. Note that the Hounsfield number curves are not parallel to the ordinate at the edges, due to the partial volume effect (thicknesses: *e* = 14 pixels = 6.8 mm; *f* = 16 pixels = 7.8 mm)

Volume of encrustations

A considerable amount of encrustations adheres to the fossilized cranium. Its volume is 31 cm^3; the total volume of fossilized bone of the specimen is 619 cm^3. Consequently, the volume of encrustations is ~5 % of the fossilized bone of the fossil.

Endocranial volume

The method of flood-filling (Prossinger et al., 2003) can be used to estimate the endocranial volume, a surrogate for brain volume. All foramena must be "sealed" (see Prossinger et al., 2003, for details) and then all voxels inside are counted. Multiplying this (counted) number with the volume of one voxel (0.116415 mm^3) gives the estimate for the endocranial volume: 1575 cm^3. This method is far superior to methods using linear dimensions of the skull. In the latter, one calculates the volume of a rectangular prism with dimensions of something close to a skull breadth with a likewise dimension correlating with skull length with a third correlating with skull height. The endocranial volume, being less than the volume of the prism, is obtained by using "regression factors", derived from a sample of "reference" skulls (their suitability is argued in several publications). Flood-filling the endocranial cavity of the CT-scan does not rely on reference specimens. Szombathy (1925) lists the results of several endocranial volumes determined by the use of regression factors: 1615 cm^3 (method of Welcker), 1698 cm^3 (method of Manouvrier), 1565 cm^3 (method of Lee-Wacker), 1618 cm^3 (method of Froriep). Szombathy extracts, from these results, a putative endocranial volume of 1629 cm^3. Wolpoff (1999) and Wolpoff et al. (this volume) use the regression factors published by Billy (1972) to estimate the endocranial volume, obtaining a value of 1540 cm^3. We emphasize that our result cannot be construed to support the method of Lee-Wacker, although that method yields a result close to ours (obtained by flood-filling and correcting for restoration work and incompleteness of the fossil specimen).

The method of counting voxels implies a simplicity that belies several technical challenges. In addition to the most obvious – namely, where to define the limiting surface across the foramen magnum –, we note that there exist not only several cracks in the cranium, but also the reconstructed gypsum parietals. Careful assessment of these two issues is necessary. Not only do we need to reconstruct the putative surface when plugging cracks and gaps, but we also had to deal with the thickness of gypsum – much thicker than the fossilized bone it reconstructs (Fig. 3; presumably, the gypsum needed to be thicker in order to ensure mechanical stability). We had to electronically erode the internal gypsum wall, as well as plug the cracks before flood-filling the endocranial cavity. Using arcane mustard seed methods cannot deal with any of these issues; the regression fac-

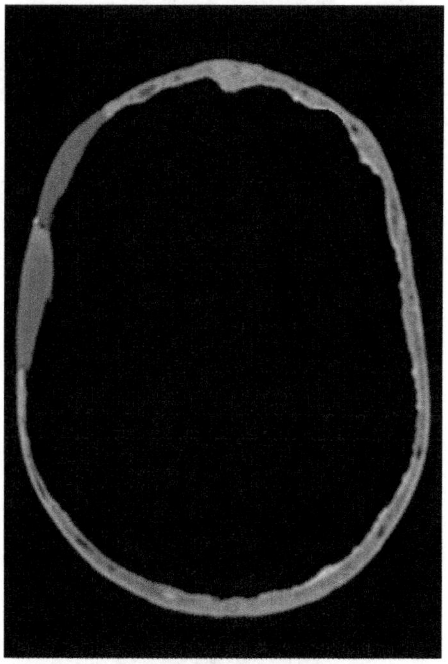

Fig. 3. A sample slice (transverse slice no. 227) showing the thickness of the gypsum used to reconstruct the parietal vault. It is obvious that the gypsum is far thicker than the bone it replaces; electronic erosion is needed before flood-filling can be performed. Note that the fossilized bone missing from this large area was reconstructed in two steps: the two batches of gypsum that have been used (and had dried at different times) can be clearly identified – as well as their sequence of application

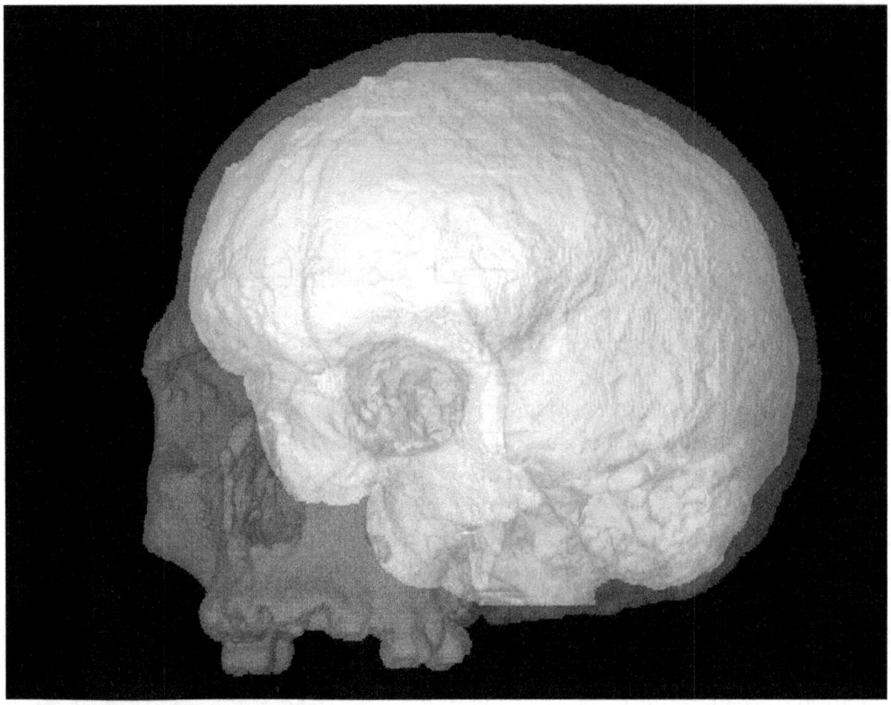

a

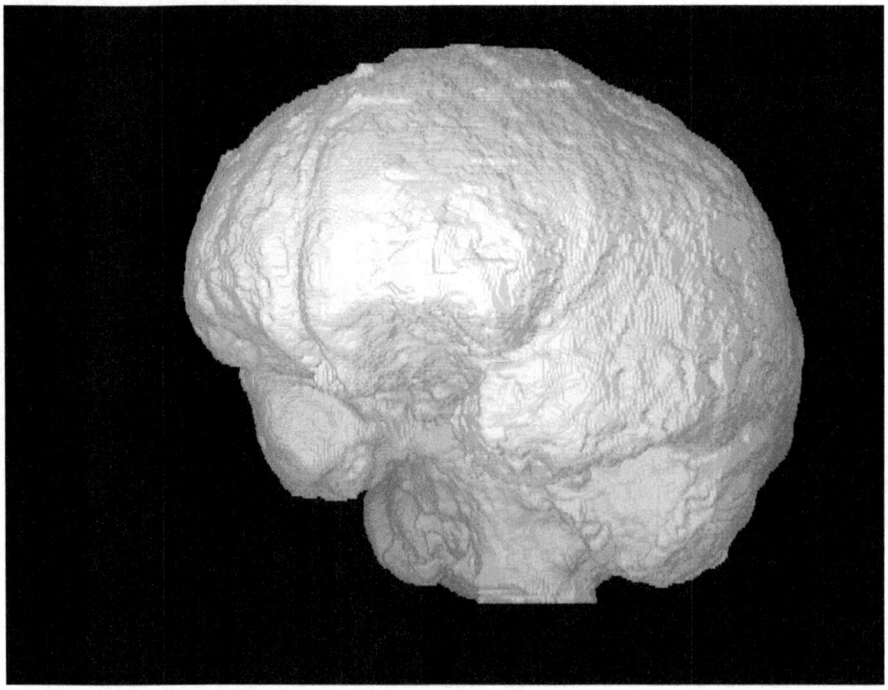

b

Fig. 4. An electronic endocast of the Mladeč I endocranium. **(a)** The endocast embedded inside the cranium, showing its relative position. **(b)** The isolated endocast, showing numerous surface features tors so often popular among paleoanthropologists are actually mustard seed methods in disguise: the volumes of the "reference" crania were also determined with mustard seeds. A further advantage of the flood-filling technique: one can image the endocast; it is shown in Fig. 4.

Discussion

Preservation status of the cranium

Several authors (among others: Szombathy, 1925 and Wolpoff et al., this volume) have noted the presence of the encrustations. Most of these were evident either because of their coloring or their slightly different surface texture. A comparison of our identified encrustations with those mentioned by other authors shows that many have remained unobserved in the course of visual inspection. We also stress that the thickness of the encrustations must be determined using CT-scans and electronic segmentation methods (Fig. 2).

There are difficulties, however, which we note here. There are many regions of the cranium where bone has been replaced by the same material as the sinter (or: with minerals having the same x-ray signature - Hounsfield numbers - as the sinter; see Fig. 1). In such cases, we had to decide whether the regions appeared to be intact bone that had fossilized or missing bone that had been filled with sinter. Such issues arose primarily on the ectocranial surface of the neurocranium (Plates I and II). Overall, the regions where we deemed an encrustation (rather than a mineral replacement of bone during fossilization) to be present were very few and of small extent. On the other hand, there are many regions of fossilized bone that have the same x-ray signatures as the sinter.

Taphonomy issues

The identification of encrustations and gypsum reconstruction has two implications. First, assessing the preservation status of the Mladeč human remains is of primary interest to the paleoanthropological community. Second, encrustations that are due to staligmatic material directly relate to the site where Mladeč 1 was found, while sinter encrustations contribute to the overall discussion of the taphonomy of the site. For instance, it is intriguing, in our opinion, why the maxillary sinuses have not been *completely* filled with encrustations, and there are no encrustations on the palate, whereas the region between the medial and lateral pterygoid plates have been filled on both the right and left sides. The formation of staligmatic material on the skull in a few, well-defined areas introduce a discussion about the orientation (more likely: several orientations) of the cranium during its (long) deposition in the cave.

No drawing of the Mladeč 1 cranium *in situ* has survived. We have Szombathy's notes of the discovery, but they do not clarify - let alone answer - basic questions about the taphonomy of this specimen.

The cranium was discovered at a depth of 20–30 cm, with cranial base upwards (Szombathy, 1925); attached to the basiocciput was a cylindrical staligmatic bulge of 2–3 cm diameter; the cylinder was attached to a plate-sized disc of calcaneous material horizontally embedded in the clayey loam. Because the staligmatic bulge was removed before the excavators were aware of the presence of the cranium, the latter was damaged (notably: parts of the right parietal were crushed). A reindeer rib of ~10 cm length and a small, unidentified piece of bone were attached, via an encrustation, to one zygomatic arch. This encrustation a restorer (Szombathy?) had carefully removed, but, in so doing, the mastoid process below the bulge was destroyed.

Szombathy points out that the cranium could have not been contemporaneous with the 35 cm deeper layer of charcoal, ascribed to a former hearth, as a layer of limestone 30 cm below the cranium separated the two and must have taken some time to form.

These field notes raise as many questions as they propose answers: (1) If the sinter derives via water action from the surrounding clayey loam, then what prevented the encrustation of the palate, which was above the base of the cranium, according to the field notes? Also: why are the tooth sockets not completely filled with encrustations? (2) How can the encrustations inside the maxillary sinus "grow" from proximal to distal in an upside-down cranium? (3) The encrustations inside the neurocranium are localized in the region of the left anterior fossa - how so?

We propose that the cranium had changed its orientation of deposition (at least) three times during the encrustation process(es).

Initially, the encrustations within the maxillary sinuses were the result of stalagmite-forming action (as evidenced by the geometry of their deposition), perhaps from minerals that precipitated out of the water seeping through the bases of the orbitae. During this first stalagmite-forming process, the cranium must have been almost upright (bregma uppermost). Later, while many of the sinter encrustations in the endocranium (Plate II, *right,* top) were formed, the cranium was perhaps lying on its side, with its face downwards. Finally, the cranium shifted into the upside-down position - the one in which the cranium was ultimately found. In this position, the staligmatic bulge connecting the basiooccipt with other deposits must have formed. We are unable to explain why the palate did not become encrusted (and the tooth sockets only a little), but speculate it may have been covered with some deposition layer that prevented encrustation. To summarize: the cranium seems to have "rolled" from an upright to an upside-down position, in at least three stages, resting long enough between them for the observed encrustation processes to take place.

There are, alas, other (minor) encrustations; their distributions are not easily explained with these postulated orientations. In any case, the description of finding this cranium upside-down does not imply that it had lain this way during the whole 31 thousand years of its fossilization.

Bayer (1925) postulated that the woman whose cranium has fossilized lived in the layer of the hearth - an interpretation fiercely denied by Szombathy (1925). However, the issue as to whether she was buried (an associated necklace had also been recovered) remains unresolved. After all, our postulated reorientation(s) of the cranium *could* be compatible with a burial. We do not wish to speculate as to how the woman (or her remains) came to be in the cave nor do we endorse the view that she was buried; we do, however, postulate that the encrustation patterns are evidence that the cranium did not lie undisturbed in one sole orientation.

Conclusions

The value of Mladeč 1 as a national heritage monument precludes any manipulative restoration work on the original. Attempting to mechanically remove the shellac, for example, will also remove parts of the outer table of the cranium. No mechanical removal of endocranial encrustations is possible, unless one attempts to remove (and destroy) the gypsum reconstructions first. Even if one were to risk damages to the specimen in the course of such undertakings, one cannot rest assured that a systematic removal will be successful. We do not know how easily the outer table of the fossilized bone can break or otherwise be damaged when encrustations and/or shellac are mechanically removed, but we suspect the risk of damage is indeed very high. Mechanical removal of encrustations on the inner table is practically impossible, in our opinion, primarily because of accessibility issues, let alone guessing the thickness of the various encrustations. Even access to them via the (removed!) gypsum portions of the parietal is limited to only small regions of the endocranial surface. The crista galli, for example, will remain encrusted and the encrustations inside the maxillary sinuses will remain inaccessible.

A further difficulty is the inability of a human restorer to see or otherwise identify all encrustations: many do not appear sufficiently different from the fossilized bone. A restorer - no matter how diligent, patient, conscientious and competent – is bound to miss some encrustations or – worse, perhaps – misjudge their thickness.

Electronic removal using a CT-scan overcomes all the aforementioned difficulties. First, all segmentation and removal steps (which are image-editing processes) are reversible and easily recoverable via documentation in a protocol. Second, encrustations that are one voxel thick (only 0.488 mm thin!) can be identified in the CT-scan. In fact, any thicker encrustations can be electronically removed to the nearest 0.488 mm. Third: even encrustations that are visually indiscernible on the original due to the lack of texture differences can be identified (via their Hounsfield numbers). Fourth, removal of encrustations covering the pterygoid plates is mechanically impossible. These plates are so thin (usually between 0.5 mm and 1 mm) that any mechanical removal attempt will unavoidably result in their breaking off.

Apart from contributing to the taphonomy discussion of one of the most prominent specimens of the Mladeč human fossil trove, the application of electronic segmentation methodology enables researchers to investigate the taphonomy and assess all morphologies of this fossil for the first time since its discovery.

References

Bayer, J. (1925) Die ältere Steinzeit in den Sudetenländern. Reichenberg, Anstalt für Sudetendeutsche Heimatforschung, Vorgeschichtliche Abteilung 2, 73–77

Billy, G. (1972) L'évolution humaine au paléolithique supérieur. *Homo* 23, 2–12

Hounsfield, G. N. (1973) Computerised transverse axial scanning tomography. Part I: Description of the system. *British Journal of Radiology* 46, 1016–1022

Jelínek, J. (1983) The Mladeč finds and their evolutionary importance. *Anthropologie* (Brno) 21, 57–64

Oliva, M. (1989) Mladopaleolitické nálezy z Mladečských jeskyní. *Časopis Moravského muzea zemského* 74, 35–54

Prossinger, H., Wicke, L., Seidler, H., Weber, G. W., Recheis, W. and Müller, G. B. (2000) The CT-scans of fossilized crania with encrustations removed allow morphological and metric comparisons of para-nasal sinuses. *American Journal of Physical Anthropology* 30 (Suppl.), 254

Prossinger, H., Seidler, H., Wicke, L., Weaver, D., Recheis, W. Stringer, C. and Müller, G. B. (2003) Electronic removal of encrustations inside the Steinheim cranium reveals paranasal sinus features and deformations, and provides a revised endocranial volume estimate. *Anatomical Record (New Anat.)* 273B, 132–142

Svoboda, J. (2000) The depositional context of the Early Upper Paleolithic human fossils from the Koněprusy (Zlatý kůň) and Mladeč Caves, Czech Republic. *Journal of Human Evolution* 38, 523–536

Szombathy, J. (1882) Fünfter Bericht der prähistorischen Commission der mathematisch-naturwissenschaftlichen Classe der kaiserlichen Akademie der Wissenschaften über die Arbeiten im Jahre 1881, vorgelegt von Ferdinand von Hochstetter in der Sitzung am 9. Februar, 1882. *Sitzungsber. der math.-naturwiss. Classe der kaiserl. Akad. der Wissenschaften* 85, [90]–[107]

Szombathy, J. (1925) Die diluvialen Menschenreste aus der Fürst-Johanns-Höhle bei Lautsch in Mähren. *Die Eiszeit* 2, 1–34, 73–95

Udupa, J. K. (1999) 3DVIEWNIX 1.2. Department of Radiology, University of Pennsylvania Medical Center, Philadelphia, PA

Wild, E., Teschler-Nicola, M., Kutschera, W., Steier, P., Trinkaus, E. and Wanek, W. (2005) Direct dating of Early Upper Palaeolithic human remains from Mladeč. *Nature* 435, 332–335

Wolpoff, M. H. (1999) Paleoanthropology. New York: McGraw-Hill

Wolpoff, M. H., Frayer, D. W., Trinkaus, E. and Teschler-Nicola, M. (2006) Inventory and photo-documentation of the Maldeč human remains. In (M. Teschler-Nicola, Ed.) *Early modern humans at the Moravian Gate. The Mladeč Caves and their remains*, this issue, pp. 159–184

Plates

Plate I.

Left: The frontal view of the Mladeč I cranium CT-scan. top: with gypsum (blue) and encrustations (orange) color-coded; bottom: without gypsum and encrustations. For further descriptions, see text
Right: The basal view of the Mladeč I cranium CT-scan. top: with gypsum (blue) and encrustations (orange) color-coded; bottom: without gypsum and encrustations. For further descriptions, see text

Plate II.

Left: The right lateral view of the Mladeč I cranium CT-scan. top: with gypsum (blue) and encrustations (orange) color-coded; below: without gypsum and encrustations. For further descriptions, see text
Right: The endocranial view of the Mladeč I cranium CT-scan towards the anterior cranial fossa. top: with gypsum (blue) and encrustations (orange) color-coded; bottom: without gypsum and encrustations. For further descriptions, see text

Plate III.

Left: The endocranial view of the Mladeč I cranium CT-scan towards the posterior cranial fossa. top: with gypsum (blue) and encrustations (orange) color-coded; bottom: without gypsum and encrustations. For further descriptions, see text
Right: The endocranial view of the Mladeč I cranium CT-scan towards the cranial base. top: with encrustations (orange) color-coded; bottom: without encrustations. For further descriptions, see text

Plate IV.

The endocranial view of the Mladeč I cranium CT-scan towards the right parietal. top: with gypsum (blue) and encrustations (orange) color-coded; bottom: without gypsum and encrustations. For further descriptions, see text

PLATE I

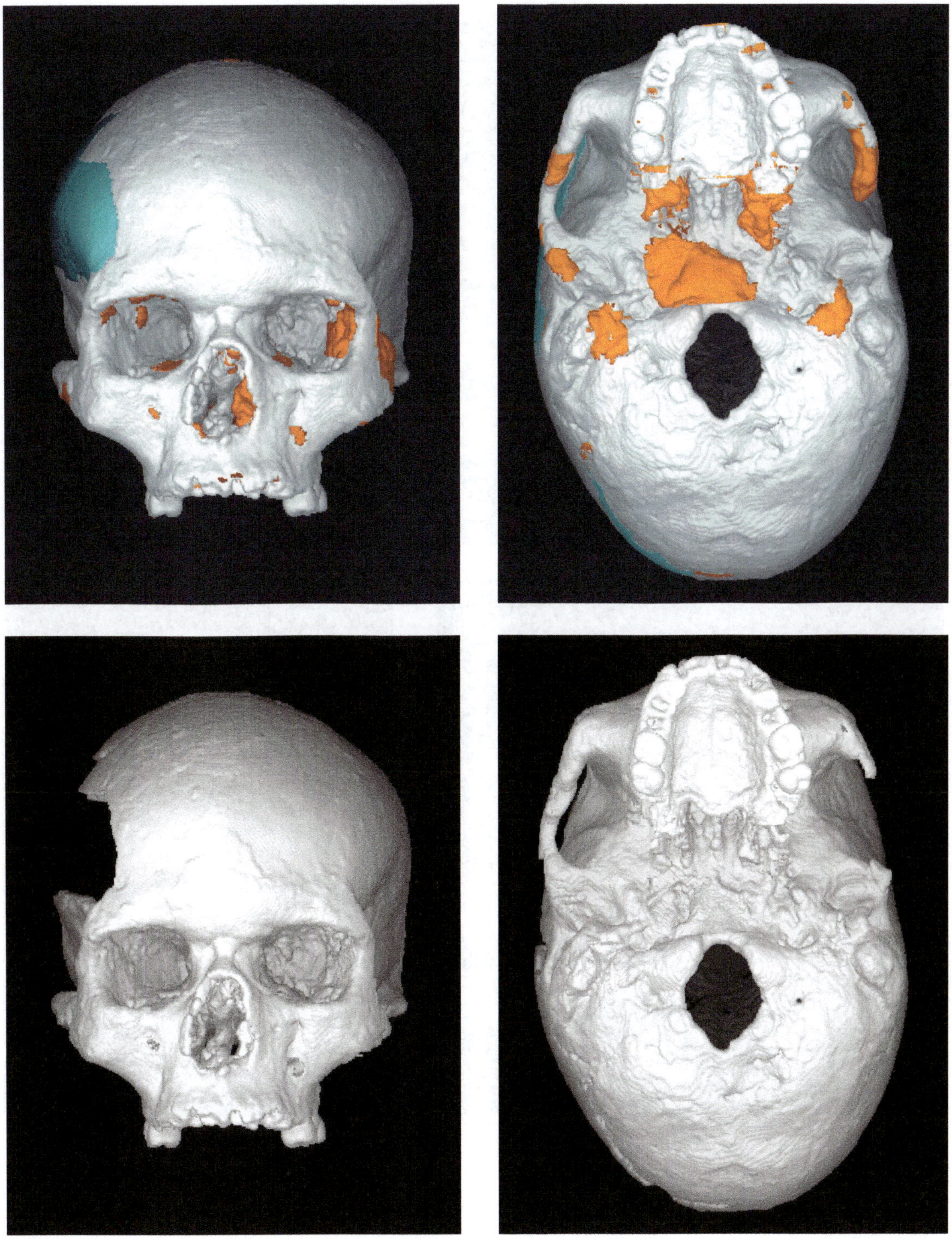

PLATE II

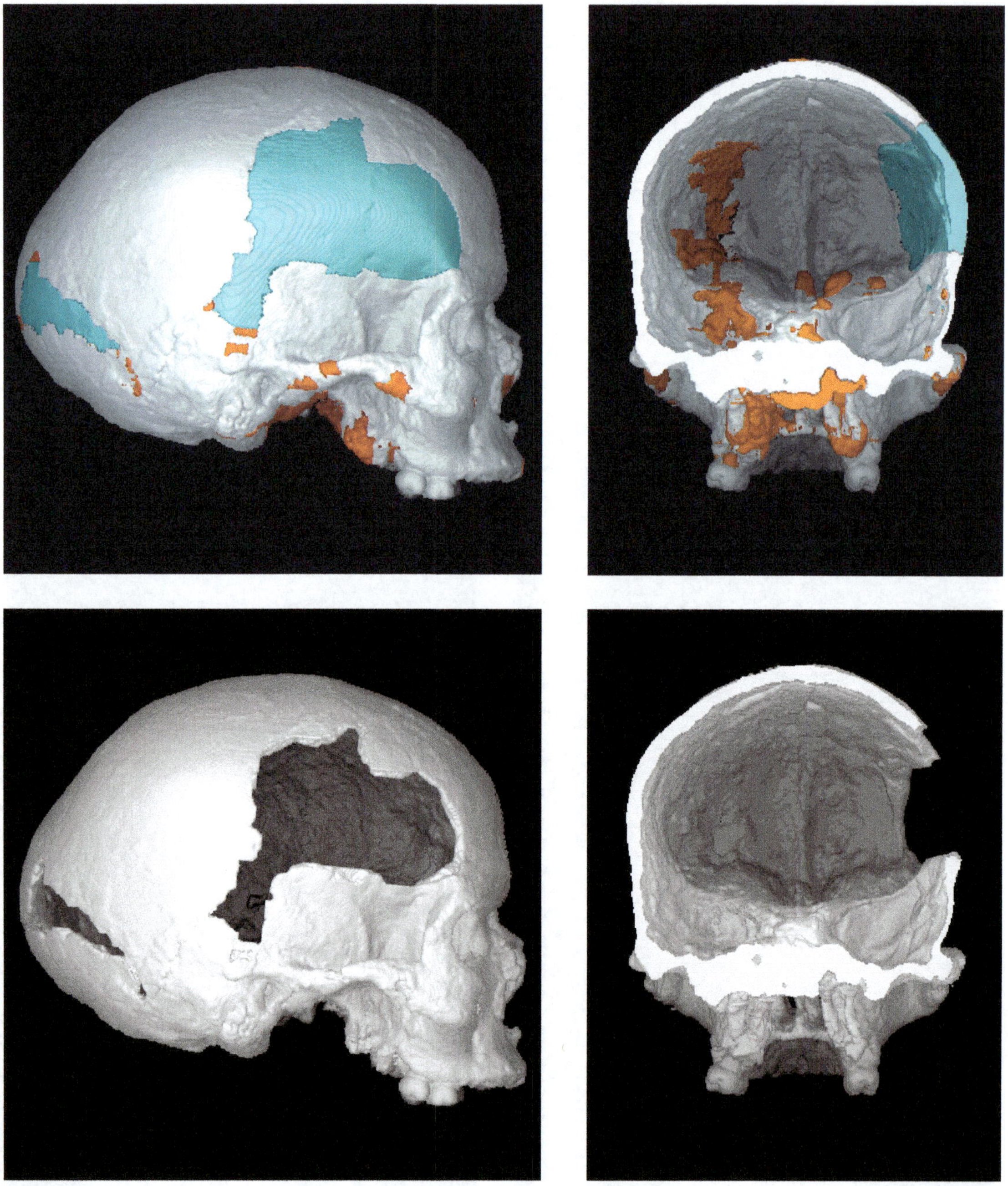

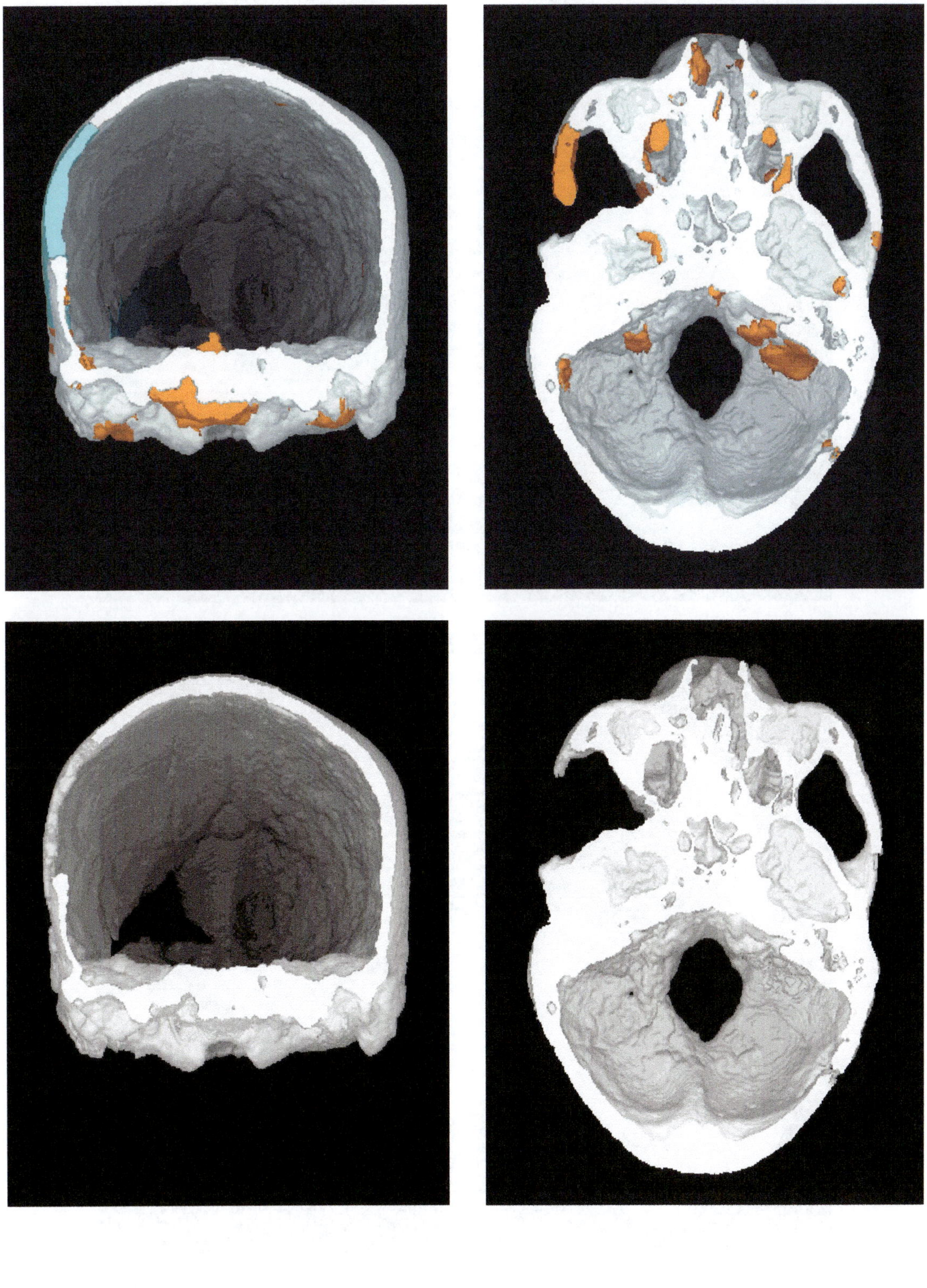

PLATE IV

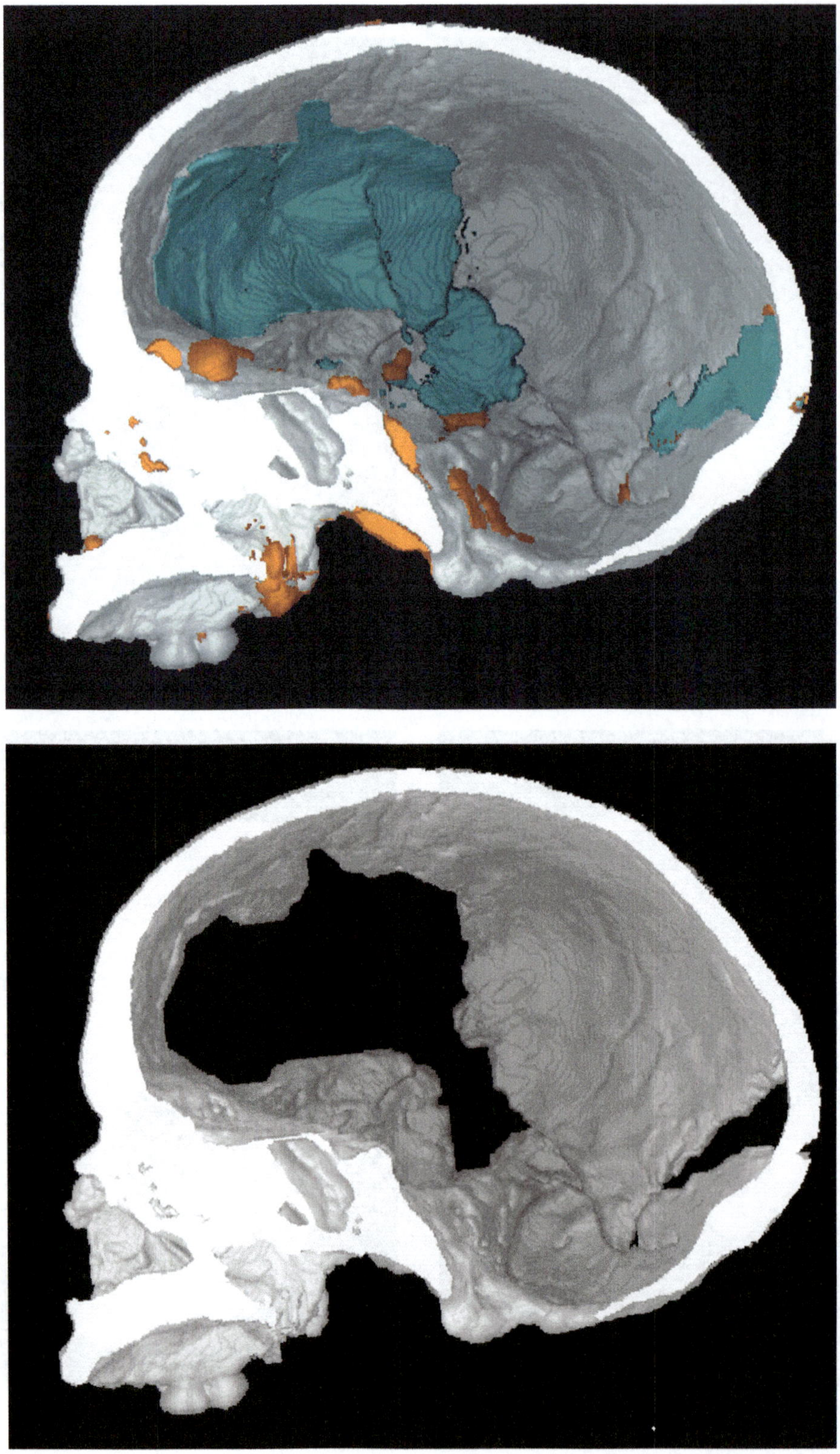

Chapter 12

THE MLADEČ 3 INFANT

Nancy Minugh-Purvis, Th. Bence Viola and Maria Teschler-Nicola

Introduction

During the course of excavations between 1881 and 1922, caves at the western end of the town of Mladeč yielded skeletal material from several children. Six of these individuals included cranial remains, of which only two survive today: Mladeč 40, excavated between 1903 and 1911, and until recently in the possession of the Knies family, and Mladeč 3, excavated in 1882 and taken back to Vienna by Szombathy for study and curation (Szombathy, 1925).

It has been established, with some certainty, that all of the Mladeč children were excavated from early to middle Central European Aurignacian deposits (Valoch, 1968). In addition to Mladeč 3 and 40 described in this volume (see Frayer et al., this volume; and chap. 8, plate XVII), Mladeč 46, comprised of numerous portions of a skeleton approximately 10 years of age at the time of death (Knies, 1905), was recovered in 1904. Three additional juveniles, Mladeč 37, 44, and 45 were unearthed in 1922 (Smith, 1982; Jelínek, 1976). Unfortunately, Mladeč 37, 44, 45 and 46, together with numerous other invaluable late Pleistocene hominid specimens, were housed at the museum at Mikulov when it was destroyed by fire in 1945 during the final days of World War II in Europe. As no casts were made of these lost Mladeč children, and their unpublished documentation perished with them, they must forever remain essentially unknown. Their loss is particularly great, considering that so few immature remains have been recovered from the early Aurignacian.

Since 1882, Mladeč 3 has been curated at the Naturhistorisches Museum Wien, Austria. As part of the Mladeč sample, the specimen provides important insight into the development of the adult cranial morphology within the important early to middle Aurignacian Mladeč population. In the broader sense, however, Mladeč 3 is presently the only known example of an infant from the Central European Aurignacian and thus provides a rare snapshot of early growth and development in a population overlapping temporally with Neandertals, potentially offering valuable clues regarding the similarities and differences in ontogenetic patterning in these groups. This is of considerable interest as, in attempting to understand the evolutionary transitions of late Upper Pleistocene peoples, our ability to recognize the significance of complex phenotypic changes in the fossil record may well depend upon our ability to interpret them within a detailed life history context.

Szombathy's 1925 publication of numerous Mladeč skeletal remains included a competent but cursory description of Mladeč 3. The present study is intended to supplement his work by analyzing the Mladeč 3 infant with the benefit of numerous comparative data available from the many Upper Pleistocene juveniles discovered since 1925 and within the context of analytical frameworks developed beginning in the latter half of the 20th Century by investigators of paleoanthropological ontogeny (Vlček, 1969; Tillier, 1982; 1983a; 1983b; 1987; 1988; 1999; Heim, 1982; Minugh-Purvis, 1988; 1993; 1998; 2000; Minugh-Purvis et al., 2000; Minugh-Purvis and McNamara, 2002; Coqueugniot, 1999; 2000; Zilhão and Trinkaus, 2002; Akazawa and Muhesen, 2003; Thompson et al. 2003; Antón, 2002; and many others).

Materials and methods

This study compares Mladeč 3 to five samples: the Mladeč adults (see Wolpoff et al., this volume, and Frayer et al., this volume); European and Middle Eastern Neandertals of comparable developmental age; the Mousterian-associated Levantine children from Skhul and Qafzeh; immature remains from other European Upper Paleolithic sites; and cranial remains of recent modern children of similar developmental age at death. Part of the recent immature sample is curated at the Department of Anatomy, Vienna Medical University, and is the same reference collection used by Szombathy (1925). This recent sample from Vienna Medical University was measured by BV. Data on all other recent modern children (see Minugh-Purvis, 1988 for details) and on the fossils were collected by NM-P from original specimens unless otherwise noted.

Measurements of the Mladeč 3 child were made with Helios needle-point sliding calipers (NM-P), a Mitutoyo digital caliper (BV), and linen measuring tape. X-rays taken prior to matrix removal using conventional radiographic techniques were obtained through the Anthropologische Abteilung at the Naturhistorisches Museum, Wien, in 1980. CT-scans were recorded at the Department of Radiology, Vienna Medical University[1] using a Philips MX8000 CT scanner. The slice thickness was 0.4 mm and a matrix of 512 x 512 was used, resulting in a reconstructed voxel size of 0.24 x 0.24 x 0.4 mm. The data analysis and visualization was done in Amira® 3.1.

Preservation and reconstruction

Mladeč 3, stored at the Naturhistorisches Museum Wien, Anthropologische Abteilung (inv. no. 5456), is known only from cranial remains. The partial calvarium consists of a numbered incomplete frontal, incomplete occipital, partial right temporal, one right and another probably right parietal fragment. In addition, a small, uninscribed calvarial fragment, covered with matching travertine matrix and of the same thickness as the Mladeč 3 vault, is kept with these remains. This piece was not mentioned by Szombathy in his 1925 description and the original inventory slip accompanying Mladeč 3 lists only the 5 labeled pieces of skull as belonging to the specimen. However, a fresh break along one border suggests this piece may have broken from a larger section of the Mladeč 3 calvarium since Szombathy's publication. Unfortunately, it was not possible to reposition it on any of the other Mladeč 3 fragments to verify this notion so that its association with the specimen has not been authenticated at this time.

At excavation, the Mladeč 3 remains were completely encrusted in a thin layer of travertine matrix (Szombathy, 1925). This is visible in this volume, chap. 8, Plate III. Most edges of the various bones were covered with this matrix, clearly indicating that the vault was broken prior to fossilization although it is clear from fresh breaks that additional damage occurred during excavation (Szombathy, 1925). Yet despite the delicate nature of infantile craniofacial remains, the few recovered portions of Mladeč 3 are, in many respects, fairly well preserved. Particularly notable, and fortunate, is the fact that the surviving fragments exhibit no apparent deformation or distortion.

Mladeč 3 was curated for many years with its hard, adhering matrix intact. This material surrounded the fossil in a thin, continuous layer, which probably added 1.0–0.5 mm to the thickness of the specimen at most locations. It was so thinly and uniformly distributed over most surfaces that many, but not all, morphological features were easily observed through the thin crust. In the 1980's a decision was made to attempt surface matrix removal despite the obvious risks to the fragile re-

1 CT-scans were possible through the courtesy of Professors Imhof and Kainberger. We thank Ms. Plischke for her technical help during the scanning.

mains and the matrix film was cleaned away by Hubert Poxleitner of the Anthropologische Abteilung of the Naturhistorisches Museum Wien in 1988. Although this resulted in some inevitable scoring of the bone surfaces and perforation of some very thin regions of the temporal and occipital squamae, visualization of other aspects of surface morphology was enhanced and several new observations possible as a result. The present study is based on study of the original at the Naturhistorisches Museum Wien, beginning in 1978, both before and after matrix removal.

The frontal is approximately 60% complete and comprised of two large fragments which have been glued together to the right of the midline. The larger of these two includes the central region of the frontal from the frontal-nasal suture to a position just short of bregma. The second, slightly smaller fragment is a sizeable portion of the right squama. Along the glue joint between these two pieces a large triangular defect is missing from the superomedial portion of the right frontal squama due to postmortem damage. Anteriorly, nasion and the fronto-nasal articulation are preserved. Anteriorly, the left dacryon and a small portion of the left medial orbital wall, continuous with a fragment of left medial orbital roof, are also preserved. Except for this tiny shelf of the left orbital roof, which extends posteriorly some 12 mm from the anterior edge of the orbital margin at its greatest depth, the orbital processes of the frontal are almost completely broken away with the lateral squama. None of the right orbit is preserved, with the bone broken through the region where the right dacryon would have been located.

From the fronto-nasal suture, the preserved area of frontal bone extends posteriorly in the midline to a position perhaps 8–10 mm anterior to the estimated location of bregma. On both the left and right, however, the lateral portions of the frontal are extensively damaged and much of the bone is missing. On the right posterior squama a portion of coronal suture begins at a point 21 mm lateral to the midline. This extends 52 mm to the right until reaching the broken lateral edge of the bone. A weak inferior temporal line, set off by the angulated roughened surface of the temporal fossa below it, arches 25 mm anteriorly from the coronal suture at stephanion. Damage to the left lateral frontal is more extensive. The bone is broken away beginning 11 mm lateral to the medial orbital margin. Posterior to this point, 66 mm of central squama are preserved but include neither the coronal suture nor temporal line. The zygomatic processes of the frontal are completely absent bilaterally. At several locations, breakage reveals the internal and external bony tables and diploë, although thickness of the diploë varies considerably in different regions of the bone.

Two small fragments remain of the Mladeč 3 right parietal. One, a triangular piece with a height of 60 mm and a base breadth of 39 mm, includes 24.3 mm of coronal suture which articulates neatly with the frontal from 30–54 mm to the right of the midline. It bears some faint cerebral impressions and a vascular groove. What is likely an adjoining portion of the right parietal, measuring 69 x 36 mm, bears no endocranial features. This fragment was broken at some point in the 1980's, and now consists of several small glued parts. Although these fragments suggest continuity in the vault curvature of the parietal region, it is not possible to establish a good bone to bone fit between them due to surface erosion and damage from repeated gluing and ungluing in previous reconstruction attempts.[2] Nor is it possible to position the 69 x 36 mm parietal fragment on the vault with confidence. As a result, we have omitted it from our CT generated total vault reconstruction.

The right temporal consists of the incomplete petromastoid portion fused to a partial squama. Their union is a solid one, although the line of fusion remains clearly visible endocranially. The bone is preserved from the sphenotemporal suture anteriorly to the lambdoid suture posteriorly, and includes both the parietal notch and asterion. Around the fragile squamosal margin there is considerable breakage, making the exact superior outline of the temporal squama difficult to estimate

2 At present, these fragments remain glued together in an erroneous position, a situation which necessitated "cutting away" the rectangular piece from our CT generated image of our total vault reconstruction.

although it could not have extended more than a few mm above the preserved border of the bone. The zygomatic process of the temporal is broken off anteriorly, but the most distal portion of its root is intact where it courses above the glenoid fossa.

Endocranially, considerable detail of the squama and petrous are preserved, but the tip of the petrous portion is broken away so that all features medial to the internal acoustic meatus are lost. The total distance from the internal wall of the squamous portion to the preserved most antero-medial petrous tip measures 36.2 mm; from floor of sigmoid sinus to the anterior petrous tip: 37.4 mm. On the inferior aspect of the petrous, considerable detail was originally visible through the thin covering of matrix, although some of these delicate features were damaged during cleaning of the bone.

The Mladeč 3 occipital is represented by the complete squamous portion; only the lateral and basilar elements are missing. The lateral portions had not yet fused with the squama as indicated by the immature, billowed surfaces of the posterior intraoccipital synchondroses on either side of the posterior margin of the foramen magnum. Internally, grooves for the dural venous sinuses are well-marked. Unfortunately, this thin bone was perforated and cracked during matrix removal, and further broken during the 1990s.

Two quite different approaches are currently used to reconstruct fossils from computerized tomographic (CT) data. Zollikofer and colleagues (Ponce de Leon and Zollikofer, 1999; Zollikofer et al., 2002; Zollikofer et al., 2005; Zollikofer et al., 1995) rely on anatomical features and broken edges to first position parts, then use mirror imaging to complete missing portions. Parts that cannot be reconstructed this way are interpolated using a reference sample. Further, taphonomic deformations are corrected using "predefined diagenetic scenarios" (Zollikofer et al., 2002, 330). Gunz and colleagues (2004) employ a different approach, using geometric morphometrics with both conventional and geometrically homologous semilandmarks to reconstruct crania. In the case of Mladeč 3, we used a methodology similar to Zollikofer and colleagues (Zollikofer et al., 2002). As mentioned above, CT data was visualized and manipulated using Amira® 3.1 software.

As a first step, we refitted the anatomically connected occipital and right temporal along their chipped but otherwise fairly well-preserved junction along the right occipito-mastoid suture. The left temporal was created by isolating the right temporal and mirroring the resulting bone. After this, it was joined to the occipital along the left occipitomastoid suture. Next, the frontal was completed using mirror-imaging. First, the bone was cut into four pieces consisting of left and right glabellar/interorbital segments and left and right portions of the squama. The interorbital area was mirror-imaged using the better-preserved left side as a template, while mirror imaging of the squama was based on the right side. As mentioned prevously, the parietal fragments are currently glued together incorrectly. Nor is it possible to be certain of the position of one of these fragments. Thus, rather than subject the original fossil to further manipulation at this time, we utilized imaging to omit this fragment from the reconstruction. The parietal fragment of certain position was fitted to the frontal along the preserved part of the right coronal suture. This same portion of the left parietal was generated by mirror-imaging, and positioned along the computer reconstructed left coronal suture.

This process produced two separate portions of reconstructed calvarium: one comprised of frontal and parietals, and the other of occipital and temporals. As no direct anatomical connections are possible between these portions, we utilized a previously scanned template specimen of comparable developmental age and very similar dimensions from the collections of the Department of Anatomy, Vienna Medical University to position them into a suggested reconstruction (Fig. 1).[3]

3 To achieve optimal fit of the Mladeč 3 fragments to the template, we used the LandmarkSurfaceWarp module of Amira. This procedure conducts an affine transformation, that is only rotation and translation, and results in a fit that minimizes squared distances of homologous landmarks. We used standard osteometric landmarks on the frontal, occipital and both temporals.

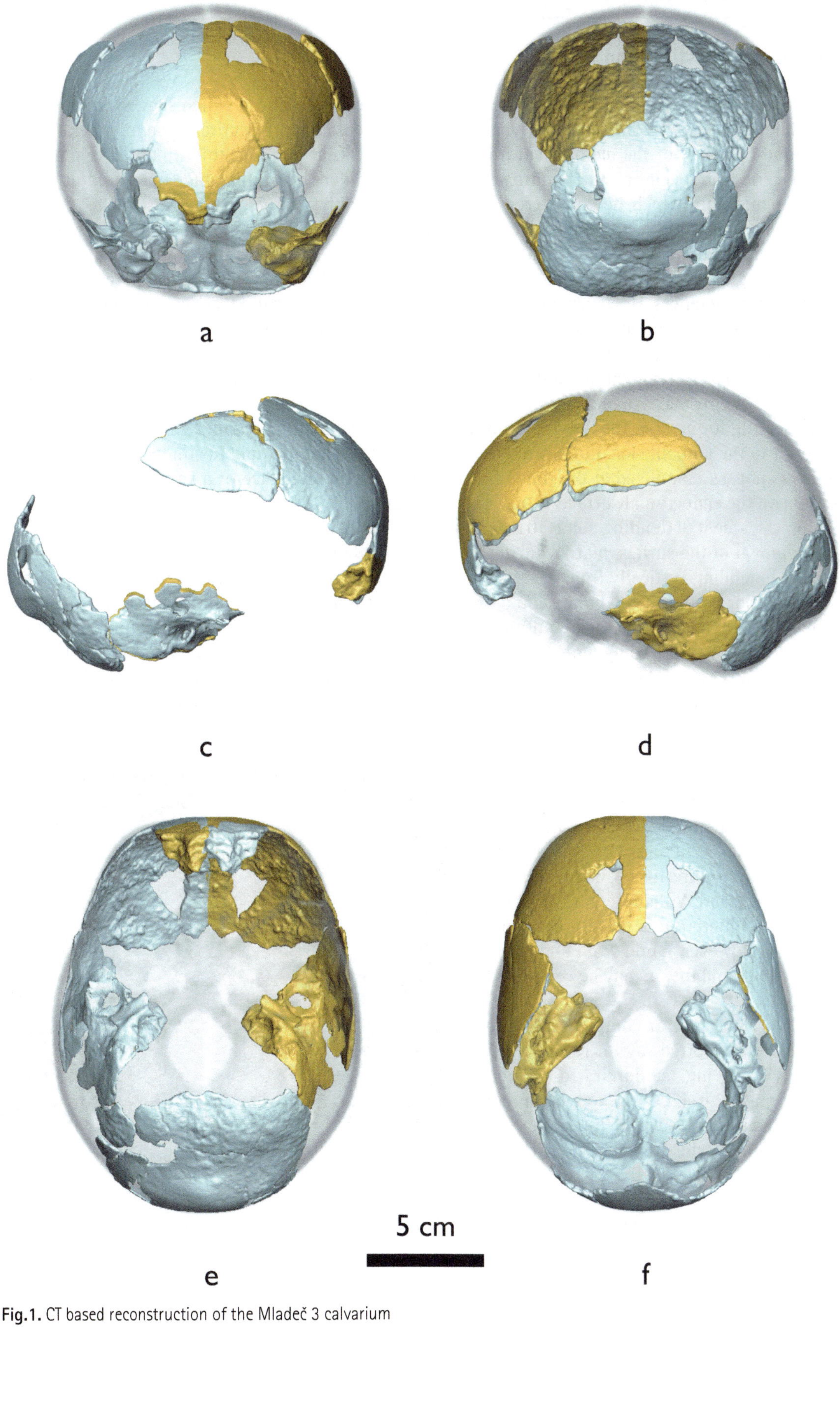

Fig. 1. CT based reconstruction of the Mladeč 3 calvarium

Age at death estimation

Szombathy (1925, 21) noted several features of the Mladeč 3 calvarium which suggested to him that the child died between two and four years of developmental age. His criteria were: general size of the bones; patency of the metopic suture at either end; recent petromastoid union; and patency of the posterior intraoccipital synchondroses. He based the older end of this range on the status of temporal maturation, which he considered indicative of a four year old, in contrast to his other age indicators suggesting two years or less. Ultimately, he settled on the average of these different determinations to arrive at an estimated age at death of three years for the Aurignacian child.

Working without the benefit of any preserved dentition, Szombathy's age estimate is consistent with classic references. However, closer examination of his aging criteria, assisted by removal of matrix from the fossil and more studies of skull maturation, suggests that Mladeč 3 was perhaps considerably younger than previously suspected.

Prior to matrix removal, a subtle, discontinuous midline groove was visible on the ectocranial surface of the frontal squama. This extended approximately 10 mm above the fronto-nasal suture, disappearing for most of the A-P length of the frontal squama, then reappearing several millimeters anterior to the probable location of bregma. This suggested that metopic fusion had begun centrally, but was not yet completed at the time of death. However, no evidence of a comparable groove was visible on the endocranial surface. X-ray examination (see Fig. 2) revealed a possible partially patent metopic segment, extending some 10 mm above the fronto-nasal suture.[4]

Removal of the surface matrix during the 1980's further confirmed the presence of this midline groove extending vertically from nasion but suggested that the midline crack terminating at the broken posterior border of the bone should be ruled out as a metopic remnant. Hauser and DeStefano (1989) caution against confusing an anterior metopic remnant for a supranasal suture, noting that the latter is characterized by more complexity than usually seen in the typically straight metopic suture. In the case of Mladeč 3, the simple supranasal midline scar appears to represent a true supranasal metopic remnant and it seems reasonable to concur with Szombathy's findings that metopic fusion had commenced but that the sutural scar was incompletely obliterated at the time of death.

Szombathy considered the status of metopic suture closure in Mladeč 3 indicative of an age of two years at death. Warwick and Williams (1975) reported that metopic suture fusion begins during the 2nd year and Williams et al. (1995) report that obliteration is usually complete by age 6. However, Ford (1958) and Sperber (2001) found fusion may be completed by age 1.

A younger age for complete metopic fusion is supported by a recent MRI study of living children in which Weinzweig et al. (2003) report complete metopic suture closure in 100% of their sample of 76 healthy, normal North

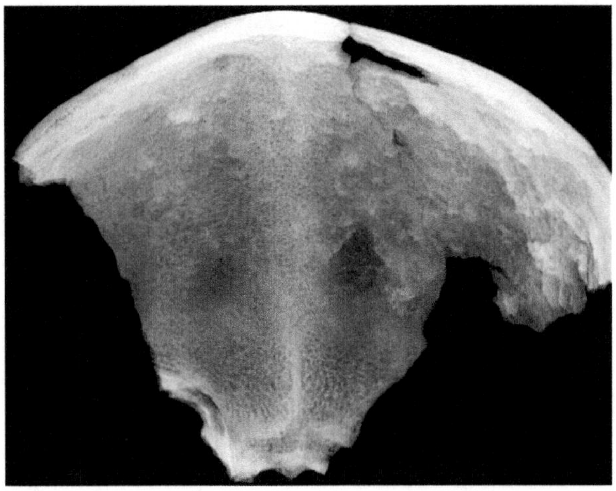

Fig. 2. A-P X-ray of the Mladeč 3 frontal

4 This feature was not visible in our more recently obtained CT images probably due to the lower resolution of this method.

American boys by 8 months of age. Moreover, they found that normal closure begins as early as 3 months. Curiously, in every case, Weinzweig at al. found that metopic fusion commenced at nasion, then progressed posteriorly along the frontal vault towards the bregmatic frontanelle. This finding disagrees with a large body of traditional literature in which metopic fusion is usually described as initiating centrally. Very possibly, closure as visualized by MRI and scar obliteration observed in dry, osteological specimens are not strictly synonymous, a matter which merits future investigation in order to permit accurate application of findings based on live samples to the study of skeletal remains. Assuming, nevertheless, that data on modern human North American children provide an appropriate reference for estimating the timing of metopic suture closure in a central European Aurignacian child, Weinzweig et al.'s study suggests that Szombathy's and other previous age determinations (Minugh, 1983; Minugh-Purvis, 1988) likely overerestimated the age at death for Mladeč 3 which should almost certainly be revised to two years or less.

Another feature of the frontal bone helpful in estimating age at death is the dramatic thinning of the posterior midline squama. While thickness at the Mladeč frontal tubers ranges from 4.0–2.0 mm, near the posterior, broken edge approaching bregma the bone thins to a mere 1 mm. Here the internal and external bony tables are separated only by a rudimentary diploë. The gradual reduction of diploë approaching bregma is also clearly visible in the CT images (see Fig. 3). Both the minimal thickness and incipient nature of the spongy layer at this location suggest that if anterior fontanelle closure was completed in this child, the event had occurred only a brief time before death. In modern children, complete closure of the anterior fontanelle occurs around the age of 2–2.5 years, also suggesting an estimated age at death of approximately 2 years for Mladeč 3.

The right temporal bone provides two criteria of assistance in estimating maturational status: the degree of petromastoid union and tympanic region development. In Mladeč 3, it is difficult to ascertain whether the ring surrounding the external auditory meatus had fused to the outside of the tympanic plate by the time of death, although some evidence of a partial tympanic ring is suggested by the configuration of the anterior external auditory meatus border. Weaver's (1978) study of temporal bone maturation found that tympanic ring fusion begins, *on average*, by approximately 6 months

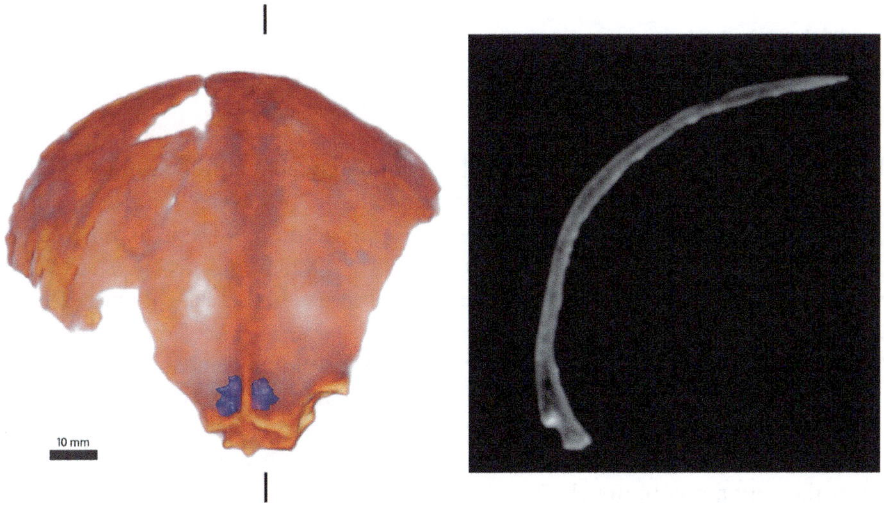

Fig. 3. Frontal view of the segmented Mladeč 3 frontal sinuses (blue) and parasagittal section through the frontal bone

of postnatal life. This agrees with the very young age suggested by the status of metopic suture closure. Another complex used by Szombathy to calculate the age of Mladeč 3 was the union of the petromastoid and squamous temporal elements. Despite the presence of matrix, he correctly described these as solidly fused together, although their junction remains clearly evident, and he considered their status indicative of a four year old child. Most modern references report petromastoid-squamous fusion as occurring at around 1 year of age (Warwick and Williams, 1975) but emphasize a high degree of variability in this timing (Williams et al., 1995). Thus, the state of petromastoid and squamous temporal fusion in Mladeč 3 provides only an estimated age at death of >1 year. Finally, we note that no clear evidence of the tympanic dehiscence, or Foramen of Huschke, is apparent in Mladeč 3. The timing of tympanic plate closure is highly variable (Ars, 1989; Coqueugniot, 1999; Trinkaus, 2002; Hauser and De Stefano, 1989) and in our comparative sample from Vienna, the tympanic dehiscence was closed in four out of six children in the age range of 2 to 2.5 years so that closure of this feature is not necessarily inconsistent with an age at death of 2 years, *contra* Schwartz and Tattersall (2002, 275).

In the occipital, the widely open posterior intraoccipital synchondroses suggest an age of two years or less as well. Warwick and Williams (1975) report these synchondroses fused by the age of two. However, Hoyte (1997) states that although the posterior intraoccipital synchondroses may fuse by two years they are known to remain patent until age three in modern children. Another feature, the *sutura mendosa*, which intervenes between the upper and lower occipital scales, is consistently fused centrally by birth. Obliteration of its lateral portions, which terminate at asterion, often occurs during the first year although in some individuals patency is evident ectocranially up to two years (Heim, 1982; Hauser and De Stefano, 1989). The *sutura mendosa* was not mentioned by Szombathy (1925), nor have we observed it before or after matrix removal from the specimen, leading to the conclusion that it was completely obliterated by the time of death. This reaffirms an age for Mladeč 3 around 2 years, but leaves open the possibility of a younger age at death as well.

In conclusion, despite its lack of dentition, Mladeč 3 preserves numerous features of skeletal growth useful for assessing its developmental age at death. Collectively, these strongly suggest that the child died at no more than two years of age. This represents a considerable revision over previous estimates, and requires that morphological comparisons of Mladeč 3 with other immature remains be revised to accommodate this younger ontogenetic context.

Developmental morphology

Frontal

In general configuration, the frontal is a smoothly rounded, well-vaulted and bossed bone, with an overall gracility consistent with its young developmental age. In *norma frontalis,* a midsagittal keel is clearly visible extending from the broken edge approaching bregma to nasion. As with the Mladeč adults, the anterior surface of the forehead is rather steep and nearly vertical, while vaulting of the longer metopion-bregma segment is much more gradual. A slight eminence is present at metopion.

In norma lateralis, the frontal squama angles markedly from an almost vertical anterior nasion-metopion plate to a nearly horizontal, > 41 mm long segment which extends from metopion posteriorly towards the broken coronal border of the bone. In this metopion-bregma segment, the keel is well-formed. More anteriorly, in the nasion-metopion segment, it is merely a low ridge. This anterior portion of the squama is flatter than the segment above metopion except where interrupted at the midline by the low keel. These morphological details seem to preview the morphology of the Mladeč 6 adult male.

Glabella forms a smooth, rounded eminence symmetrically continuous with the medial left and right supraorbital regions. A visibly subtle but readily palpable immature browridge projects slightly from the frontal squama. This incipient browridge is more completely preserved on the left where its medial portion reaches a maximum height of 13.4 mm before projecting smoothly across the midline to merge with the right browridge at glabella. It is accentuated by bilateral, slightly depressed areas extending about 23 mm to either side of the keel just above the supraorbital borders where a postsupraorbital sulcus was developing. The presence of a browridge in Mladeč 3 provides an interesting contrast with the Skhul I Mousterian child in which a vertically steep forehead and moderate frontal bossing are accompanied by a more typically modern lack of medial superciliary arch development with a thin, pinched area over the lateral orbit. Unfortunately, as the lateral supraorbital margins are not preserved in Mladeč 3, comparisons of this diagnostic region with other Upper Pleistocene children is not possible. However, in the medial supraorbital region, Mladeč 3 most closely resembles the evenly rounded relief of the developing browridges in the 2.5 year old Pech de l'Azé and Subalyuk Neandertal children, although neither Neandertal possesses as prominent a glabella as Mladeč 3. A survey of other Upper Paleolithic associated juveniles reveals a hint of browridge development in the photograph of the Předmostí 6 Gravettian infant published by Matiegka (1934), and an incipient browridge is also present on the Cro-Magnon 5 infant frontal. In contrast, among the young European Neandertals sample, strong glabellar development is not seen in extremely young individuals, but is evident in the 3 year old Roc de Marsal child, and is well-developed in the Devil's Tower 4.5–5 year old. However, the 2 year old Dederiyeh 2 Neandertal from Syria appears to have a prominent glabella (Ishida and Kondo, 2002, 285, Fig. x–19). The strong medial expression of supraorbital and glabellar relief in Mladeč 3 is interesting in that both of these features are also particularly well-developed in the Mladeč 5 and 6 adult males, and, to a lesser extent, in the Mladeč 2 adult female. Thus, just as this region in Mladeč adult frontal morphology is reminiscent of Neandertals in several respects, the same is true of the Mladeč 3 child.

Metrically, estimated frontal sagittal length in Mladeč 3 cannot be distinguished from comparably aged Neandertals, Upper Paleolithic associated children, or the slightly older Skhul 1 specimen (see Table 1, Figs. 4 and 5). Frontal chord length for Mladeč 3 falls just above the mean for recent modern 2 year olds and only slightly below the huge value of this dimension in the Předmostí 6 2 year old Gravettian child. Interestingly, in Mladeč 3 this dimension is larger than in the Dederiyeh 2 Syrian 2 year old Neandertal. The Mladeč 3 value also exceeds the reconstructed measurement of Pap et al. (1996) for the Subalyuk 2 Neandertal. However, all older Neandertal measurements are consistent with the values for recent modern children and other Upper Pleistocene and Levantine juveniles between 2.5 -7 years. Frontal arc length in Mladeč 3 (Fig. 5) falls into the upper portion of the recent modern range and above the value of most comparably aged Neandertals. Other Upper Pleistocene children of this developmental age plot close to the recent mean with a few Neandertals falling below the recent range and two of the Předmostí children plotting above the recent range. Although collectively the young Neandertals show a tendency towards shorter frontal arcs than the *Homo sapiens sapiens,* the considerable overlap between these groups is, nevertheless, quite evident. Compared with the Mladeč adults (see Wolpoff et al., this volume; Frayer et al., this volume), the Mladeč 3 sagittal frontal length is approximately 82% of both the Mladeč 5 and 6 male nasion-bregma arc, 83% of the Mladeč 5 and 80% of the Mladeč 6 nasion-bregma chord. Comparisons with the Mladeč adult female frontals show the Mladeč 3 nasion-bregma arc as approximately 86% of the Mladeč 1 and 95% of the Mladeč 2 arc, while the Mladeč 3 nasion-bregma chord is 84% the length of that reported for Mladeč 1 and 93% of the Mladeč 2 nasion-bregma chord. The index of sagittal frontal curvature (calculated as the nasion-bregma chord/arc index) shows no differences between Mladeč 3 and the immature comparative samples. These findings are consistent with previous observations that this aspect of neurocranial morphology is quite similar in most young Upper Pleistocene and recent children (Minugh-Purvis, 1988; 2002).

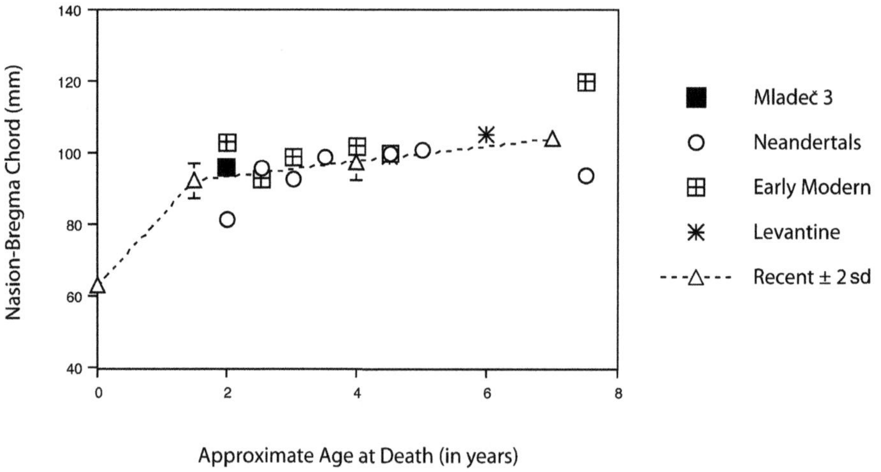

Fig. 4. Growth in frontal chord length from birth to mid-childhood in *Homo sapiens*

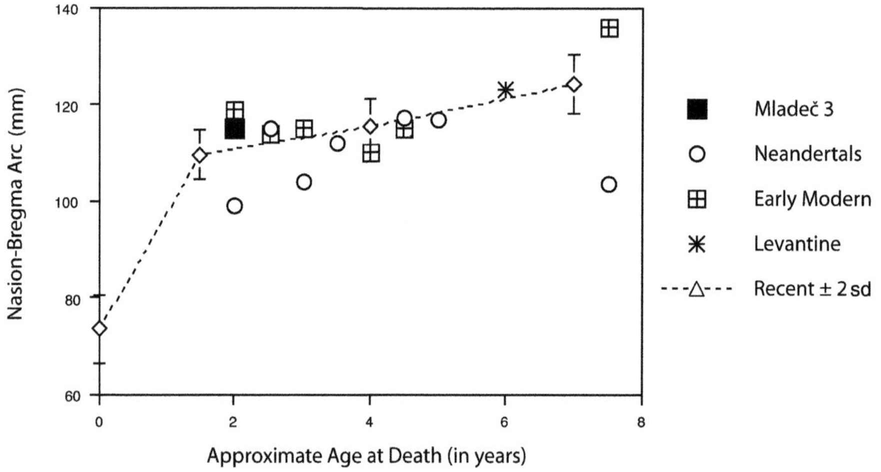

Fig. 5. Growth in frontal arc length from birth to mid-childhood in *Homo sapiens*

Not surprisingly, given its young developmental age, the bone of the Mladeč 3 frontal squama is very thin. Thickness was measured laterally on the left frontal boss at 1.95 mm, slightly below the values of 2.5–3.1 for comparably aged Neandertal children, but quite similar to values for the early Upper Paleolithic associated children and 2.0 for Skhul 1. Differentiation of the vault into external and internal bony tables with an intervening diploë is visible at several places along broken edges. This differentiation is well developed in the central squama, particularly in the region of the frontal boss, but reduces posteriorly approaching the vicinity of bregma where the frontal squama thins to 1 mm, and CT images reveal an absence of diploë, due to a recently or incompletely closed bregmatic fontanelle as mentioned previously (see also Fig. 3).

Separation of the external and internal bony tables is strikingly evident over the left orbit where frontal thickness expands from the 2 mm squamous average to 5.4 mm. Laterally, where breakage exposes the internal architecture of the left superior orbital margin, it is obvious that the diploë is responsible for most of this expanded thickness. More medially, however, just internal to the supero-

Table 1. Mladeč 3 and comparative frontal dimensions

Specimen	Approx. age at death (years)	Nasion-Bregma chord (M29)	Nasion-Bregma arc (M26)	Nasion-Bregma index (M29/M26)	Interorbital breadth (M50)
Mladeč 3					
this study	≤2	**(96)**	**(115)**	**(83)**	**18.5**
Szombathy, 1925	2-4	-	(110)	-	-
Upper Paleolithic Associated					
Pataud 3[1]	3-6 mos	-	-	-	22.0
Předmostí 6[2]	2.0-2.5	103.0	119.0	86.5	-
Le Figuier 2[3]	2-3	93.0	114.0	81.5	-
La Madeleine 4	3[4]	99.0	115.0	86.1	(22.5)
Předmostí 8[2]	4.0-4.5	102.0	110.0	92.7	-
Lagar Velho 1	4.5-5[5]	(100)[6]	(115)[6]	(87)	(20.4)[7]
Předmostí 2[2]	7-8	(120)	(136)	(88)	-
Levantine Early Anatomically Moderns					
Skhul 1	4.5	(99)	(115)	(86)	(24)
Qafzeh 10[8]	6	(105)	(123)	(85)	(23.4)[9]
Neandertals					
Dederiyeh 2[10]	2	81.5	99	82.3	-
Subalyuk 2	2.5	(96)	(115)	(83)	-
	3.0[11]	(78)[11]	(95)[11]	(82)	-
Pech de l'Azé	2.5-3.0	(93)	(104)	(89)	23.5
Roc de Marsal	3	99	112	88.4	22.2
Engis 2	4.0-5.0	100.1	117.5	(86.0)	22.4
Devil's Tower	4.5-5.0	101.0	117.0	86.3	24.1
La Quina 18	7.5	94.0	103.5	90.8	16.0
	Range, approx. age at death (years)	**Nasion-Bregma chord (M29)**	**Nasion-Bregma arc (M26)**	**Nasion-Bregma index (M29/M26)**	**Interorbital breadth (M50)**
Recent Children					
	0-0.24	63.2 ± 3.5 (N=10)	73.4 ± 7.1 (N=10)	86.4 ± 4.2 (N=10)	15.2 ± 1.4 (N=9)
	0.25- 2.4	92.2 ± 4.9 (N=16)	109.5 ± 5.1 (N=15)	84.6 ± 3.0 (N=15)	18.3 ± 1.8 (N=18)
	2.5-5.9	97.6 ± 5.3 (N=22)	115.3 ± 6.0 (N=22)	84.9 ± 2.5 (N=22)	18.5 ± 1.7 (N=23)
	6.0-7.9	103.9 ± 3.2 (N=9)	124.1 ± 6.1 (N=9)	83.8 ± 3.5 (N=9)	19.8 ± 1.8 (N=13)

[1] Billy (1975)
[2] Matiegka (1934)
[3] Billy (1979)
[4] Heim (1991)
[5] Trinkaus (2002)
[6] Zollikofer et al. (2002)
[7] Franciscus (2002)
[8] Tillier (1999)
[9] Franciscus (1995)
[10] Ishida and Kondo (2002)
[11] Pap et al. (1996); these measurements are based on the Pap et al. reconstruction of Subalyuk 2, a dramatically different interpretation of the fossil's shape than that available when measured this same specimen in 1980.

medial corner of the left orbit, a thin, concave shell of bone, indicates that pneumatization by the frontal air sinus had commenced.

This finding is also clearly visible in the computer tomographic images (see Fig. 3). Using a semi-automatic segmentation tool in Amira® we segmented the frontal sinuses which revealed two cavities, separated by a median septum, beginning slightly above the level of the fronto-maxillary sutures bilaterally and extending superiorly to a position behind glabella. The right lobe is irregularly shaped and about 8 mm high, 5 mm wide and 1.5 mm deep. The left lobe is more triangular, about 6.5 mm high, 5 mm wide and 2 mm deep. We estimate their combined volume at 0.082 cm^3.

The literature reports considerable variation in the onset of frontal sinus development in modern humans. According to Hasselwander (1927) and the extensive study by Szilvássy (1981) frontal sinus formation begins in the medial supraorbital region around the twelfth month of postnatal life although a contemporary German sample found the frontal sinus already present at birth in 1.5% of individuals (Spaeth et al., 1997). Far less commonly, later ages are reported for the onset of frontal sinus formation. Barghouth et al. (2002) described frontal sinus initiation at the age of 6, and Vlček (1967) associates their appearance with the time of permanent canine eruption.

Data on frontal sinus development in fossil hominids is sparse. A first detailed study of frontal sinus morphology and development in Neandertals was done by Vlček (1967). He found no frontal sinuses in the 2 year old Pech de l'Azé 1 and the 4.5 year old Devil's Tower individuals, but observed small (ethmoidofrontal) cellulae in the 8 year old La Quina 15 and 10 year old Teshik-Tash specimens. Tillier (1999) reports some ethmoido-frontal cellulae in the 6 year old Qafzeh 10 child, while the 3–4 year old Qafzeh 12 has ethmoido-frontal pneumatization with "limited vertical development" (Tillier, 1999, 66).

Prossinger and colleagues (2005), reported a volume of 5.3 cm^3 for the frontal sinuses of the Le Moustier 1 adolescent, and 0.064 cm^3 for the 2 year old Subalyuk child. However, after examining the same CT scans utilized by Prossinger and colleagues, we question the presence of a real frontal sinus in Subalyuk 2. Rather, the area above the Subalyuk 2 frontonasal suture is, in our opinion, filled by diploë. Nor does it present the classic bilobate shape characterizing a true frontal sinus from its earliest inception.

The left orbit preserves the medial orbital angle and some 8 mm of the superior orbital margin. The medial orbital angle is dramatic, changing abruptly from vertical to horizontal and suggesting a rectangular-shaped eye socket as is the usual case for modern Europeans (Szilvássy, 1986). This

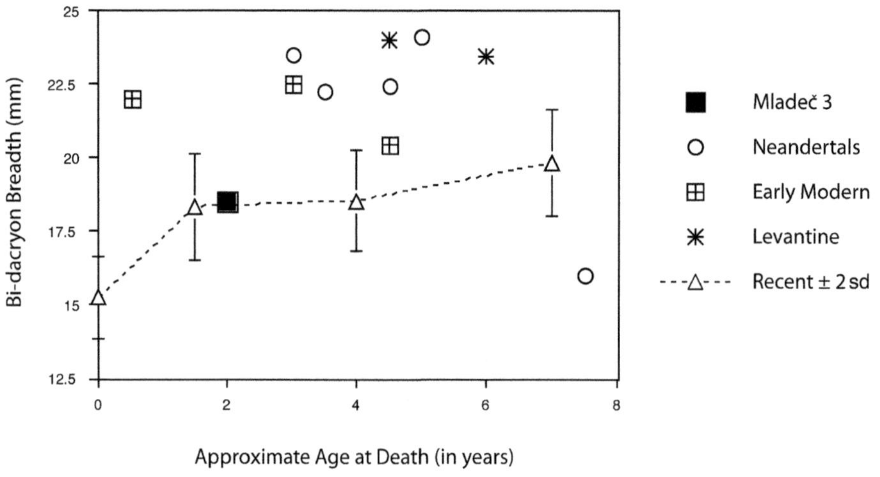

Fig. 6. Growth in biorbital breadth from birth to mid-childhood in *Homo sapiens*

contrasts with the more rounded orbital aperture of European Neandertals, including young individuals, in whom orbital angles transition gradually from their vertical to horizontal borders. Between the orbits, the frontonasal suture forms a large inverted "V" where the now missing nasal bones previously articulated. Mladeč 3 exhibits an extremely wide superior nasal breadth (17.0 mm). Interorbital breadth (as measured by bi-dacryon breadth) is estimated at a minimum of 18.5 mm. This interorbital breadth is quite narrow, falling directly at the recent human mean for 2 year olds but considerably lower than this value for similarly aged Neandertal and other Upper Paleolithic associated children (Table 1, Fig. 6).

Endocranially, a few faint cerebral cortical impressions are present but overall, the bone is quite smooth and lacking in details of taxonomic interest. A strong, well-preserved frontal crest, 22 mm in length, is found along the endocranial midline anteriorly, with a foramen caecum at its junction with the anterior cranial fossa floor. As it runs superiorly, the frontal crest becomes trough-like, incorporating the groove for the superior sagittal venous sinus. This groove, which reaches a maximum width of 7.0 mm, remains visible for approximately 1/3 the length of the frontal, then disappears as it merges imperceptibly into the inner bony table of the squama.

Occipital

Despite its very young age, the Mladeč 3 occipital squama was already a large, strongly sculpted bone, previewing the morphology seen in adult specimens such as the Mladeč 5 male. From the midsagittal plane, the superior nuchal line sweeps superolaterally to the highest point in its arc, some 12 mm from the midline, to create a downward pointing, subtle inion triangle. Occipital plane length in Mladeč 3 is considerably shorter than in Skhul 1, but given the rapid growth of the brain during infancy and early childhood, comparisons with the slightly older Skhul 1 specimen may not be appropriate. Interestingly, the specimen closest to Mladeč 3 in occipital plane length is La Ferrassie 8, a Neandertal but an individual of approximately the same developmental age as Mladeč 3.

In *norma occipitalis*, the left and right superior nuchal lines can be seen forming an prominent external occipital protuberance. This protuberance is flanked on either side and accentuated by the superior nuchal lines where they mark the upper limits for the attachment sites of strongly developed semi-spinalis capitis muscles. The arc formed by the superior nuchal line continues laterally and inferiorly from its summit, but becomes so weak laterally that exactly where it crosses the lambdoid suture cannot be determined. Bilateral, bulges are a prominent feature of the occipital. These are a common characteristic of immature Würm specimens, where the bone of the posterior cranial fossa conformed to the shape of the underlying posterior poles of the left and right cerebellar lobes. Prior to matrix removal, a groove for the occipital artery was clearly visible on the right where it marked the superior course of that vessel approaching the superior nuchal line.

Prior to matrix removal, a concentration of pits defined a shallow, ovoid suprainiac fossa, approximately 19.6 mm wide and 13.9 mm high located on the occipital plane just above the superior nuchal line. Unfortunately, this interesting feature was destroyed during cleaning. The Mladeč 3 suprainiac pitting was typical of that common in *Homo sapiens sapiens* of similar developmental age (Heim, 1982), as distinct from the deep fossa typical of juvenile as well as adult Neandertals. As noted by Tillier (1999), such suprainiac pitting is also weakly expressed in the 12 year old Qafzeh 11 and Sungir' 2 specimens and is not an unexpected finding given its occurrence in a number of Upper Paleolithic associated adults, including Mladeč 6 (Caspari, 1991; Frayer et al., this volume).

Thirty-five millimeters posterior to lambda, the flattened occipital plane merges into the nuchal plane. The superior nuchal line forms the inferior boundary of this junction where the occipital and nuchal planes meet at an approximately 135° angle when viewed in *norma lateralis*. Laterally, angulation between the occipital and nuchal planes is reduced. As with the Mladeč 5 adult male, the nuchal plane is strongly depressed about the midline. The remarkable extent of lambdoid flattening in Mladeč 3 is perhaps best visualized in *norma verticalis*. A strongly protruding area between the flattened superior portion of the occipital plane and the inferior angulation at the superior nuchal line forms the distinct projection of an unmistakable occipital bun. This is further accentuated by a bilateral flattening of the occipital plane some 25 mm to either side of the midline. The presence of an occipital bun in Mladeč 3 is not surprising, given the presence of this feature in the Mladeč adults of both sexes and it is strikingly similar, in a number of elements, to that of the Mladeč 5 adult male. Both individuals exhibit significant lambdoid flattening which transitions abruptly to a vertical posterior face of the occipital bun itself.

Caspari (1991) observed that the Mladeč 5 occipital bun is more medially restricted than in Neandertals and is accentuated by a side-to-side compression which flattens the occipital plane between the bun and asterion. This same morphology describes the Mladeč 3 occipital bun and, in part, differentiates it from those of young Neandertal children. Comparison of the occipital in Mladeč 3 with early central European Upper Paleolithic children such as Předmostí 6, 2, and 22 reveals a wide range of variation in this feature. Předmostí 6, the specimen closest in developmental age to Mladeč 3, has no evidence of a bun. The 8–9 year old Předmostí 22, which Matiegka (1934) considered a female, seems to have a slight expression of this feature while the 7–8 year old Předmostí 2 has a very distinct bun. This morphology differs from that of the Skhul 1 Mousterian child which lacks an occipital bun as do the juveniles from Qafzeh, Israel (Tillier, 1999).

Below the inferior nuchal line, the attachment sites of the *m. rectus capitis posterior minor* and *major* are strongly sculpted and separated by a well-defined median occipital crest, indicating that these muscles were well-developed. Posterior foramen magnum breadth, between the open posterior intraoccipital synchondroses measures 6.1 mm. Nuchal plane length in Mladeč 3 exceeds nuchal plane chord and arc dimensions for the La Ferrassie 8 two year old Neandertal, but is less than those of the slightly older La Madeleine 4 Upper Paleolithic child. Interestingly, curvature of the nuchal plane, calculated using the inion-opisthion chord/arc index, is greater in both Mladeč 3 and La Madeleine 2 than in the La Ferrassie 8 Neandertal.

Maximum occipital width, measured by a biasterionic breadth of 91.4 mm in Mladeč 3, is equivalent to the recent human mean for children of comparable developmental age (see Fig. 7). The finding that the Mladeč 3 bi-asterionic measurement is narrower than most of the comparably aged Neandertal children, greater than that of the only other available 2 year old Upper Paleolithic child (Le Figuier) and quite close to the Lagar Velho Gravettian 4.5–5 year old is consistent with mean values

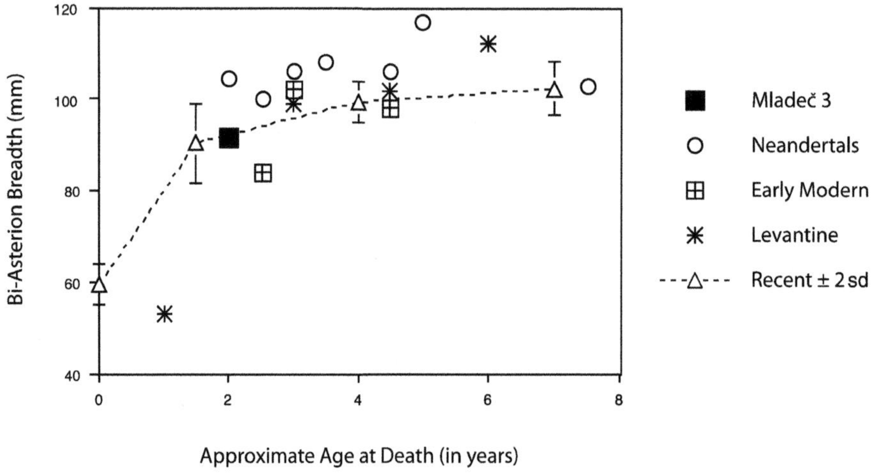

Fig. 7. Growth in occipital breadth from birth to mid-childhood in *Homo sapiens*

Table 2. Mladeč 3 and comparative occipital dimensions

Specimen	Approx. age at death (years)	Lambda-Opisthion chord (M31)	Lambda-Opisthion arc (M28)	Lambda-Opisthion index (M31/M28)	Lambda-Inion chord (M31.1)	Lambda-Inion arc (M28.1)	Lambda-Inion index (M31.1/M28.1)	Inion-Opisthion chord (M31.2)	Inion-Opisthion arc (M28.2)	Inion-Opisthion index (M31.2/M28.2)	Bi-Asterion breadth chord	Bi-Asterion breadth arc
Mladeč 3												
this study	≤ 2	**80.6**	**100.0**	**79.8**	**50.5**	**56**	**90.1**	**39.5**	**44**		**89.8**	**91.5**
Szombathy (1925)	2–4	81.0	103	78.6	52	62	83.9	40	41		97.6	92.0
Upper Paleolithic Associated												
Předmostí 6[1]	2.0–2.5	101.0	121.0	83.4	–	–	–	–	–		–	–
Le Figuier[2]	2–3	–	–	–	–	–	–	–	–		–	84
La Madeleine 4	3.0[3]	105	121	87	66.7	75	88.9	43.3	48		90.2	102
Lagar Velho 1[4]	4.5-5	94	115	81.7	62	–	–		43		–	98
Kostenki 3	6.0–7.0	–	(117)	–	–	–	–	–	–		–	–
Předmostí 2[1]	7.0–8.0	(91)	(115)	(79)	–	–	–	–	–		–	–
Levantine Early Anatomically Moderns												
Qafzeh 13[5]	10mos.	55	–	–	–	–	–	–	–		–	(53)
Qafzeh 12[5]	3	91	109	83.4	–	–	–	–	–		–	(99)
Skhul 1	4.5	(87)[6]	105[6]	(83)	–	76[6]	–	–	29[6]		–	101.7
Qafzeh 10[5]	6	(92)	(112)	(82)	(57.8)	65	(89)	–	–		–	(112)
Neandertals												
La Ferrassie 8[7]	2.0	79.7	92	86.6	52.2	57	91.5	35.3	36		98	(95)
Dederiyeh 1[8]	2.0		–	–	–	–	–	(36.7)[9]	–		–	–
Dederiyeh 2[10]	2.0	–	–	–	–	–	–	–	–		–	(114)
Subalyuk 2	2.5	(85)	(110)	(86)	–	–	–	–	–		–	(100)
	3.0[11]	84[10]	102[11]	82.3	–	–	–	–	–		–	112[11]
Pech de l'Azé	2.5–3.0	–	–	–	–	–	–	–	–		–	(106)
Roc de Marsal	3.0	(88)	(109)	(81)		–	–	–	–		–	108
Engis 2	4.0–5.0	83[12]	106[12]	78.3	61.5[12]	69[12]	89.1	46.4	50		92.8	106.0
La Quina 18	7.5	–	–	–	–	–	–	–	–		–	(103)

	Range, approx. age at death (years)	Lambda-Opisthion chord (M31)	Lambda-Opisthion arc (M28)	Lambda-Opisthion index (M31/M28)	Bi-Asterion Breadth chord (M12)
Recent children					
	0–0.24	58.3 ± 4.7 (N=9)	71.9 ± 5.7 (N=9)	81.0 ± 1.6 (N=9)	59.5 ± 4.4 (N=9)
	0.25–2.4	81.3 ± 7.0 (N=12)	97.9 ± 9.4 (N=12)	83.1 ± 2.8 (N=12)	90.4 ± 8.6 (N=14)
	2.5–5.9	87.5 ± 5.7 (N=18)	106.3 ± 8.3 (N=18)	82.3 ± 2.7 (N=18)	99.3 ± 4.5 (N=22)
	6.0–7.9	92.1 ± 4.3 (N=7)	111.1 ± 7.2 (N=7)	82.9 ± 2.7 (N=7)	102.3 ± 5.8 (N=7)

1 Matiegka (1934)
2 Billy (1979)
3 Heim (1991)
4 Trinkaus (2002)
5 Tillier (1999)
6 McCown and Keith (1939)
7 From Heim (1983)
8 Dodo, Kondo and Nara (2002).
9 Note that this measurement was taken slightly off the midline, "from the medial end of the posterior intraoccipital synchondrosis to the superior nuchal line"(Dodo et al., 2002, 103).
10 From Ishida and Kondo (2002)
11 Pap et al. (1996); these measurements, are based on the Pap et al. reconstruction of Subalyuk 2, a dramatically different interpretation of the fossil's shape than that available when NM-P measured this same specimen in 1980.
12 Tillier (1983b)

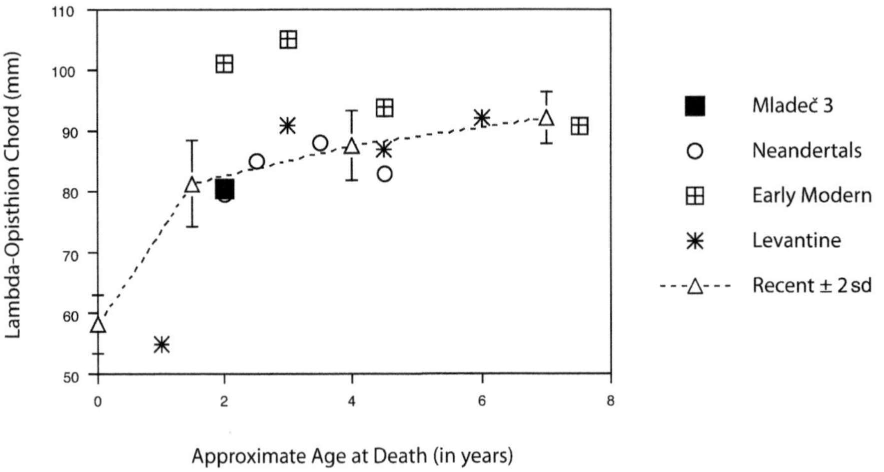

Fig. 8. Growth in occipital height from birth to mid-childhood in *Homo sapiens*

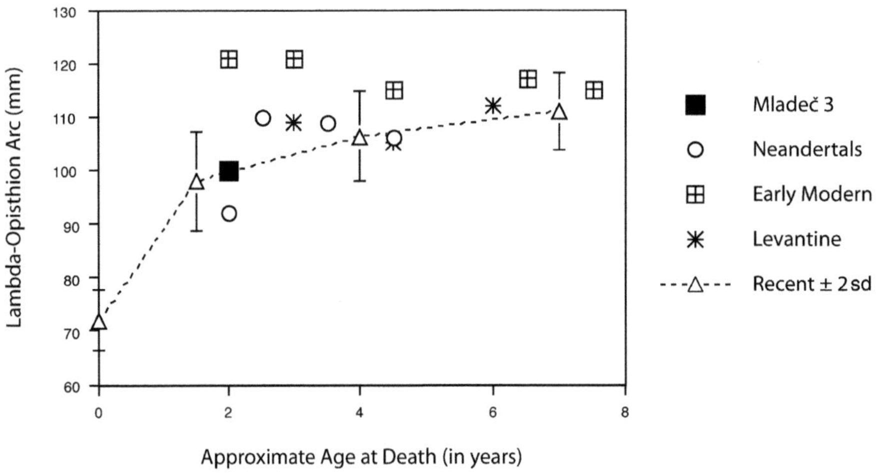

Fig. 9. Growth in occipital sagittal arc from birth to mid-childhood in *Homo sapiens*

for adult specimens from these populations (see Wolpoff et al., this volume; Frayer et al., this volume). It is broader than that of the Le Figuier 2–3 year old Upper Paleolithic child, similar to the 95 mm bi-asterionic breadth of the La Ferrassie 8 Neandertal, but considerably narrower than the 2 year old Dederiyeh 2 Syrian Neandertal child. In the slightly older, 2–2.5 year old Subalyuk 2 and 2.5–3 year old Pech de l'Azé Neandertals, this dimension is considerably wider than in Mladeč 3 (see Table 2 and Fig. 7). In assessing the significance of these comparisons, it is important to note that bi-asterion breadth is essentially a measurement of cerebellar breadth. At 1.5 years, the modern human cerebellum has attained only 50% of its adult size which is usually reached by age 3 (Blinkov and Glezer, 1968). Thus, Mladeč 3 bi-asterion breadth is age appropriate considering that, by the time of death, it was 76% of the bi-asterionic breadth of the Mladeč 5, and 72% of the Mladeč 6 males and was 83% of the Mladeč 1 adult female dimension (see see Wolpoff et al., this volume; Frayer et al., this volume).

Total occipital height, measured from lambda-opisthion is 100 mm (arc) and 80.6 mm (chord). For both these values, Mladeč 3 falls around the mean for recent modern children of comparable developmental age. However, compared with other Upper Pleistocene juveniles, Mladeč 3 occipital height is most similar to La Ferrassie 8 and other young Neandertals, and considerably less than the Předmostí 6, La Madeleine 4, Lagar Velho 1 and Skhul 1 children (see Table 2, Figs. 8 and 9). This is an interesting finding, as among the adult male and female means for these samples, Neandertals consistently have the lowest occipital height values. The similarity between the Mladeč 3 and La Ferrassie occipitals, and their contrast with the 2 year old Gravettian associated Předmostí 6 and 4.5–5 year old Lagar Velho 1, suggest that, in infancy, vertical occipital growth at Mladeč more closely resembled the Neandertal pattern. When compared with the only available Mladeč adult values for this dimension (Wolpoff, et al., this volume), Mladeč 3 occipital height is 75% of the Mladeč 1 female lambda-opisthion arc and 76% of her lambda-opisthion chord.

Relative to the occipital plane, the Mladeč 3 nuchal plane is fairly long. Comparing it with adults from the site (Frayer et al., this volume; Wolpoff et al., this volume) reveals that Mladeč 3 occipital plane length had attained 76% of the Mladeč 5 adult male lambda-inion arc and 75% of his lambda-inion chord values. Comparisons with the Mladeč 6 adult male reveals the Mladeč 3 value being 81% of his lambda-inion arc and 81% of his lambda-inion chord values. Not surprisingly, comparisons of these Mladeč 3 metrics with the extremely high Mladeč 1 female adult show Mladeč 3 having reached only 65% of her lambda-inion arc and 68% of her lambda inion chord values.

Occipital squama thickness is fairly uniform in Mladeč 3 and resembles that of the frontal, averaging about 1.9 mm around both lambda and asterion. At the position of the external occipial protuberance, the squama thickens slightly to 4.0 mm. These thicknesses are well within the ranges for European Neandertal children of the same developmental age as Mladeč 3 and similar to the Skhul 1 Levantine specimen.

The endocranial surface of the Mladeč 3 occipital is well preserved, and reveals an unusual venous drainage pattern. Grooves for both transverse venous sinuses are well-defined. The right is somewhat larger with a span of up to 10 mm between the summits defining the margins of the groove. It is joined centrally at the confluence by the vertically descending impression for the superior sagittal sinus which deviates more markedly to the right than is usual before contributing to the formation of the transverse sinus. Slightly further to the left, a groove for an unnamed, accessory dural sinus also descends to merge with the confluence. Brash (1951) and Anson and Maddock (1952) reported that the superior sagittal sinus may bifurcate in which case it contributes to both the left and right transverse sinuses and this appears to have been the case for Mladeč 3. Endinion, therefore, is not a definite point, but rather extends along the confluence of sinuses for more than 18 mm. Inferior to the confluence, the slightly fainter occipital crest, bears a narrow but well-defined occipital sinus. This channel descends, then divides into short left (16.2 mm) and right (14.76 mm) arms on approaching the posterior border of the foramen magnum where it is truncated at the unfused edge of the squamous occipital. According to Browning (1953) large occipital sinuses are common in late fetal and neonatal specimens, so that the obvious development of this dural sinus in Mladeč 3 is not surprising. A number of small, emissary foramina perforate the Mladeč 3 dural sinus impressions.

The posterior cranial fossa displays some additional interesting asymmetries. The left cerebral fossa is about double the size of the right, having a transverse breadth of 44 mm compared with 23 mm on the right side. The cerebellar fossae are less assymetrical. The right is the larger of the two, but this difference is not great and this internal asymmetry results from a displacement of the superior sagittal sinus as described above. As in Mladeč 3, the left cerebral fossa of Skhul 1 is considerably larger than the right (McCown and Keith, 1939) while Heim (1982) reports asymmetrical left and right cerebellar fossae in the La Ferrassie 8 Neandertal as well.

Temporal

For unknown reasons, Szombathy included neither a description nor measurements of the Mladeč 3 temporal in his 1925 study. Nevertheless, the bone is an excellent specimen and despite Szombathy's

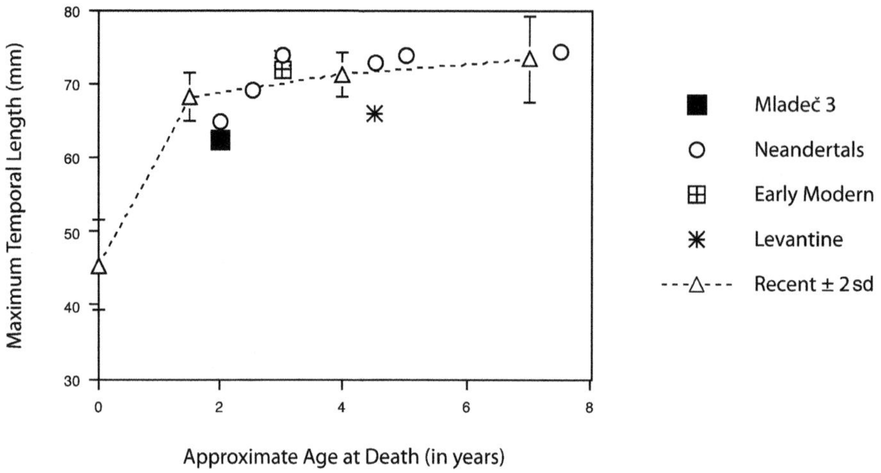

Fig. 10. Growth in maximum temporal length from birth to mid-childhood in *Homo sapiens*

slight consternation regarding its accelerated development compared with the frontal and occipital (see section "Age at death estimation"), it fits well with the other Mladeč 3 remains leaving no doubt regarding their association as a single individual.

Although its edges are chipped a reasonable estimate of maximum temporal length, from spheno-temporal articulation to occipitomastoid suture, is 62.5 mm. This total temporal length falls below the recent modern mean and into the small end of the immature European Neandertal range (see Table 3 and Fig. 10). Compared with the Mladeč adults (Frayer et al., this volume; Wolpoff et al., this volume), Mladeč 3's total temporal length is 85% of the Mladeč 5 male's dimension and 86% and 91% of this dimension in the Mladeč 1 and 2 females, respectively. The squamous portion alone, measured from the most anterior squamous margin to the parietal notch, is an estimated 50 mm in length. This is long, representing some 80% of Mladeč 3's total temporal length. This falls above the recent modern mean as do other Upper Paleolithic associated children, while Neandertals and the Skhul 1 child tend to fall below the recent mean or even more than 2sd below the range for our recent comparative sample (Fig. 11). Such a proportioning of squamous to total temporal length resembles that of the Upper Paleolithic associated La Madeleine 4 3-year old (85%), but is considerably greater than in the 2 year old Dederiyeh 2 (73%), the 2.5–3 year old Pech de l'Azé (73%), and the 3 year old Roc de Marsal (65%) Neandertals. In the 4.5 year old Skhul 1 child, squamous temporal length is approximately 76% of total temporal length, thus exceeding young Neandertal values, but falling below those of the known Upper Paleolithic associated children.

As with the Mladeč adults, the temporal squama is flat. This configuration is apparent in other European Upper Paleolithic young children as well as in the Skhul sample, but contrasts markedly with the young European Neandertals who usually present more vertical curvature in their squamous temporal morphology, a pattern consistent with their lower vaulted crania. The estimated height of the Mladeč 3 squamous temporal falls below the recent modern range for 2 year olds, (see Fig.12), grouping with comparably aged Neandertals as does Skhul 1. In contrast, the other Upper Paleolithic associated children examined fall at or above the recent mean for this dimension. Although this low vertical height of Mladeč 3 might, in part, be due to a slight underestimate in measuring given the damage to the squamous margins, it is interesting to note that the Mladeč 2 adult female temporal squamae also exhibit a low vertical height which falls within the Neandertal range (see Wolpoff et al., this volume).

Table 3. Mladeč 3, comparative temporal dimensions

Specimen	Approx. age at death (years)	Maximum length of temporal	Squamous length (to incisure)	Squamous height (above porion)	Squamous height/length index	Mastoid process height (to porion)	Mastoid process breadth
Mladeč 3							
this study[1]	≤ 2	**R:(62.5)**	**R:(50)**	**R:(26)**	**R:(52)**	**R:8.10**	**R:10.2**
Upper Paleolithic Associated							
La Madeleine 4	3.0[2]	L:72.0	L:(61)	L:33.5	L:(55)	L:13.6	L: 19.9
						R:17.2	R:20.9
Lagar Velho 1[3]	4.5-5	-	R: 53.5	R:(35)	R:(65)	R:18.0	-
Kostenki 3	6.0-7.0	-	R:59.0	L:44.2	R:74.9	R:12.1	R:14.2
				R:44.6			
Levantine Early Anatomically Moderns							
Qafzeh 12[4]	3	L:73.5	-	-	-	-	-
Skhul 1	4.5	R:(66)[4]	R:(50)[5]	R:(30)	R:(60)	R:13.4	R:14.1
Qafzeh 10[4]	6	-	-	-	-	15.5[3]	-
Neandertals							
Krapina 38.22	0.5-1.5	-	-	-	-	R:10.4	R:10.8
Krapina 17	1-2.5	-	-	R:(37.5)	-	R:13.1	R:12.0
Krapina 39.3	1-2.5	L: 73.5	L: 57	-	-	L:14.7	L:12.8
Dederiyeh 2[6]	2.0	R:65	R:47.4	R:28.4	R:59.9	R:13.1	R:12.7
Subalyuk 2	2.5	-	-	R:(30)	-	L:13.6	L:12.6
		-	-	-	-	R:(11)	R:(12)
Amud 3	2.5-3.0	-	-	-	-	R:14.0	R:17.0
Pech de l'Azé	2.5-3.0	R:(65)[7]	R:46.5	R:(26)	R:(56)	L:13.3	L:11.4
						R:13.0	R:12.2
Roc de Marsal	3.0	R:74.0[7]	R:48.4	R:33.5	R:69.2	-	-
Engis 2	4.0-5.0	R:73	R:54.2	R:31.3	R:72.6	L:14	L:14.7
		-	-	-	-	R:13.7	R:15.6
Devil's Tower	4.5-5.0	R:74	R:48.8	R:(35)	R:(72)	R:13.4	R:16.1
La Quina 18	7.5	L:74.5	L:51.0	L:(33)	L:64.7	L:13.3	L:16.6
		R:74.2	R:51.2	-	-	R:17.6	R:17.6

	Range, approx. age at death (years)	Maximum length of temporal	Squamous length (to incisure)	Squamous height (above porion)	Squamous height/length index
Recent children					
	0-0.24	45.4 ± 6.1 (N=10)	33.1 ± 5.2 (N=10)	21.1 ± 2.6 (N=10)	64.9 ± 8.8 (N=10)
	0-2.4	68.1 ± 3.2 (N=4)	46.2 ± 4.2 (N=10)	31.4 ± 3.5 (N=9)	68 ± 7.8 (N=9)
	2.5-5.9	71.2 ± 3.0 (N=14)	51.8 ± 2.9 (N=21)	35.2 ± 4.0 (N=21)	67.6 ± 4.6 (N=21)
	6.0-7.9	73.3 ± 5.8 (N=3)	54.9 ± 3.0 (N=10)	37.0 ± 2.9 (N=10)	66.5 ± 4.6 (N=10)

[1] Szombathy provided no metric data for the Mladeč 3 temporal in his 1925 description.
[2] Heim (1991)
[3] Trinkaus (2002)
[4] Tillier (1999)
[5] Measured from a cast
[6] Ishida and Kondo (2002)
[7] Madre-Dupouy (1992)

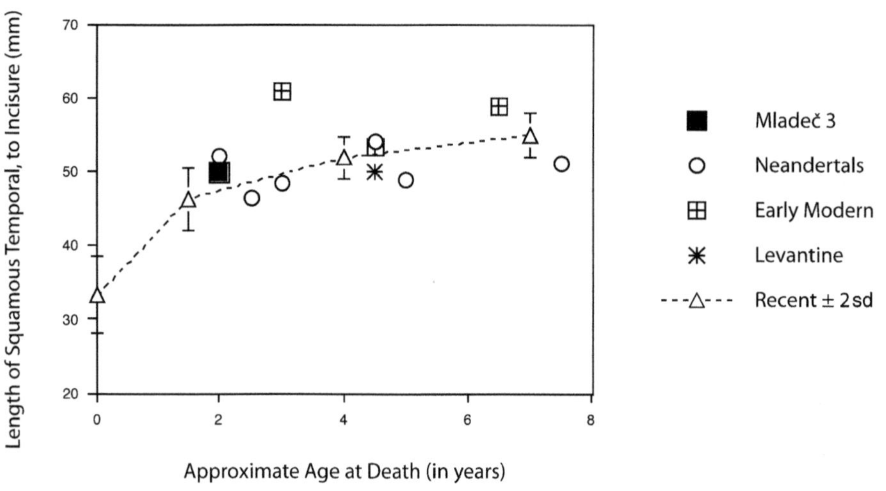

Fig. 11. Growth in squamous length from birth to mid-childhood in *Homo sapiens*

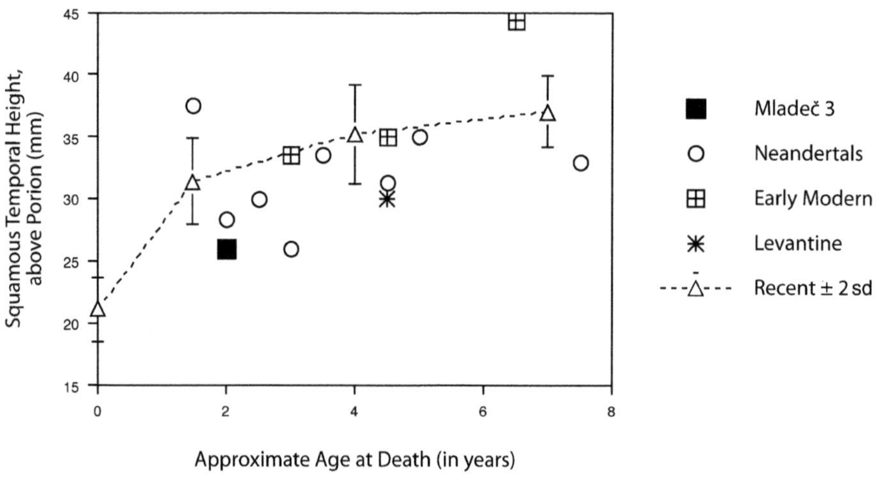

Fig. 12. .Growth in squamous height from birth to mid-childhood in *Homo sapiens*

The terminal stump of the Mladeč 3 zygomatic process of the temporal merges with the squamous portion superior to a broad, shallow glenoid fossa. Prior to cleaning, a slight linear elevation marked the position where a supraglenoid crest, extending posteriorly beyond the fossa and above the external auditory meatus to the root of the mastoid had started to develop. Unfortunately, this feature was lost in the process of matrix removal.

The glenoid fossa is shallow, as is typical for all very young *Homo sapiens*. Its articular surface faces infero-laterally, at an angle of 78° to the sagittal plane, a posture within the range of variation for young *Homo sapiens* up to the age of 6–8 years (Minugh-Purvis, 1988; Minugh-Purvis et al., 2000). Behind it, abutting against the anterior tympanic plate, is a post-glenoid tubercle. Despite its youth, the region immediately anterior to the glenoid fossa had already clearly developed into an articular eminence. It thus unambiguously resembles the morphology most commonly found in *Homo sapiens sapiens*, consistent with adult Mladeč temporal morphology. This arrangement contrasts with that usually described as typical of Neandertals in which the articular eminence is lacking and the glenoid fossa and location for the protracted mandibular condyle are situated entirely in one plane. However, it should be noted that the numerous temporals included in the Krapina Neandertal sample display a considerable range of variation in glenoid fossa morphology including that described above as typifying modern humans (Smith, 1976; Minugh-Purvis et al., 2000).

Because of the specimen's immaturity, the final shape of the external auditory meatus, had Mladeč 3 survived to adulthood, is a matter of conjecture. This region was incompletely ossified and it is difficult to determine whether the tympanic ring had partially fused at the time of death. However, it appears that the external auditory meatus was elliptical in form with the longest (7.2 mm) axis angling strongly antero-superiorly. This differs from Neandertal children of similar developmental age, in whom the form of the external auditory meatus often begins to assume a posteriorly oriented ellipse early in childhood (Trinkaus, 1983; 2002). No evidence of the tympanic dehiscence is present in Mladeč 3.

As is to be expected in such a young individual, the Mladeč 3 mastoid process is small, although some very slight development and pneumatization had commenced. Its breadth measures 11.4 mm at its base and its height measures 12.7 mm at the level of porion. However, because mastoid size is highly dependent upon the mechanical pull of attaching muscles, such as the sternocleidomastoid, which alter with head posture and activity patterns, this feature changes dramatically during ontogeny and is highly variable in both recent and fossil adult human samples. With that proviso, we note that Mladeč 3 mastoid size is comparable to the Neandertal mean for children of similar developmental age, but smaller than in the slightly older 4.5 year old Skhul 1 Levantine child (see Table 3). Medial to the base of the mastoid process is a broad, shallow, single digastric groove, in contrast to the double furrow typical of Neandertals. A strong juxamastoid eminence is present as a raised area medial to the digastric groove. This projects 1.3 mm inferior to the the mastoid tip and effectively dwarfs it. Prior to matrix removal, but obscured since cleaning, a deep pit forming the vaginal foramen was visible where the root of the styloid process, presumably still cartilaginous and unfused within its socket, was anchored in life. Inferolateral to the vaginal process, the stylomastoid foramen is situated in a more lateral than inferior position as is typical of infants and very young children.

Above the mastoid process, the mastoid notch is well separated from asterion by about 15 mm. The bone thickens at the notch to a maximum of 2.8 mm, considerably more than the average thickness (0.9 mm) of the temporal squama.

Endocranially, the area of the jugular notch is broken but the path of the sigmoid sinus is well-defined, presenting a wide (approx. 7 mm), deep channel in the bone. The superior petrosal sinus is also well-marked but the region where the groove for the inferior petrosal sinus would normally be expected is missing, so that this feature cannot be observed. The arcuate eminence, subarcuate fossa, and vestibular aqueduct are strongly marked. The arcuate eminence is also well-developed in La Ferrassie 3, Pech de l'Azé, Devil's Tower, and several Krapina young European Neandertals temporals. Tillier (1999) reports that in the Qafzeh 13 neonate, the arcuate eminence is quite pronounced and describes the subarcuate fossa as small but well-defined. She also reports similarly well-developed arcuate eminences in the Qafzeh 21 3 year old and the Qafzeh 12 3–4 year old children. The internal acoustic meatus of the Mladeč 3 child is quite large (5.7 x 3.8 mm) relative to petrous size. Examination of CT images reveal a spacious tympanic antrum within the petrous portion, typical of a young child. The antrum is partly filled with matrix and no auditory ossicles are visible.

Total vault

Utilizing CT reconstruction, a limited number of observations concerning the overall configuration of the Mladeč 3 vault are possible. In this regard, it is particularly interesting to note the correspondence between CT images of Mladeč 3 and the modern 2 year old reference skull over which it was superimposed (see Fig. 1).

Mladeč 3's frontal betrays an interesting mosaic in which a very steep forehead, as typical of a modern child, is found in combination with browridge development, a more archaic character at this young developmental age but consistent with the already evident browridges of the Mladeč adult phenotype. The frontal appears to have led to a well-vaulted parietal region clearly seen in normal lateralis.

Viewed in *norma verticalis,* the natural integration of the temporal segment is clearly evident - further supporting the association of this bone with the rest of the remains. From this view it is obvious that the vault is a bit long and narrow, strongly resembling the Předmostí 6 2 year old. In *norma occipitalis,* it is clear that Mladeč 3 possesses a fairly modern braincase morphology with its greatest breadth positioned high on the parietals. Integrated into its anatomical position in this way, it becomes evident that the flat temporal squama (Fig. 1) of Mladeč 3 is canted superolaterally rather than having a strictly vertical orientation. This pattern is typical of modern humans of this developmental age, and results from differential growth between the rapidly expanding neurocranium above a cranial base which is expanding in breadth more slowly during development. Other features, such as lambdoidal flattening and low occipital height clearly evident in *norma verticalis* and *norma lateralis* (Fig. 1), together with a well-developed occipital bun, are reminiscent of Neandertals as is true of several features in the Mladeč adults, particularly the Mladeč 5 and 6 males.

One of the most rapidly evolving regions of the neurocranium during the Upper Pleistocene, the parietal segment - has not been preserved in the Mladeč 3 child. However, CT reconstruction now permits limited observations on the parietal vault. Although the Mladeč 3 parietal sagittal arc length cannot be reconstructed, lambda is present and the location of bregma easy to estimate, so that it is possible to estimate Mladeč 3 sagittal parietal chord length as between 98 and 103 mm. Unexpectedly, this dimension is quite short, falling into the Neandertal range, well below that of other Upper Paleolithic associated children, including Předmostí 6 and Lagar Velho. Among the Mladeč adults, this proportioning most closely resembles the Mladeč 2 female - an individual with which Mladeč 3 also shares in common a low temporal squamous height - another trait more characteristic of Neandertals. In contrast, the Mladeč 3 midvault is proportionately shorter than either of the Mladeč adult males or the Mladeč 1 female (see Frayer et al., this volume; Wolpoff et al., this volume), in whom parietal lengths exceed those of Neandertals as is typical in Upper Paleolithic associated humans.

Conclusions

Many students of human evolution have long argued that the Central European fossil record, where large sites such as Krapina and Předmostí have provided considerable evidence for a Neandertal range of morphological variability overlapping with that of early Upper Paleolithic associated humans, provides evidence for biological continuity between Neandertals and *Homo sapiens sapiens* (see Frayer et al., this volume; Wolpoff et al., this volume). More recently, Trinkaus and Zilhão (2002) have convincingly argued that numerous skeletal and dental characteristics of the Lagar Velho Gravettian child illustrate genetic continuity between the late Neandertals and early modern humans of Iberia as well. Yet although suggestive of such continuity, the juvenile material from these Gravettian sites are considerably more gracile, and their overall morphology more modern than Neandertals - a situation which phenotypically positions them closer to modern populations and thus leaves the door open for continued arguments by skeptics who would omit Neandertals from any participation in modern European ancestry.

Without question, however, the morphology of Mladeč 3 is truly intermediate between a Neandertal and modern human neurocranial phenotype. In the frontal bone, the suggested shape of the orbits and the degree of frontal vaulting strongly resemble the morphology of *Homo sapiens sapi-*

ens, while the thick, well-developed medial browridges and prominent glabella suggest the form found in young Neandertals. The occipital also presents a mixture of traits. Its strongly developed occipital bun and associated features such as low occipital height, conspicuous lambdoid flattening, and short occipital plane length align Mladeč 3 with young Neandertals, yet in nuchal plane length and occipital breadth the specimen is intermediate between a more modern and Neandertal morphology. The temporal is short and low with a prominent juxtamastoid eminence as in Neandertals, but possesses numerous modern features, including antero-superior orientation of the external auditory meatus; a strongly concave glenoid fossa with a well-developed articular eminence; presence of a post-glenoid tubercle; single digastric groove; and a long squamous portion relative to maximum anterior-posterior temporal length. Also modern is the flatness of the squamous temporal and its angulation to the sagittal plane, suggesting a modern pattern of cranial vault to base growth.

Perhaps the most intriguing finding to result from the present study is the presumptive pattern of parietal growth inferred from the Mladeč 3 CT generated vault reconstruction. During the European Upper Pleistocene, the parietal vault is characterized by a dramatic evolutionary shift in development (Minugh-Purvis, 2002), resulting in an accelerated rate of midvault growth in most Upper Paleolithic associated infants over that seen in Neandertals. Thus, the presence of a short midvault segment at age 2 strongly suggests that, had it survived to adulthood, Mladeč 3 would have possessed a parietal segment of Neandertal length. It is particularly noteworthy that Mladeč 3 possesses this Neandertal trait, as all other known Upper Paleolithic associated children exhibit a midvault of, or greater than, modern length.

In light of this and other evidence from Mladeč, the possibility of biological continuity between Neandertals and early modern Europeans is extremely difficult to refute. As is true of the Mladeč adults, Mladeč 3 morphology exhibits far more Neandertal attributes than many other Upper Paleolithic associated Europeans who followed them in time. However, the abundant similarities in *ontogenetic patterning* which Mladeč 3 shared both with Neandertals and modern humans, strongly argues that the genetic underpinnings of key growth phenomena persisted from a Neandertal gene pool into the Mladeč population. There they were expressed in the Mladeč phenotype as an amalgam of features reminiscent of Neandertals and those later common in modern *Homo sapiens*. The remains of the Mladeč people, from cradle to grave, leave little doubt that minimally, some early Aurignacian populations of Central Europe carried Neandertal alleles. As a bridge between Neandertals and later Gravettian associated Europeans, such as Předmostí and Lagar Velho, the Mladeč people also provide evidence for continuity of the ever-more diluted Neandertal gene pool across generations of Europeans.

Acknowledgements

NM-P thanks Maria Teschler-Nicola for her kind invitation to contribute to this volume. Thanks also to H. Kritscher, H. Poxleitner, Maria Teschler, and Johann Szilvássy of the Anthropologische Abteilung, Naturhistorisches Museum, Wien for their kind assistance and access to the Mladeč 1, 2, and 3 material over the years and to Milford Wolpoff, David Frayer, and Fred Smith for first encouraging this study of the Mladeč 3 remains. NM-P is especially grateful to the late Jan Jelínek, Ustav Anthropos, Moravske Muzeum, Brno, Czech Republic, for access to the Mladeč 5 remains and to the Mladeč 6 primary cast as well as for many hours of valuable and stimulating discussion regarding the site and people of Mladeč. Finally, special thanks to Doug Purvis, without whose photographs, technical assistance, critical discussion, and patience, this work would not have been possible. NM-P was supported by grants from the Department of Anthropology, University of Pennsylvania; Foundation for Research into the Origin of Man, LSB Leakey Foundation, Richard D. Irwin Foundation, and Sigma Xi, and the Dept. of Neurobiology, Drexel University School of Medicine (formerly Medical College of Pennsylvania – Hahnemann University School of Medicine). BV was supported during the research for this paper by the Austrian Federal Ministry of Education, Science and Culture, Grant No. GZ 200.093/I-vI/2004 (PI Horst Seidler). BV would also like to thank H. Seidler, K. Schäfer, P. Gunz, P. Mitteröcker, S. Neubauer, A. Stadlmayr, G. Weber and especially L. Blundell for discussions and support. We thank Professors H. Imhof and F. Kainberger for the CT scans of the Mladeč material, and P. Brugger from the Institute for Anatomy, Vienna Medical University for the immature comparative material.

References

Akazawa, T. and Muhesen, S. (Eds.) (2002) *Neanderthal burials: Excavations of the Dederiyeh Cave, Afrin, Syria.* Kyoto: International Research Center for Japanese Studies

Anson, B. A. and Maddock, W. G. (1952) *Callender's surgical anatomy.* 3rd Edition. Philadelphia: W. B. Saunders

Antón, S. (2002) Cranial growth in *Homo erectus*. In (N. Minugh-Purvis and K. J. McNamara, Eds.) *Human evolution through developmental change.* Baltimore: Johns Hopkins University Press, pp. 349–380

Ars, B. (1989) Organogenesis of the middle ear structures. *Journal of Laryngology and Otology (London)* 103, 16–21

Barghouth, G., Prior, J. O., Lepori, D., Duvoisin, B., Schnyder, P. and Gudinchet, F. (2000) Paranasal sinuses in children: size evaluation of maxillary, sphenoid, and frontal sinuses by magnetic resonance imaging and proposal of volume index percentile curves. *European Radiology* 12, 1451–58

Bennett, K. S. (1965) The etiology and gentics of wormian bones. *American Journal of Physical Anthropology* 23, 255–260

Billy, G. (1975) Étude anthropologique des restes humains de l'Abri Pataud. In (J. L. Movius, Jr., Ed.) *Excavation of the Abri Pataud: Les Eyzies (Dordogne).* Harvard: Peabody Museum (Bulletin of the American School of Prehistoric Research 30), pp. 201–261

Billy, G. (1979) L'enfant magdalénien da la grotte du Figuier (Ardèche). *L'Anthropologie* 83, 223–252

Blinkov, S. M. and Glezer, I. I. (1968) *The human brain in figures and tables.* New York: Plenum Press

Brash, J. C. (Ed.) (1951) *Cunningham's textbook of anatomy.* 9th ed. New York: Oxford University Press

Browning, H. (1953) The confluence of dural venous sinuses. *American Journal of Anatomy* 93, 307–329

Buxton, L. D. H. (1928) Human remains. In (D. A. E. Garrod, L. D. H. Buxton, G. E. Smith and D. M. A. Bate, Eds.) *Excavation of a Mousterian rock-shelter at Devil's Tower, Gibraltar.* London: Royal Anthropological Institute of Great Britain and Ireland 58, 57–91

Caspari, R. (1991) *The evolution of the posterior cranial vault in the central European Upper Pleistocene.* PhD Dissertation, University of Michigan. Ann Arbor: University Microfilms

Coqueugniot, H. (1999) *Le Crâne d'*Homo sapiens *en Eurasie: Croissance et variation depuis 100 000 ans.* BAR Int. Series 822. Oxford: British Archaeological Reports

Coqueugniot, H. (2000) La position du foramen mentionnier chez l'enfant: Révision ontogénetique et phylogénetique. *Bulletins et Mémoires de la Société d'Anthropologie (Paris), n.s.* 12 (3–4), 227–246

Dean, M. C., Stringer, C. B. and Bromage, T. G. (1985) A new age at death for the Neanderthal child from Devil's Tower, Gibraltar and the implications for studies of general growth and development in Neanderthals. *American Journal of Physical Anthropology* 70, 301–309

Dodo, Y., Kondo, O. and Nara, T. (2002) The skull of the Neanderthal child of burial no. 1. In (T. Akazawa and S. Muhesen, Eds.) *Neanderthal burials: Excavations of the Dederiyeh Cave, Afrin, Syria.* Kyoto: International Research Center for Japanese Studies, pp. 93–137

Ford, E. H. R. (1958) Growth of the human cranial base. *American Journal of Orthodontics* 44, 498

Franciscus, R. (1995) *Nasal morphology in the Western Old World Later Pleistocene and the origins of modern humans.* PhD Thesis, University of New Mexico

Franciscus, R. (2002) The midfacial morphology. In (J. Zilhão and E. Trinkaus, Eds.) *Portrait of the artist as a child: The Gravettian human skeleton from the Abrigo do Lagar Velho and its archaeological context.* Trabalhos de Arqueologia 22, pp. 297–311

Frayer, D. W., Jelínek, J., Oliva, M. and Wolpoff, M. H. (2006) Aurignacian male crania, jaws and teeth from the Mladeč Caves, Moravia, Czech Republic. In (M. Teschler-Nicola, Ed.) *Early modern humans at the Moravian Gate: The Mladeč Caves and their remains,* this issue, pp. 185–272

Gorjanović, D. (1906) *Der diluviale Mensch von Krapina in Kroatien: Ein Beitrag aus Paläoanthropologie.* Wiesbaden: Kreidels

Gunz, P., Mitteröcker, P., Bookstein, F. and Weber, G. W. (2004) Computer aided reconstruction of incomplete human crania using statistical and geometrical estimation methods. In (Stadtarchäologie, Ed.): *Enter the past: Computer applications and quantitative methods in Archaeology.* BAR International Series 1227. Oxford: Archaeopress, pp. 92–94

Hasselwander, A. (1927) Der Schädel. In (H. Peter, G. Wetzek and H. Heiderich, Eds.) *Handbuch der Anatomie des Kindes.* Vol. 2. München: J. F. Bergman

Hauser, G. and De Stefano, G. F. (1989) *Epigenetic variants of the human skull.* Stuttgart: Schweizerbart

Heim, J.-L. (1982) *Les enfants néandertaliens da La Ferrassie.* Foundation Singer Polignac. Paris: Masson

Heim, J.-L. (1991) L'enfant Magdalénien de La Madeleine. *L'Anthropologie* 2/3, 611–638

Hoyte, D. (1997) Growth of the cranial base. In (A. D. Dixon, D. A. N. Hoyte and O. Rönning, Eds.) *Fundamentals of craniofacial growth.* New York: CRC Press, pp. 257–333

Isihida, H. and Kondo, O. (2003) The skull of the Neanderthal child of burial no. 2. In (T. Akazawa and S. Muhesen, Eds.) *Neanderthal burials: Excavations of the Dederiyeh Cave, Afrin, Syria.*). Kyoto: International Research Center for Japanese Studies, pp. 271–297

Jelínek, J. (1969) Neanderthal man and *Homo sapiens* in Central and Eastern Europe. *Current Anthropology* 10, 475–503

Jelínek, J. (1976) The *Homo sapiens neanderthalensis* and *Homo sapiens sapiens* relationship in Central Europe. *Anthropologie (Brno)* 14, 79–81

Jelínek, J. (1978) Earliest *Homo sapiens sapiens* from Central Europe (Mladeč, Czechoslovakia). Paper presented at the Xth International Congress of Anthropological and Ethnological Sciences, Delhi, India

Knies, J. (1905) Nový nález diluviálního člověka u Mladče na Moravě. *Věstnik Klubu přírovědeckého* 9, 3–19

Krovitz, G. E. (2003) Shape and growth differences between Neandertals and modern humans: Grounds for a species-level distinction? In (J. L. Thompson, G. E. Krovitz and A. J. Nelson, Eds.) *Patterns of growth and development in the genus Homo.* Cambridge: University Press, pp. 320–342

Madre-Dupouy, M. (1992) *L'Enfant du Roc de Marsal.* Paris: CNRS

Matiegka, J. (1934) *Homo předmostensis: Fosilni čelovek ž Předmostí na Morave.* II. Prag: Nkladem Česke Akademie Ved a Umani

McCown, T. D. and Keith, A. (1939) *The stone age of Mount Carmel II: The fossil human remains from the Levallois-Mousterian.* Oxford: Clarendon

Minugh, N. S. (1983) The Mladeč 3 child: Aspects of cranial ontogeny in early anatomically modern Europeans. Paper presented at the 52nd Annual Meeting of the American Association of Physical Anthropologists, Indianapolis

Minugh-Purvis, N. (1988) *Patterns of craniofacial growth and development in Upper Pleistocene hominid.* Ann Arbor: University Microfilms

Minugh-Purvis, N. (1993) Reexamination of the immature hominid maxilla from Tangier, Morocco. *American Journal of Physical Anthropology* 92 (4), 449–61

Minugh-Purvis, N. (1998) The search for the earliest modern Europeans: a comparison of the Krapina 1 and es-Skhul 1 juveniles. In (T. Akazawa, K. Aoki and O. Bar-Yosef, Eds.) *Neanderthals and modern humans in West Asia.* New York: Plenum Press, pp. 339–352

Minugh-Purvis, N. (2000) Ontogeny and morphology of the child's mandible from Šipka (Moravia), Czech Republic. *Anthropologie* (Brno) 38, 71–82

Minugh-Purvis, N. (2002) Heterochronic change in the neurocranium and the emergence of modern humans. In (N. Minugh-Purvis and K. J. McNamara, Eds.) *Human evolution through developmental change.* Baltimore: Johns Hopkins University Press, pp. 479–498

Minugh-Purvis, N. and McNamara, K. J. (Eds.) (2002) *Human evolution through developmental change.* Baltimore: Johns Hopkins University Press

Minugh-Purvis, N., Radovčić, J. and Smith, F. H. (2000) Krapina 1: a juvenile Neandertal from the early late Pleistocene of Croatia. *American Journal of Physical Anthropology* 111 (3), 393–424

Minugh-Purvis, N., Radovčić, J. and Tobias, P. V. (in prep.) Variation in dural venous sinus drainage and population continuity in the Central European Late Pleistocene

Musil, R. and Valoch, K. (1966) Beitrag zur Gliederung des Würms in Mitteleuropa. *Eiszeitalter und Gegenwart* 17, 131–138

Pap, I, Tillier, A.-M., Arensburg, B. and Chech, M. (1996) The Subalyuk Neanderthal remains (Hungary): A re-examination. *Annales Historico-Naturales Musei Nationalis Hungarici* 88, 233–270

Ponce de Leon, M. S. and Zollikofer, C. P. (1999) New evidence from Le Moustier 1: computer-assisted reconstruction and morphometry of the skull. *Anatomical Record* 254, 474–489

Prossinger, H., Seidler, H., Teschler-Nicola, M., Schaefer, K., Weber, G. W., Recheis, W., Szikossy, I. and zur Nedden, D. (2005) Neanderthal versus modern frontal sinuses: preliminary considerations of ontogenetic aspects of their morphologies and volumes. In (H. Ullrich, Ed.) *The Neandertal adolescent Le Moustier 1: New aspects, new results*. Berlin: Staatliche Museen zu Berlin – Preußischer Kulturbesitz, pp. 134–148

Radovčić, J., Smith, F. H., Trinkaus, E. and Wolpoff, M. H. (1988) *The Krapina hominids: An illustrated catalog of the skeletal collection*. Zagreb: Mladost

Schwartz, J. H. and Tattersall, I. (2002) *The human fossil record – Volume 1: Terminology and craniodental morphology of genus* Homo *(Europe)*. New York: Wiley-Liss Inc.

Smith, F. H. (1976) *The Neanderthal remains from Krapina: A descriptive and comparative study*. Report of Investigations No. 15. Knoxville: Department of Anthropologie, University of Tennessee

Smith, F. (1982) Upper Pleistocene hominid evolution in South-Central Europe: A review of the evidence and analysis of trends. *Current Anthropology* 23, 667–703

Spaeth, J., Krugelstein, U. and Schlondorff, G. (1997) The paranasal sinuses in CT-imaging: development from birth to age 25. *International Journal of Pediatric Otorhinolaryngology* 39, 25–40

Sperber, G. (2001) *Craniofacial development*. Hamilton, Canada: BC Decker Inc.

Szilvássy, J. (1981) Die Entwicklung der Stirnhöhlen. *Anthropologischer Anzeiger* 39, 138–49

Szilvássy, J. (1986) Eine neue Methode zur intraserialen Analyse von Gräberfeldern. *Mitteilungen der Berliner Gesellschaft für Anthropologie, Ethnologie und Urgeschichte* 7, 49–62

Szombathy, J. (1925) Die diluvialen Menschenreste aus der Fürst-Johanns-Höhle bei Lautsch in Mähren. *Die Eiszeit* 2, 1–34; 73–95

Thompson, J. L., Krovitz, G. E. and Nelson, A. J. (Eds.) (2003) *Patterns of growth and development in the genus* Homo. Cambridge: University Press

Tillier, A.-M. (1982) Les enfants Néanderthaliens de Devil's Tower (Gibraltar). *Zeitschrift für Morphologie und Anthropologie* 73, 125–148

Tillier, A.-M. (1983a) A new Mousterian child from Qafzeh (Israel): Qafzeh 41. *Bulletins et Mémoires de la Société d'Anthropologie (Paris)*, y. 10, série XIII, 61–69

Tillier, A.-M. (1983b) Le crâne d'enfant d'Engis 2: un example de distribution des caractères juvéniles, primitifs et néanderthaliens. *Bulletin de la Société Royal Belge d'Anthropologie et de Préhistoire* 94, 51–75

Tillier, A.-M. (1987) L'enfant de la Quina H18 et l'ontogénie des Néanderthaliens. In (B. Vandermeersch, Ed.) *Préhistoire de Poitou-Charentes, problèms actuels*. Paris: Éditions du CTHS, pp. 201–206

Tillier, A.-M. (1988) La place des restes de Devil's Tower (Gibraltar) dans l'ontogenése des Néanderthaliens. *Bulletins et Mémoires de la Société d'Anthropologie* 514, 257–266

Tillier, A.-M. (1999) *Les enfants Moustériens de Qafzeh*. Paris: CNRS

Trinkaus, E. (1983) *The Shanidar Neanderthals*. New York: Academic Press

Trinkaus, E. (2002) The cranial morphology. In (J. Zilhão and E. Trinkaus, Eds.) *Portrait of the artist as a child: The Gravettian human skeleton from the Abrigo do Lagar Velho and its archaeological context*. Trabalhos de Arqueologia 22, pp. 256–286

Trinkaus, E. and Zilhão, J. (2002) Phylogenetic implications. In (J. Zilhão and E. Trinkaus, Eds.) *Portrait of the artist as a child: The Gravettian human skeleton from the Abrigo do Lagar Velho and its archaeological context*. Trabalhos de Arqueologia 22, pp. 497–518

Valoch, K. (1968) Evolution of the Paleolithic in Central and Eastern Europe. *Current Anthropology* 9, 351–390

Vlček, E. (1967) Die Sinus frontalis bei europäischen Neandertalern. *Anthroplogischer Anzeiger* 30, 166–189

Vlček, E. (1969) *Neandertaler der Tschechoslowakei*. Prag: Böhau

Vlček, E. (1970) Étude comparative onto-phylogénetique de l'enfant du Pech-de-l'Azé par rapport a d'autres enfants Néanderthaliens. In (D. Ferembach, P. Legoux, R. Fenart, R. Empereur-Buisson and E. Vlček, Eds.) *L'Enfant du Pech-de-l'Azé*. Paris: Masson, pp. 149–178

Vlček, E. (1971) Czechoslovakia. In (K. P. Oakley, B. G. Campbell and T. I. Molleson, Eds.) *Catalogue of fossil hominids. Part II: Europe*. London: Trustees of the British Museum (Natural History), pp. 47–64

Warwick, R. and Williams, P. L. (Eds.) (1975) *Gray's anatomy*. 35th British edition. Philadelphia: Saunders

Weaver, D. (1978) Age estimation using the infant and child temporal. Paper presented at the 47th Annual Meeting of the American Association of Physical Anthropologists, Toronto, April, 1978

Weinzweig, J., Kirschner, R. E., Farley, A., Reiss, P. M. S., Hunter, J., Whitaker, L. A. and Bartlett, S. P. (2003) Metopic synostosis: Defining the temporal sequence of normal suture fusion and differentiating it from synostosis on the basis of computed tomography images. *Plastic and Reconstructive Surgery* 112 (5), 1211–1218

Williams, P. L., Bannister, L. H., Berry, M. M., Collins, P., Dyson, M., Dussek, J. E. and Ferguson, M. W. J., (Eds.) (1995) *Gray's anatomy: The anatomical basis of medicine and surgery*. 38th British edition. New York: Churchill Livingstone

Wolpoff, M. H. (1999) *Paleoanthropology*. New York: McGraw Hill

Wolpoff, M. H., Frayer, D. W. and Jelínek, J. (2006) Aurignacian female crania and teeth from the Mladeč Caves, Moravia, Czech Republic. In (M. Teschler-Nicola, Ed.) *Early modern humans at the Moravian Gate: The Mladeč Caves and their remains*, this issue, pp. 273–340

Zollikofer, C. P., Ponce De Leon, M. S., Esteves, F., Silva, T. F. and Dias, R. P. (2002) The Computer-assisted reconstruction of the skull. In (J. Zilhão and E. Trinkaus, Eds.) *Portrait of the artist as a child: The Gravettian human skeleton from the Abrigo do Lagar Velho and its archaeological context*. Trabalhos de Arqueologia 22, pp. 326–341

Zollikofer, C. P., Ponce de Leon, M. S., Lieberman, D. E., Guy, F., Pilbeam, D., Likius, A., Mackaye, H. T., Vignaud, P. and Brunet, M. (2005) Virtual cranial reconstruction of *Sahelanthropus tchadensis*. *Nature* 434, 755–759

Zollikofer, C. P., Ponce de Leon, M. S., Martin, R. D. and Stucki, P. (1995) Neanderthal computer skulls. *Nature* 375, 283–285

Chapter 13

THE HUMAN POSTCRANIAL REMAINS FROM MLADEČ

Erik Trinkaus, Fred H. Smith, Trenton C. Stockton and Laura L. Shackelford

Introduction

The paleontological analysis of Late Pleistocene human postcranial remains from Europe has highlighted a series of changes in morphology associated with the emergence of modern humans and the eventual replacement of a late archaic human (specifically Neandertal) morphological pattern with one of robust early modern humans. This has involved aspects of body proportions (Trinkaus, 1981; Holliday, 1997; 2005a), upper limb diaphyseal strength and muscularity (Trinkaus, 1983; 1997; 2000; 2005a; Churchill, 1994), lower limb diaphyseal shape (especially femoral) (Trinkaus, 1976; 2005b; Trinkaus et al., 1999a), and vertebral spinous process robusticity (Matiegka, 1938; Heim, 1976; Trinkaus, 1983). Other aspects of lower limb robusticity, such as diaphyseal robusticity, articular hypertrophy, knee moment arms and femoral curvature, largely disappear once appropriately analyzed and/or scaled to body mass and limb length (Trinkaus et al., 1999a, b; 2005b; Trinkaus and Rhoads, 1999; Trinkaus, 2000; Shackelford and Trinkaus, 2002).

Despite these observations, it is also apparent that most of the postcranial comparisons are between Middle Paleolithic Neandertals and earlier Upper Paleolithic modern humans dating to ≤ 28 kyr BP (Gravettian *sensu lato*). The rare exceptions to this are the Châtelperronian Saint Césaire Neandertal partial skeleton (Vandermeesch, 1984; Trinkaus et al., 1999a), the Fontana Nuova 4 talus (Chilardi et al., 1996), isolated phalanges from Brassempouy (Henry-Gambier et al., 2004), and the collection of human remains from Mladeč Cave. Indeed, although distinctive Neandertal remains have been dated to ca. 30 kyr BP in at least southwestern Europe (Hublin et al., 1995), diagnostic and well-dated early modern human remains ≥ 30 kyr BP in Europe are poorly known. Indeed, only the remains from the Peştera Muierii (Nicolăescu-Plopşor, 1968), the teeth and phalanges from Brassempouy (Henry-Gambier et al., 2004), and the craniofacial remains from the Peştera cu Oase (Trinkaus et al., 2003a, b; 2005) provide secure paleontological information on modern humans ≥ 30 kyr BP in Europe, and all of them except for an incomplete scapula and fibula from the Peştera Muierii and phalanges from Brassempouy are cephalic. The remainder of the European early modern human remains are immature mandibles, isolated teeth and/or insecurely dated to this time period.

The only other human limb bone which has been considered in this sample, the Vogelherd 3 humerus (Churchill and Smith, 2000), is now known to be Holocene in age (Conard et al., 2004). Similarly, the Cro-Magnon and La Rochette human remains (including postcrania) are now dated to the succeeding Gravettian of Western Europe (Orschiedt, 2003; Henry-Gambier, 2003).

As a result, the key sample of early modern human postcrania for providing current information on the postcranial morphology of the earliest European modern humans are the disassociated human axial and limb remains from Mladeč. Incompletely described (Szombathy, 1925) and only partially integrated into appropriate analyses (e.g., Wolpoff, 1989; Liston et al., 1989; Churchill, 1994; Chilardi et al., 1996; Stockton, 1997; Trinkaus, 2005a, b), the postcranial remains from the karstic cave system at Mladeč appear to represent the oldest currently known sample (as opposed to isolated elements) of early modern humans limbs in Europe. Directly associated with early modern human craniofacial remains, these postcrania provide important information on the early modern human biology in Europe. We therefore provide here a detailed description and assessment of these elements.

Inventory

This description of the Mladeč human postcranial remains is based principally on those specimens excavated by Josef Szombathy in 1881 and 1882 (Szombathy, 1881; 1882; 1904) in Chamber D, originally described by him (1925) and currently curated in the Naturhistorisches Museum Wien. Other human postcranial remains have been found in the Mladeč Caves and briefly described (e.g., Maška, 1905; Knies, 1906); the majority were destroyed in 1945 in Mikulov Castle or their whereabouts is not known, but several hand bones excavated by Jan Knies in Chamber E between 1903 and 1911 are preserved in the Moravské zemské muzeum in Brno. Given the apparent depositional complexity of the Mladeč karstic cave system and the fact that different Chambers, "Sites," or "Halls," may have accumulated archeological and human remains at different times (Oliva, 1989; Svoboda, 2000; Svoboda, this volume, chap. 3), the focus here is principally on those remains which are currently available and appear to derive from one discrete depositional event, the Chamber D remains. The Chamber E remains, essentially the aforementioned hand bones, are also included but considered as a separate sample.

The Mladeč human postcranial remains that are no longer available for analysis (see Wolpoff et al., this volume, chap. 8 (inventory)) are listed and briefly described in chap. 14 based on the original publications of them (Wolpoff et al., this volume). Since it cannot be adequately assessed whether they are of the same geological age as the remains from Szombathy's excavations in Chamber D, they are not considered further in this presentation.

The Mladeč human postcranial remains in the Naturhistorisches Museum Wien were originally inventoried in a hand written document by Josef Szombathy (December 12, 1922) as presented in Table 1. Comments and notes follow each of his original entries. The inventory of currently available human postcranial remains in the Naturhistorisches Museum Wien and the Moravské zemské muzeum in Brno are presented by catalog number in Table 2. Although the identifications of the

Table 1. Szombathy's handwritten December 12, 1922 inventory of Mladeč human postcranial remains in the Naturhistorisches Museum Wien, with comments on current identifications

1 Halswirbel, etwa Nr V oder VI	Mladeč 11
1 Körper eines ähnlichen Halswirbels	Mladeč 34, probably non-human
16 Rippenfragmente	13 are currently present; Mladeč 12 to 19 and 20a to 20e; Mladeč 13 is a clavicle
1 Schlüsselbein eines Säuglings (?)	Whereabouts unknown
4 Beckenfragmente	Mladeč 21 and 22; one is non-human, one is missing
2 Oberarmfragmente	Mladeč 23 and 24
1 prox. Ulnafragment	Mladeč 25c
4 Radiusfragmente	Mladeč 25a, 25b and 26. The fourth is not known, and it may be the piece of immature proximal femur (Mladeč 102)
1 fragliches Radiusfragment, kindlich	Non-human postcranial element
2 Oberschenkelknochenstücke	Mladeč 27 and 28
3 Metatarsalknochen III, IV, V l.	Mladeč 32 (metatarsal 3), Mladeč 31 (metacarpal 3, mislabeled "IV"), the bone identified as a metatarsal 5 is absent.
1 Fingerknochen I	Whereabouts unknown
1 Sprungbein l.	Mladeč 30
1 distale Epiphyse des Schienbeins r.	Mladeč 29

Table 2. Summary list, by specimen number, of the human postcranial remains from Mladeč currently known and available in the Naturhistorisches Museum Wien (Mladeč 11 to 32 and 102) and the Moravské zemské muzeum in Brno (Mladeč 88 to 91)

Specimen number	Identification	Specimen number	Identification
11	Cervical vertebra C3 to C6	24	Humerus right
12	Rib 1 left	25a	Radius right
13	Clavicle right immature	25b	Radius right
14	Rib 2 or 3 left	25c	Ulna right
15	Middle rib left	26	Radius left
16	Middle to caudal rib right	27	Femur right
17	Caudal rib right	28	Femur left
18	Rib 12 left	29	Tibia immature right
19	Rib 11 or 12 right	30	Talus left
20a	Rib section, side indeterminate	31	Metacarpal 3 immature right
20b	Rib section, side indeterminate	32	Metatarsal 3 immature left
20c	Rib section, side indeterminate	88	Manual proximal phalanx 2 left?
20d	Rib section, side indeterminate	89	Manual proximal phalanx 3 left?
20e	Rib section, side indeterminate	90	Metacarpal 3 right
21	Os coxae left	91	Metacarpal 4 right
22	Os coxae immature right	102	Femur immature left
23	Humerus right		

larger postcranial elements are the same as those of Szombathy, several of the smaller pieces have been reidentified based on further considerations. The current identifications, based on our work and that of M. Teschler-Nicola, are presented in Tables 1 and 2.

It should be noted that most of the human remains from Mladeč in the Naturhistorisches Museum Wien bear the museum catalog number 5459. This same number is on the section of a proximal immature human femur, a piece which was not originally included in the comprehensive Mladeč catalog (chap. 8, Inventory). Since it is now recognized as both human and deriving from the same collection as the remainder of the Mladeč postcrania, based on museum association and preservation, including similar adhering matrix, it has been given the number of Mladeč 102.

There is no indication in Szombathy's inventory of the possible associations between these bones, but it is apparent that at least three and probably more individuals are represented by the postcrania. This is based on duplication of the Mladeč 23 and 24 right humeri, the size differences between the Mladeč 25b and 26 radial diaphyses, the size and maturity contrasts between the Mladeč 21 and 22 pelvic pieces, and the morphological contrasts between the Mladeč 27 and 28 femora. A series of associations are possible, and they are discussed after the descriptions of the individual elements.

The hand remains from Chamber E include Mladeč 88 to Mladeč 91. Mladeč 88 and 89 are proximal hand phalanges, whereas Mladeč 90 and 91 are right metacarpal bones.

Comparative materials

The description of human fossil remains entails, by its very nature, comparisons to relevant samples of human remains. Since the Mladeč fossils derive from the central European earlier Upper Paleolithic and are cranially attributable to early modern humans (Szombathy, 1925; Smith, 1984; Frayer, 1986; Wolpoff et al., 2001; see this volume, chaps. 9 and 10, male and female crania), the most relevant sample is that of earlier Upper Paleolithic northwestern Old World early modern humans. Comparative data as appropriate and available have therefore been assembled for European and western Asian earlier (> 18,000 years BP) Upper Paleolithic early modern humans. These specimens derive from the sites of Arene Candide, Barma Grande, Cro-Magnon, Dolní Věstonice I and II, Fanciulli (Grotte-des-Enfants), Fontana Nuova, Lagar Velho, Nahal-Ein-Gev, Ohalo II, Paglicci, Pataud, Paviland, Pavlov I, Předmostí, La Rochette, Sunghir, Veneri (Parabita) and Willendorf. The majority of them date to after 30,000 years BP, and the sample is therefore predominantly Gravettian (or Middle Upper Paleolithic) in age. Several of the specimens, as noted above, were originally attributed to the Aurignacian (e.g., Cro-Magnon and La Rochette), but reassessment and dating (Orschiedt, 2002; Henry-Gambier, 2003) have placed them within the Gravettian. The only Aurignacian specimen within this postcranial comparative sample is the Fontana Nuova 4 talus, despite uncertainties as to its association with the Aurignacian of that site (Chilardi et al., 1996).

At the same time, given that the Mladeč fossils represent one of the earlier populations of early modern humans in central Europe, overlapping the time period of the latest Neandertals and probably resulting in part from a dispersal of early modern humans from southeast Europe and eventually southwest Asia (Smith and Trinkaus, 1991; Smith et al., 1999; Trinkaus et al., 2003b), the sample may well have close affinities to the earliest modern humans from southwest Asia, the Middle Paleolithic Qafzeh-Skhul sample. Data for the adult and immature remains from those two sites are therefore included in the comparisons.

In addition, it has also been argued (Smith, 1984; Frayer, 1986; Smith and Trinkaus, 1991; Wolpoff et al., 2001) that the Mladeč remains exhibit evidence of some degree of genetic continuity between central European Neandertals and their early modern human temporal successors. It is therefore necessary to compare the Mladeč postcranial remains to those of last glacial Neandertal specimens. Unfortunately, such Neandertal postcranial remains are rare and fragmentary in central Europe, consisting principally of those from the Middle Paleolithic of Subalyuk (Pap et al., 1996) and Vindija G_3 (Wolpoff et al., 1981; Malez and Ullrich, 1982; Ahern et al., 2004) and the initial Upper Paleolithic of Vindija G_1 (Ahern et al., 2004). The Subalyuk 1 and Vindija G_3 postcranial elements are small and/or fragmentary, and only the probably immature Vindija 228, 253 and 266 distal humeri, the Vindija 13.8 proximal radial shaft, and the metatarsal 3 of Subalyuk 1 anatomically match those of the Mladeč sample. Given that the Mladeč remains preserve several relatively complete portions of major long bones, the comparative Neandertal sample consists principally of last glacial remains from western European and southwestern Asian Neandertals. These are from the sites of Amud, La Chapelle-aux-Saints, Cova Negra, Dederiyeh, La Ferrassie, Font-de-Forêt, Kebara, Lezetxiki, La Quina, Neandertal, Regourdou, Roc de Marsal, Rochers-de-Villeneuve, Saint Césaire, Shanidar and Spy. The Krapina remains predate this time period (Rink et al., 1995), as do the Tabun Layers B and C human remains (Mercier et al., 1995; Grün and Stringer, 2000); they are therefore not included in the comparative samples.

Measurements affected by pathological lesions or post-traumatic remodeling [e.g., Neandertal 1 humeral diaphyses (Trinkaus et al., 1994)] are not included. The comparative data were assembled from personal investigation of the original specimens and the primary published descriptions of specimens, supplemented by data particularly from Churchill (1994, pers. comm.), Hollliday (1995), and Holt (1999).

Comparative methods

The comparative assessment of the Mladeč postcranial remains involves three forms of data. The material are qualitatively described and compared to other samples. A full set of comparative linear and angular metrics are provided; most of these measurements follow the Martin system (Bräuer, 1988), but additional measurements are provided and defined in the notes to the tables. The majority of the Mladeč postcranial measurements were taken by three of us (FHS, TCS and ET) on separate occasions; in 1998 they were all rechecked on the originals by ET to resolve differences in measurement techniques. The presented data represent the result of that combined effort.

In addition, cross-sectional geometric parameters (cross-sectional areas and second moments of area) are provided for the Mladeč long bone and metapodial diaphyses. All of them were reconstructed using polysiloxane molding putty (Cuttersil Putty Plus) to transcribe the anatomically oriented subperiosteal contours and biplanar radiography to determine parallax corrected cortical thicknesses. The endosteal contours were then interpolated and the resultant cross sections were projected enlarged onto a Summagraphics III digitizing tablet and digitized twice. Cross-sectional parameters were computed using a PC-DOS version (Eschman, 1992) of SLICE (Nagurka and Hayes, 1980), and the results of the repeated digitizing of each section averaged.

None of the Mladeč diaphyses is complete, and the sections are located at percentages of bone lengths; it was therefore necessary to estimate the positions of the cross sections. For the two most complete long bones (Mladeč 24 and 27) plus the two metapodials (Mladeč 31 and 32), lengths were estimated, and the sections were located at the appropriate calculated percentages of bone biomechanical length. For the other diaphyses, the positions of the sections were estimated using anatomical landmarks (including proximity to epiphyses and diaphyseal muscle markings); their locations are therefore more approximate. The orientations of the sections relative to anatomical planes were assessed principally using diaphyseal shape indicators, in particular the anterior crests on the humeri, the interosseus crests on the radii, and the linea aspera on the femora. Only for the Mladeč 25c proximal ulna, the Mladeč 31 metacarpal and the Mladeč 32 metatarsal could articulations be employed for anatomical orientation. For this reason, the orientations of the maximum second moments of area (theta) are not provided, and the anatomically oriented second moments of area (I_x and I_y) should be considered approximate.

Since the humeral diaphyses of Late Pleistocene humans frequently exhibit pronounced levels of asymmetry (Trinkaus et al., 1994; Churchill and Formicola, 1997), and since the dominant arm in almost all of the individuals providing data on asymmetry is the right one, comparisons should be made solely to the same side bone for those portions of humeral remains which exhibit significant levels of asymmetry. For this reason, the Mladeč humeral diaphyses, both of which are right, are compared only to right humeri. The few forearm and hand bones are compared to both right and left sides, even though there may be some systematic asymmetry in at least diaphyseal features related to handedness (Sakaue, 1999). The lower limbs may exhibit asymmetry, but it is generally lower in magnitude and random with respect to side (Ruff and Hayes, 1983; Trinkaus et al., 1994; Anderson and Trinkaus, 1998). The forearm, hand and lower limb comparisons therefore pool right and left specimens in the comparative samples, averaging the right and left values for specimens which preserve both sides prior to any sample calculations or graph composition.

Given the incompleteness of the Mladeč postcranial remains and the need to have length measurements for long bones to scale their diaphyseal and articular properties, lengths have been estimated for several of the Mladeč bones. For the Mladeč 24 humerus and Mladeč 27 femur, lengths were estimated from landmarks adjacent to the epiphyses. For the Mladeč 31 metacarpal and the Mladeč 32 metatarsal, both of which lack their unfused head epiphyses, lengths were estimated based on the distance from the dorsal tubercles to the proximal facet using mature metacarpals. In each case, the reference samples were mature recent human bones; given the immature status of the Mladeč 31 and 32 bones and the absence of the epiphyses on Mladeč 24 and 27, the resultant lengths may slightly underestimate the original mature lengths of these bones. Given that this would have the effect of overestimating the robusticities of these bones, and that one of the concerns is whether they exhibit the reduction in robusticity seen in some earlier Upper Paleolithic human remains (Trinkaus, 2000), this possible bias should be conservative for the comparisons.

In the comparisons, assessments of the degrees of difference between the three comparative samples were done using parametric tests (ANOVA and t-tests) for linear and angular measurements, and non-parametric (Kruskal-Wallis and Wilcoxon) tests for ratios (indices) and discrete data. Metric comparisons which are not amenable to ratios are presented graphically. The differences between the samples are assessed with linear residuals relative to the reduced major axis lines through the pooled comparative sample, which are then compared parametrically.

The Mladeč 11 cervical vertebra and the Mladeč 34 vertebral body

Preservation, identification and maturity

Mladeč 11 is a largely complete adult cervical vertebra, almost entirely covered with a thin layer of carbonate crust (see chap. 8, Plate VIII). The crust, however, is sufficiently thin to permit assessment of most aspects of its preserved morphology and proportions. The superior (cranial) margins of the uncinate processes, as well as the anterocaudal margin of the vertebral body, are slightly abraded. The inferior and superior articular facets are complete, with only slight damage to the superolateral aspect of the left superior facet and the inferior aspects of the right and left inferior articular facets.

Moderate damage has occurred to the lateral and posterior (dorsal) portions of the vertebra. The most lateral portions of the posterior tubercles and the entire anterior tubercles of the transverse processes are absent. The right costotransverse bar (intertubercular lamella) is complete while only a very thin bridge of bone remains on the left side. The transverse foramina are complete and unobscured, as are the laminae, pedicles, neural arch and neural canal. The spinous process, however, is damaged and only the base remains. The preserved length of the spinous process is only 11 mm. The vertebra provides a total preserved transverse diameter of 57.3 mm and a preserved dorsoventral maximum dimension of 42.4 mm.

The annular rings are fully fused to the centrum, and there are no traces of pathological degenerations on the bone. These suggest a minimum age-at-death at the end of the second decade (Scheuer and Black, 2000) and probably not an advanced age for the individual.

The presence of foramina in each transverse process, superior articular facets that face superiorly and posteriorly, posterolateral lips (uncinate processes) on the superior surface of the centrum, and a triangular neural (vertebral) canal indicate that Mladeč 11 is a cervical vertebra. Determining the position of Mladeč 11 within the cervical vertebral sequence has proven to be difficult, however, since no other vertebrae in the sequence were recovered and damage to the spinous process makes it impossible to determine the length or degree of bifidity. It is clearly not a C1 or C2, and the absence of any thoracic-like morphology (e.g., the inferior surface of the centrum is not flat and the transverse processes do not appear large) precludes it from being a C7. Therefore, based on morphology alone, it is equally likely that Mladeč 11 is a C3, C4, C5 or C6.

Szombathy (1925, 24) identified this specimen as a C6 but did not make explicit his reasons for doing so. This is unfortunate because the single feature that can be used to differentiate between C6 and C3–C5 (a large anterior tubercle of the transverse process known as the carotid tubercle) (McMinn et al., 1993) is not markedly large on the Mladeč 11 specimen. Therefore, although only a single vertebra is preserved in the Mladeč sample and carotid tubercle development is variable (Stockton, pers. observ.), the absence of a large anterior tubercle of the transverse process of Mladeč 11 makes it unlikely (but not impossible) that this specimen is a C6.

Metric data do not help to resolve the issue. This is in part due to variation within populations that occurs naturally in the vertebral columns of fossil and recent humans, but it is also caused by overlapping ranges of morphometric variation between different (not necessarily consecutive) vertebrae in the cervical vertebral column. For example, despite the well-supported assumption that the more caudal vertebral bodies are larger than the cranial ones as a virtue of weight transmission in orthograde primates (Taylor and Twomey, 1984), the considerable ranges of overlap in within-sample and between-sample vertebral body size from one vertebra to the next preclude accurate positional assessment from a single isolated vertebra (Hasebe, 1912; Lanier, 1939; Pal and Routal, 1986).

One feature which may help is the essentially horizontal orientation of the Mladeč 11 spinous process. Cervical spinous processes do become more horizontal as they go caudally, including among Late Pleistocene humans (Matiegka, 1938; Heim, 1976; Trinkaus, 1983), making it more likely that Mladeč 11 is a C5 or a C6.

For these reasons, the Mladeč 11 vertebra is compared to available data for C3 to C6 vertebrae from the Late Pleistocene comparative samples. Yet, given the limited sample sizes available for even C5 and C6 (Tables 4 and 5), it is principally to the more caudal two vertebrae that proportional metrics are compared.

The Mladeč 34 specimen is a vertebral body, which was originally identified as human (see chap. 8, Plate VIII). It is partially obscured by a massive carbonate encrustation on its cranial body surface and its ventrolateral surface. It is abraded along its dorsal surface. The strong dorsoventral concavity of its caudal body surface with the caudally turned ventral edge indicates that it is a cervical vertebra. Its number is unknowable, but its small size suggests that it derives from the middle of the cervical column. The caudal annular ring is partially fused. It is completely fused along the ventral midline, but partially fused bilaterally ventrolaterally. This suggests a late adolescent age for the individual (Scheuer and Black, 2000).

The unusually small size of Mladeč 34 (the body dorsoventral diameter is 13.4 mm and its transverse diameter is 19.5 mm; the dorsal and ventral body heights are 11.7 and 10.9 mm respectively) bring into question whether this specimen is indeed human. Moreover, the carbonate encrustation obscures most of the relevant morphology. It is therefore put into a suspense account and is not considered further.

Morphology

The Mladeč 11 vertebra exhibits large foramina transversaria for the vertebral arteries and a small foramen on the dorsal side of each one. The right articular facets are concave craniocaudally and appear to be slightly displaced dorsally relative to the left ones. The spinous process is incomplete, but the preserved portions of it suggest that it tapered markedly from its base towards its dorsal tip. For example, the craniocaudal height of the process is 10.6 mm at its juncture with the laminae, but is reduced to 7.5 mm near its dorsal break, only 12.4 mm from the vertebral canal. At the same time, viewed in *norma lateralis,* the spinous process appears to have been largely horizontal (estimated angle of 0°) relative to the planes defined by the cranial and caudal body surfaces.

Relatively horizontal cervical spinous processes, although once considered as a Neandertal or "archaic" characteristic (e.g., Boule, 1911–13), have been shown repeatedly (e.g., Kleinschmidt, 1938; Arambourg, 1955; Straus and Cave, 1957; Toerien, 1957; Stewart, 1962; Heim, 1976; Trinkaus, 1983; Arensburg, 1991) to be present in variable frequencies among recent humans (see also Cunningham, 1886). They are also present in the Předmostí sample (Matiegka, 1938) and on Cro-Magnon 2 (Trinkaus, pers. observ.). The apparently horizontal orientation of the Mladeč 11 spinous process is therefore unremarkable.

At the same time, what appears to contrast more consistently between late archaic and early modern human cervical spinous processes is the degree of tapering of the process, in which the cranial and caudal margins of Neandertal lower cervical processes remain largely parallel, especially close to the laminae, while those of early modern humans become progressively thinner dorsally (Matiegka, 1938; Piveteau, 1963–66; Heim, 1976; Trinkaus, 1983, pers. observ.). There is nonetheless some overlap in this feature between the two samples. In this aspect, the Mladeč 11 spinous process, especially if it represents a C5 or a C6, is closer to the pattern seen in the few earlier Upper Paleolithic specimens available for comparison.

Table 3. Osteometric dimensions of the Mladeč 11 cervical vertebra (in mm and degrees). Values in parentheses are estimated

Body ventral height (M-1)	9.5
Body dorsal height (M-2)	12.2
Body median height (M-3)	9.3
Body cranial dorsoventral diameter (M-4)	16.7
Body caudal dorsoventral diameter (M-5)	(17.0)
Body cranial transverse diameter (M-7)	25.5
Body caudal transverse diameter (M-8)	24.6
Cranial external transverse diameter[1]	(53.2)
Cranial internal transverse diameter[2]	22.1
Caudal external diameter[1]	53.6
Caudal internal transverse diameter[2]	20.5
Spinal canal cranial dorsoventral diameter (M-10)	11.9
Spinal canal caudal dorsoventral diameter	13.6
Spinal canal transverse diameter (M-11)	22.4
Pedicle dorsoventral diameter - right	6.3
Pedicle dorsoventral diameter - left	6.8
Pedicle mediolateral diameter - right	5.6
Pedicle mediolateral diameter - left	7.0
Cranial facet angle - right[3]	62°
Cranial facet angle - left[3]	59°
Spinous process angle[4]	(0°)

[1] Maximum transverse dimension measured to the lateral margins of the superior or inferior articular facets.
[2] Transverse diameter measured to the medial margins of the cranial or caudal facets.
[3] Angle between the facet and the mid-sagittal plane in the horizontal plane of the cranial body.
[4] Angle between the cranial body and the midline of the spinous process.

The Mladeč 11 measurements (Table 3) fall near the mean values of the comparative samples. Additionally, the inability to distinguish between samples based on the vertebral metrics presented here holds true at each level of the spinal sequence: C3, C4, C5 and C6. There is, however, one interesting aspect of these data that deserves mention.

While the overall vertebral morphology of European Neandertals and of more recent humans is very similar, the cervical bodies of European Neandertals have been described as relatively low and broad (McCown and Keith, 1939; Piveteau, 1963–66; Heim, 1976). Relatively low and broad centra are also found in southwest Asian late archaic human vertebrae from Shanidar (Trinkaus, 1983), Kebara (Arensburg, 1991), and the Mousterian associated third cervical vertebra from Hayonim (Arensburg and Nathan, 1980). Wolpoff (1989) observed that the low and broad vertebral body of Mladeč 11 resembled European Neandertals more than Skhul 5 or the small sample (N = 3) of early modern humans from Předmostí and argued that the Mladeč 11 proportions provided evidence for Neandertal to early modern human morphological continuity in central Europe. Wolpoff, however, regarded Mladeč 11 as a C3 and therefore only compared it to other third cervical vertebrae.

To reevaluate the body heights of Mladeč 11 in light of additional data, absolute heights are provided in Table 4 and indices comparing it to C5 and C6 proportions (samples of the paired measurements are too small to be meaningful for C3 and C4) are in Table 5. From these, it is apparent that the ventral body height of Mladeč 11 is relatively small. Only the Skhul 5 C4 and the Regourdou 1 and Shanidar 1 C5s have ventral heights as small as Mladeč 11. However, its dorsal height is well within the ranges of variation of all of the samples. Moreover, although the earlier Upper Paleolithic

Table 4. Comparative cervical vertebral body heights for Mladeč 11. Mean and standard deviation (N) provided for samples greater than 4; individual values provided for smaller samples. T-test P-values for the Neandertal versus earlier Upper Paleolithic samples. * P < 0.05 with a multiple comparison correction (α/8)

	Mladeč 11	Neandertals	Skhul 5	Earlier Upper Paleolithic	Nean/EUP P
C3					
Ventral height	9.5	11.0, 11.0, 12.2	–	11.9 ± 1.0 (6)	0.414
Dorsal height	12.2	10.0, 12.5, 13.0	10.5	12.4 ± 1.2 (5)	0.608
C4					
Ventral height	9.5	11.1 ± 0.2 (5)	8.5	12.1 ± 0.8 (5)	0.033
Dorsal height	12.2	12.3 ± 0.8 (4)	9.5	12.4 ± 1.3 (8)	0.797
C5					
Ventral height	9.5	10.5 ± 1.3 (7)	10.0	12.0 ± 1.0 (6)	0.037
Dorsal height	12.2	11.8 ± 1.0 (6)	10.0	13.3 ± 1.0 (8)	0.013
C6					
Ventral height	9.5	10.9 ± 0.9 (7)	12.2	11.9 ± 1.0 (7)	0.062
Dorsal height	12.2	12.3 ± 0.7 (7)	13.5	13.7 ± 0.9 (9)	0.003*

Table 5. Comparative body proportions of the Mladeč 11 cervical vertebra. Mean and standard deviation (N) provided for samples greater than 3; individual values provided for smaller samples. T-test P-values provided for the Neandertal versus earlier Upper Paleolithic samples. ** P < 0.01 with a multiple comparison correction (α/8)

	Mladeč 11	Neandertals	Skhul 5	Earlier Upper Paleolithic	Nean/EUP P
C5					
Ventral height / Cranial body breadth	37.3	46.9 ± 6.4 (5)	45.5	47.8 ± 3.5 (5)	0.794
Ventral height / Canal breadth	42.4	40.4 ± 5.4 (6)	40.0	50.5 ± 5.7 (5)	0.015
Dorsal height / Cranial body breadth	47.8	54.3 ± 6.1 (5)	45.5	56.2 ± 2.0 (5)	0.532
Dorsal height / Canal breadth	54.5	44.8 ± 4.7 (6)	40.0	58.3 ± 3.8 (6)	0.001**
C6					
Ventral height / Cranial body breadth	37.3	47.4 ± 11.9 (4)	52.0	45.6 ± 5.2 (6)	0.789
Ventral height / Canal breadth	42.4	37.1, 41.0, 50.6	54.2	46.3 ± 4.8 (4)	0.501
Dorsal height / Cranial body breadth	47.8	52.2 ± 9.5 (4)	40.0	52.3 ± 5.9 (6)	0.844
Dorsal height / Canal breadth	54.5	43.9, 48.5, 54.9	41.7	53.4 ± 5.1 (4)	0.280

sample has consistently higher mean values than the Neandertals for these heights, only the C6 dorsal value is significantly greater. When the Mladeč 11 heights are compared to body cranial breadth or canal breadth for C5s and C6s (Table 5), only its ventral height to body breadth value is even moderately low (being approached or matched by values for La Chapelle-aux-Saints 1 and Shanidar 1). Moreover, contrary to previous impressions, only the C5 dorsal height to canal breadth index is significantly different between the Neandertal and earlier Upper Paleolithic samples, and Skhul 5 exhibits relatively low heights as well.

Therefore, morphologically the Mladeč 11 cervical vertebra falls well within the ranges of variation of Late Pleistocene C3 to C5 vertebrae. Its moderate ventral body height (but not dorsal body height) suggests closer affinities to the Mousterian late archaic and early modern human remains, whereas its strongly tapering spinous process aligns it closer to early modern human vertebrae.

The Mladeč costal remains

Preservation, identification and maturity

Twelve rib pieces are associated with the Mladeč human postcrania, all of them heavily encrusted with dark carbonate (see chap. 8, Plates IX–XI). Only the Mladeč 12 left first rib is essentially complete; the others are missing significant portions. Their identifications as to side are based on their patterns of proximodistal torsion. Their attributions to rib number are based on morphology for the cranial three ribs and the eleventh and twelfth rib; the remainder which have been assigned numbers (even approximate) have been so identified based on general degrees of size and curvature. Their preservations and inferred numbers are as follows.

Mladeč 12

An essentially complete left first rib, with slight abrasion to the costal cartilaginous surface. The head epiphysis appears to have been unfused, with a portion of the metaphyseal surface preserved cranially. Maximum length is 70.2 mm.

Mladeč 14

A left second or third rib with the head metaphysis, the articular facet, the tubercle and the angle area. Portions of the middle of the head metaphysis are preserved, indicating that the head epiphysis was unfused. The attribution to the second or third rib given preservation depends upon the assessment of the degree of curvature, and it appears more likely that it derives from a second rib. Maximum preserved length is 62.5 mm.

Mladeč 15

A left middle rib, probably from the fourth, fifth or sixth one, from the distal margin of the articular facet around to the region of the angle. Its maximum preserved length is 50.6 mm.

Mladeč 16

A right middle or more caudal rib; it may be from the sixth to the ninth rib. It is present from the curve near the angle to the distal tapering for the costal cartilage, and its maximum preserved length is 106.8 mm.

Mladeč 17

A right middle to caudal rib preserved from the head (or head metaphysis) to the middle of the body, for a preserved length of 127.5 mm. It could be a ninth or a tenth one. The head metaphysis is preserved across the caudal half of the surface, indicating that it was unfused.

Mladeč 18

A twelfth rib from the left side, preserved from the distal articular facet most of the way to its distal end. Maximum preserved length is 111.2 mm. It is encrusted similarly to the other ribs, and the distocaudal margin is abraded.

Mladeč 19

A piece of a right rib, preserved from the distal facet margin towards the angle. It should derive from an eleventh or twelfth rib. Maximum preserved length is 42.5 mm.

Mladeč 20a

A piece of a middle to more cranial right rib, with a preserved length of 75.7 mm.

Mladeč 20b

A piece probably of a middle to more cranial rib, with an original preserved length of 69.0 mm. It was shortened to 64.2 mm in 1998 through the removal of a sample for accelerator mass spectrometry radiocarbon dating; the dating attempt failed due to low carbon yield and contamination.

Mladeč 20c

A piece probably of a middle to more caudal rib, with a preserved length of 65.6 mm.

Mladeč 20d

A piece of a probably distal rib of indeterminate side. Preserved maximum length is 29.6 mm.

Mladeč 20e

A polished piece of proximal rib, possibly from the left side. Maximum preserved length is 35.2 mm.

It is not possible to determine how many individuals are represented by these ribs. Given the apparent lack of duplication, they could derive from one thorax. However, three of the ribs (Mladeč 12, 14 and 17) have unfused head epiphyses. This indicates a late second decade to young third decade age-at-death, and they may represent the same individual.

Morphology

As an essentially complete left first rib, Mladeč 12 provides some indications of overall morphology. However, among Late Pleistocene humans, largely intact first ribs are rare, being preserved mainly for Dolní Věstonice 13, 14 and 15 plus Skhul 4 among early modern humans and for Kebara 2 and Regourdou 1 for the Neandertals (McCown and Keith, 1939; Arensburg, 1991; Kuželka, 1992; Sládek et al., 2000; Holliday, 2005b); fragments are preserved for a number of others (Trinkaus, 1983; Franciscus and Churchill, 2002).

The Mladeč 12 rib, as far as can be determined through the carbonate encrustation, is gently convex caudally and gently concave cranially with a modest development of a cranial ridge externally near the distal end. It narrows dorsally for the neck relative to the dorsal projection of the angle, and it presents a modest torsion proximodistally. In terms of overall dimensions, the rib measures 69 mm from the ventral costal cartilage surface to the dorsal head, 44 mm from the dorsal cartilage surface to the ventral head, 25 mm from the dorsal head to the dorsal angle, and 73 mm from the dorsal angle to the ventral costal cartilage surface. The internal margin is maximally 25 mm from a line drawn between the ventral head and the dorsal cartilage surface.

The Mladeč 14 left second or third rib has its margins generally rounded along the length of the preserved rib.

The Mladeč 16 right middle or more caudal rib (possibly number 8) has a strong flange of bone for the intercostal muscles and connective tissue near the angle, which then tapers to nothing distally. The rib arcs caudally as it goes distally, in addition to the normal axial torsion.

Table 6. Osteometric dimensions of the Mladeč ribs (in mm). Estimated values are in parentheses

Number	Side	Neck height[1]	Neck thickness	Facet height[2]	Facet thickness	Angle height[3]	Angle thickness
12	1 - Lt	6.1	8.1	6.6	12.6		
14	2/3 - Lt	10.0	5.4	10.7	8.6	12.4	8.0
15	Mid - Lt	-	-	9.4	8.3	-	-
16	Mid/Caud - Rt	-	-	(13.4)	(7.4)	18.3	9.2
17	Caudal - Rt	(15.5)	7.0	10.0	9.0	16.4	9.1
18	12 - Lt	-	-	10.6	6.8	14.2	6.3
19	11/12 - Rt	-	-	11.3	6.6	-	-

[1] Diameters in the middle of the costal neck.
[2] Diameters taken just distal of the articular facet.
[3] Diameters taken in the middle of the angle.

Table 7. Minimum to maximum diameter indices from diameters taken at the posterior angle for Mladeč and comparative Late Pleistocene and recent Euroamerican male right ribs (N = 18–19). Comparative data from Trinkaus (1983), Arensburg (1991), Sládek et al. (2000) and Franciscus and Churchill (2002). Right and left indices, when available for the fossils, are averaged. For Mladeč 14 and 17, given uncertainties in rib number, comparative data for the adjacent probable ribs are provided

<table>
<tr><th></th><th>Rib 2</th><th>Rib 3</th><th>Rib 8?</th><th>Rib 9?</th><th>Rib 10?</th><th>Rib 12</th></tr>
<tr><td>Mladeč 14</td><td colspan="2">64.5</td><td></td><td></td><td></td><td></td></tr>
<tr><td>Mladeč 16</td><td></td><td></td><td>50.3</td><td></td><td></td><td></td></tr>
<tr><td>Mladeč 17</td><td></td><td></td><td></td><td colspan="2">55.5</td><td></td></tr>
<tr><td>Mladeč 18</td><td></td><td></td><td></td><td></td><td></td><td>44.4</td></tr>
<tr><td>Neandertals</td><td></td><td></td><td></td><td></td><td></td><td></td></tr>
<tr><td>Kebara 2</td><td>46.4</td><td>58.4</td><td>57.9</td><td>52.4</td><td>60.0</td><td>62.3</td></tr>
<tr><td>Shanidar 3</td><td>66.3</td><td>55.8</td><td>52.1</td><td>53.8</td><td>62.2</td><td>56.5</td></tr>
<tr><td>Shanidar 4</td><td>70.8</td><td>-</td><td>-</td><td>-</td><td>-</td><td>52.8</td></tr>
<tr><td>Shanidar 5</td><td>71.0</td><td>-</td><td>-</td><td>-</td><td>-</td><td>-</td></tr>
<tr><td>Qafzeh-Skhul</td><td></td><td></td><td></td><td></td><td></td><td></td></tr>
<tr><td>Skhul 4</td><td>-</td><td>48.1</td><td>65.8</td><td>39.6</td><td>65.5</td><td>55.8</td></tr>
<tr><td>Skhul 5</td><td>50.0</td><td>57.3</td><td>68.5</td><td>56.3</td><td>95.2</td><td>-</td></tr>
<tr><td>European earlier Upper Paleolithic</td><td></td><td></td><td></td><td></td><td></td><td></td></tr>
<tr><td>Dolní Věstonice 13</td><td>41.6</td><td>71.5</td><td>57.1</td><td>53.1</td><td>63.6</td><td>-</td></tr>
<tr><td>Dolní Věstonice 14</td><td>-</td><td>-</td><td>-</td><td>-</td><td>-</td><td>58.8</td></tr>
<tr><td>Dolní Věstonice 15</td><td>54.9</td><td>57.3</td><td>53.6</td><td>-</td><td>48.2</td><td>40.7</td></tr>
<tr><td>Euroamericans males – right</td><td>58.0
± 5.4</td><td>71.7
± 14.2</td><td>57.2
± 9.8</td><td>47.8
± 8.1</td><td>45.6
± 8.5</td><td>54.9
± 7.9</td></tr>
</table>

The Mladeč 17 more caudal right rib (probably 9 or 10) has a convex articular facet, a large pit dorsally between the head and the articular facet, a rounded shaft between the articular facet and the angle, and then a modest intercostal crest and sulcus distal of the angle.

The Mladeč 18 left twelfth rib appears to be relatively robust with rounded margins. In addition to its smooth proximodistal spiral, it is caudally convex proximally and then cranially convex distally.

There are no notable features on the Mladeč 15, 19, and 20a to 20e rib pieces.

The available rib diameters at the neck, articular facet and angle are provided in Table 6. Comparisons of the index of the minimum to maximum diameters (generally the breadth versus the height, except for rib 2) are provided in Table 7. For each of Mladeč 14 and 17, comparisons are made to the adjacent ribs given uncertainties in rib number. Although it has long been stated that Neandertals have very round ribs compared to recent humans (Schaaffhausen, 1858; Boule, 1911–13), recent analyses (e.g., Heim, 1976; Arensburg, 1991; Franciscus and Churchill, 2002) have documented considerable within-sample variation in this feature. However, Franciscus and Churchill (2002) have argued that Neandertal ribs are distinctive mostly in the robusticity of their more caudal ribs, and this is supported by the data in Table 7, particularly for the eleventh rib. In this, only Kebara 2 among the Neandertals has a relatively thin eleventh rib, although Dolní Věstonice 13 has relatively robust ones. The Mladeč 18 twelfth (or eleventh) rib is one of the thinnest of these Late Pleistocene specimens exceeded in this feature only by Dolní Věstonice 15 and matched by some recent Euroamericans. However, the Mladeč 16 and 17 ribs are also relatively thin for Late Pleistocene humans. Moreover, the Mladeč 14 second or third rib falls in the middle of variable samples.

The Mladeč 13 clavicle

Mladeč 13 was originally identified as a proximal rib fragment, but reconsideration suggests that it is the distal portion of an immature right clavicle (see chap. 8, Plate IX, Fig. b). Its maximum preserved length is 60.7 mm. Its fragmentary condition provides little morphological information, reflected in difficulties in determining its identification.

The Mladeč 23 and 24 humeri

Preservation and maturity

The Mladeč sample preserves portions of two right humeri, Mladeč 23 and Mladeč 24. The former retains the proximal epiphysis and proximal third of the diaphysis, whereas the latter retains most of the diaphysis (see chap. 8, Plate XIII).

Mladeč 23 is an almost mature right proximal humerus, broken distally at the level of the deltoid tuberosity, approximately where the tuberosity begins its superoinferior course from the lateral to the anterior (ventral) aspects of the diaphysis. The proximal shaft is complete distal to the level of the surgical neck. The proximal epiphysis, however, has endured some damage. The lesser tubercle lacks only the most anterior projection, and the intertubercular sulcus is completely preserved, but nothing remains of the greater tubercle. The damage associated with the loss of the greater tubercle has exposed cancellous bone from the lateral surgical neck to the superior margin of the humeral head. Posteriorly, a band (ca. 10 mm wide) of mostly cortical bone is missing along the posterior aspect of the anatomical neck. While preservation of the bone is excellent, the bone surface is obscured by encrusting matrix on the medial head and shaft, and slight surface erosion is visible on the anteromedial border of the shaft.

Mladeč 24 is a virtually complete right humeral diaphysis, preserved from slightly proximal of the level of the surgical neck to the distal epiphyseodiaphyseal region. Nothing remains of the humeral head, anatomical neck, or greater and lesser tubercles, but the insertion for *M. pectoralis major* is visible slightly inferior to the level of the lesser tubercle, as are the insertions for *M. latissimus dorsi* and *M. teres major.* Distally, the complete olecranon fossa is preserved, and a lateral portion of the trochlea remains attached to the inferior aspect of the olecranon fossa. The bulk of the trochlea, the capitulum, and the medial and lateral epicondyles are missing. On the anterior surface, the coronoid fossa is completely preserved with a small shelf of bone directly inferior of it, where a fragment of the trochlea remains attached. The radial fossa is also preserved, but it is broken along its inferior margin. Overall, the specimen exhibits good preservation, although surface matrix is found on the proximal shaft and in the olecranon, radial and coronoid fossae and surrounding areas. In the mid-distal diaphysis there is a fossilization break which has been reassembled with minimal separation of the two pieces; in 1998 the join was opened, a sample of bone was taken from the middle of the cortical bone for accelerator mass spectrometry radiocarbon dating of the specimen, the adjoining surfaces cleaned of matrix, and the two pieces reassembled. Dating of the specimen failed due to contamination and low carbon yield.

The Mladeč 23 humerus exhibits radiographically an indication of its proximal (head) epiphyseal line, extending from the medial side about two-thirds of the distance laterally, even though externally it is not apparent on the cortical bone anteriorly and medially or on the exposed trabeculae posteriorly and laterally. This provides a general adolescent (13 to 20 years) age-at-death for Mladeč 23, although the degree of fusion suggests an age toward the higher end of that range. The Mladeč 24 humerus exhibits no trace externally or radiographically of the distal epiphyseal fusion line, indicating a late adolescent or adult status for it.

Mladeč 24 length estimation

Since it is necessary to scale detailed measurements of the humerus to a measure of overall size whenever possible, the anatomical lengths of the more complete Mladeč 24 humerus were estimated; the Mladeč 23 humerus is insufficiently preserved to allow an adequate estimation of its original length. In order to do this, the bone was placed on a metric grid and the locations of the middle of the proximal medial diaphyseal curvature (for the medial deviation of the head), the proximal end of the *M. pectoralis major* tuberosity, the posteroproximal trochlea and the distal coronoid fossa were located. The average of each pair of proximal or distal landmarks was noted, and the distance between them determined (296.5 mm for Mladeč 24). The same measurements were taken on a geographically mixed sample of recent human humeri, and least squares regressions for maximum length, articular length and biomechanical length were determined (see notes to Table 8). The resultant values for Mladeč 24 are provided in Table 8; the standard errors of the estimates for Mladeč 24 are between 3.0 and 3.5 mm, all of which are less than 1% of the predicted lengths.

This approach only assumes that the relative overall dimensions of the epiphyses are similar to those of the recent human sample; given the apparently modest dimension of at least one aspect of the distal epiphysis (see olecranon fossa discussion below), this assumption appears to be reasonable. If the dimensions of the epiphyses have been underestimated, then the errors will be conservative since they will tend to make diaphyseal and epiphyseal aspects of the Mladeč 24 humerus appear more robust than they originally were.

Morphology

The Mladeč 23 humeral head is evenly rounded with the mediolateral diameter probably (given damage) greater than the anteroposterior diameter (see Table 8). The head-neck angle is 142°. The lesser

Table 8. Osteometric dimensions of the Mladeč 23 and 24 right humeri (in mm and degrees). Estimated values are in parentheses

	Mladeč 23	Mladeč 24
Maximum length[1]	–	(373.5)
Articular length[2]	–	(366.0)
Biomechanical length[3]	–	(366.5)
Head mediolateral diameter (M-9)	44.1	–
Head anteroposterior diameter (M-10)	(40.0)	–
Head mediolateral arc[3]	63.0	–
Head depth	17.5	–
Head-neck angle (M-17)	(142°)	–
Surgical neck maximum diameter	27.0	22.6
Surgical neck minimum diameter	22.0	19.9
Surgical neck circumference	78.0	69.0
Lesser tubercle length	22.0	–
Lesser tubercle breadth	12.5	–
Intertubercular sulcus depth (S-23)[5]	(3.8)	–
Intertubercular sulcus breadth (S-24)	7.9	–
Midshaft maximum diameter (M-5)	–	21.9
Midshaft minimum diameter (M-6)	–	15.5
Midshaft circumference (M-7a)	–	61.0
Deltoid diameter (M-6a)	19.7	21.3
Deltoid circumference	–	63.0
Distal minimum circumference (M-7)	–	59.0
Distal diaphyseal AP diameter	–	18.5
Distal diaphyseal ML diameter	–	16.7
Maximum deltoid tuberosity breadth[6]	10.8	9.0
Pectoralis major breadth[6]	7.9	4.7
Latissimus dorsi / teres major breadth[6]	6.3	–
Supraolecranon AP diameter	–	14.6
Supraolecranon ML diameter	–	28.8
Medial pillar thickness (S-12)	–	7.5
Lateral pillar thickness (S-13)	–	(16.0)
Olecranon fossa breadth (M-14)	–	28.0
Olecranon fossa depth (M-15)	–	11.6
Coronoid fossa breadth	–	16.2
Septal aperture	–	absent
Maximum preserved length	147.2	313.0

[1] Maximum length estimated from the mean distance between the proximal medial diaphyseal curvature and the proximal pectoralis major tuberosity proximally and the posteroproximal trochlea and the distal coronoid fossa distally (PresLen = 296.5 mm), using a least squares regression based on recent humans (MaxLen = 1.30 × PresLen – 14.1, r^2 = 0.945, N = 29). SE_{est} = 3.4 mm.

[2] Articular length estimated from the mean distance between the proximal medial diaphyseal curvature and the proximal pectoralis major tuberosity proximally and the posteroproximal trochlea and the distal coronoid fossa distally (Pres Len = 296.5 mm), using a least squares regression based on recent humans (ArtLen = 1.22 × PresLen + 2.0, r^2 = 0.949, N = 29). SE_{est} = 3.1 mm.

[3] Biomechanical length (proximal head to distal lateral trochlear margin) estimated from the mean distance between the proximal medial diaphyseal curvature and the proximal pectoralis major tuberosity proximally and the posteroproximal trochlea and the distal coronoid fossa distally (Pres Len = 296.5 mm), using a least squares regression based on recent humans (BiomLen = 1.25 × PresLen – 4.9, r^2 = 0.950, N = 29). SE_{est} = 3.1 mm.

[4] Maximum arcs (AP and ML relative to head) of the subchondral bone (Vandermeersch and Trinkaus, 1995).

[5] Measurement definitions from Senut (1981).

[6] Maximum distinct breadth of the muscular insertion rugosity.

tubercle is obliquely oriented with a peak distally; there is a large muscle tendon facet (for *M. subscapularis*) on it with a broad sulcus between it and the head articular margin.

The Mladeč 23 proximal anterior diaphysis exhibits a broad (maximum width 15.6 mm) area of mildly rugose surface bone for the combined attachments of *M. pectoralis major, M. latissimus dorsi* and *M. teres major.* It is possible to largely discern the attachment for *M. pectoralis major* from that for the other two muscles, and the maximum breadths of these attachments are in Table 8. The preserved portion of the deltoid tuberosity is a broad, roughened area that is minimally raised from the adjacent diaphyseal surface and exhibits minimal rugosity.

On the Mladeč 24 proximal diaphysis, there is a clear but narrow and minimally rugose insertion for *M. pectoralis major.* However, the attachments for *M. latissimus dorsi* and *M. teres major* cannot be discerned; although the region for their insertions is partly obscured with matrix, it is unlikely that their attachments were more than minimally developed. The deltoid tuberosity is discernable as a gentle swelling rather than as a distinct tuberosity. One can discern longitudinal striations on the tuberosity, but there is none of the rugosity frequently associated with the osteological insertion of the deltoid muscle. Distally, the supracondylar crests are minimally present.

The size of the *M. pectoralis major* tuberosity in particular, as reflected in its breadth relative to humeral length, is one of the features that have been shown to largely separate the Neandertals and early modern humans (Trinkaus, 2000). As illustrated in Fig. 1, the relative *M. pectoralis major* breadths of the Neandertals are above those of all early modern humans. Among the Neandertals, only those of La Ferrassie 1 and Lezetxiki 1 approach the early modern human distribution. It is not possible to appropriately compare the Mladeč 23 *M. pectoralis major* breadth, since humeral length is not available for it; however its breadth of 7.9 mm is in the middle of the early modern human distribution and among the Neandertals only the La Ferrassie 2 and Lezetxiki 1 humeri have lower absolute values. The Mladeč 24 breadth of 4.7 mm, however, is among the smaller of the early modern human values, and it is the lowest value relative to humeral length of any of the Late Pleistocene specimens for which data are available. Only the Dolní Věstonice 14 humerus approaches it.

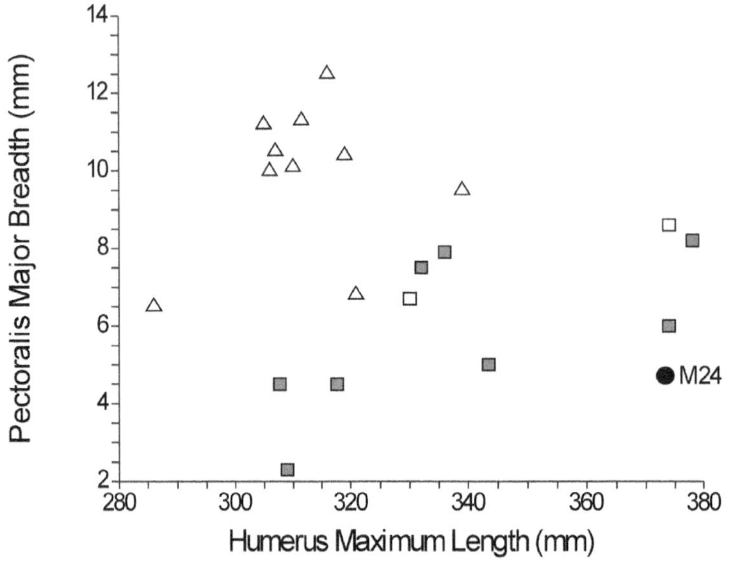

Fig. 1. *M. pectoralis major* tuberosity breadth versus humeral length for Mladeč 24 (black circle – M24), earlier Upper Paleolithic humans (gray squares), Neandertals (open triangles) and Qafzeh-Skhul humans (open squares)

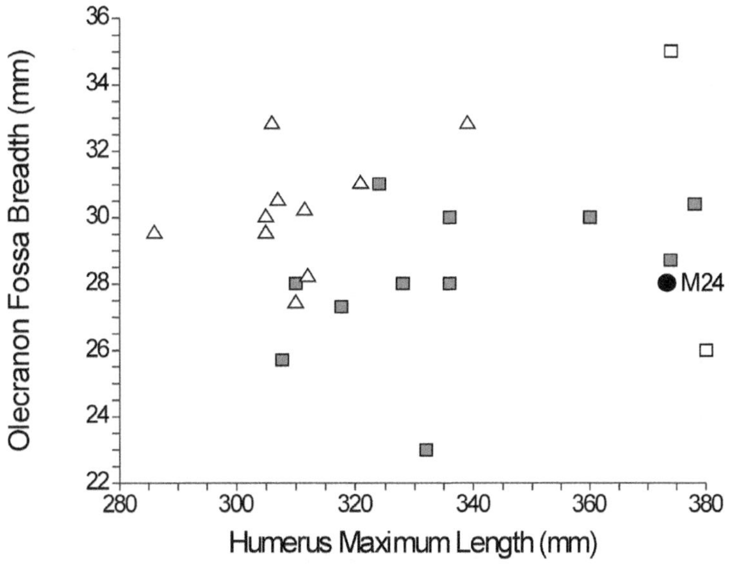

Fig. 2. Olecranon fossa breadth versus humeral length for Mladeč 24 (black circle – M24), earlier Upper Paleolithic humans (gray squares), Neandertals (open triangles) and Qafzeh-Skhul humans (open squares)

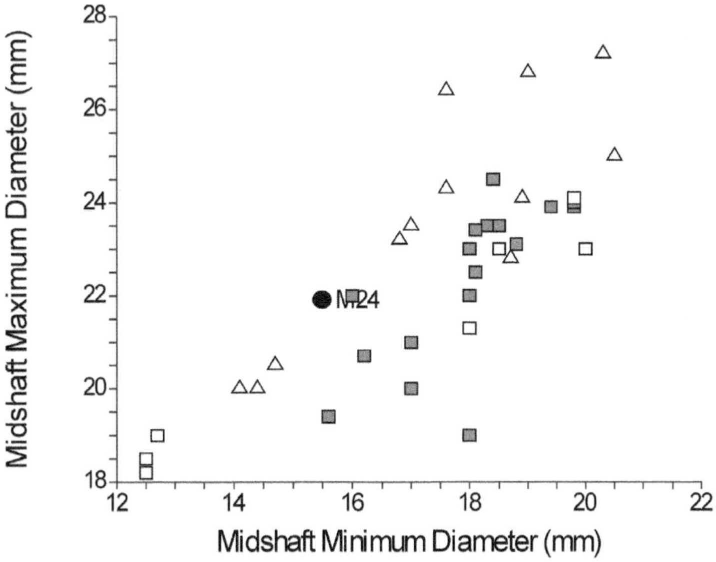

Fig. 3. Humeral midshaft maximum versus minimum subperiosteal diameters, for Mladeč 24 (black circle – M24), earlier Upper Paleolithic humans (gray squares), Neandertals (open triangles) and Qafzeh-Skhul humans (open squares)

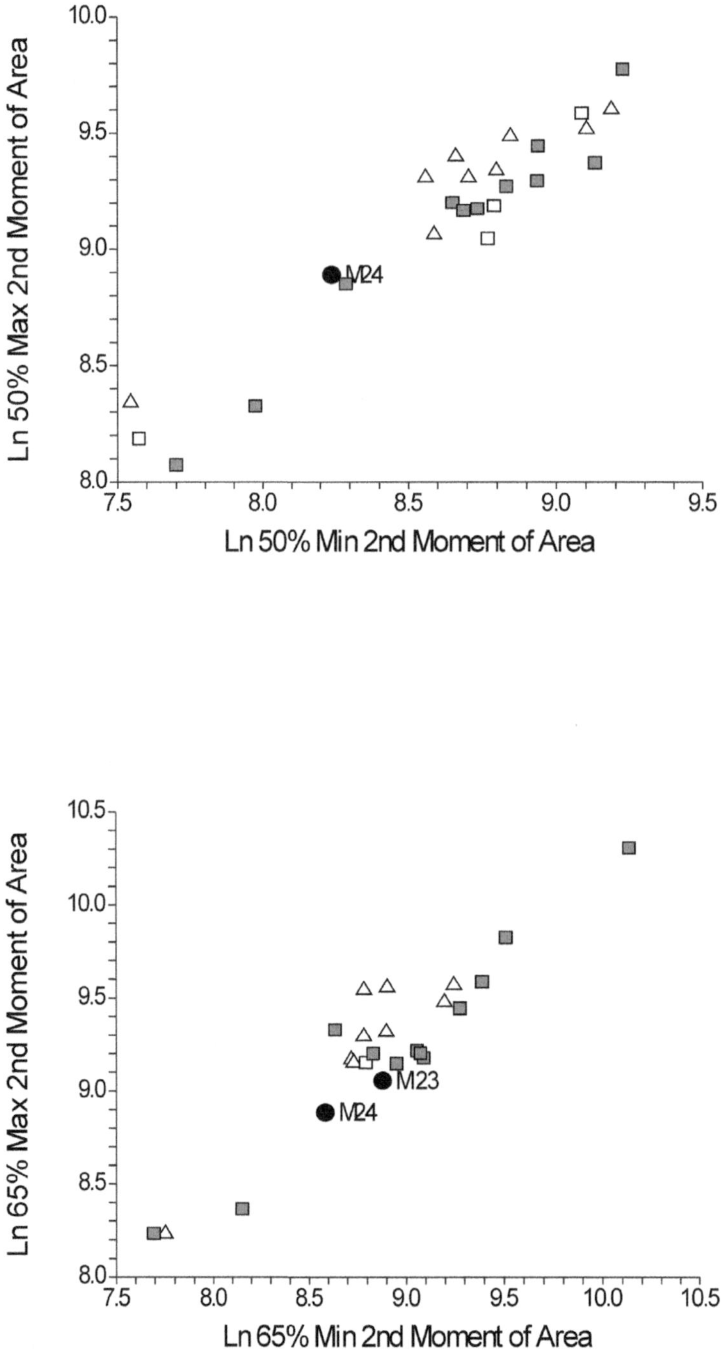

Fig. 4. Maximum versus minimum second moments of area of the humeral mid-shaft (50%) and mid-proximal shaft (65%), for Mladeč 23 and 24 (black circles – M23 and M24), earlier Upper Paleolithic humans (gray squares), Neandertals (open triangles) and Qafzeh-Skhul humans (open squares)

The other measure of muscular hypertrophy available on the Mladeč humeri is an indirect one, the breadth of the olecranon fossa (as reflecting olecranon breadth and hence *M. triceps brachii* tendon size) relative to humeral length. The Neandertals again have higher relative values than the early modern humans, although the degree of overlap is greater than with the *M. pectoralis major* tuberosity breadth (Table 10; Fig. 2). The Mladeč 24 humerus is among the most gracile of the available specimens. It is close to that of Dolní Věstonice 14 and only Dolní Věstonice 16 and Skhul 5 have lower relative values.

The diaphyseal shape has been assessed using midshaft maximum and minimum external diameters, to maximize sample size, and using 50% (midshaft) and 65% (mid-proximal diaphysis) maximum versus minimum second moments of area. The latter provide more accurate measurements of diaphyseal cortical bone distribution and permit the inclusion of the Mladeč 23 humerus (for the 65% section), but available comparative data are more limited.

The midshaft subperiosteal diameter comparison (Fig. 3) shows a general trend, with the Neandertals and the smaller Qafzeh-Skhul specimens having more ellipsoid diaphyseal proportions and the other early modern humans having rounder (or, more accurately, less ellipsoid) midshafts. Since a more ellipsoid shaft would provide for greater rigidity to bending in the plane of the maximum diameter, this appears to suggest greater robusticity of the Neandertal and some Qafzeh-Skhul humeri. In this, Mladeč 24 is the most ellipsoid of the earlier Upper Paleolithic specimens; although it is close to the middle of the Neandertal range of variation, it is not significantly different from either early modern human sample (Table 10).

The second moments of area comparisons provide less separation of the Late Pleistocene samples, although they continue the general pattern of more ellipsoid humeral diaphyses among the Neandertals (Fig. 4; Table 10). In the 50% cross section, Mladeč 24 remains in the middle of the Neandertals but close to other early modern humans, whereas in the 65% cross section, both of the Mladeč humeri are among the early modern humans but Mladeč 24 remains close to the Neandertal distribution (Fig. 4; Table 10).

Diaphyseal robusticity

These aspects of muscular hypertrophy and diaphyseal shape can be, for Mladeč 24 in particular, associated with an assessment of diaphyseal robusticity using especially cross sectional geometry (Table 9). Relative cortical area at the mid-distal (35%) diaphysis provides little separation of the comparative samples, and Mladeč 24 falls in the middle of the Late Pleistocene distribution (Table 10; Fig. 5). The more proximal diaphyseal sections provide some separation of the late archaic and early modern human samples despite considerable overlap and the high relative val-

Table 9. Cross-sectional geometric properties of the Mladeč 23 and 24 humeri. Areas in mm^2, second moments of area in mm^4. 0% is distal

	Mladeč 23 65%	Mladeč 24 35%	Mladeč 24 50%	Mladeč 24 65%
Total area (TA)	350.8	263.4	271.9	294.0
Cortical area (CA)	193.1	201.9	181.1	201.5
Medullary area (MA)	157.7	61.5	90.8	92.5
AP 2nd moment of area (I_x)	8337	6350	6240	6366
ML 2nd moment of area (I_y)	7387	4516	4802	6188
Max 2nd moment of area (I_{max})	8525	6753	7248	7184
Min 2nd moment of area (I_{min})	7199	4113	3794	5370
Polar moment of area (J)	15724	10866	11042	12554

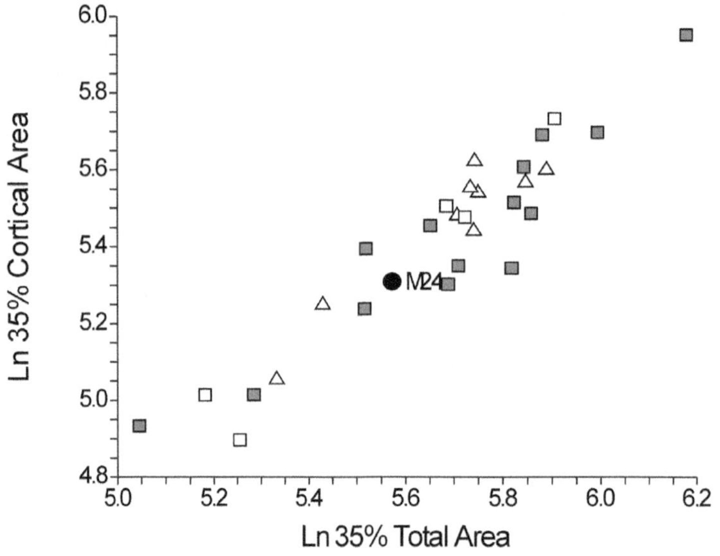

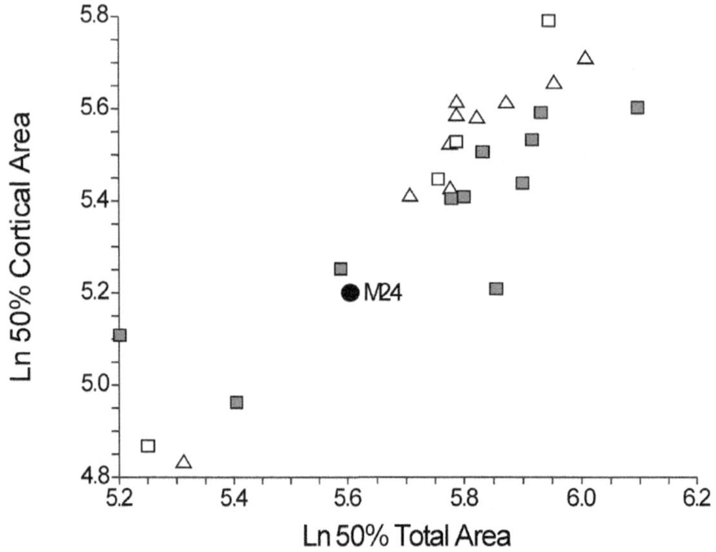

Fig. 5. Cortical area versus total subperiosteal area for the mid-distal (35%) and midshaft (50%) humerus for Mladeč 24 (black circles – M24), earlier Upper Paleolithic humans (gray squares), Neandertals (open triangles) and Qafzeh-Skhul humans (open squares)

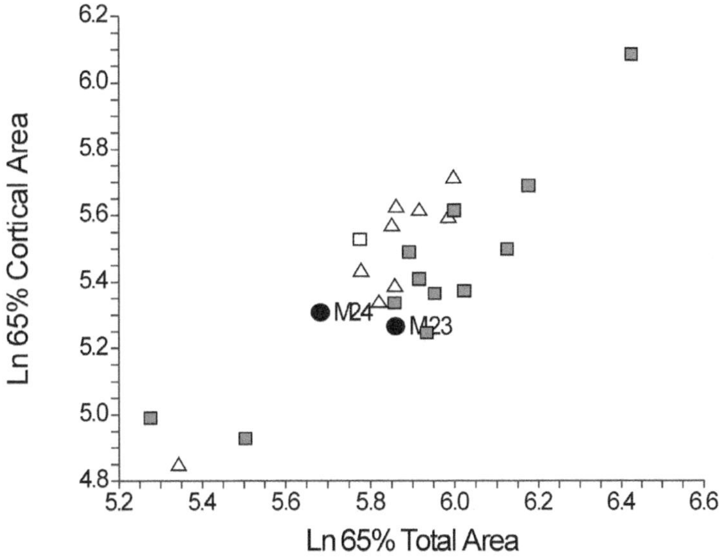

Fig. 6. Cortical area versus total subperiosteal area for the mid-proximal (65%) humerus for Mladeč 23 and 24 (black circles – M23 and M24), earlier Upper Paleolithic humans (gray squares), Neandertals (open triangles) and Qafzeh-Skhul humans (open squares)

ues of the very small Nahal-Ein-Gev 1 specimen (Fig. 6). In the 50% comparisons, Mladeč 24 is among the earlier Upper Paleolithic specimens but close to the Middle Paleolithic ones, and in the 65% section Mladeč 24 has moderately higher relative cortical area whereas Mladeč 23 remains with the majority of the earlier Upper Paleolithic specimens and distinct from the Neandertals (Fig. 6).

It is debatable to what extent relative cortical area reflects diaphyseal robusticity rather than the life history of endosteal resorption versus subperiosteal deposition of the humerus in question. Comparisons of diaphyseal measures of bone quantity to bone length, however, provide better assess-

Table 10. Mladeč 23 and 24 humeral metric comparisons using linear residuals from the reduced major axis line through the pooled comparative sample. The P-value derives from the ANOVA comparison across the three reference samples, and the Mladeč values are Z-scores [(Mladeč values – sample mean)/standard deviation]. Z-scores are provided only relative to samples ≥ 5. I_{max} and I_{min}: maximum and minimum second moments of area; CA and TA: cortical and total subperiosteal areas; J: polar moment of area.* $P < 0.05$ with a multiple comparison correction ($\alpha/9$)

	P-Value	Neandertals		Qafzeh-Skhul	Earlier Upper Paleolithic	
		Mladeč 23	Mladeč 24	Mladeč 24	Mladeč 23	Mladeč 24
Olec. br. / Length	<0.001*	-	5.04	-	-	1.44
Midshaft Max. / Min.	0.004*		0.26	1.18	-	1.73
50% I_{max} / I_{min}	0.011		0.19	-	-	1.74
65% I_{max} / I_{min}	0.069	2.01	1.40	-	0.73	0.20
35% CA / TA	0.372	-	0.55	0.46	-	0.10
50% CA / TA	0.045	-	1.33		-	0.59
65% CA / TA	0.013	2.38	0.08	-	0.44	1.05
Distal circ. / Length	0.001*	-	3.43	-	-	3.35
J / Length	<0.001*	-	3.66	0.40	-	2.54

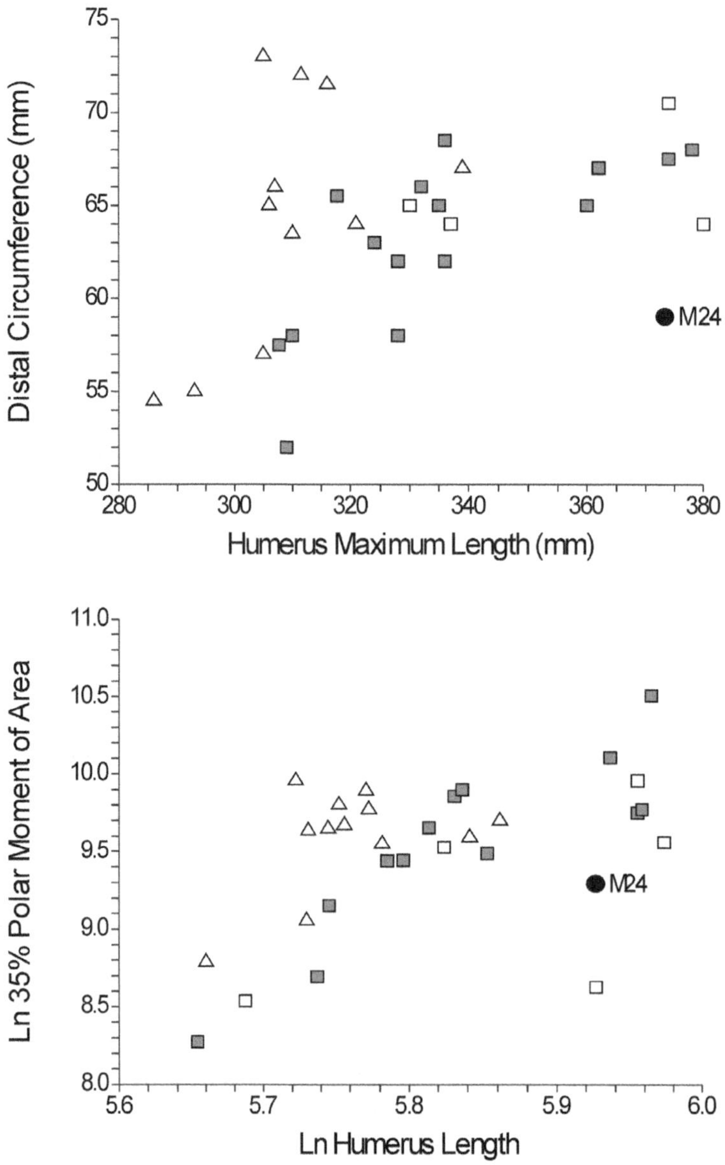

Fig. 7. Measures of humeral robusticity for Mladeč 24. Distal minimum circumference (above) and mid-distal (35%) polar moment of area (below) for Mladeč 24 (black circle – M24), earlier Upper Paleolithic humans (gray squares), Neandertals (open triangles) and Qafzeh-Skhul humans (open squares).

ments of the bone's robusticity. Previous assessments of humeral diaphyseal scaling (e.g., Ruff et al., 1993; Trinkaus and Churchill, 1999; Ruff, 2000a) have varied as to whether humeral length alone or humeral length combined with body mass should be used to scale humeral robusticity. However, since a number of recent humans with fully functional upper limbs are incapable of supporting their body weights on their upper limbs (in either suspensory or supportive modes) (Trinkaus, personal experience), body weight should not necessarily be used to scale humeral robusticity. Therefore, measures of human humeral diaphyseal size are scaled here solely against an estimate of beam length, humeral length.

To maximize sample size, initially distal minimum diaphyseal circumference relative to humeral length was compared (Table 10; Fig 7). Several Neandertals provide high values, several early modern humans have more gracile humeri, and the two samples are significantly different (Table 10); there is nonetheless some overlap between the samples. However, regardless of the degree of Neandertal to early modern human similarity or difference in relative distal circumference, none of them is as gracile as the Mladeč 24 humerus. The Mladeč 24 humerus is approached only by the Skhul 5 and, to a lesser extent, the Dolní Věstonice 3 humeri.

The patterns of sample tendencies hold for the more appropriate comparison of 35% polar moment of area to humeral length (Fig. 7). The reference samples are significantly different, Mladeč 24 remains very gracile compared to the Neandertal and earlier Upper Paleolithic samples, although in this comparison it is close to the values for Dolní Věstonice 14, Pavlov 1 and Skhul 4 and above the low value for Skhul 2. Mladeč 24 is completely separate from even the most gracile Neandertal humeri.

The interpretation of the relative gracility of the Mladeč 24 humerus depends in part on the appropriate scaling employed for humeral diaphysis. As discussed above, it is most appropriate to use humeral length alone, as beam length, to scale human humeral diaphyseal and articular dimensions. If humeral length alone is employed, then Mladeč 24 is clearly a very gracile humerus, among the most gracile humeri known from the Middle Paleolithic and earlier Upper Paleolithic, suggesting relatively light habitual loads on the upper limb. If, however, it were argued that some percentage of body mass should be combined with humeral length, then the relatively small dimensions of the Mladeč 24 diaphysis could be the product of a very linear body (low mass relative to arm length), or a combination of a linear body and a gracile bone.

The Mladeč 25c ulna

Preservation and maturity

The Mladeč 25c right ulna preserves most of the proximal end of the bone plus some of the adjacent diaphysis (chap. 8, Plate XIV). For the portions preserved, most of the bone is intact. The primary damage is to the volar margins of the olecranon and coronoid processes, although the coronoid damage is mostly abrasion to the margin with little bone loss. The proximal olecranon epiphysis appears to have been fully fused at the time of death, but it has been abraded and gives the superficial appearance of the olecranon metaphysis. This indicates an age-at-death of at least ca. 15 years (Scheuer and Black, 2000).

Morphology

The Mladeč 25c ulna is a large and rugose bone (Tables 11 and 12). The various muscular and ligamentous attachments are strongly marked on the bone, and the diaphyseal crests appear to have been fully formed. The proximal diaphysis is triangular in cross section, and there is a strong development of a posterolateral sulcus. The proximal end of the interosseus crest is thick (4.0 mm) and prominent. There is a strong marking for the *M. brachialis* insertion with a distinct concavity formed on the distal coronoid process for it. However, the supinator crest is poorly developed.

The abrasion of the coronoid process and particularly erosion of the olecranon process prevent metrical assessment of the trochlear notch orientation. However, it is evident in lateral view from the

Table 11. Osteometric dimensions of the Mladeč 25c right proximal ulna (in mm and degrees). Estimated values are in parentheses

Proximal shaft AP diameter (M-14)	24.5
Proximal shaft ML diameter (M-13)	24.0
Olecranon breadth (M-6)	(25.0)
Olecranon thickness (MCH-7)[1]	21.3
Olecranon length (M-8; MCH-12)	(20.5)
Coronoid height (MCH-8)	43.0
Tuberosity position (MCH-11)	33.0
Proximal trochlear angle[2]	19°
Radial facet height	(20.0)
Radial facet breadth	(17.0)
Anterior breadth of coronoid-radial half (M-9)	14.0
Posterior breadth of coronoid-radial half (M-10)	16.6
Radial breadth of the coronoid[3]	10.7
Ulnar breadth of the coronoid[3]	18.3
Maximum preserved length	114.2

[1] Measurement definition from McHenry et al. (1976).
[2] Taken in the coronal plane of the bone.
[3] From the mid-coronoid crest to the radial or ulnar margin.

Table 12. Cross-sectional geometric properties of the Mladeč 25a ulna, Mladeč 25a, 25b and 26 radii, and Mladeč 31 metacarpal 3. Areas in mm^2, second moments of area in mm^4. 0% is distal. Proximodistal positions are based on the morphology and are therefore approximate

	Mladeč 25c 80%	Mladeč 25a 80%	Mladeč 25b 35%	Mladeč 25b 50%	Mladeč 26 35%	Mladeč 26 50%	Mladeč 31 50%
Total area (TA)	246.6	180.0	144.1	163.4	114.1	110.2	55.9
Cortical area (CA)	172.4	117.0	116.0	139.0	97.0	85.9	48.5
Medullary area (MA)	74.2	63.0	28.1	24.4	17.1	24.3	7.4
AP 2nd moment of area (I_x)	4097	2531	1443	1724	1040	858	220
ML 2nd moment of area (I_y)	4985	2049	1774	2608	994	1004	276
Max 2nd moment of area (I_{max})	5057	2592	1775	2610	1096	1039	284
Min 2nd moment of area (I_{min})	4025	1988	1442	1723	938	824	212
Polar moment of area (J)	9083	4580	3217	4332	2034	1863	496

large size of the coronoid process and the modest development of what is preserved of the volar olecranon that the orientation of the trochlear notch was predominantly volar and proximal. In this, it contrasts with the more strictly volar orientation of the trochlear notch seen in Neandertals and archaic *Homo* generally (Trinkaus, 1983; Churchill et al., 1996).

The trochlear notch also turns strongly laterally relative to the diaphysis, or vice versa. The angle between the midline crest of the trochlear notch and the midline of the proximal diaphysis is ca. 19°.

The coronoid process is notable primarily for its strong medial flare, indicating a large medial portion to the humeral trochlea. As an indication of the degree of medial flare, the distance from the mid-trochlear notch crest to the radial side is 10.7 mm, whereas it is 18.3 mm to the medial side. The radial facet is large and broad, and in terms of curvature fits well with the head of the Mladeč 25a right radius.

The Mladeč 25a, 25b and 26 radii

Preservation and maturity

The radial remains from Mladeč include a proximal right radius with the head, the neck, the tuberosity, and a small portion of the proximal diaphysis to the beginning of the interosseus crest (Mladeč 25a), a right midshaft section from the proximal interosseus crest to the beginning of the flare for the distal epiphysis (Mladeč 25b), and a smaller left radial midshaft section (Mladeč 26) (chap. 8, Plate XIV). The last has its proximal break dorsally near the maximum development of the interosseus crest but closer to midshaft elsewhere, and distally it is broken near the minimum circumference.

The first two specimens, Mladeč 25a and 25b, are very similar in diaphyseal size, both are right, and they could very well belong to the same bone. If Mladeč 25a derives from the same elbow as the Mladeč 25c ulna, then the three pieces could derive from the same forearm. Mladeč 26 is clearly smaller (Tables 12 and 13), and it is therefore unlikely to be from the left arm of the same individual. If they are assumed to be from the same individual, their distal minimum circumferences provide an asymmetry value [(right – left) / (right + left) / 2] of 15.0. A sample of 14 earlier Upper Paleolithic paired radii provides a median of 4.2 and a maximum value of 12.2 (Barma Grande 2), small Neandertal and Qafzeh-Skhul samples provide ranges of 0.0 to 8.7 (N = 4) and 2.4 to 3.1 (N = 3) respectively, and a robust recent human sample (Ohba, 1935) has a median of 2.7, a 95% value of 10.2, and a maximum outlier of 16.0. It is therefore possible, but unlikely, that the Mladeč 25b and 26 derive from the same individual.

The Mladeč 25a proximal radius gives the appearance of having a partially, but largely, fused head epiphysis. However, close inspection of the purported epiphyseal fusion line indicates that it is a postmortem break and not a fusion line. This is indicated by the exposed trabeculae within the break, rather than a metaphyseal surface, and by its location within the head subchondral bone rather than between the head and the neck. This indicates an age-at-death of at least ca. 15 years (Scheuer and Black, 2000).

Morphology

The Mladeč 25a proximal radius is notable especially for its absolutely large head, neck and tuberosity (Tables 12 to 14). The sagittal diameter of the head (25.2 mm) is at the upper end of the range of variation for earlier Upper Paleolithic humans, being approached only by that of Paviland 1 (24.6 mm); among Middle Paleolithic humans, only Skhul 4 (24.8 mm) and Shanidar 3 (25.0 mm) come close. In concert with the large size of its head, the sagittal neck diameter is also large; its diameter of 15.6 mm is above those of all earlier Upper Paleolithic specimens except Cro-Magnon 4303 and 4305a (16.3 and 15.5 mm, respectively), and among Middle Paleolithic specimens it is exceeded only by that of Skhul 4 (16.3 mm).

The combination of head and neck diameters provides it with an index of 161.5, which is close to the mean of a variable earlier Upper Paleolithic sample (Table 14) and relatively low for a Neandertal specimen. Given the shift, albeit non-significant, for an increase in relative head diameter between Neandertals and early modern humans (Table 14), Mladeč 25a clusters with the early modern humans and only a minority of the Neandertals. In the context of this, the Mladeč 25a neck anteroposterior to mediolateral diameter proportions are similar to those of the comparative samples, which differ little among themselves.

Since Neandertals have long radial necks relative to radial length (Trinkaus, 2000), reflecting the mechanical advantage of *M. biceps brachii* for elbow flexion, an index was devised using radial head diameter as a surrogate for overall radial size, given the similarities in the radial head to length proportions of later Pleistocene and recent humans (Trinkaus, 1983). The resultant indices reinforce the

Table 13. Dimensions of the Mladeč 25a right proximal radius, 25b right radial midshaft and 26 left radial midshaft (in mm and degrees).

	Mladeč 25a	Mladeč 25b	Mladeč 26
Head AP diameter (M-5[1])	25.2	–	–
Head depth (mid-proximal fossa)	3.4	–	–
Prox. ulnar prox.-dist. dia.: lat. surf.[1]	11.0	–	–
Prox. ulnar prox.-dist. dia.: maximum[1]	12.5	–	–
Neck AP diameter (M-5[2])	15.6	–	–
Neck ML diameter (M-4[2])	14.2	–	–
Neck circumference (M-5[4])	51.0	–	–
Head-neck length (M-1a)	38.2[2]	–	–
Neck-shaft angle (M-7)	(14°)	–	–
Tuberosity length (S-1)	29.4	–	–
Tuberosity breadth (S-4)	17.8	–	–
Tuberosity projection (S-8)	19.0	–	–
Shaft diameter at tuberosity	16.2	–	–
Tuberosity position[3]	2	–	–
Proximal AP diameter[4]	13.7	–	–
Proximal ML diameter[4]	14.6	–	–
Proximal circumference[4]	48.0	–	–
Midshaft AP diameter (M-5a)[5]	–	13.6	11.7
Midshaft ML diameter (M-4a)	–	17.1	14.1
Distal minimum circumference (M-3)	–	43.0	37.0
Maximum preserved length	82.0	127.0	93.8

[1] The lateral surface measurement includes only the proximodistally flat portion of the surface, whereas the maximum dimension extends to the plane of the proximal head.
[2] This is the distance to the proximodistal middle of the tuberosity; the measurement to the most prominent point on the tuberosity provides a head-neck length of 36.0 mm.
[3] Position 2 indicates that the interosseus crest is in line with the dorsal third of the tuberosity (Trinkaus and Churchill, 1988).
[4] Diameters and circumference taken at the midpoint between the tuberosity and the proximal extent of the interosseus crest.
[5] The positions of midshaft estimated based on interosseus crest morphology.

pattern of greater *M. biceps brachii* power for Neandertals in general, although the differences between the Late Pleistocene samples are non-significant. The Mladeč 25a value, however, is well above the comparative sample means, exceeded only by those for Dolní Věstonice 14, La Ferrassie 2, and Shanidar 6 and 8. The three Neandertals with high values are all small females with small radial heads; however, both Dolní Věstonice 14 and especially Mladeč 25a have rather large radial heads, emphasizing their combinations of long radial necks and especially large radial tuberosities.

The estimated Mladeč 25a neck-shaft angle is close to the Neandertal mean value and about a standard deviation above the earlier Upper Paleolithic mean. Among the latter sample, it is exceeded by Cro-Magnon 4303 and Paviland 1, and it therefore remains within the earlier Upper Paleolithic distribution.

The Mladeč 25a radial tuberosity is large and prominent, especially on its dorsal margin. It is moderately rotated anteriorly, such that its dorsal third is in line with the interosseus crest (position 2 of Trinkaus and Churchill, 1988). This arrangement is found in all three comparative samples, but there is a highly significant shift between Neandertals and earlier Upper Paleolithic humans in their frequency distributions of radial tuberosity orientation. The Neandertals have mainly the more directly medial orientation, including the initial Upper Paleolithic Vindija 13.8 radius but not the Saint Cé-

Table 14. Comparative osteometrics for the Mladeč radii. For comparative samples with N > 4, the mean, standard deviation and N are provided. For continuous variables, the ANOVA P-value is provided between the comparative samples for head diameter and neck-shaft angle; for indices and tuberosity position, the Kruskal-Wallis P-value between these samples is provided. * P < 0.05 with a multiple comparison correction (α/7)

	Mladeč 25	Mladeč 26	Neandertals	Qafzeh-Skhul	Earlier Upper Paleolithic	Comparative P-value
Head AP diameter (mm)	25.2	-	21.6 ± 2.4 (13)	20.2, 22.0, 24.8	23.1 ± 1.4 (14)	0.160
Head/neck AP index	161.5	-	178.2 ± 13.4 (13)	152.3, 165.6, 184.9	166.0 ± 16.5 (8)	0.297
Neck AP/ML index	109.9	-	114.4 ± 7.6 (14)	90.2, 115.1, 119.5	110.6 ± 8.1 (11)	0.349
Tuberosity/ head index[1]	107.7	-	99.3 ± 8.2 (12)	93.5, 99.4	94.0 ± 8.1 (12)	0.228
Neck-shaft angle	14°	-	14.4° ± 1.1° (12)	-	11.1° ± 3.1° (8)	0.003*
Tuberosity position	2	-	1: 3.6% 2: 28.6% 3: 67.9% (14)	1: 25.0% 2: 50.0% 3: 25.0% (4)	1: 21.2% 2: 78.8% 3: 0.0% (26)	<0.001*
Midshaft AP/ML index	79.5	83.0	77.1 ± 3.3 (13)	78.9, 92.5, 92.9	79.1 ± 5.4 (12)	0.091

[1] Tuberosity / head index = (tuberosity length x tuberosity breadth x head-neck length)$^{1/3}$ / head AP diameter (x 100).

saire 1 specimen. All of the earlier Upper Paleolithic ones are anteromedial to some degree (Table 14); the Qafzeh-Skhul sample is intermediate. The position of the Mladeč 25a tuberosity is not diagnostic of one group or the other, even though it is found more frequently among the early modern humans.

Both the Mladeč 25b and 26 diaphyses are tear-dropped shaped with small sulci on either side of the interosseus crest; there are no concavities elsewhere on the diaphyses. As such, they contrast with some early modern human radial diaphyses and are within the range of overlap between Neandertal and early modern humans in radial diaphyseal cross-sectional shape. There is little difference between the comparative samples in relative anteroposterior versus mediolateral midshaft proportions (except for a couple of high values in the Qafzeh-Skhul sample), and the two Mladeč radial diaphyses fall comfortably within the distributions (Table 14).

The Mladeč 31 metacarpal bone

Preservation and age-at-death

The Mladeč 31 metacarpal bone presents most of a right immature third metacarpal (chap. 8, Plate XVI). The shaft is complete, the base is largely covered with carbonate crust and sustained abrasion to the palmar surface and the styloid process, and the distal metaphysis is preserved dorsally and palmar-ulnarly. The maximum preserved length is 69.1 mm. The complete absence of head epiphysis fusion indicates an age-at-death less than 14 to 16 years (Scheuer and Black, 2000).

Morphology

The diaphysis presents very weakly marked lines for the dorsal interosseus muscles. They are gentle angles proximally and become flat distally. The extent of the projection of the styloid process is unknown, but the capitate facet curves clearly up onto the base of the styloid process, suggesting that it was projecting beyond the plane of the capitate facet. The capitate facet is turned about 10° ulnarly relative to the perpendicular to the diaphyseal axis. The adjacent metacarpal facets are partially obscured by matrix. However, the metacarpal 2 facet is large and dorsopalmarly concave. Palmarly it is 7.0 mm proximodistally, and dorsally it is ca. 5.0 mm proximodistally. Ulnarly, there are two facets for the metacarpal 4, one dorsal and one palmar with a clear sulcus between them.

Base and midshaft dimensions are provided in Tables 12 and 15. Articular length (mid capitate surface to the most distal point on the head) was estimated using a least squares regression based on recent human mature third metacarpals from the preserved length (65 mm) from the mid-capitate facet to the dorsal epiphyseal margin ($ArtLen = 1.15 \times PresLen - 0.75$; $r^2 = 0.952$, $N = 34$). The resultant value is 74.0 mm. On the basis of this, a robusticity index (geometric mean of the midshaft diameters/articular length) was computed as 11.3 for Mladeč 31. This value is relatively low compared to other Late Pleistocene sample values, which do not differ significantly between them (Kruskal-Wallis $P = 0.135$) (Table 16). Among the earlier Upper Paleolithic specimens, it is approached only by the value of 11.5 for the female Dolní Věstonice 3. The modest value for Mladeč 31 may be influenced by its relatively long articular length, which is 3.3 standard deviations from the Neandertal mean but only 1.5 standard deviations from the earlier Upper Paleolithic mean length. However, the similarly long Barma Grande 2 third metacarpals (75.4 mm) have an average robusticity index of 13.9, suggesting that either Mladeč 31 is simply gracile or the immature status of the bone is affecting its degree of diaphyseal hypertrophy.

Table 15. Osteometric measurements of the Mladeč 31 metacarpal 3 (in mm and degrees)

Proximal dorsal articular breadth	12.7
Proximal palmar articular breadth	8.4
Horizontal base angle (ulnar deviation)	80°
Midshaft height	8.1
Midshaft breadth	8.6

Table 16. Osteometric comparisons for the Mladeč 31 and 90 third metacarpals. Mean and standard deviation (N) are provided for samples greater than five; individual values provided for smaller samples. Robusticity index = (geometric mean of midshaft diameters / articular length) x 100

	Articular length (mm)	Robusticity index
Mladeč 31	74.0	11.3
Mladeč 90	77.8	12.8
Neandertals	61.9 ± 3.7 (6)	13.3 ± 1.0 (6)
Qafzeh-Skhul	63.1, 65.6	13.7, 16.7
Earlier Upper Paleolithic	67.0 ± 4.7 (8)	13.0 ± 0.9 (7)

The Mladeč 88 to 91 hand remains

Preservation

Since they derive from Chamber E, rather than Chamber D, and may therefore not be associated with, or the same age as, the other Mladeč postcranial remains, the Mladeč 88 to 91 hand bone (see chap. 8, Plate XVIII) are described here, separately from the Mladeč 31 metacarpal bone. Measurements are in Table 17.

Mladeč 90 is a right metacarpal 3, which is complete except for the radial half of the head and the adjoining shaft. Its preserved length equals its anatomical maximum length of 81.9 mm. The articular surfaces are normal, and there is no evidence of abnormalities. The Mladeč 91 metacarpal bone is from the right fourth ray. Both articular ends and the associated metaphyseal areas are missing, and thus it cannot be determined if this is an adult specimen. However, the flares for both epiphyses and the complete diaphysis are present. The total preserved length is 44.5 mm.

The Mladeč 88 phalanx is a partial proximal hand phalanx, which lacks part of the distal shaft and all of the head. Other damage includes postmortem chipping on the marginal ridges, damage to the dorsal left base, and slight damage to the dorsal right surface adjacent to the metacarpal facet. Total length of the preserved specimen is 45 mm. It probably derives from the second ray, based on the relatively broad expansion of the base for the interosseus muscles. Given the greater expansion of the base on the right side (presumably for the first dorsal interosseus muscle), the specimen is probably left.

The Mladeč 89 proximal hand phalanx is complete except for some minor postmortem chipping along the well-developed palmar marginal ridges and slightly more damage on the dorsal aspect of the proximal articular rim. There are no abnormalities. Preserved maximum length is 49 mm. The specimen is most likely from the third digit, given the largely, but not entirely, symmetrical development of the radial and ulnar proximal tubercles and its relatively broad shaft. The moderately greater right basal tubercle suggests that it is left.

The Mladeč 88, 89 and 90 bones are generally similar in relative overall size and musculoligamentous markings, and they may derive from the same individual. The Mladeč 91 metacarpal 4 is smaller and more gracile and therefore probably represents a second individual. An assessment of the possible association of the first three hand bones can be made by comparing the lengths

Table 17. Osteometric measurements of the Mladeč 88 to 91 metacarpals and proximal phalanges (in mm). Estimated values are in parentheses

	Mladeč 90 Metacarpal 3	Mladeč 91 Metacarpal 4	Mladeč 88 Proximal Phalanx 2	Mladeč 89 Proximal Phalanx 3
Articular length	77.8	–	–	47.5
Midshaft height	10.5	7.8	8.8	8.1
Midshaft breadth	9.4	7.3	10.9	10.0
Proximal max. height	19.6	–	14.1	(13.0)
Proximal max. breadth	15.9	–	18.2	16.9
Proximal artic. height	–	–	12.0	11.2
Proximal artic. breadth	15.9	–	14.5	13.4
Styloid projection	4.1			
Distal height	16.4	–	–	9.1
Distal maximum breadth	–	–	–	13.0
Distal articular breadth	–	–	–	12.0

of the Mladeč 89 and 90 third ray proximal phalanx and metacarpal. The resultant index is 61.1; this value is low for a recent European sample (68.1 ± 2.3, N = 38), being 3.04 standard deviations from the recent human mean. Yet, it is only close to the lower limit of a European Gravettian sample (65.8 ± 3.6, N = 11); it is 1.31 standard deviations below that mean, matched by the value of 61.0 for Dolní Věstonice 13, and above the index of 59.6 for Paglicci 25. A small Neandertal sample has similar values (66.0 ± 2.6, N = 5), and the Qafzeh 9 index (71.6) is near the top of the Late Pleistocene range. It may be that the Mladeč remains, like Neandertals (Villemeur, 1994) and apparently also earlier Upper Paleolithic humans, have more abbreviated ulnar proximal phalanges than at least recent Europeans. Consequently, the combination of Mladeč 89 and 90 based on their lengths would be moderately unusual for a recent human, but well within the ranges of variation of Late Pleistocene humans. The hypothesis that these three bones represent one individual is thus not rejected.

Morphology

As with the Mladeč 31 metacarpal, the Mladeč 90 bone is long. Its articular length of 77.8 mm is 2.3 standard deviations above an earlier Upper Paleolithic mean (Table 16) and 4.3 standard deviations from a Neandertal mean. It is therefore slightly longer than those of Mladeč 31 and even Barma Grande 2. Its diaphysis has a relatively smooth dorsal surface, but the palmar crest is strongly marked. A robusticity index, between the geometric mean of the midshaft diameters and the articular length, is 12.8, which is very close to the earlier Upper Paleolithic mean and within one standard deviation of a Neandertal sample. Therefore, despite its length, it is not particularly gracile.

Proximally, Mladeč 90 has a minimally transversely concave capitate facet and a moderately projecting styloid process. The ratio of its styloid projection from the mid-carpal surface (4.1 mm) to its articular length of 77.8 mm is 5.27. This index is similar to the values of a Neandertal sample (4.94 ± 1.83, N = 8) and well within the range of variation of a small earlier Upper Paleolithic sample (4.09 ± 1.22, N = 5); the Qafzeh specimens have very small styloid processes (Qafzeh 9: 1.49 mm). Neandertal third metacarpals have relatively small styloid processes compared to recent humans (recent Euroamericans: 7.76 ± 1.71 mm, N = 30; see Niewoehner et al., 1997); it is apparent from these comparisons that early modern humans share this archaic pattern. Mladeč 90 is similar to both of these Late Pleistocene samples.

The Mladeč 91 fourth metacarpal retains essentially only its diaphysis, which has weakly developed dorsal interosseus crests and a sharp palmar crest. There is little else of note on it.

The Mladeč 88 and 89 proximal hand phalanges are notable for the degree of hypertyrophy of the palmar crests for the flexor tendon sheaths. They are prominent on both bones, but especially on the Mladeč 88 proximal phalanx 2. Both of their bases turn slightly palmarly, and the head of Mladeč 89 is deviated slightly radially (assuming that it is left). The completeness of the Mladeč 89 phalanx permits a robusticity index (geometric mean of the midshaft diameters/articular length) comparison. The value is 18.9 for Mladeč 89, which is close to those of Qafzeh 3 and 9 (19.2 and 19.6) and an earlier Upper Paleolithic mean (19.4 ± 1.3, N = 7). It is, however, 2.13 standard deviations below the mean of a Neandertal sample (22.1 ± 1.5, N = 9) and below the range of that sample. Given the similarities of third ray proximal phalanx to metacarpal length proportions across these samples, this suggests a decrease in phalangeal robusticity between these late archaic and early modern human samples (Kruskal-Wallis P-value without Mladeč 89 = 0.002; P = 0.001 with Mladeč 89 in the earlier Upper Paleolithic sample).

The Mladeč 21 ands 22 pelvic remains

Preservation, age and sex

The Mladeč 21 and 22 pelvic remains consist of two partial os coxae (innominate bones). Mladeč 21 is a mature partial left ilium and ischium, and Mladeč 22 is an immature partial right ischium. They derive from two individuals, based on size, morphology and developmental status.

Mladeč 21 is a partially preserved adult left os coxae (chap. 8, Plate XII). It retains the posterior half of the acetabulum, including most of the lunate articular surface (except for the anterosuperior portion) and the acetabular notch. The proximal portion of the ischial body and tuberosity, the greater sciatic notch, and the sacroiliac surface are present and undistorted.

The acetabulum is broken away superiorly and anteriorly, and the adjacent iliopectineal line and the entire pubis are missing. An ischial segment extends from the inferior rim of the acetabulum for about 40 mm and includes a 29.5 mm segment of ischial tuberosity. The superior and posterolateral borders of the ischial tuberosity are the only ones remaining. A segment of the obturator foramen rim is preserved from the inferior extent of the acetabular rim to the most superior extent of the root of the ischiopubic ramus. The posterior surface of the ischium is not preserved, except for the inferior base of the ischial spine. The lesser sciatic notch is not preserved; however, a wide greater sciatic notch is present and the medial (internal) aspect of the ilium preserves the arcuate buttress and line extending toward the sacroiliac articular surface. Inferior to the sacroiliac articular surface lies a deep and wide preauricular sulcus. The edges of the preauricular sulcus and the entire sacroiliac articular surface are obscured by the encrusting matrix, which adheres to most of this specimen. The posterior inferior iliac spine and the base of the iliac tubercle are present, but the posterior superior iliac spine and all of the iliac crest are missing in addition to the entire ilium anterosuperior of the level of the arcuate line and sacroiliac articular surface.

Although nothing remains of the iliac crest, it is apparent that Mladeč 21 had attained an adult skeletal age. No traces of epiphyseal lines occur in the acetabulum, indicating that the primary elements of the os coxae are completely fused, and the ischial tuberosity is fused to the body of the ischium. According to recent human standards, fusion of the ischial tuberosity occurs approximately between 19 and 20 years of age (Johnston and Zimmer, 1989; Scheuer and Black, 2000). Therefore, it is unlikely that the age of Mladeč 21 is younger than this.

Two features strongly suggest that Mladeč 21 is a female based on the sexually dimorphic characteristics of modern human pelves (Brůžek, 2002). First, the greater sciatic notch of Mladeč 21 is wide and open (e.g., the distance from the posterior inferior iliac spine to the ischial spine is relatively great) and, second, there is a wide, deep preauricular sulcus. The first is documented by measurements (Table 18), in which the subtense from the chord between the posterior inferior iliac spine and the dorsal ischium is only 25.5% of that chord, and the point of the maximum subtense is 72.3% of the distance from the posterior inferior iliac spine to the ischium. The precise form of the Mladeč 21 preauricular sulcus is obscured by encrustations, but it appears to be relatively shallow superiorly and then becomes a deeper groove that rounds inferiorly adjacent to the posterior inferior iliac spine. Its form may occur in males or females as a result of tension on the sacroiliac ligaments (a paraglenoid sulcus), but the depth of the sulcus suggests the bony resorption through pregnancy that is diagnostic of its being female. However, the degree of openness of the greater sciatic notch would be very unusual in a male pelvis.

Mladeč 22 is a right os coxae preserving a mostly complete acetabulum, most of the ischial body, and an anteroinferior segment of the ilium. The acetabulum is complete except for the anterosuperior-most segment of the lunate articular surface and the anterior-most margin of the acetabular notch. The rim of the acetabulum is well preserved except for its posterosuperior margin, which is abraded. The specimen is broken anterior to the acetabulum, and nothing remains of the iliopubic eminence, the pubic body and symphysis, or the ischiopubic ramus.

Inferior to the acetabulum, the ischial body is nearly complete and lacks only a small postero-medial portion including the tip of the ischial spine, although the lesser sciatic notch is visible just below it. The ischial tuberosity metaphysis is well preserved with only slight surface abrasion to the rims, but none of the epiphysis remains. The posterior surface of the obturator foramen rim and the root of the ischiopubic ramus are present.

Superior to the acetabulum, a small portion of the ilium is present including the anterior inferior iliac spine. The ilium extends cranially from the superior rim of the acetabulum for 55 mm. A well-developed iliopsoas groove separates the anterior inferior iliac spine from the arcuate buttress on the medial surface of the bone. The preserved segment of arcuate buttress and line is 44 mm long. Posterior to the preserved segment of arcuate buttress and line and superior to the anterior inferior iliac spine, nothing remains of the ilium.

The subadult age of Mladeč 22 is apparent in its epiphyseal union status. The ilium, ischium and pubis have undergone complete fusion, and there are no signs of epiphyseal lines in the acetabulum. The ridges and furrows present on the Mladeč 22 ischium indicate that the ischial tuberosity epiphysis had not united with the ischial body. According to recent human standards (Johnston and Zimmer, 1989), the age of Mladeč 22 would be between 13 to 15 (based on the fused primary elements of the os coxae) and 19 to 20 years of age (based on the unfused ischial tuberosity epiphysis). Unlike Mladeč 21, there are no morphological indicators of sex preserved on the Mladeč 22 specimen. However, the large size of the acetabulum, ischial body and anterior inferior iliac spine strongly suggest that this specimen is male.

Morphology

Despite the series of measurements that are possible on the Mladeč 21 and 22 os coxae (Table 18), there is little Late Pleistocene comparative data available for these measurements. Moreover, the partially encrusted state of the bones obscures many of the finer surface details.

Overall size is best indicated by the acetabular heights of these two specimens. To provide a comparative sample, acetabular height was measured (or taken from the literature) for thirteen earlier Upper Paleolithic specimens, three Qafzeh-Skhul specimens, and four Neandertals. To maximize the sample size, acetabular height was predicted for an additional fourteen earlier Upper Paleolithic specimens, two Qafzeh-Skhul individuals, and five Neandertals from their sagittal femoral head diameters. This was done using a least squares regression based on the pooled Late Pleistocene sample of associated femoral heads and acetabulae ($AcetHt = 1.30 \times FemHd - 6.5$, $r^2 = 0.932$, $N = 13$). The resultant samples (Table 19) provide a modest difference across the three pooled-sex comparative samples (ANOVA $P = 0.079$), which is in agreement with the slightly higher predicted body masses for the Neandertals (Ruff et al., 1997).

The articular height of Mladeč 21 (ca. 53 mm) falls very close to the pooled sex means of the two early modern human samples, being slightly below them. It is slightly above the mean for the earlier Upper Paleolithic female sample, and at the top of the very small Middle Paleolithic female range, being matched by La Ferrassie 2. However, it is exceeded by the predicted values (55.4 mm each) for the Předmostí 4 and 10 Gravettian female remains.

The higher value of 59.8 mm for Mladeč 22 is slightly above all of the male comparative means for acetabular height, but it is within one standard deviation of the Neandertal and earlier Upper Paleolithic mean values and is only relatively high compared to the Qafzeh-Skhul sample. Among earlier Upper Paleolithic males, it is matched by the value predicted for Sunghir 1 (59.9 mm) and exceeded by the predicted value of Barma Grande 2 (61.8 mm) and the measured diameter of Fanciulli 4 (65.3 mm). Moreover, the Neandertal and earlier Upper Paleolithic male means are not significantly

Table 18. Osteometric dimensions of the Mladeč 21 left os coxae and the Mladeč 22 right os coxae (in mm). Values in parentheses are estimated

	Mladeč 21	Mladeč 22
Acetabulum height (articular)	(53.0)	59.8
Acetabulum height (capsule)	-	61.0
Acetabulum depth (articular)	-	27.5
Acetabulum depth (capsule)	-	30.5
Lunate surface superior minimum diameter	-	28.0
Lunate surface lateral minimum diameter	20.0	22.9
Lunate surface lateral maximum diameter	(23.0)	-
Lunate surface inferior minimum diameter	22.0	22.7
Lunate surface inferior maximum diameter	23.2	26.1
Acetabulum center to closest point on auricular surface	(70.0)	-
Acetabulum center to tip of anterior inferior iliac spine	-	62.6
Minimum inferior iliac thickness	22.2	-
Iliac arcuate line chord[1]	56.6	-
Iliac arcuate line subtense	9.1	-
Greater sciatic notch: dorsal ischium to post. inf. iliac spine	47.0	-
Greater sciatic notch: superior notch to post. inf. iliac spine	34.0	-
Greater sciatic notch: notch chord to superior notch	12.0	-
Acetabulum center to ischial tuberosity angle	(65.5)	77.0
Acetabulum center to ischial tuberosity center	-	63.0
Acetabular rim to closest point on ischial tuberosity	10.5	14.7
Ischial tuberosity breadth	(28.0)	(29.3)
Ischial spine to posterior inferior iliac spine	(71.0)	-
Acetabulosciatic breadth	34.5	36.0
Auricular surface breadth	(53.0)	-
Maximum preserved dimension	199.0	160.3

[1] Chord distance and maximum subtense along the iliac arcuate line from the anterior auricular surface to the point opposite the anterior inferior iliac spine (Ruff, 1995).

Table 19. Comparative acetabular height measurements. Mean and standard deviation (N) for samples greater than four; individual measurements for smaller samples

	Pooled sex	Male	Female
Neandertals	58.6 ± 3.5 (9)	59.3 ± 2.9 (8)	53.0
Qafzeh-Skhul	54.7 ± 2.4 (5)	54.1, 55.3, 56.3, 57.2	50.8
Earlier Upper Paleolithic	54.8 ± 4.7 (27)	57.8 ± 3.2 (15)	51.5 ± 3.4 (10)

Male earlier Upper Paleolithic versus Neandertal P = 0.279

different (t-test P = 0.279). However, the acetabular height for Mladeč 22 would be unusual for a Late Pleistocene female, since it is 2.44 standard deviations from the earlier Upper Paleolithic female mean; this supports its diagnosis as male.

Consequently, even though the acetabular heights of Mladeč 21 and 22 are moderately large, they are by no means unusual for either a Neandertal or an earlier Upper Paleolithic individual of the same sex. These acetabular dimensions, contra Wolpoff (1989), do not serve to align the Mladeč 21 and 22 with either the late archaic or the early modern human Late Pleistocene European samples.

The *M. obturator internus* sulcus is not preserved on Mladeč 21, but the ischial tuberosity is strongly rotated ventrally. The Mladeč 22 *M. obturator internus* groove is not evident, but the ischial spine is at the same vertical level as the superior margin of the superior ischial tuberosity metaphysis. Its ischial tuberosity is also strongly rotated ventrally.

The Mladeč 22 anterior inferior iliac spine is prominent and vertically straight, with only a slight internal concavity. The adjacent area of the acetabular rim for the insertion of the reflected head of *M. rectus femoris* is crushed and provides no evidence of its original configuration.

The break of the Mladeč 22 right ilium exposes the base of the iliac pillar (or acetabulocristal buttress), which has a total internal-external thickness of 14.1 mm and internal and external cortical thicknesses of 2.3 mm and 3.8 mm, respectively. These values are very close to the mean values for a recent European sample (2.4 ± 0.5 mm and 3.9 ± 0.6 mm, respectively, N = 23) provided by Stringer (1986). Comparative Late Pleistocene data are unavailable, but radiographic observations and inspection of fossilization breaks of both Neandertal and earlier Upper Paleolithic ilia indicate that they generally have iliac cortical thicknesses similar to those of recent humans.

On the dorsal surface of the posterior Mladeč 21 ilium, there is a tubercle opposite the auricular surface, which extends as a raised crest to the area of the iliac tuberosity.

The Mladeč 27 and 28 femora

Preservation and maturity

Mladeč 27 is a well preserved right femoral diaphysis (chap. 8, Plate XV). Proximally, the distal base of the lesser trochanter is observable, but the head, neck, greater trochanter, intertrochanteric region and lesser trochanter are absent. Much of the popliteal surface is visible at the distal metaphysis, but nothing remains of the condyles, epicondyles or patellar articular surface. The most distal extent is posterolateral, just proximal of the capsular attachment above the dorsal lateral condyle. The entire length of the diaphysis is complete and in excellent condition, but much of the bone is covered by matrix that is quite thick and, in some areas, globular in form. Despite this encrustation, all of the major muscle attachment sites are visible, and the original subperiosteal contour can be discerned without difficulty. There are no rejoined breaks in the bone and no evidence of distortion. There are no indications as to the maturity of the specimen, since all epiphyses/metaphyses are absent, but the size and general morphology of the diaphysis (see below) suggest minimally a late adolescent age and probably a fully mature status.

The Mladeč 28 femur is less complete but less encrusted (chap. 8, Plate XIV). The specimen consists of a 198 mm long proximal segment of an adult left femur. Unlike the other fossil remains in the Mladeč sample, no matrix adheres to this specimen. The head is absent, as are most of the anterior, superior and posterior aspects of the medial femoral neck. The inferior contour of the lateral neck is preserved and extends uninterrupted into the medial surface of the proximal shaft. Superiorly, a short (ca. 10 mm) portion of the lateral neck is preserved along with the adjacent greater trochanter and trochanteric fossa. The greater trochanter, however, is not completely preserved, since much of the lateral and posterior surfaces are broken away.

The intertrochanteric crest is missing and only the base of the lesser trochanter is preserved. This leaves a band of exposed trabecular bone extending superiorly and laterally from the lesser trochanter to the anterosuperior margin of the lateral greater trochanter and continuing medially across the anterior surface of the bone to the femoral neck. There is slight loss of cortical bone on the third trochanter, and the diaphysis is preserved for 102 mm below the distal base of the lesser trochanter.

The distal break is within the proximal half of the diaphysis, clearly proximal of midshaft. The specimen therefore does not provide midshaft measurements, and those provided by Szombathy (1925) are inaccurate. The cortical bone and medullary canal are visible at the distal fracture, which is flat and horizontally disposed anteriorly, thus forming a right angle with the anterior surface of the diaphysis. The fracture is somewhat more irregular posteriorly and exhibits a slight posteroinferior bevel.

The full maturity of Mladeč 28 is indicated by the complete obliteration of the epiphyseal fusion lines for the trochanteric epiphyses.

Femoral length estimation

The Mladeč 27 femur is sufficiently complete to provide a reasonable estimate of its original biomechanical and interarticular lengths (see notes to Table 20). The distance between the distal margin of the lesser trochanter and the posterolateral supracondylar margin is 322 mm. This value may slightly

Table 20. Osteometric dimensions of the Mladeč 27 right femur, Mladeč 28 left femur and Mladeč 102 immature left femur (in mm)

	Mladeč 27	Mladeč 28	Mladeč 102
Maximum length (M-1)[1]	485.7	–	–
Bicondylar length (M-2)[2]	483.3	–	–
Biomechanical length[3]	454.7	–	–
Proximal sagittal diameter (M-10)	25.5	23.5	12.6
Proximal transverse diameter (M-9)	30.0	33.8	15.4
Proximal circumference	89.0	89.0	–
Midshaft sagittal diameter (M-6)	28.3	–	(13.4)[4]
Midshaft transverse diameter (M-7)	24.0	–	(14.0)[4]
Midshaft circumference (M-8)	83.0	–	–
Gluteal tuberosity breadth	8.0	12.5	–
Hypotrochanteric fossa	absent	absent	–
Third trochanter	–	present	–
Anterior curvature chord	301.0	–	–
Anterior curvature subtense	12.0	–	–
Anterior curvature position (to proximal)	123.0		
Maximum preserved length	338.0	198.0	53.0

[1] Maximum length estimated from the bicondylar length using a least squares regression based on recent humans (MaxLen = 0.98 × BicLen + 9.9, r^2 = 0.996, N = 50).

[2] Bicondylar length estimated from the biomechanical length using a least squares regression based on earlier Upper Paleolithic humans (BicLen = 1.02 × BiomLen + 21.0, r^2 = 0.992, N = 13).

[3] Biomechanical length (average distance parallel to the diaphyseal axis between each distal condyle and the proximal neck just medial of the greater trochanter) estimated from the distance between the distal lesser trochanter and the lateral supracondylar margin (BiomLen = 1.23 × PresLen + 57.3, r^2 = 0.945, N = 50), SEest = 1.2 mm.

[4] The Mladeč 102 "midshaft" diameters may be taken slightly proximal of midshaft. This is unlikely to greatly affect the values, given the near circularity of immature femora and the absence of a pilaster. Contrary to previous assessments, the Mladeč 28 femur is not sufficiently preserved toward midshaft to provide reliable midshaft diameters.

underestimate the length of the bone, since the distal landmark may be 1–2 mm from the actual supracondylar margin; any such underestimation is minimal, and it is conservative in the comparisons below, since it will tend to accentuate, trivially, the perceived robusticity of the diaphysis. This value was used to estimate biomechanical length using a least squares regression based on recent human femora, from which bicondylar and maximum length were then estimated (see Table 20).

The length of the Mladeč 28 femur cannot be reliably estimated for more than general size comparisons. However, it is possible to locate the 80% diaphyseal cross section morphologically near the middle of the proximal posterolateral gluteal buttress. This position is 87 mm distal of the proximal end of the biomechanical length (on the superior neck just medial of the greater trochanter), which provides a very approximate biomechanical length of 435 mm (and a bicondylar length of 465 mm). Note that placing the 80% section 5 mm more distal would increase the estimated biomechanical length to 460 mm (and the bicondylar length to 490 mm). The differences in these length estimates have little effect on the positions and the biomechanical implications of the 80% and 65% diaphyseal cross sections compared below, but they are sufficient to provide a general indication of the original length of the Mladeč 28 femur; it was probably close to that of Mladeč 27.

Length comparisons

The estimated bicondylar length of Mladeč 27 of ca. 483 mm places it well above the female values for all three of the comparative samples, and it is 2.33 standard deviations above the mean of the earlier Upper Paleolithic female mean (Table 21). However, it falls well within the male ranges of variation of both the earlier Upper Paleolithic and the Qafzeh-Skhul samples. It remains just over two standard deviations (2.04) from the Neandertal male mean; it is approximately the same as the high outlier in femoral length among the Neandertals, Amud 1. The roughly estimated values for the original length of the Mladeč 28 femur would make it either a reasonably average early modern human male in size or a relatively large female.

Table 21. Comparisons of femoral bicondylar lengths. Mean and standard deviation (N) for samples greater than four; individual measurements for smaller samples

	Pooled sex	Males	Females
Neandertals	436.8 ± 26.1 (12)	445.0 ± 18.8 (10)	384.0, 407.0
Qafzeh-Skhul	474.0 ± 29.9 (7)	475.0, 486.0, 490.0, 515.0	417.8, 464.8, 469.5
Earlier Upper Paleolithic	456.7 ± 33.6 (25)	478.6 ± 21.1 (13)	428.3 ± 23.6 (10)

Body mass estimation for Mladeč 27

The appropriate scaling of weight-bearing limbs for assessments of diaphyseal robusticity and muscle hypertrophy requires an estimation of body mass for each specimen, since the baseline load on the diaphysis or muscle is body mass times the beam length around which the load is operating. For this reason, measures of cortical area, which reflect resistance to axial loading, should be compared to body mass, and measures of muscle hypertrophy or diaphyseal bending strength should be compared to body mass times bone length (Ruff et al., 1993; Trinkaus and Ruff, 2000).

Such considerations of the effects of body mass are especially important for comparisons across samples which vary in body proportions, as did the Neandertals and early modern humans of the northwestern Old World (Trinkaus, 1981; Ruff, 1994; Holliday, 1997; 2000). In cases in which there is significant variation in body shape and relative limb length, comparisons using only bone length will overestimate hypertrophy for stockier individuals and underestimate skeletal hypertrophy for linear individuals. In fact, although earlier analyses using only bone length for size standardization "documented" a significant reduction in femoral robusticity with the emergence of modern humans (e.g., Twiesselmann, 1961; Trinkaus, 1976), assessments combining the effects of differential body mass to lower limb bone length proportions (e.g., Ruff et al., 1993; Trinkaus and Ruff, 1999a, b; Trinkaus et al., 1999b; Trinkaus and Rhoads, 1999; see Trinkaus, 2000; Ruff et al., 2000) have shown that there was little change in femoral (or tibial) diaphyseal, muscular or articular hypertrophy through the Late Pleistocene.

Body mass for the comparative samples was estimated following Ruff et al. (1997) in which body mass was predicted from recent human samples using estimated stature and bi-iliac breadth. Stature was estimated using ecogeographically appropriate, and sex-specific when known, formulae from Trotter and Gleser (1952); male and female estimates were averaged for individuals of unknown sex. Bi-iliac breadth was measured when available (see Holliday, 1995). Otherwise, it was estimated from either a pooled-sex sample from the same group when available (for the earlier Upper Paleolithic sample) or from the most complete specimen providing limb length and bi-iliac breadth (Kebara 2 for the Neandertals and Skhul 4 for the Qafzeh-Skhul sample).

To provide a body mass estimate for Mladeč 27, whose body proportions are unknown and whose inferred body proportions could provide information on its morphological affinities (Holliday, 1997), it was modeled as an average Neandertal, Qafzeh-Skhul individual and earlier Upper Paleolithic individual. The resultant values for body mass are 69.4 kg for the earlier Upper Paleolithic model, 69.7 kg for the Qafzeh-Skhul one, and 83.5 kg for the Neandertal one. Since the Qafzeh-Skhul model provides almost the same value as the earlier Upper Paleolithic one, it was not included in the comparisons. Consequently, in the comparisons of femoral properties of Mladeč 27 that involve body mass estimation, two data points are provided, one for it with stocky Neandertal body proportions and one with it having earlier Upper Paleolithic linear body proportions.

Morphology

A portion of the Mladeč 28 anterolateral greater trochanteric region is preserved, and the *M. gluteus minimus* insertion is moderately rugose. The medial aspect of the greater trochanter exhibits a well-developed trochanteric fossa for the insertion of *M. obturator externus* and *internus*. The anterior surface of the proximal shaft is smoothly concave.

On the posterolateral surface of Mladeč 28, lateral to the preserved base of the lesser trochanter, there is a well-developed and distinct gluteal tuberosity, for the insertion of *M. gluteus maximus*. At the superior extent of the gluteal tuberosity, there is an oval protuberance of bone, a well-developed third trochanter (*sensu* Hrdlička, 1937), which marks the superior extent of the gluteal tuberosity. From the third trochanter, a gluteal ridge extends mediodistally for a distance of ca. 81 mm until it merges with the lateral margin of the *linea aspera*. The gluteal tuberosity is broad (see below) but not at all rugose, in contrast to those seen on most Neandertal femora and the Cro-Magnon ones. It is bordered laterally by a prominent, but blunt, proximolateral gluteal buttress, or flange. There is a clear sulcus between the gluteal tuberosity and the full lateral extent of the gluteal buttress, and the buttress is separated from the anterior diaphyseal convexity by a broad and shallow sulcus. The pectineal area is smooth, and there is a faint spiral line anteriorly, that is evident only adjacent to the lesser trochanter.

The Mladeč 27 proximolateral femoral diaphysis exhibits far less development of the gluteal buttress and a more modest gluteal tuberosity. There is no anterior sulcus between the buttress and the anterior diaphysis, and the posterior one is modest. Its gluteal tuberosity is partially obscured by matrix, but it is clearly modest in size (see below) and rugosity. There is a clear pectineal crest and a faint but apparent spiral line. These two femora from Mladeč contrast strongly in their subtrochanteric configurations, in terms of both diaphyseal cross-sectional shape (reflecting in large part the relative development of the gluteal buttress) and gluteal tuberosity size.

The relative development of the gluteal buttress is reflected in the cross-sectional diaphyseal proportions at the subtrochanteric (or meric) level (Table 20). In the comparison of the external diaphyseal diameters [through the meric index (Tables 20 and 24) and graphically (Fig. 8)], there is a general pattern in which the earlier Upper Paleolithic specimens have proportionately greater mediolateral diameters than most of the Neandertals, and the Qafzeh-Skhul specimens fall generally with the rounder diaphyses of the Neandertals (the high outlier is Skhul 5). In this comparison, the measurements are normally taken such that the "mediolateral diameter" is the maximum external diameter of the cross section across the gluteal buttress; as a result of the variable but ubiquitous anteversion (or torsion) of the femoral head and neck of these femora (Twiesselmann, 1961; Sládek et al., 2000), the "mediolateral diameter" is actually slightly anteromedial to posterolateral on most specimens. The "anteroposterior diameter" is then taken perpendicular to the mediolateral one.

The proportions of the subtrochanteric region were also compared using second moments of area at the 80% level (Tables 22 and 23; Fig. 8). However, it is not appropriate to simply compare maximum versus minimum diameters at this level, since there are several Neandertal and early modern human femora in which the maximum second moment of area is oriented largely anteroposteriorly, whereas in most of the earlier Upper Paleolithic femora the maximum second moment of area is predominantly mediolateral (similar to the maximum external diameter). Consequently, even though they do not take into account the effects of anteversion on the proximal femoral diaphysis, anteroposterior and mediolateral second moments of area (oriented with respect to the midshaft and distal femur) are compared in Fig. 8. The resultant distribution provides less separation of the reference samples, although all of the Neandertals and all of the Qafzeh-Skhul specimens except Qafzeh 8 are along the upper (rounder with less of a gluteal buttress) half of the earlier Upper Paleolithic distribution. Mladeč 27 falls in at the edge of the Neandertal distribution and close to the early Upper Paleolithic mean (Table 25). Mladeč 28 is also close to the middle of the earlier Upper Paleolithic distribution, very distinct from the Neandertals and separate from most of the Qafzeh-Skhul specimens (Table 25). It is probable that, should the 80% second moment of area be oriented with respect to the anteversion angles of these femora, which is possible for neither the Mladeč femora nor many of the other Late Pleistocene femora given damage to their necks, that a greater separation of the Middle and Upper Paleolithic samples, and of Mladeč 27 and 28, similar to that seen in the external diameters, would be evident.

Table 22. Cross-sectional geometric properties of the Mladeč 27 femur. Areas in mm², second moments of area in mm⁴. 0% is distal

	20%	35%	50%	65%	80%
Total area (TA)	879.8	562.3	497.7	504.3	597.3
Cortical area (CA)	311.4	309.5	390.7	397.0	389.8
Medullary area (MA)	568.4	252.8	107.0	107.3	207.5
AP 2nd moment of area (I_x)	32413	25065	23588	20056	21088
ML 2nd moment of area (I_y)	40825	15758	15493	18636	29263
Max 2nd moment of area (I_{max})	41359	25077	24169	20676	29325
Min 2nd moment of area (I_{min})	31880	15746	14912	18015	21027
Polar moment of area (J)	73239	40823	39081	38692	50352

Table 23. Cross-sectional geometric properties of the Mladeč 28 and 102 femora and the Mladeč 32 metatarsal 3. Areas in mm², second moments of area in mm⁴. 0% is distal

	Mladeč 28 65%	Mladeč 28 80%	Mladeč 102 50%[1]	Mladeč 102 80%	Mladeč 32 50%
Total area (TA)	495.5	574.4	135.6	148.9	71.0
Cortical area (CA)	409.3	426.9	101.4	106.3	60.4
Medullary area (MA)	86.2	147.5	34.2	42.6	10.6
AP 2nd moment of area (I_x)	16469	21049	1375	1292	381
ML 2nd moment of area (I_y)	21937	32743	1398	2019	452
Max 2nd moment of area (I_{max})	22443	37190	1541	2021	499
Min 2nd moment of area (I_{min})	15963	16602	1232	1289	334
Polar moment of area (J)	38406	53792	2773	3310	833

[1] The midshaft cross-section of Mladeč 102 may be slightly proximal of midshaft.

Table 24. Diaphyseal indices for the Mladeč and comparative sample femora. Mean and standard deviation (N) for samples greater than four; individual measurements for smaller samples

	Meric index	Pilastric index
Mladeč 27	85.0	117.9
Mladeč 28	69.5	--
Neandertals	82.0 ± 3.3 (13)	103.3 ± 9.2 (13)
Qafzeh-Skhul	80.6, 83.8, 86.1, 103.5	124.2 ± 11.9 (8)
Earlier Upper Paleolithic	73.4 ± 5.7 (25)	116.6 ± 11.2 (23)
Kruskal-Wallis P-values	<0.001	0.001

Table 25. Mladeč 27 and 28 femoral metric comparisons using linear residuals from the reduced major axis line through the pooled comparative sample. The P-value derives from the ANOVA comparison across the three reference samples, and the Mladeč values are Z-scores [(|Mladeč values – sample mean|) / standard deviation]. Z-scores are provided only relative to samples > 5. I_x, I_y, I_{max} and I_{min}: anteroposterior, mediolateral, maximum and minimum second moments of area; CA and TA: cortical and total subperiosteal areas; J: polar moment of area; Z_p: polar section modulus; BMxLen.: estimated body mass times length. For the gluteal tuberosity breadth and polar section modulus comparisons, values are provided for Mladeč 27 modeled with the body proportions of an earlier Upper Paleolithic human (UP) and of a Neandertal (N). * $P < 0.05$ with a multiple comparison correction (α/12)

	P-value	Neandertals		Qafzeh-Skhul		Earlier Upper Paleolithic	
		Mladeč 27	Mladeč 28	Mladeč 27	Mladeč 28	Mladeč 27	Mladeč 28
Proximal AP/ML	<0.001*	1.03	3.74	-	-	1.66	0.59
80% I_x/I_y	0.034	2.60	4.06	0.78	1.17	0.06	0.44
Gluteal Br./BMxLen.	0.060	1.41 (UP) 2.47 (N)	-	-	-	0.90 (UP) 2.32 (N)	-
Midshaft AP/ML	<0.001*	2.63	-	0.27	-	0.27	-
50% I_x/I_y	<0.001*	3.70	-	0.18	-	0.20	-
65% I_x/I_y	<0.001*	3.58	0.16	0.57	5.21	0.24	1.97
50% CA/TA	0.656	0.78	-	0.95	-	0.61	-
65% CA/TA	0.923	0.62	0.07	0.25	0.10	0.50	0.13
80% CA/TA	0.559	0.71	0.39	1.50	0.12	1.41	0.14
PseudoJ/Length	<0.001*	4.48	-	2.78	-	2.70	-
50% J/Length	<0.001*	5.10	-	2.05	-	1.97	-
50% Z_p/BMxLen	0.553	2.27 (UP) 4.03 (N)	-	1.89 (UP) 3.41 (N)	-	1.78 (UP) 3.78 (N)	-

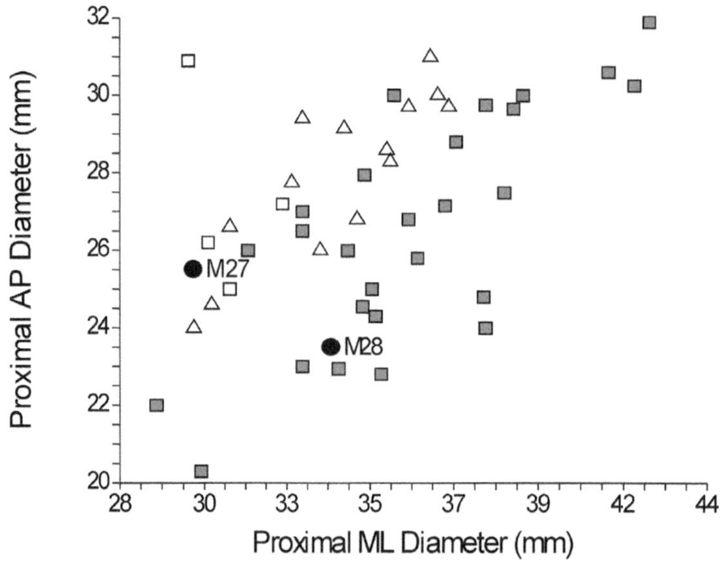

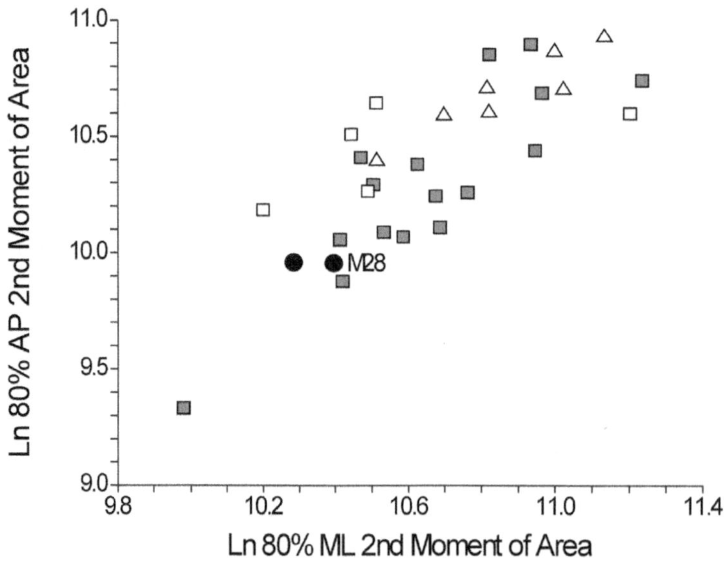

Fig. 8. Subtrochanteric femoral diaphyseal proportions. Anteroposterior versus posterior subperiosteal diameters (above) and anteroposterior versus mediolateral second moments of area (below) for Mladeč 27 and 28 (black circles – M27 and M28), earlier Upper Paleolithic humans (gray squares), Neandertals (open triangles) and Qafzeh-Skhul humans (open squares). In the second moment of area comparison, Mladeč 27 is the value to the left

Table 26. Comparisons of femoral osteometric values. Mean and standard deviation (N) for samples greater than four; individual measurements for smaller samples

	Gluteal tuberosity breadth (mm)	Neck-shaft angle (°)	Anterior curvature subtense (mm)
Neandertals	12.9 ± 2.0 (10)	121.0° ± 4.7° (9)	15.5 ± 3.4 (6)
Qafzeh-Skhul	6.6, 8.5, 9.5, 11.4	133.2° ± 2.6° (6)	10.0, 17.0, 21.0
Earlier Upper Paleolithic	9.7 ± 2.1 (12)	121.5° ± 8.0° (16)	14.1 ± 3.1 (10)
ANOVA P-values	0.001	0.002	0.631

In the comparison of the gluteal tuberosity size, measured as the breadth of the tuberosity given difficulties in discerning the distal end of the tuberosity in many femora, the Mladeč femora are also dichotomous. There is a significant difference across the Late Pleistocene samples (Table 26; ANOVA P = 0.001), with the Neandertals having generally broader tuberosities. The tuberosity breadth of 12.5 mm of Mladeč 28 is at the top of the early modern human range of variation, being matched only by that of Cro-Magnon 4322/4323A, but it is in the middle of the Neandertal range of variation. In contrast, the value of 8.0 mm for Mladeč 27 is below the means of all of the samples, and only the female Dolní Věstonice 3 and Qafzeh 9 have narrower tuberosity breadths. It is not possible to scale the Mladeč 28 gluteal tuberosity breadth to body size, but this can be done for Mladeč 27. Despite the significant difference in absolute tuberosity breadths across the Late Pleistocene samples (Table 26), scaling them to body mass time's femoral length (Table 25; Fig. 9) provides only a modest separation of the samples; it is principally the small individuals in both samples that have proportionately large tuberosities. The earlier Upper Paleolithic body mass for Mladeč 27 places it within the ranges of variation of Late Pleistocene humans, if in a relatively gracile position, whereas the Neandertal body mass would make it the most gracile of the known Middle or earlier Upper Paleolithic specimens and significantly different from the reference samples (Table 25).

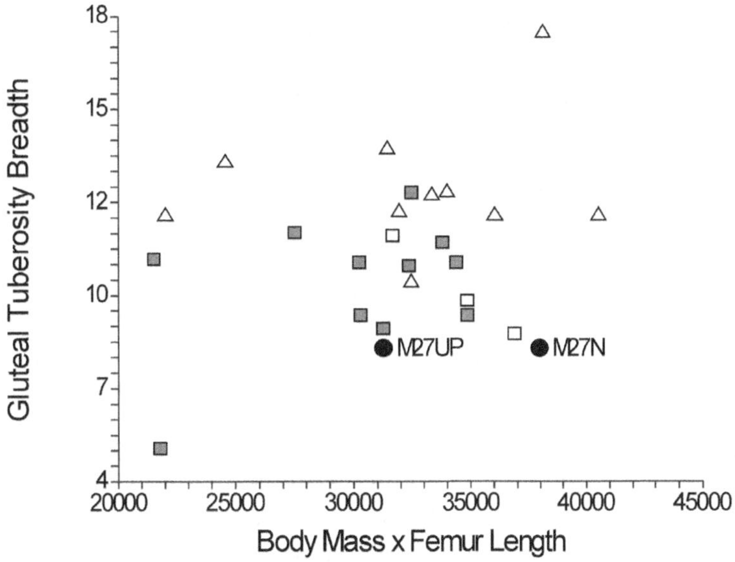

Fig. 9. Gluteal tuberosity breadth versus body mass times femoral length for Mladeč 24 (black circles), earlier Upper Paleolithic humans (gray squares), Neandertals (open triangles) and Qafzeh-Skhul humans (open squares). The left value for Mladeč 27 (M27UP) is based on a body mass estimate using earlier Upper Paleolithic body proportions, whereas the right value for Mladeč 24 (M27N) employed a body mass estimate based on Neandertal body proportions

It is possible to estimate the neck-shaft angle of Mladeč 28 at about 123° to 125°, despite the damage to the head and neck region. This value is unexceptional for a Neandertal or an earlier Upper Paleolithic human, falling minimally above the essentially identical means for the two samples (Table 26). It is, however, well below the mean of the Qafzeh-Skhul sample, whose femoral neck-shaft angles are anomalously high for a Pleistocene human sample, being closest to those of recent urban human populations (Trinkaus, 1993; Anderson and Trinkaus, 1998). The three Late Pleistocene samples are significantly different (ANOVA P = 0.002), but this difference is driven entirely by the Qafzeh-Skhul sample, since the Neandertal and earlier Upper Paleolithic angles are very close (t-test P = 0.836).

The midshaft of Mladeč 27 is dominated by a clear pilaster and its associated *linea aspera*. The *linea aspera* has a single narrow angle, reaching a maximum breadth of 3.0 mm. Laterally along the pilaster the bone is flat, and there is no sulcus adjacent to the *linea aspera*. Medially, there is a small ridge which is ca. 6.2 mm anteromedial from the *linea aspera* at midshaft, which parallels the *linea aspera*, only to blend in with it distally. The Mladeč 28 *linea aspera*, where preserved proximally, is smooth and relatively narrow, being 3.4 mm wide by the distal break, or moderately proximal of midshaft.

The cross-sectional shape of the Mladeč 27 femur at midshaft can be quantified by both its external diameters [the pilastric index and a graph of the data (Table 24 and Fig. 10)] and using anteroposterior versus mediolateral second moments of area (Table 22; Fig. 10). In the external diameters, but especially in the second moments of area, the Neandertals and the early modern humans show little overlap and are highly significantly different (Table 25). The low earlier Upper Paleolithic values are several Gravettian specimens from Pavlov and Předmostí; the highest Neandertal value is the late Saint Césaire 1 (see Trinkaus et al., 1999a), and the slightly lower Neandertal ones are the slightly older Neandertal 1 and Rochers-de-Villeneuve 1. In both comparisons, Mladeč 27 falls clearly with the early modern humans and significantly distinct from the Neandertals.

It is not possible to compare the Mladeč 28 diaphysis to other femora using the midshaft; however, it preserves the mid-proximal diaphysis, and it is possible to estimate the position of the 65% diaphyseal cross-section (Table 23). This was done by using morphological criteria to locate the 80% (subtrochanteric section), measuring the distance from the proximal neck to the 80% section (ca. 20% of biomechanical length), and then placing the 65% section 15% of that estimated biomechanical length distal of the 80% one. Its position is unlikely to be more than a few millimeters proximal or distal of the original position. The resultant plot of the 65% anteroposterior versus mediolateral second moments of area (Fig. 11) provides a high degree of separation of the Neandertal and early modern human samples (Table 25). The one high Neandertal value is Spy 2, and the one low earlier Upper Paleolithic value is Dolní Věstonice 41, an isolated short shaft section whose location of the 65% section may be too proximal (Trinkaus et al., 2000). Therefore, even though the pilaster (when present) is not fully developed in this mid-proximal section of the femoral diaphysis, there is sufficient development of it in these early modern human femora to provide separation between the Neandertals and the early modern humans. The Mladeč 27 position is consistent with the midshaft comparisons in falling in the middle of the pilastric early modern human femora. Mladeč 28, however, is in line with the Neandertal distribution and largely separate from the earlier Upper Paleolithic sample and small errors in the location of the section could not account for its diaphyseal mechanical proportions. However, its cross-sectional shape indicates the beginning of a pilaster, in that the posteromedial and especially posterolateral diaphyseal surfaces adjacent to the *linea aspera* exhibit some degree of flattening. Moreover, it is not known how it would compare to some of the now-lost non-pilastric Gravettian Předmostí femora.

The Mladeč 27 femur is sufficiently complete to assess its anterior diaphyseal curvature. Its point of maximum curvature is located slightly proximal of midshaft, 41% of the distance from the prox-

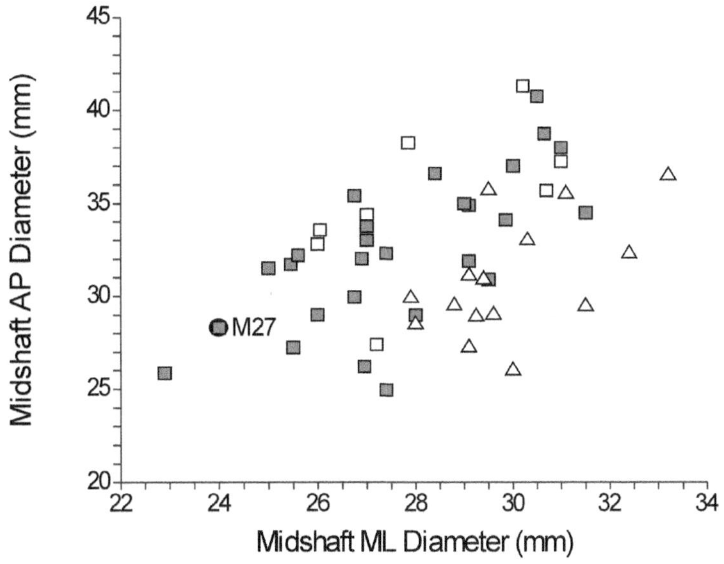

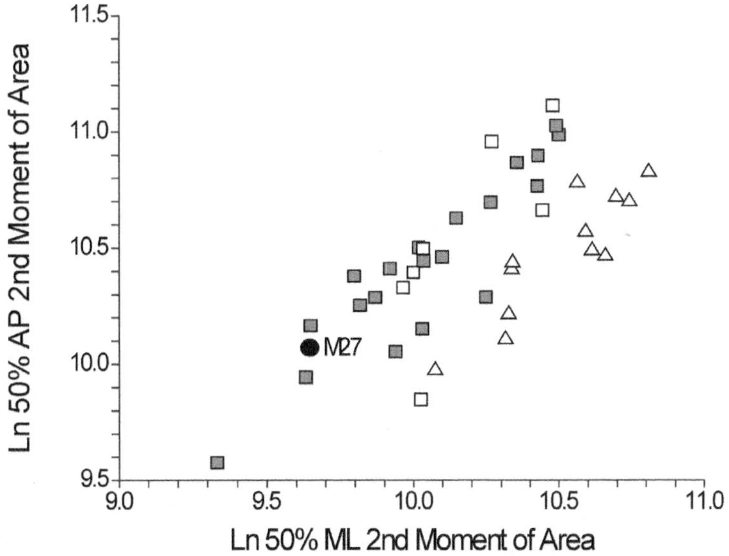

Fig. 10. Midshaft femoral proportions. Anteroposterior versus mediolateral subperiosteal diameters (above) and second moments of area (below) for Mladeč 27 (black circle – M27), earlier Upper Paleolithic humans (gray squares), Neandertals (open triangles) and Qafzeh-Skhul humans (open squares). The value for Mladeč 27 in the subperiosteal diameter comparison is partly obscured by identical values for Willendorf 2

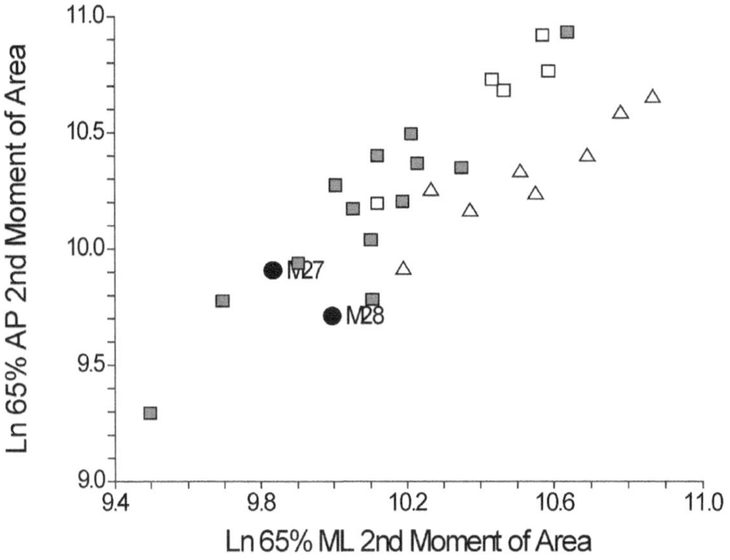

Fig. 11. Mid-proximal (65%) anteroposterior versus mediolateral second moments of area for Mladeč 27 and 28 (black circles – M27 and M28), earlier Upper Paleolithic humans (gray squares), Neandertals (open triangles) and Qafzeh-Skhul humans (open squares)

imal end of the chord. Since it is the absolute subtense, rather than the subtense scaled to any measure of bone length, which is of relevance (Shackelford and Trinkaus, 2002), comparisons are made solely to the absolute subtenses of the comparative samples. The value of 12 mm for Mladeč 27 is slightly below the means for those samples but well within their ranges of variation (Table 26). In this, it is at the top of a later Upper Paleolithic sample and those of more recent humans (Shackelford and Trinkaus, 2002).

Diaphyseal robusticity

Even though it is commonly considered a reflection of overall robusticity, the relative proportion of the diaphyseal cross section made up of cortical bone (or percent cortical area) is a better reflection of differential endosteal resorption versus subperiosteal deposition through the life cycle than of the relative strength of the diaphysis (Ruff and Hayes, 1983; Ruff et al., 1994). The relative cortical to total subperiosteal areas of the Mladeč femora (Figs. 12 and 13) align them with the Late Pleistocene comparative samples, which are little different despite a couple of outliers in the distributions (Table 25). In the 65% and especially the 80% sections, the Mladeč 27 femur has a moderately lower percent cortical area than Mladeč 28, but both of them are within the Late Pleistocene distributions.

Given the absence of a reliable length estimate and midshaft, it is not possible to assess the overall robusticity of the Mladeč 28 femur. It is possible, however, to assess it indirectly for the Mladeč 27 femur using several approaches.

Initially, to maximize sample size (and to be able to include the Předmostí femora), a "pseudo polar moment of area" was calculated, by modeling the midshaft as a solid ellipse and using the external diameters to calculate second moments of area. Given the similarities across the samples in percent cortical area, modeling them as solid beams is not likely to distort the data greatly, although treating the highly pilastric femoral diaphyses as ellipses does increase their relative

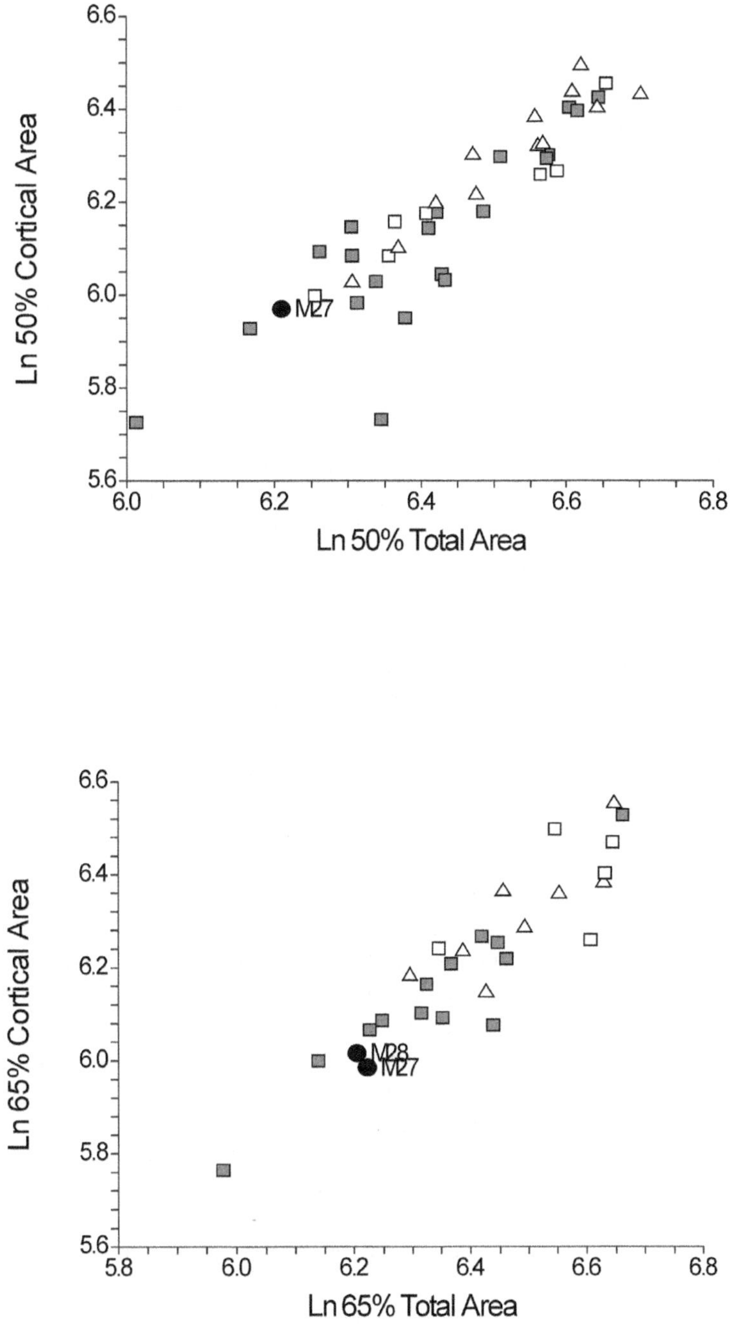

Fig. 12. Comparisons of femoral midshaft (50%) and mid-proximal (65%) diaphyseal cortical versus total subperiosteal area for Mladeč 27 and 28 (black circles – M27 and M28), earlier Upper Paleolithic humans (gray squares), Neandertals (open triangles) and Qafzeh-Skhul humans (open squares)

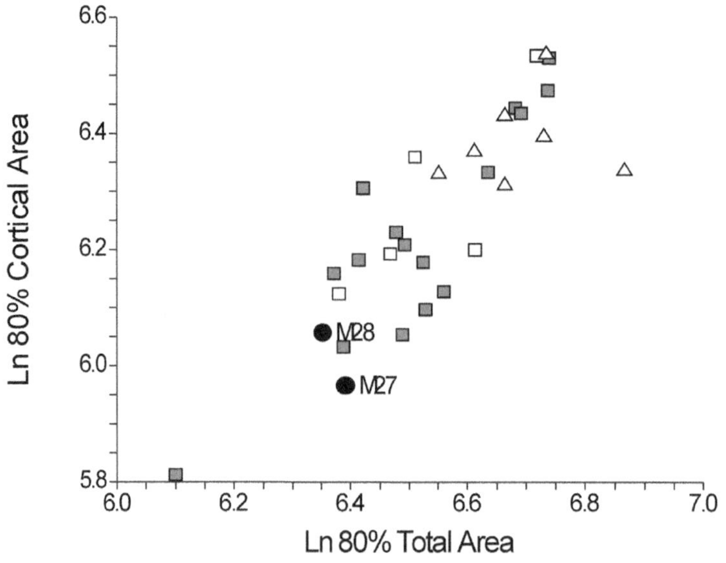

Fig. 13. Comparisons of femoral subtrochanteric (80%) diaphyseal cortical versus total subperiosteal area for Mladeč 27 and 28 (black circles – M27 and M28), earlier Upper Paleolithic humans (gray squares), Neandertals (open triangles) and Qafzeh-Skhul humans (open squares)

values. The resultant plot of it to femoral length (Fig. 14) provides a significant degree of separation between the Neandertals and early modern humans, with the latter appearing more gracile (Table 25). Mladeč 27 falls at the gracile end of the early modern human distribution. A similar result is obtained when the cross-sectionally measured polar moment of area is compared to femoral length (Fig. 14). However, both of these comparisons ignore the effects of body proportions on femoral robusticity and largely reflect the stockier proportions of the Neandertals compared to early modern humans.

To correct for variance in body proportions, the midshaft diaphyseal strength was plotted against femoral length time body mass (Fig. 15). The result is little separation of any of the Late Pleistocene samples, as has been previously documented (Ruff et al., 1993; Trinkaus and Ruff, 1999a; Trinkaus, 2000; see Table 25). The two models of Mladeč 27 body proportions provide contrasting results. The earlier Upper Paleolithic model places it at the gracile margin of the Late Pleistocene distribution, relatively gracile but close to several early modern human specimens. The Neandertal model for its proportions makes it exceptionally gracile for a Late Pleistocene human, being about four standard deviations from the Neandertal and earlier Upper Paleolithic samples. Given its similarity to other early modern human femora in its midshaft anteroposterior to mediolateral proportions [which reflect mobility patterns (Ruff, 2000b)] and relative cortical area, it is unlikely that Mladeč 27 was an abnormally gracile early modern human. More likely, as suggested by its relative gluteal tuberosity breadth, it was moderately gracile for a Late Pleistocene human and had the linear body proportions of a European early modern human.

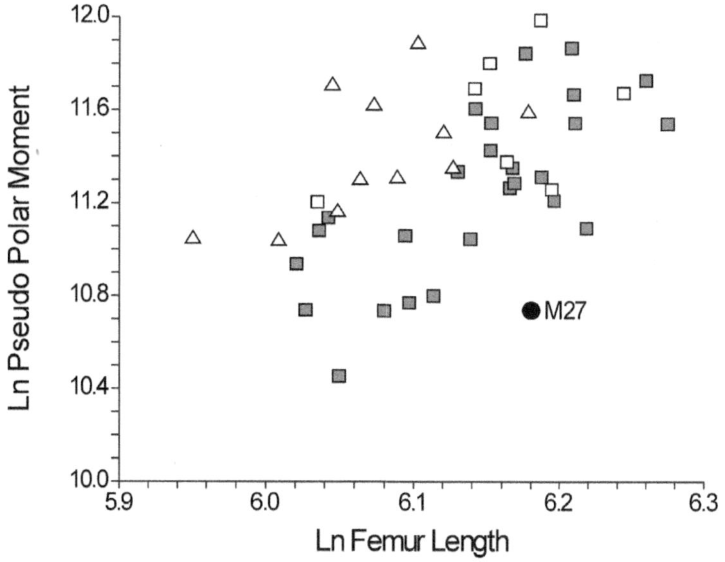

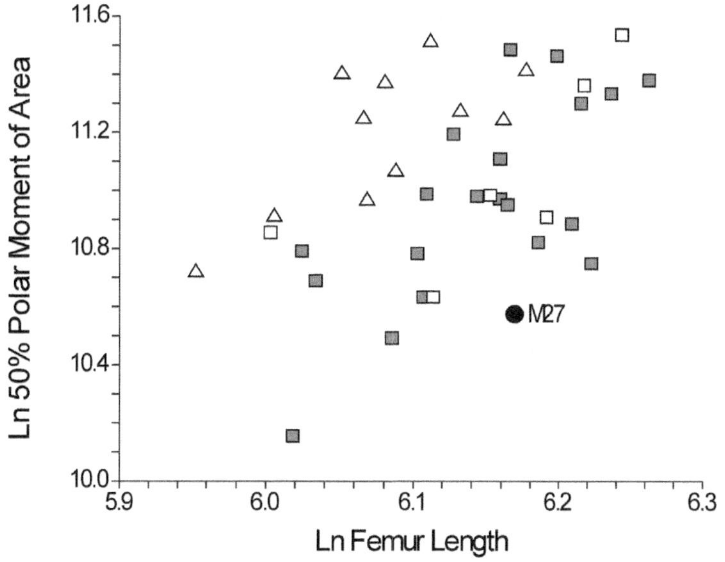

Fig. 14. Comparisons of femoral midshaft dimensions versus length, using an estimate of strength using external diameters (above) and cross-sectional geometry (the polar moment of area) (below), for Mladeč 27 (black circle – M27), earlier Upper Paleolithic humans (gray squares), Neandertals (open triangles) and Qafzeh-Skhul humans (open squares)

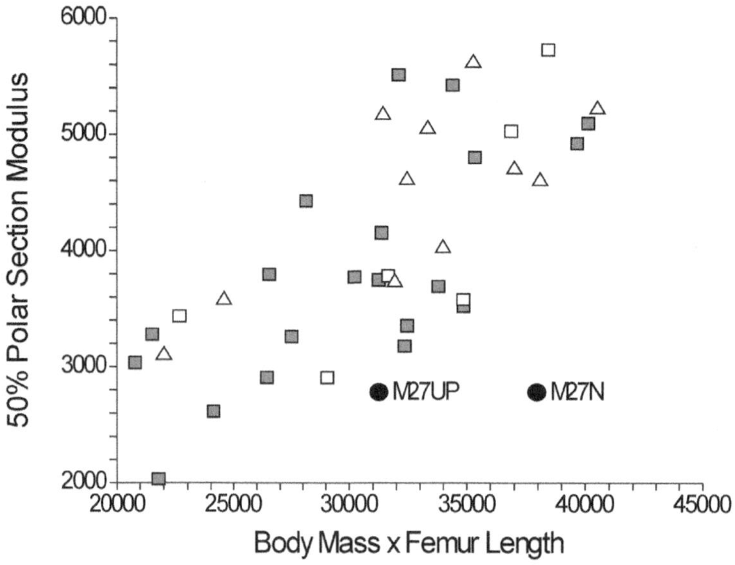

Fig. 15. Femoral robusticity of Mladeč 27, comparing the midshaft polar section modulus to body mass times femoral length. The left value for Mladeč 27 (M27UP) is based on a body mass estimate using earlier Upper Paleolithic body proportions, whereas the right value for Mladeč 27 (M27N) employed a body mass estimate based on Neandertal body proportions

The Mladeč 102 immature femur

Preservation

Curated with the Mladeč human remains in the Naturhistorisches Museum Wien is a previously undescribed proximal immature human femoral diaphyseal section (chapter 8, Plate XVIII). It exhibits the same bone preservation and carbonate encrustation as the other Mladeč remains and therefore is considered to be part of the sample. It also exhibits the museum catalog number of 5459 present on the other Mladeč human remains and in the same style of handwriting. It is given the number of Mladeč 102, adding on to the comprehensive catalog list of the Mladeč human remains. It may be the specimen listed by Szombathy in 1922 as the fourth radial fragment (Table 1).

Mladeč 102 is a left juvenile femur from the middle of the gluteal buttress to the proximal *linea aspera*, 53 mm long. The shaft is complete for the distal 38.3 mm, but proximally the gluteal buttress is partially absent and crushed inwards. There is carbonate concretion anteriorly and posterolaterally, the latter obscuring the gluteal tuberosity. The age-at-death of the specimen cannot be determined precisely, but it resembles the size and degree of diaphyseal development seen in earlier juvenile (ca. 3–6 year old) human femora. It could derive from the same individual as the Mladeč 3 cranium or be from a slightly older individual.

Morphology

There is little of note on this small specimen (Tables 20 and 23). The proximal diaphysis provides a meric index of 81.8. This value is in the middle of the variation of other Late Pleistocene juvenile specimens, including the Neandertal Cova Negra 3 (82.9) and La Ferrassie 6 (77.0) specimens, the Qafzeh 10 (88.4) and Skhul 1 (83.3) femora, and the earlier Upper Paleolithic Lagar Velho 1 (83.0) skeleton. The distal break, which is probably slightly proximal of midshaft, provides a pilastric index of 95.7. This value is similarly within the ranges of variation of Late Pleistocene juvenile femora, since the Neandertal Cova Negra 3, La Ferrassie 6 and Roc de Marsal 1 provides indices of 93.5, 93.9 and 96.2 respectively, the younger Dederiyeh 1 and 2 Neandertal femora provide indices of 98.1 and 103.0, and the Skhul 1 (112.9), Qafzeh 10 (105.6) and Skhul 8 (94.7) femora span most of the range. The earlier Upper Paleolithic Lagar Velho 1 and the east Asian Yamashita-cho 1 early modern human juveniles have indices of 94.5 and 100.6 respectively.

The Mladeč 29 tibia

Preservation and maturity

This specimen is the lateral portion of a right distal tibial epiphysis (chap. 8, Plate XVI). The fibular notch is preserved, as well as the lateral aspects of the anterior and posterior epiphyseal surfaces. The preserved anterior breadth of the specimen is 22 mm; the preserved posterior breadth is 18 mm. The specimen has been broken since its recovery, since Szombathy's (1925) description indicates that it was a complete right distal tibial epiphysis. This is confirmed by his illustration of the specimen, description of the medial malleolus and malleolar articular surface, and his reported distal epiphyseal breadth of 58 mm (Szombathy, 1925, 26). Most of the specimen is covered by surface matrix except for the medial broken edge, where the internal trabeculae are clearly visible and undistorted.

The proximal surface of the bone, although covered by a thin layer of matrix, is clearly that of an unfused epiphysis. This is confirmed by the presence of a thin layer of cortical bone along the proximal edge of the current break through the bone. Given the fusion of this epiphysis in later adolescence (15 to 18 years) (Scheuer and Black, 2000), this bone therefore represents at most an individual about 16 years old at death.

Morphology

The Mladeč 29 tibial epiphysis is notable for two features, its large size and its squatting facet.

To provide a measure of its size (Table 27), the "area" of the distal articular facet for the talar trochlea was computed using the formula for a trapezoid as 1040 mm^2. This is the largest value known among Late Pleistocene late archaic and early modern humans [data from Holliday (1995)

Table 27. Osteometric dimensions of the Mladeč 29 right distal tibia (in mm)

Distal epiphyseal breadth (M-6)[1]	58.0
Distal epiphyseal sagittal length (M-7)	43.5
Distal articular breadth[1]	32.0
Medial articular depth[1]	30.0
Lateral talar articulation sagittal depth	34.5
Distal fibular articulation sagittal length	27.9
Lateral squatting facet height	≥ 9.4
Lateral squatting facet breadth	> 15.5

[1] Dimension not preserved on specimen, and measurement is taken from Szombathy (1925). It is presumed to have been broken after Szombathy's analysis.

and Sládek et al. (2000)], although Barma Grande 2 (948.6 mm²), Caviglione 1 (904.8 mm²) and Fanciulli 4 (956.8 mm²) are close. It is well above the means of the three references samples: Neandertals (765.2 ± 115.7 mm², N = 6), Qafzeh-Skhul (738.2 ± 84.8 mm², N = 4) and earlier Upper Paleolithic (806.7 ± 102.1 mm², N = 14). That makes it, respectively, 2.38, 3.56 and 2.29 standard deviations above those sample means.

This large talar trochlear facet is accompanied by a prominent lateral squatting facet, which extends vertically up to the epiphyseal cartilage margin. Its medial margin was lost in the recent breakage. Similar squatting facets are commonly found on other Late Pleistocene distal tibiae (Trinkaus, 1975a; Sládek et al., 2000).

The Mladeč 30 talus

Preservation

Mladeč 30 is a large left talus and represents one of the most complete specimens in the Mladeč postcranial sample (chap. 8, Plate XVI). The specimen is nearly completely preserved, lacking only the anteroinferior corner of the lateral malleolar surface and adjacent posterior calcaneal surface. Minor surface abrasion occurs elsewhere on the bone, including the inferomedial border of the head and much of the medial neck, the anterior-most portion of the posterior calcaneal facet, and the medial margin of the medial talar tubercle. Most of the specimen is covered by a thin layer of matrix, except for portions of the trochlea and the lateral malleolar articular surface.

Comparison of talar trochlear articular surface size and curvature suggests that Mladeč 30 may derive from the same individual as the Mladeč 29 distal tibia, even though they derive from opposite sides. The complete fusion of the posterior tubercles suggests an age-at-death greater than 10 to 13 years (Scheuer and Black, 2000).

Morphology

Given its completeness, the Mladeč 30 talus provides a largely complete set of osteometrics and discrete trait observations, with only those requiring reconstruction of the lateral tips of the lateral malleolar surface and posterior calcaneal facet having estimation (Tables 28 and 29). The talus is notable primarily for its large size, since all of its overall and trochlear dimensions are well above the means of the Late Pleistocene comparative samples (Table 30). However, they all remain within two standard deviations of those means, and its length is within one standard deviation of the mean of the earlier Upper Paleolithic sample. In fact, its length is matched or exceeded by those of Barma Grande 2, Cro-Magnon 4377, Fancuilli 4 and Veneri 1.

Despite its large overall size, its head-neck length is average for a Late Pleistocene human, being below the mean of the earlier Upper Paleolithic sample. To assess its relative head-neck length, which is an indirect measure of relative trochlear size (Rhoads and Trinkaus, 1977), it was compared to talar length. The resultant index (33.3) is below the means of the comparative samples (Table 31) but well within their ranges of variation. There is a trend for the earlier Upper Paleolithic tali to have longer necks relative to Middle Paleolithic specimens, and the difference approaches significance. The same pattern is evident in the comparison of trochlear size to talar length (Table 31), in which Mladeč 30, in this comparison, is more than two standard deviations from the earlier Upper Paleolithic sample but similar to the two Middle Paleolithic samples. The Mladeč 30 talus therefore exhibits the relative expansion of the trochlea seen in the earlier Late Pleistocene samples, as noted by Wolpoff (1989), but it is equally close to the Neandertals and

Table 28. Osteometric dimensions of the Mladeč 30 left talus (in mm and degrees). Estimated values are in parentheses

Length (M-1)	57.0
Medial maximum length	57.4
Lateral maximum length	62.0
Maximum height	34.5
Maximum articular height (to trochlear margins) (M-3b)	30.0
Minimum articular height (to trochlear sulcus)	28.2
Articular breadth (M-2b)	(53.0)[1]
Trochlear length (M-4)	38.7
Anterior trochlear breadth	35.5
Middle trochlear breadth (M-5)	34.0
Posterior trochlear breadth	26.2
Trochlear height (M-6)	10.8
Lateral malleolar breadth (M-7a)	(10.0)[1]
Lateral malleolar height	(29.5)[1]
Lateral malleolar oblique height (M-7)	(30.0)[1]
Medial malleolar breadth	9.0
Head-neck length (M-8)	19.0
Head length (M-9)	37.2
Head breadth (M-10)	26.5
Posterior calcaneal length (M-12)	37.0
Posterior calcaneal breadth (M-13)	27.0
Trochlear angle[2]	8°
Neck angle (M-16)	26°
Trochlear torsion angle (M-17)	35°
Basal torsion angle (M-17a)	35°
Posterior calcaneal angle (M-15)	34°
Subtalar angle[3]	44°

[1] The lateral tip of the lateral malleolar process is absent, and it was necessary to model it in plasticene using the adjacent contours of the lateral malleolar surface and the posterior calcaneal surface. These measurements are dependent in part on that restoration.

[2] The angle between the two dorsal trochlear margins (Trinkaus, 1975b).

[3] The angle between the proximo-distal trochlear axis and the "subtalar" line across the posterior and medial calcaneal surfaces (Trinkaus, 1975b).

Table 29. Discrete traits of the Mladeč 30 left talus. See Barnett (1954) and Trinkaus (1975a) for definitions

Anterior extension of medial malleolar surface	present (minimal)
Anterior extension of medial trochlea	absent
Anterior extension of lateral trochlea	present
Lateral squatting facet	present
Sulcus tali margin rounding	present
Sulcus tali facet	present
Anterior-medial calcaneal facet fusion	complete

Table 30. Osteometric comparisons of the Mladeč 30 talus to Late Pleistocene comparative samples. Mean and standard deviation (N) provided. All in millimeters. ANOVA P-values provided for the comparative samples

	Length	Articular breadth	Trochlear length	Trochlear breadth	Head-neck length
Mladeč 30	57.0	(53.0)	38.7	34.0	19.0
Neandertals	52.0 ± 3.2 (11)	49.2 ± 2.4 (11)	35.9 ± 2.3 (11)	28.4 ± 1.8 (11)	18.8 ± 2.1 (11)
Qafzeh-Skhul	52.6 ± 4.2 (6)	44.7 ± 4.1 (7)	34.5 ± 1.9 (7)	28.7 ± 2.7 (7)	18.1 ± 3.2 (6)
Earlier Upper Paleolithic	53.2 ± 4.3 (23)	46.2 ± 5.6 (13)	34.4 ± 2.8 (23)	29.3 ± 2.7 (20)	19.9 ± 2.5 (15)
ANOVA P-value	0.715	0.095	0.238	0.599	0.311

Table 31. Comparative talar proportions. Mean and standard deviation (N) and ANOVA P-values for the comparative samples provided

	Head-neck length/Length	Trochlea/Length	Neck Angle	Torsion Angle
Mladeč 30	33.3	63.6	26°	35°
Neandertals	36.1 ± 3.1 (11)	61.6 ± 2.0 (11)	26.0° ± 4.0° (11)	40.4° ± 4.8° (11)
Qafzeh-Skhul	34.2 ± 3.7 (6)	60.7 ± 4.2 (6)	25.8° ± 2.3° (6)	38.4° ± 9.0° (5)
Earlier Upper Paleolithic	38.2 ± 3.8 (16)	59.7 ± 1.8 (21)	24.2° ± 3.9° (20)	34.6° ± 2.4 (8)
ANOVA P-value	0.058	0.099	0.369	0.841

the Middle Paleolithic early modern humans in this respect. Interestingly, the one other European Aurignacian talus, that from Fontana Nuova, provides indices of 37.6 and 56.7 which contrast with those of Mladeč 30.

The angular orientations of the Mladeč 30 articulations are similar to those of other Late Pleistocene humans and recent human samples (Table 31; Trinkaus, 1975b). The articular discrete traits of the Mladeč 30 specimen mostly suggest frequent hyperdorsiflexion of the talocrural articulation, presumably through the assumption of a squatting position (Trinkaus, 1975a). The one exception to this is the minimal anterior extension of the medial malleolar surface, contrasting with the ubiquitous anterior extension of this facet among Late Pleistocene human tali. The anterior and medial calcaneal surfaces are completely fused, and they have a projection into the middle of the sulcus tali. This results in a narrow sulcus tali, minimally 4.9 mm wide. Finally, the lateral posterior tubercle is much more strongly developed than the medial one, and it borders a broad sulcus for the *M. flexor hallucis longus* tendon (ca. 8.5 mm wide).

The Mladeč 32 metatarsal 3

Preservation and age-at-death

The Mladeč 32 left metatarsal 3 is a complete immature bone, lacking only the unfused head epiphysis (chap. 8, Plate XVI). There is a thin layer of carbonate crust over the entire bone, and there was minor crushing to the plantar base. Since metatarsal heads normally fuse between the ages of 12 and 16 years (Scheuer and Black, 2000), this bone likely represents an individual no older than the middle of the second decade of life.

Morphology

The Mladeč 32 metatarsal 3 has a triangular diaphysis with rounded angles. There is a strong twist to the bone, which provides the torsion angle of 11° between the base and the head metaphysis, indicating a well-formed pedal arch. The base has a large triangular facet for the lateral cuneiform bone. The metatarsal 2 facets are double, with a 6 mm gap between them (Table 32). There is a single and larger metatarsal 4 facet. The plantar tubercle for the short plantar ligaments is moderately large.

Using a least squares regression based on mature recent human third metatarsals, the articular length of the bone has been estimated to be 80.9 mm (ArtLen = 0.99 × PresLen + 8.5, r^2 = 0.964, N = 44). This value, as with the Mladeč 31 metacarpal 3, is large for a Late Pleistocene human, being approached by Barma Grande 2 (79.8 mm), Skhul 4 (80.3 mm) and Veneri 1 (79.8 mm) (Table 33). Its robusticity index (geometric mean of the midshaft diameters / articular length) was computed as 12.0. This value is similar to other Late Pleistocene sample values, which do not differ significantly between them (Table 33).

Table 32. Osteometric dimensions of the Mladeč 32 left metatarsal 3 (in mm). Values in parentheses are estimated

Midshaft height	10.8
Midshaft breadth	8.7
Proximal maximum height	22.1
Proximal articular height	17.8
Dorsal (maximum) proximal articular breadth	14.0
Plantar (minimum) proximal articular breadth	7.2
Metatarsal 2 dorsal facet height	7.2
Metatarsal 2 dorsal facet breadth	10.7
Metatarsal 2 plantar facet breadth	(7.7)
Metatarsal 4 facet height	9.9
Metatarsal 4 facet breadth	14.3
Plantar tuberosity proximodistal length	17.0
Preserved length	73.4

Table 33. Osteometric comparisons for the Mladeč 32 third metatarsal. Mean and standard deviation (N) are provided for samples greater than five; individual values provided for smaller samples. ANOVA P-value for the comparative samples provided for length and Kruskal-Wallis P-value provided for the index

	Articular length (mm)	Robusticity index
Mladeč 32	80.9	12.0
Neandertals	68.3 ± 3.9 (6)	12.3 ± 0.9 (6)
Qafzeh-Skhul	67.6, 73.7, 80.3	11.3, 12.5, 14.3
Earlier Upper Paleolithic	72.1 ± 6.4 (10)	11.8 ± 0.8 (10)
Comparative P-values	0.324	0.322

Paleopathology of the Mladeč postcranial elements

The Mladeč postcranial remains in the Naturhistorisches Museum Wien exhibit no macroscopic evidence of lesions externally. It is possible that there are minor ones obscured by the ubiquitous carbonate encrustations, but they are likely to be very minor if present. There is also no evidence of transverse (Harris) lines radiographically on the Mladeč 23 proximal humerus, the Mladeč 24 distal humerus, the Mladeč 25a proximal radius, the Mladeč 31 metacarpal or the Mladeč 32 metatarsal. Although they are (probably) not from the same individuals, this absence of transverse lines is in agreement with the absence of dental enamel hypoplasias on Mladeč 1 and 2 maxillary teeth and the Mladeč 8, 9 and 10 posterior teeth; there are minor pits on the Mladeč 8 and 9 maxillary canines. The hand remains from Chamber E also lack pathological lesions.

Associations by individual

The Mladeč postcranial remains from Chamber D therefore include the remains of both mature individuals and immature individuals. It is likely that these do not each represent separate individuals but can be associated based on size, morphology, and age-at-death. The following associations are considered to be reasonable, but they remain hypothetical in the absence of DNA fingerprinting of the remains.

The Mladeč 102 femoral diaphysis, as that of a younger juvenile, may come from the same individual as the Mladeč 3 cranial remains, depending largely on the age estimations for each.

There is a series of immature remains, whose ages-at-death provide a range between 13 and 16 years. These include three rib pieces with unfused head epiphyses (Mladeč 12, 14 and 17; age-at-death < 20–25 years), the Mladeč 23 proximal humerus (age-at-death 16–20 years), the Mladeč 31 metacarpal 3 (age-at-death < 14–16 years), the Mladeč 22 os coxae (between 13–15 years and 19–20 years), the Mladeč 29 distal tibia (age-at-death < ca. 15 years), and the Mladeč 32 metatarsal 3 (age-at-death < 12–16 years). In addition, the Mladeč 30 talus has an age-at-death greater than 10–13 years, and it is a good match morphologically and dimensionally for the Mladeč 29 distal tibia. The Mladeč 24 humerus, whose epiphyseal line is fused where preserved, may be developmentally too old to be included within this individual and, in any case, would duplicate the Mladeč 23 humerus. In addition, the Mladeč 25a and 25c proximal radius and ulna, which appear on the basis of size and morphology to derive from the same arm, have a minimum age of ca. 15 years based on complete fusion of their proximal epiphyses.

It is therefore possible that these remains represent a single individual, although one would have to argue for a relatively late fusion of the metacarpal 3 head, distal tibial and metatarsal 3 head epiphyses and a relatively early fusion of the proximal humeral, radial and ulnar epiphyses. It is more likely that more than one individual is represented by these remains, and that the mid-adolescent remains are restricted to those that show clear immature status (the ribs, the metacarpal 3, the os coxae, the distal tibia and the metatarsal 3). The remainder of the immature pieces (the proximal humerus and possibly the immature ribs), perhaps joined by other late adolescent or skeletally mature remains, represent an additional individual.

Most of these adolescent or possibly adolescent remains are moderately (but not exceptionally) large in size compared to other earlier Upper Paleolithic Europeans. These include especially the Mladeč 25a and 25c radius and ulna, the Mladeč 31 metacarpal, the Mladeč 22 os coxae, the Mladeč 29 and 30 tibia and talus, and the Mladeč 32 metatarsal. It is therefore likely that these remains all derive from large, mid-adolescent and late adolescent individuals. The two more complete crania from Chamber D of the Mladeč Cave, Mladeč 1 and 2, are late adolescent in age based on dental attrition (especially the absence of distal interproximal facets on the Mladeč 1 M^2s), and it is possible that one of them is the same individual as the late adolescent or young adult postcrania. However, these crania have been considered to be female (chap. 10), since they contrast with the marked cranial superstructures of Mladeč 5 and 6 and the large palate of Mladeč 8, whereas the late adolescent based on these postcrania is likely, on the basis of size, to be male. Therefore, either there is greater variation in male cranial superstructure development than has been assumed for the Mladeč sample [such as is seen in the Dolní Věstonice 11/12, 13, 14 and 16 and Pavlov 1 males (Vlček, 1991; Franciscus and Vlček, 2005)], or the Mladeč 1 and 2 cranial remains do indeed represent different individuals than these apparently male postcrania. The additional Mladeč Chamber D skull remains, the Mladeč 8 maxilla and the Mladeč 9 and 10 isolated teeth, are fully mature.

The remainder of the Mladeč postcrania are either undiagnostic as to age (beyond being adolescent or adult) and could belong to these Mladeč adolescents (e.g., additional rib fragments, the Mladeč 27 femoral diaphysis), clearly belong to other individuals based on duplication of elements, size and/or morphology (e.g., Mladeč 24 humerus, Mladeč 21 os coxae, Mladeč 26 radius), and/or they are fully mature (Mladeč 11 vertebra, Mladeč 24 humerus, Mladeč 21 os coxae, Mladeč 28 femur). There are no reliable criteria available to assess whether these mature remains might be associated with each other or with the mature craniodental remains from Chamber D.

As noted above, it is likely that three of the Chamber E hand remains, Mladeč 88 to 90, derive from one mature individual.

Morphological affinities of the Mladeč human postcranial remains

The basic morphological pattern of the Mladeč human postcranial elements is that of relatively robust early modern humans. There is a suite of characteristics in the remains that align them principally with the early modern human remains known from the earlier portion of the Upper Paleolithic (generally between 20,000 and 28,000 years BP) of Europe. These characteristics include (1) the tapering of the lower cervical spinous process, (2) the relative thinness of several of the rib remains, (3) the modest *M. pectoralis major* tuberosities, (4) the small size of the olecranon fossa, (5) the gracility of the humeral diaphysis, (6) the more proximal orientation of the ulnar trochlear notch, (7) the proximal phalangeal robusticity, (8) the smooth gluteal tuberosities, and (9) the modest femoral hypertrophy that is best seen as a reflection of linear body proportions in the context of robust Late Pleistocene femora.

At the same time, there are several features which are either intermediate or variable between late archaic and early modern humans. The ventral height of the Mladeč 11 vertebral body appears relatively low, similar to those of the Neandertals, but it is also matched by those of Skhul 5. The more anterior radial tuberosity position is closely aligned with early modern humans, but it overlaps the Neandertal range of variation. The metacarpal 3 robusticity is variable. The marked gluteal buttress/flange of Mladeč 28 is proportionately and morphologically similar to those of earlier Upper Paleolithic humans and distinct from those of the Neandertals, but the Mladeč 27 rounder subtrochanteric femur is close to the Neandertals in proportions; however, in this feature Mladeč 27 is also close to the Qafzeh-Skhul sample. The gluteal tuberosity of Mladeč 27 is very modest similar to those of many earlier Upper Paleolithic humans, but the one of Mladeč 28 (which cannot be scaled to body size) is one of the largest known for early modern humans and, in absolute terms, most similar to those of the Neandertals. The femoral midshaft of Mladeč 27 exhibits a distinctively modern human pilaster, whereas the mid-proximal shaft (to the extent preserved) of Mladeč 28 is broader, mechanically similar to Neandertal femora, and has only a suggestion of the proximal formation of a pilaster. And finally, the relative trochlear size of the Mladeč 30 talus separates it from earlier Upper Paleolithic humans and places it among the Neandertals, but it is also indistinguishable from the Qafzeh-Skhul sample.

Consequently, although there are several features which might be taken to provide evidence of morphological affinities of the Mladeč postcranial remains to those of the Neandertals, most of them are variable within the small Mladeč sample and/or they align the Mladeč remains as much with the Middle Paleolithic Qafzeh-Skhul early modern humans as with the Neandertals. It becomes difficult to assess whether the "archaic" features of the Mladeč human postcrania are the result of some degree of Neandertal ancestry, affinities to Middle Paleolithic early modern humans, or aspects of robusticity reflecting earlier Upper Paleolithic human behavioral patterns in Europe.

It should be emphasized that these conclusions do not necessarily mean that the Mladeč sample provides no evidence of Neandertal ancestry for Central European early modern human samples (see Smith and Trinkaus, 1991; Frayer, 1993; Wolpoff et al., 2001). It only means that it is difficult to use those portions of the postcranial remains preserved for these early Upper Paleolithic humans to substantiate such a hypothesis.

Acknowledgments

We would like to thank Maria Teschler-Nicola for inviting us to write up a description of the Mladeč human postcranial remains and for her assistance in identifying smaller pieces of the collection, and both her and the curators of the Moravské zemské muzeum (the late Jan Jelínek, Marta Dočklová and Martin Oliva) for access to the Mladeč human remains. Comparative data has been collected through the permissions of multiple curators of specimens, including B. Arensburg, A. Czarnetzki, Y. Coppens, H. Delporte, I. Hershkovitz, J. J. Hublin, H. Joachim, W. J. Kennedy, the late B. Klíma, A. Langaney, A. Leguebe, B. Maureille, Muyaid Sa'id al-Damirji, R. Orban, J. Radovčić, Y. Rak, C. B. Stringer, J. Svoboda, F. Twiesselmann, B. Vandermeersch, J. Zias, and J. Zilhão, in addition to M. Teschler-Nicola, J. Jelínek, M. Dočkalová and M. Oliva. This work has been funded by the National Science Foundation, the Wenner-Gren Foundation (IRG-14), the L.S.B. Leakey Foundation, Alexander von Humboldt Foundation, Northern Illinois University and Washington University.

References

Ahern, J. C. M., Karavanić, I., Paunović, M., Janković, I. and Smith, F. H. (2004) New discoveries and interpretations of hominid fossils and artifacts from Vindija Cave, Croatia. *Journal of Human Evolution* 46, 27–67

Anderson, J. Y. and Trinkaus, E. (1998) Patterns of sexual, bilateral and inter-populational variation in human femoral neck-shaft angles. *Journal of Anatomy* 192, 279–285

Arambourg, C. (1955) Sur l'attitude, en station verticale, des néanderthaliens. *Comptes rendus de l'Académie des Sciences* 240D, 804–806

Arensburg, B. (1991) The vertebral column, thoracic cage and hyoid bone. In (O. Bar-Yosef and B. Vandermeersch, Eds.) *Le squelette Moustérien de Kébara 2*. Paris: C.N.R.S. Éditions, pp. 113–146

Arensburg, B. and Nathan, H. (1980) A Mousterian third cervical vertebra from Hayonim Cave, Israel. *Journal of Human Evolution* 9, 193–195

Barnett, C. H. (1954) Squatting facets on the European talus. *Journal of Anatomy* 88, 509–513

Boule, M. (1911–1913) L'homme fossile de La Chapelle-aux-Saints. *Annales de Paléontologie* 6, 111–172; 7, 21–56, 85–192; 8, 1–70

Bräuer, G. (1988) Osteometrie. In (R. Knussmann, Ed.) *Anthropologie: Handbuch der vergleichenden Biologie des Menschen, Bd I: Wesen und Methoden der Anthropologie*. Stuttgart: Gustav Fischer, pp. 160–192

Brůžek, J. (2002) A method for visual determination of sex, using the human hip bone. *American Journal of Physical Anthropology* 117, 157–168

Chilardi, S., Frayer, D. W., Giola, P., Macchiarelli, R. and Mussi, M. (1996) Fontana Nuova di Ragusa (Sicily, Italy): Southernmost Aurignacian site in Europe. *Antiquity* 70, 553–563

Churchill, S. E. (1994) *Human upper body evolution in the Eurasian Later Pleistocene*. Ph. D. Thesis, University of New Mexico

Churchill, S. E. and Formicola, V. (1997) A case of marked bilateral asymmetry in the upper limb of an Upper Palaeolithic male from Barma Grande (Liguria), Italy. *International Journal of Osteoarcheology* 7, 18–38

Churchill, S. E., Pearson, O. M., Grine, F. E., Trinkaus, E. and Holliday, T. W. (1996) Morphological affinities of the proximal ulna from Klasies River Mouth Main Site: Archaic or modern? *Journal of Human Evolution* 31, 213–237

Churchill, S. E. and Smith, F. H. (2000) A modern human humerus from the early Aurignacian of Vogelherdhöhle (Stetten, Germany). *American Journal of Physical Anthropology* 112, 251–273

Conard, N. J., Grootes, P. M. and Smith, F. H. (2004) Unexpectedly recent dates for human remains from Vogelherd. *Nature* 430, 198–201

Cunningham, D. J. (1886) The neural spines of the cervical vertebrae as a race-character. *Journal of Anatomy and Physiology* 20, 637–640

Eschman, P. N. (1992) *SLCOMM Version 1.6*. Albuquerque: Eschman Archeological Services

Franciscus, R. G. and Churchill, S. E. (2002) The costal skeleton of Shanidar 3 and a reappraisal of Neandertal thoracic morphology. *Journal of Human Evolution* 42, 303–356

Franciscus, R. G. and Vlček, E. (2005) The cranial remains. In (E. Trinkaus and J. A. Svoboda, Eds.) *Early modern human evolution in Central Europe. The people of Dolní Věstonice and Pavlov*. New York: Oxford University Press, pp. 63–152

Frayer, D. W. (1986) Cranial variation at Mladeč and the relationship between Mousterian and Upper Paleolithic hominids. *Anthropos* 23, 243–256

Frayer, D. W. (1993) Evolution at the European edge: Neandertal and Upper Paleolithic relationships. *Préhistoire Européene* 2, 9–69

Grün, R. and Stringer, C. (2000) Tabun revisited: revised ESR chronology and new ESR and U-series analyses of dental material from Tabun C1. *Journal of Human Evolution* 39, 601–612

Hasebe, K. (1912) Die Wirbelsäule der Japaner. *Zeitschrfit für Morphologie und Anthropologie* 15, 259–308

Heim, J. L. (1976) Les hommes fossiles de La Ferrassie I: Le gisement. Les squelettes adultes (crâne et squelette du tronc). *Archives de l'Institut de Paléontologie Humaine* 35, 1–331

Henry-Gambier, D. (2003) Les fossiles de Cro-Magnon (Les Eyzies-de-Tayac, Dordogne): Nouvelles données sur leur position chronologique et leur attribution culturelle. *Bulletins et Mémoires de la Société d'Anthropologie de Paris* ns14, 89–112

Henry-Gambier, D., Maureille, B. and White, R. (2004) Vestiges humains des niveaux de l'Aurignacien ancien du site de Brassempouy (Landes). *Bulletins et Mémoire de la Société d'Anthropologie de Paris* ns 16, 49–87

Holliday, T. W. (1995) *Body size and proportions in the Late Pleistocene Western Old World and the origins of modern humans*. Ph. D. Thesis, University of New Mexico

Holliday, T. W. (1997) Body proportions in Late Pleistocene Europe and modern human origins. *Journal of Human Evolution* 32, 423–447

Holliday, T. W. (2000) Evolution at the crossroads: Modern human emergence in western Asia. *American Anthropologist* 102, 54–68

Holliday, T. W. (2005a) Body proportions. In (E. Trinkaus and J. Svoboda, Eds.) *Early modern human evolution in central Europe. The people of Dolní Věstonice and Pavlov*. New York: Oxford University Press, pp. 224–232

Holliday, T. W. (2005b) The costal skeletons. In (E. Trinkaus and J. Svoboda, Eds.) *Early modern human evolution in central Europe. The people of Dolní Věstonice and Pavlov*. New York: Oxford University Press, pp. 295–326

Holt, B. M. (1999) *Biomechanical evidence of decreased mobility in Upper Paleolithic and Mesolithic Europe*. Ph. D. Thesis, University of Missouri-Columbia

Hrdlička, A. (1937) The gluteal ridge and the gluteal tuberosities (3rd trochanter). *American Journal of Physical Anthropology* 23, 127–198

Hublin, J. J., Barroso Ruiz, C., Medina Lara, P., Fontugne, M. and Reyss, J. L. (1995) The Mousterian site of Zafarraya (Andalucia, Spain): Dating and implications on the palaeolithic peopling processes of Western Europe. *Comptes rendus de l'Académie des Sciences Paris Série IIa* 321, 931–937

Johnston, F. E. and Zimmer, L. O. (1989) Assessment of growth and age in the immature skeleton. In (M. İşcan and K. A. R. Kennedy, Eds.) *Reconstruction of life from the skeleton*. New York: Alan R. Liss, pp. 11–21

Kleinschmidt, O. (1938) Unausrottbare falsche Behauptungen. II. Die Halswirbel des Neandertalmenschen. *Beilage zu Falco* 34, 1–4

Knies, J. (1906) Nový nález diluviálního člověka u Mladeč na Moravé. *Věstník klubu přírodovědného v Prostějově* 8, 1–19

Kuželka, V. (1992) Osteologické zhodnoceni žeber mladopaleolitických jedinců z nálezů v Dolních Věstonicích. In (E. Vlček, Ed.) *Lovci Mamutů z Dolních Věstonice. Acta Musei Nationalis Pragae* 48B, 110–129

Lanier, R. R. Jr. (1939) The presacral vertebrae of American white and negro males. *American Journal of Physical Anthropology* 25, 341–420

Liston, M., Falsetti, A. B. and Smith, F. H. (1989) Morphometric analysis of the Mladeč postcranial remains (abstract). *American Journal of Physical Anthropology* 78, 305

Malez, M. and Ullrich, H. (1982) Neuere paläoanthropologische Untersuchungen am Material aus der Höhle Vindija (Kroatien, Jugoslawien). *Palaeontologia Jugoslavica* 29, 1–44

Maška, J. (1905) Poznámky k diluviálním nálezům v jeskyních mladečských a stopám glaciálním na severovýchodní Moravě. *Časopis Moravského zemského musea* 5, 263–266

Matiegka, J. (1938) *Homo předmostensis. Folsilní člověk z Předmostí na Moravě II.* Prague: Česká Akademie Věd a Umění

McCown, T. D. and Keith, A. (1939) *The Stone Age of Mount Carmel II. The fossil human remains from the Levalloiso-Mousterian.* Oxford: Clarendon Press

McHenry, H. M., Corruccini, R. S. and Howell, F. C. (1976) Analysis of an early hominid ulna from the Omo Basin, Ethiopia. *American Journal of Physical Anthropology* 44, 295–304

McMinn, R. M. H., Hutchings, R. T., Pegington, J. and Abrahams, P. (1993) *Color atlas of human anatomy.* 3rd ed. St. Louis: Mosby Year Book

Mercier, N., Valladas, H., Valladas, G., Reyss, J. L., Jelínek, A., Meignan, L. and Joron, J. L. (1995) TL dates of burnt flints from Jelínek's excavations at Tabun and their implications. *Journal of Archaeological Science* 22, 495–509

Nagurka, M. L. and Hayes, W. C. (1980) An interactive graphics package for calculating cross-sectional properties of complex shapes. *Journal of Biomechanics* 13, 59–64

Nicolăescu-Plopşor, D. (1968) Les hommes fossiles en Roumanie. *Proceedings of the VIIth International Congress of Anthropological and Ethnological Sciences (Moscow)* 3, 381–386

Niewoehner, W. A., Weaver, A. H. and Trinkaus, E. (1997) Neandertal capitate-metacarpal articular morphology. *American Journal of Physical Anthropology* 103, 219–233

Ohba, H. (1935) Anthropologische Untersuchungen über das Skelett der Yoshiko-Steinzeitmenschen aus der Prov. Mikawa, Japan. IV. Teil. Über die oberen Extremitäten. *Journal of the Anthropological Society of Nippon* 50, 1–189

Oliva, M. (1989) Mladopaleolitické nálezy z Mladečských Jeskyní (The Upper Paleolithic finds from the Mladeč Cave). *Časopis Moravského Muzea* 74, 35–54

Orschiedt, J. (2002) Datation d'un vestige humain provenant de La Rochette (Saint Léon-sur-Vézère, Dordogne) par la méthode du carbone 14 en spectrométrie de masse. *Paléo* 14, 239–240

Pal, G. P. and Routal, R. V. (1986) A study of weight transmission through the cervical and upper thoracic regions of the vertebral column in man. *Journal of Anatomy* 148, 245–261

Pap, I., Tillier, A. M., Arensburg, B. and Chech, M. (1996) The Subalyuk Neanderthal remains (Hungary): a re-examination. *Annales Historico-Naturales Musei Nationalis Hungarici* 88, 233–270

Piveteau, J. (1963–66) La grotte de Regourdou. *Annales de Paléontologie* 49, 285–305; 50, 155–194; 52, 163–194

Rhoads, J. G. and Trinkaus, E. (1977) Morphometrics of the Neandertal talus. *American Journal of Physical Anthropology* 46, 29–44

Rink, W. J., Schwarcz, H. P., Smith, F. H. and Radovčić, J. (1995) ESR dates for Krapina hominids. *Nature* 378, 24

Ruff, C. B. (1994) Morphological adaptation to climate in modern and fossil hominids. *Yearbook of Physical Anthropology* 37, 65–107

Ruff, C. B. (1995) Biomechanics of the hip and birth in early *Homo. American Journal of Physical Anthropology* 98, 527–574

Ruff, C. B. (2000a) Body size, body shape, and long bone strength in modern humans. *Journal of Human Evolution* 38, 269–290

Ruff, C. B. (2000b) Biomechanical analyses of archaeological human skeletal samples. In (M. A. Katzenberg and S. R. Saunders, Eds.) *Biological anthropology of the human skeleton.* New York: Alan R. Liss. pp. 71–102

Ruff, C. B. and Hayes, W. C. (1983) Cross-sectional geometry of Pecos Pueblo femora and tibiae - A biomechanical investigation. II. Sex, age, and side differences. *American Journal of Physical Anthropology* 60, 383–400

Ruff, C. B., Trinkaus, E., Walker, A. and Larsen, C. S. (1993) Postcranial robusticity in *Homo*. I: Temporal trends and mechanical interpretations. *American Journal of Physical Anthropology* 91, 21–53

Ruff, C. B., Trinkaus, E. and Holliday, T. W. (1997) Body mass and encephalization in Pleistocene *Homo*. *Nature* 387, 173–176

Ruff, C. B., Trinkaus, E. and Holt, B. (2000) Lifeway changes as shown by postcranial skeletal robustness (abstract). *American Journal of Physical Anthropology (Suppl.)* 30, 266

Ruff, C. B., Walker, A. and Trinkaus, E. (1994) Postcranial robusticity in *Homo*. III: Ontogeny. *American Journal of Physical Anthropology* 93, 35–54

Sakaue, K. (1999) Bilateral asymmetry of the first metacarpal bone in the Jomon shellmound males (in Japanese). *Anthropological Science* 107, 21–30

Schaaffhausen, H. (1858) Zur Kenntnis der ältesten Rassenschädel. *Archiv für Anatomie, Physiologie und wissenschaftliche Medicin* 25, 453–478

Scheuer, L. and Black, S. (2000) *Developmental juvenile osteology*. London: Academic Press

Senut, B. (1981) *L'Humérus et ses Articulations chez les Hominidés Plio-Pleistocènes*. Paris: C.N.R.S.

Shackelford, L. L. and Trinkaus, E. (2002) Late Pleistocene human femoral diaphyseal anterior curvature. *American Journal of Physical Anthropology* 118, 359–370

Sládek, V., Trinkaus, E., Hillson, S. W. and Holliday, T. W. (2000) *The people of the Pavlovian: Skeletal catalogue and osteometrics of the Gravettian fossil hominids from Dolní Věstonice and Pavlov. Dolní Věstonice Studies* 5. Brno: Akademie věd Česke republicky

Smith, F. H. (1984) Fossil hominids from the Upper Pleistocene of central Europe and the origin of modern Europeans. In (F. H. Smith and F. Spencer, Eds.) *The origins of modern humans*. New York: Alan R. Liss, pp. 137–209

Smith, F. H. and Trinkaus, E. (1991) Les origines de l'homme moderne en Europe centrale: Un cas de continuité. In (J. J. Hublin and A. M. Tillier, Eds.) *Aux origines d'*Homo sapiens. *Nouvelle Encyclopédie Diderot*. Paris: Presses Universitaires de France, pp. 251–290

Smith, F. H., Trinkaus, E., Pettitt, P. B., Karavanić, I. and Paunović, M. (1999) Direct radiocarbon dates for Vindija G_1 and Velika Pećina Late Pleistocene hominid remains. *Proceedings of the National Academy of Science USA* 96, 12281–12286

Stewart, T. D. (1962) Neanderthal cervical vertebrae with special attention to the Shanidar Neanderthals from Iraq. *Bibliotheca Primatologica* 1, 130–154

Stockton, T. C. (1997) *Upper Pleistocene hominid postcrania from Mladeč Cave, Moravia: A descriptive and comparative analysis with respect to current models of modern human emergence*. M. A. Thesis, Northern Illinois University

Straus, W. L. Jr. and Cave, A. J. E. (1957) Pathology and the posture of Neanderthal man. *Quarterly Review of Biology* 32, 348–363

Stringer, C. B. (1986) An archaic character in the Broken Hill innominate E. 719. *American Journal of Physical Anthropology* 71, 115–120

Svoboda, J. (2000) The depositional context of the Early Upper Paleolithic human fossils from the Koněprusy (Zlatý Kůň) and Mladeč Caves, Czech Republic. *Journal of Human Evolution* 38, 523–536

Svoboda, J. (2006) The structure of the cave, stratigraphy, and depositional context. In (M. Teschler-Nicola, Ed.) *Early modern humans at the Moravian Gate: The Mladeč Caves and their remains*, this issue, pp. 27–40

Szombathy, J. (1881) Die Lautscher Höhle. *Fünfter Bericht der prähistorischen Commission der math.-naturwiss. Classe der kaiserliche Academie der Wissenschaften* 85, 17

Szombathy, J. (1882) Über Ausgrabungen in den mährischen Höhlen im Jahre 1881. *Sitzungsberichte der Kaiserlichen Akademie der Wissenschaften* 85, 90–107

Szombathy, J. (1904) Neue diluviale Funde von Lautsch in Mähren. *Jahrbuch der k. k. Zentral-Kommission für Kunst- und historische Denkmäler* 2, 9–16

Szombathy, J. (1925) Die diluvialen Menschenreste aus der Fürst-Johanns-Höhle bei Lautsch in Mähren. *Die Eiszeit* 2, 1–34, 73–95

Taylor, F. R. and Twomey, L. T. (1984) Sexual dimorphism in human vertebral body shape. *Journal of Anatomy (London)* 138, 281–286

Toerien, M. J. (1957) Note on the cervical vertebrae of the La Chapelle man. *South African Journal of Science* 53, 447–449

Trinkaus, E. (1975a) Squatting among the Neandertals: A problem in the behavioral interpretation of skeletal morphology. *Journal of Archaeological Science* 2, 327–351

Trinkaus, E. (1975b) *A functional analysis of the Neandertal foot.* Ph. D. Thesis, University of Pennsylvania

Trinkaus, E. (1976) The evolution of the hominid femoral diaphysis during the Upper Pleistocene in Europe and the Near East. *Zeitschrift für Morphologie und Anthropologie* 67, 291–319

Trinkaus, E. (1981) Neanderthal limb proportions and cold adaptation. In (C. B. Stringer, Ed.) *Aspects of human evolution.* London: Taylor and Francis, pp.187–219

Trinkaus, E. (1983) *The Shanidar Neandertals.* New York: Academic Press

Trinkaus, E. (1993) Femoral neck-shaft angles of the Qafzeh-Skhul early modern humans, and activity levels among immature Near Eastern Middle Paleolithic hominids. *Journal of Human Evolution* 25, 393–416

Trinkaus, E. (1997) Appendicular robusticity and the paleobiology of modern human emergence. *Proceedings of the National Academy of Sciences USA* 94, 13367–13373

Trinkaus, E. (2000) The "Robusticity Transition" revisited. In (C. B. Stringer, R. N. E. Barton and C. Finlayson, Eds.) *Neanderthals on the edge.* Oxford: Oxbow Books, pp. 227–236

Trinkaus, E. (2005a) The upper limb remains. In (E. Trinkaus and J. Svoboda, Eds.) *Early modern human evolution in Central Europe. The people of Dolní Věstonice and Pavlov.* New York: Oxford University Press, pp. 327–372

Trinkaus, E. (2005b) The lower limb remains. In (E. Trinkaus and J. Svoboda, Eds.) *Early modern human evolution in Central Europe. The people of Dolní Věstonice and Pavlov.* New York: Oxford University Press, pp. 380–418

Trinkaus, E. and Churchill, S. E. (1988) Neandertal radial tuberosity orientation. *American Journal of Physical Anthropology* 75, 15–21

Trinkaus, E. and Churchill, S. E. (1999) Diaphyseal cross-sectional geometry of Near Eastern Middle Paleolithic humans: The humerus. *Journal of Archaeological Science* 26, 173–184

Trinkaus, E., Churchill, S. E. and Ruff, C. B. (1994) Postcranial robusticity in *Homo*, II: Humeral bilateral asymmetry and bone plasticity. *American Journal of Physical Anthropology* 93, 1–34

Trinkaus, E., Churchill, S. E., Ruff, C. B. and Vandermeersch, B. (1999a) Long bone shaft robusticity and body proportions of the Saint-Césaire 1 Châtelperronian Neandertal. *Journal of Archaeological Science* 26, 753–773

Trinkaus, E., Moldovan, O., Milota, Ş., Bîlgăr, A., Sarcina, L., Athreya, S., Bailey, S. E., Rodrigo, R., Gherase, M., Higham, T., Bronk Ramsey, C. and van der Plicht, H. (2003a) An early modern human from the Peştera cu Oase, Romania. *Proceedings of the National Academy of Sciences USA* 100, 11231–11236

Trinkaus, E., Milota, Ş., Rodrigo, R., Gherase, M. and Moldovan, O. (2003b) Early modern human cranial remains from the Peştera cu Oase, Romania. *Journal of Human Evolution* 45, 245–253

Trinkaus, E., Moldovan, O., Zilhão, J., Rougier, H., Rodrigo, R., Milota, Ş., Gherase, M., Sarcina, L., Băltean, I., Codrea, V., Bailey, S. E., Franciscus, R. G., Ponce de León, M. and Zollikofer, C. P. E. (2005) The Peştera cu Oase and early modern humans in southeastern Europe. In (N. Conard, Ed.) *Neanderthals and modern humans meet?* Tübingen Publications in Prehistory (in press)

Trinkaus, E. and Rhoads, M. L. (1999) Neandertal knees: Power lifters in the Pleistocene? *Journal of Human Evolution* 37, 833–859

Trinkaus, E. and Ruff, C. B. (1999a) Diaphyseal cross-sectional geometry of Near Eastern Middle Paleolithic humans: The femur. *Journal of Archaeological Science* 26, 409–424

Trinkaus, E. and Ruff, C. B. (1999b) Diaphyseal cross-sectional geometry of Near Eastern Middle Paleolithic humans: The tibia. *Journal of Archaeological Science* 26, 1289–1300

Trinkaus, E. and Ruff, C. B. (2000) Comment on: O. M. Pearson, "Activity, climate, and postcranial robusticity. Implications for modern human origins and scenarios of adaptive change." *Current Anthropology* 41, 598

Trinkaus, E., Stringer, C. B., Ruff, C. B., Hennessy, R. J., Roberts, M. B. and Parfitt, S. A. (1999b) Diaphyseal cross-sectional geometry of the Boxgrove 1 Middle Pleistocene human tibia. *Journal of Human Evolution* 37, 1–25

Trinkaus, E., Svoboda, J., West, D. L., Sládek, V., Hillson, S. W., Drozdová, E. and Fišáková, M. (2000) Human remains from the Moravian Gravettian: Morphology and taphonomy of isolated elements from the Dolní Věstonice II site. *Journal of Archaeological Science* 27, 1115–1132

Trotter, M. and Gleser, G. C. (1952) Estimation of stature from long bones of American whites and negroes. *American Journal of Physical Anthropology* 10, 463–514

Twiesselmann, F. (1961) Le fémur Néanderthalien de Fond-de-Forêt (Province de Liège). *Mémoire de l'Institut Royal des Sciences Naturelles de Belgique* 149, 1–164

Vandermeersch, B. (1984) Á propos de la découverte du squelette Néandertalien de Saint-Césaire. *Bulletins et Mémoires de la Société d'Anthropologie de Paris Série* XIV, 1, 191–196

Vandermeersch, B. and Trinkaus, E. (1995) The postcranial remains of the Régourdou 1 Neandertal: The shoulder and arm remains. *Journal of Human Evolution* 28, 439–476

Villemeur, I. (1994) *La Main des Néandertaliens*. Paris: C.N.R.S.

Vlček, E. (1991) Die Mammutjäger von Dolní Věstonice. *Archäologie und Museum* 22

Wolpoff, M. H. (1989) The place of the Neandertals in human evolution. In (E. Trinkaus, Ed.) *The emergence of modern humans*. Cambridge: Cambridge University Press, pp. 97–143

Wolpoff, M. H., Frayer, D. W. and Jelínek, J. (2006) Lost, destroyed or misidentified postcranial specimens from Mladeč. In (M. Teschler-Nicola, Ed.) *Early modern humans at the Moravian Gate: The Mladeč Caves and their remains*, this issue, pp. 447–452

Wolpoff, M. H., Frayer, D. W., Trinkaus, E. and Teschler-Nicola, M. (2006) Inventory and photo-documentation of the Mladeč hominid remains. In (M. Teschler-Nicola, Ed.) *Early modern humans at the Moravian Gate: The Mladeč Caves and their remains*, this issue, pp. 159–184

Wolpoff, M. H., Hawks, J., Frayer, D. W. and Hunley, K. (2001) Modern human ancestry at the peripheries: A test of the replacement theory. *Science* 291, 293–297

Wolpoff, M. H., Smith, F. H., Malez, M., Radovčić, J. and Rukavina, D. (1981) Upper Pleistocene human remains from Vindija Cave, Croatia, Yugoslavia. *American Journal of Physical Anthropology* 54, 499–545

Chapter 14

LOST, DESTROYED OR MISIDENTIFIED POSTCRANIAL SPECIMENS FROM MLADEČ

Milford H. Wolpoff, David W. Frayer and Jan Jelínek

Introduction

Along with the postcranial material covered in the previous chapter, there are 23 other specimens described or mentioned by Szombathy which have been lost, were destroyed at Mikulov or were misidentified. Despite the limited details we have for these specimens, they provide some additional information about skeletal variation at the site and potential associations of certain skeletal parts into individuals. In some cases we can only list the elements since Szombathy published neither data nor photographs for many of the more fragmentary specimens. However, he provided some information, if sketchy, for some postcrania. For example, the strongly curved femoral shafts, catalogued by us as Mladec 78, were proposed by Szombathy to be associated with the Mladec 5 cranium in the side cave. If so, this would indicate that several whole cadavers were likely deposited in the both caves, as we argued in previous chapters. Likewise, Szombathy described postcranial remains associated with the child vault (lost) from the Quarry Cave. Thus, there is some evidence from the Quarry and Main Cave that more than just parts of cadavers were deposited in these burial sites. If so, for a variety of reasons, not all the bones were preserved or collected. Besides the lost details about the nature of the burial sites in the Main and Quarry Caves, the missing skeletal parts described here would have been extremely helpful in sorting out some of the taphonomic details from the Main and Quarry Caves.

Here, we list the additional specimens and describe them from Szombathy's text or, in some cases, his plates and tables. We have included comparative data for Neandertal and the Skhul / Qafzeh samples. For these data, we used male measurements as complied from our personal dataset. In fact, for this material, most of the specimens for which Szombahy provided measurements, have very large dimensions and we consider them to be most likely male.

Mladeč 34 (misidentified)

In addition to Mladeč 11, only two other vertebrae are known to have been found at Mladeč. Mladeč 34 is described by Szombathy (1925, 24) as consisting of a vertebral body "...originating from an individual smaller than Mladeč 11." This specimen was excavated in 1882 and is listed by Vlček (1971) as being present at the Naturhistorisches Museum, Wien. It was not located in the inventory in 1961 and has been considered lost. However, among the Mladeč remains we found a non-hominid specimen that fits the description of being slightly smaller than Mladeč 11 and believe it is also possible that the fragment was misidentified as hominid. The other vertebra was Mladeč 60.

Mladeč 35 (lost)

Although several phalanges are reported to have been excavated at Mladeč (Table 1; Szombathy, 1925, 75, 77), only one is described. This was Mladeč 35, a proximal row phalanx lacking its proximal articular end. Szombathy (1925, 26) suggests that this slender bone is probably from the left third finger. He gives no metric data, stating basically that there are none worth giving. This specimen was excavated by Szombathy but can no longer be located in the collections at the Naturhistorisches Museum, Wien.

Mladeč 36 (lost)

This specimen was a proximal articular surface and shaft of a 5th metatarsal. This specimen had a preserved length of 61 mm. Unfortunately it cannot be located in the Naturhistorisches Museum, Wien collections and is presumed lost.

Mladeč 46 (destroyed or lost)

The postcranial remains associated with this child from the Quarry Cave include most or all of a humerus, ulna, and radius, and fragments of a fibula and two pieces of radius.

Mladeč 60 (destroyed)

Mladeč 60 is described by Szombathy (1925, 74) as a multiply damaged lumbar vertebra. No other information is given, and this specimen perished at Mikulov.

Mladeč 62–65 (destroyed)

Four clavicles (Mladeč 62, 63, 64 and 65) were recovered from the cave between 1903 and 1922. Szombathy (1925, 74) notes that one (Mladeč 62) was rather fragile, strongly curved, and lacked either end. A second specimen (Mladeč 65) is described as robust, with particularly strong curvature and extensive damage on both ends (1925, 77).

Mladeč 66 (destroyed)

Only two scapulae were apparently excavated from Mladeč. Mladeč 66 is listed only as a fragment of scapula (Szombathy, 1925, 77). There is no description of Mladeč 101, the other scapular specimen.

Absence of information on these scapular pieces is particularly frustrating given the patterns of scapular axillary border morphology which tend to differentiate Neandertals from recent humans (Gorjanovič-Kramberger, 1914; Smith, 1976) and the intermediate nature of the known European Upper Paleolithic scapular sample (Trinkaus, 1977; 1983).

Mladeč 68 (destroyed)

Mladeč 68 is listed as a 130 mm long, slender midshaft fragment of a humerus which had been heavily battered.

Mladeč 69 (destroyed)

Mladeč 69 was evidently a very robustly developed distal end of a left humerus, with a broken or perforated olecranon fossa. Szombathy's text is unclear as to whether the break was postmortem or was a normally perforated septal aperture. Szombathy remarked that the medial epicondyle was "especially very strongly developed". He provides a 65 mm dimension for what we presume was the biepicondylar breadth. This is quite large in the context of the Upper Paleolithic (only the massive Pavlov humerus is larger), but close to the means of the males from Skhul/Qafzeh (66.5, n=3) and Neandertal males (65.0, n=5). Because of its massiveness, we regard the specimen as a male.

Mladeč 70 and 71 (destroyed)

Szombathy described these as "2 diaphyses of the upper arm". It is not clear whether he believed that they were humeri of the same individual. The diaphyses were long and slender, with areas of strong muscle attachments.

Mladeč 73 (destroyed)

The only information we have about this specimen is a short description by Szombathy, who stated it was the proximal end of an ulna with a robust articular area.

Mladeč 74 (destroyed)

Szombathy describes this right femoral diaphysis as "strong with weak muscle attachments." Strong indeed, its midshaft circumference of 90 mm (the only measurement he provides) is greater than any other Mladeč femur, even exceeding the robustly developed Mladeč 78 (see Table 1) it also contrasts with Mladeč 78 in possessing a straight shaft. For these reasons, it seems likely Mladeč 74 was a male, in spite of the "weak muscle attachments."

Table 1. Femur dimensions (mm)

	Mladeč 78	Skhul/Qafzeh		Neandertal	
		Mean (n)	Range	Mean (n)	Range
Midshaft					
Anteroposterior (M6)	37.0	33.1 (11)	24.6–41.3	29.2 (9)	26.0–31.7
Mediolateral (M7)	26.0	27.4 (11)	24.0–31.9	30.0 (9)	28.6–31.9
Circumference (M8)	86.0	96.7 (10)	85.0–112.0	93.0 (7)	90.0–99.9
Pilastric index	142.3	120.8 (13)	96.5–143.8	97.1 (9)	86.7–103.6

While the midshaft circumference of Mladeč 74 fits within the range of the comparative samples, the other four Mladeč femora with circumferences (27, 28, 75 and 78 – some of which are likely females) are at or below the Skhul/Qafzeh minimum and well below the Neandertal range. Although no complete Mladeč femur was preserved, it is clear from these data that the Mladeč specimens were almost certainly built in a slender form relative to the earlier samples, similar to the other Upper Paleolithic Europeans. Mladeč 74 probably was quite tall.

Mladeč 75 (destroyed)

Szombathy described this right femoral diaphysis as being "extremely curved [with] a very projecting *linea aspera*". While the curvature and linea development, which we assume reflects a prominent pilaster, may resemble Mladeč 78 (below), the size of this specimen does not. The minimum shaft circumference (the only dimension Szombathy recorded) was 84 mm, as small as that of Mladeč 27, a female, and also Mladeč 28 if we can assume that minimum and midshaft circumferences are close to the same. (This seems to be the case with virtually all Upper Pleistocene specimens we have studied). Mladeč 75 thereby appears to be one of the smaller of the earlier Upper Paleolithic folk.

Knowledge of its age at death would be useful in establishing whether the size difference between the smaller and larger femora reflected sexual dimorphism.

Mladeč 77 (destroyed)

The specimen was a 250 mm long distal half of a left femur. It exhibited a well-developed *linea aspera* and pilaster. Szombathy reports that the distal end was 78 mm in breadth and damaged. We are unclear about whether this breadth is biepicondylar or articular because comparisons with other European early Upper Paleolithic femora suggest it could be either. Consequently, his metric is not usuable.

Mladeč 78 (destroyed)

Ths specimen comprises the proximal end and shaft, a detached damaged distal portion of the left femur of an adult, and a right diaphyseal fragment that may be from the same individual. Szombathy does not discuss the right specimen further, except to state that it is identical to the left one. On the other hand, he describes and discusses the significance of the left femur in some detail and he provides a photograph of it (1925, 77, reproduced as Fig. 1).

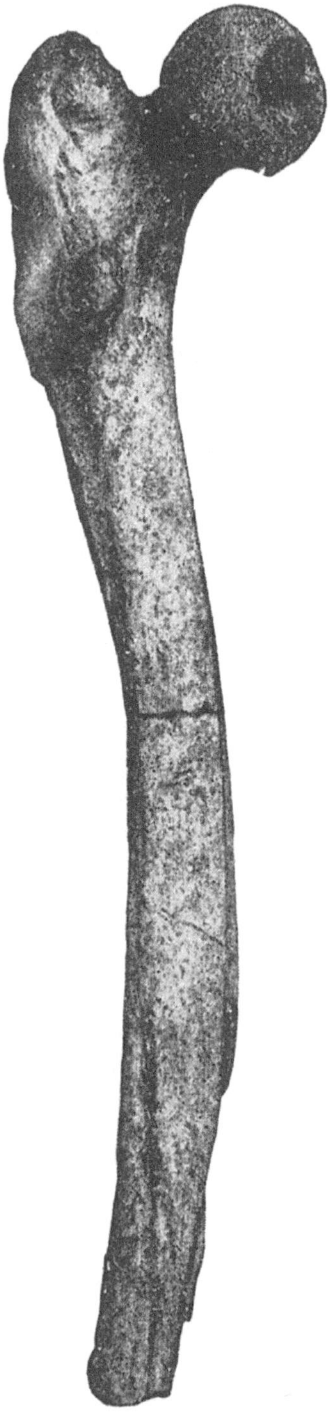

Fig. 1. Mladeč 78 (destroyed), left femur (reproduced from Szombathy, 1925)

According to Szombathy, the distal part of the bone was not attachable, so that the exact length of the bone could not be given. It would have been, in his opinion, a long femur. The femoral neck was short and according to Szombathy it formed almost a right angle with the shaft. The greater and lesser trochanters were very strongly developed. Szombathy also describes a well-developed third trochanter for this specimen, which he described as a 35 mm long and 16 mm broad rugosity. According to Szombathy the third trochanter

> *"begins on the external margin at the same level as the lesser trochanter, and reaches a point that is more inferior. From this roughly wrinkled area the linea aspera stretches downwards as a coarsely wrinkled ledge. It is more than 1 centimeter wide; in its upper portion a furrow accompanies the line. In the mid-region of the bone the linea turns into a strongly projecting pilaster."*

Szombathy stressed the strong anteroposterior curvature of this specimen and gives estimates for a radius and index of curvature (518 and 8.5, respectively). He reports that if the main portion of the bone (i.e. without the disarticulated distal end) is put on a table on its dorsal size, the shaft curves 27 mm above the surface. Unfortunately, given the orientation of Szombathy's illustration (Fig. 1), it is not possible to compare this curvature with other specimens. Interestingly, Szombathy (1925, 89) noted that the strongly curved Mladeč 78 femora were found in association with Mladeč 5, which he described as a "... particularly low skull, reminiscent of Neandertal form". He interpreted these as members of a presumably Neandertal-like "race" of the Upper Paleolithic, generally then referred to as the Brux race.

The large anteroposterior midshaft dimensions (and related prominent pilaster) of this specimen suggest it is male. In fact, the anteroposterior diameter is one of the largest in the European early Upper Paleolithic – only two specimens from Cro-Magnon are larger. The mediolateral diameter is much less, and consequently the circumference at midshaft is at the low end of the Skhul / Qafzeh range. Related to this, the Mladeč 78 pilastric index (Table 1) is extremely high (142.3 mm). This value is matched by only a single Skhul / Qafzeh femur and confirm Szombathy's description of a "marked pilaster."

Unfortunately, Szombathy gives no femur head dimensions, nor does he note if a hypertrochanteric fossa or proximolateral femoral flange is found on the specimen. No judgment regarding these last two features can be made from the only photograph. The femoral head, however, appears to be relatively small and is clearly fused to the shaft.

Mladeč 79 (destroyed)

The second largest tibial fragment from Mladeč was a 150 mm long right proximal end fragment, "worn" around the edge of the joint. Szombathy (1925, 75) further described this specimen as possessing a well-developed tibial tuberosity, a triangular shaft shape, and a 97 mm circumference at its distal break.

Mladeč 80 (destroyed)

This is described as portion of tibial epiphysis. Szombathy gives dimensions of 37 mm by 25 mm and a cnemic index of 67.5; it is uncertain how long the bone was, or whether it extended to the midshaft and thereby is comparable with the other remains. It appears to have been a rather large.

Mladeč 81 (destroyed)

According to Szombathy this was the robust diaphysis of a left tibia. Its midshaft circumference is very large (99 mm) and we regard it as a male. This circumference is above the Neandertal male range, but lies within the range of the Skhul / Qafzeh males. The Mladeč 81 index fits the European early Upper Paleolithic pattern of a smaller index (a less rounded bone) and is below the Skhul / Qafzeh and Neandertal ranges. This low index is mainly a consequence of the marked anteroposterior midshaft diameter, which is above the ranges of the comparative samples, and is only matched by Mladeč 84.

Mladeč 84 (destroyed)

The specimen is a 400 mm long portion of a right tibia with badly damaged articular areas. Szombathy describes the muscle attachments as being "fairly strong". At the approximate midshaft the cross-section is moderately platycnemic, with dimensions of 40 mm by 28 mm and a cnemic index of 70, within the ranges of the comparative samples. The cnemic index is virtually identical to the Neandertal mean, although the average Neandertal dimensions are quite a bit smaller. Based on the muscle attachments and the size of this specimen we regard it as a male (Table 2). At midshaft, both the anteroposterior and mediolateral diameters exceed all other specimens. The closest to it is Skhul 5, which also has a very similar cnemic index.

Table 2. Tibia dimensions (mm)

	Mladeč		Skhul/Qafzeh		Neandertal[1]	
	81	84				
			Mean (n)	Range	Mean (n)	Range
Midshaft						
Anteroposterior (M8)	39.0	40.0	34.2 (4)	30.4–38.8	33.4 (4)	30.5-35.6
Mediolateral	24.0	28.0	24.7 (4)	23.0–26.0	23.9 (4)	23.1–24.7
Cnemic index	61.5	70.0	72.6 (4)	65.7–78.9	71.8 (4)	65.2–81.0
Circumference	99.0		94.2 (3)	86.5–100.0	90.0 (3)	89.0–91.0

[1] includes Krapina 219

These two male tibia fragments from the Quarry Cave, Mladeč 81 and 84, support the contention that at least some of the Mladeč males were unusually large and robust individuals, even by standards set by the large males from the later Upper Paleolithic such as the Cro-Magnon and Pavlov males. They are, for the most part, as large as the largest specimens of Neandertals and from Skhul/Qafzeh, more often than not exceeding their ranges.

Conclusion

A limited amount of information can be gleaned from Szomathy's descriptions of numerous, now lost or destroyed postcranial fragments from the Main and Quarry Cave. The most interesting details from these bones are further indications that the Quarry Cave contained the burial of three individuals. It will always be a tragedy of paleoanthropology that these materials were not more systematically collected and retained for subsequent analyses.

References

Gorjanovič-Kramberger, D. (1914) Der Axillarand des Schulterblattes des Menschen von Krapina. *Glasnik Hrvatskoga Prirodoslovnog Društva* 26, 231–257

Smith, F. H. (1976) The Neandertal remains from Krapina: A descriptive and comparative study. *University of Tennessee Department of Anthropology Reports of Investigations* 15, 1–359

Steele, D. G. (1976) The estimation of sex on the basis of the talus and calcaneus. *American Journal of Physical Anthropology* 45, 581–588

Szombathy, J. (1925) Die diluvialen Menschenreste aus der Fürst-Johanns-Höhle bei Lautsch in Mähren. *Die Eiszeit* 2, 1–34, 73–95

Trinkaus, E. (1977) A functional interpretation of the axillary border of the Neandertal scapulae. *Journal of Human Evolution* 6, 231–234

Trinkaus, E. (1983) *The Shanidar Neandertals.* New York: Academic Press

Vlček, E. (1971) Czechoslovakia. In (K. P. Oakley, B. G. Campbell and T. I. Molleson, Eds.) *Catalogue of fossil hominids, Part II: Europe.* London: British Museum (Natural History), pp. 47–64

Chapter 15

EXTERNAL GEOMETRY OF MLADEČ NEUROCRANIA COMPARED WITH ANATOMICALLY MODERN HUMANS AND NEANDERTALS

Gerhard W. Weber, Philipp Gunz, Philipp Mitteröcker, Andrea Stadlmayr, Fred L. Bookstein and Horst Seidler

Introduction

Because of their age and their Central European location, the Upper Paleolithic Mladeč human remains (also known as "Lautsch") have played a central role in the discussion of modern human origins ever since the first work with them began in 1881 (Szombathy, 1925). The fossil material consists of more than 100 specimens from the large Main Cave (e.g., Mladeč 1 and 2) and from the Quarry Cave (e.g., Mladeč 5 and 6). Both sites seem to be accumulations of human remains that fell through chimneys or fissures into the karstic cave system (Frayer, 1986; Svoboda, 2000). The morphological differences of Mladeč 5 and 6 vs. 1 and 2 (interestingly each group originating from a different cave) are attributed to sexual dimorphism by most authors (Jelínek, 1983; Frayer, 1986; Wolpoff et al., this volume). In spite of substantial morphological variation within sex (Frayer, 1986), the specimens considered to be male (Mladeč 5 and 6) are very robust, with a low braincase, thick cranial bones, Neandertal-like occipital bunning, marked spongy bone development, thick projecting supraorbitals shaped differently than in Neandertals, and large cranial capacities, whereas the female specimens (Mladeč 1 and 2) are more gracile than the Mladeč males but robust nevertheless in comparison with recent *Homo sapiens.*

Since 1900 (Anthropology-Congress Halle a. S., IX/1900), the Mladeč remains have been generally considered to be modern *Homo sapiens* ("Cro-Magnon race"; Szombathy, 1925) or, according to the discoverer, as "definitely belonging to the *Homo sapiens* fossilis" (ibid.). Most anthropologists today accept Mladeč as modern *Homo sapiens,* but the prominent supraorbitals, the well-developed occipital bun reminiscent of the Neandertal "chignon", and the distinct nuchal ridge have stimulated speculations about possible anatomical links to the Neandertals who preceded them. The relation of this assemblage (one of the earliest in Central Europe) to other groups of fossil hominids is crucial and likewise the question of hybridisation (Bräuer, 1980; Duarte et al., 1999; Tattersall and Schwartz, 1999) because of the chronological and regional overlapping of anatomically modern humans and Neandertals in Europe. That relation has therefore been the focus of diverse studies, including comparisons of gross morphology (Jelínek, 1951; Smith, 1982; Frayer, 1986; Bräuer and Broeg, 1998, Wolpoff et al., 2001; Frayer et al., this volume; Wolpoff et al., this volume) and of mtDNA (Serre et al., 2004) and new radiometric datings of hominid and faunal remains (Wild et al., 2005).

Especially Mladeč 5 (see Fig. 1.) was described as Neandertal-like in several aspects by Smith (1982), Jelínek (1983), and Wolpoff (1999), and also as resembling Jebel Irhoud-like or archaic *Homo* (Stringer et al., 1984). Frayer (1986) saw Mladeč as "good evidence in support of the gradualist model," with "a greater number of archaic features shared with Neandertals" (see also Frayer et al., this volume). Others have reached quite opposite conclusions: for instance, Bräuer and Broeg (1998) via a non-metric analysis. Klein (2003), too, noted that the typical cranial traits of Neandertals preclude a major Neandertal contribution to living human populations.

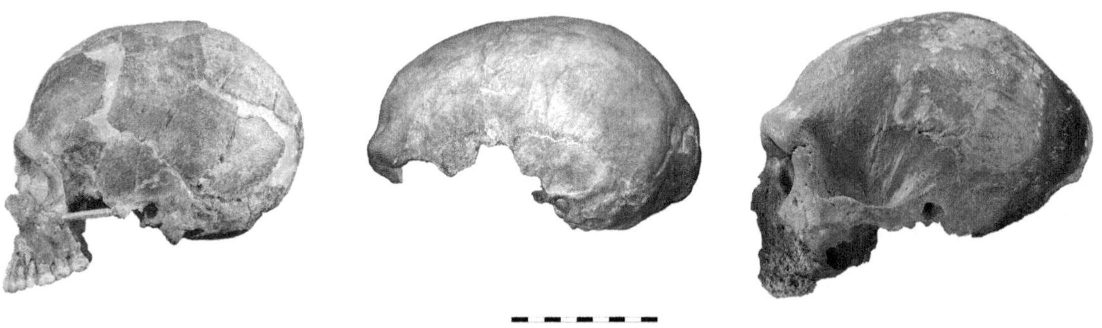

Fig. 1. Lateral views of an anatomically modern *Homo sapiens* (Qafzeh 9, left), Mladeč 5 (center) and a classic Neandertal (Guattari 1, right). Image of Qafzeh 9 mirrored for comparison, from Schwartz and Tattersall (2002). Image of Mladeč 5 by Inst. for Anthropology, University of Vienna. Image of Guattari 1 from Piperno and Scichilone (1991). All to scale

In 2001, Wolpoff et al. argued that pairwise difference analysis of non-metric characters of Mladeč crania, Neandertals and Levantine specimens weighed against the Out of Africa hypothesis. They found no evidence that the Mladeč crania could be grouped with the Levantines (early modern humans) to the exclusion of European Neandertals, and thus argued that a dual-ancestry model could not be rejected. Neandertal anatomy in later populations is an indication of sufficient Neandertal ancestry to refute the replacement theory, they noted. Predictably, serious criticisms were immediately raised against these arguments regarding relevance to the replacement theory, choice of material, choice of characters, assignment of characters, and method (Bräuer et al., 2004).

The debate between proponents of the replacement hypothesis and proponents of the regional continuity hypothesis evidently continues. In fact, there is generally little agreement about systematics of Mid- to Late-Pleistocene hominid fossils, owing partly to the subtle differences in morphological traits that distinguish putative taxa and partly to temporal and spatial gaps within the fossil record.

In light of all this disagreement, the Mladeč assemblage becomes a particularly valuable case study inasmuch as it provides evidence for modern humans from the same region inhabited by Neandertals contemporaneously or nearly contemporaneously (Smith et al., 1999; Churchill and Smith, 2001; Conard et al., 2003). Assuming that modern humans evolved in Africa between 160 and 250 kyr and successfully migrated into Eurasia (Klein, 2003; White et al., 2003), we need to think about complete replacement of the indigenous Neandertals between about 40 kyr and 30 kyr, versus the possibility of gene flow between the two groups as also less rigorous proponents of the replacement theory would accept (Bräuer, 1984; Stringer, 1992).

One direct assault upon this question is via the comparison of DNA between fossils and moderns. Most such studies (Krings et al., 1997; Krings et al., 1999; Ingman et al., 2000; Ovchinnikov et al., 2000; Krings et al., 2000; Serre et al., 2004, Cooper et al., 2004) argue for significant differences between the genotypes of Neandertals and modern humans, with little evidence of persisting sequences from the former. Nevertheless, DNA of unequivocally anatomically modern fossils can itself be strongly divergent from that of living humans (Adcock et al., 2001; Relethford, 2001). Templeton (2002), analysing human haplotype trees, concludes that there has been continual substantial interbreeding over at least the past 1.7 myr, and methodological challenges (Maddison, 1991; Templeton, 1992) that have become particularly serious recently (Pusch and Bachmann, 2004) call the DNA comparative method itself into question.

An indirect approach would study the phenotypic variation – the hard evidence – that is preserved in fossils. Of course loud disputes always arise wherever more than one paleanthropologist is considering such a question. Working backwards from phenotype to genotype (or inherited characters), we have to be careful to distinguish between those traits that are related to the genome and such that reflect environmental stimuli. Lovejoy et al. (1999) have compiled a classification system for morphological trait types that ranges from traits whose expression is under genetic influence and who are exposed to selection, to traits that are epigentic and not pleiotropic. The adult skull can be seen as the result of genetic and envrionmental influences acting upon shape and size. These interactions are between anatomical units and the rest of the organism as well as the environment (for example, mechanical loading) which makes it difficult to infer ancestry from morphological similarity alone. But generally, there is agreement that the skull of anatomically modern *Homo sapiens* (AMHS) is characterized by a number of autapomorphies such as longer and more flexed anterior cranial base, an overall diminution of facial size, and an increased globularity of the neurocranium (Lieberman et al., 2004). The greater the phyletic distances of specimens, the more prominent a part of the comparison these derived features become. But use of characters so global for discriminating closely related phenotypes within a single genus is more problematic. Moreover, underlying data can be only qualitative, thereby greatly depending on observer's experience, or data can be quantitative but insufficient to describe the form variation under examination. We are convinced that there are better quantitative tools to be exploited, tools built to account for sufficient geometric detail that inference can proceed in breadth and in depth simultaneously.

Classic quantitative description and analysis of cranial elements of the Mladeč assemblage was conducted by Jelínek (1951), Frayer (1986), Frayer et al. (in this volume) and Wolpoff et al. (in this volume) as well as others. But new techniques have been adopted for paleoanthropological questions within the last years that overcome the flaws of traditional morphometric approaches (Bookstein, 1991; 1997; Dryden and Mardia, 1998; O'Higgins, 2000; Weber et al., 2001). The present chapter will provide a state-of-the-art multivariate analysis of the geometry of the more complete Mladeč crania.

Three out of four Mladeč crania preserve sufficient parts of the neurocranium, as well as do many other fossil specimens known from the Late Pleistocene. For these reasons, our study focuses upon overall geometry of the Mladeč neurocranium in comparison with the variability of other populations such as modern people, upper paleolithic people, Neandertals, and some archaic forms of the genus *Homo*. The neurocranium reflects evolutionary changes in anatomy on the way to modern humans very well. Of the major changes, facial reduction and braincase restructuring both pertain to neurocranial form.

The special type of analysis conducted here provides several advantages compared with earlier approaches: (a) a dense data matrix of three-dimensional coordinates, (b) quantitative data in previously unattended regions, (c) independence of traditional orientation problems, (d) separate analysis of size and shape, (e) visualisation of regional shape differences.

Material

The excavations at the Fürst-Johanns-Höhle were begun in 1881 by Josef Szombathy and later continued by Jan Knies and others. The specimens are named after a nearby village, Mladeč, 4km from Litovel in central Moravia, what is now the Czech Republic. As the cranial materials are described in detail in Frayer et al. (this volume) and Wolpoff et al. (this volume), we focus only on the information most relevant for our own computations. For historical reasons, Mladeč 1 and 2 are housed at the Naturhistorisches Museum in Vienna under the curation of Maria Teschler-Nicola, along with a cast of Mladeč 6, which was lost in World War II. Mladeč 5 is at the Moravské zemské muzeum in Brno under the curation of Martin Oliva. Mladeč 1, 2, and 8 are consistently dated to ~ 31 ^{14}C kyr BP (Wild et al., 2005), and, both in the Main Cave and the adjacent Small Cave, seem associated with the Aurignacian stone and bone industry. The non-hominin remains, while quite variable, represent mostly a Würm II fauna (see Pacher, this volume).

Mladeč 1

The first to be found by Josef Szombathy, is the most complete cranium of an almost adult individual, characterized as female by others (Jelínek, 1983; Frayer, 1986; Wolpoff et al., this volume). It is weathered but largely intact. It is missing only parts of the right frontal, parietal and squamosal and teeth except all M1-2. The cranium is long and narrow (dolichocephal). Orbits are low and rectangular, and the supraorbital region is bipartite with supraorbital swellings that fuse across glabella. The occipital region appears to be swollen and rounded (Schwartz and Tattersall, 2002).

Mladeč 2

Mladeč 2 is an adult partial cranium (usually interpreted as female) lacking the occipital, sphenoid, malars and parts of squamosals. It is less dolichocephal than Mladeč 1, with gracile, thin bones. The supraorbital region is bipartite with a weakly developed "butterfly" swelling across glabella. The occipital would not have had protruded posteriorly as in Mladeč 1, and the parietals curve in much more steeply. This cranium was not used in the current analysis because it was too incomplete.

Mladeč 5

Is an adult calvaria (thought to be of male sex) with healed wounds. The braincase is very large, broad and robust. The supraciliary arch is again bipartite, and the occipital is well rounded and bulging. The specimen was slightly damaged by fire during World War II (Schwartz and Tattersall, 2002).

Mladeč 6

Is an incomplete adult calotte of massive bone (usually described as beeing male). The original was lost in 1945 in the Mikulov Castle fire, so that only a cast is available. The cranium is reminiscent of Mladeč 5 but not as broad. The supraciliary region is well developed but not protruding, with a glabellar butterfly. The occiput is only moderately protruding (ibid.).

Other specimens in the study (s. Table 1)
For geometrical comparison of the neurocranium, we distinguish between two groups:

- AMHS – anatomically modern *Homo sapiens* (from Late Pleistocene to Holocene)
- AH – more archaic forms of *Homo* (from Middle to Late Pleistocene)

The AMHS sample can be further divided into:

- AMHS recent – modern *Homo sapiens* on the one end of the timescale (end of Holocene)
- AMHS fossil – anatomically modern *Homo sapiens* from the Middle Paleolithicum to the border of Late Upper Paleolithicum/Mesolithicum
- AMHS /archaic – some crania that were included based on the distinctivness of specimens described as "morphologically archaic" (Grün and Stringer, 1991 – Jebel Irhoud 1), as "evolved *Homo erectus*" (Santa Luca, 1980 – Ngandong 7).

Table 1. Specimens (109) and grouping in the study

Group	Specimens
AMHS recent (50)	50 modern *Homo sapiens* crania of both sexes from Africa, Asia, Polynesia, South America, Europe, and Australia
AMHS fossil (30)	Bruenn III, Comb Capelle, Cro Magnon I, Cro Magnon 3, Grotte des Enfants IV, Oberkassel, Předmostí III, Předmostí IV, Skhul 5, Wadjak I, Bruenn II, Dolní Věstonice 2, Mladeč 5, Mladeč 6, Mladeč 1, Fish Hoek, Hohlenstein 2, Hohlenstein 1, Kaufertsberg, Oberkassel 2, Oberkassel 1, Paderborn, Pavlov, Wahlwies, Qafzeh 9, Liujiang, Qafzeh 6, Singa, Kelsterbach, Kow Swamp 5
AMHS/archaic (7)	Jebel Irhoud 1, Jebel Irhoud 2, Dali, Ngandong 14, Ngandong 7, LH 18, Omo 2
Homo neanderthalensis (11)	Amud 1, La Chapelle aux Saints, La Ferrassie 1, La Quina H5, Spy 1, Spy 2, Guattari 1, Le Moustier, Shanidar 1, Ehringsdorf, Tabun 1
Homo heidelbergensis (3)	Kabwe 1, Petralona, Atapuerca SH5
Homo ergaster/erectus (8)	KNMER 3733, Dmanisi I, Dmanisi II, Sangiran 17, Trinil 2, Zhoukoutien 1, Zhoukoutien 11, Zhoukoutien 12

The AH sample consists of:

- *Homo neanderthalensis*
- *Homo heidelbergensis*
- *Homo ergaster/erectus.*

Most measurements were taken from high quality casts at the Naturhistorisches Museum Wien, at the Natural History Museum London, and at the Johann Wolfgang Goethe-University in Frankfurt/Main, except that the original of Mladeč 1 was measured in Vienna, while Atapuerca SH 5, Petralona and Guattari 1 was measured from high-resolution CT-scans. The reconstructions of specimens from Frankfurt were those made by F. Weidenreich. The geographically diverse sample of recent *Homo* was taken from the Naturhistorisches Museum and the Institute for Anthropology in Vienna, with all measurements from original specimens. The specimens Australier G33, G34, G35, G54 and K72 were casts from the Institute for Anthropology in Vienna made and donated in the early days of the Institute by Hermann Klaatsch, who collected recent Australian aboriginals during his expeditions 1904–1907 along the coast regions.

Methods

Size and shape of all specimens are based on a dataset of Cartesian coordinates either digitised by a Microscibe 3DX digitizer or extracted from surfaces detected in 3D CT-scans. We restricted the sample to specimens that preserved the mid-sagittal profile from Nasion to Inion, a landmark-based curve along the supraorbital rim (from glabella both sides to frontomalare temporale), another such curve from the inion both sides along the nuchal crest, and surface form for at least one side of the neurocranium (Fig. 2). Landmarks or semilandmarks missing on one side of the cranium were mirrored along the midsagittal plane. When landmarks were missing on both sides of the cranium or could not be located reliably this missing data was estimated by thin plate spline relaxation against the Procrustes average (Gunz et al, 2004). Note that Mladeč 2 was too incomplete to be used here.

Landmarks and semilandmarks

The analysis is of 16 landmark points (Table 2) along with 470 geometrically homologous semilandmarks in regions that are complete in all hemispecimens. None of the analyses to follow are of the landmarks alone. The method of semilandmarks (Bookstein, 1997) allows homologous curves and surfaces to be studied with the existing statistical toolkit. An earlier morphometric practice uses some non-landmark points from curves or surfaces as if they were landmarks: the extremal points (Type III of Bookstein, 1991) that have definitions like "most anterior" or "widest point". These locations, however useful for traditional distance measurements, are ambiguous regarding the one or two coordinates "perpendicular to the ruler". Semilandmarks eliminate the confounding influence of the deficient coordinates by computing them solely using the part of the data that is not deficient. To be specific, they are treated as missing data and estimated, all at once, in order to minimize the net bending energy of the data set as a whole around its own Procrustes average by allowing the semilandmarks to slide along the curvature (Gunz et al., 2005). All computations and statistical analyses were done in Mathematica 5.0 (Wolfram Research) using routines programmed by the authors (PG and PM). Figure 2 shows the computer reconstruction of the CT scan of Mladeč 1 with the full point set.

Regions

In addition to the overall analysis of the full 486-point shape we carried out three separate regional analysis (Fig. 2): the frontal region (120 points), the parieto-occipital region (90 points), and the midsagittal curve (26 points). The parieto-occipital region reaches down only to the linea nuchae superior, excluding the basal part of the occipital bone, which is often missing in fossils.

Procrustes shape analysis of the full 486 points and of these three subsets was carried out by the standard methods of the current geometric morphometrics toolkit (Bookstein, 1991; 1996; Marcus et al., 1996; O'Higgins, 2000; Mitteröcker et al., 2004; Gunz et al., 2005). In addition, all four analyses were duplicated in size-shape space (Mitteröcker et al., 2004), in which the usual Procrustes shape coordinates are supplemented by log Centroid Size prior to principal components analysis and principal coordinates analyses. These four size-shape spaces involved four different log size variables, each corresponding to the specific subset of landmarks and semilandmarks involved.

In conventional biometrics, there are two different ways of reporting one familiar analysis. The same computation that results in "principal components" (in this context, relative warps) that "explain fractions of variance" also results in "principal coordinates" that "account for fractions of interspecimen distance" (Marcus et al., 1996; Reyment and Joreskog, 1992). While yielding the same ordinations of the original data, the two methods differ in their actual computation: relative warp analysis is a decomposition of the variance-covariance matrix of Procrustes coordinates and princi-

Table 2. Classic landmarks in the study

Abbreviation	Landmarks	Abbreviation	Landmarks
N	Nasion	STL	Stephanion left
G	Glabella	STR	Stephanion right
B	Bregma	FMTL	Frontomalare temporale left
I	Inion	FMTR	Frontomalare temporale right
AUL	Auriculare left	FMOL	Frontomalare orbitale left
AUR	Auriculare right	FMOR	Frontomalare orbitale right
MSL	Mastoidale left	FTL	Frontotemporale left
MSR	Mastoidale right	FTR	Frontotemporale right

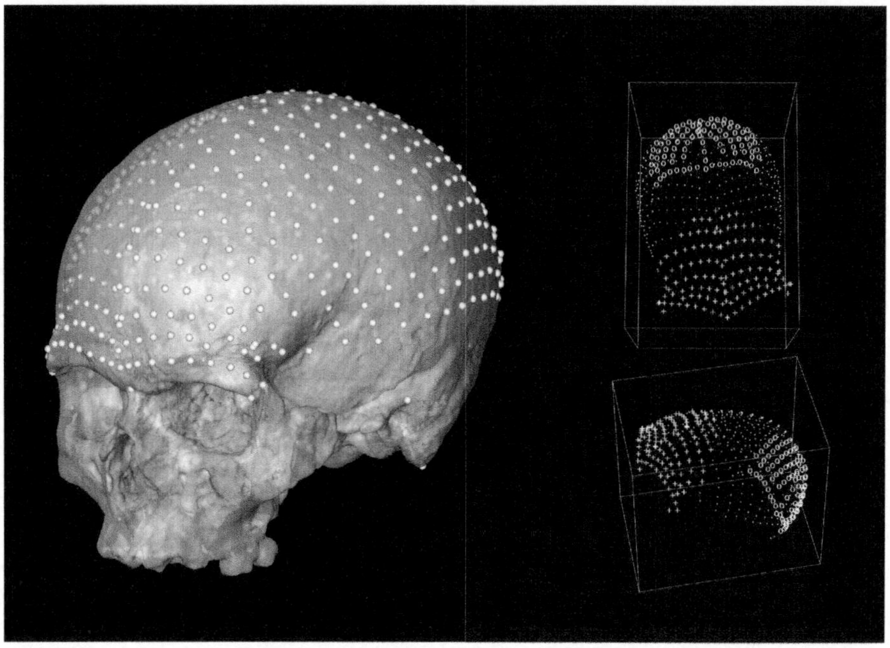

Fig. 2. Landmarks and semilandmarks indicated upon a rendering of Mladeč 1. Left – all 486 points. Right – subsets for the regional analyses: crosses – frontal points; circles – parieto-occipital points

pal coordinate analysis is a decomposition of the inter-specimen Procrustes distance matrix. In the current study, we will often use the language of distances, because each ordination of the principal coordinates will be overlaid by a representation of the underlying Procrustes distances, the Minimum Spanning Tree.

Minimum spanning tree

The shape coordinates of the standard Procrustes analysis are in fact a representation of a more fundamental underlying geometric structure, the Procrustes distance matrix among a set of specimen forms. When such distances exist a priori, there is one classic graphical enhancement of ordinations that often affords greater insight than scatterplots alone: the overlay of a Minimum Spanning Tree (MST) that connects all the forms up into one graph (usually with branches) in the way that minimum distance according to the distance measure. We use this graphic device here particularly because all of the ordinations (Figs. 3–5) below are projections into two dimensions, whereas the spaces in which Procrustes shape distance or size-shape distance is computed range up to 1452 dimensions and thus convey considerable additional information. Two specimens A and B, for example, might be much closer to each other in a particular principal component plot relative to a third form C, even though the relative positions are opposite in full space (AC and BC are smaller than AB). The MST computed from the full space will then represent the true relationship (connecting AC and BC, the two forms that are near each other in the PCA with the third form that appears more distant in the projection) and thus adding important information not present in the PCA ordination.

MST's can be drawn by utility software in a variety of popular packages, and there are no free parameters (i.e., all competent computations of MST's give identical trees). The MST is more interesting here than the more usual graphic, the display of relative warps (axes of the principal coordinate space) as deformations, because the main thrust of these analyses is the ordination of individual Mladeč specimens, not the representation of dimensions of variability per se.

In applications to ordinations among multiple groups, which is the main concern here, one usually inspects the tree for a variety of interpretively useful features: *dead ends,* which are terminal branches of the tree that connect one specimen to precisely one other form instead of the more usual two; *partial MST's,* which are restrictions of the full MST to links among members of the same subgroup (e.g., recent humans); and *bridges,* which are the remaining links of the MST, those connecting specimens of different subgroups. In the color figures below, partial MST's are drawn in distinct colors, and bridges are drawn in gray. Dead ends are not graphically enhanced, but are often referred to in our text.

Conventionally, systematic applications of morphometrics end with the production of ordinations like those of Figs. 3–5. For the present scientific context, it is more helpful to ask about the *actual strength of evidence* supporting classification of the Mladeč specimens into one of two subsamples, the AMHS or the Neandertals. We will ask this question, in fact, eight different times: separately for shape space and for size-shape space for the full set of 486 semilandmarks and for the three parcellations in Fig. 2. The question is asked in the standard loglikelihood framework (Mardia et al., 1979): in some reasonable multivariate feature space, the AMHS data are used to specify a Gaussian distribution contingent on AMHS membership, the Neandertal data are used to specify a Gaussian distribution contingent on Neandertal status, and the ratio of probability densities of the two distributions at each of the three Mladeč forms is taken (in the form of its likelihood) as a measure of the strength of shape evidence (or size-shape evidence) for membership in one of the taxa vis-a-vis the other (cf. Jeffreys, 1961). The computations do not report either of these likelihoods separately, only their ratio; so a form that is 10 S.D. out from one mean but 15 S.D. out from the other is "assigned" to the former group, as the less unlikely of the two alternatives.

For each of the eight versions of this computation, the Gaussian models require inversion of the corresponding covariance structures. As the smaller of our two groups, the Neandertals, comprises only nine specimens, we have limited the dimensionality of the descriptor spaces within which we are working to five: the first five relative warps of the pool of all 91 AMHS and Neandertal forms together, for any of the three regions or for the whole, either in shape space or in size-shape space. Logodds reported below correspond to formal log likelihood ratios for these eight separate computations. These eight versions need not necessarily be consistent – in fact in the discussion we will focus on the implications of their discrepancies. Logodds of the specimens pre-assigned to one taxon or the other are of course likely to confirm those assignments; our interest is only in the properties of the specimens regarding which our community has not yet arrived at a taxonomic consensus.

Results

A useful initial summary of the data from the 109 forms for 486 landmarks and semilandmarks is the ordination in shape space displayed in Fig. 3. The illustration represents the space of the first three principal components (a total of 65% of the summed squared Procrustes distances) but is shown here in a particularly helpful projection onto two dimensions only. The overall minimum spanning tree is not shown, but several of its subtrees are highlighted: that for all recent *H. sapiens* (red), that for all fossil *H. sapiens* (blue), that for all Neandertals (green), and that for all other AH (orange). Inter-taxon links of the MST that will be discussed in this text are drawn as dashed black lines. The two groups of AMHS overlap quite thoroughly in this space (not only in the projection here), but the Neandertals and the even older specimens are distinctly separate from the AMHS and from each other. There is only one region of substantial "bridging," a two-

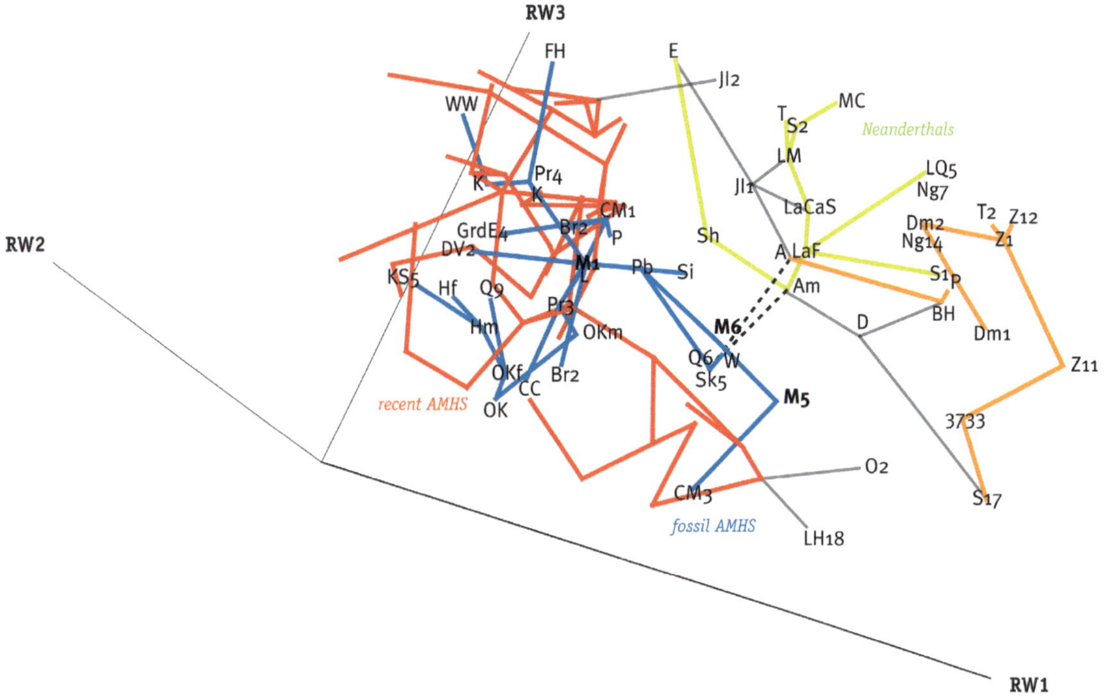

Fig. 3. A projection of the first three principal components in shape space (relative warps)

lane bridge connecting Qafzeh 6 (Q6) and Amud 1 (Am) on the one hand, and Shkul V (Sk5) and Atapuerca SH 5 (At) on the other. The gray link of a modern female to Jebel Irhoud 2 (JI2) is a "dead end."

We recommend that this Fig. 3 be supplanted by the corresponding Fig. 4. in size-shape space. These are convenient projections of a three-dimensional principal components (= principal coordinates) analysis including size, one that accounts for 75% of summed squared interspecimen distances (where now, because this is a size-shape space, squared distance incorporates squared log ratio of Centroid Sizes as well as squared Procrustes shape distance: Mitteröcker et al., 2004). It is now much clearer that the fossil *H. sapiens* group lies at the bifurcation for a pair of allometric sequences, one for the recent humans and one for the AH subsample. Please note that the modern and fossil AMHS are separated better by the analysis in which size was left in. The size-shape MST has three bridges between AMHS and the archaic forms: a central bridge from Qafzeh 6 to Amud 1, again, along with two side bridges, one from Jebel Irhoud 1 to a modern Indonesian male and one from LH 18 to Sangiran 17 (S17).

A summary classification of the Mladeč forms can be carried out via likelihood ratio test (s. Methods) using all of the semilandmarks in either shape space or size-shape space. In shape space, Mladeč 1 (M1) is distinctly AMHS in affinity (logodds 14.1) and Mladeč 5 (M5) and Mladeč 6 (M6) show logodds of about 4.5 for AMHS in analyses for which the logodds of the actual Neandertal skulls average about. In size-shape space, the modernity of all these forms is slightly less persuasive: Mladeč 5 and Mladeč 6 are categorised as AMHS at logodds of 1.46 and 3.97, respectively, while Mladeč 1 is anatomically modern at logodds of 9.0. As Fig. 4. indicates, considered as an AMHS, Mladeč 5 is somewhat too large for its shape, as is Cro-Magnon 3 (CM3).

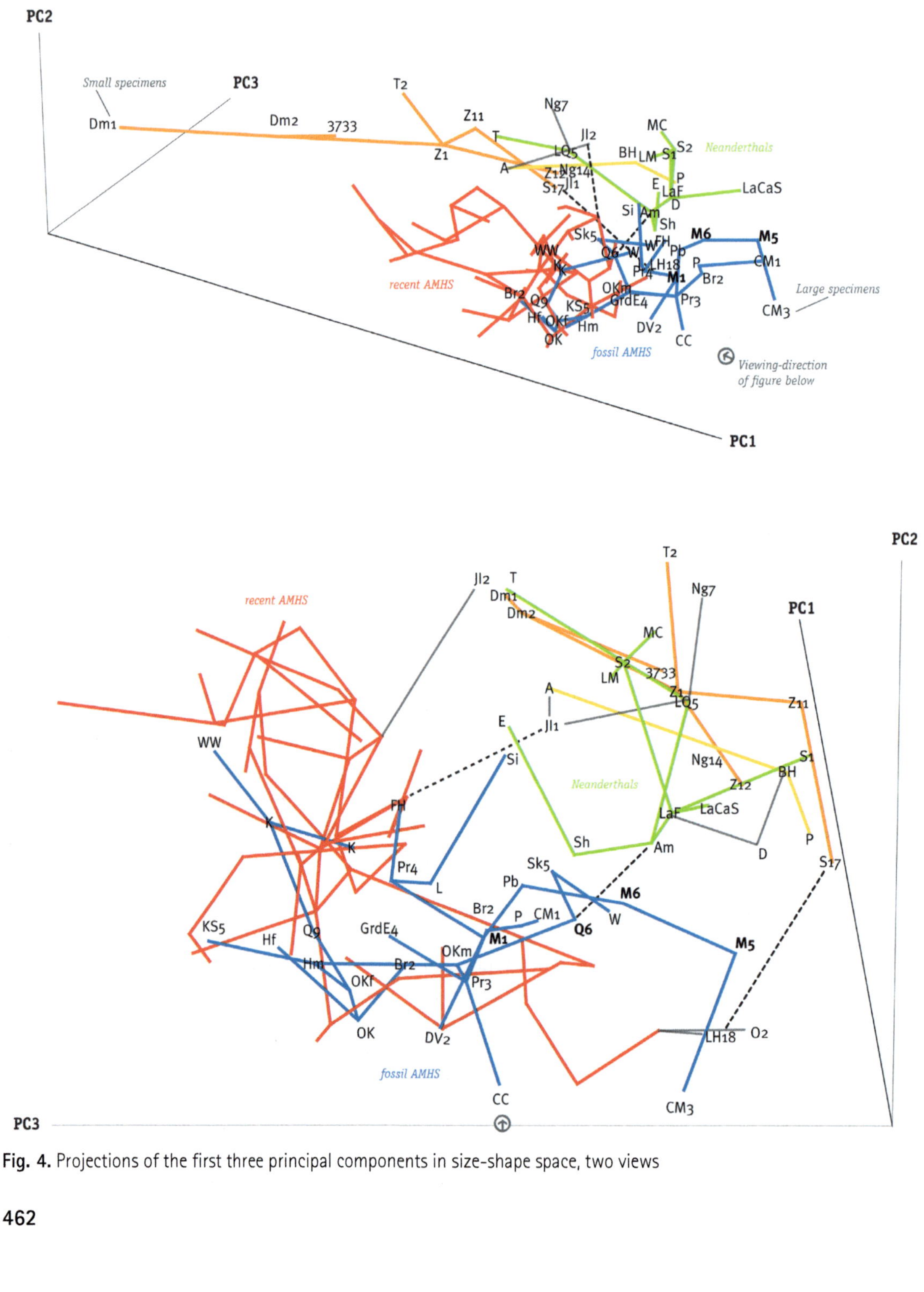

Fig. 4. Projections of the first three principal components in size-shape space, two views

Regional analysis

Because the regional analyses are aimed at the likelihood-ratio tests for the taxon of the Mladeč specimens, they are carried out in the space of principal components of the Neandertal and AMHS subsample only; and because the tests require matrix inversion, we limit the principal components to the first five of this pooled covariance structure.

Table 3. Logodds for all specimens in size-shape space. Modern *Homo sapiens* specimens are printed in gray. The red positive numbers are pointing towards Neandertal affinity, the negative numbers towards AMHS (higher numbers mean "more likely"). Note that Mladeč 5 lies right at the boundary (black line) – its ratio of −1.46 means that it is 4.3 (e^1.46) times more likely to be a AMHS than a Neandertal when all 486 landmarks and semilandmarks of the neurocranium are taken into account

LABEL	NAME	FULL	MIDSAG	PAR-OCCIP	FRONTAL
S1	Spy I	15.73	11.83	2.01	19.76
LQ5	La Quina H5	14.72	15.92	5.76	21.31
LM	Le Moustier	14.05	6.91	7.39	18.38
P	Petralona	12.74	9.45	-5.18	5.67
MC	Guattari	12.39	11.14	5.84	8.91
T	Tabun 1	12.29	11.14	2.33	7.44
S2	Spy II	11.05	5.37	7.46	9.18
LaCaS	La Chapelle aux Saints	10.65	9.31	7.29	7.43
BH	Broken Hilll	9.09	4.83	1.55	-2.07
D	Dali	8.10	9.27	2.04	-0.74
JI1	Jebel Irhoud I	7.60	2.78	1.68	6.23
JI2	Jebel Irhoud 2	7.13	-18.64	-0.45	-4.11
LaF	La Ferrassie I	6.93	9.50	1.58	6.98
Am	Amud I	6.51	8.76	2.17	7.05
A	Atapuerca	3.92	5.01	2.39	-1.99
Sh	Shanidar 1	2.24	4.95	-7.75	5.21
W	Wadjak I	1.27	-7.74	-11.56	1.12
M5	**Mladeč 5**	**-1.46**	**-7.71**	**-8.79**	**-9.07**
Pb	Paderborn	-3.80	-10.48	-8.13	-6.23
M6	**Mladeč 6**	**-3.97**	**-7.94**	**-0.56**	**-18.92**
Q6	Quafzeh VI	-5.32	-4.50	-5.23	-4.66
Sk5	Skhul V	-6.11	-4.54	-6.99	3.37
O2	Omo II	-6.20	9.00	-5.12	-11.58
CM1	Cro-Magnon I	-6.79	-8.63	-1.64	-7.55
	Inuit	-7.32	-2.86	-9.07	-7.13
LH18	LH 18	-7.50	8.53	0.55	-30.22
Br2	Bruenn II	-8.85	-0.96	-10.24	-12.22
M1	**Mladeč 1**	**-9.02**	**-13.21**	**-4.32**	**-6.84**
	Europe	-9.60	-3.08	-9.07	-7.65
	M.Australia	-9.75	4.35	-3.02	-3.47
	Europe	-9.82	-21.96	1.17	-15.30
	F.Europe	-10.46	-15.52	-5.21	-16.83
P	Pavlov	-10.52	-3.13	1.75	-7.16
	Australia34	-10.85	0.74	-11.54	-7.45
	AustraliaK72	-11.37	0.73	-13.48	-11.05
	F.Europe	-12.57	-21.64	-0.35	-6.09
	M.Indonesia	-12.71	-8.00	-3.05	-7.01
Pr3	Předmostí III	-13.15	-5.14	-4.37	-6.12
	M.Europe	-13.31	-22.31	-1.48	-9.64
OKm	Oberkassel.m	-14.40	-6.16	-7.65	-6.57
Pr4	Předmostí IV	-14.62	-10.94	-3.45	-9.47
	AustraliaG35	-14.75	-0.92	-6.88	0.42
	M.Java	-14.89	-9.82	-8.58	-9.63
L	Liujiang	-15.11	-7.48	-2.38	-8.41
	F.Europe	-15.25	-15.27	-2.66	-11.81
CM3	Cromagnon 3	-15.83	-9.40	-10.27	-16.61
	Australia54	-16.34	-0.89	-9.12	-4.36
	M.Europe	-17.04	-23.60	-11.67	-8.86
	M.Afrika	-17.24	-4.26	-4.13	-8.00
FH	Fish Hoek	-17.67	-15.23	-3.13	-16.46
	M.SouthAmerika	-18.12	-15.06	-11.92	-11.72
	F.Europe	-18.38	-13.60	-5.51	-12.52
CC	Comb Capelle	-18.74	-7.52	-14.24	-6.38
WW	Wahlwies	-19.06	-22.74	-27.11	-11.00
	F.Afrika	-19.24	-9.18	-8.91	-9.56
GrdE4	Grotte des enfants IV	-19.70	-15.47	-5.21	-20.87
	F.Europe	-19.77	-8.84	0.63	-10.44
	Europe	-20.07	-26.59	-10.71	-9.77
	F.Europe	-20.24	-4.86	-2.11	-14.29
Br2	Bruenn III	-20.57	-6.69	-1.70	-10.56
	F.Malaysia	-21.02	-13.23	-8.04	-11.43
	M.Philippines	-21.38	-7.69	-13.45	-10.28
K	Kelsterbach	-21.47	-10.66	-5.43	-8.68
	F.Europe	-21.49	-9.02	-3.48	-12.66
	M.SouthAmerika	-22.26	-14.30	-14.29	-13.42
	M.Afrika	-22.28	-10.66	-12.14	-8.68
	F.Polynesia	-22.60	-16.61	-5.61	-14.17
Q9	Qafzeh 9	-25.30	-5.75	-9.62	-8.88
	PNeuGuinea	-25.57	-15.30	-8.36	-11.25
OKf	Oberkassel f	-26.43	-13.85	-8.00	-12.85
	F.Europe	-26.56	-40.65	1.32	-20.58
Hm	Hohlestein m	-27.02	-16.06	-9.01	-15.16
K	Kaufertsberg	-27.02	-14.29	-3.82	-13.34
	F.Europe	-27.43	-10.60	-6.82	-6.58
	F.Timor	-27.51	-9.74	-14.24	-12.89
	M.Sumatra	-27.62	-9.42	-20.46	-9.82
	M.SouthAmerika	-28.08	-14.69	-26.05	-11.60
OK	Oberkassel	-28.16	-11.52	-7.00	-9.19
	F.Europe	-28.35	-22.96	-8.99	-13.31
Hf	Hohlestein f	-28.46	-13.64	-7.84	-16.22
	F.Europe	-31.70	-28.35	-2.39	-22.76
	M.Afrika	-31.77	-8.18	-11.98	-8.64
	M.Afrika	-32.07	-20.11	-6.76	-16.32
	M.Afrika	-32.20	-15.22	-29.80	-11.37
DV2	Dolni Vestonice 2	-32.79	-13.69	-12.96	-12.66
	M.Indonesia	-33.45	-19.53	-11.78	-16.01
Si	Singa	-33.83	-22.10	2.47	-28.74
KS5	Kow Swamp 5	-33.96	-28.51	-17.40	-6.16
	F.Asia	-35.16	-28.00	-23.30	-13.10
	F.Java	-35.64	-26.88	-33.66	-16.14
	F.NewGuinea	-37.52	-12.16	-16.78	-13.47
	M.India	-38.64	-11.94	-8.47	-12.54
	M.Malaysia	-38.79	-23.46	-39.64	-13.81
	M.Indonesia	-40.89	-25.96	-37.90	-11.64
	M.Indonesia	-42.37	-28.18	-29.94	-12.36
	M.Euro.Malaysia	-45.73	-23.72	-16.91	-17.14
	F.Europe	-47.43	-30.43	-35.14	-15.93
	F.Indonesia	-54.31	-20.84	-16.26	-21.79

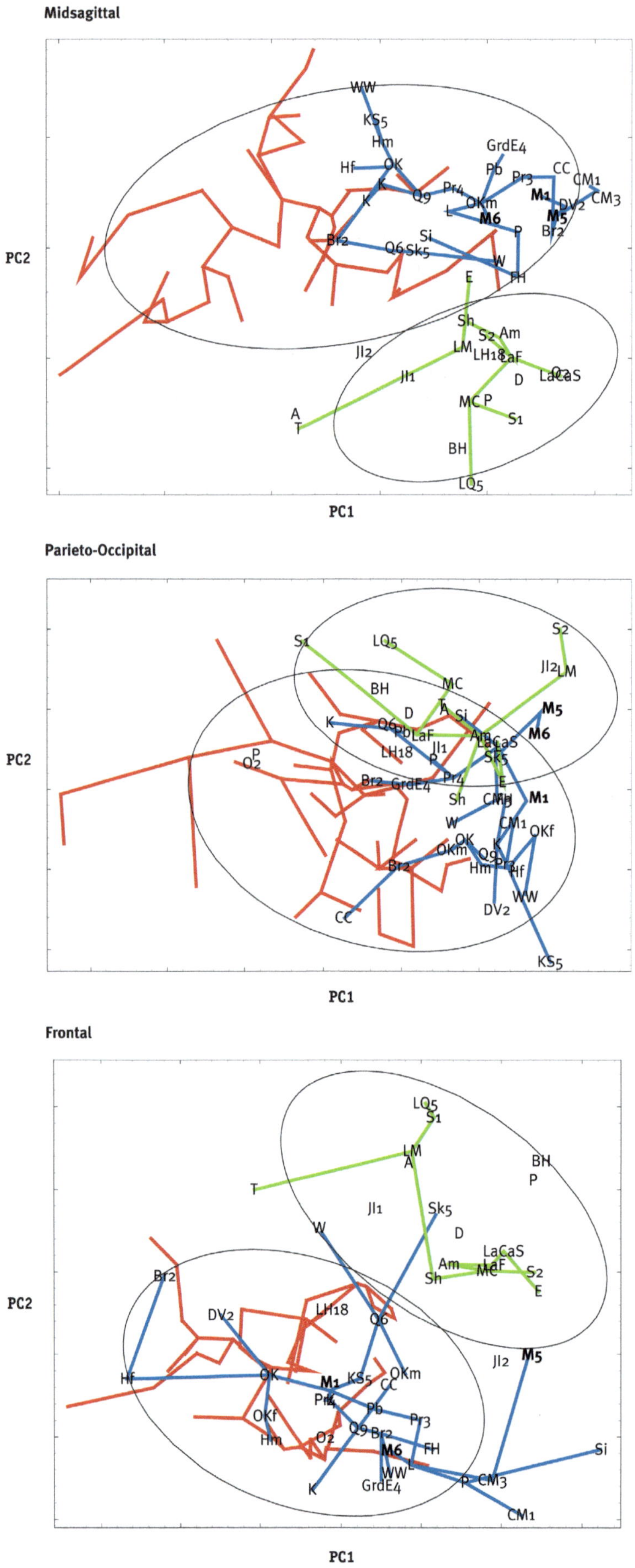

Fig. 5. Principal components of size-shape space for the mid-sagittal profile, parieto-occipital and frontal regions of the AMHS and Neandertal group. Logodds based on five dimensions in Table 3

When only the parieto-occipital region is analysed in shape space, the loglikelihoods confirm the fact that the classification of the Mladeč specimens is controversial: they cluster quite closely around the 0.0 value for even odds: for Mladeč 1, logodds of 1 for Neandertal status, 1.7 for Mladeč, and -0.7 for Mladeč 6. When considered in the context of size, all of these shift radically (Table 3): Mladeč 5's parieto-occipital region is distinctly modern for its size, with logodds 8.8, and Mladeč 1 almost as distinctly so (logodds 4.32 in favour of AMHS). Only Mladeč 6 is ambiguous as regards taxon (logodds of modernity, 0.56, corresponding to a factor of 1.75:1 only). Fig. 5 shows the first two principal components of this parieto-occipital size-shape space with the two $|z|=2$ ellipses for AMHS and Neandertals. In the first two dimensions these two ellipses overlap meaning that the occipital region warrants no classification between AMHS and Neandertals.

The analogous statistics for the frontal region indicate that all of the Mladeč forms have a distinct affinity for the AMHS sub-sample in shape space (logodds 21.3 M1, 16.3 M5, 42.2 M6) and only somewhat less support of the AMHS affinity in size-shape space (logodds 6.8 M1, 9.1 M5, 18.9 M6) in Fig. 5 and Table 3.

The situation is the same in both analyses of the midsagittal curve (Fig. 5) as well. All three Mladeč specimens are distinctly more likely to be AMHS than Neandertal - the minimum of the six logodds involved is 7.7 translating into a factor of 2200:1.

In summary, all parts of all three Mladeč specimens are clearly anatomically modern except the shapes of Mladeč 5 and Mladeč 6 in the parieto-occipital region.

Discussion

Phenetics and cladistics

The two main groups of hominin crania in our study (AMHS and AH) that were distinguished based on common classification are well separated by our quantitative approach. In other words, the analysis here, which relied on Procrustes distances for a very dense information grid along the external neurocranial surface, generally supports the traditional classification based on cladistics instead of aggregated geometry. The first three components in both spaces assort the AMHS group along a grade from more gracile and rounded neurocranial forms to robust and elongated shapes (Fig. 3–5), whereas the archaic forms clearly sequester themselves in another corner of the plots. If specimens like Wahlwies and the modern Australian G33 are taken to mark the extremes of the AMHS variation, Mladeč 1 could be taken as a good intermediate sample (Fig. 6)

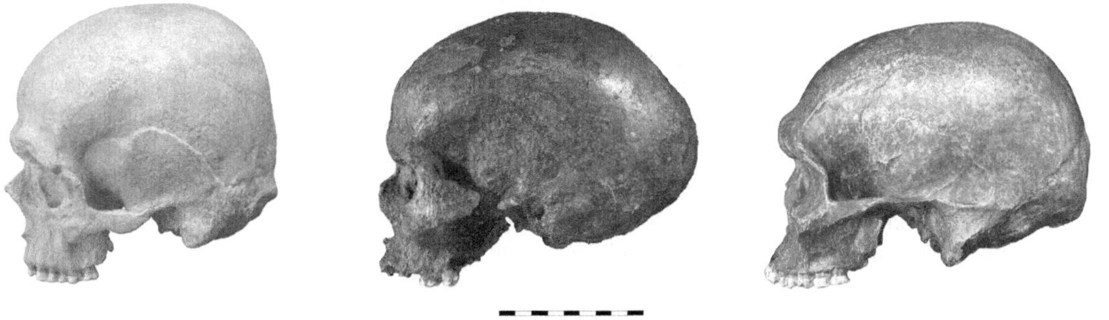

Fig. 6. Left lateral views of Wahlwies, Mladeč 1 and a modern Australian (G33). Images copyright of Inst. for Anthropology, University of Vienna (Wahlwies, Australian) and Naturhistorisches Museum Wien (Mladeč 1). All to scale

We have presented eight separate analyses, two each for 90 parieto-occipital semilandmarks, 26 midsagittal landmarks, 120 frontal landmarks, or the full configuration of 486. All of them reduce to analysis of matrices of interspecimen distances (not, for instance, distances between points): four in the classic Procrustes shape space and the other four in the relatively newer Procrustes size-shape space, where squared distance equals squared Procrustes shape distance plus squared log ratio of Centroid Size measures. As large-scale summary distance measures, these eight alternate analyses share several characteristics with the wide variety of other summary distance methods (for example, principal components analysis of measured lengths, measured volumes or areas, or "image distances"). Perhaps the most important of these shared characteristics is the a-priori refusal of the algebra to make any particular use of the relatively more informative parts of the data, where "informative" means informative for the hypothesis under examination. Formulas used in the distance methods are constructed prior to any actual data, and cannot be weighted toward (or against) the patterns the investigators wishes they would convey.

This is in striking contrast to the methods of cladistics, which are based on the insights of generations of scholars regarding the characteristics of organisms that, regardless of scale, prove informative for predicting other relationships, and thus represent, each one, a hypothesis all its own. Distance measures overturn that logic. They represent characteristics that are born as averages over large ranges of a single data source, and they are weighted by a semantics of "completeness" and "equal weighting" in a mostly mathematical sense, not by any criterion of faithfulness or consistency for answering the question under study. The tools we are using here are likewise tools that apply generally to measures of phenetic distance, not specialised for application to evolutionary hypotheses. Principal components and minimum spanning trees are part of the standard statistical toolkit for every field from market research through bioinformatics. Like the distance measures themselves, they embody no hypothesis per se, and when interpreted correctly (as we have tried to do in these pages), they are as much a comment on the prophylaxis afforded by redundancy of biometrical quantification as they are on any particular scientific hypothesis.

No distance-based analysis can therefore be seen as authoritative or canonical; all must be judged in reference to findings by alternative formulas and in reference to evidence completely outside of the algebraic framework of the data from which the distances are computed. A good example of this prophylaxis is our check for relative consistency of the classifications of the Mladeč regions vis-a-vis the two Procrustes spaces. Addition of one additional huge quantity known to be selectively important, namely Centroid Size, does not change the ordination as radically as the reader might have expected. Many of the forms bridging AMHS to the others remain bridging, and others remain near the springs of the bridge in spite of the change of space.

This paper is not the place to delve more deeply into the paradox of "objective" systematics, which has been the subject of the often-bitter debate between the pheneticists and the phylogeneticists since the dawn of computed taxonomy. We note only that when summary distance measure analyses like ours agree with cladistic findings, it is not typically because they are based on the same information flows, but instead because the different channels are truly tapping consilient (redundant) representations of the same true phenomenon. Conclusions supported by both phenetic and cladistic methods are evidently more persuasive than those supported by either method separately, and far more persuasive than those supported by one but contradicted by the other. If the summary morphometric methods mostly agree with the other approaches in their areas of substantive overlap, both gain credibility in the extension to domains of their own: for the Procrustes methods, this includes analysis of fully quantitative representations of form changes; for the cladistic methods, extensions to broader time frames affording far fewer corresponding geometric points.

MST bridges

The Mladeč crania 5 and 6 appear near the central MST bridge from AMHS to AH in the relative warps but the tree does not connect them directly. The bridge itself is interesting because it emerges in shape space as well as in size-shape space and separates the sample into two large groups. The two bridging specimens both come from sites in Israel that are geographically about 30 km apart. Amud 1 has been described as a progressive Neandertal (Suzuki and Takai, 1970), as a Neandertal with unusual small brow ridge and teeth (Trinkhaus, 1984), or as "peculiarly eclectic" Neandertal (Rak et al., 1994). It is not surprising that Amud 1 lies at the end of our Neandertal variation in the direction towards modern humans. The Qafzeh sample, on the other hand, while recognized as anatomically modern *Homo sapiens* by most authors (Vandermeersch, 1981; Howells, 1974; Trinkaus, 1984), has nevertheless been pushed into the Neandertal corner by others (Brose and Wolpoff, 1971).

The Mousterian association of Qafzeh raised additional questions. In their paper dealing with characteristics of mandibles, Schwartz and Tattersall (2000) point out that within the Qafzeh sample, only Qafzeh 8, 9 and 11 show the characteristic *Homo sapiens* chin structure in contrast to the other specimens from the same site. In shape space and size-shape space, we find a wide separation of the two Qafzeh crania 6 and 9, and no connection of these specimens via the MST, which is in agreement with the latter hypothesis. The few other bridges in MSTs vary from space to space. Except for one bridge, they lie at the extremes of the MST and are difficult to interpret. The affinity in shape space between Skhul V and Atapuerca SH 5 is interesting but will be not treated here.

The situation of Mladeč

In shape space, the two Mladeč crania 5 and 6, usually classified as male on account of their robustness, are close to each other but widely separated from their counterpart Mladeč 1, which is usually classified as female. Mladeč 1 itself seems entirely unexceptionable within the modern human variation if only shape is tapped here. In size-shape space however, the two very robust and the one more gracile Mladeč crania are represented on the MST as a subtree requiring only one intermediate (Paderborn). Generally, fossil and recent AMHS are better separated if size is included (Figs. 3 and 4), as the fossil AMHS group seem to have the larger crania. One might thus argue, that size difference actually matters more than shape difference over the form change from the Upper Paleolithic AMHS to recent humans. Not surprisingly, the Mladeč crania cluster with most of their European contemporaries. Within the Mladeč sample, the remaining size differences between M5 and 6 and M1 as well as the existing shape differences, could be an expression of sexual dimorphism.

Those Mladeč specimens commonly sexed as male are exceeded in size only by Cro-Magnon 3 (CR3) that is directly connected to Mladeč 5 (see Fig. 4) and from there to Mladeč 6, which itself is linked to Mladeč 1 via Paderborn. Mladeč 1 is also linked by the MST with Dolní Věstonice II, Předmostí III and Předmostí IV, all specimens that originate from sites that are not more than 100 km from the Mladeč Caves and that are dated not more than 3–5 kyr younger. Thus this group provides a glimpse of morphological variation in a narrow geographical area at roughly the same time. It also highlights the somehow unusual position of the large Mladeč crania 5 and 6 – Mladeč 1 indeed represents a more generalised form in the AMHS fossil sample.

Neurocranial regional analysis

The regional likelihood analyses in shape space and size-shape space confirm that the frontal region as well as the midsagittal plane of all Mladeč specimens in the study may be unequivocally classified as AMHS. In contrast, Mladeč parieto-occipital regions are neither obviously Neandertal-like nor obviously modern. Note that there is much more overlap of the two ellipses in Fig. 5 for the parieto-occipital region than for the frontal region and the mid-sagittal profile, which means that classification based solely on the posterior neurocranium is ambiguous for many specimens, not only the Mladeč crania. Note, too, that the frontal region of Skhul V and Wajak 1 is classified as completely archaic (Table 3).

Other aspects

Variation in the younger group of fossil AMHS dated on the onset or early in the Holocene (Brno 3, Fish Hoek, Hohlenstein female and male, Kaufertsberg, Kow Swamp 5, Oberkassel 1 and 2, Paderborn 1, Wahlwies) seems wholly consistent with variation in extant humans, except for Wajak 1, the only Asian specimen of the group. Note also that in shape space Omo 2 and LH 18, both dated to around 120–130 kyr, directly connect to one end of the extant human variation represented by Australian specimens (G33, G34, G35, G54 and K72) in the direction of *H. sapiens* allometry. Thus they may embody an extreme of the same shape robustness dimension that already characterises the rest of this group. Both have been considered to be late archaic *Homo sapiens* (Bräuer and Leakey, 1986) and thus possible candidates for the emergence of modern human morphology in Africa, and its later dispersal into Eurasia. Omo 2 and LH 18 are clearly separated from all other fossil hominids of the same time period, such as Qafzeh, Skhul, Jebel Irhoud, Singa, Liujiang, Dali, Petralona, Kabwe and Atapuerca. According to the MST, Singa and Liujiang connect also with the AMHS group, but through other bridges, Liujiang being virtually indistinguishable from moderns and Singa seemingly deviating in principal component 2.

In sum, Neandertals form a good connected group in the minimum spanning tree, while the *Homo ergaster* and *Homo erectus* specimens from Africa and Asia form a separated group. Sangiran 17 and likewise Trinil 2 take an unusual position that is not further treated here. The Ngangdong specimens are not homogeneous but clearly both belong in the group of archaic Homo (AH) and not to AMHS as their dating (~40 kyr) would have suggested. The archaic morphology of the Jebel Irhoud (JI) specimens, previously inferred by Grün and Stringer (1991), is confirmed in our morphometric analysis of the neurocranium only for JI 1 whereas JI 2 is connected to the AMHS sample.

The neurocrania of the three Mladeč specimens are distinctly more likely to be AMHS than Neandertal. In the statistic framework of log-likelihood ratios, we measure the support for different hypotheses of group affinity. The specimen that is closest to the Neandertal group is Mladeč 5 and its logodds of 4.03 in shape space and 1.46 in size-shape space translate into factors of 56:1 and 4:1 respectively (i.e., means that it is 56 times/4 times more likely to be a member of the AMHS group than one of the Neandertal group). These numbers are of course sensitive functions of the selection of landmarks and sample. With a dense mesh of hundreds of semilandmarks, we have tried to capture every aspect of shape variation in the exterior neurocranial vault. The availability of fossil material that is sufficiently complete to be used here, is of course limited but our sample of recent *Homo sapiens* embraces the full range of modern shape variability. We concur with previous metric and morphological analyses that the occipital region of the Mladeč crania is ambiguous with respect to group affinity, however the considerable overlap of the two ellipses in Fig. 5 shows that overall form of the posterior cranium does not distinctly separate Neandertals and AMHS.

Acknowledgements

We like to thank Maria Teschler-Nicola for the invitation to contribute to the important monography on the Mladeč material. We also thank her team at the Naturhistorisches Museum in Vienna, Chris Stringer and Rob Kruszynski at the Dept. of Palaeontology of the Natural History Museum in London, and Rainer Protsch v. Zieten and Stefan Flohr at the Institute of Anthropology of the Johann Wolfgang Goethe-University in Frankfurt/Main for access to their collection of fossils and casts and Markus Bernhard for sharing his data of modern humans. We thank Juan-Luis Arsuaga, George Koufos, Luca Bondioli and Roberto Macchiarelli for the permission to scan fossil specimens and Wolfgang Recheis and Dieter zur Nedden at the Dept. for Radiology II from the University Hospital Innsbruck, and Herwig Imhof and Franz Kainberger at the Institute of Radiodiagnostics from the General Hospital Vienna for technical support. We thank Thomas Bence Viola and Simon Neubauer for support and discussion. This work was funded by the Austrian Science Foundation (P14738) and the Austrian Council for Science and Technology and the Austrian Federal Ministry for Education, Science and Culture (AD 387/25-30, GZ200.093/I/VI/2004).

References

Adcock, G. J., Dennis, E. S., Easteal, S., Huttley, G. A., Jermiin, L. S., Peacock, W. J. and Thorne, A. (2001) Mitochondrial DNA sequences in ancient Australians: Implications for modern human origins. *Proceedings of the National Academy of Science USA* 98, 537–542

Bookstein, F. L. (1991) *Morphometric tools for landmark data: Geometry and biology.* Cambridge: Cambridge University Press

Bookstein, F. L. (1996) Biometrics, biomathematics and the morphometric synthesis. *Bulletin of Mathematical Biology* 58, 313–365

Bookstein, F. L. (1997) Landmark methods for forms without landmarks: Morphometrics of group differences in outline shape. *Medical Image Analysis* 1, 225–243

Bräuer, G. (1980) The morphological affinities of the early Pleistocene frontal bone from the estuary area of the Elbe River near Hahnofersand. *Zeitschrift für Morphologie und Anthropologie* 71, 1–42

Bräuer, G. (1984) Presapiens hypothesis or Afro-European sapiens hypothesis? *Zeitschrift für Morphologie und Anthropologie* 75, 1–25

Bräuer, G. and Leakey, R. E. (1986) A new archaic *Homo sapiens* cranium from Eliye Springs, West Turkana, Kenya. *Zeitschrift für Morphologie und Anthropologie* 76, 245–252

Bräuer, G. and Broeg, H. (1998) On the degree of Neandertal-modern continuity in the Earliest Upper Palaeolithic crania from the Czech Republic: Evidence from non-metrical features. In (K. Omoto and P. V. Tobias, Eds.) *The origins and past of modern humans: Towards reconciliation.* Singapore: World Scientific, pp. 106–125

Bräuer, G., Collard, M. and Stringer, C. (2004) On the reliability of recent tests of the out of Africa hypothesis for modern human origins. *Anatomical Record* 279 A, 701–707

Brose, D. S. and Wolpoff, M. H. (1971) Early Upper Paleolithic man and late Middle Paleolithic tools. *America Anthropologist* 73 (5), 1156–1194

Churchill, S. E. and Smith, F. H. (2001) Makers of the Early Aurignacian of Europe. *Yearbook of Physical Anthropology.* 43, 61–115

Conard, N. J. and Bolus, M. (2003) Radiocarbon dating the appearance of modern humans and timing of cultural innovations in Europe: New results and new challenges. *Journal of Human Evolution* 44, 331–371

Cooper, A., Drummond, A. J. and Willerslev, E. (2004) Ancient DNA: Would the real Neandertal please stand up? *Current Biology* 14, R431–R433

Dryden, I. L. and Mardia, K. V. (1998) *Statistical shape analysis.* New York: Wiley

Duarte, C., Mauricio, J., Pettitt, P. B., Souto, P., Trinkaus, E., van der Plicht, H. and Zilhão, J. (1999) The early Upper Paleolithic human skeleton from the Abrigo do Lagar Velho (Portugal) and modern human emergence in Iberia. *Proceedings of the National Academy of Science USA* 96, 7604–7609

Frayer, D. W. (1986) Cranial variation at Mladeč and the relationship between Mousterian and Upper Paleolithic hominids. In (V. V. Novotný and A. Mizerová, Eds.) *Fossil man – new facts, new ideas. Papers in Honor of Jelínek's Life Anniversary.* Brno: Anthropos Institute – Moravian Museum, 23, pp. 243–256

Frayer, D. W., Jelínek, J., Oliva, M. and Wolpoff, M. H. (2006) Aurignacian males from the Mladeč Caves, Moravia, Czech Republic. In (M. Teschler-Nicola, Ed.) *Early modern humans at the Moravian Gate: The Mladeč Caves and their remains*, this issue, pp. 185–272

Grün, R. and Stringer, C. (1991) Electron spin resonance dating and the evolution of modern humans. *Archaeometry* 33, 153–199

Gunz, P., Mitteröcker, P., Bookstein, F. L. and Weber, G. W. (2004) Computer aided reconstruction of incomplete human crania using statistical and geometrical estimation methods. In (K. Fischer Ausserer, W. Börner, M. Goriany and L. Karlhuber-Vöckl, Eds.) [Enter the Past] *The E-way into the four dimensions of cultural heritage.* CAA 2003. *Computer Applications and Quantitative Methods in Archaeology, Proceedings of the 31st Conference, Vienna, Austria, April 2003.* Oxford: Archaeopress, pp. 92–95

Gunz, P., Mitteröcker, P. and Bookstein, F. L. (2005) Semilandmarks in three dimensions. In (D. E. Slice, Ed.) *Modern morphometrics in physical anthropology.* New York: Kluwer Academic/Plenum Publishers, pp. 73–98

Howells, W. (1974) Neanderthals: Names, hypotheses and scientific method. *American Anthropologist* 76, 24–38

Ingman, M., Kaessmann, H., Paabo, S. and Gyllensten, U. (2000) Mitochondrial genome variation and the origin of modern humans. *Nature* 408, 708–713

Jeffreys, H. (1961) *Theory of probability.* Oxford: The Clarendon Press

Jelínek, J. (1951) A contribution to the classification of the Moravian (Czechoslovakia) Upper Paleolithic man. *Acta Musei Moraviae* 36, 1–12

Jelínek, J. (1983) The Mladeč finds and their evolutionary importance. *Anthropologie* 21, 57–64

Klein, R. G. (2003) Whither the Neanderthals? *Science* 299, 1525–1527

Krings, M., Stone, A., Schmitz, R. W., Krainitzki, H., Stoneking, M. and Pääbo, S. (1997) Neandertal DNA sequences and the origin of modern humans. *Cell* 90, 19–30

Krings, M., Geisert, H., Schmitz, R. W., Krainitzki, H. and Pääbo, S. (1999) DNA sequence of the mitochondrial hypervariable region II from the Neandertal type specimen. *Proceedings of the National Academy of Science USA* 96, 5581–5585

Krings, M., Capelli, C., Tschentscher, F., Geisert, H., Meyer, S., von Haeseler, A., Grossschmidt, K., Possnert, G., Paunovic, M. and Pääbo, S. (2000) A view of Neandertal genetic diversity. *Nature Genetics* 26, 144–146

Lieberman, D. E. (2004) Humans and primates: New model organisms for evolutionary developmental biology? *J. Exp. Zoolog. Part B. Mol. Dev. Evol.* 302, 195

Lovejoy, C. O., Cohn, M. J. and White, T. D. (1999) Morphological analysis of the mammalian postcranium: A developmental perspective. *Proceedings of the National Academy of Science USA* 96, 13247–13252

Maddison, D. R. (1991) African origin of human mitochondrial DNA reexamined. *Systematic Zoology* 40, 355–363

Marcus, L. F., Corti, M., Loy, A., Naylor, G. and Slice, D. E. (1996) *Advances in morphometrics.* New York: Plenum Press

Mardia, K. V., Kent, J. T. and Bibby, J. M. (1979) *Multivariate analysis.* London: Academic Press

Mitteröcker, P., Gunz, P., Bernhard, M., Schaefer, K. and Bookstein, F. L. (2004) Comparison of cranial ontogenetic trajectories among great apes and humans. *Journal of Human Evolution* 46, 679–697

O'Higgins, P. (2000) Quantitative approaches to the study of craniofacial growth and evolution: Advances in morphometric techniques. In (P. O'Higgins and M. Cohn, Eds.) *Development, growth and evolution: Implications for the study of the hominid skeleton.* San Diego: Academic Press, pp. 163–185

Ovchinnikov, I. V., Gotherstrom, A., Romanova, G. P., Kharitonov, V. M., Liden, K. and Goodwin, W. (2000) Molecular analysis of Neanderthal DNA from the northern Caucasus. *Nature* 404, 490–493

Pacher, M. (2006) Large mammal remains from the Mladeč Caves and their contribution to site formation processes. In (M. Teschler-Nicola, Ed.) *Early modern humans at the Moravian Gate: The Mladeč Caves and their remains,* this issue, pp. 99–148

Pusch, C. M. and Bachmann, L. (2004) Spiking of contemporary human template DNA with ancient DNA extracts induces mutations under PCR and generates non-authentic mitochondrial sequences. *Molecular Biology and Evolution* 21, 957–964

Rak, Y., Kimbel, W. H. and Hovers, E. (1994) A Neandertal infant from Amud Cave, Israel. *Journal of Human Evolution* 26, 313–324

Relethford, J. H. (2001) Ancient DNA and the origin of modern humans. *Proceedings of the National Academy of Science USA* 98, 390–391

Reyment, R. A. and Joreskog, K. G. (1993) *Applied factor analysis in the natural sciences.* Cambridge: The University Press

Santa Luca, A. P. (1980) The Ngandong fossil hominids: A comparative study of a Far Eastern *Homo erectus* group. *Yale Univ. Publ. Anthropol.* 78, 1–175

Schwartz, J. H. and Tattersall, I. (2000) The human chin revisited: What is it and who has it? *Journal of Human Evolution* 38, 367–409

Schwartz, J. H. and Tattersall, I. (2002) *Terminology and craniodental morphology of genus Homo (Europe).* New York: Wiley-Liss, pp. 273–280

Serre, D., Langaney, A., Chech, M., Teschler-Nicola, M., Paunovic, M., Mennecier, P., Hofreiter, M., Possnert, G. and Pääbo, S. (2004) No evidence of Neandertal mtDNA contribution to early modern humans. *Public Library of Science Biology* 2, 313–317

Smith, F. H. (1982) Upper Pleistocene hominid evolution in south-central Europe: A review of the evidence and analysis of trends. *Current Anthropology* 23, 667–703

Smith, F. H., Trinkaus, E., Pettitt, P. B., Karavanic, I. and Paunovic, M. (1999) Direct radiocarbon dates for Vindija G_1 and Velika Pećina Late Pleistocene hominid remains. *Proceedings of the National Academy of Science USA* 96, 12281–12286

Stringer, C. B., Hublin, J. J. and Vandermeersch, B. (1984) The origin of anatomically modern humans in Western Europe. In (F. H. Smith and F. Spencer, Eds.) *The origins of modern humans: A world survey of the fossil evidence.* New York: Alan R. Liss, Inc., pp. 51–135

Stringer, C. B. (1992a) Reconstructing recent human evolution. *Philos. Trans. R. Soc. Lond. B. Biol. Sci.* 337, 217–224

Suzuki, H. and Takai, F. (Eds.) (1970) *The Amud man and his cave site.* Tokyo: Tokyo University Press

Svoboda, J. (2000) The depositional context of the Early Upper Paleolithic human fossils from the Koneprusy (Zlatý kůň) and Mladeč Caves, Czech Republic. *Journal of Human Evolution* 38, 523–536

Szombathy, J. (1925) Die diluvialen Menschenreste aus der Fürst-Johanns-Höhle bei Lautsch in Mähren. *Eiszeit* 2, 1–34, 73–95

Tattersall, I. and Schwartz, J. H. (1999) Hominids and hybrids: The place of Neanderthals in human evolution. *Proceedings of the National Academy of Science USA* 96, 7117–7119

Templeton, A. R. (1992) Human origins and analysis of mitochondrial DNA sequences. *Science* 255, 737

Templeton, A. R. (2002) Out of Africa again and again. *Nature* 416, 45–51

Trinkaus, E. (1984) Western Asia. In (F. H. Smith and F. Spencer, Eds.) *The origins of modern humans: A world survey of the fossil evidence.* New York: Alan R. Liss, pp. 251–325

Vandermeersch, B. (1981) *Les hommes fossiles de Qafzeh (Israel).* Paris: C.N.R.S.

Weber, G. W., Schäfer, K., Prossinger, H., Gunz, P., Mitteröcker, P. and Seidler, H. (2001) Virtual anthropology: The digital evolution in anthropological sciences. *J. Physiol. Anthropol. Appl. Human Sci.* 20, 69–80

White, T. D., Asfaw, B., De Gusta, D., Gilbert, H., Richards, G. D., Suwa, G. and Howell, F. C. (2003) Pleistocene *Homo sapiens* from Middle Awash, Ethiopia. *Nature* 423, 742–747

Wild, E. M., Teschler-Nicola, M., Kutschera, W., Steier, P., Trinkhaus, E. and Wanek, W. (2005) Direct dating of Early Upper Palaeolithic human remains from Mladeč. *Nature* 435, 332–335

Wolpoff, M. H. (1999). The systematics of Homo. *Science* 284, 1774–1775

Wolpoff, M. H., Hawks, J., Frayer, D. W. and Hunley, K. (2001) Modern human ancestry at the peripheries: A test of the replacement theory. *Science* 291, 293–297

Wolpoff, M. H., Frayer, D. W. and Jelínek, J. (2006) Aurignacian female crania from the Mladeč Caves, Moravia, Czech Republic. In (M. Teschler-Nicola, Ed.) *Early modern humans at the Moravian Gate: The Mladeč Caves and their remains*, this issue, pp. 273–340

Chapter 16

PATHOLOGICAL ALTERATIONS AND TRAUMAS IN THE HUMAN SKELETAL REMAINS FROM MLADEČ

Maria Teschler-Nicola, Christian Czerny, Martin Oliva, Doris Schamall and Michael Schultz

Introduction

There are numerous descriptions of pathological features of Paleolithic (Straus and Cave, 1957; Dastugue, 1962; Trinkaus and Zimmerman, 1982; Walker et al., 1982; Trinkaus, 1985; Tappen, 1985; Crubézy and Trinkaus, 1992; Rothschild et al., 1995; Antón, 1997; Tillier et al., 2001; Lebel et al., 2001; DeGusta, 2002; Zollikofer et al. 2002; Bräuer et al., 2003), in particular Middle and Late Upper Paleolithic associated finds presented so far (Dastugue, 1967; Thillaud, 1981–1982; Bukhman, 1984; Klíma, 1987; 1991; Frayer et al., 1988; Jelínek, 1992; Formicola, 1995; Oliva, M., 2000; Formicola et al., 2001). Some of the alterations noted are mentioned and discussed within monographic presentations of site complexes and include a wide spectrum of dental pathologies, ante mortem traumatic lesions and other pathologies, which shed light on the degree of physiological stress to which Pleistocene hunter-gatherer populations were exposed (e.g., Svoboda, 1997; Tillier, 1999; Roebroeks et al., 2000; Zilhão and Trinkaus, 2002; Alexeeva and Bader, 2000; Trinkaus and Svoboda, 2006). Several recorded anomalies enable us to reconstruct working activities and to get an idea of the social system and care of severely incapacitated individuals within these early societies (e.g., traumatic injuries or congenital diseases, Formicola et al., 2001; Formicola and Buzhilova, 2004).

Interestingly, the most extensive human fossil assemblage from the early Upper Paleolithic of central Europe represented at Mladeč, including cranial, mandibular, dental and postcranial remains has never been systematically investigated in this respect. Other than the limited paleopathology remarks by Szombathy (1925), Jelínek (1969), Frayer et al. (this volume) and Trinkaus et al. (this volume) little has been documented in this sample. Here we present results of the first systematic screening by using non-invasive methods. To avoid destruction of the original material we rely primarily upon macroscopic analysis (cf. Aufderheide and Rodríguez-Martín, 1998; Ortner, 2003; Steinbock, 1976). Our survey of paleopathology in the Mladeč skeletal remains is listed in numeric order of the bone specimens.

Materials and methods

All cranial and postcranial remains were investigated macroscopically, radiologically and by multidetector computed tomography. The CT-scans were performed at the Department of Radiology, Vienna Medical University and the Department of Medical Imaging, St. Anna University Hospital, Brno, Czech Republic, by using a PHILIPS MX8000 CT scanner. The vestiges of pathological processes were scored systematically by the pattern suggested by Schultz (1988). According to this procedure, special sheets for the documentation of the pathological findings were used to note, if possible, every alteration. Endocranial bone surfaces were examined by using a small dentist's mirror and a battery lamp. Tooth eruption was diagnosed by the method of Ubelaker (1978).

Results

Mladeč 1: cranium

As a thin matrix almost completely covers the cranium's surface (see Wolpoff et al., this volume, chap. 10; Prossinger and Teschler-Nicola, this volume, chap. 11), only very few morphological features could be observed on actual bone. The left external and the right internal auditory meatus are pathology free. As already noted by Szombathy, the right external acoustical porous seems to be slightly deformed (Szombathy, 1925, 18) due to a small exostosis, which originates in the mastoideal portion. Although the internal structures of the nasal cavity are encrusted by matrix as well, the position of the nasal septum is severely deviated from the mid-sagittal plane (see Wolpoff et al., this volume, chap. 10, Plate I). Such a displacement may be congenital or caused by an injury and could have led to a blockage of the nostril and frequent sinus infections or chronic sinusitis (Ballenger, 1991; Casano, 2002). Unfortunately the latter cannot be verified due to the matrix accretion. Finally, there is no evidence of dental caries (0 out of 4), dental calculus (0 out of 4) or transverse linear enamel hypoplasia (0 out of 4) nor of periodontitis (0 out of 16).

Mladeč 2: cranium

The posterior third of the parietal bones (between the tuber and the sagittal suture) exhibit porotic abnormalities (probably porotic hyperostosis caused by malnutrition). All the other parts of the incomplete preserved skull vault and the facial portion show no macroscopic evidence of pathological changes (we observed in detail a section of the sagittal sinus and the right and left sigmoid sinus, the right and the left external auditory meatus, the left mastoid process, the right nasolacrimal duct and the mandibular fossae). The maxillary sinuses as well as the floor of the nasal cavity are exposed but covered with matrix and cannot be observed in detail. This applies to the hard palate as well. Although the latter shows a slightly remodeled, plate-like surface which might be interpreted as the result of a healed chronic process of the gum, it is more likely that these structures are caused by different chemical preservatives and hardeners which were used for restoration and conservation of the fragmentary remains.

High-resolution multi-detector-spiral CT images of the right temporal bone region delineated an inner ear malformation (see Casselman et al., 1993; Czerny et al., 1998; 2000; Som and Curtin, 2003), consisting of a severe hypoplasia of the vestibular system: compared to the left side, the right vestibule is very small and hypoplastic. Also the system of the semicircular canals on the right side is not well delineated compared with the normal one on the left. Moreover, the internal auditory canal seems to be narrower, and the facial nerve leaves the internal auditory canal a bit posterior-laterally to its normal exit. The geniculate ganglion and the course of the facial nerve in its tympanic segment are also malformed (see Fig. 1a–c).

Despite the low age of death of this individual (a young adult according to Wolpoff et al., this issue, chap. 10), there are vestiges characteristic of periodontitis (tooth sockets right M^1 [degree II], right P^3 and M^1 [degree I] = 3 out of 16)) and traces of dental calculus (right and left M^1 [degree 1] = 2 out of 4). No dental caries (0 out of 4), and no transverse linear enamel hypoplasia (0 out of 4) were observed.

Mladeč 3: cranial fragments

The external surface of the fragments of the right and left parietal bones and the right temporal bone show no conspicuous irregularities. Only the occipital bone, below the lambda and along the lambdoidal sutures, exhibits porosities, but these originate most probably in normal skeletal growth processes. No sign of diseases could be observed in the right external auditory meatus. The external

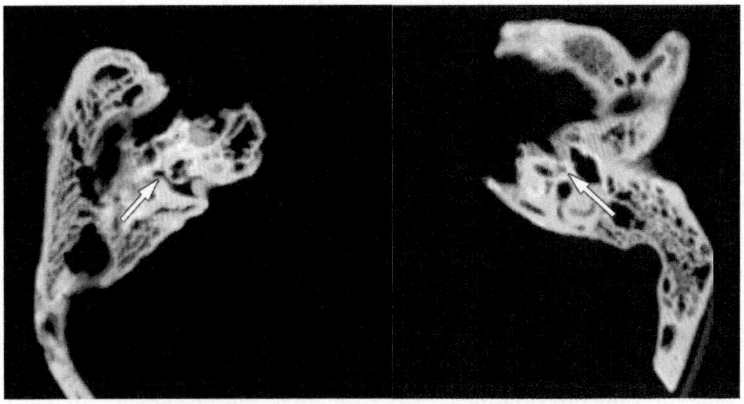

a

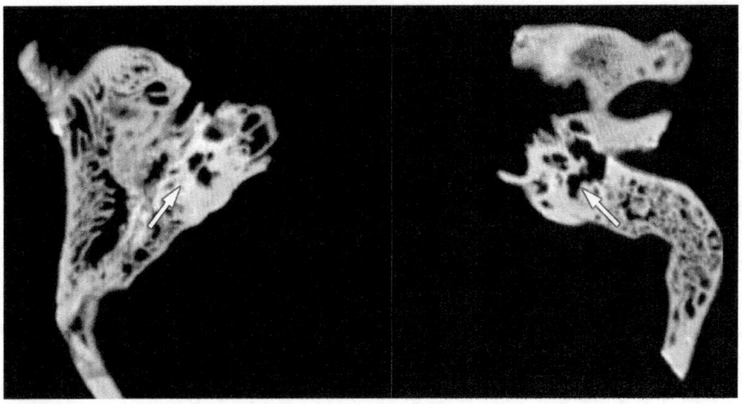

b

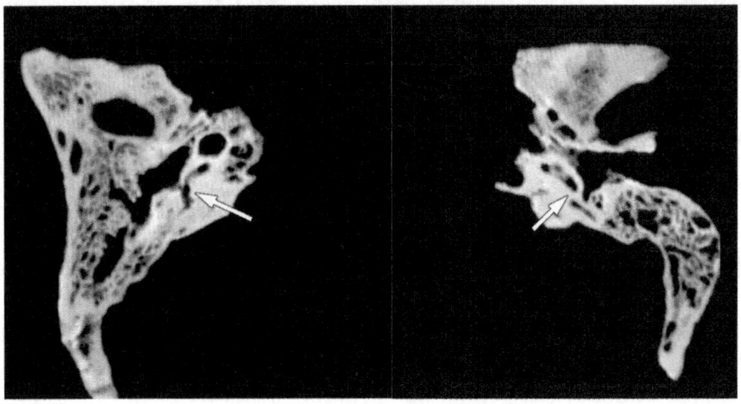

c

Fig. 1. CT-scans of Mladeč 2: **(a)** On the right side a malformed cochlea-vestibule region is shown compared to the normal one left (arrows). **(b)** No normal vestibular system is depicted on the left while the right one is normal (arrows). **(c)** Abnormal extrusion of the facial nerve of the internal auditory canal is depicted on the right while the left one is normal (arrows) layer of the frontal bone exhibits several scratches and, about 30 mm anterior from bregma, there is a small irregular-shaped loss of bony substance. These are most probably postmortem, related to attempts to remove the thin matrix coating over the bone.

The internal layer, although covered in large parts by thin matrix, show two noticeable features: on the right parietal digital impressions seem to be more pronounced. This also applies to the right temporal bone. Furthermore, the ground of the impressions have – in a few very restricted areas – an irregular morphology, as can be observed in slightly cicatrized and porotic surfaces. A similar irregular porosity is visible in the lateral quarter of the right sigmoid sinus and in the lateral angle of the right middle cerebral fossa of the temporal bone; this is observable even in the occipital bone, namely the left cerebellar fossa and in the right and left cerebral fossae. Furthermore, the internal occipital crest bears a depression (probably an occipito-marginal sulcus) where similar, very tiny porosities can be identified as well. All these changes are partly originating in postmortem alterations or may even represent physiological structures (e.g., blood vessel canals). In principle these findings are not really very unusual and frequently well developed in infants and young children of recent populations.

The most conspicuous internally located feature in the occipital bone, which was mentioned by Szombathy as well (Szombathy, 1925, 22), is the asymmetry of the superior cerebral fossae, so that the left fossa is strikingly broader than the right. Accordingly, the internal longitudinal crest turns extensively to the right portion of the occipital bone. This feature, which may represent an anatomical variation, is discussed in more detail by Minugh-Purvis et al. (this volume, chap. 12).

Mladeč 5: calotte

The external layer of the vault exhibits several traumatic lesions. The most conspicuous is located in the frontal, approximately 21 mm above and in the middle of the left orbital rim (see also Frayer et al., this volume, chap. 9). The defect is oval-shaped, 12 mm in sagittal and 8 mm in transversal dimension (Plate I, Figs. a and b). Computed tomography in the sagittal plane (Plate II, Fig. c) shows that this defect is limited to the external layer and superficial diplöe. It tapered towards the center of the impression. The ground of this lesion exhibits irregular structures and the adjacent area of the external table shows remarkable "pitting", most likely caused by a reactive inflammation of the skin. The advanced remodeling of these features emphasizes that this injury originated from a blunt trauma and was completely healed.

A second minor, oval depression (40 x 25 mm) is noticeable in the right parietal bone, at a distance of about 20 mm from the sagittal suture and close to the bregma (Plate II, Figs. a and d). According to Szombathy (1925, 75) this depression pervaded the internal layer and left a sagittal oriented 35 mm long cleft, most probably caused by a blow.[1] Although the internal fracture line cannot yet be verified macroscopically because of recent renovation efforts, the CTs show that the defect seems to pertain to the outer and inner layers of the cranial vault without any features of healing and remodeling. It is not possible to make a clear diagnosis, if the fracture originated ante- or post-mortem.

A third very small defect, most probably also caused by an injury, is located in the right parietal bone, about 45 mm right of the lambda and approximately 30 mm behind the parietal tuber. The defect encloses an area of about 6 x 9 mm and is characterized by a slight loss of substance, which pertains to the external layer only. Despite a slight porosity at the lesion's base, it appears to be an old, remodeled and more or less completely healed wound (Plate I, Figs. e and f).

1 „Am rechten Scheitelbein, nahe dem Bregma, eine seichte ovale Depression von 4,2 cm Länge und 2,6 cm Breite, die den Knochen durchdringt und an der Innenseite eine sagittal verlaufende, 3,5 cm lange Bruchspalte erzeugt; wahrscheinlich durch einen dem Lebenden zugefügten Schlag hervorgebracht" (Szombathy, 1925, 75)

One striking impression fracture could be observed as well. In contrary to the above, this defect is most probably caused post-mortem by a blunt object: on the frontal bone between the left frontal boss and the median sagittal line an oval shaped (28 x 17 mm) impression with a few irregular radial cracks can be identified (Plate II, Figs. b and e). One indirect argument in support of post-mortem origin (likely related to treatment of the vault in 1945) is given by the fact that this lesion was obviously not visible in 1925, when Szombathy investigated and discussed the specimen. In addition, several small superficially located defects of unclear genesis (most probably post-mortem alterations) could be identified (Plate II, Fig. f).

Finally, for the ectocranial surface a, what appears to be an osteoma can be observed on the frontal bone, 10 mm left of the median sagittal plane and 40 mm anterior of the coronal suture. This pathologic alteration is circular, approximately 4 mm in diameter and of compact nature (Plate I, Figs. c and d).

The internal layer of the cranium exhibits not only very complex and reticulated meningeal arteries (Plate I, Fig. g; see also Frayer et al., this volume, chap. 9), but also alterations within the sagittal sulcus of the frontal bone. The latter exhibits irregular bony structures and porosities along the anterior part of the sagittal sulcus, originating directly above the frontal crest and expanding over a distance of about 40 mm posteriorly. This more or less superficial structural change is most probably caused by an inflammatory process of the superior sagittal sinus (perisinusitis) (Plate I, Fig. h).

Mladeč 6: calotte (cast)

Since only a cast of the cranial vault has been preserved, it is not possible to differentiate small pathological changes from erosions and inaccuracies resulting from the plaster material, but there are at least three areas where substance loss or conspicuous porosities can definitely be observed: on the frontal, approximately 22 mm left of the median sagittal line and about 32 mm anterior of the coronal suture, a small (13 x 6 mm) depression is detectable (see Frayer et al., this issue, chap. 9). The adjacent region, located posterior to the defect, seems porotic and uneven; similar porosities are noticeable in the anterior part of the right parietal bone, between the temporal crest and the sagittal suture as well as within the posterior part of the right parietal bone, just above the asterion. Such severe porosities (porotic hyperostosis) are frequently caused by malnutrition, but could originate from a tissue inflammation (e.g., after a blunt trauma) as well.

Mladeč 8: maxilla with teeth

Since the preserved facial portion is completely covered by matrix (see Wolpoff et al., this volume, chap. 8, Plate VI) superficially located signs such as newly built bone structures and porosities caused by inflammation or chronic processes cannot be reliably diagnosed. Very probably, there is no pathology on the rim of the right nasal aperture and the hard palate. There is evidence of slight periodontal disease (tooth sockets of the right P^3, C, I^2 and I^1 and the left I^1, C, P^3, P^4 and M^1 ([degree I] = 9 out of 14). There is no dental caries (0 out of 4), and no transverse linear enamel hypoplasia (0 out of 4). Brownish calculus [degree I]) is preserved on the buccal and lingual side of the cervix of the left M^2; moreover, a small remnant of calculus is noticeable on the mesial interproximal tooth surface of the left M^1 and the buccal and distal tooth surface of the canine as well (3 out of 4). Dental attrition is pronounced, in particular in the frontal teeth; the right I^2 exhibits a skewed attrition facet, which is lingually oriented, in contrast to other Upper Paleolithic finds (Pavlov 1, Dolní Věstonice 16 and Brno 2; see Vlček, 1997). The huge canine is horizontally abraded and reaches the dentine as well. A similar wear stage is visible in the first molar. Despite this advanced degree of attrition, the pulp chamber is still intact in all teeth due to the formation of secondary dentine. This secondary dentine is dark-brownish discolored (probably by manganese, which could be detected in postcranial bone samples recovered from this site; see Teschler-Nicola, this volume, chap. 5). Moreover, there are small congenital enamel defects (oval shaped pits) on the labial side of the left canine, a crushing of enamel along the mesial crown edge and several mesio-distally oriented fracture lines. These alterations most probably relate in brief local traumatic incidents, whereas the molars show multiple small superficial cracks and slight substance losses, most likely caused by a postmortem phenomenon.

Mladeč 9: right maxillary C (9a) and P^3 (9b)

The upper right canine (9a) shows no sign of dental caries and dental calculus. A large number of very fine and discrete transverse linear enamel defects, covering approximately half of the crown (5 years ±16 months) and some labial pits due to enamel hypoplasia could be identified. The latter are congenital and identical to those identified in Mladeč 8. The isolated crown of the right P^3 (9b; most probably the adjacent tooth) shows no pathological features.

Mladeč 11: cervical vertebra (C3, C4 or C5)

All articular faces of this well-preserved vertebra are observable. The cranial plate of the vertebral body shows signs of slightly degenerative changes (rim: degree A III, face: degree A I–III). The caudal plate is affected in a similar way, although the diagnosis is hampered by postmortem loss of bone around the ventral margin. Also the four joints of the vertebral arches are slightly marked by degenerative joint disease: a) upper right joint: rim: degree A 0–II, face: degree A 0–III, b) upper left joint: rim: degree A I–II, face: degree A 0–III, c) lower right joint: rim: degree A I–II, face: degree A 0–III, d) lower left joint: rim: degree A 0–I, face: degree A 0–III).

The height of the vertebral body in its ventral region is slightly reduced. This reduction is probably caused by chronic physical strain and correlates with the changes observed in the plates of the vertebral body.

Mladeč 14: left vertebral fragment of rib 2 or 3

None to moderate degenerative changes (arthrosis) on the rim and the articular face of the tubercle of the costovertebral joint were found (degree A 0–III). Postmortem sinter makes diagnosis difficult.

Mladeč 17: right vertebral fragment of rib 9 or 10

None to moderate degenerative changes (arthrosis) on the rim and the articular face of the tubercle of the costovertebral joint (degree A 0–III) were discernable. As in Mladeč 16 postmortem sinter makes diagnosis difficult.

Mladeč 21: fragment of left hip bone (incomplete os ilium and os ischium)

The specimen is free of pathology except for some moderate degenerative changes (arthrosis) on the rim (degree B 0–III) and the articular face (degree B 0–III) of the left acetabulum were developed.

Mladeč 23: proximal half of right humerus

The right humerus is characterized by signs of degenerative joint disease (arthrosis), specifically in the superior rim of the head (degree C IV). The changes are of a proliferative nature. The articular face of the head shows only slight changes represented by *intra vitam* erosion, which (most probably) damaged not only the joint cartilage, but also the subchondral bone (degree B I–II). Furthermore, there is probably evidence of a slightly developed myotendinitis. This inflammatory process, which, as a rule, affects primarily muscle and its tendon, has disturbed the insertion area of the subscapularis muscle. All changes were healed at time of death of the individual which can be seen from the morphology of the relatively unobtrusive lesions. In detail, these changes can be observed very well at the lesser tuberosity. There are well-organized, originally plate-like, porotic structures, representing reactive bone. The insertion area's irregular rim is surrounded by a bulgy structure due to a reactive response. All these changes are characteristic of an old process representing a chronic course of inflammatory disease.

The medial lip of the bicipital groove, where the latissimus dorsi and teres major muscles insert, as well as the lateral lip of the bicipital groove, where the pectoralis major muscle inserts, are both slightly pronounced. There are no convincing signs of disease such as the one in the lesser tuberosity. In the medial lip, the distal part (teres major muscle) is a little more pronounced than the proximal part (latissimus dorsi muscle). In sum, the muscles responsible for internal rotation in the shoulder joint were more or less normal.

At the well-pronounced lateral lip of the bicipital groove, where the pectoralis major muscle inserts, no significant morphological differences are observable.

Mladeč 24: right humerus diaphysis

This gracile shaft of the right humerus of a young adult individual shows only slightly developed muscle areas and no pathology.

Mladeč 25c: proximal third of the right ulna

This proximal fragment of a robust right ulna belonged to a muscular individual. The trochlear notch is characterized by changes of (most probably) degenerative origin (arthrosis) which affected the rim of this joint face (degree C IV). The radial notch exhibits probably similar features, but the region is damaged postmortem and this cannot be confirmed. Moreover, on the proximal shaft's medial face, approximately 70 mm below the olecranon process, a circular area of reactive compact bone formation is detectable. The apposition represents a remodeled remnant of a subperiosteal haematoma (originating in a localized blunt trauma).

Mladeč 27: right femur diaphysis

Although the femur shaft is more or less completely covered by matrix, there are several areas which exhibit linear, vertically oriented and uneven striation. They are especially developed in the middle third of the medial portion of the shaft, close to the linea aspera. At first glance these features could represent periosteal bone growth, probably caused by a (systemic) pathological process. Clinically, these are described as one of the symptoms characterizing Bamberger's disease, but this diagnosis is impaired due to postmortem alterations.

Mladeč 28: proximal two fifths of the shaft of a left femur

Along the medial and posterior side is a slight porosity with uneven areas, probably originating in local periostitis.

Mladeč 30: complete left talus

No degenerative changes (degree B 0) in the rim, however, slight to medium degenerative changes (degree B II<III) in the articular face of the superior articular surface. No degenerative changes (degree C 0) in the rim and almost no degenerative changes (degree C I) in the medial malleolar's articular face. The lateral malleolar articular surface is also free of degenerative changes. The posterior calcanear articular surface shows almost no degenerative changes to medium degenerative

changes on the rim and the articular face (degree A 0–III). Similar changes (degree B 0–III) are observable in the rim and the articular face of the medial calcanear face. The head of the talus is characterized, on its articular surface of the navicular bone, by almost no degenerative changes to medium degenerative changes on the rim (degree C 0–III) as well as its articular face (degree B 0–III).

Mladeč 102: second fourth of a shaft of a left femur

Along the complete length of this fragment, the lateral face shows (by x-ray) a thickening of the cortical bone (Fig. 2). We are uncertain as to the significance of this.

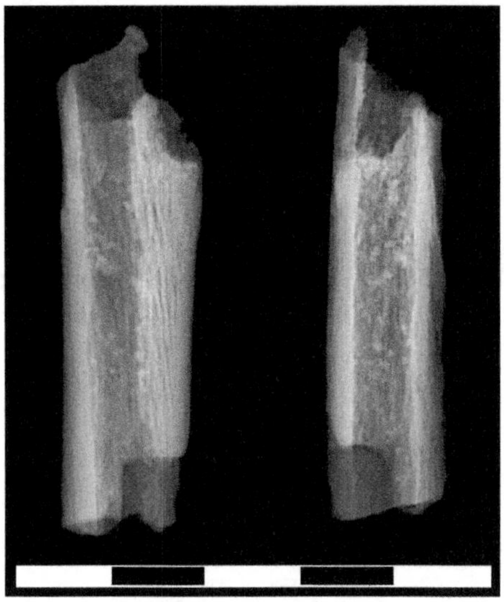

Fig. 2. Mladeč 102: X-ray of the second fourth of a shaft of a left femur showing an unclear thickening of the cortical bone along the lateral margin of the fragment: Left: anterio-posterior direction; right: medio-lateral direction

Finally, for the following specimens, we found no evidence of pathological changes:

Mladeč 10 (maxillary right third molar)
Mladeč 12 (left first rib)
Mladeč 13 (acromial half of right clavicle)
Mladeč 15 (vertebral fragment of a left rib 4, 5 or 6)
Mladeč 16 (right fragment of corpus of rib 6–9)
Mladeč 18 (vertebral fragment of a right rib 11)
Mladeč 19 (right fragmentary rib 11 or 12)
Mladeč 20 (5 rib fragments)
Mladeč 22 (right ilium and ischium fragment)
Mladeč 25a (right proximal radius fragment)
Mladeč 25b (second and third quarter of shaft of right radius)
Mladeč 26 (second third of shaft of left radius)
Mladeč 29 (distal epiphysis of right tibia)
Mladeč 31 (right third metacarpal bone)
Mladeč 32 (shaft of left third metatarsal bone)
Mladeč 34 (body of a cervical vertebra)
Mladeč 39 (right parietal fragment)
Mladeč 40 (left occipital/parietal fragment)
Mladeč 41 (left occipitotemporal fragment)
Mladeč 88 (left proximal hand phalanx 2)
Mladeč 89 (left proximal hand phalanx 3)
Mladeč 90 (right metacarpal 3)
Mladeč 91 (right metacarpal 4)

Discussion

Although most of the skeletal elements are poorly preserved and in part still covered by matrix, which limited the accurate identification of slightly expressed pathological lesions (e.g., periostitis), we could identify evidence for traumatic lesions, slightly chronic and degenerative diseases, proliferative changes (osteoma, and reactive bone caused by bacterial infection), malnutrition as well as congenital abnormalities, among them a very rare inner ear defect. As mentioned above, the investigation of these "bio historical documents" (Schultz and Kunter, 1999; Schultz, 2001; Schultz and Schmidt-Schultz, 1992; Schultz et al., 2001; 2003) was carried out by using non-invasive techniques, thereby taking curatorial concerns into account.

We have no evidence of severe infectious disease. However, there is a suspected inflammatory process in the child Mladeč 3 and the male individual Mladeč 5. In Mladeč 3 the internal surfaces of the neurocranium, including the sulci of venous brain sinuses, exhibit features which might be associated with a slight brain pressure. The latter could originate in meningeal affections, such as bacterial meningitis or a sinus inflammation. The process healed many weeks before the child died. Since these changes are very small and not determined histologically, they could also be caused by a postmortem diagenesis. The symptoms of a slightly increased intracranial pressure, represented by enlarged digital impressions, can obviously not root in diagenetic alterations, but they could represent an individual feature which has nothing to do with a pathological origin. Interestingly, Buzhilova (2000a) noted similar digital impressions on the frontal bone and partly on both temporal bones of the Sunghir 2 child, which she interpreted as evidence of high intracranial pressure as well. However, in combination with the changes observed on the internal lamina represented by the irregular (slightly healed) bone structures, a meningeal affection seems plausible in the Mladeč 3 child. A second intracranial process could be identified in Mladeč 5 calotte. The internal layer of the cranium exhibits not only very complex and reticulated meningeal arteries (Plate I, Fig. g; see also Frayer et al., this volume, chap. 9) but also alterations within the sagittal sulcus of the frontal. The latter reveals irregular reactive bony structures and porosities along the anterior part of the sagittal sulcus, originating directly above the frontal crest and expanding over a distance of about 40 mm posteriorly. This more or less superficial structural change is likely caused by an inflammatory process of the superior sagittal sinus (perisinusitis) (Plate I, Fig. h).

Reliable findings characteristic of deficiency diseases are, with the exception of four possible cases, not diagnosable. In one case (Mladeč 27), there is a possible layer on the surface of the shaft of the right femur which is suspicious of a periosteal process, such as periostosis or periostitis. Macroscopically, the differential diagnosis suggested signs of postmortem destruction. If this alteration is of *intra vitam* origin, it seems probable that it was due to a hemorrhagic process typical of chronic vitamin-C deficiency (scurvy). Eventually the slight porosities and uneven structures on the femur fragment Mladeč 28 may have the same origin. Furthermore, there are two cases which are characterized by morphological structures which might be due to healed chronic vitamin-C deficiency. The hard palate – more so in Mladeč 2 than in Mladeč 8 – has a plate-like appearance which is frequently the result of a healed scorbutic stomatitis. Additionally, in Mladeč 2, the surface of the hard palate is slightly cicatrized. Porotic hyperostosis of the skull vault, often associated with iron malnutrition, could be observed in several cases: Mladeč 2 exhibits severe porosities in the area of the parietal tubers, Mladeč 5 along the sagittal suture and Mladeč 6, although represented as (highly precise) cast, is characterized by conspicuous and clearly visible porosities.

Degenerative disease (arthrosis) is slightly developed (Mladeč 21 and 30), a result which is most probably in accordance with the low age of death of the Mladeč individuals. Only in two cases, Mladeč 23 and 25c, young mature to old adult individuals, pronounced changes of degenerative joint disease could be diagnosed. Relatively slight degenerative joint disease was also observed in the costovertebral joints (Mladeč 14 and 17). These findings correlate with slight degenerative changes (arthroses) in the vertebral column of an adult individual (Mladeč 11) according to the plates of the vertebral body (spondylosis) and the joints of the vertebral arches (spondylarthrosis). In all, the pathological changes due to degenerative joint disease are age-appropriate and do not represent severe physical strain.

Additionally, the young mature to old adult, probably male individual (Mladeč 23) who suffered from pronounced changes due to degenerative joint disease, shows well-developed signs of myotendinitis. This disease frequently co-occurs with degenerative joint disease/osteoarthritis (cf. Jaffe, 1972; Resnick and Niwayama, 1981). The changes in tendonous insertion are minor and may simply represent an example of physical overstrain in the shoulder joint.

The dental samples are not large, but typical of other Upper Paleolithic specimens (Frayer, 1989), there is no evidence of dental caries and just a slight amount of dental calculus in an older individual (Maldeč 8). Periodontal diseases are, as a rule, only slightly expressed - perhaps also in accordance with the age of death of the individuals. Enamel hypoplasias could only be detected in one case, Mladeč 9, although it is a more or less common feature observed in Upper Paleolithic finds (e.g., Dolní Věstonice 15, Formicola et al., 2001; Sunghir 1, Buzhilova, 2000b). The localization in the Mladeč 9 canine allows to date the critical period of physical stress around 5 years (±16 months), which therefore cannot represent the so end of weaning. Instead we suggest malnutrition or infectious disease as possible physical stress generators. The most conspicuous dental feature within the present sample is the pronounced attrition observed in the Mladeč 8 specimen, a mature individual. The substantial loss is noticeable specifically on the anterior teeth, the lateral incisor and the canine. The attrition of the left second incisor reaches the tooth neck lingually and leaves an intact, just 1-1.5 mm crown portion labially. The crown of the canine is reduced approximately to one third, the wear plane is horizontally oriented. A similar degress of differential wear can be documented among Middle Paleolithic and other early Upper Paleolithic, specifically the Gravettian finds from the Czech Republic. Pavlov I, Dolní Věstonice 13 and, in particular, Dolní Věstonice 16 and Brno 2 (Vlček, 1997), exhibit a similar pattern, which is generally attributed to a high functional stress caused by using the anterior teeth in non-dietary activities (e.g., skin processing).

Our survey revealed various types of trauma. It is most likely that the small, circular bone apposition on the ulna is the product of a minor trauma to the lateral elbow region. This was followed by a subperiosteal disruption, hemorrhagic process and endosteal remodeling with callus formation. There are no signs which would indicate an inflammation. Healed as well as unhealed traumas could be identified on the cranium remains of the male individual Mladeč 5. There are two healed wounds above the left orbit and in the right parietal and a bigger, perimortem defect in the right parietal. As it was documented by Trinkaus (1985), fractures are frequently observed in Neandertals and their patterns of skeletal injuries differ from early modern human groups (Berger and Trinkaus, 1995). Trinkaus and Zimmerman (1982) found that the survival of the injured individuals could be characteristic of the Neandertals (for further references see Trinkaus and Zimmerman, 1982); however, there are also several Upper Paleolithic sites where individuals with healed (and unhealed) injuries have been found (Trinkaus et al., 2000; Formicola et al., 2001). Several of the reported injuries represent indeed craniofacial trauma, among them the Gravettian Lagar Velho I child, where one trauma to the lower left face is documented (Trinkaus et al, 2002, 495) and Dolní Věstonice 16, which (probably) exhibit "a partly healed fracture on the maxilla, in the sutura palatina area" (Vlček, 1997, 74) and Dolní Věstonice 12, which shows a healed pronounced depression above the right supraorbital notch (Trinkaus et. al., 2000, 1119). Also one of the Qafzeh juveniles (Qafzeh 11) exhibit a fracture on the right side of the frontal bone (Tillier, 1999, 165) and Ullrich (1986) claimed that (unhealed) impressed skull fractures are even present in the Upper Paleolithic finds from Balla (Hungary) and Cioclovina (Romania). Referring to Mladeč 5, there is some uncertainty (because of the restorative manipulations carried out), but it is possible that the oval depression localized near the sagittal suture in the right parietal bone is caused by perimortal blunt trauma.

Finally, malformations were found in Mladeč 1 and Mladeč 2: In Mladeč 1 a nasal septum deviation, a very frequent anomaly in recent living populations, could be observed. Such a deformation may be congenital or possibly have occurred later (e.g., by a trauma). In Mladeč 1 the septum is conspicuously shifted away from the sagittal midline to the right. This condition could have caused symptoms such as a blockage of the right nostril, difficulties in breathing or sinus infections (Ballenger, 1991). By using computed tomography, a very rare malformation of the vestibular system in specimen Mladeč 2 was identified. The vestibule of the right side is very small and hypoplastic compared to the left side. With high probability, such a defect could have caused impairment of hearing or deafness.

Finally, we should mention that Szombathy (1925, 75–77) described several specimens which were excavated during the 1904 Mladeč campaign, among them a proximal portion of a left femur (Mladeč 78). According to Szombathy (1925, 77, Abb. 35) this fragment is characterized by an unusual shaft curvature.[2] The same is true for the second femur, where only part of the diaphysis is preserved. Although it is not possible to investigate this specimen thoroughly by the use of the figure alone,[3] which is less meaningful due to the reproduction quality, we should point to this feature since it has been described from other Upper Paleolithic sites as well. For example, Dolní Věstonice 15 possesses a similar deformation, where the right femur shows a severe asymmetric shortening and bowing compared to the left. Klima (1987) argued for semiparalysis of the right side, possibly resulting from encephalitis in early life. His hypothesis has been disputed by Jelínek (1992), who suggested a rachitic condition, and Formicola et al. (2001), who suggested that the most likely etiology is chondrodysplasia calcificans punctata, a rare inherited disorder. The Upper Paleolithic burial from Sunghir, Russia, the 9–10 year old girl Sunghir 3 exhibits symmetric shortening as well as marked anterior-posterior bowing of both femora (Buzhilova, 2000a). Following Formicola and Buzhilova (2004), this anomaly is comparable to the Mladeč 78 left femur and may point to a "congenital bowing of long bones."

Conclusion

We present the results of the first systematic, non-destructive investigation of pathological and traumatic conditions of the early Upper Paleolithic human fossils from Mladeč. In sum the nature and the location of the pathological changes observed are characteristic of a population which did not live under great physical strain (although we have to take into account that several remains derive from young individuals and it has been reported, e.g., that injuries tend to accumulate over an individual's lifetime, Berger and Trinkaus, 1995). While traumatic injuries are not uncommon in the Upper Paleolithic record, congenital anomalies are seldom described. The scope of diseases and anomalies diagnosed, including (very minor) chronic and degenerative diseases and proliferous changes, possible malnutrition as well as congenital abnormalities, such as in particular the rare inner ear malformation and the limb bone bowing, complete our knowledge on the diseases from which early Upper Paleolithic modern humans might have suffered.

Acknowledgement

We thank Prof. H. Imhof, Institute for Radiology, General Hospital Vienna, and Dr. J. Křístek, Dept. of Medical Imaging, St. Anna University Hospital, Brno, Czech Republic, for providing us CT scanning and x-ray analysis and Helga Plischke for her assistance in the preparation procedure. Furthermore, we thank Wolfgang Reichmann for the photographic documentation, Doris Pany for her helpful commentaries on the identification of musculo-skeletal stress markers and David Frayer for critical comments on an earlier version of the manuscript.

2 „Bei dem unvollständigen (des distalen Gelenksendes beraubten) Stück, das mit seiner dorsalen Seite auf eine Ebene aufgelegt wird, erhebt sich die Mitte des Unterrandes um 27 mm über die Unterlage" (Szombathy, 1925, 77).

3 The originals are lost and casts do not exist.

References

Alexeeva, T. I. and Bader, N. O. (2000) *Homo sungirensis. Upper Palaeolithic man: ecological and evolutionary aspects of the investigation.* Moscow: Scientific World

Antón, S. C. (1997) Endocranial hyperostosis in Sangiran 2, Gibraltar 1, and Shanidar 5. *American Journal of Physical Anthropology* 102, 111–122

Aufderheide, A. C. and Rodríguez-Martín, C. (1998) *The Cambridge encyclopedia of human paleopathology.* Cambridge, New York: Cambridge University Press

Ballenger, J. J. (1991) *Disorders of the nose, throat, ear, head, and neck.* Philadelphia: Lea & Febiger

Berger, T. D. and Trinkaus, E. (1995) Patterns of trauma among the Neandertals. *Journal of Archaeological Science* 22, 841–852

Bräuer, G., Groden, C., Delling, G., Kupczik, K., Mbua, E. and Schultz, M. (2003) Pathological alterations in the archaic *Homo sapiens* cranium from Eliye Springs, Kenya. *American Journal of Physical Anthropology* 120, 200–204

Bukhman, A. I. (1984) Roentgenological studies of the children's skeletons from the Upper Paleolithic site Sungir [in Russian]. In (A. A. Zubov and V. M. Kharitonov, Eds.) *Sungir anthropological investigations.* Moscow: Nauka, pp. 203–204

Buzhilova, A. P. (2000a) The analysis of anomalies and indicators of physiological stress in non-mature Sunghir individuals. In (T. I. Alexeeva and N. O. Bader, Eds.) *Homo Sungirensis. Upper Paleolithic man: Ecological and evolutionary aspects of the investigation.* Moscow: Scientific World, pp. 302–314

Buzhilova, A. P. (2000b) Palaeopathological analysis of the adult Sunghir man. In (T. I. Alexeeva and N. O. Bader, Eds.) *Homo Sungirensis. Upper Paleolithic man: Ecological and evolutionary aspects of the investigation.* Moscow: Scientific World, pp. 227–234

Casano, P. (2002) Deviated nasal septum. www.sinuscarecenter.com/dns.html

Casselman, J. W., Kuhweide, R., Deimling, M., Ampe, W., Dehaene, I. and Meeus, L. (1993) Constructive interference in steady state-3DFT MR imaging of the inner ear and cerebellopontine angle. *American Journal of Neuroradiology* 14, 47–57

Crubézy, E. and Trinkaus, E. (1992) Shanidar 1: A case of hyperostotic disease (DISH) in the Middle Paleolithic. *American Journal of Physical Anthropology* 89, 411–420

Czerny, C., Rand, T., Gstoettner, W., Woelfl, G., Imhof, H. and Trattnig, S. (1998) MR imaging of the inner ear and cerebellopontine angle: Comparison of three-dimensional and two-dimensional sequences. *American Journal of Roentgenology* 170, 791–806

Czerny, C., Gstoettner, W., Adunka, O., Hamzavi, J. and Baumgartner, W. D. (2000) Präoperative Bildgebung vor dem Einsetzen eines multikanalikulären Cochlea-Implantates mittels Computer- und Magnetresonanztomographie der Innenohrregion. *Wiener Klinische Wochenschrift* 112, 481–486

Dastugue, J. (1962) Pathologie de quelques Neandertaliens. *Actes Vie Congr. Int. Sci. Anthropol. Ethnol. (Paris 1960)* 1, 577–581

Dastugue, J. (1967) Pathologie des hommes fossiles de l'abri de Cro-Magnon. *L'Anthropologie* 71, 479–492

De Gusta, D. (2002) Comparative skeletal pathology and the case of conspecific care in Middle Pleistocene Hominids. *Journal of Archaeological Science* 29, 1435–1438

Formicola, V. (1995) X-linked hypophosphatemic rickets: A probable Upper Paleolithic case. *American Journal of Physical Anthropology* 98, 403–409

Formicola, V., Pontradolfi, A. and Svoboda, J. (2001) The Upper Paleolithic triple burial of Dolní Věstonice: Pathology and funerary behavior. *American Journal of Physical Anthropology* 115, 372–379

Formicola, V. and Buzhilova, A. P. (2004) Double child burial from Sunghir (Russia): Pathology and inferences for Upper Paleolithic funerary practices. *American Journal of Physical Anthropology* 124, 189–198

Frayer, D. W. (1989) Oral pathologies in the European Upper Paleolithic and Mesolithic. In (I. Hershkovitz, Ed.) *People and culture in change.* Oxford: BAR International Series (508i), pp. 255–281

Frayer, D. W., Macchiarelli, R. and Mussi, M. (1988) A case of chondrodystrophic dwarfism in the Italian late Upper Paleolithic. *American Journal of Physical Anthropology* 75, 549–565

Frayer, D. W., Jelínek, J., Oliva, M. and Wolpoff, M. H. (2006) Aurignacian male crania and teeth from the Mladeč Caves, Moravia, Czech Republic. In (M. Teschler-Nicola, Ed.) *Early modern humans at the Moravian Gate: The Mladeč Caves and their remains,* this issue, pp. 185–272

Jaffe, H. L. (1972) *Metabolic, degenerative, and inflammatory diseases of bones and joints.* München Berlin Wien: Urban & Schwarzenberg

Jelínek, J. (1969) Neanderthal man and *Homo sapiens* in central and eastern Europe. *Current Anthropology* 10, 475–503

Jelínek, J. (1992) New Upper Paleolithic burials from Dolní Věstonice. *ERAUL* 56, 207–227

Klíma, B. (1987) Das jungpaläolithische Massengrab von Dolní Věstonice. *Quartär* 37/38, 53–62

Klíma, B. (1991) Die jungpaläolithischen Mammutjäger-Siedlungen Dolní Věstonice and Pavlov in Südmähren – ČSFR. *Archäologie und Museum* 23, 1–30

Lebel, S., Trinkaus, E., Faure, M., Fernandez, P., Guérin, C., Richter, D., Mercier, N., Valladas, H. and Wagner, G. A. (2001) Comparative morphology and paleobiology of Middle Pleistocene human remains from the Bau de l'Aubesier, Vaucluse, France. *Proceedings of the National Academy of Sciences USA* 98, 11097–11102

Mednikova, M. B. (2000) X-ray morphology of children from the grave 2. In (T. I. Alexeeva and N. O. Bader, Eds.) *Homo Sungirensis. Upper Paleolithic man: Ecological and evolutionary aspects of the investigation.* Moscow: Scientific World, pp. 286–298

Minugh-Purvis, N., Bence Viola, T. and Teschler-Nicola, M. (2006) The Mladeč 3 infant. In (M. Teschler-Nicola, Ed.) *Early modern humans at the Moravian Gate: The Mladeč Caves and their remains,* this issue, pp. 357–384

Oliva, M. (2000) The Brno II Upper Paleolithic burial. In (W. Roebroeks, M. Mussi, J. Svoboda and K. Fennema, Eds.) *Hunters of the Golden Age. The Mid Upper Paleolithic of Eurasia (30,000–20,000 BP).* Leiden: University Press, pp. 143–153

Ortner, D. (2003) *Identification of pathological conditions in human skeletal remains.* Amsterdam Boston London New York: Academic Press/Elsevier Science

Prossinger, H. and Teschler-Nicola, M. (2006) Electronic segmentation methods reveal the preservation status and otherwise unobservable features of the Mladeč 1 cranium. In (M. Teschler-Nicola, Ed.) *Early modern humans at the Moravian Gate: The Mladeč Caves and their remains,* this issue, pp. 341–356

Resnick, D. and Niwayama, G. (1981) *Diagnosis of bone and joint disorders.* Philadelphia London Toronto: WB Saunders Comp

Roebroeks, W., Mussi, M., Svoboda, J. and Fennema, K. (Eds.) (2000) Hunters of the Golden Age. The Mid Upper Palaeolithic of Eurasia (30,000–20,000 BP). Leiden: University Press

Rothschild, B. M., Hershkovitz, I. and Rothschild, C. (1995) Origins of yaws in the Pleistocene. *Nature* 378, 343–344

Schultz, M. (1988) Paläopathologische Diagnostik. In (R. Knussmann, Ed.) *Anthropologie. Handbuch der vergleichenden Biologie des Menschen.* Volume 1. Stuttgart: Fischer, pp. 480–496

Schultz, M. (2001) Paleohistopathology of bone: A new approach to the study of ancient diseases. *Yearbook of Physical Anthropology* 44, 106–147

Schultz, M. and Kunter, M. (1999) Erste Ergebnisse der anthropologischen und paläopathologischen Untersuchungen an den menschlichen Skelettfunden aus den neuassyrischen Königinnengräbern von Nimrud. *Jahrbuch des Römisch-Germanischen Zentralmuseums* 45, 85–128

Schultz, M. and Schmidt-Schultz, T. H. (1992) Ergebnisse osteologischer Untersuchungen an menschlichen Skelettfunden aus Milet. *Istanbuler Mitteilungen des Deutschen Archäologischen Instituts* 41, 163–186

Schultz, M., Walker, R., Strouhal, E. and Schmidt-Schultz, T. H. (2001) Skeletal remains. Merinebti, Hefi and Iries. In (N. Kanawati and M. Abder-Raziq, Eds.) *The Teti cemetery at Saqqara: The tombs of Shepsipuptah, Mereri (Merinebti), Hefi and others.* Warminster: Aris and Phillips, pp. 65–74

Schultz, M., Walker, R., Strouhal, E. and Schmidt-Schultz, T. H. (2003) Skeletal remains. Report on the skeleton of Jj-nfrt excavated from his mastaba in the north cemetery of Unis's pyramid (5th Dynasty). In (N. Kanawati and M. Abder-Raziq, Eds.) *The Unis cemetery at Saqqara.* Warminster: Aris and Phillips, pp. 75–86

Som, P. M. and Curtin, H. D. (2003) *Head and neck imaging.* St. Louise: Mosby

Steinbock, R. T. (1976) *Paleopathological diagnoses and interpretation. Bone diseases in ancient human populations.* Springfield: CC Thomas

Straus W. L., Jr. and Cave, A. J. E. (1957) Pathology and the posture of Neanderthal man. *Quarterly Review of Biology* 32, 348–363

Svoboda, J. (Ed.) (1997) *Pavlov I – Northwest. The Upper Paleolithic burial and its settlement context* (The Dolní Věstonice Studies 4). Brno: Academy of Sciences

Szombathy, J. (1925) Die diluvialen Menschenreste aus der Fürst-Johanns-Höhle bei Lautsch in Mähren. *Die Eiszeit* 2, 1–34, 73–95

Tappen, N. C. (1985) The dentition of the "Old Man" of La Chapelle-aux-Saints and inferences concerning Neandertal behavior. *American Journal of Physical Anthropology* 67, 43–50

Teschler-Nicola, M. (2006) Taphonomic aspects of the human remains from the Mladeč caves. In (M. Teschler-Nicola, Ed.) *Early modern humans at the Moravian Gate: The Mladeč Caves and their remains*, this issue, pp. 75–98

Thillaud, P. L. (1981–1982) L'histiocytose X au paléolithique (sujet n° 1 de Cro-Magnon). Problématique du diagnostic ostéo-archéologique. *L'Anthropologie* 71, 219–239

Tillier, A.-M. (1999) Les enfants Mousteriens de Qafzeh. Interpretation phylogenetique et paleoauxologique. Paris: C.N.R.S. Editions

Tillier, A.-M., Arensburg, B., Duday, H. and Vandermeersch, B. (2001) An early case of hydrocephalus: The Middle Paleolithic Qafzeh 12 child (Israel). *American Journal of Physical Anthropology* 114, 166–170

Trinkaus, E. (1985) Pathology and the posture of the La Chapelle-aux-Saints Neandertal. *American Journal of Physical Anthropology* 67, 19–41

Trinkaus, E. and Svoboda, J. (Eds.) (2006) Early modern human evolution in Central Europe. Oxford: Oxford University Press

Trinkaus, E. and Zimmerman, M. R. (1982) Trauma among the Shanidar Neandertals. *American Journal of Physical Anthropology* 57, 61–76

Trinkaus, E., Hillson, S. W. and Santos Coelho, J. M. S. (2002) Paleopathology. In (J. Zilhão and E. Trinkaus, Eds.) *Portrait of the artist as a child. The Gravettian human skeleton from the Abrigo do Lagar Velho and its archeological context* (Trabalhos de arqueologia 22) Lisboa: Instituto Português de Arqueologia, pp. 489–495

Trinkaus, E., Smith, F. H., Stockton, C. and Shackelford, L. L. (2006) The human postcranial remains from Mladeč. In (M. Teschler-Nicola, Ed.) *Early modern humans at the Moravian Gate: The Mladeč Caves and their remains*, this issue, pp. 385–446

Trinkaus, E., Svoboda, J., West, D. L., Sládek, V., Hillson, S. W., Drozdová, E. and Fišákova, M. (2000) Human remains from the Moravian Gravettian: Morphology and taphonomy of isolated elements from the Dolní Věstonice II site. *Journal of Archaeological Science* 27, 1115–1132

Ubelaker, D. H. (1978) *Human skeletal remains. Excavations, analysis, interpretation.* Chicago: Aldine

Ullrich, H. (1986) Manipulations on human corpses, mortuary practice and burial rites in Paleolithic times. *Anthropos* (Brno) 23, 227–236

Vlček, E. (1997) Human remains from Pavlov and the biological Anthropology of the Gravettian human population of south Moravia. In (J. Svoboda, Ed.) *Pavlov I - Northwest. The Upper Paleolithic burial and its settlement context* (The Dolní Věstonice Studies 4) Brno: Academy of Sciences, pp. 53–153

Walker, A., Zimmerman, M. R. and Leakey, R. E. (1982) A possible case of hypervitaminosis A in *Homo erectus*. *Nature* 296, 248–250

Wolpoff, M. H., Frayer, D. W., Trinkaus, E. and Teschler-Nicola, M. (2006) Inventory and photo-documentation of the Mladeč hominid remains. In (M. Teschler-Nicola, Ed.) *Early modern humans at the Moravian Gate: The Mladeč Caves and their remains*, this issue, pp. 159–184

Wolpoff, M. H., Frayer, D. W. and Jelínek, J. (2006) Aurignacian female crania and teeth from the Mladeč Caves, Moravia, Czech Republic. In (M. Teschler-Nicola, Ed.) *Early modern humans at the Moravian Gate: The Mladeč Caves and their remains*, this issue, pp. 273–340

Zilhão J. and Trinkaus E. (Eds.) (2002) *Portrait of the artist as a child. The Gravettian human skeleton from the Abrigo do Lagar Velho and its archeological context* (Trabalhos de arqueologia 22). Lisboa: Instituto Português de Arqueologia

Zollikofer, Ch. P. E., Ponce de León, M. S., Vandermeersch, B. and Lévêque, F. (2002) Evidence for interpersonal violence in the St. Césaire Neanderthal. *Proceedings of the National Academy of Sciences USA* 99, 6444–6448

Plates

Plate I.
Mladeč 5: male calotte

(a) Healed traumatic lesion in the frontal bone, above and in the middle of the left orbital rim.
(b) Same defect as (a), 1.8-times magnified, showing irregular structures, a remarkable pitting and remodeling
(c) Small osteoma in the frontal bone near the median sagittal plane
(d) Same defect as (c), higher magnification
(e) Small, circular intravital substance loss in the right parietal bone, pertaining the external layer only
(f) Same defect as (e), higher magnification, showing irregular structures and porosities
(g) Endocranial view showing complex and reticulated meningeal arteries
(h) Endocranial view with irregular newly built bony structures and porosities along the anterior part of the sagittal sulcus (perisinusitis)

Plate II.
Mladeč 5: computed tomographies[4]

(a) Sagittal view: depression in the right parietal bone (most probably perimortal)
(b) Sagittal view: postmortal alteration in the frontal bone
(c) Sagittal view: blunt trauma in the frontal bone above the left orbital rim (intravital)
(d) Axial view: depression in the right parietal bone (most probably perimortal)
(e) Coronal view: postmortal impression in the left frontal bone
(f) Coronal view: small superficial (most probably postmortal) alteration in the right temporal bone; frontal sinuses with some encrustation; normal built petrosal bones

4 Copyright: Dr. J. Křístek, Dept. of Medical Imaging, St. Anna University Hospital, Brno, Czech Republic

PLATE I

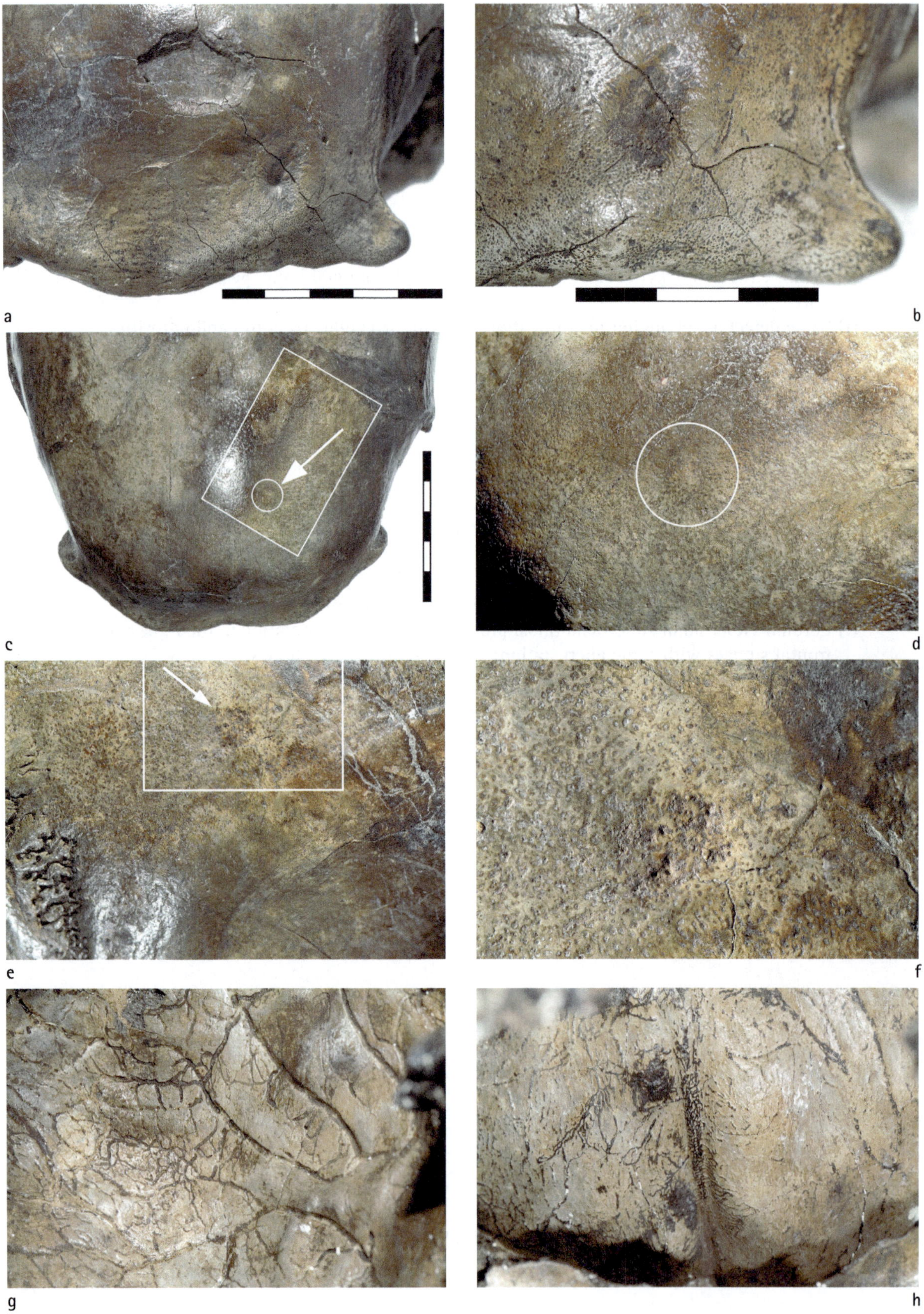

PLATE II

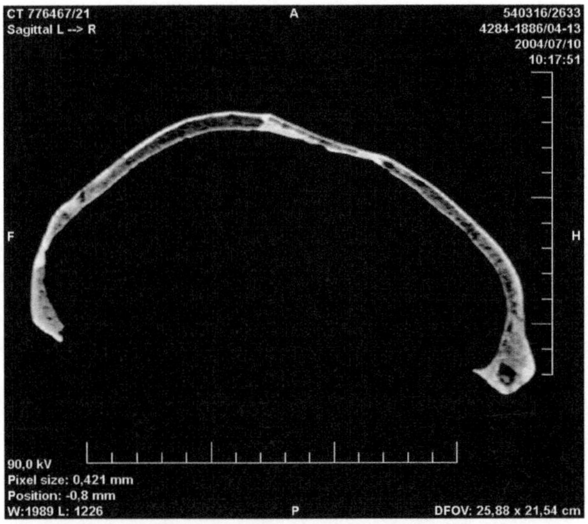

a

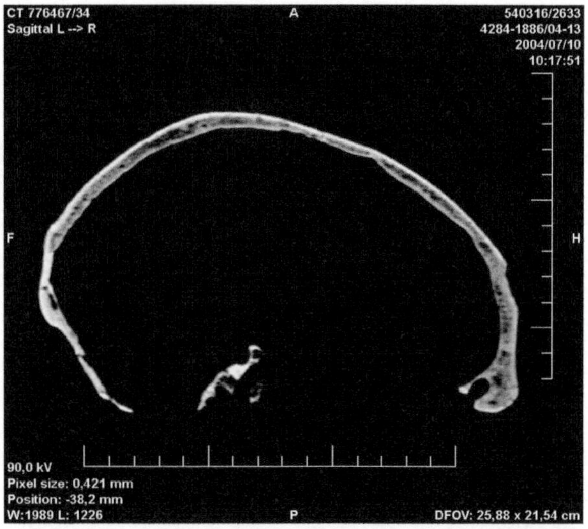

b

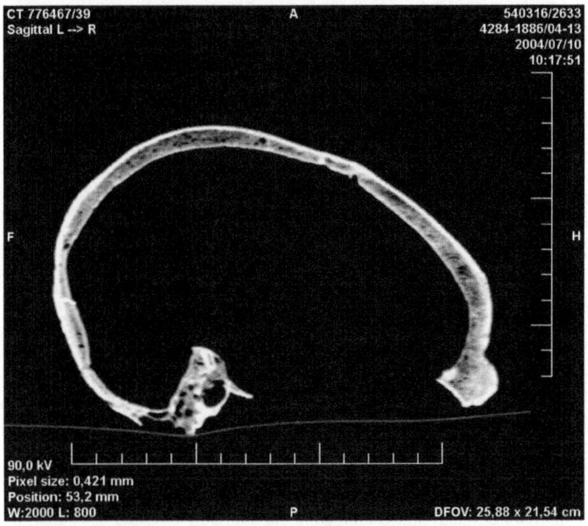

c

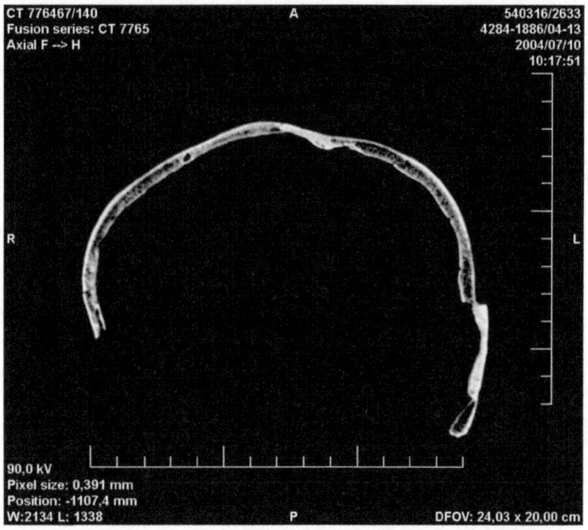

d

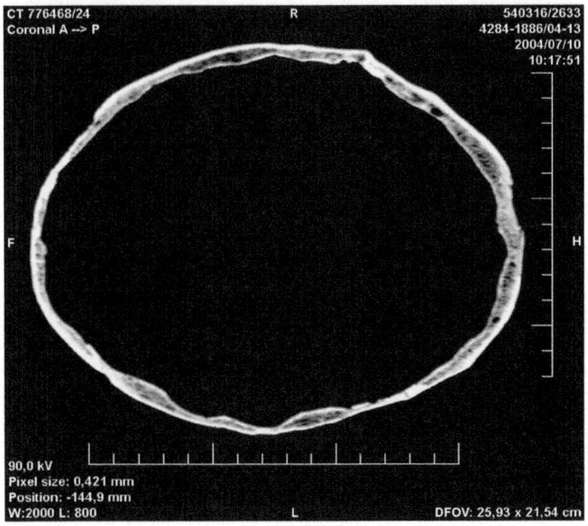

e

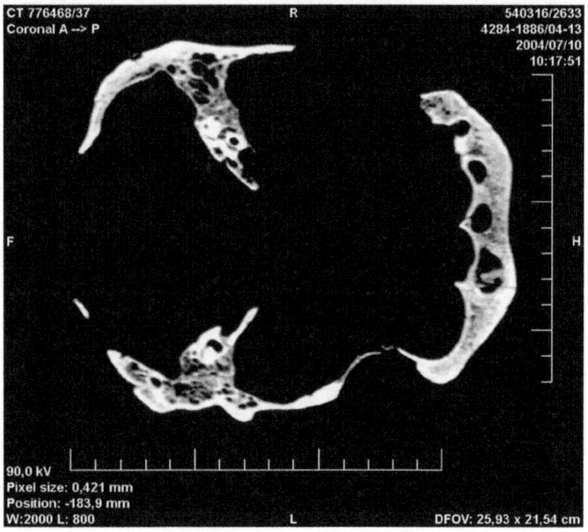

f

Chapter 17

NO EVIDENCE OF NEANDERTAL mtDNA CONTRIBUTION TO EARLY MODERN HUMANS

David Serre, André Langaney, Mario Chech, Maria Teschler-Nicola, Maja Paunovic, Philippe Mennecier, Michael Hofreiter, Göran Possnert and Svante Pääbo

Introduction

Despite intense research efforts, no consensus has been reached about the genetic relationship between early modern humans and archaic human forms such as the Neandertals. While supporters of "multiregional evolution" argue for genetic exchange or even continuity between archaic and modern humans (Weidenreich, 1943; Wolpoff et al., 1984; Duarte et al., 1999; Wolpoff et al., 2000; Hawks and Wolpoff, 2001), proponents of a "single African origin" of contemporary humans claim that negligible genetic interaction took place (Cann et al., 1987; Stringer and Andrews, 1988; Ingman et al., 2000; Underhill et al., 2000; Stringer, 2002). Mitochondrial DNA (mtDNA) sequences from early modern humans would in principle be able to resolve the question of a contribution of Neandertal mtDNA to modern humans. However, human DNA is pervasive in palaeontological and archaeological remains as well as in most laboratory environments (e.g., Krings et al., 2000; Hofreiter et al., 2001b; Wandeler et al., 2003). It is therefore currently impossible to differentiate contaminating modern DNA sequences from endogenous human DNA in humans remains. Thus, although mtDNA sequences have been reported from remains of early modern humans (Adcock et al., 2001; Caramelli et al., 2003), it is not possible to determine whether such DNA sequences indeed represent endogenous DNA sequences (Abbott, 2003). A related problem is that if a Neandertal fossil yields modern-like DNA sequences, those might be discarded as putative contaminations (Nordborg, 1998; Trinkaus, 2001), even if they may be endogenous and represent evidence for a close genetic relationship or interbreeding between the two groups.

To explore the genetic relationship between early modern humans and Neandertals in spite of these difficulties, we made use of the fact that the four Neandertal mtDNA sequences determined to date can easily be distinguished from those of modern humans (Krings et al., 1997; 2000; Ovchinnikov et al., 2000; Schmitz et al., 2002; Knight, 2003). This allowed us to ask whether all well-preserved Neandertal remains contain Neandertal-like mtDNA and whether all well-preserved early modern human remains fail to contain such DNA sequences. Thus, we did not attempt to determine DNA sequences that are similar to present-day, human mtDNA. Instead we determined whether Neandertal-like mtDNA sequences were present or absent in well-preserved remains of Neandertals and early modern humans.

Methods

Amino acid preservation

About 10 mg of bone were removed from each specimen and analyzed as in Schmitz et al. (2002) with minor modifications. In brief, proteins are hydrolyzed and amino acids labeled with *o*-phtaldialdehyde/*N*-acetyl-L-cystein and analyzed by high performance liquid chromatography (Schimadzu, Japan) under conditions that separate the different amino acids as well as their stereoisomers. Eight amino acids are analyzed and their respective concentration measured: D- and L-alanine, glycine, D- and L-aspartic acid, serine, glutamic acid, valine, D- and L-leucine and isoleucine.

DNA extraction and amplification

DNA extractions were performed in a laboratory dedicated to ancient DNA work. In this laboratory, positive air pressure is maintained with filtered air at all times, and all areas and equipment are treated with UV light when the laboratory is not used. A maximum of six bone or teeth samples were processed together with two blank extractions. Neandertal samples were always processed together with early modern human samples or cave bear samples. For each extraction, the samples were ground and between 30 mg and 120 mg of bone powder were extracted as in Krings et al. (1997). mtDNA sequences were amplified by the polymerase chain reaction (PCR) using 5 µl of extract and 60 cycles. In addition, a minimum of four blank PCRs were performed together with each amplification from extracts. The "Neandertal-specific" amplification was carried out using the primers NL16230/NH16262 (Krings et al., 1997) and an annealing temperature of 60°C. We consider it highly unlikely that the Neandertal-specific mtDNA fragments represent contaminations from other Neandertals, given that none of the extracts of modern humans or cave bears processed in parallel with the Neandertal remains yielded such products. The "hominoid" amplification was performed with the primers L16022/H16095 (Krings et al., 1997) and an annealing temperature of 54°C. PCR products were cloned into *Escherichia coli*kit (Invitrogen, Leek, The Netherlands), and ten or 30 clones of each amplification were sequenced on a ABI 3700 (Applied Biosystems, Forster City, California, United States).

Estimation of admixture

Given that previous analyses of mtDNA sequences have rejected a model of complete panmixia between Neandertals and early modern humans (Nordborg, 1998), we focused on the estimation of the level of admixture between Neandertals and early modern humans that can be excluded. For this purpose, we considered a population of early modern humans that merged at *Tm* with a (genetically different) population of Neandertal individuals (see Fig. 4) from which point the fused population was panmictic. The probability of picking K individuals by chance in the merged population that all carry a modern human mtDNA sequence is $(1 - c)^K$, where c represents the Neandertal genetic contribution to the merged population. If none of n mtDNA sequences sampled in the merged population is Neandertal-like, we can exclude (at the 5% level) contributions that give a probability smaller than 0.05 of observing only modern human sequences, i.e., $(1 - c)^K < 0.05$. The number of ancestors of n samples at the time t is represented by a probability distribution, $An(t)$. Thus, the probability of observing only one kind of sequences in n samples becomes:

$$\Pr(\text{only human sequences observed})=\sum(\Pr(An(t)=K)\times(1-c)^K),$$

where K vary from 1 to n. For a population of constant size over time, $\Pr(An(t)=K$ has been derived in Tavare (1984). We estimated the number of ancestors of n samples at time t as the expected value of $An(t)$, $E(An(t))$, according to this model and calculate the probability of observing only human sequences for different values of c.

Results and discussion

The preservation of endogenous DNA in fossils is correlated with the amount, composition, and chemical preservation of amino acids (Poinar et al., 1996). We find that endogenous DNA can be amplified from Pleistocene remains when the amino acid content is more than 30,000 parts per million (ppm), the ratio of glycine to aspartic acid between two and ten, and the aspartic acid racemization (i.e., the stereoisomeric D-/L-ratio) less than 0.10 (Poinar et al., 1996; Krings et al., 1997; 2000; Schmitz et al., 2002; data not shown). We analyzed the amino acid preservation of 24 Neandertal and 40 early modern human fossils (see Table 1). Several important Neandertal fossils, such as La

Table 1. Results of the amino acid analyses of 40 human and 24 Neandertal remains.
The bones were analyzed by high performance liquid chromatography for their amino acid content (see Materials and Methods). The extent of racemization of aspartic acid (D-/L-Asp), the ratio glycine to aspartic acid (Gly/Asp) and the total amount of the eight amino acid analyzed (ppm) are given for each specimen. Zero indicates values below detection level. The five human and four Neandertal specimens from which DNA extraction were performed are displayed in gray

Specimen/Provenience	Taxon	D/L Asp	Gly/Asp	PPM
Abri Pataud, France	Modern Humans	0,086	3,18	14380
Abri Pataud, France	Modern Humans	0	3,87	24817
Abri Pataud, France	Modern Humans	0	0	0
Abri Pataud, France	Modern Humans	0,068	4,61	24743
Abri Pataud, France	Modern Humans	0	3,77	17582
Abri Pataud, France	Modern Humans	0,062	7,37	56119
Cro-Magnon, France	Modern Humans	0	6,98	6438
Cro-Magnon, France	Modern Humans	0,011	2,11	2287
Cro-Magnon, France	Modern Humans	0	1,9	1784
Cro-Magnon, France	Modern Humans	0,095	1,7	2060
Cro-Magnon, France	Modern Humans	0,117	2,97	13532
Cro-Magnon, France	Modern Humans	0,081	7,59	56684
La Madeleine, France	Modern Humans	0	5,02	14384
La Madeleine, France	Modern Humans	0	5,34	18880
La Madeleine, France	Modern Humans	0	0	0
La Madeleine, France	Modern Humans	0,096	1,17	4985
La Madeleine, France	Modern Humans	0,095	1,75	7688
La Madeleine, France	Modern Humans	0,086	0,78	5256
La Madeleine, France	Modern Humans	0,091	7,21	29901
Mladeč 2, Czech Republic	Modern Humans	0,080	8,25	87436
Mladeč 21, Czech Republic	Modern Humans	0,330	13,71	4895
Mladeč 25a, Czech Republic	Modern Humans	0,003	8,22	14565
Mladeč 25c, Czech Republic	Modern Humans	0,009	9,06	53718
Mladeč 26, Czech Republic	Modern Humans	0	9,89	3253
Mladeč 27, Czech Republic	Modern Humans	0	113,48	24156
Mladeč 27, Czech Republic	Modern Humans	0	11,81	3883
Mladeč 28, Czech Republic	Modern Humans	0	8,53	3343
Mladeč 28, Czech Republic	Modern Humans	0	9,69	2151
Mladeč 33, Czech Republic	Modern Humans	0	11,37	4001
Mladeč 8, Czech Republic	Modern Humans	0,133	5,21	3488
Mladeč child, Czech Republic	Modern Humans	0	115,01	9308
Sandalja, Croatia	Modern Humans	0,115	0,97	10848
Sandalja, Croatia	Modern Humans	0,12	0,95	22256
Sandalja, Croatia	Modern Humans	0,111	1,19	10950
Veternica, Croatia	Modern Humans	0,089	3,7	9429

Table 1. (continued)

Specimen/Provenience	Taxon	D/L Asp	Gly/Asp	PPM
Veternica, Croatia	Modern Humans	0,092	1	11886
Veternica, Croatia	Modern Humans	0,066	1,09	12218
Veternica, Croatia	Modern Humans	0,082	6,7	23509
Veternica, Croatia	Modern Humans	0,076	1,37	22340
Veternica, Croatia	Modern Humans	0,07	1,11	13161
Monte Circeo, Italia	Neandertal	0,244	4,57	371
Amud, Israel	Neandertal	0,171	1,79	164
Dederiyeh, Syria	Neandertal	0,000	4,48	200
Dederiyeh, Syria	Neandertal	0,000	0	0
El Sidron, Spain	Neandertal	0,053	9,92	9924
Engis 2, Belgium	Neandertal	0,086	7,55	31561
Krapina, Croatia	Neandertal	0,223	3,77	7044
Krapina, Croatia	Neandertal	0	0	0
Krapina, Croatia	Neandertal	0	0	0
La Ferrassie 2, France	Neandertal	0,033	9,80	1419
La Ferrassie 1, France	Neandertal	0,251	7,12	40353
La Ferrassie 1, France	Neandertal	0,296	2,06	85510
La Ferrassie 1, France	Neandertal	0,263	3,13	30150
La Ferrassie 1, France	Neandertal	0,165	6,07	36405
La Ferrassie 1, France	Neandertal	0,141	4,28	42145
La-Chapelle-aux-Saints, France	Neandertal	0	7,49	2190
La-Chapelle-aux-Saints, France	Neandertal	0,068	4,89	19650
La-Chapelle-aux-Saints, France	Neandertal	0,065	7,97	55128
LaQuina, France	Neandertal	0	0	0
LaQuina, France	Neandertal	0,081	1	2931
LaQuina, France	Neandertal	0,225	0	194
LaQuina, France	Neandertal	0	0	0
Vindija Vi-80, Croatia	Neandertal	0,074	5,95	45212
Vindija Vi-77, Croatia	Neandertal	0,071	3,01	74596

Ferrassie and Krapina, as well as important modern human fossils, such as Veternica, proved to be too poorly preserved to be likely to allow DNA retrieval. Thus, further destructive sampling of these specimens was not considered justified. However, four Neandertal and five early modern human fossils fulfilled the above criteria for amino acid preservation and were thus expected to contain endogenous DNA (Fig. 1; Table 2). These samples were geographically well distributed across Europe (Fig. 2) and included remains whose morphology is typical of Neandertals (e.g., La Chapelle-aux-Saints) and of modern humans (La Madeleine, Cro-Magnon). They also included samples that have sometimes been considered "transitional" between Neandertals and modern humans based on their morphological features: Vindija (Smith, 1984) and Mladeč (Frayer, 1986; 1992; Wolpoff, 1999).

If low amounts of DNA are preserved in a specimen, some extracts will fail to contain DNA molecules by chance (Hofreiter et al., 2001a). Therefore, except in the case of Mladeč 2, in which the amount of material available permitted only two extractions, we extracted each of the four Neandertal and the five early modern human samples three times. For each extraction, amplifications were performed using two primer pairs: (i) "hominoid primers" that amplify homologous mtDNA sequences from the previously determined Neandertals, contemporary modern humans as well as African great apes; (ii) "Neandertal primers" that under the conditions used amplify only Neandertal mtDNAs even in the presence of a large excess of modern human DNA (Krings et al., 2000;

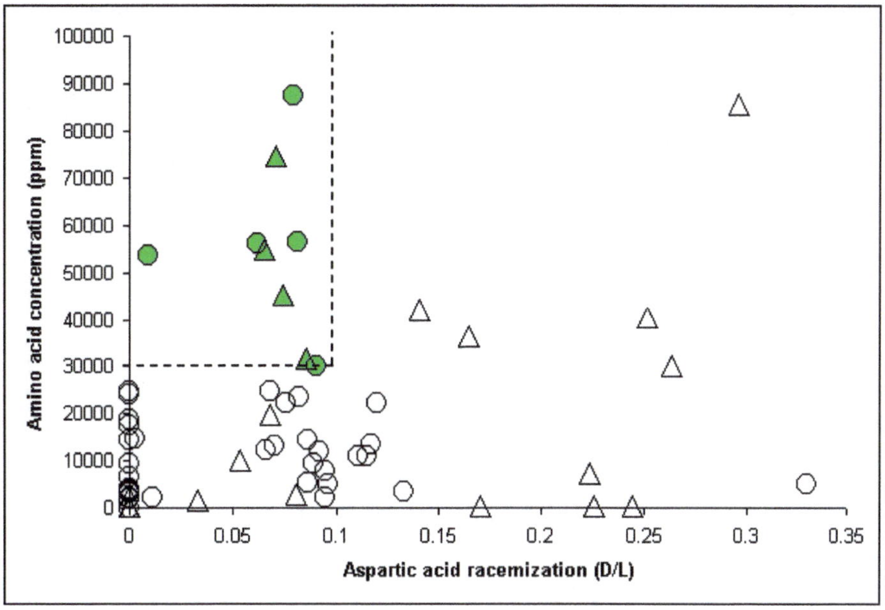

Fig. 1. Amino acid analyses of 64 hominid remains. For each bone, the extent of aspartic acid racemization (D/L) and the amino acid concentration (ppm) is given. The dash lines delimit the area of amino acid preservation compatible with DNA retrieval. Circles and triangles represent early modern humans and Neandertals, respectively. The samples from which DNA extractions were performed are green. (see also Table 1)

Table 2. DNA retrieved from Late Pleistocene fossils in this study

Specimen	Primers used and products obtained[a]	
	"Hominoid"	"Neandertal"
Neandertal remains		
Vindija 77 (Croatia)	3/3	2/3
Vindija 80 (Croatia)	3/3	1/1[b]
Engis 2 (Belgium)	2/3	2/3
La Chapelle aux Saints (France)	3/3	2/3
Early modern human remains		
Mladeč 25c (Czech Republic)	3/3	0/3
Mladeč 2 (Czech Republic)	2/2	0/2
Cro-Magnon (France)	3/3	0/3
Abri Pataud (France)	3/3	0/3
La Madeleine (France)	2/3	0/3
6 cave bears	13/18	0/18

[a] For each specimen and primer pair, the number of amplifications yielding a specific product is given followed by the total number of amplification attempted.

[b] A single amplification using the indicated "Neandertal" primers was attempted. The sequence was confirmed by amplification of larger overlapping fragments (cf. Fig. S1).

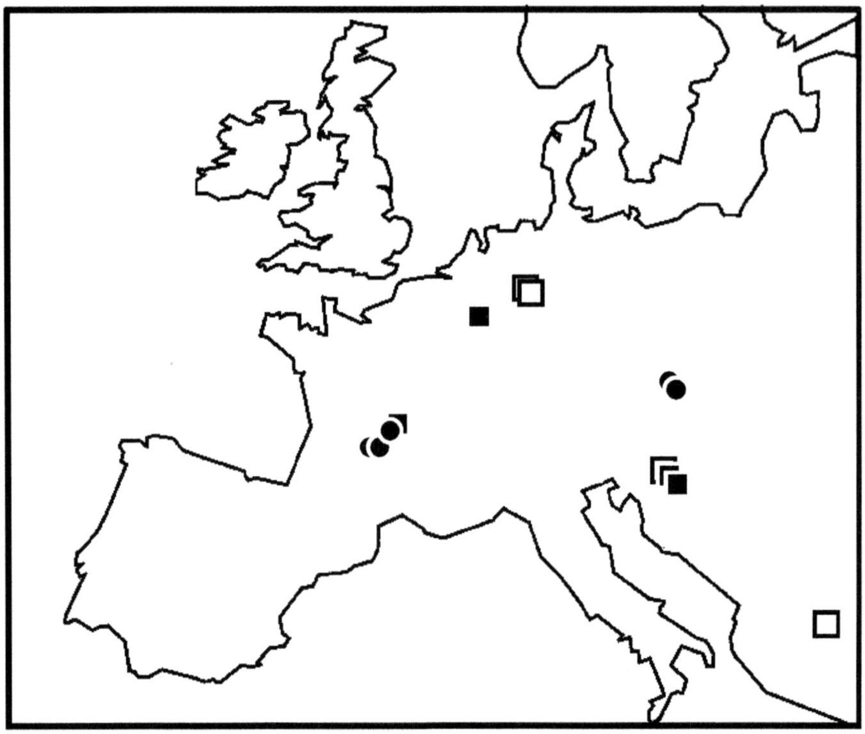

Fig. 2. Geographical origin of Neandertal and early Modern Human samples from which mtDNA sequences have been analyzed. Filled squares and filled circles represent Neandertal and early modern human remains, respectively, analyzed in this study. The four Neandertal remains formerly analyzed are represented by empty squares

Schmitz et al., 2002). Since authentic ancient DNA is typically highly degraded both primer pairs were designed to amplify short mtDNA fragments (72 and 31 bp, respectively, excluding primers). In each of these fragments, two substitutions allow the discrimination of previously determined Neandertal mtDNA sequences from contemporary modern human sequences. The sensitivity of both primer pairs is similar, as shown by the fact that they are both able to amplify single template molecules as judged from nucleotide misincorporation patterns (Hofreiter et al., 2001a). In order to determine the nature of the DNA sequences amplified, each amplification product was cloned and approximately 30 clones were sequenced for each "hominoid product" and ten clones for each "Neandertal product".

When amplifyed with the hominoid primers, all Neandertal and all early modern human remains yielded modern human DNA sequences (see Table 2). In addition, five cave bear teeth from Vindija, Croatia, and one from Gamssulzen, Austria, extracted in parallel with the hominid samples, all yielded human sequences. This confirms previous results in showing that most, if not all, ancient remains yield human DNA sequences when amplification conditions that allow single DNA molecules to be detected are used (Hofreiter et al., 2001b). For three Neandertal and all five modern human remains, several different mtDNA sequences were retrieved from individual extractions and in the case of one Neandertal and one modern human at least two of the sequences were also found in an independent extraction from the same specimen. Additionally, one of the cave bear teeth yielded a human sequence found in two independent extracts. Thus, the fact that a DNA sequence is found in two independent extracts is a necessary, but not sufficient, criterion of authenticity

```
Cambridge Reference Sequence    TCACACATCAACTGCAACTCCAAAGCCACCC
Vi80                            ...T.........A...........A.G...
Vi77                            ...T.........A...........A.G...
Engis 2                         ...T.........A...........A.....
La Chapelle-aux-Saints          ...T.........A...........A.....

Feldhofer                       ...T.........A...........A.G...
Mezmaiskaya                     ...T.........A...........A.....
Vi75                            ...T.........A...........A.G...
Feldhofer II                    ...T.........A...........A.....
```

Fig. 3. Sequences obtained from the Neandertal remains using one of the two sets of "Neandertal primers". Dots indicate identity to the human reference sequence (Anderson et al., 1981) given above. The four upper DNA sequences were determined in this study. Previously determined DNA sequences are shown below

when human remains are analyzed. This implies that in absence of further technical improvements, it is impossible to produce undisputable human mtDNA sequences from ancient human remains. In addition to DNA sequences identical to those previously amplified from present-day humans, the Neandertal bones Vi-77 and Vi-80 from Vindija yielded four out of 89 and 73 out of 85 mtDNA sequences, respectively, that were identical to previously determined Neandertal sequences. Thus, these two specimens contain a proportion of Neandertal-like mtDNA sequences (i.e., sequences that carry two substitutions that differentiate Neandertal mtDNA sequences from modern human mtDNA sequences as described above) that is high enough to detect using primers that amplify also modern human DNA.

When amplified with Neandertal-specific primers, Neandertal-like mtDNA sequences were amplified from two independent extractions from all Neandertal fossils (Table 2; Fig. 3). For one of these, Vi-80 from Vindija, DNA preservation was sufficient to allow the retrieval of longer fragments and thus the reconstruction of 357 bp of the hypervariable region I (see Appendix and Fig. S1). This mtDNA sequence was identical to that retrieved from another bone from the same locality (Vi-75; Krings et al., 2000). In contrast to the Neandertal remains, none of the early modern human extracts yielded any amplification products with the Neandertal primers, although these remains are similar in chemical preservation to the Neandertal remains (see Fig. 1).

Thus, all Neandertal remains analyzed yield mtDNA sequences that are not found in the human mtDNA gene pool today but are similar to those found in four previously published Neandertals (Krings et al., 1997; 2000; Ovchinnikov et al., 2000; Schmitz et al., 2002) (see Fig. 3). This is compatible with results suggesting that the extent of Neandertal mtDNA diversity was similar to that of current humans and lower than that of the great apes (Krings et al., 2000; Schmitz et al., 2002). It is noteworthy that this result is not an artifact created by discarding "modern-like" mtDNA sequences amplified from Neandertals (Trinkaus, 2001), since all Neandertal remains with good biomolecular preservation yield "Neandertal-like" mtDNA sequence. Furthermore, none of the five early modern humans yields "Neandertal-like" mtDNA sequences in spite of the fact that these remains are as well preserved in terms of amino acids as the Neandertal remains. Thus, we fail to detect any evidence of mtDNA gene flow from Neandertals to early modern humans, or from early modern humans to Neandertals.

However, a relevant question is what extent of gene flow between Neandertals and early modern humans the current data allow us to exclude. In this regard, it is of relevance that the five early modern humans analyzed lived much closer in time to the Neandertals than do contemporary individuals. The probability that mtDNA sequences potentially contributed to modern hu-

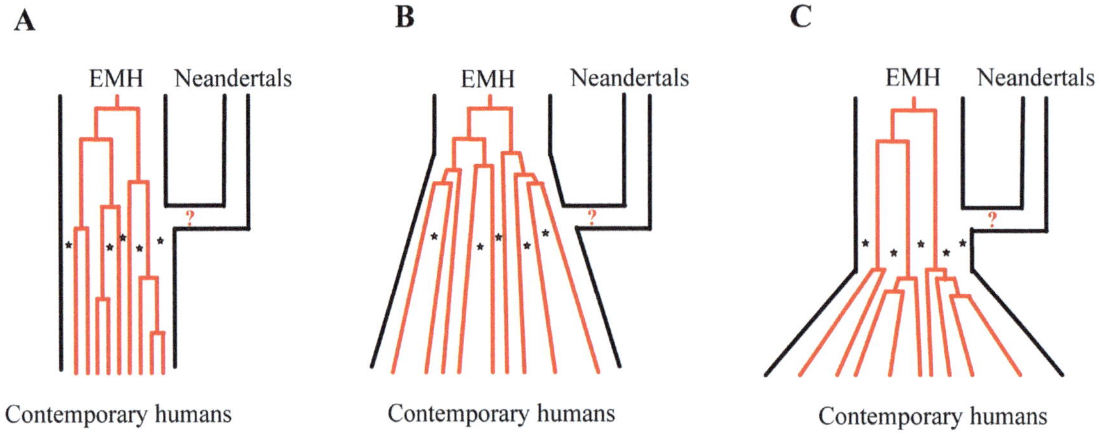

Fig. 4. Schematic model of putative contribution of Neandertal mtDNA to the gene pool of modern humans. (**A**) Under the assumption of a constant effective population size of 10,000 for modern humans, contemporary mtDNAs trace back to approximately five mtDNA lineages 25,000 years ago. The modern human fossils represent five additional samples from around the time of putative admixture (red stars). The contemporary and early modern human (EMH) samples reject a Neandertal contribution of 25% or more to modern humans about 30,000 years ago ($p < 0.05$). (**B**) Under the more realistic scenario of an expansion of the human population size during and after the colonization of Europe, a smaller Neandertal contribution can be excluded because the number of ancestors of the current human gene pool was larger 30,000 years ago. However, the contribution that can be excluded would depend on when and how the expansion occurred. (**C**) Under the scenario that population size was constant before a putative merging with the Neandertal population and expanded only thereafter, the Neandertal contribution could have been larger but similarly depends on how the expansion occurred

mans by Neandertals were lost by drift (Nordborg, 1998) or swamped by continuous influx of modern human mtDNAs (Enflo et al., 2001) in the Neandertal gene pool is therefore much smaller than when contemporary humans are analyzed (e.g., Relethford, 1999). In fact, the five early modern humans analyzed almost double the amount of information about the Upper Pleistocene mtDNA gene pool since, under a model of constant effective population size, all contemporary humans trace their mtDNA ancestors back to only four to seven mtDNA lineages 20,000 to 30,000 years ago (Fig. 4A; Fig. 5), while all other mtDNA sequences present in the gene pool at that time have been lost by random genetic drift. Since the probability is very low ($p < 0.007$) that one or more of the five early modern humans analyzed here are among these few ancestors of current humans, the five Upper Pleistocene individuals can be added to the ancestors of the current mtDNA gene pool to allow us to ask what extent of Neandertal mtDNA contribution to early modern humans can be statistically excluded using the coalescent. Under the model of a constant human effective population size (Tavare, 1984; Nordborg, 1998) of 10,000 over time (Fig. 4A), any contribution of Neandertal mtDNA to modern humans 30,000 years ago larger than 25% can be excluded at the 5% level (Fig. 6). A more realistic scenario may be that the spread of modern humans was accompanied by an increase in population size before and during their migration out of Africa and subsequent colonization of western Eurasia (see Fig. 4B). In that case, the Neandertal contribution that can be excluded is smaller (i.e., less gene flow could have taken place), but that depends critically on when and how the expansion occurred. Finally, under the unlikely scenario that population size was constant during the migration out of Africa and colonization of Europe and expanded only after a putative merging with Neandertals, the Neandertal contribution could have been larger, but this also depends on the nature of the growth (see Fig. 4C).

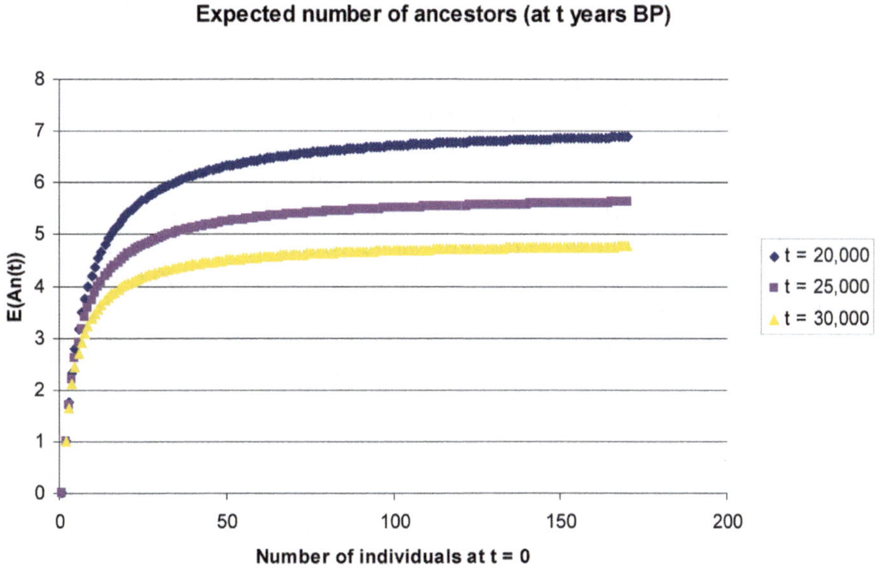

Fig. 5. Expected number of ancestors E(A*n*(t)) of n individuals under a model of constant population size of Ne = 10,000. The number of ancestors of *n* individuals is estimated at 20,000, 25,000, and 30,000 years ago. For example, 150 humans living today have approximately seven ancestors 20,000 years ago

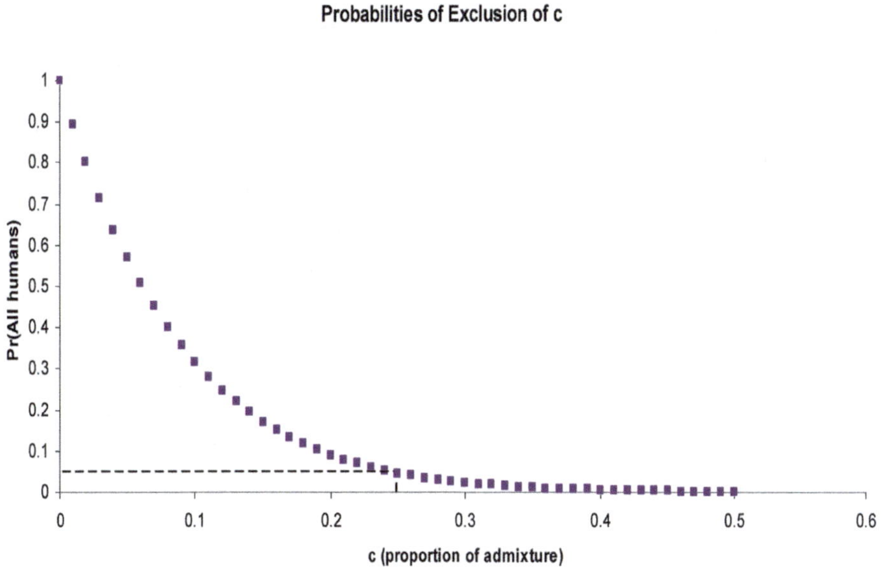

Fig. 6. Probability of observing only modern human mtDNA sequences in both five early human remains and the current mtDNA gene pool given different proportion of Neandertal contribution c (X-axis) under a model of constant population size (see text and *"Materials and methods")*. For example, the probability of observing only human mtDNA sequences given a Neandertal contribution of 25% or more is smaller than 0.05 (dotted line)

Concluding remarks

It is noteworthy that under the model of constant population size, about 50 early modern human remains would need to be studied to exclude a Neandertal mtDNA contribution of 10%. To exclude a 5% contribution, one would need to study more early modern human remains than have been discovered to date. Thus, definitive knowledge of the extent of a putative contribution of Neandertals to the modern human gene pool will not be possible, although extensive studies of variation in the current human gene pool may clarify this question (Wall, 2000). It is, however, worthwhile to note that samples considered as anatomically "transitional" between modern humans and Neandertals, such as Vindija (Smith, 1984; Wolpoff, 1999) and Mladeč (Frayer, 1986; 1992; Wolpoff, 1999), analyzed here, fail to show any evidence of mtDNA admixture between the two groups. Thus, while it cannot be excluded that Neandertals contributed variants at some genetic loci to contemporary humans, no positive evidence of any such contribution has yet been detected.

Acknowledgements

We are grateful to the editors of PLoS (Public Library of Science) for their permission to reprint the original article (Serre et al., 2004). We are indebted to J.-J. Hublin, M. Nordborg, M. Przeworski, M. Stoneking and L. Vigilant for helpful discussions and comments; to many persons and institutions that allowed access to fossils; and to the Max-Planck-Gesellschaft and the Deutsche Forschungsgemeinschaft for financial support.

Appendix

Determination of the mtDNA sequence of Vi-80 from Vindija, Croatia

The entire hypervariable region I sequence was determined from this specimen using amplifications and clones given in Figure S1. Its sequence is identical to the sequence previously determined from individual Vi-75 from Vindija (Krings et al., 2000). We could exclude cross-contamination from the old extract to this bone because different primers were used and some of the fragments of mtDNA amplified Vi-80 were longer than those used to determine the sequence of Vi-75. Morphological analyses do not exclude that these two fragmentary bones (Vi-75 and Vi-80) may come from a single individual. Carbon-14 accelerator mass spectrometry dating, conducted in the Ångstrom Laboratory (Uppsala University, Sweden), yielded a date for Vi-80 of 38,310 ± 2,130 BP (before present). Since Vi-75 has been previously dated to over 42,000 years BP (Krings et al., 2000) the possibility exists that the dates overlap since 42,000 BP is within two standard deviations of the Vi-80 date. Therefore, the bone labeled Vi-80 that yields the new mtDNA sequence could either be (i) a fragment of the same skeleton (individual) that was already successfully extracted, (ii) a bone from another individual maternally related to the first individual amplified or (iii) another unrelated individual having by chance the same mtDNA sequence which is not unlikely given the apparently low mtDNA diversity of Neandertals (Krings et al., 2000; Schmitz et al., 2002).

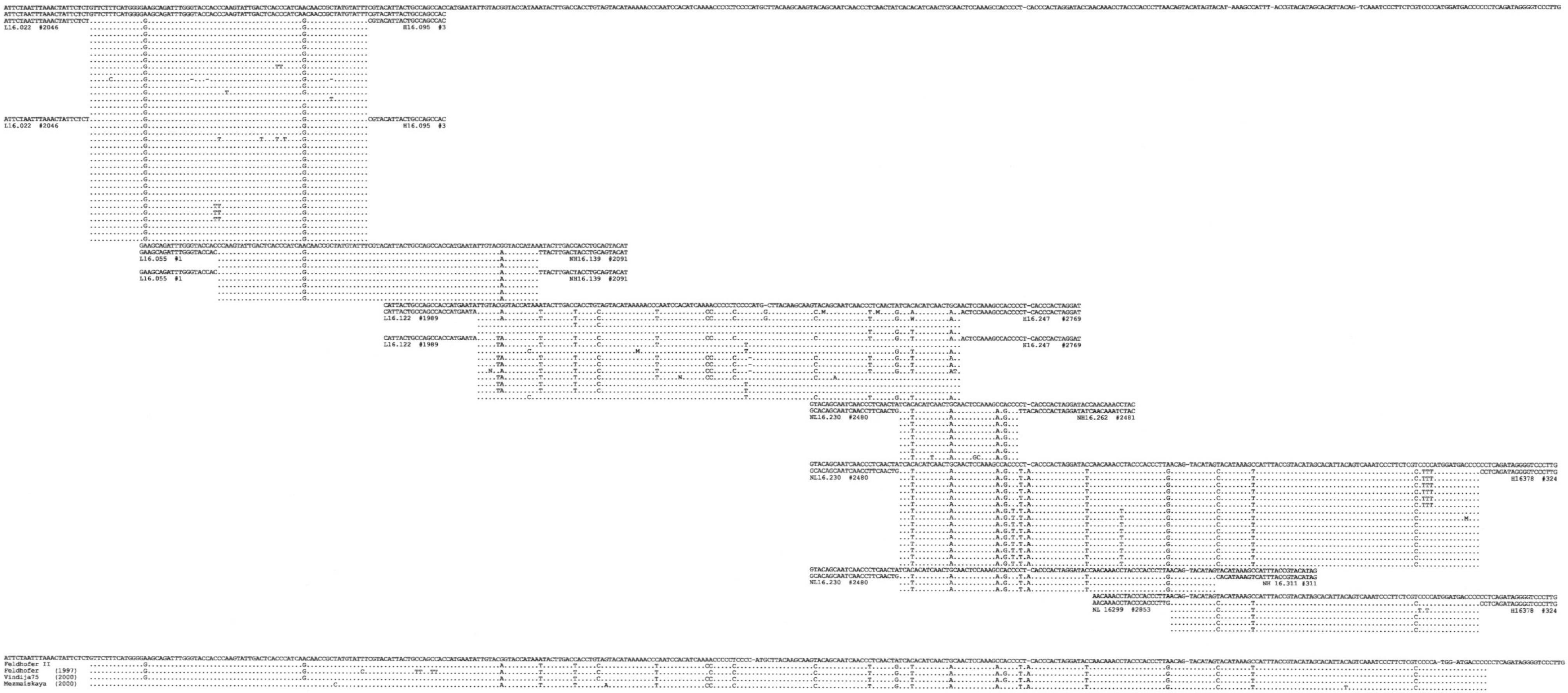

Fig. S1. The DNA sequences of the clones used to reconstruct the sequence of the mitochondrial hypervariable region I from the bone Vi-80

References

Abbott, A. (2003) Anthropologists cast doubt on human DNA evidence. *Nature* 423, 468

Adcock, G. J., Dennis, E. S., Easteal, S., Huttley, G. A., Jermiin, L. S., et al. (2001) Mitochondrial DNA sequences in ancient Australians: Implications for modern human origins. *Proceedings of the National Academy of Sciences USA* 98, 537–542

Anderson, S., Bankier, A. T., Barrell, B. G., de Bruijn, M. H., Coulson, A. R., et al. (1981) Sequence and organization of the human mitochondrial genome. *Nature* 290, 457–465

Cann, R. L., Stoneking, M. and Wilson, A. C. (1987) Mitochondrial DNA and human evolution. *Nature* 325, 31–36

Caramelli, D., Lalueza-Fox, C., Vernesi, C., Lari, M., Casoli, A., et al. (2003) Evidence for a genetic discontinuity between Neandertals and 24,000-year-old anatomically modern Europeans. *Proceedings of the National Academy of Sciences USA* 100, 6593–6597

Duarte, C., Maurício, J., Pettitt, P. B., Souto, P., Trinkaus, E., van der Plicht, H. and Zilhão, J. (1999) The early Upper Paleolithic human skeleton from the Abrigo do Lagar Velho (Portugal) and modern human emergence in Iberia. *Proceedings of the National Academy of Sciences USA* 96, 7604–7609

Enflo, P., Hawks, K. and Wolpoff, M. (2001) A simple reason why Neanderthal ancestry can be consistent with current DNA information. *American Journal of Physical Anthropology* 114, 62

Frayer, D. W. (1986) Cranial variation at Mladeč and the relationship between Mousterian and Upper Paleolithic hominids. *Anthropos* 23, 243–256

Frayer, D. W. (1992) Evolution at the European edge: Neanderthal and Upper Paleolithic relationships. *Prehistoire Europeenne* 2, 9–69

Hawks, J. D. and Wolpoff, M. H. (2001) The accretion model of Neandertal evolution. *Evolution* 55, 1474–1485

Hofreiter, M., Jaenicke, V., Serre, D., von Haeseler, A. and Pääbo, S. (2001a) DNA sequences from multiple amplifications reveal artifacts induced by cytosine deamination in ancient DNA. *Nucleic Acids Research* 29, 4793–4799

Hofreiter, M., Serre, D., Poinar, H. N., Kuch, M. and Pääbo, S. (2001b) Ancient DNA. *Nature Reviews Genetics* 2, 353–359

Ingman, M., Kaessmann, H., Pääbo, S. and Gyllensten, U. (2000) Mitochondrial genome variation and the origin of modern humans. *Nature* 408, 708–713

Knight, A. (2003) The phylogenetic relationship of Neandertal and modern human mitochondrial DNAs based on informative nucleotide sites. *Journal of Human Evolution* 44, 627–632

Krings, M., Stone, A., Schmitz, R. W., Krainitzki, H., Stoneking, M., et al. (1997) Neandertal DNA sequences and the origin of modern humans. *Cell* 90, 19–30

Krings, M., Capelli, C., Tschentscher, F., Geisert, H., Meyer, S., et al. (2000) A view of Neandertal genetic diversity. *Nature Genetics* 26, 144–146

Nordborg, M. (1998) On the probability of Neanderthal ancestry. *American Journal of Human Genetics* 63, 1237–1240

Ovchinnikov, I. V., Gotherstrom, A., Romanova, G. P., Kharitonov, V. M., Liden, K., et al. (2000) Molecular analysis of Neanderthal DNA from the northern Caucasus. *Nature* 404, 490–493

Poinar, H. N., Hoss, M., Bada, J. L. and Pääbo, S. (1996) Amino acid racemization and the preservation of ancient DNA. *Science* 272, 864–866

Relethford, J. H. (1999) Models, predictions, and the fossil record of modern human origins. *Evolutionary Anthropology* 8, 7–10

Schmitz, R. W., Serre, D., Bonani, G., Feine, S., Hillgruber, F., et al. (2002) The Neandertal type site revisited: Interdisciplinary investigations of skeletal remains from the Neander Valley, Germany. *Proceedings of the National Academy of Sciences USA* 99, 13342–13347

Serre, D., Langaney, A., Chech, M., Teschler-Nicola, M., Paunovic, M., Mennecier, P., Hofreiter, M., Possnert, G. and Pääbo, S. (2004) No evidence of Neandertal mtDNA contribution to early modern humans. *Public Library of Science Biology* 2 (3), 313–317

Smith, F. H. (1984) Fossil hominids from the Upper Pleistocene of Central Europe and the origin of modern Europeans. In (F. Spencer, Ed.) *The origins of modern humans: A world survey of the fossil evidence.* New York: Alan R. Liss., pp. 137–210

Stringer, C. (2002) Modern human origins: Progress and prospects. *Philosophical Transactions of the Royal Society of London B: Biological Sciences* 357, 563–579

Stringer, C. B. and Andrews, P. (1988) Genetic and fossil evidence for the origin of modern humans. *Science* 239,1263–1268

Tavare, S. (1984) Line-of-descent and genealogical processes, and their applications in population genetics models. *Theoretical Population Biology* 26,119–164

Trinkaus, E. (2001) The Neandertal paradox. In (C. Finlayson, Ed.) *Neanderthals and modern humans in late Pleistocene Eurasia*. Gibraltar: The Gibraltar Museum, pp. 73–74

Underhill, P. A., Shen, P., Lin, A. A., Jin, L., Passarino, G., et al. (2000) Y-chromosome sequence variation and the history of human populations. *Nature Genetics* 26, 358–361

Wall, J. (2000) Detecting ancient admixture in humans using sequence polymorphism data. *Genetics* 154, 1271–1279

Wandeler, P., Smith, S., Morin, P. A., Pettifor, R. A. and Funk, S. M. (2003) Patterns of nuclear DNA degeneration over time: A case study in historic teeth samples. *Molecular Ecology* 12, 1087–1093

Weidenreich, F. (1943) The "Neanderthal man" and the ancestors of "Homo sapiens". *American Anthropologist* 45, 39–48

Wolpoff, M. H. (1999) *Paleoanthropology*. Boston: McGraw-Hill

Wolpoff, M., Wu, X. and Thorne, A. G. (1984) Modern *Homo sapiens* origins: A general theory of hominid evolution involving the fossil evidence from East Asia. In (F. Spencer, Ed.) *The origins of modern humans: A world survey of the fossil evidence*. New York: Alan R. Liss, pp. 411–483

Wolpoff, M. H., Hawks, J. and Caspari, R. (2000) Multiregional, not multiple origins. *American Journal of Physical Anthropology* 112, 129–136

Chapter 18

NON-DESTRUCTIVE DETERMINATION OF $^{87}SR/^{86}SR$ ISOTOPE RATIOS IN EARLY UPPER PALEOLITHIC HUMAN TEETH FROM THE MLADEČ CAVES – PRELIMINARY RESULTS

Thomas Prohaska, Maria Teschler-Nicola, Patrick Galler, Antonin Přichystal, Gerhard Stingeder, Monika Jelenc and Urs Klötzli

Introduction

Strontium isotope ratio analysis was applied successfully within the last decade as one method of choice to reconstruct migration events in past human populations (Runia, 1985; Price et al., 1994, 2000; 2002; Grupe et al., 1997; Ezzo et al., 1997; 2002; Latkoczy et al., 1998; Sillen et al., 1998; Budd et al., 2000; Teschler-Nicola et al., 1999; 2001; Chiaradia et al., 2003; Bentley et al., 2002; Kutschera and Müller 2003; Schweissing and Grupe, 2003; Hodell et al., 2004). $^{87}Sr/^{86}Sr$ ratios are functions of the local environment. Sr incorporation in human bones and dentine undergoes a periodical turnover, since these tissues equilibrate with the individual metabolism. Therefore a later stage of an individual's life is recorded herein. Dental enamel, on the other hand, does not undergo a turnover and can be considered an "archive of childhood" (Grupe et al., 1997; Lee et al., 1999), since it is formed during the early years of the life of an individual (see Hillson, 1996). Under favorable preservation conditions a significant difference in the Sr isotope signature between enamel and dentine (or bone) can point to a residential change of an individual (Lee et al., 1999; Beard and Johnson, 2000). But the preservation status of human skeletal remains, diagenetic changes or contamination have to be taken into consideration in order to avoid conclusions drawn from artifacts (Sillen, 1986; Sillen and Le Geros, 1991; Budd et al., 2000; Latkoczy et al., 2001; Prohaska et al., 2002; Chiaradia et al., 2003; Trickett et al., 2003). Most of the original biogenic Sr is preserved in enamel (Lee-Thorp and Sponheimer, 2003). Investigations of solubility distributions of hydroxyapatite lead to the same conclusion, namely that biogenic Sr is sufficiently conserved in enamel over long term periods (Shellis and Wilson, 2004). This is an important prerequisite for the present pilot study of Sr-isotope ratios of the Mladeč Upper Paleolithic specimen.

The strontium isotope ratio technique has rarely been used for identifying locals and non-locals in the Paleolithic (Sillen et al., 1995). So far, most investigations have focused on tooth/bone pair analysis of prehistoric and historic populations (see above). One of the reasons might be that the common technologies for the determination of Sr-isotopes have been of an invasive nature. For the present pilot study, we applied the laser ablation technique (Prohaska et al., 2002) in combination with a multi collector inductively coupled plasma mass spectrometer (LA-MC-ICP-MS). The coupling of laser ablation enhances the isotope ratio analysis capabilities of ICP-MS by introducing a direct solid sampling system with high spatial resolution (spot diameter down to a few µm), which enables direct "quasi non-destructive" isotope ratio analysis of solid surfaces. Single collector ICP-MS in combination with laser ablation has so far been limited for its achievable total combined uncertainties on the final ratio (Prohaska et al., 2002). MC-ICP-MS instruments allow the assessment of Sr isotope ratios with an improved instrumental precision up to three orders of magnitude. Several publications report the capabilities of resolving Sr isotopic variations within different samples on a

10–100 µm scale by LA-MC-ICP-MS (Bizzarro et al., 2003; Schmidberger et al., 2003; Ramos et al., 2004) and a correction for systematic bias which has to be taken into account (Waight et al., 2002). Laser ablation in combination with MC-ICP-MS is therefore the method of choice for direct isotope ratio determination of valuable human remains.

Within this study, we analyzed $^{87}S/^{86}Sr$ isotope ratios of small teeth fragments of the most representative specimens of Mladeč 1, Mladeč 2 and Mladeč 8. The results contribute to shed light on whether early modern humans of Mladeč changed their residence or stayed as autochthones in this area.

Material and site description

Material

Analysis was performed on teeth of individuals Mladeč 1, Mladeč 2 and Mladeč 8 (Table 1). The samples were excavated by Josef Szombathy at the end of the 19th century and have since then been inventorised in the osteological collection of the Anthropological Department at the Naturhistorisches Museum Wien (Szombathy, 1925). We used fragments of about 1–3 mm^3 of enamel and/or dentine. The majority of the fragments were taken in the course of the radiocarbon dating procedure carried out recently (Wild et al., 2005).

From the specimen of Mladeč 1, most probably a female, we used a small section of the right maxillary second molar (M^2), consisting mainly of enamel. Because of the insufficiently preserved dentine in this fragment, we also took 2 mm^3 of dentine from the partly preserved left first maxillary premolar (P^3). In the latter, the tooth crown is more or less completely destroyed post mortally and dentine is exposed in the whole area. We had to follow curatorial concerns, since all other teeth are well and completely preserved or restored.

In case of Mladeč 2, also a young female, an approximately 3 mm^3 sample was taken from the left third maxillary molar (M^3), containing both a very small enamel section as well as the enclosing dentine section.

From the Mladeč 8 specimen, a male individual, we took 2–2,5 mm^3 of dentine from the root of the left maxillary second molar (M^2). A further sample was taken from the right second maxillary incisor (I^2), which exhibited fracture cracks and minimal post mortem damage, allowing an accurate sampling of approximately 2 mm^3 of tooth material, mainly composed of dentine. The adjacent enamel was not adequate for subsequent analysis. Additional sampling of enamel has not be performed so far due to curatorial concerns.

All samples had fresh and uncontaminated cleavage areas, which were used for further LA-ICP-MS analysis.

Table 1. Specification of the enamel and dentine samples from Mladeč 1, Mladeč 2 and Mladeč 8 taken for LA-ICP-MS analysis of Sr-isotope ratios

Specimen no.	Sex	Sample material
Mladeč 1	Female	Enamel/right M^2
Mladeč 1	Female	Dentine/left P^3
Mladeč 2	Female	Enamel/left M^3
Mladeč 2	Female	Dentine/left M^3
Mladeč 8	Male	Dentine (1)/ right I^2
Mladeč 8	Male	Dentine (2)/ left M^2

Since the Mladeč Caves are located in an area of variable geology and unknown whole rock, soil and water Sr-isotope ratios, the local signal of Sr isotopes has first to be determined for the different potential Sr reservoirs. Within this pilot study, two drop water samples obtained from the "Witch Cave" and the "Virgin Cave" have been analyzed so far. The sampling sites are indicated in the map given in Fig. 1. It might be relevant that both samples are from dripping water; there is no running water within the cave system today.

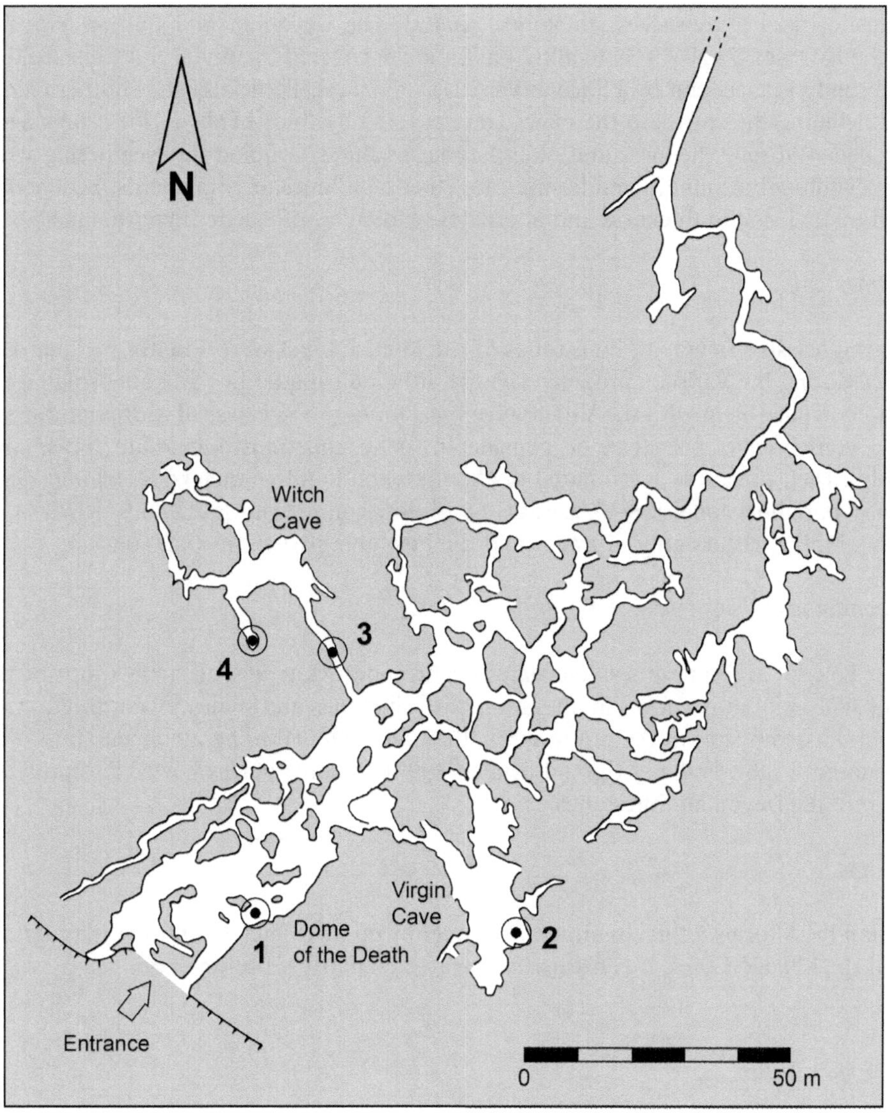

Fig. 1. Map of the Mladeč Cave system showing the sampling sites (2 and 4 = water samples; 1 and 3 = soil samples)

Geomorphology of the site

The Mladeč Caves have developed in a small area (2.4 x 1.5 km) of Devonian limestones near the town of Litovel (central/northern Moravia). They comprise an intricate maze of fissure passages, domes, chimneys filled with sediments and large boulders. From the geomorphological point of view, limestones form the Třesín hill are situated at the eastern margin of the Zábřeh Highland. The highest point (345 m) is more than 105 m above the large flat fluvial plain of the river Morava and its branches (the Upper Moravian Basin). The limestone area represents a fault block surrounded by Lower Carboniferous siliciclastic rocks (graywackes, siltstones, shales). The Devonian and Carboniferous rocks are sunken in the East along a NW – SE tending fault and are covered by fluvial Holocene sandy clays, by Pleistocene sandy gravels and by Pliocene varied sandy clays. Especially the southern slopes of the Třesín hill including the entries to the caves were covered by thick banks of loess prevalently in the Late Pleistocene (Würm). The loess and older laterite products of limestone weathering were washed into the caves through chimneys and fissures, together with limestone fragments. Sedimentary filling was formed up to 3–4 m in thickness and alternates with layers of calcite travertine in several places.

Devonian limestone

The light-gray massive Devonian limestones of the Mladeč Karst were slightly metamorphosed and recrystalized during the Variscan orogeny. General lithological qualities and one isolated fossil finding allow to compare them with the Vilémovice limestones of the classical Moravian Karst development. As a consequence, they can be considered to be uppermost Devonian (Frasnian, roughly 385–375 My). Their thickness is estimated to about several hundred meters. Štelcl and Zimák (1999) reported very pure limestones (98–99% $CaCO_3$) with low content of $MgCO_3$ (0.52–0.73%) and $FeCO_3$ (up to 0.1%). $MnCO_3$ occurs only in traces and the insoluble part forms 0.27–0.65%.

Lower Carboniferous siliciclastics

The nearest Lower Carboniferous siliciclastics occur about 2 km west or 1 km south of the Mladeč Caves. They consist of graywackes in alteration with siltstones and shales. According to the study of heavy mineral assemblages in the graywackes, they are assumed to be a continuation of the Protivanov Formation in the Drahany Upland (Otava, 1997). In some places Lower Carboniferous shales are folded into the Devonian limestones.

Würmian loess

The Würmian loess forms a thick sedimentary cover of up to 15 meters on the Devonian limestones just around the Mladeč Caves. It is distinctly calcareous, partly with lime nodules.

Analytical setup

Analysis of tooth samples was performed at the test site of NU Instruments by using a NU plasma multi collector inductively coupled plasma mass spectrometer (MC-ICP-MS) (NU Instruments Ltd., Wrexham, UK) in combination with a 213 nm UV laser (New Wave Research Co. Ltd., USA). Laser ablation and MC-ICP-MS parameters are summarized in Table 2.

Five replicate single spot analyses using a beam diameter of 100 µm were performed on each sample. Laser parameters were adjusted in order to obtain a constant signal of about 2 minutes per shot. Data evaluation was performed on the transient signal after the Sr signal had reached a stable

Table 2. Laser ablation and MC-ICP-MS instrumentation parameters

MC-ICP-MS and laser ablation parameters	
RF Power	1300 W
Argon gas flow rates	
Cooling	13 L/min
Auxiliary	0.35 L/min
Cones	Ni
Collector arrangement	
Sr	^{89}Sr:H5 ^{88}Sr:H4 ^{87}Sr:H2 ^{86}Sr:Ax ^{84}Sr:L3
Rb	^{85}Rb:L2
Kr	^{83}Kr:L4 ^{82}Kr:L5
Data acquisition mode	Time resolved analysis (TRA)
Dwell time	0.2 sec
Laser ablation parameters	
Wavelength	213 nm
Beam diameter	100 µm
Ablation mode	Single spot
Pulse energy	75% (≈ 27,5 J/cm^2)
Repetition rate	10 Hz
Ablation gas flow rates	
Mixing gas 1 He	0.7 L/min
Mixing gas 2 Ar	0.5 L/min

maximum. All intensities were corrected for blank including a Kr correction. Therefore, a gas blank from the purging gas of the ablation cell was measured for 20 seconds prior to each ablation. ^{87}Sr was corrected for minor ^{87}Rb interferences via the ^{85}Rb signal. The ^{87}Rb contribution was calculated via the ^{85}Rb-signal intensity by using the natural abundances as recommended by the IUPAC (Coplen, 2001). All raw ratios were corrected for mass bias using the $^{88}Sr/^{86}Sr$ signal applying a power law mass bias correction. The total combined uncertainty (calculated via propagation of errors to the final result) was 0.04% RSU for $^{87}Sr/^{86}Sr$.

The Sr isotopic composition of the drop water sample from the "Witch Cave" and the "Virgin Cave" of the Mladeč Cave system was analyzed by static MC-TIMS (Triton, Finnigan, Bremen, Germany) after complete Rb/Sr separation and purification using conventional ion exchange procedures (Klötzli et al., 2001).

Results and discussion

In this pilot study, we obtained a stable distribution of LA-ICP-MS data within the investigated areas, which covered a field of approximately 2 mm^2. The final mean $^{87}Sr/^{86}Sr$ ratios, polled from five individual/replicate spot analyses, are summarized in Table 3 and Fig. 2.

The results show a noticeable variability: Mladeč 1 reveals a statistically significant difference between the isotope ratios of dentine of the first left premolar (0.7108) and the enamel (0.7119) of the second right molar. The latter represents the highest value within the Mladeč samples. Mladeč 2 possesses a similar pattern: The Sr-isotope ratio of the enamel of the third left molar (0.7113) is statistically significantly higher than the ratio obtained from dentine of the same tooth (0.7106). The Mladeč 8 dentine samples of the second permanent incisor (0.7104) and the left second molar (0.7106) show

Table 3. $^{86}Sr/^{87}Sr$ isotope ratios of enamel and dentine samples of individuals Mladeč 1, Mladeč 2 and Mladeč 8 (mean of five replicate analyses) and the water samples of the "Witch Cave" and the "Virgin Cave"

Individual	Tooth	Tissue	Analytical technique	$^{87}Sr/^{86}Sr$	SU
Mladeč 1	right M^2	Enamel	LA-MC-ICP-MS	0.7119	0.0003
Mladeč 1	left P^3	Dentine	LA-MC-ICP-MS	0.7108	0.0003
Mladeč 2	left M^3	Enamel	LA-MC-ICP-MS	0.7113	0.0003
Mladeč 2	left M^3	Dentine	LA-MC-ICP-MS	0.7106	0.0003
Mladeč 8	right I^2	Dentine (1)	LA-MC-ICP-MS	0.7104	0.0003
Mladeč 8	left M^2	Dentine (2)	LA-MC-ICP-MS	0.7106	0.0003
Sample name	**Cave**	**Type**			
Mladeč W01	Witch Cave	cave water	TIMS	0.710545	0.000003
Mladeč W02	Virgin Cave	cave water	TIMS	0.71056	0.00008

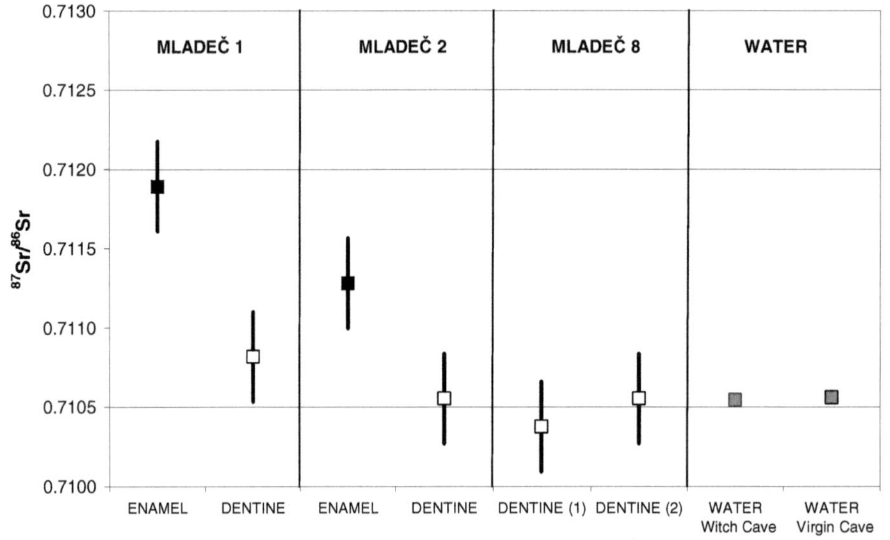

Fig. 2. $^{86}Sr/^{87}Sr$ isotope ratios of enamel and dentine samples of Mladeč 1, Mladeč 2 and Mladeč 8 and the water samples of the "Witch Cave" and the "Virgin Cave"

no intra-individual variation and exhibit nearly identical, very low Sr isotope ratios which are consistent with the dentine ratios of the other Mladeč specimen, in particular Mladeč 2.

The Mladeč 1 and Mladeč 2 specimens show an analogue Sr isotope ratio pattern, namely a higher ratio in the enamel samples and a statistically significantly lower ratio in the dentine (Fig. 2). According to theoretical considerations and the examined potential of the Sr isotope ratio technique for the study of prehistoric migration (Price et al., 2001; 2004; Bentley et al., 2002; 2003; 2004), it seems plausible at first glance that both individuals could represent non-locals. But one of the problems arising in such investigation is the determination of the local Sr reservoirs and the corresponding Sr isotope ratio signals as a necessary prerequisite for distinguishing migrants from locals. Although it has been suggested to investigate local animals for that purpose (Price et al., 2002), the study of Neolithic animal teeth by Bentley et al. (2003) demonstrated that this approach cannot solve the problem in a satisfactory way. Moreover, we do not have a sufficient number of recommended 'ideal' faunal remains within the preserved fossil record of the Mladeč Caves. In particular, teeth of stationary animals are missing (see Pacher, this volume, chap. 6). Another problem is due to the fact that the Mladeč Caves are situated in an area where Devonian limestone, Lower Carboniferous siliciclastics and Würmian Loess occur. Since Sr isotope ratio mapping of that area does not exist, we have used as proxy one water sample from the "Witch Cave", located in an cave area which is not accessible to the general public. The second water sample is from the 'Virgin Cave' of the Mladeč Cave, where public access is allowed (nonetheless, this fact did not influence the Sr isotopic composition, even if tourists are used to throw coins into the water). The Sr isotope ratios of the "Witch Cave" and the "Virgin Cave" water are 0.710545 and 0.71056, respectively. These values are in accordance with all dentine values of the investigated specimens. Moreover, one of the Mladeč 8 dentine samples from the right maxillary second incisor shows a (very low) Sr isotope ratio, which is practically identical to the value obtained in a single sample from the Bell Beaker period site Moravská Nová Ves, located in South Moravia on "Quaternary deposits of loess, fluviatile sediments and Cretaceous deposits". Price et al. (2004) reported a Sr ratio of about 0.7103 in the bone and 0.71009 in the enamel sample, "suggesting a locally born individual". The results given by Price at al. (2004) seem to underline the assumption that the dentine values of the Mladeč samples represent the local signal.

It is established that biogenic Sr is well preserved in enamel (Lee-Thorp and Sponheimer, 2003; Shellis and Wilson, 2004), whereas dentine or bone can exhibit remarkable diagenetic alterations. Therefore, even if the Sr isotope ratios of the dentine from the Mladeč samples correspond to the local signal, we cannot assume a priori that this Sr was incorporated *intra vitam*. Post mortem diagenetic alterations have to be taken into account as well. In this context it might be of interest that analytical scanning electron microscopy (JEOL JSM6400 equipped with an energy dispersive system) of embedded small block preparates of several human postcranial elements, including Mladeč 20 (a rib fragment), Mladeč 25a (a right proximal radius) and Mladeč 28 (a left proximal femur) have provided evidence for the presence of calcite ($CaCO_3$) with minor amounts of Mg, Mn and Fe (see Teschler-Nicola, this volume, chap. 5). A similar composition was reported by Štelcl and Zimák (1999) for the almost pure limestone of the Mladeč Karst. Moreover, SiO_2, typically for siliciclastics, which represent a further component of the Mladeč geomorphology, was also detected in parts of the human bone block preparates.

Interestingly, the high enamel values obtained from the Mladeč1 and 2 specimens correspond to ratios which Price et al. (2004) determined for individuals recovered from Velke Prilepy, a multi-component site in Bohemia located more north and westward (Price et al. 2004).

In sum, the results of our non-invasive Strontium isotope ratio investigation on the early Upper Paleolithic specimens from the Maldeč Cave could provide first direct measures of the assumed mobility in hunter-gatherer populations. But they must be seen as preliminary: not only because of the insufficient number of samples, but also because of our inadequate knowledge on the variability of the local bedrock and water signals, the sparse reference dates, specifically the lack of supra-regional Sr isotope ratio signals, and the problems arising from diagenetic alterations in fossil human remains.

Future investigations must include further samples of the local site, water samples of the immediate vicinity, faunal remains unearthed from the cave as well as enamel and dentine samples of two other human tooth samples (Mladeč 8, Mladeč 9b). The nature and degree of post mortem alterations will be of additional concern. Referring to the highly valuable material, it will not be possible to solve this question by using leaching procedures and therewith differentiate biogenic and geogenic apatite in the nearest future. In this context, multi-elemental pattern analysis might be of interest. For this reason, we are currently investigating the elemental distribution and diffusion profiles of major and trace elements in tooth tissues in comparison to the elemental pattern of the adjacent repository material.

Acknowledgement

We highly acknowledge Nu Instruments Ltd, especially Andrew Burrows, and New Wave Research Co. Ltd for help in obtaining the analytical results. We thank Franz Branstätter and Robert Seemann, Department of Mineralogy, NHM Vienna for their comments on an earlier version of the manuscript and Wolfgang Reichmann, Ronald Mühl and August Walch for their technical support. Further acknowledge is given to the FWF (START project VIRIS, Y267-N11 and project 12480-CHE).

References

Beard, B. L. and Johnson C. M. (2000) Strontium isotope composition of skeletal material can determine the birth place and geographic mobility of humans and animals. *Journal of Forensic Sciences* 45, 1049–1061

Bentley, R. A., Price, T. D., Lüning, J., Gronenborn, D., Wahl, J. and Fullagar, P. D. (2002) Human migration in early Neolithic Europe. *Current Anthropology* 43, 799–804

Bentley, R. A., Krause, R., Price, T. D. and Kaufmann, B. (2003) Human mobility at the early Neolithic settlement of Vaihingen, Germany: Evidence from strontium isotope analysis. *Archaeometry* 45, 481–496

Bentley, R. A., Price, T. D. and Stephan, E. (2004) Determining the local $^{87}Sr/^{86}Sr$ range for archaeological skeletons: a case study from Neolithic Europe. *Journal of Archaeological Science* 31, 365–375

Bizzarro, M., Simonetti, A., Stevenson, R. K. and Kurszlaukis, S. (2003) In situ $^{87}Sr/^{86}Sr$ investigation of igneous apatites and carbonates using laser ablation MC-ICP-MS. *Geochimica et Cosmochimica Acta* 67, 289–302

Budd, P., Montgomery, J., Barreiro, B. and Thomas, R. G. (2000) Differential diagenesis of strontium in archaeological human dental tissues. *Applied Geochemistry* 15, 687–694

Chiaradia, M., Gallay, A. and Todt, W. (2003) Different contamination styles of prehistoric human teeth at a Swiss necropolis (Sion, Valais) inferred from lead and strontium isotopes. *Applied Geochemistry* 18, 353–370

Coplen, T. B. (2001) Atomic weights of the elements 1999. *Pure and Applied Chemistry* 73 (4), 667–683

Ezzo, J. A., Johnson, C. M. and Price, T. D. (1997) Analytical perspectives on prehistoric migration: A case study from East-Central Arizona. *Journal of Archaeological Science* 24, 447–466

Ezzo, J. A. and Price, T. D. (2002) Migration, regional reorganization, and spatial group composition at Grasshopper Pueblo, Arizona. *Journal of Archaeological Science* 29, 499–520

Grupe, G., Price, T. D., Schröter, P., Söllner, F., Johnson, C. M. and Beard, B. L. (1997) Mobility of Bell Beaker people revealed by Sr isotope ratios of tooth and bone: A study of southern Bavarian skeletal remains. *Applied Geochemistry* 12, 517–525

Hillson, S. (1996) *Dental anthropology.* Cambridge: University Press, p. 123

Hodell, D. A., Quinn, R. L, Brenner, M. and Kamenov, G. (2004) Spatial variation of strontium isotopes ($^{87}Sr/^{86}Sr$) in the Maya region: A tool for tracking ancient human migration. *Journal of Archaeological Science* 31, 585–601

Klötzli, U. S., Koller, F., Scharbert, S. and Höck, V. (2001) Cadomian lower crustal contributions to Variscan granite petrogenesis (South Bohemian pluton, Austria): Constraints from zircon typology and geochronology, whole-rock, and feldspar Pb-Sr isotope systematics. *Journal of Petrology* 42 (9), 1621–1642

Kutschera, W. and Müller, W. (2003) 'Isotope Language' of the Alpine iceman investigated with AMS and MS. *Nuclear Instruments and Methods in Physics Research* B 24, 705–719

Latkoczy, C., Prohaska, T., Stingeder, G. and Teschler-Nicola, M. (1998) Strontium isotope ratio measurements in prehistoric human bone samples by means of HR-ICPMS. *Journal of Analytical Atomic Spectrometry* 13 (6), 561–566

Latkoczy, C., Prohaska, T., Watkins, M., Stingeder, G. and Teschler-Nicola, M. (2001) Strontium isotope ratio determination after on-line matrix separation by coupling ion chromatography (HPIC) to an inductively coupled plasma sector field mass spectrometer (SF-ICP-MS). *Journal of Analytical Atomic Spectrometry* 16, 806–811

Lee, K. M., Appleton, J., Cooke, M., Keenan, F. and Sawicka-Kapusta, K. (1999) Use of Laser ablation inductively coupled plasma mass spectrometry to provide element versus time profiles in teeth. *Analitica Chimica Acta* 395, 179–185

Lee-Thorp, J. and Sponheimer, M. (2003) Three case studies used to reassess the reliability of fossil bone and enamel isotope signals for paleodietary studies. *Journal of Anthropological Archaeology* 22, 208–216

Otava, J. (1997) Geology of Paleozoic of the Litovelské Pomoraví Protected Area, Central Moravia. *Acta Musei Moraviae, Scientiae Geologicae* 82, 93–103

Pacher, M. (2006) Large mammal remains from the Mladeč Caves and their contribution to side formation processes. In (M. Teschler-Nicola, Ed.) *Early modern humans at the Moravian Gate: The Mladeč Caves and their remains*, this issue, pp. 99–148

Price, T. D., Grupe, G. and Schroeter, P. (1994) Reconstruction of migration patterns in the Bell Beaker period by stable strontium isotope analysis. *Applied Geochemistry* 9, 413–417

Price, T. D., Manzanilla, L. and Middleton, W. D. (2000) Immigration and the ancient city of Teotihuacan in Mexico: A study using strontium isotope ratios in human bone and teeth. *Journal of Archaelogical Science* 27, 903–913

Price, T. D., Bentley, R. A., Lüning, J., Gronenborn, D. and Wahl, J (2001) Prehistoric human migration in the Linearbandkeramik of Central Europe. *Antiquity* 75 (289), 593–603

Price, T. D., Burton, J. H. and Bentley, R. A. (2002) The characterization of biologically available strontium isotope ratios for the study of prehistoric migration. *Archaeometry* 44, 117–135

Price, T. D., Knipper, C., Grupe, G. and Smrcka, V. (2004) Strontium isotopes and prehistoric human migration: The bell beaker period in central Europe. *European Journal of Archaeology* 7 (1), 9–40

Prohaska, T., Schultheis, G., Latkoczy, C., Teschler-Nicola, M. and Stingeder, G. (2002) Investigation of Sr isotope ratios in prehistoric human bones and teeth using Laser ablation ICP-MS and ICP-MS after Rb/Sr separation. *Journal of Analytical Atomic Spectrometry* 17, 887–891

Ramos, F. C., Wolff, J. A. and Tollstrup, D. L. (2004) Measuring $^{87}Sr/^{86}Sr$ in minerals and groundmass from basalts using LA-MC-ICP-MS. *Chemical Geology* 211, 135–158

Runia, L. T. (1985) Gebruik van strontium, andere sporenelementen en stabile isotopen als voedingsindicatoren in de archeologie. *Voeding* 46, 368–375

Schmidberger, S. S., Simonetti, A. and Francis, D. (2003) Small scale Sr isotope investigation of clinopyroxenes from peridotite xenoliths by laser ablation MC-ICP-MS. Implications for metasomatism. *Chemical Geology* 199, 317–329

Schweissing, M. M. and Grupe, G. (2003) Stable Sr isotopes in human teeth and bone: A key to migration events of the late Roman period in Bavaria. *Journal of Archaeological Science* 30, 1373–1383

Shellis, R. P. and Wilson, R. M. (2004) Apparent solubility distributions of hydroxyapatite and enamel apatite. *Journal of Colloid and Interface Science* 278, 325–332

Sillen, A. (1986) Biogenic and diagenetic Sr/Ca in Plio-Pleistocene fossils of the Omo Shungura Formation. *Paleobiology* 12, 311–323

Sillen, A. and Le Geros, R. (1991) Solubility profiles of synthetic apatites and of modern and fossil bones. *Journal of Archaeological Science* 18, 385–397

Sillen, A., Hall, G., Richardson, S. and Armstrong, R. (1995) Strontium calcium ratios (Sr/Ca) and strontium isotopic ratios ($^{87}Sr/^{86}S$) of *Australopithecus robustus* and *Homo* sp. from Swartkrans. *Journal of Human Evolution* 28, 277–285

Sillen, A., Hall, G., Richardson, S. and Armstrong, R. (1998) $^{87}Sr/^{86}Sr$ ratios in modern and fossil food-webs of the Sterkfontein Valley: Implications for early hominid habitat preference. *Geochimica et Cosmochimica Acta* 62, 2463–2473

Štelcl, J. and Zimák, J. (1999) Results of geological, petrographical and geochemical study in the speleotherapeutic sanatory at Mladeč. *Geologické výzkumy na Moravě a ve Slezsku* VI, 157–161

Szombathy, J. (1925) Die diluvialen Menschenreste aus der Fürst-Johanns-Höhle bei Lautsch in Mähren. *Die Eiszeit* 2, 1–34, 73–95

Teschler-Nicola, M., Gerold, F., Bujatti-Narbeshuber, M., Prohaska, T., Latkoczy, C., Stingeder, G. and Watkins, M. (1999) Evidence of genocide 7000 BP. Neolithic paradigm and geo-climatic reality. *Collegium Antropologicum* 23, 437–450

Teschler-Nicola, M., Prohaska, T., Gerold, F., Watkins, M., Latkoczy, C. and Stingeder, G. (2001) $^{87}Sr/^{86}Sr$ Isotopenverhältnis in (prä)historischen menschlichen Skelettresten. Indikator für individuellen Ortswechsel. *Archäologie Österreichs* 12 (1–2), 70–74

Teschler-Nicola, M. (2006) Taphonomic aspects of the human skeletal remains from Mladeč. In (M. Teschler-Nicola, Ed.) *Early modern humans at the Moravian Gate: The Mladeč Caves and their remains*, this issue, pp. 75–98

Trickett, A. M., Budd, P., Montgomery, J. and Evans, J. (2003) An assessment of solubility profiling as decontamination procedure for the $^{87}Sr/^{86}Sr$ analysis of archeological human skeletal tissue. *Applied Geochemistry* 18, 653–658

Waight T., Baker J. and Peate, D. (2002) Sr isotope ratio measurements by double-focusing MC-ICPMS: Techniques, observations and pitfalls. *International Journal of Mass Spectrometry* 221, 229–244

Wild, E., Teschler-Nicola, M., Kutschera, W., Steier, P., Trinkaus, E. and Wanek, W. (2005) Direct dating of Early Upper Palaeolithic human remains from Mladeč. *Nature* 435, 332–335

LIST OF CONTRIBUTORS

Antl-Weiser Walpurga, Dr., Natural History Museum Vienna, Department of Prehistory, Burgring 7, 1010 Vienna, Austria
E-mail: walpurga.antl@nhm-wien.ac.at

Bookstein Fred L., Prof. Ph.D., Department of Anthropology, University Vienna, Althanstrasse 14, 1090 Vienna, Austria
E-mail: fred.bookstein@univie.ac.at

Chech Mario, Dr., Laboratoire d'Anthropologie Biologique, Musée de L'Homme, 17, Place du Trocadéro, 75116 Paris, France
E-mail: mario.chech@free.fr

Czerny Christian, Univ.-Prof. Dr., Department of Radiology, AKH Wien, Währinger Gürtel 18–20, 1090 Vienna, Austria
E-mail: christian.czerny@univie.ac.at

Frayer David W., Prof. Ph.D., Department of Anthropology, University of Kansas, Lawrence, KS 66044-2110, USA
E-mail: frayer@ukans.edu

Galler Patrick, Dipl.-Ing., Department für Chemie, University of Natural Resources and Applied Life Sciences, Muthgasse 18, 1190 Vienna, Austria
E-mail: patrick.galler@bku.ac.at

Gunz Philipp, Ph.D., Department of Human Evolution, Max Planck Institute for Evolutionary Anthropology, Deutscher Platz 6, 04103 Leipzig, Germany
E-mail: gunz@eva.mpg.de

Hofreiter Michael, Dr., Max Planck Institute for Evolutionary Anthropology, Deutscher Platz 6, 04103 Leipzig, Germany
E-mail: hofreiter@eva.mpg.de

Jelenc Monika, Labor für Geochronologie, Institut für Geologische Wissenschaften, Universität Wien, Althanstrasse 14, 1090 Wien, Austria

Jelínek Jan †, Prof. Dr., Moravian Museum, Anthropos Institute, Zelný trh 6, 659 37 Brno, Czech Republic

Klötzli Urs, Univ.-Prof. Dr., Labor für Geochronologie, Institut für Geologische Wissenschaften, Universität Wien, Althanstrasse 14, 1090 Wien, Austria
E-mail: urs.kloetzli@univie.ac.at

Kutschera Walter, Univ.-Prof. Dr., Istitut für Isotopenforschung und Kernphysik, Universität Wien, Währingerstraße 17, 1090 Vienna, Austria
E-mail: walter.kutschera@univie.ac.at

Langaney André, Prof. Ph.D., Laboratoire D'Anthropologie Biologique, Musée de L'Homme, 17, Place du Trocadéro, 75116 Paris, France
E-mail: langaney@mnhn.fr

Mennecier Philippe, Dr., Laboratoire d'Anthropologie Biologique, Musée de L'Homme, 17, Place du Trocadéro, 75116 Paris, France
E-mail: phm@mnhn.fr

Minugh-Purvis Nancy, Dr., Department of Anatomy and Cell Biology, University of Pennsylvania School of Dental Medicine, 422 Levy, 420 South 40th Street, Philadelphia, PA 19104-6002, USA
E-mail: minugh@biochem.dental.upenn.edu

Mitteröcker Philipp, Mag., Department of Anthropology, University Vienna, Althanstrasse 14, 1090 Vienna, Austria
E-mail: philipp.mitteroecker@univie.ac.at

Oliva Martin, Doz. Ph.D., Anthropos Institute Moravian Museum, Zelný trh 6, 659 37 Brno, Czech Republic
E-mail: moliva@mzm.cz

Pacher Martina, Dr., Institute of Paleontology, University of Vienna, Althanstraße 14, 1090 Vienna, Austria
E-mail: martina.pacher@univie.ac.at

Pääbo Svante, Prof. Ph.D., Max Planck Institute for Evolutionary Anthropology, Deutscher Platz 6, D-04103 Leipzig, Germany
E-mail: paabo@eva.mpg.de

Paunovic Maja †, Dr., Institue of Quaterny Paleontology and Geology, Croatian Academy of Sciences and Arts, A. Kovacica 5, 10000 Zagreb, Croatia

Přichystal Antonin, Prof. RNDr.CSc., Masaryk University Brno, Faculty of Science, Institute of Geological Science, CSc. Kotlářská 2, 611 37 Brno, Czech Republik
E-mail: prichy@sci.muni.cz

Possnert Göran, Dr., Ångström Laboratory, Uppsala University, 751 05 PO Box 256, 751 05 Uppsala, Sweden
E-mail: Goran.Possnert@Angstrom.uu.se

Prohaska Thomas, ao. Univ.-Prof. Dipl.-Ing. Dr.techn., Department für Chemie, University of Natural Resources and Applied Life Sciences, Muthgasse 18, 1190 Vienna, Austria
E-mail: thomas.prohaska@boku.ac.at

Prossinger Hermann, Univ.-Doz. Dr., Department of Anthropology, University Vienna, Althanstrasse 14, 1090 Vienna, Austria
E-mail: hermann.prossinger@univie.ac.at

Schamall Doris, Mag., Natural History Museum Vienna, Department of Anthropology, Burgring 7, 1010 Vienna, Austria
E-mail: doris.schamall@nhm-wien.ac.at

Schultz Michael, Prof. Dr. Dr., Universität Göttingen, Bereich Humanmedizin, Abt. Morphologie, Kreuzbergring 36, 37075 Göttingen, Germany
E-mail: mschult1@gwdg.de

Seidler Horst, o. Univ.Prof. Dr., Department of Anthropology, University Vienna, Althanstrasse 14, 1090 Vienna, Austria
E-mail: horst.seidler@univie.ac.at

Serre David, Dr., McGill University and Genome Quebec Innovation Center, Montreal, Quebec H3A 1A4, Canada
E-mail: david.serre@mail.mcgill.ca

Shackelford Laura L., Ass. Prof. Ph.D., Department of Anthropology, Campus Box 1114, Washington University, St. Louis, MO 63130, USA
E-mail: llshacke@artsci.wustl.edu

Smith Fred H., Prof. Ph.D., Department of Sociology and Anthropology, Loyola University Chicago, Lake Shore Campus, 6525 North Sheridan Road, Chicago, IL 60626, USA
E-mail: fsmith3@luc.edu

Stadlmayr Andrea, Department of Anthropology, University Vienna, Althanstrasse 14 A-1090 Vienna, Austria
E-mail: andrea.stadlmayr@univie.ac.at

Steier Peter, Mag. Dr., Institut für Isotopenforschung und Kernphysik, Universität Wien, Währingerstraße 17, 1090 Vienna, Austria
E-mail: peter.steier@univie.ac.at

Stingeder Gerhard, Univ.Prof. Dipl.-Ing. Dr.techn., Department für Chemie, University of Natural Resources and Applied Life Sciences, Muthgasse 18, 1190 Vienna, Austria
E-mail: gerhard.stingeder@boku.ac.at

Stockton Trenton C., Ph.D., Office of Communications and Public Affairs, Johns Hopkins School of Medicine, Baltimore, MD 21205, USA
E-mail: tstockt1@jhmi.edu

Svoboda Jiři A., Dr.Sc., Institute of Archaeology, Academy of Sciences of the Czech Republic, Královopolská 147, 612 00 Brno, Czech Republic
E-mail: svoboda@iabrno.cz

Teschler-Nicola Maria, HR Univ.-Prof. Dr., Natural History Museum Vienna, Department of Anthropology, Burgring 7, 1010 Vienna, Austria
E-mail: maria.teschler@univie.ac.at; maria.teschler@nhm-wien.ac.at

Trinkaus Erik, Prof. (Mary Tileston Hemenway), Ph.D., Department of Anthropology, Campus Box 1114, Washington University, St. Louis, MO 63130, USA
E-mail: trinkaus@artsci.wustl.edu

Viola Th. Bence, Mag., Department of Anthropology, University Vienna, Althanstrasse 14, 1090 Vienna, Austria
E-mail: bence.viola@univie.ac.at

Wanek Wolgang, Ass.Prof. Dr., Department für Chemische Ökologie und Ökosystemwissenschaften, Universität Wien, Althanstrasse 14, 1090 Vienna, Austria
E-mail: wolfgang. wanek@univie.ac.at

Weber Gerhard, Univ.Prof. Dr., Department of Anthropology, University Vienna, Althanstrasse 14, 1090 Vienna, Austria
E-mail: gerhard.weber@univie.ac.at

Wild Eva M., Univ.Prof. Dr., Istitut für Isotopenforschung und Kernphysik, Universität Wien, Währingerstraße 17, 1090 Vienna, Austria
E-mail: eva.maria.wild@univie.ac.at

Wolpoff Milford H., Prof. Ph.D., Paleoanthropology Laboratory, Department of Anthropology, 231 West Hall, University of Michigan, Ann Arbor, MI 48109-1382, USA
E-mail: wolpoff@umich.edu

INDEX

GPSR Compliance

The European Union's (EU) General Product Safety Regulation (GPSR) is a set of rules that requires consumer products to be safe and our obligations to ensure this.

If you have any concerns about our products, you can contact us on ProductSafety@springernature.com

In case Publisher is established outside the EU, the EU authorized representative is:

Springer Nature Customer Service Center GmbH
Europaplatz 3
69115 Heidelberg, Germany

Batch number: 10731592

Printed by Printforce, the Netherlands